QUANTUM INFORMATION AND COMPUTATION FOR CHEMISTRY

ADVANCES IN CHEMICAL PHYSICS

VOLUME 154

EDITORIAL BOARD

QUANTUM INFORMATION AND COMPUTATION FOR CHEMISTRY

ADVANCES IN CHEMICAL PHYSICS

VOLUME 154

Edited by

SABRE KAIS

Purdue University
QEERI, Qatar
Santa Fe Institute

Series Editors

STUART A. RICE

Department of Chemistry
and
The James Franck Institute
The University of Chicago
Chicago, Illinois

AARON R. DINNER

Department of Chemistry
and
The James Franck Institute
The University of Chicago
Chicago, Illinois

WILEY

Published by John Wiley & Sons, Inc., Hoboken, New Jersey.
Published simultaneously in Canada.

Library of Congress Cataloging-in-Publication Data:

ISBN: 978-1-118-49566-7

CONTRIBUTORS TO VOLUME 154

ALÁN ASPURU-GUZIK, Department of Chemistry and Chemical Biology, Harvard University, 12 Oxford Street, Cambridge, MA 02138, USA

JONATHAN BAUGH, Institute for Quantum Computing, University of Waterloo, Waterloo, Ontario, N2L 3G1, Canada; Departments of Physics and Astronomy, University of Waterloo, Waterloo, Ontario, N2L 3G1, Canada; Department of Chemistry, University of Waterloo, Waterloo, Ontario, N2L 3G1, Canada

JACOB BIAMONTE, ISI Foundation, Via Alassio 11/c, 10126, Torino, Italy; Centre for Quantum Technologies, National University of Singapore, Block S15, 3 Science Drive 2, Singapore 117543, Singapore

SERGIO BOIXO, Department of Chemistry and Chemical Biology, Harvard University, 12 Oxford Street, Cambridge, MA 02138, USA; Google, 340 Main St, Venice, CA 90291, USA

KENNETH R. BROWN, School of Chemistry and Biochemistry, School of Computational Science and Engineering, School of Physics, Georgia Institute of Technology, Ford Environmental Science and Technology Building, 311 Ferst Dr, Atlanta, GA 30332-0400, USA

GARNET KIN-LIC CHAN, Department of Chemistry and Chemical Biology, Cornell University, Ithaca, NY 14850, USA

ROBIN CÔTÉ, Department of Physics, U-3046, University of Connecticut, 2152 Hillside Road, Storrs, CT 06269-3046, USA

BEN CRIGER, Institute for Quantum Computing, University of Waterloo, Waterloo, Ontario, N2L 3G1, Canada; Departments of Physics and Astronomy, University of Waterloo, Waterloo, Ontario, N2L 3G1, Canada

BORIVOJE DAKIĆ, Faculty of Physics, University of Vienna, Boltzmanngasse 5, A-1090 Vienna, Austria

R. DE VIVIE-RIEDLE, Department Chemie, Ludwig-Maximilians-Universität, Butenandt-Str. 11, 81377 München, Germany

FRANK GAITAN, Laboratory for Physical Sciences, 8050 Greenmead Drive, College Park, MD 20740-4004, USA

C. GOLLUB, Department Chemie, Ludwig-Maximilians-Universität, Butenandt-Str. 11, 81377 München, Germany

SABRE KAIS, Department of Chemistry and Physics, Purdue University, 560 Oval Drive, West Lafayette, IN 47907, USA; Qatar Environment & Energy Research Institute (QEERI), Doha, Qatar; Santa Fe Institute, Santa Fe, NM 87501, USA

GRAHAM KELLS, Dahlem Center for Complex Quantum Systems, Fachbereich Physik, Freie Universität Berlin, Arminallee 14, D-14195 Berlin, Germany

JESSE M. KINDER, Department of Chemistry and Chemical Biology, Cornell University, Ithaca, NY 14850, USA

M. KOWALEWSKI, Department Chemie, Ludwig-Maximilians-Universität, Butenandt-Str. 11, 81377 München, Germany

DANIEL A. LIDAR, Departments of Electrical Engineering, Chemistry, and Physics, and Center for Quantum Information Science & Technology, University of Southern California, Los Angeles, CA 90089, USA

PETER J. LOVE, Department of Physics, Haverford College, 370 Lancaster Avenue, Haverford, PA 19041, USA

XIAO-SONG MA, Faculty of Physics, University of Vienna, Boltzmanngasse 5, A-1090 Vienna, Austria; Institute for Quantum Optics and Quantum Information (IQOQI), Austrian Academy of Sciences, Boltzmanngasse 3, A-1090 Vienna, Austria

DAVID A. MAZZIOTTI, Department of Chemistry and the James Franck Institute, The University of Chicago, Chicago, IL 60637, USA

J. TRUE MERRILL, School of Chemistry and Biochemistry, School of Computational Science and Engineering, School of Physics, Georgia Institute of Technology, Ford Environmental Science and Technology Building, 311 Ferst Dr, Atlanta, GA 30332-0400, USA

SEBASTIAN MEZNARIC, Clarendon Laboratory, University of Oxford, Parks Road, Oxford OX1 3PU, UK

FRANCO NORI, CEMS, RIKEN, Saitama 351-0198, Japan; Physics Department, University of Michigan, Ann Arbor, MI 48109-1040, USA

A. PAPAGEORGIOU, Department of Computer Science, Columbia University, New York, NY 10027, USA

DANIEL PARK, Institute for Quantum Computing, University of Waterloo, Waterloo, Ontario, N2L 3G1, Canada; Departments of Physics and Astronomy, University of Waterloo, Waterloo, Ontario, N2L 3G1, Canada

JIŘÍ PITTNER, J. Heyrovský Institute of Physical Chemistry, Academy of Sciences of the Czech Republic, v.v.i., Dolejškova 3, 18223 Prague 8, Czech Republic

CLAIRE C. RALPH, Department of Chemistry and Chemical Biology, Cornell University, Ithaca, NY 14850, USA

GEHAD SADIEK, Department of Physics, King Saud University, Riyadh, Saudi Arabia; Department of Physics, Ain Shams University, Cairo 11566, Egypt

NOLAN SKOCHDOPOLE, Department of Chemistry and the James Franck Institute, The University of Chicago, Chicago, IL 60637, USA

DAVID G. TEMPEL, Department of Chemistry and Chemical Biology, Harvard University, 12 Oxford Street, Cambridge, MA 02138, USA; Department of Physics, Harvard University, 17 Oxford Street, Cambridge, MA 02138, USA

J. F. TRAUB, Department of Computer Science, Columbia University, New York, NY 10027, USA

U. TROPPMANN, Department Chemie, Ludwig-Maximilians-Universität, Butenandt-Str. 11, 81377 München, Germany

JIŘÍ VALA, Department of Mathematical Physics, National University of Ireland Maynooth, Maynooth, Co. Kildare, Ireland; School of Theoretical Physics, Dublin Institute for Advanced Studies, 10 Burlington Road, Dublin 4, Ireland

LIBOR VEIS, J. Heyrovský Institute of Physical Chemistry, Academy of Sciences of the Czech Republic, v.v.i., Dolejškova 3, 18223 Prague 8, Czech Republic; Department of Physical and Macromolecular Chemistry, Faculty of Science, Charles University in Prague, Hlavova 8, 12840 Prague 2, Czech Republic

P. VON DEN HOFF, Department Chemie, Ludwig-Maximilians-Universität, Butenandt-Str. 11, 81377 München, Germany

PHILIP WALTHER, Faculty of Physics, University of Vienna, Boltzmanngasse 5, A-1090 Vienna, Austria; Institute for Quantum Optics and Quantum Information (IQOQI), Austrian Academy of Sciences, Boltzmanngasse 3, A-1090 Vienna, Austria

PAUL WATTS, Department of Mathematical Physics, National University of Ireland Maynooth, Maynooth, Co. Kildare, Ireland, School of Theoretical Physics, Dublin Institute for Advanced Studies, 10 Burlington Road, Dublin 4, Ireland

JAMES D. WHITFIELD, Department of Chemistry and Chemical Biology, Harvard University, 12 Oxford Street, Cambridge, MA 02138, USA; NEC Laboratories America, 4 Independence Way, Princeton, NJ 08540, USA; Department of Physics, Columbia University, 538 West 120th Street, New York, NY 10027, USA

QING XU, Department of Chemistry, Purdue University, 560 Oval Drive, West Lafayette, IN 47907, USA

MAN-HONG YUNG, Center for Quantum Information, Institute for Interdisciplinary Information Sciences, Tsinghua University, Beijing 100084, P. R. China; Department of Chemistry and Chemical Biology, Harvard University, 12 Oxford Street, Cambridge, MA 02138, USA

FOREWORD

Quantum mechanics and information theory were key new areas for scientific progress in the 20th century. Toward the end of the 1900s, when both of these fields were generally regarded as mature, a small community of researchers in physics, computer science, mathematics, and chemistry began to explore the fertile ground at the intersection of these two areas. Formulation of a quantum version of information theory and analysis of computational and communication protocols with quantum, rather than classical states, led to a powerful new paradigm for computation. Dramatic theoretical results were achieved in the mid-1990s with the presentation of both a quantum algorithm for solving factorization problems with an exponential speedup relative to the best-known classical algorithm and a protocol for performing fault-tolerant quantum computation (i.e., computation allowing for errors but in which these errors are guaranteed not to propagate). These two key results, both due to Peter Shor, motivated the launch of experimental efforts to realize physical implementations of quantum computations. Initially restricted to a small set of candidate physical systems, experimental studies have since multiplied to encompass an increasingly wide range of physical qubit systems, ranging from photons to solid-state quantum circuits. As the study of physical candidates for viable qubit arrays for computation has grown and as attention turns to more complex quantum systems that promise scalability, many quantum theorists active in the field are exploring the use of information theoretic concepts to bring new insights and understanding to the study of quantum systems. Theorists and experimentalists alike are also increasingly heavily focused on quantum simulation, the art of making one Hamiltonian emulate another, according to Feynman's original proposal (1982) for a quantum computer.

Today, the field of quantum information science covers an increasingly broad range of physical systems in addition to theoretical topics in mathematics, computer science, and information theory. As a result, the community of scientists working in this field is quite diverse and interdisciplinary. The unifying feature of the community is interest in a fully quantum description of both information processing and the physical systems that might enable it to be realized in a controllable and programmable fashion. Chemical physics, rooted in quantum physics and the intellectual home of molecular quantum mechanics, is centrally located in this interdisciplinary community. The chemical physicist's interest and expertise in analysis and control of molecular systems bring key tools and perspectives to the twin challenges of experimentally realizing quantum computation and of using quantum information theory for analysis of chemical problems in the full quantum regime. I use the latter term instead of the simpler and oft-used term *quantum chemistry*, which has unfortunately become associated almost exclusively with the specific subdiscipline of electronic structure of atoms and molecules. For quantum information science however, the field quantum chemistry covers much more, namely, the set of molecular phenomena that require a full quantum description for energetics, structure, and dynamics, whether electronic or nuclear, local or collective in nature.

This volume brings together chapters from the quantum information science community that describe topics in quantum computation and quantum information related to or overlapping with key topics in chemical physics. The motivation for this volume may be summarized by two questions. First, what can chemistry contribute to quantum information? Second, what can quantum information contribute to the study of chemical systems? The contributions in this volume address both perspectives, while surveying theoretical and experimental quantum information-related research within chemical physics.

In Chapter 1, Sabre Kais introduces the fields of quantum information, quantum computation, and quantum simulation, outlining their relevance and impact for chemistry. In Chapter 2, Peter Love presents ideas from electronic structure theory that might facilitate quantum computation and quantum simulation. Alan Aspuru-Guzik and Joseph Traub describe quantum algorithms that are relevant for efficient solution of a range of problems in both physics and chemistry, in Chapters 3 and 6, respectively. Libor Veis and Jiří Pittner discuss the solution for both nonrelativistic and relativistic electronic energies via quantum computations in Chapter 4. Several contributions specifically address the relation between quantum computation and electronic structure calculations in terms of the two motivating questions stated earlier. Two contributions examine the use and impact of electronic structure calculations for quantum computation. In Chapter 5, Frank Gaitan and Franco Nori review an approach based on use of density functional theory to generate efficient calculation of energy gaps for adiabatic quantum computation. In Chapter 7, Jesse Kinder, Claire Ralph, and Garnet Chan discuss the use of matrix product states for understanding the dynamics of complex electronic states. The understanding offered by quantum information theoretic concepts for realizing correlations in complex quantum systems is also the focus of Gehad Sadiek and Sabre Kais's discussion of entanglement for spin systems in Chapter 15.

Other authors address the realization of high-fidelity quantum operations, a key requirement of implementation of all quantum information processing. In Chapter 10, True Merrill and Ken Brown discuss the use of compensating pulse sequences from NMR for correcting unknown errors in the quantum bits (*qubits*). In Chapter 11, Daniel Lidar compares the benefits of passive versus active error correction, with analysis of decoherence-free subspaces, noiseless subsystems, and dynamical decoupling pulse sequences. Fault tolerance, the ability to effectively compute despite a bounded rate of error, is also a powerful driver for development of the topological materials that Jiří Vala describes in Chapter 16. The exotic "topological" phases of these materials are characterized by massive ground-state degeneracy, an energy gap for all local excitations and anyonic excitations—properties that mathematicians have shown guarantee a remarkable natural fault tolerance for quantum computation.

Several experimentalists have contributed chapters describing specific physical implementations of qubits and outlining experimental schemes for the realization of quantum computation, quantum simulation, or quantum communication. Ben Criger, Daniel Park, and Jonathan Baugh describe spin-based quantum information processing in Chapter 8. Xiao-Song Ma, Borivoje Drakić, and Philip Walther describe the use of photonic systems for quantum simulation of chemical phenomena in Chapter 9. Regina de Vivie-Riedle describes how information may be transferred through molecular chains by taking advantage of vibrational energy transfer (Chapter 13), and Robin Côté summarizes the application of the rapidly growing field of ultracold molecules to quantum information processing (Chapter 14). Finally, David Mazziotti examines the roles of electron correlation, entanglement, and redundancy in the energy flow within the pigment–protein structure of a light-harvesting complex in Chapter 12.

The contributions to this volume represent just a partial overview of the synergy that has developed over the past 10 years between quantum information sciences and chemical physics. Many more areas of overlap have not been addressed in detail here, notably coherent quantum control of atoms and molecules. However, we hope this selection of chapters will provide a useful introduction and perspective to current directions in quantum information and its relation to the diverse set of quantum phenomena and theoretical methods that are central to the chemical physics community.

University of California, Berkeley BIRGITTA WHALEY

PREFACE TO THE SERIES

Advances in science often involve initial development of individual specialized fields of study within traditional disciplines, followed by broadening and overlap, or even merging, of those specialized fields, leading to a blurring of the lines between traditional disciplines. The pace of that blurring has accelerated in the past few decades, and much of the important and exciting research carried out today seeks to synthesize elements from different fields of knowledge. Examples of such research areas include biophysics and studies of nanostructured materials. As the study of the forces that govern the structure and dynamics of molecular systems, chemical physics encompasses these and many other emerging research directions. Unfortunately, the flood of scientific literature has been accompanied by losses in the shared vocabulary and approaches of the traditional disciplines. Scientific journals are exerting pressure to be ever more concise in the descriptions of studies, to the point that much valuable experience, if recorded at all, is hidden in supplements and dissipated with time. These trends in science and publishing make this series, *Advances in Chemical Physics*, a much needed resource.

Advances in Chemical Physics is devoted to helping the reader obtain general information about a wide variety of topics in chemical physics, a field that we interpret broadly. Our intent is to have experts present comprehensive analyses of subjects of interest and to encourage the expression of individual points of view. We hope this approach to the presentation of an overview of a subject will both stimulate new research and serve as a personalized learning text for beginners in a field.

Stuart A. Rice
Aaron R. Dinner

CONTENTS

INTRODUCTION TO QUANTUM INFORMATION AND COMPUTATION FOR CHEMISTRY

SABRE KAIS

Department of Chemistry and Physics, Purdue University, 560 Oval Drive, West Lafayette, IN 47907, USA; Qatar Environment & Energy Research Institute (QEERI), Doha, Qatar; Santa Fe Institute, Santa Fe, NM 87501, USA

Advances in Chemical Physics, Volume 154: Quantum Information and Computation for Chemistry, First Edition. Edited by Sabre Kais.

I. INTRODUCTION

The development and use of quantum computers for chemical applications has the potential for revolutionary impact on the way computing is done in the future [1–7]. Major challenge opportunities are abundant (see next fifteen chapters). One key example is developing and implementing quantum algorithms for solving chemical problems thought to be intractable for classical computers. Other challenges include the role of quantum entanglement, coherence, and superposition in photosynthesis and complex chemical reactions. Theoretical chemists have encountered and analyzed these quantum effects from the view of physical chemistry for decades. Therefore, combining results and insights from the quantum information community with those of the chemical physics community might lead to a fresh understanding of important chemical processes. In particular, we will discuss the role of entanglement in photosynthesis, in dissociation of molecules, and in the mechanism with which birds determine magnetic north. This chapter is intended to survey some of the most important recent results in quantum computation and quantum information, with potential applications in quantum chemistry. To start with, we give a comprehensive overview of the basics of quantum computing (the gate model), followed by introducing quantum simulation, where the phase estimation algorithm (PEA) plays a key role. Then we demonstrate how PEA combined with Hamiltonian simulation and multiplicative inversion can enable us to solve some types of linear systems of equations described by $A\vec{x} = \vec{b}$. Then our subject turns from gate model quantum computing (GMQC) to adiabatic quantum computing (AQC) and topological quantum computing, which have gained increasing attention in the recent years due to their rapid progress in both theoretical and experimental areas. Finally, applications of the concepts of quantum information theory are usually related to the powerful and counter intuitive quantum mechanical effects of superposition, interference, and entanglement.

Throughout history, man has learned to build tools to aid computation. From abacuses to digital microprocessors, these tools epitomize the fact that laws of physics support computation. Therefore, a natural question arises: "Which physical laws can we use for computation?" For a long period of time, questions such as this were not considered relevant because computation devices were built exclusively based on classical physics. It was not until the 1970s and 1980s when Feynmann [8], Deutsch [9], Benioff [10], and Bennett [11] proposed the idea of using quantum mechanics to perform calculation that the possibility of building a quantum computing device started to gain some attention.

What they conjectured then is what we call today a quantum computer. A quantum computer is a device that takes direct advantage of quantum mechanical phenomena such as superposition and entanglement to perform calculations [12]. Because they compute in ways that classical computers cannot, for certain problems quantum algorithms provide exponential speedups over their classical counterparts. As an example, in solving problems related to factoring large numbers [13] and simulation of quantum systems [14–28], quantum algorithms are able to find the answer exponentially faster than classical algorithms. Recently, it has also been proposed that a quantum computer can be useful for solving linear systems of equations with exponential speedup over the best-known classical algorithms [29]. In the problem of factoring large numbers, the quantum exponential speedup is rooted in the fact that a quantum computer can perform discrete Fourier transform exponentially faster than classical computers [12]. Hence, any algorithm that involves Fourier transform as a subroutine can potentially be sped up exponentially on a quantum computer. For example, efficient quantum algorithms for performing discrete sine and cosine transforms using quantum Fourier transform have been proposed [30]. To illustrate the tremendous power of the exponential speedup with concrete numbers, consider the following example: the problem of factoring a 60-digit number takes a classical computer 3×10^{11} years (about 20 times the age of universe) to solve, while a quantum computer can be expected to factor a 60-digit number within 10^{-8} seconds. The same order of speedup applies for problems of quantum simulation.

In chemistry, the entire field has been striving to solve a number of "Holy Grail" problems since their birth. For example, manipulating matter on the atomic and molecular scale, economic solar splitting of water, the chemistry of consciousness, and catalysis on demand are all such problems. However, beneath all these problems is one common problem, which can be dubbed as the "Mother of All Holy Grails: exact solution of the Schrödinger equation. Paul Dirac pointed out that with the Schrödinger equation, "the underlying physical laws necessary for the mathematical theory of a large part of physics and the whole of chemistry are thus completely known and the difficulty is only that the exact application of these laws leads to equations much too complicated to be soluble" [31]. The problem of solving the Schrödinger equation is fundamentally hard [32,33] because as the number of particles in the system increases, the dimension of the corresponding Hilbert space increases exponentially, which entails exponential amount of computational resource.

Faced with the fundamental difficulty of solving the Schrödinger equations exactly, modern quantum chemistry is largely an endeavor aimed at finding approximate methods. *Ab initio* methods [34] (Hartree–Fock, Moller–Plesset, coupled cluster, Green's function, configuration interaction, etc.), semi-empirical methods (extended Huckel, CNDO, INDO, AM1, PM3, etc.), density functional methods [35] (LDA, GGA, hybrid models, etc.), density matrix methods [36], algebraic methods [37] (Lie groups, Lie algebras, etc.), quantum Monte Carlo methods [38] (variational, diffusion, Green's function forms, etc.), and dimensional scaling methods [39] are all products of such effort over the past decades. However, all the methods devised so far have to face the challenge of unreachable computational requirements as they are extended to higher accuracy to larger systems. For example, in the case of full CI calculation, for N orbitals and m electrons there are $C_N^m \approx \frac{N^m}{m!}$ ways to allocate electrons among orbitals. Doing full configuration interaction (FCI) calculations for methanol (CH_3OH) using 6-31G (18 electrons and 50 basis functions) requires about 10^{17} configurations. This task is impossible on any current computer. One of the largest FCI calculations reported so far has about 10^9 configurations (1.3 billion configurations for Cr_2 molecules [40]).

However, due to exponential speedup promised by quantum computers, such simulation can be accomplished within only polynomial amount of time, which is reasonable for most applications. As we will show later, using the phase estimation algorithm, one is able to calculate eigenvalues of a given Hamiltonian H in time that is polynomial in $O(\log N)$, where N is the size of the Hamiltonian. So in this sense, quantum computation and quantum information will have enormous impact on quantum chemistry by enabling quantum chemists and physicists to solve problems beyond the processing power of classical computers.

The importance of developing quantum computers derives not only from the discipline of quantum physics and chemistry alone, but also from a wider context of computer science and the semiconductor electronics industry. Since 1946, the processing power of microprocessors has doubled every year simply due to the miniaturization of basic electronic components on a chip. The number of transistors on a single integrated circuit chip doubled every 18 months, which is a fact known as Moore's law. This exponential growth in the processing power of classical computers has spurred revolutions in every area of science and engineering. However, the trend cannot last forever. In fact, it is projected that by the year 2020 the size of a transistor would be on the order of a single atom. At that scale, classical laws of physics no longer hold and the behavior of the circuit components obeys laws of quantum mechanics, which implies that a new paradigm is needed to exploit the effects of quantum mechanics to perform computation, or in a more general sense, information processing tasks. Hence, the mission of quantum computing is to study how information can be processed with quantum mechanical devices as well as what kinds of tasks beyond the capabilities of classical computers can be performed efficiently on these devices.

Accompanying the tremendous promises of quantum computers are the experimental difficulties of realizing one that truly meets its above-mentioned theoretical potential. Despite the ongoing debate on whether building a useful quantum computer is possible, no fundamental physical principles are found to prevent a quantum computer from being built. Engineering issues, however, remain. The improvement and realization of quantum computers are largely interdisciplinary efforts. The disciplines that contribute to quantum computing, or more generally quantum information processing, include quantum physics, mathematics, computer science, solid-state device physics, mesoscopic physics, quantum devices, device technology, quantum optics, optical communication, and nuclear magnetic resonance (NMR), to name just a few.

A. Qubits and Gates

In general, we can think of information as something that can be encoded in the state of a physical system. If the physical system obeys classical laws of physics, such as a classical computer, the information stored there is of "classical" nature. To quantify information, the concept of *bit* has been introduced and defined as the basic unit of information. A *bit* of information stored in a classical computer is a value 0 or 1 kept in a certain location of the memory unit. The computer is able to measure the bit and retrieve the information without changing the state of the bit. If the bit is at the same state every time it is measured, it will yield the same results. A bit can also be copied and one can prepare another bit with the same state. A string of bits represents one single number.

All these properties of bits seem trivial but in the realm of quantum information processing, this is no longer true (Table I). The basic unit of quantum information is a *qubit*. Physically, a qubit can be represented by the state of a two-level quantum system of various forms, be it an ion with two accessible energy levels or a photon with two states of polarization. Despite the diverse physical forms that a qubit can take, for the most part the concept of "qubit" is treated as an abstract mathematical object. This abstraction gives us the freedom to construct a general theory of quantum computation and quantum information that does not depend on a specific system for its realization [12].

Unlike classical bits, a qubit can be not only in state $|0\rangle$ or $|1\rangle$, but also a superposition of both: $\alpha|0\rangle + \beta|1\rangle$. If a qubit is in a state of quantum superposition, a measurement will collapse the state to either one of its component states $|0\rangle$ or $|1\rangle$, which is a widely observed phenomenon in quantum physics. Suppose we repetitively do the following: prepare a qubit in the same state $\alpha|0\rangle + \beta|1\rangle$ and then measure it with respect to the basis state $\{|0\rangle, |1\rangle\}$. The measurement outcomes would most probably be different—we will get $|0\rangle$ in some measurements and $|1\rangle$ in the others—even the state of the qubit that is measured is identical each time. Furthermore, unlike classical bits that can be copied, a qubit cannot be copied due to the *no-cloning* theorem, which derives from a qubit's quantum mechanical nature

TABLE I
Comparison Between Classical Bits and Qubits

Classical Bit	Qubit
State 0 or 1	$|0\rangle$, $|1\rangle$, or superposition
Measurement does not change the state of the bit	Measurement changes the system
Deterministic result	Obtain different results with the same system
Can make a copy of bit (eavesdrop)	Cannot clone the qubit (security)
One number for a string bit	Store several numbers simultaneously due to superposition

(see Ref. [12], p. 24 for details). Such no-cloning property of a qubit has been used for constructing security communication devices, because a qubit of information is impossible to eavesdrop. In terms of information storage, since a qubit or an array of qubits could be in states of quantum superposition such as $\alpha_{00}|00\rangle + \alpha_{01}|01\rangle + \alpha_{10}|10\rangle + \alpha_{11}|11\rangle$, a string of qubits is able to store several numbers $\alpha_{00}, \alpha_{01}, \ldots$ simultaneously, while a classical string of bits can only represent a single number. In this sense, n qubits encode not n bits of classical information, but 2^n numbers. In spite of the fact that none of the 2^n numbers are efficiently accessible because a measurement will destroy the state of superposition, this exponentially large information processing space combined with the peculiar mathematical structure of quantum mechanics still implies the formidable potential in the performance of some of computational tasks exponentially faster than classical computers.

Now that we have introduced the basic processing units of quantum computers—the qubits, the next question is: How do we make them compute? From quantum mechanics we learned that the evolution of any quantum system must be *unitary*. That is, suppose a quantum computation starts with an initial state $|\Psi_{\text{initial}}\rangle$, then the final state of the computation $|\Psi_{\text{final}}\rangle$ must be the result of a unitary transformation U, which gives $|\Psi_{\text{final}}\rangle = U|\Psi_{\text{initial}}\rangle$. In classical computing, the basic components of a circuit that transforms a string $\{0, 1\}^n$ to another string $\{0, 1\}^m$ are called *gates*. Analogously, in quantum computing, a unitary transformation U that transforms a system from $|\Psi_{\text{initial}}\rangle$ to $|\Psi_{\text{final}}\rangle$ can also be decomposed into sequential applications of basic unitary operations called *quantum gates* (Table II). Experimentally, the implementation of a quantum gate largely depends on the device and technique used for representing a qubit. For example, if a qubit is physically represented by the state of a trapped ion, then the quantum gate is executed by an incident laser pulse that perturbs the trapped atom(s) and alters its state; if the qubit states are encoded in the polarization states of photons, then a quantum gate consists of optical components that interact with photons and alter their polarization states as they travel through the components.

If we use vectors to describe the state of a qubit, that is, using $|0\rangle$ to represent $(1, 0)^{\mathrm{T}}$ and $|1\rangle$ to represent $(0, 1)^{\mathrm{T}}$, a single-qubit quantum gate can be represented

TABLE II
Comparison Between Classical and Quantum Gates

Classical Logic Gates	Quantum Gates
Each gate corresponds to a mapping $\{0, 1\}^m \rightarrow \{0, 1\}^n$	Each quantum gate corresponds to a transformation $|\Psi\rangle \rightarrow |\Psi'\rangle$ or a rotation on the surface of Bloch sphere ([12], p. 15)
Nonunitary	Unitary
Irreversible	Reversible

using a 2×2 matrix. For example, a quantum NOT gate can be represented by the Pauli X matrix

$$U_{\text{NOT}} = X = \begin{pmatrix} 0 & 1 \\ 1 & 0 \end{pmatrix} \tag{1}$$

To see how this works, note that $X|0\rangle = |1\rangle$ and $X|1\rangle = |0\rangle$. Therefore, the sheer effect of applying X to a qubit is to flip its state from $|0\rangle$ to $|1\rangle$. This is just one example of single-qubit gates. Other commonly used gates include the Hadamard gate H, Z rotation gate, phase gate S, and $\frac{\pi}{8}$ gate T:

$$H = \frac{1}{\sqrt{2}} \begin{pmatrix} 1 & 1 \\ 1 & -1 \end{pmatrix}, \quad S = \begin{pmatrix} 1 & 0 \\ 0 & i \end{pmatrix}, \quad T = \begin{pmatrix} 1 & 0 \\ 0 & e^{i\pi/4} \end{pmatrix} \tag{2}$$

If a quantum gate involves two qubits, then it is represented by a 4×4 matrix. The state of a two-qubit system is generally in form of $\alpha_{00}|00\rangle + \alpha_{01}|01\rangle + \alpha_{10}|10\rangle + \alpha_{11}|11\rangle$, which can be written as a vector $(\alpha_{00}, \alpha_{01}, \alpha_{10}, \alpha_{11})^{\text{T}}$. In matrix form, the CNOT gate is defined as

$$U_{\text{CNOT}} = \begin{pmatrix} 1 & 0 & 0 & 0 \\ 0 & 1 & 0 & 0 \\ 0 & 0 & 0 & 1 \\ 0 & 0 & 1 & 0 \end{pmatrix} \tag{3}$$

It is easy to verify that applying CNOT gate to a state $|\Psi_0\rangle = \alpha_{00}|00\rangle + \alpha_{01}|01\rangle + \alpha_{10}|10\rangle + \alpha_{11}|11\rangle$ results in a state

$$U_{\text{CNOT}}|\Psi_0\rangle = \alpha_{00}|00\rangle + \alpha_{01}|01\rangle + \alpha_{10}|11\rangle + \alpha_{11}|10\rangle \tag{4}$$

Hence, the effect of a CNOT gate is equivalent to a conditional X gate: If the first qubit is in $|0\rangle$, then the second qubit remains intact; on the other hand, if the first qubit is in $|1\rangle$, then the second qubit is flipped. Generally, the first qubit is called the *control* and the second is the *target*.

In classical computing, an arbitrary mapping $\{0, 1\}^n \to \{0, 1\}^m$ can be executed by a sequence of basic gates such as AND, OR, NOT, and so on. Similarly in quantum computing, an arbitrary unitary transformation U can also be decomposed as a product of basic quantum gates. A complete set of such basic quantum gates is a *universal gate set*. For example, Hadamard, phase, CNOT, and $\pi/8$ gates form a universal gate set ([12], p. 194).

Now that we have introduced the concepts of qubits and quantum gates and compared them with their classical counterpart, we can see that they are the very building blocks of a quantum computer. However, it turns out that having qubits and executable universal gates is not enough for building a truly useful quantum computer that delivers its theoretical promises. So what does it really take to build such a quantum computer? A formal answer to this question is the following seven criteria proposed by DiVincenzo [41]:

- A scalable physical system with well-characterized qubits.
- The ability to initialize the state of the qubits to a simple fiducial state.
- Long (relative) decoherence times, much longer than the gate operation time.
- A universal set of quantum gates.
- A qubit-specific measurement capability.
- The ability to inter convert stationary and flying qubits.
- The ability to faithfully transmit flying qubits between specified locations.

For a detailed review of state-of-the-art experimental implementation based on the preceding criteria, refer to Ref. [42]. The take-home message is that it is clear that we can gain some advantage by storing, transmitting, and processing information encoded in systems that exhibit unique quantum properties, and a number of physical systems are currently being developed for quantum computation. However, it remains unclear which technology, if any, will ultimately prove successful in building a scalable quantum computer.

B. Circuits and Algorithms

Just as in classical computing, logic gates are cascaded to form a circuit. A *quantum circuit* is a sequence of quantum gates. When an algorithm needs to be implemented with a quantum computer, it must first be translated to a quantum circuit in order to be executed on the quantum hardware (qubits). Figure 1 is an example of a quantum circuit. Each horizontal line represents a qubit and every box on the line is a quantum gate applied on that qubit. If the box is connected with a vertical line that joins it to the line(s) with solid circles, then the box is a controlled gate operation and the qubit(s) that it is joined to are the control qubit(s). Just like a CNOT gate, only when the control qubit (s) is (are all) in $|1\rangle$ state will the controlled operation be applied onto the target qubit.

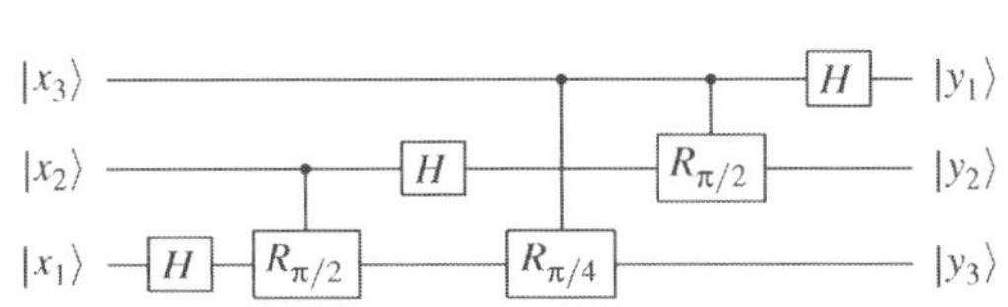

Figure 1. Quantum circuit for quantum Fourier transform on the quantum state $|x_1, x_2, x_3\rangle$. Starting from the left, the first gate is a Hadamard gate H that acts on the qubit in the state $|x_1\rangle$, and the second gate is a $|x_2\rangle$-controlled phase rotation $R_\theta(a|0\rangle + b|1\rangle) \rightarrow (a|0\rangle + be^{i\theta}|1\rangle)$ on qubit $|x_1\rangle$, where $\theta = \pi/2$. The rest of the circuit can be interpreted in the same fashion.

C. Teleportation

Quantum teleportation exploits some of the most basic and unique features of quantum mechanics, which is quantum entanglement, essentially implies an intriguing property that two quantum correlated systems cannot be considered independent even if they are far apart. The dream of teleportation is to be able to travel by simply reappearing at some distant location. Teleportation of a quantum state encompasses the complete transfer of information from one particle to another. The complete specification of a quantum state of a system generally requires an infinite amount of information, even for simple two-level systems (qubits). Moreover, the principles of quantum mechanics dictate that any measurement on a system immediately alters its state, while yielding at most one bit of information. The transfer of a state from one system to another (by performing measurements on the first and operations on the second) might therefore appear impossible. However, it was shown that the property of entanglement in quantum mechanics, in combination with classical communication, can be used to teleport quantum states. Although teleportation of large objects still remains a fantasy, quantum teleportation has become a laboratory reality for photons, electrons, and atoms [43–52].

More precisely, quantum teleportation is a quantum protocol by which the information on a qubit A is transmitted exactly (in principle) to another qubit B. This protocol requires a conventional communication channel capable of transmitting two classical bits, and an entangled pair (B, C) of qubits, with C at the location of origin with A and B at the destination. The protocol has three steps: measure A and C jointly to yield two classical bits; transmit the two bits to the other end of the channel; and use the two bits to select one of the four ways of recovering B [53,54].

Efficient long-distance quantum teleportation is crucial for quantum communication and quantum networking schemes. Ursin and coworkers [55] have performed a high-fidelity teleportation of photons over a distance of 600 m across the River Danube in Vienna, with the optimal efficiency that can be achieved using linear optics. Another exciting experiment in quantum communication has also been done with one photon that is measured locally at the Canary Island of La Palma, whereas the other is sent over an optical free-space link to Tenerife, where the Optical Ground Station of the European Space Agency acts as the receiver [55,56]. This exceeds previous free-space experiments by more than an order of magnitude in distance, and is an essential step toward future satellite-based quantum communication.

Recently, we have proposed a scheme for implementing quantum teleportation in a three-electron systems [52]. For more electrons, using Hubbard Hamiltonian, in the limit of the Coulomb repulsion parameter for electrons on the same site $U \to +\infty$, there is no double occupation in the magnetic field; the system is reduced to the Heisenberg model. The neighboring spins will favor the anti parallel configuration for the ground state. If the spin at one end is flipped, then the spins on the whole chain will be flipped accordingly due to the spin–spin correlation. Such that the spins at the two ends of the chain are entangled, a spin entanglement can be used for quantum teleportation, and the information can be transferred through the chain. This might be an exciting new direction for teleportation in molecular chains [57].

II. QUANTUM SIMULATION

A. Introduction

As already mentioned, simulating quantum systems by exact solution of the Schrödinger equation is a fundamentally hard task that the quantum chemistry community has been trying to tackle for decades with only approximate approaches. The key challenges of quantum simulation include the following (see next five chapters) [28]:

1. Isolate qubits in physical systems. For example, in a photonic quantum computer simulating a hydrogen molecule, the logical states $|0\rangle$ and $|1\rangle$ correspond to horizontal $|H\rangle$ and vertical $|V\rangle$ polarization states [58].
2. Represent the Hamiltonian H. This is to write H as a sum of Hermitian operators, each to be converted into unitary gates under the exponential map.
3. Prepare the states $|\psi\rangle$. By direct mapping, each qubit represents the fermionic occupation state of a particular orbital. Fock space of the system is mapped onto the Hilbert space of qubits.
4. Extract the energy E.
5. Read out the qubit states.

A technique to accomplish challenge 2 in a robust fashion is presented in Section II.B.2. Challenge 4 is accomplished using the phase estimation quantum algorithm (see details in Section II.B). Here, we can mention some examples of algorithms and their corresponding quantum circuits that have been implemented experimentally: (a) the IBM experiment, which factors the number 15 with nuclear magnetic resonance (NMR) (for details see Ref. [59]); (b) using quantum computers for quantum chemistry [58].

B. Phase Estimation Algorithm

The phase estimation algorithm (PEA) takes advantage of quantum Fourier transform ([12], see chapter by Gaitan and Nori) to estimate the phase φ in the eigenvalue $e^{2\pi i\varphi}$ of a unitary transformation U. For a detailed description of the algorithm, refer to Ref. [60]. The function that the algorithm serves can be summarized as the following: Let $|u\rangle$ be an eigenstate of the operator U with eigenvalue $e^{2\pi i\varphi}$. The algorithm starts with a two-register system (a *register* is simply a group of qubits) in the state $|0\rangle^{\otimes t}|u\rangle$. Suppose the transformation U^{2^k} can be efficiently performed for integer k, then this algorithm can efficiently obtain the state $|\tilde{\varphi}\rangle|u\rangle$, where $|\tilde{\varphi}\rangle$ accurately approximates φ to $t - \lceil\log(2 + \frac{1}{2\epsilon})\rceil$ bits with probability of at least $1 - \epsilon$.

1. General Formulation

The generic quantum circuit for implementing PEA is shown in Fig. 2. Section 5.2 of Ref. [12] presents a detailed account of how the circuit functions mathematically to yield the state $|\tilde{\varphi}\rangle$, which encodes the phase φ. Here we focus on its capability of finding the eigenvalues of a Hermitian matrix, which is of great importance in quantum chemistry where one often would like to find the energy spectrum of a Hamiltonian.

Suppose we let $U = e^{iAt_0/2^t}$ for some Hermitian matrix A, then $e^{iAt_0}|u_j\rangle = e^{i\lambda_j t}|u_j\rangle$, where λ_j and $|u_j\rangle$ are the j-th eigenvalue and eigenvector of matrix A. Furthermore, we replace the initial state $|u\rangle$ of register b (Fig. 2) with an arbitrary vector $|b\rangle$ that has a decomposition in the basis of the eigenvectors of A: $|b\rangle = \sum_j^n \beta_j|u_j\rangle$. Then the major steps of the algorithm can be summarized as the following.

1. Transform the t-qubit register C (Fig. 2) from $|0\rangle^{\otimes t}$ to $\frac{1}{\sqrt{2^t}}\sum_{\tau=0}^{2^t-1}|\tau\rangle$ state by applying Hadamard transform on each qubit in register C.
2. Apply the U^{2^k} gates to the register b, where each U^{2^k} gate is controlled by the $(k-1)$th qubit of the register C from bottom. This series of controlled operations transforms the state of the two-register system from $\frac{1}{\sqrt{2^t}}\sum_{\tau=0}^{2^t-1}|\tau\rangle \otimes |b\rangle$ to $\frac{1}{\sqrt{2^t}}\sum_{\tau=0}^{2^t-1}|\tau\rangle\sum_{j=1}^{n} e^{i\lambda_j\tau t/2^t}\beta_j|u_j\rangle$.
3. Apply inverse Fourier transform $\mathrm{FT}^\dagger$ to the register C. Because every basis state $|\tau\rangle$ will be transformed to $\frac{1}{\sqrt{2^t}}\sum_{k=0}^{2^t-1} e^{-2\pi i\tau k/2^t}|k\rangle$ by $\mathrm{FT}^\dagger$, the final

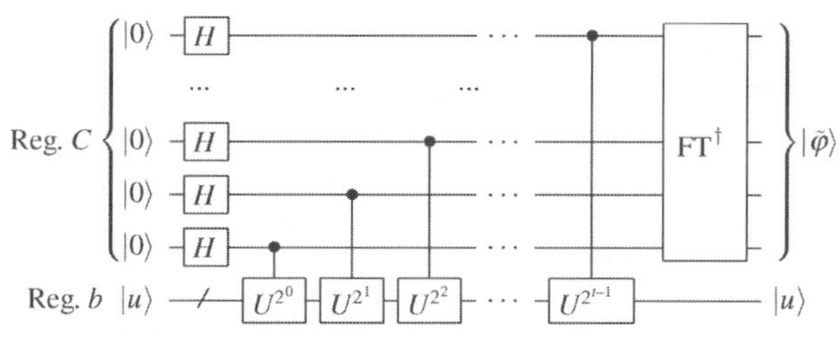

Figure 2. Schematic of the quantum circuit for phase estimation. The quantum wire with a "/" symbol represents a register of qubits as a whole. $\mathrm{FT}^\dagger$ represents inverse Fourier transform, whose circuit is fairly standard ([12], see chapter by Gaitan and Nori).

state of the PEA is proportional to $\sum_{k=0}^{2^t-1}\sum_{j=1}^{n} e^{i(\lambda_j t_0 - 2\pi k)\tau/2^t}\beta_j|k\rangle|u_j\rangle$. Due to a well-known property of the exponential sum, in which sums of the form $\sum_{k=0}^{N-1}\exp(2\pi ik\frac{r}{N})$ vanish unless $r = 0 \bmod N$, the values of k are concentrated on those whose value is equal or close to $\frac{t_0}{2\pi}\lambda_j$. If we let $t_0 = 2\pi$, the final state of system is $\sum_j \beta_j|\tilde{\lambda}_j\rangle|u_j\rangle$ up to a normalization constant.

In particular, if we prepare the initial state of register b to be one of matrix A's eigenvector $|u_i\rangle$, according to the procedure listed above, the final state of the system will become $|\tilde{\lambda}_i\rangle|u_i\rangle$ up to a constant. Hence, for any $|u_i\rangle$ that we can prepare, we can find the eigenvalue λ_j of A corresponding to $|u_i\rangle$ using a quantum computer. Most importantly, it has been shown that [17] quantum computers are able to solve the eigenvalue problem significantly more efficiently than classical computers.

2. *Implementation of Unitary Transformation U*

Phase estimation algorithm is often referred to as a *black box* algorithm because it assumes that the unitary transformation U and its arbitrary powers can be implemented with basic quantum gates. However, in many cases U has a structure that renders finding the exact decomposition $U = U_1U_2...U_m$ either impossible or very difficult. Therefore, we need a robust method for finding approximate circuit decompositions of unitary operators U with minimum cost and minimum fidelity error.

Inspired by the optimization nature of the circuit decomposition problem, Daskin and Kais [61,62] have developed an algorithm based on group leader optimization technique for finding a circuit decomposition $U = U_1U_2...U_m$ with minimum gate cost and fidelity error for a particular U. Hence, there are two factors that need to be optimized within the optimization: the error and the cost of the circuit. The costs of a one-qubit gate and a control gate (two-qubit gate) are defined as 1 and 2, respectively. Based on these two definitions, the costs of other quantum gates can be deduced. In general, the minimization of the error to an acceptable level is more important than the cost in order to get more reliable results in the optimization process. The circuit decompositions for $U = e^{iAt}$ presented in Fig. 3b for the particular instance of A in Eq. (6) are found by the algorithm such that the error $||U' - U||$ and the cost of U' are both minimized.

3. *Group Leaders Optimization Algorithm*

The group leaders optimization algorithm (GLOA) described in more detail in Refs [61,62] is a simple and effective global optimization algorithm that models the influence of leaders in social groups as an optimization tool. The algorithm starts with dividing the randomly generated solution population into several disjunct groups and assigning for each group a leader (the best candidate solution inside

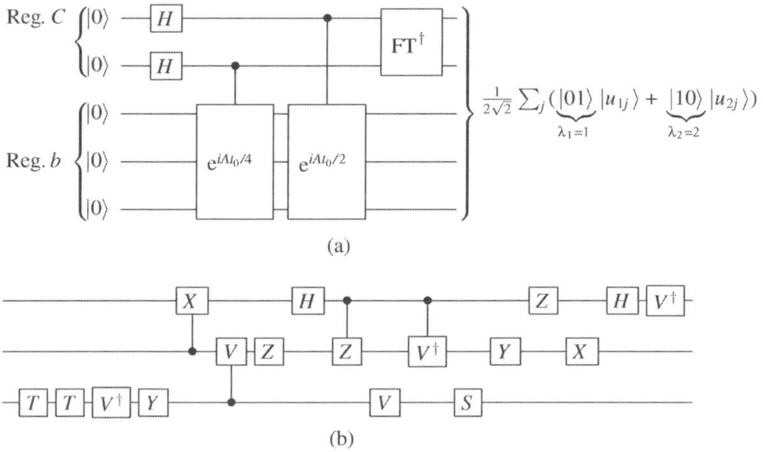

Figure 3. Quantum circuit for estimating the eigenvalues of A, which is the 8×8 matrix shown in Eq. (6). (**a**) The overall circuit implementing the phase estimation algorithm. (**b**) Decomposition of the gate $e^{iA\frac{t_0}{4}}$ in terms of basic gates.

the group). The algorithm basically is built on two parts: mutation and parameter transfer. In the mutation part, a candidate solution (a group member that is not leader) is mutated by using some part of its group leader, some random part, and some part of this member itself. This mutation is formulated as

$$\begin{aligned}\text{new member} &= r_1 \text{ part of the member}\\ &\cup\ r_2 \text{ part of its leader}\\ &\cup\ r_3 \text{ part of random}\end{aligned} \tag{5}$$

where r_1, r_2, and r_3 determine the rates of the member, the group leader, and the newly created random solution into the newly formed member, and they sum to 1. The values of these rates are assigned as $r_1 = 0.8$ and $r_2 = r_3 = 0.1$. The mutation for the values of the all angles in a numerical string is done according to the arithmetic expression: $\text{angle}_{\text{new}} = r_1 \times \text{angle}_{\text{old}} + r_2 \times \text{angle}_{\text{leader}} + r_3 \times \text{angle}_{\text{random}}$, where $\text{angle}_{\text{old}}$, the current value of an angle, is mutated: $\text{angle}_{\text{new}}$, the new value of the angle, is formed by combining a random value and the corresponding leader of the group of the angle and the current value of the angle with the coefficients r_1, r_2, and r_3. The mutation for the rest of the elements in the string means the replacement of its elements by the corresponding elements of the leader and a newly generated random string with the rates r_2 and r_3. In the second part of the algorithm, these disjoint groups communicate with each other by transferring some parts of their members. This step is called parameter transfer. In this process, some random part of a member is replaced with the equivalent part of a random member from a different group. The amount of this communication process is limited with some parameter that is set to $\frac{4\times \max_{\text{gates}}}{2} - 1$, where the numerator is the number of variables forming a numeric string in the optimization. During the optimization, the replacement criterion between a newly formed member is and an existing member is defined as follows: If a new member formed by a mutation or a parameter transfer operation gives less error-prone solution to the problem than the corresponding member, or they have the same error values but the cost of the new member is less than this member, then the new member takes the former one's place as a candidate solution; otherwise, the newly formed member is disregarded.

4. *Numerical Example*

In order to demonstrate how PEA finds the eigenvalues of a Hermitian matrix, here we present a numerical example. We choose A as a Hermitian matrix with the degenerate eigenvalues $\lambda_i = 1, 2$ and corresponding eigenvectors $|u_{11}\rangle = |+++\rangle$, $|u_{12}\rangle = |++-\rangle$, $|u_{13}\rangle = |+-+\rangle$, $|u_{14}\rangle = |-++\rangle$, $|u_{21}\rangle = |---\rangle$, $|u_{22}\rangle = |--+\rangle$, $|u_{23}\rangle = |-+-\rangle$, $|u_{24}\rangle = |+--\rangle$:

$$A = \begin{pmatrix} 1.5 & -0.25 & -0.25 & 0 & -0.25 & 0 & 0 & 0.25 \\ -0.25 & 1.5 & 0 & -0.25 & 0 & -0.25 & 0.25 & 0 \\ -0.25 & 0 & 1.5 & -0.25 & 0 & 0.25 & -0.25 & 0 \\ 0 & -0.25 & -0.25 & 1.5 & 0.25 & 0 & 0 & -0.25 \\ -0.25 & 0 & 0 & 0.25 & 1.5 & -0.25 & -0.25 & 0 \\ 0 & -0.25 & 0.25 & 0 & -0.25 & 1.5 & 0 & -0.25 \\ 0 & 0.25 & -0.25 & 0 & -0.25 & 0 & 1.5 & -0.25 \\ 0.25 & 0 & 0 & -0.25 & 0 & -0.25 & -0.25 & 1.5 \end{pmatrix} \tag{6}$$

Here $|+\rangle = \frac{1}{\sqrt{2}}(|0\rangle + |1\rangle)$ and $|-\rangle = \frac{1}{\sqrt{2}}(|0\rangle - |1\rangle)$ represent the Hadamard states. Furthermore, we let $\vec{b} = (1, 0, 0, 0, 0, 0, 0, 0)^{\mathrm{T}}$. Therefore, $|b\rangle = |000\rangle = \sum_j \beta_j |u_j\rangle$ and each $\beta_j = \frac{1}{2\sqrt{2}}$. Figure 3 shows the circuit for solving the 8×8 linear system. The register C is first initialized with Walsh–Hadamard transformation and then used as the control register for Hamiltonian simulation e^{iAt_0} on the register B. The decomposition of the two-qubit Hamiltonian simulation operators in terms of basic quantum circuits is achieved using group leader optimization algorithm [61,62]. The final state of system is $\sum_j \beta_j |\lambda_j\rangle |u_j\rangle = \frac{1}{2\sqrt{2}} \sum_{i=1}^{4} (|01\rangle |u_{1i}\rangle + |10\rangle |u_{2i}\rangle)$, which encodes both eigenvalues 1 (as $|01\rangle$) and 2 (as $|10\rangle$) in register C (Fig. 3).

5. *Simulation of the Water Molecule*

Wang et al.'s algorithm [19] can be used to obtain the energy spectrum of molecular systems such as water molecule based on the multiconfigurational self-consistent field (MCSCF) wave function. By using a MCSCF wave function as the initial guess, the excited states are accessible. The geometry used in the calculation is near the equilibrium geometry (OH distance $R = 1.8435a_0$ and the angle HOH = $110.57°$). With a complete active space type MCSCF method for the excited-state simulation, the CI space is composed of 18 CSFs, which requires the use of five qubits to represent the wave function. The unitary operator for this Hamiltonian can be formulated as

$$\hat{U}_{\mathrm{H_2O}} = \mathrm{e}^{i\tau(E_{\max}-\mathcal{H})t} \tag{7}$$

where τ is given as

$$\tau = \frac{2\pi}{E_{\max} - E_{\min}} \tag{8}$$

$E_{\max}$ and $E_{\min}$ are the expected maximum and minimum energies. The choice of $E_{\max}$ and $E_{\min}$ must cover all the eigenvalues of the Hamiltonian to obtain the correct results. After finding the phase ϕ_j from the phase estimation algorithm, the corresponding final energy E_j is found from the following expression:

$$E_j = E_{\max} - \frac{2\pi\phi_j}{\tau} \tag{9}$$

Because the eigenvalues of the Hamiltonian of the water molecule are between $-80 \pm \epsilon$ and $-84 \pm \epsilon$ ($\epsilon \leq 0.1$), taking $E_{\max} = 0$ and $E_{\min} = -200$ gives the following:

$$\hat{U} = \mathrm{e}^{\frac{-i2\pi H}{200}t} \tag{10}$$

Figure 4 shows the circuit diagram for this unitary operator generated by using the optimization algorithm and procedure as defined. The cost of the circuit is

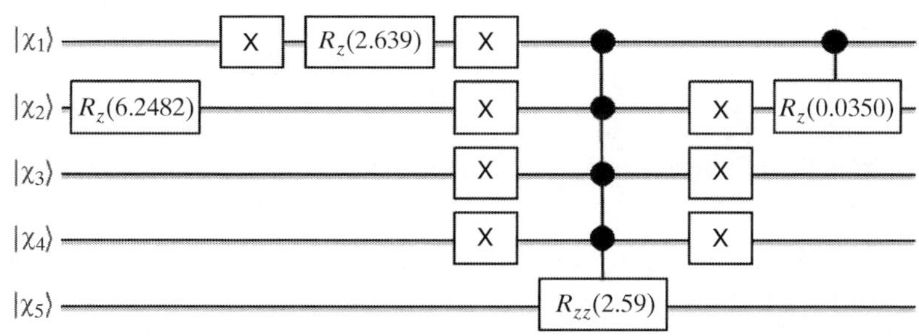

Figure 4. The circuit design for the unitary propagator of the water molecule.

TABLE III
Energy Eigenvalues of the Water Molecule

Phase	Found Energy	Exact Energy
0.4200	−84.0019	−84.0021
0.4200	−84.0019	−83.4492
0.4200	−84.0019	−83.0273
0.4200	−84.0019	−82.9374
0.4200	−84.0019	−82.7719
0.4200	−84.0019	−82.6496
0.4200	−84.0019	−82.5252
0.4200	−84.0019	−82.4467
0.4144	−82.8884	−82.3966
0.4144	−82.8884	−82.2957
0.4144	−82.8884	−82.0644
0.4144	−82.8884	−81.9872
0.4144	−82.8884	−81.8593
0.4144	−82.8884	−81.6527
0.4144	−82.8884	−81.4592
0.4144	−82.8884	−81.0119
0.4122	−82.4423	−80.9065
0.4122	−82.4423	−80.6703

44, which is found by summing up the cost of each gates in the circuit. Because we take $E_{\max}$ as zero, this deployment does not require any extra quantum gate for the implementation within the phase estimation algorithm. The simulation of this circuit within the iterative PEA results in the phase and energy eigenvalues given in Table III: The left two columns are, respectively, the computed phases and the corresponding energies, while the rightmost column of the matrix is the eigenvalues of the Hamiltonian of the water molecule (for each value of the phase, the PEA is run 20 times).

III. ALGORITHM FOR SOLVING LINEAR SYSTEMS $A\vec{x} = \vec{b}$

A. General Formulation

The algorithm solves a problem where we are given a Hermitian s-sparse $N \times N$ matrix A and a unit vector $\vec{b}$ (Fig. 5). Suppose we would like to find $\vec{x}$ such that $A\vec{x} = \vec{b}$. The algorithm can be summarized as the following major steps [29]:

1. Represent the vector $\vec{b}$ as a quantum state $|b\rangle = \sum_{i=1}^{N} b_i |i\rangle$ stored in a quantum register (termed *register b*). In a separate quantum register (termed *register C*) of t qubits, initialize the qubits by transforming the register to state $|\Psi\rangle^C$ from $|0\rangle$ up to error ϵ_Ψ.

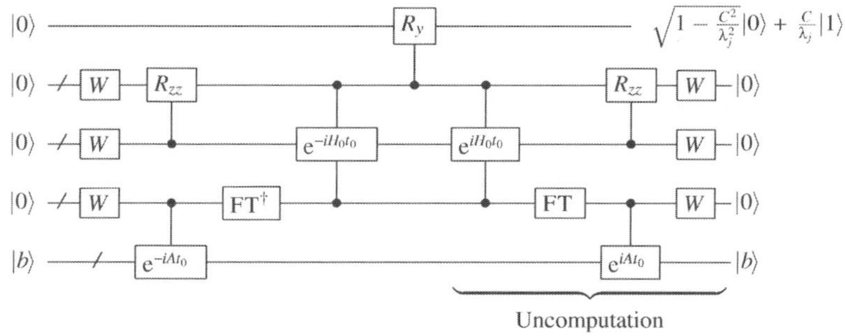

Figure 5. Generic quantum circuit for implementing the algorithm for solving linear systems of equations. The registers from the bottom of the circuit diagram upwards are, respectively, registers b, C, m, and l. The qubit on the top of the figure represents the ancilla bit.

2. Apply the conditional Hamiltonian evolution $\sum_{\tau=0}^{T-1} |\tau\rangle\langle\tau|^C \otimes e^{iA\tau t_0/T}$ up to error ϵ_H.
3. Apply quantum inverse Fourier transform to the register C. Denote the basis states after quantum Fourier transform as $|k\rangle$. At this stage in the superposition state of both registers, the amplitudes of the basis states are concentrated on k values that approximately satisfy $\lambda_k \approx \frac{2\pi k}{t_0}$, where λ_k is the kth eigenvalue of the matrix A.
4. Add an ancilla qubit and apply conditional rotation on it, controlled by the register C with $|k\rangle \approx |\lambda_k\rangle$. The rotation transforms the qubit to $\sqrt{1-\frac{C^2}{\lambda_j^2}}|0\rangle + \frac{C}{\lambda_j}|1\rangle$. This key step of the algorithm involves finding the reciprocal of the eigenvalue λ_j quantum mechanically, which is not a trivial task on its own. Now we assume that we have methods readily available to find the reciprocal of the eigenvalues of matrix A and store them in a quantum register.
5. Uncompute the register b and C by applying Fourier transform on register C followed by the complex conjugates of same conditional Hamiltonian evolution as in step 2 and Walsh–Hadamard transform as in the first step.
6. Measure the ancilla bit. If it returns 1, the register b of the system is in the state $\sum_{j=1}^{n} \beta_j {\lambda_j}^{-1}|u_j\rangle$ up to a normalization factor, which is equal to the solution $|x\rangle$ of the linear system $A\vec{x} = \vec{b}$. Here $|u_j\rangle$ represents the jth eigenvector of the matrix A and let $|b\rangle = \sum_{i=1}^{n} \beta_j |u_j\rangle$.

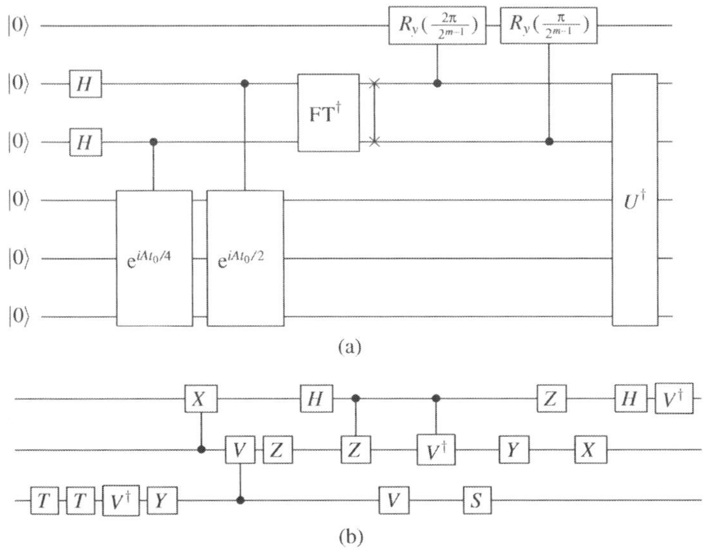

Figure 6. Quantum circuit for solving $A\vec{x} = \vec{b}$ with $A_{8\times 8}$ being the matrix shown in Eq. (6). (**a**) The overall circuit. From bottom up are the two qubits for $|b\rangle$, zeroth qubit in register C for encoding the eigenvalue, first qubit in register C for eigenvalue, and ancilla bit. $U^\dagger$ represents uncomputation. (**b**) Decomposition of the gate $e^{iA\frac{t_0}{4}}$ in terms of basic gates.

B. Numerical Example

For this example we choose A as the same Hermitian matrix as the one in Eq. (6). Furthermore, we let $\vec{b} = (1, 0, 0, 0, 0, 0, 0, 0)^T$. Therefore, $|b\rangle = |000\rangle = \sum_j \beta_j |u_j\rangle$ and each $\beta_j = \frac{1}{2\sqrt{2}}$. To compute the reciprocals of the eigenvalues, a quantum *swap* gate is used (Fig. 6) to exchange the values of the zeroth and first qubit. By exchanging the values of the qubits, one inverts an eigenvalue of A, say 1 (encoded with $|01\rangle$), to $|10\rangle$, which represents 2 in binary form. In the same way, the eigenvalue 2 ($|10\rangle$) can be inverted to 1 $|01\rangle$.

Figure 6 shows the circuit for solving the 8×8 linear system. The register C is first initialized with Walsh–Hadamard transformation and then used as the control register for Hamiltonian simulation e^{iAt_0} on the register B. The decomposition of the two-qubit Hamiltonian simulation operators in terms of basic quantum circuits is achieved using group leader optimization algorithm [61,62].

The final state of system, conditioned on obtaining $|1\rangle$ in the ancilla bit, is $\frac{1}{2\sqrt{10}}(6|000\rangle + |001\rangle + |010\rangle + |100\rangle - |111\rangle)$, which is proportional to the exact solution of the system $\vec{x} = (0.75, 0.125, 0.125, 0, 0.125, 0, 0, -0.125)^T$.

IV. ADIABATIC QUANTUM COMPUTING

The model of adiabatic quantum computation (AQC) was initially suggested by Farhi, Goldstone, Gutman, and Sisper [63] for solving some classical optimization problems. Several years after the proposition of AQC, Aharonov, Dam, Kempe, Landau, Lloyd, and Regev [64] assessed its computational power and established that the model of AQC is polynomially equivalent to the standard gate model quantum computation. Nevertheless, this model provides a completely different way of constructing quantum algorithms and reasoning about them. Therefore, it is seen as a promising approach for the discovery of substantially new quantum algorithms.

Prior to the work by Aharonov et al. [64], it had been known that AQC can be simulated by GMQC [65,66]. The equivalence between AQC and GMQC is then proven by showing that standard quantum computation can be efficiently simulated by adiabatic computation using 3-local Hamiltonians [64]. While the construction of three-particle Hamiltonians is sufficient for substantiating the theoretical results, it is technologically difficult to realize. Hence, significant efforts have been devoted to simplifying the universal form of Hamiltonian to render it feasible for physical implementation [64,67–70].

From the experimental perspective, current progress [71,72] in devices based on superconducting flux qubits has demonstrated the capability to implement Hamiltonian of the form $\Sigma_i h_i \sigma_i^z + \Sigma_i \Delta_i \sigma_i^x + \Sigma_{i,j} J_{ij} \sigma_i^z \sigma_j^z$. However, this is not sufficient for constructing a universal adiabatic quantum computer [71]. It is shown [68] that this Hamiltonian can be rendered universal by simply adding a tunable 2-local transverse $\sigma^x\sigma^x$ coupling. Once tunable $\sigma^x\sigma^x$ is available, all the other 2-local interactions such as $\sigma^z\sigma^x$ and $\sigma^x\sigma^z$ can be reduced to sums of single σ^x, σ^z spins and $\sigma^x\sigma^x$, $\sigma^z\sigma^z$ couplings via a technique called *Hamiltonian gadgets*. In Section IV.3, we will present a more detailed review of this subject.

A. Hamiltonians of n-Particle Systems

In the standard GMQC, the state of n qubits evolves in discrete time steps by unitary operations. Physically, however, the evolution is continuous and is governed by the *Schrödinger equation*: $-i\frac{\mathrm{d}}{\mathrm{d}t}|\psi(t)\rangle = H(t)|\psi(t)\rangle$, where $|\psi(t)\rangle$ is the state of n qubits at time t and $H(t)$ is a Hermitian $2^n \times 2^n$ matrix called the *Hamiltonian* operating on the n-qubit system; it governs the dynamics of the system. The fact that it is Hermitian derives from the unitary property of the discrete time evolution of the quantum state from t_1 to a later time t_2. In some context, the eigenvalues of Hamiltonians are referred to as *energy levels*. The *ground-state energy* of a Hamiltonian is its lowest eigenvalue and the corresponding eigenvector(s) are the *ground state(s)*. The *spectral gap* $\Delta(H)$ of a Hamiltonian H is defined as the difference between lowest eigenvalue of H and its second lowest eigenvalue.

We say that a Hamiltonian H is k-local if H can be written as $\sum_A H^A$ where A runs over all subsets of k particles. In other words, H^A is a tensor product of a Hamiltonian on A with identity on the particles outside A. Note that although a k-local Hamiltonian H operating on n qubits dwells in the Hilbert space of dimension 2^n, it can be described by $2^{2k}n^k = \text{poly}(n)$ numbers.

B. The Model of Adiabatic Computation

To perform useful computations, the model of AQC hinges on a well-known principle called adiabatic theorem [73,74]. Consider a system with a time-dependent Hamiltonian $H(s)$, where $s \in [0, 1]$ is the normalized time parameter. The system is initialized at $t = 0$ in the ground state of $H(0)$ (assuming that for any s the ground state of $H(s)$ is unique). Then we let the system evolve according to the Hamiltonian $H(t/T)$ from time $t = 0$ to T. We refer to such process as an *adiabatic evolution* according to H for time T. The adiabatic theorem ensures that for T sufficiently large, the final state of the system is very close to the ground state of $H(1)$. The minimum T required for this process is a function of the minimum spectral gap $\Delta(H(s))$, as is stated in the adiabatic theorem:

Theorem 1 (The Adiabatic Theorem [75]) *Let H_{init} and H_{final} be two Hamiltonians acting on a quantum system and consider the time-dependent Hamiltonian $H(s) := (1 - s)H_{\text{init}} + sH_{\text{final}}$. Assume that for all s, $H(s)$ has a unique ground state. Then for any fixed $\delta > 0$, if*

$$T \geq \Omega\left(\frac{||H_{\text{final}} - H_{\text{init}}||^{1+\delta}}{\epsilon^\delta \min_{s\in[0,1]}\{\Delta^{2+\delta}(H(s))\}}\right) \tag{11}$$

then the final state of an adiabatic evolution according to H for time T (with an approximate setting of global phase) is ϵ-close in l_2-norm to the ground state of H_{final}. The matrix norm is the spectral norm $||H|| := \max_w ||Hw||/||w||$.

Based on Eq. (11), a reasonable definition of the running time of the adiabatic algorithm is $T \cdot \max_s ||H(s)||$, because we must take into account the physical trade-off between time and energy [64]. (The solution to the Schrödinger equation remains the same if time is divided by some factor and at the same time the Hamiltonian is multiplied by the same factor.) Hence, in order to show that an adiabatic algorithm is efficient, it is enough to use Hamiltonian of at most poly(n) norm, and show that for all $s \in [0, 1]$ the spectral gap $\Delta(H(s))$ is at least inverse polynomial in n.

C. Hamiltonian Gadgets

A perturbation gadget or simply a gadget Hamiltonian refers to a Hamiltonian construction invented by Kempe, Kitaev, and Regev [67] first used to approximate the ground states of k-body Hamiltonians using the ground states of two-body Hamiltonians. Gadgets have been used and/or extended by several authors including Oliveira and Terhal [76–79]. Recent results have been reviewed in the article by Wolf [80]. Gadgets have a range of applications in quantum information theory, many-body physics, and are mathematically interesting to study in their own right. They have recently come to occupy a central role in the solution of several important and long-standing problems in quantum information theory. Kempe, Kitaev, and Regev [81] introduced these powerful tools to show that finding the ground state of a 2-local system (i.e., a system with at most two-body interactions) is in the same complexity class QMA as finding the ground-state energy of a system with k-local interactions. This was done by introducing a gadget that reduced 3-local to 2-local interactions. Oliveira and Terhal [76] exploited the use of gadgets to manipulate Hamiltonians acting on a 2D lattice. The work in [76,81] was instrumental in finding simple spin models with a QMA-complete ground-state energy problem [78]. Aside from complexity theory, Hamiltonian gadget constructions have important application in the area of adiabatic quantum computation [76,77,81]. The application of gadgets extends well beyond the scope mentioned here.

V. TOPOLOGICAL QUANTUM COMPUTING

Topological quantum computation seeks to exploit the emergent properties of many-particle systems to encode and manipulate quantum information in a manner that is resistant to error. This scheme of quantum computing supports the gate model of quantum computation. Quantum information is stored in states with multiple quasi particles called *anyons*, which have a topological degeneracy and are defined in the next section. The unitary gate operations that are necessary for quantum computation are carried out by performing *braiding* operation on the anyons and then measuring the multiquasipartite states. The fault tolerance of a topological quantum computer arises from the nonlocal encoding of the quasipartite states, which render them immune to errors by local perturbations.

A. Anyons

Two-dimensional systems are qualitatively different [82] from three-dimensional ones. In three-dimensional space, only two symmetries are possible: the time-dependent wave function of bosons is symmetric under exchange of particles while that of fermions is antisymmetric. However, in a two-dimensional case when two particles are interchanged twice in a clockwise manner, their trajectory in space-time involves a nontrivial winding, and the system does not necessarily come back to the same state. The first realization of this topological difference dates back to the 1980s [83,84] and it leads to a difference in the possible quantum mechanical properties for quantum systems when particles are confined to two dimensions. Suppose we have two identical particles in two dimensions. When one particle is exchanged in a counterclockwise manner with the other, the wave function $\psi(\vec{r}_1, \vec{r}_2)$ can change by an arbitrary phase: $\psi(\vec{r}_1, \vec{r}_2) \rightarrow e^{i\phi}\psi(\vec{r}_1, \vec{r}_2)$. The special cases where $\phi = 0, \pi$ correspond to bosons and fermions, respectively. Particles with other values of ϕ are called *anyons* [85].

B. Non-Abelian Braid Groups

In three-dimensional space, suppose we have N indistinguishable particles and we consider all possible trajectories in the space-time (or four-dimensional worldliness), which take these N particles from initial positions $\vec{r}_1, \vec{r}_2, \ldots, \vec{r}_N$ at time t_0 to final positions $\vec{r}'_1, \vec{r}'_2, \ldots, \vec{r}'_N$ at time t_f. Then the different trajectories fall into topological classes corresponding to the elements of the permutation group S_N, with each element specifying how the initial positions are permuted to obtain the final positions. The way the permutation group acts on the states of the system defines the quantum evolution of such a system. Fermions and bosons correspond to the only two one-dimensional irreducible representations of the permutation group of N identical particles.

In two-dimensional space, the topological classes of the trajectories that take these particles from initial positions $\vec{r}_1, \vec{r}_2, \ldots, \vec{r}_N$ at time t_0 to final positions $\vec{r}'_1, \vec{r}'_2, \ldots, \vec{r}'_N$ at time t_f are in one-to-one correspondence with the elements of the braid group $\mathcal{B}_N$. An element of the braid group can be visualized by considering the trajectories of particles as world lines in $(2 + 1)$-dimensional space-time originating at initial positions and terminating at final positions (Fig. 7a). The multiplication of two elements of the braid group is the successive execution of the corresponding trajectories (i.e., the vertical stacking of the two drawings). As shown in Fig. 7b, the order in which they are multiplied is important because the group is *non-abelian*, which means multiplication is not commutative. Algebraically, the braid group can be represented in terms of elementary braid operations, or generators σ_i, with $1 \leq i \leq N - 1$. σ_i is a counterclockwise exchange of the ith and $(i + 1)$th particles. σ_i^{-1} is therefore a clockwise exchange of the ith and the $(i + 1)$th particles. The σ_i's satisfy the defining relations (Fig. 7c)

$$\begin{aligned} \sigma_i\sigma_j &= \sigma_j\sigma_i && \text{if } |i - j| \geq 2 \\ \sigma_i\sigma_{i+1}\sigma_i &= \sigma_{i+1}\sigma_i\sigma_{i+1} && \text{if } 1 \leq i \leq n - 1 \end{aligned} \tag{12}$$

The richness of the braid group is that it supports quantum computation. To define the quantum evolution of a system, we specify how the braid group acts on the states of the system. An element of the braid group, say σ_1, which exchanges particles 1 and 2, is represented by a $g \times g$ unitary matrix $\rho(\sigma_1)$ acting on these

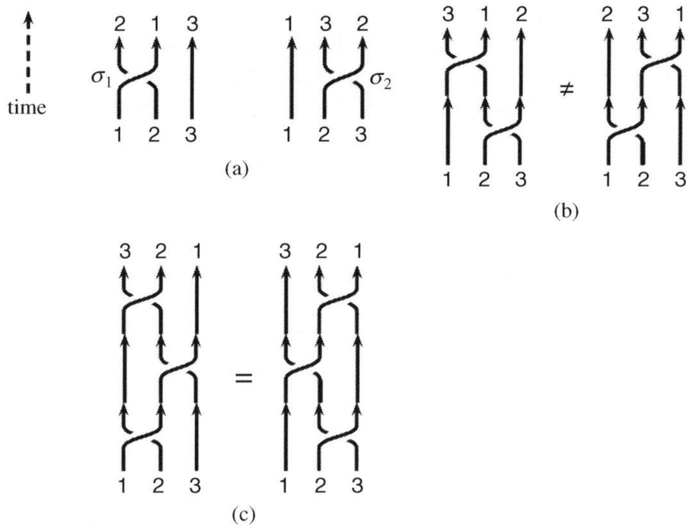

Figure 7. Graphical representation of elements of the braid group. (**a**) The two elementary braid operations σ_1 and σ_2 on three particles. (**b**) Because $\sigma_2\sigma_1 \neq \sigma_1\sigma_2$, the braid group is non-abelian. (**c**) The braid relation.

states,

$$\psi_\alpha \to [\rho(\sigma_1)]_{\alpha\beta}\psi_\beta \tag{13}$$

Clearly, if $\rho(\sigma_1)$ and $\rho(\sigma_2)$ do not commute, the particles obey *non-abelian braiding statistics*. In this case, braiding quasiparticles will cause nontrivial rotations within the degenerate many-quasiparticle Hilbert space. Furthermore, it will essentially be true at low energies that the only way to make nontrivial unitary operations on this degenerate space is by braiding quasiparticles around each other. Hence, no local perturbation can have nonzero matrix elements within this degenerate space.

C. Topological Phase of Matter

The non-abelian braiding statistics discussed in Section V.B indicates a theoretical possibility, but not any information on the occasions where such braiding statistics may arise in nature. Electrons, photons, and atoms are either fermions or bosons in two-dimensional space. However, if a system of many electrons (or bosons, atoms, etc.) confined to a two-dimensional plane has excitations that are localized disturbances of its quantum mechanical ground state, known as *quasiparticles*, then these quasiparticles can be anyons. When a system has anyonic quasiparticle excitations above its ground state, it is in a *topological phase of matter*.

Topological quantum computation is predicated on the existence in nature of topological phases of matter. Topological phases can be defined as (i) degenerate ground states, (ii) gap to local excitations, (iii) abelian or non-abelian quasiparticle excitations. Because these topological phases occur in many-particle physical systems, field theory techniques are often used to study these states. Hence, the definition of topological phase may be stated more compactly by simply saying that a system is in a topological phase if its low-energy effective field theory is a topological quantum field theory (TQFT), that is, a field theory whose correlation functions are invariant under diffeomorphisms. For a more detailed account of recent development in TQFT and topological materials, see the work of Vala and coworkers (see chapter by Watts et al.).

D. Quantum Computation Using Anyons

The braiding operation $\rho(\sigma_i)$ defined in Eq. (13) can be cascaded to perform quantum gate operation. For example, Georgiev [86] showed that a CNOT gate can be executed on a six-quasiparticle system (for details refer to Ref. [86] or a comprehensive review in Ref. [87]):

$$\rho(\sigma_3^{-1}\sigma_4\sigma_3\sigma_1\sigma_5\sigma_4\sigma_3^{-1}) = \begin{pmatrix} 1 & 0 & 0 & 0 \\ 0 & 1 & 0 & 0 \\ 0 & 0 & 0 & 1 \\ 0 & 0 & 1 & 0 \end{pmatrix} \tag{14}$$

In the construction by Georgiev, quasiparticles 1 and 2 are combined to be qubit 1 via an operation called *fusion*. Similarly, quasiparticles 5 and 6 are combined to be qubit 2. We will not be concerned about the details of fusion in this introduction. Interested readers can refer to Ref. [87], Section II for more information. Apart from CNOT, single-qubit gates are also needed for universal quantum computation. One way to implement those single-qubit gates is to use nontopological operations. More details on this topic can be found in Ref. [88].

VI. ENTANGLEMENT

The concept of *entanglement* can be defined based on a postulate of quantum mechanics. The postulate states that the state space of a composite physical system is the tensor product of the states of the component physical systems. Moreover, if we have systems numbered 1 through n, and the system number i is prepared in the state $|\psi_i\rangle$, then the joint state of the total system is $|\psi\rangle = |\psi_1\rangle \otimes |\psi_2\rangle \otimes \ldots \otimes |\psi_n\rangle$. However, in some cases $|\psi\rangle$ cannot be written in the form of a tensor product of states of individual component systems. For example, the well-known Bell state or EPR pair (named after Einstein, Podolsky, and Rosen for their initial proposition of this state [89]) $|\psi\rangle = \frac{1}{\sqrt{2}}(|00\rangle + |11\rangle)$ cannot be written in form of $|a\rangle \otimes |b\rangle$. States such as this are called *entangled states*. The opposite case is a *disentangled state*.

In fact, entanglement in a state is a physical property that should be quantified mathematically, which leads to a question of defining a proper expression for calculating entanglement. Various definitions have been proposed to mathematically quantify entanglement (for a detailed review see Refs. [5,90]). One of the most commonly used measurement for pairwise entanglement is *concurrence* $C(\rho)$, where ρ is the density matrix of the state. This definition was proposed by Wootters [91]. The procedure for calculating concurrence is the following (for more examples illustrating how to calculate concurrence in detail, refer to Refs. [5,92,93]):

- Construct the density matrix ρ.
- Construct the flipped density matrix, $\tilde{\rho} = (\sigma_y \otimes \sigma_y)\rho^*(\sigma_y \otimes \sigma_y)$.
- Construct the product matrix $\rho\tilde{\rho}$.
- Find eigenvalues λ_1, λ_2, λ_3, and λ_4 of $\rho\tilde{\rho}$.
- Calculate concurrence from square roots of eigenvalues via

$$C = \max\left(0, \sqrt{\lambda_1} - \sqrt{\lambda_2} - \sqrt{\lambda_3} - \sqrt{\lambda_4}\right) \tag{15}$$

Physically, $C = 0$ means no entanglement in the two-qubit state ρ and $C = 1$ represents maximum entanglement. Therefore, any state ρ with $C(\rho) > 0$ is an entangled state. An entangled state has many surprising properties. For example, if one measures the first qubit of the EPR pair, two possible results are obtained: 0 with probability $1/2$, where postmeasurement the state becomes $|00\rangle$, and 1 with probability $1/2$, where postmeasurement the state becomes $|11\rangle$. Hence, a measurement of the second qubit always gives the same result as that of the first qubit. The measurement outcomes on the two qubits are therefore correlated in some sense. After the initial proposition by EPR, John Bell [94] proved that the measurement correlation in the EPR pair is stronger than could ever exist between classical systems. These results (refer to Ref. [12], Section 2.6 for details) were the first indication that laws of quantum mechanics support computation beyond what is possible in the classical world.

Entanglement has also been used to measure interaction and correlation in quantum systems. In quantum chemistry, the correlation energy is defined as the difference between the Hartree–Fock limit energy and the exact solution of the nonrelativistic Schrödinger equation. Other measures of electron correlation exist, such as the statistical correlation coefficients [95] and, more recently, the Shannon entropy [96]. Electron correlations strongly influence many atomic, molecular, and solid properties. Recovering the correlation energy for large systems remains one of the most challenging problems in quantum chemistry. We have used the entanglement as a measure of the electron–electron correlation [5,92,93] and show that the configuration interaction (CI) wave function violates the Bell inequality [97]. Entanglement is directly observable via macroscopic observable called entanglement witnesses [98]. Of particular interest is how entanglement plays a role in conical intersections and Landau–Zener tunneling and whether ideas from quantum information such as teleportation can be used to understand spin correlations in molecules [99].

Since the original proposal by DeMille [100], arrays of ultracold polar molecules have been counted among the most promising platforms for the implementation of a quantum computer [101–103]. The qubit of such an array is realized by a single dipolar molecule entangled via its dipole–dipole interaction with the rest of the array's molecules. Polar molecule arrays appear as scalable to a large number of qubits as neutral atom arrays do, but the dipole–dipole interaction furnished by polar molecules offers a faster entanglement, one resembling that mediated by the Coulomb interaction for ions. At the same time, cold and trapped polar molecules exhibit similar coherence times as those encountered for trapped atoms or ions. The first proposed complete scheme for quantum computing with polar molecules was based on an ensemble of ultracold polar molecules trapped in a one-dimensional optical lattice, combined with an inhomogeneous electrostatic field. Such qubits are individually addressable, thanks to the Stark effect, which is different for each qubit in the inhomogeneous electric field. In collaboration with Wei, Friedrich, and Herschbach [93], we have evaluated entanglement of the pendular qubit states for two linear dipoles, characterized by pairwise concurrence, as a function of the molecular dipole moment and rotational constant, strengths of the external field and the dipole–dipole coupling, and ambient temperature. We also evaluated a key frequency shift, $\delta\omega$, produced by the dipole–dipole interaction. Under conditions envisioned for the proposed quantum computers, both the concurrence and $\delta\omega$ become very small for the ground eigenstate. In principle, such weak entanglement can be sufficient for operation of logic gates, provided the resolution is high enough to detect the $\delta\omega$ shift unambiguously. In practice, however, for many candidate polar molecules it appears a challenging task to attain adequate resolution. Overcoming this challenge, small $\delta\omega$ shift, will be a major contribution to implementation of the DeMille proposal. Moreover, it will open the door for designing quantum logical gate: one-qubit gate (such as the rotational X, Y, Z gates and the Hadamard gate) and two-qubit quantum gates (such as the CNOT gate) for molecular dipole arrays. The operation of a quantum gate [104] such as CNOT requires that manipulation of one qubit (target) depends on the state of another qubit (control). This might be characterized by the shift in the frequency for transition between the target qubit states when the control qubit state is changed. The shift must be kept smaller than the differences required to distinguish among addresses of qubit sites. In order to implement the requisite quantum gates, one might use algorithmic schemes of quantum control theory for molecular quantum gates developed by de Vivie-Riedle and coworkers [105–107], Herschel Rabitz, and others [108–112].

Recent experimental discoveries in various phenomena have provided further evidence of the existence of entanglement in nature. For example, photosynthesis is one of the most common phenomena in nature. Recent experimental results show that long-lived quantum entanglement are present in various photosynthetic complexes [113–115]. One such protein complex, the Fenna–Matthews–Olson (FMO) complex from green sulfur bacteria [116,117], has attracted considerable experimental and theoretical attention due to its intermediate role in energy transport. The FMO complex plays the role of a molecular wire, transferring the excitation energy from the light-harvesting complex (LHC) to the reaction center (RC) [118–120]. Long-lasting quantum beating over a timescale of hundreds of femtoseconds has been observed [121,122]. The theoretical framework for modeling this phenomenon has also been explored intensively by many authors [123–145].

The FMO complex, considered as an assembly of seven chromophores, is a multipartite quantum system. As such, useful information about quantum correlations is obtained by computing the bipartite entanglement across any of the cuts that divide the seven chromophores into two subsystems, seen in Fig. 8. Similarly, if we take the state of any subsystem of the FMO complex, we can compute the entanglement across any cut of the reduced state of that subsystem [128].

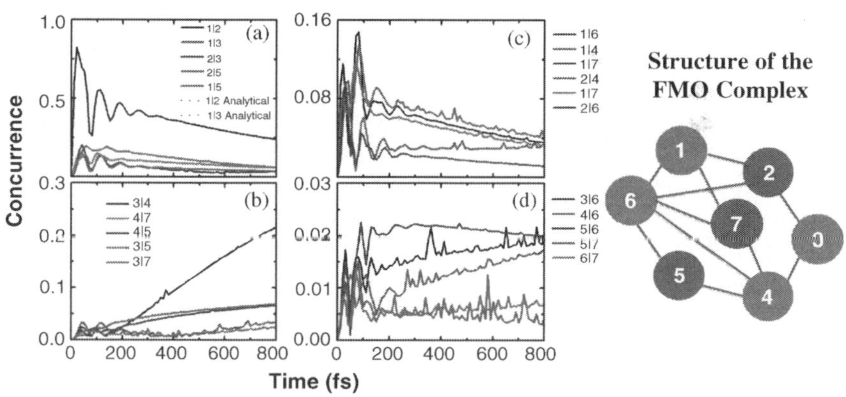

Figure 8. The quantum entanglement evolution for the pairwise entanglement in the FMO complex with the site 1 initially excited. The left panel shows the entanglement with all pairs. Based the amplitude of the concurrence, all pairs had been divided into four groups, from the largest pairs (**a**) to the smallest pairs (**d**). The solid line indicates the entanglement computed via the convex roof approach [147], while the dotted line shows the evolution calculated through the concurrence method. The right panel is the geometry structure of the FMO complex. (*See the color version of this figure in the Color Plates section.*)

Pairwise entanglement plays the dominant entanglement role in the FMO complex. Because of the saturation of the monogamy bounds, the entanglement of any chromophore with any subset of the other chromophores is completely determined by the set of pairwise entanglements. For the simulations in which site 1 is initially excited, the dominant pair is sites 1 and 2, while in the cases where 6 is initially excited sites 5 and 6 are most entangled. This indicates that entanglement is dominant in the early stages of exciton transport, when the exciton is initially delocalized away from the injection site. In addition, we observe that the entanglement mainly happens among the sites involved in the pathway. For the site 1 initially excited case, the entanglement of sites 5, 6, and 7 is relatively small compared with the domain pairs.

Although the final state is the same for both initial conditions, the role of sites 3 and 4 during the time evolution is different. For the initial condition where site 1 is excited, the entanglement is transferred to site 3 and then from site 3 to site 4. While for the site 6 initially excited case, sites 4 and 5 first become entangled with site 6 and then sites 3 and 4 become entangled. This is due to the fact that site 3 has strong coupling with sites 1 and 2, while site 4 is coupled more strongly to sites 5, 6, and 7. The initial condition plays an important role in the entanglement evolution, the entanglement decays faster for the cases where site 6 is initially excited compared with cases where the site 1 is initially excited. Increasing the temperature unsurprisingly reduces the amplitude of the entanglement and also decreases the time for the system to go to thermal equilibrium. Recently [146], using the same formalism, we have calculated the pairwise entanglement for the LH2 complex, as seen in Fig. 9.

Apart from photosynthesis, other intriguing possibilities that living systems may use nontrivial quantum effects to optimize some tasks have been raised, such as natural selection [148] and magnetoreception in birds [149]. In particular, *magnetoreception* refers to the ability to sense characteristics of the surrounding magnetic field. There are several mechanisms by which this sense may operate [150]. In certain species, the evidence supports a mechanism called *radical pair* (RP). This process involves the quantum evolution of a spatially separated pair of electron spins [151,152]. The basic idea of the RP model is that there are molecular structures in the bird's eye that can each absorb an optical photon and give rise to a spatially separated electron pair in a singlet spin state. Because of the different local environments of the two electron spins, a singlet–triplet evolution occurs. This evolution depends on the inclination of the molecule with respect to Earth's magnetic field. Recombination occurs from either the singlet or triplet state, leading to different chemical products. The concentration of these products constitutes a chemical signal correlated to Earth's field orientation. Such a model is supported by several results from the field of spin chemistry [153–156]. An artificial chemical compass operating according to this principle has been demonstrated experimentally [157], and the presence of entanglement has been examined by a theoretical study [158].

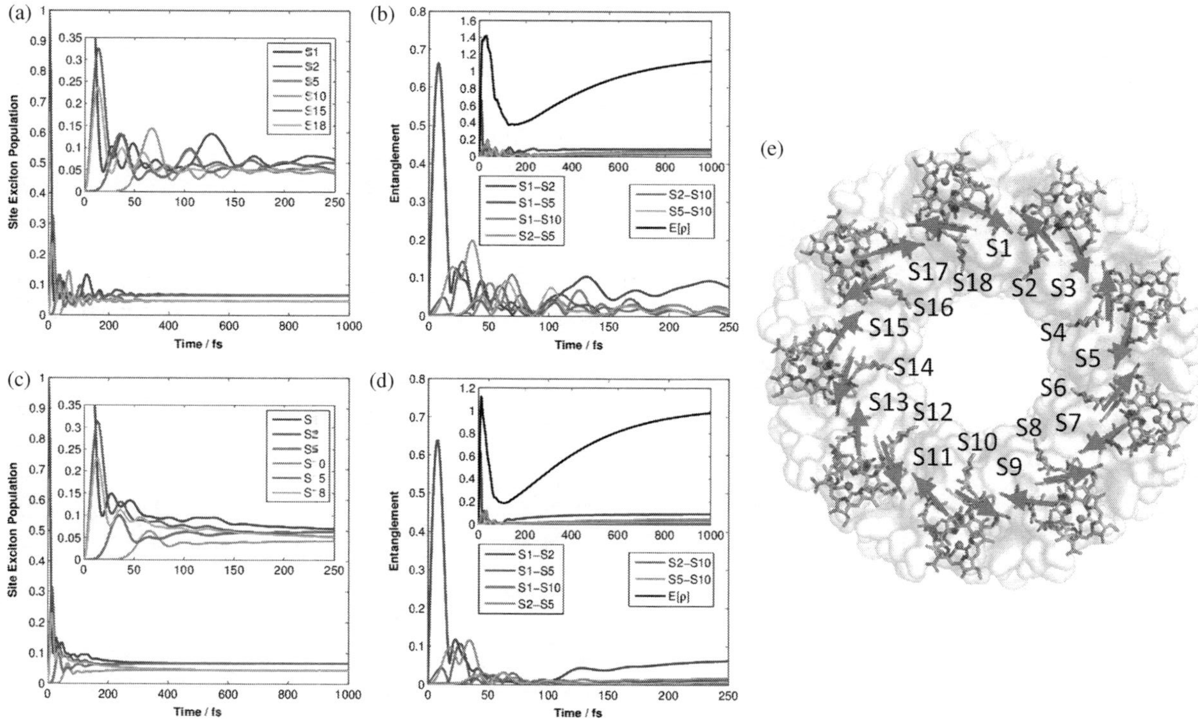

Figure 9. (**a**) The excitation population dynamics of LH2 B850 18 sites at 77 K. (**b**) The concurrence and global entanglement (black line in the inset) in the 18 sites at 77 K. (**c**) The excitation population dynamics of LH2 B850 18 sites at 300 K. (**d**) The concurrence and global entanglement (black line in the inset) in the 18 sites at 300 K. (**e**) Top view (from cytoplasmic to periplasmic side) of 2.0 Å resolution LH2 structure of purple bacteria (*Rhodopseudomonas acidophila*). The 9 B800 and 18 B850 bacteriochlorophylls (BChls) are shown in green and cyan, respectively. The red arrows represent the transition dipoles of B850 BCHls Q_Y excitation. Top view (from cytoplasmic to periplasmic side) of 2.0 Å resolution LH2 structure of purple bacteria. The 9 B800 and 18 B850 bacteriochlorophylls (BChls) are shown in green and cyan, respectively. The red arrows represent the transition dipoles of B850 BCHls Q_Y excitation [146]. (*See the color version of this figure in the Color Plates section.*)

Not only is entanglement found in nature, but it also plays a central role in the internal working of a quantum computer. Consider a system of two qubits 1 and 2. Initially, qubit 1 is in the state $|\psi_1\rangle = |+\rangle = \frac{1}{\sqrt{2}}(|0\rangle + |1\rangle)$ and qubit 2 is in the state $|\psi_2\rangle = |0\rangle$. Hence, the initial state of the system can be written as $|\Psi_0\rangle = |\psi_1\rangle \otimes |\psi_2\rangle = \frac{1}{\sqrt{2}}(|00\rangle + |10\rangle)$. Now apply a CNOT on qubits 1 and 2 with qubit 1 as the control and qubit 2 as the target. By definition of CNOT gate in Section I.A, the resulting state is $\frac{1}{\sqrt{2}}(|00\rangle + |11\rangle)$, which is the EPR pair. Note that the two qubits are initially disentangled. It is due to the CNOT operation that the two qubits are in an entangled state. Mathematically U_{CNOT} cannot be represented as $A \otimes B$, because if it could, $(A \otimes B)(|\psi_1\rangle \otimes |\psi_2\rangle) = (A|\psi_1\rangle \otimes B|\psi_2\rangle)$ would still yield a disentangled state.

CNOT gate is indispensable for any set of universal quantum gates. Therefore, if a particular computation process involves n qubits that are initially disentangled, most likely all the n qubits need to be entangled at some point of the computation. This poses a great challenge for experimentalists because in order to keep a large number of qubits entangled for an extended period of time, a major issue needs to be resolved—decoherence.

VII. DECOHERENCE

So far in our discussion of qubits, gates, and algorithms we assume the ideal situation where the quantum computer is perfectly isolated when performing computation. However, in reality this is not feasible because there is always interaction between the quantum computer and its environment, and if one would like to read any information from the final state of the quantum computer, the system has to be open to measurements at least at that point. Therefore, a quantum computer is in fact constantly subject to environmental noise, which corrupts its desired evolution. Such unwanted interaction between the quantum computer and its environment is called *decoherence*, or *quantum noise*.

Decoherence has been a main obstacle to building a quantum computing device. Over the years, various ways of suppressing quantum decoherence have been explored [159]. The three main error correction strategies for counteracting the errors induced by the coupling with the environment include the following (see chapter by Lidar):

- Quantum error correction codes (QECCs), which uses redundancy and an active measurement and recovery scheme to correct errors that occur during a computation [160–164]
- Decoherence-free subspaces (DFSs) and noiseless subsystems, which rely on symmetric system–bath interactions to end encodings that are immune to decoherence effects [165–170]

- Dynamical decoupling, or "bang-bang" (BB) operations, which are strong and fast pulses that suppress errors by averaging them away [171–175]

Of these error correction techniques, QECCs require at least a five physical qubit to one logical qubit encoding [163] (neglecting ancillas required for fault-tolerant recovery) in order to correct a general single-qubit error [162]. DFSs also require extra qubits and are most effective for collective errors, or errors where multiple qubits are coupled to the same bath mode [170]. The minimal encoding for a single qubit undergoing collective decoherence is three physical qubits to one logical qubit [168]. The BB control method requires a complete set of pulses to be implemented within the correlation time of the bath [171]. However, it does not necessarily require extra qubits.

Although the ambitious technological quest for a quantum computer has faced various challenges, as of now no fundamental physical principle has been found to prevent a scalable quantum computer from being built. Therefore, if we keep this prospect, one day we will be able to maintain entanglement and overcome decoherence to a degree such that scalable quantum computers become reality.

VIII. MAJOR CHALLENGES AND OPPORTUNITIES

Many of the researchers in the quantum information field have recognized quantum chemistry as one of the early applications of quantum computing devices. This recognition is reflected in the document "A Federal Vision for Quantum Information Science," published by the Office of Science and Technology Policy (OSTP) of the White House. As mentioned earlier, the development and use of quantum computers for chemical applications has potential for revolutionary impact on the way computing is done in the future. Major challenges and opportunities are abundant; examples include developing and implementing quantum algorithms for solving chemical problems thought to be intractable for classical computers. To perform such quantum calculations, it will be necessary to overcome many challenges in experimental quantum simulation. New methods to suppress errors due to faulty controls and noisy environments will be required. These new techniques would become part of a quantum compiler that translates complex chemical problems into quantum algorithms. Other challenges include the role of quantum entanglement, coherence, and superposition in photosynthesis and complex chemical reactions.

Many exciting opportunities for science and innovation can be found in this new area of quantum information for chemistry including topological quantum computing (see chapter by Watts et al.), improving classical quantum chemistry methods (see chapter by Kinder et al.), quantum error corrections (see chapter by Lidar), quantum annealing with D-Wave machine [176], quantum algorithms for solving linear systems of equations, and entanglement in complex biological systems (see chapter by Mazziotti and Skochdopole). With its 128 qubits, the

D-Wave computer at USC (the DW-1) is the largest quantum information processor built to date. This computer will grow to 512 qubits (the chip currently being calibrated and tested by D-Wave Inc. (D. A. Lidar, private communication, USC, 2012)), at which point we may be on the cusp of demonstrating, for the first time in history, a quantum speedup over classical computation. Various applications of DW-1 in chemistry include, for example, the applications in cheminformatics (both in solar materials and in drug design) and in lattice protein folding, and the solution of the Poisson equation and its applications in several fields. While the DW-1 is not a universal quantum computer, it is designed to solve an important and broad class of optimization problems—essentially any problem that can be mapped to the NP-hard problem of finding the ground state of a general planar Ising model in a magnetic field.

This chapter focused mainly on the theoretical aspects of quantum information and computation for quantum chemistry and represents a partial overview of the ongoing research in this field. However, a number of experimental groups are working to explore quantum information and computation in chemistry: using ion trap (see chapter by Merrill and Brown) [177], NMR (see chapter by Criger et al.) [178–182], trapped molecules in optical lattice (see chapter by Côté) [183], molecular states (see chapter by Gollub et al.) [184,185], and optical quantum computing platforms (see chapter by Ma et al.) [186,187], to name just a few. For example, Brown and coworkers [177] proposed a method for laser cooling of the AlH^+ and BH molecules. One challenge of laser cooling molecules is the accurate determination of spectral lines to 10^{-5} cm^{-1}. In their work, the authors show that the lines can be accurately determined using quantum logic spectroscopy and sympathetic heating spectroscopy techniques, which were derived from quantum information experiments. Also, Dorner and coworkers [188] perform the simplest double-slit experiment to understand the role of interference and entanglement in the double photoionization of H_2 molecule. Moreover, many more areas of overlap have not been reviewed in detail here, notably coherent quantum control of atoms and molecules. However, we hope this chapter provides a useful introduction to the current research directions in quantum information and computation for chemistry.

Finally, I would like to end this chapter by quoting Jonathan Dowling [189]: "We are currently in the midst of a second quantum revolution. The first one gave us new rules that govern physical reality. The second one will take these rules and use them to develop new technologies. Today there is a large ongoing international effort to build a quantum computer in a wide range of different physical systems. Suppose we build one, can chemistry benefit from such a computer? The answer is a resounding YES!"

Acknowledgments

I would like thank Birgitta Whaley for her critical reading of this chapter and my students Yudong Cao, Anmer Daskin, Jing Zhu, Yeh Shuhao, Qi Wei, Qing Xu, and Ross Hoehn for their major contributions to the materials discussed in this chapter. Also, I benefited from collaborating and discussing different aspects of this chapter with my colleagues: Alan Aspuru-Guzik, Daniel Lidar, Ken Brown, Peter Love, V. Ara Apkarian, Anargyros Papageorgiou, Bretislav Friedrich, Uri Peskin, Gehad Sadiek, and Dudley Herschbach.

This work is supported by NSF Center for Quantum Information for Chemistry (QIQC), http://web.ics.purdue.edu/ kais/qc/ (award number CHE-1037992).

REFERENCES

1. M. Sarovar, A. Ishizaki, G. R. Fleming, and K. B. Whaley, *Nat. Phys.* **6**, 462 (2010).
2. K. B. Whaley, *Nat. Phys.* **8**, 10 (2012).
3. I. Kassal, J. D. Whitfield, A. Perdomo-Ortiz, M. Yung, and A. Aspuru-Guzik, *Annu. Rev. Phys. Chem.* **62**, 185 (2011).
4. W. J. Kuo and D. A. Lidar, *Phys. Rev. A* **84**, 042329 (2011).
5. S. Kais, in *Reduced-Density-Matrix Mechanics: With Application to Many-Electron Atoms and Molecules, Advances in Chemical Physics*, Vol. 134, D. A. Mazziotti, ed., Wiley-Interscience, 2007, p. 493.
6. S. Lloyd, *Science* **319**, 1209 (2008).
7. J. Elbaz, O. Lioubashevski, F. Wang, F. Remacle, R. D. Levine, and I. Willner. *Nat. Nanotechnol.* **5**, 417–422 (2011).
8. R. P. Feynmann, *Int. J. Theor. Phys.* **21**(6–7), 467–488 (2008).
9. D. Deutsch, *Proc. R. Soc. Lond.* **400**, 97–117 (1985).
10. P. Benioff, *J. Stat. Phys.* **29**, 515–546 (1982).
11. C. H. Bennett and G. Brassard, *Proceedings of IEEE International Conference on Computers Systems and Signal Processing*, 1984, pp. 175–179.
12. M. A. Nielsen and I. L. Chuang, *Quantum Computation and Quantum Information*, Cambridge University Press, Cambridge, UK, 2000.
13. P. W. Shor, in *Proceedings of the 35th Annual Symposium on Foundations of Computer Science*, S. Goldwasser, ed., IEEE Computer Society Press, New York, 1994, pp. 124–134.
14. S. Lloyd, *Science* **273**, 1073–1078 (1996).
15. D. Abrams and S. Lloyd, *Phys. Rev. Lett.* **83**(24), 5162–5165 (1999).
16. D. Abrams and S. Lloyd, *Phys. Rev. Lett.* **79**(13), 2586–2589 (1997).
17. A. Papageorgiou, I. Petras, J. F. Traub, and C. Zhang, *Mathematics of Computation* **82**, 2293–2304 (2013).
18. A. Papageorgiou and C. Zhang, arXiv:1005.1318v3 (2010).
19. H. Wang, S. Kais, A. Aspuru-Guzik, and M. R. Hoffmann, *Phys. Chem. Chem. Phys.* **10**, 5388–5393 (2008).
20. A. Aspuru-Guzik, A. D. Dutoi, P. J. Love, and M. Head-Gordon, *Science* **379**(5741), 1704–1707 (2005).
21. D. Lidar and H. Wang, *Phys. Rev. E* **59**, 2429 (1999).
22. H. Wang, S. Ashhab, and F. Nori, *Phys. Rev. A* **85**, 062304 (2012).
23. J. Q. You and F. Nori, *Nature* **474**, 589 (2011).

24. J. Q. You and F. Nori, *Phys. Today* **58**(11), 42–47 (2005).
25. I. Buluta and F. Nori, *Science* **326**, 108 (2009).
26. J. P. Dowling, *Nature* **439**, 919 (2006).
27. L. Veis and J. Pittner, *J. Chem. Phys.* **133**, 194106 (2010).
28. L. Veis, J. Visnak, T. Fleig, S. Knecht, T. Saue, L. Visscher, and J. Pittner, arXiv:1111.3490v1 (2011).
29. A. W. Harrow, A. Hassidim, and S. Lloyd, *Phys. Rev. Lett.* **15**, 150502 (2009).
30. A. Klappenecker and M. Roetteler, arXiv:quant-ph/0111038 (2001).
31. P. A. M. Dirac, *Proc. Roy. Soc.* **123**, 714 (1929).
32. S. Aaronson, *Nat. Phys.* **5**, 707 (2009).
33. N. Schuch and F. Verstraete, *Nat. Phys.* **5**, 732 (2009).
34. A. Szabo and N. S. Ostlund, *Modern Quantum Chemistry—Introduction to Advanced Electronic Structure Theory*, Dover Publications Inc., Mineola, NY, 1982.
35. R. G. Parr and W. Yang, *Density-Functional Theory of Atoms and Molecules*, Oxford Science Publications, New York, 1989.
36. D. A. Mazziotti, *Adv. Chem. Phys.* **134**, 1 (2007).
37. I. Iachello and R. D. Levine, *Algebraic Theory of Molecules*, Oxford Univesity Press, 1995.
38. M. P. Nightingale and C. J. Umrigar, *Quantum Monte Carlo Methods in Physics and Chemistry*, *NATO Science Series*, Vol. 525, Springer, 1998.
39. D. R. Herschbach, J. Avery, and O. Goscinski, *Dimensional Scaling in Chemical Physics*, Kluwer Academic Publishers, Dordrecht, The Netherlands, 1993.
40. H. Dachsel, R. J. Harrison, and D. A. Dixon, *J. Phys. Chem. A* **103**, 152–155 (1999).
41. D. P. DiVincenzo, arXiv:quant–ph/0002077 (2000).
42. T. D. Ladd, F. Jelezko, R. Laflamme, Y. Nakamura, C. Monroe, and J. L. O'Brien, *Nature* **464**(4), 45 (2010).
43. R. L. de Visser and M. Blaauboer, *Phys. Rev. Lett.* **96**, 246801 (2006).
44. P. Jian-Wei, B. Dik, D. Matthew, W. Harald, and A. Zeilinge, *Nature* **403**, 515 (2000).
45. P. Chen, C. Piermarocchi, and L. J. Sham, *Phys. Rev. Lett.* **87**, 067401 (2001).
46. F. de Pasquale, G. Giorgi, and S. Paganelli, *Phys. Rev. Lett.* **93**, 12052 (2004).
47. J. H. Reina and N. F. Johnson, *Phys. Rev. A* **63**, 012303 (2000).
48. D. Bouwmeester, J. Pan, K. Mattle, M. Eibl, H. Weinfurter, and A. Zeilinger, *Nature* **390**, 575 (1997).
49. C. W. J. Beenakker and M. Kindermann, *Phys. Rev. Lett.* **95**, 056801 (2004).
50. M. D. Barrett, J. Chiaverinl, and T. Schaetz, et al., *Nature* **429**, 737 (2004).
51. M. Riebe, et al., *Nature* **429**, 734 (2004).
52. H. Wang and S. Kais, *Chem. Phys. Lett.* **421**, 338 (2006).
53. C. H. Bennett, G. Brassard, C. Crepeau, R. Jozsa, A. Peres, and W. Wootters, *Phys. Rev. Lett.* **70**, 1895 (1993).
54. C. P. Williams. *Explorations in Quantum Computing*, 2nd ed., Springer, 2011, p. 483.
55. R. Ursin, T. Jennewein, M. Aspelmeyer, R. Kaltenbaek, M. Lindenthal, P. Walther, and A. Zeilinge, *Nature* **430**, 849 (2004).
56. R. Ursin, F. Tiefenbacher, T. Schmitt-Manderbach, et al., *Nat. Phys.* **3**, 481 (2007).
57. H. Wang and S. Kais, in *Handbook of Nanophysics*, K. Sattler, ed., Taylor & Francis, 2012.

58. B. P. Lanyon, J. D. Whitfield, G. G. Gillet, M.l E. Goggin, M. P. Almeida, I. Kassal, J. D. Biamonte, M. Mohseni, B. J. Powell, M. Barbieri, A. Aspuru-Guzik, and A. G. White, *Nat. Chem.* **2**, 106–111 (2009).

59. L. M. K. Vandersypen, M. Steffen, G. Breyta, C. S. Yannoni, M. H. Sherwood, and I. L. Chuang, *Nature* **413**, 883–887 (2001).

60. M. Mosca, Quantum computer algorithms, Ph.D. thesis, University of Oxford, 1999.

61. A. Daskin and S. Kais, *J. Chem. Phys.* **134**, 144112 (2011).

62. A. Daskin and S. Kais, *Mol. Phys.* **109**(5), 761–772 (2011).

63. E. Farhi, J. Goldstone, S. Gutmann, and M. l Sisper, arXiv:quant-ph/0001106v1 (2000).

64. D. Aharonov, W. van Dam, J. Kempe, Z. Landau, S. Lloyd, and O. Regev, *SIAM J. Comput.* **37**(1), 166–194 (2007).

65. W. van Dam, M. Mosca, and U. Vazirani, *Proceedings of the 42nd Symposium on Foundations of Computer Science*, 2001, pp. 279–287.

66. E. Farhi, J. Goldstone, S. Gutmann, J. Lapan, A. Lundgren, and D. Preda, *Science* **292**(5516), 472–476 (2001).

67. J. Kempe, A. Kitaev, and O. Regev, *SIAM J. Comput.* **35**, 1070 (2006).

68. J. D. Biamonte and P. J. Love, *Phys. Rev. A* **78**, 012352 (2008).

69. S. Bravyi, D. P. DiVincenzo, R. I. Oliveira, and B. M. Terhal, *Quant. Inf. Comput.* **8**(5), 0361–0385 (2008).

70. A. Kitaev, A. Shen, and M. Vyalyi, *Classical and Quantum Computation: Graduate Studies in Mathematics*, Vol. 47, American Mathematical Society, Providence, RI, 2002.

71. R. Harris, A. J. Berkley, M. W. Johnson, P. Bunyk, S. Govorkov, M. C. Thom, S. Uchaikin, A. B. Wilson, J. Chung, E. Holtham, J. D. Biamonte, A. Yu. Smirnov, M. H. S. Amin, and A. M. van den Brink, *Phys. Rev. Lett.* **98**, 177001 (2007).

72. R. Harris, M. W. Johnson, T. Lanting, A. J. Berkley, J. Johansson, P. Bunyk, E. Tolkacheva, E. Ladizinsky, N. Ladizinsky, T. Oh, F. Cioata, I. Perminov, P. Spear, C. Enderud, C. Rich, S. Uchaikin, M. C. Thom, E. M. Chapple, J. Wang, B. Wilson, M. H. S. Amin, N. Dickson, K. Karimi, B. Macready, C. J. S. Truncik, and G. Rose, *Phys. Rev. B* **82**, 024511 (2010).

73. T. Kato, *J. Phys. Soc. Jap.* **5**, 435–439 (1951).

74. A. Messiah, *Quantum Mechanics*, Wiley, New York, 1958.

75. B. Reichardt, *Proceedings of the 36th Symposium on Theory of Computing*, 2004, pp. 502–510.

76. R. Oliveira and B. Terhal, *Quant. Inf. and Comp.* **8**, 900–924 (2008).

77. J. D. Biamonte, *Phys. Rev. A* **77**(5), 052331 (2008).

78. J. D. Biamonte and P. J. Love, *Phys. Rev. A* **8**(1), 012352 (2008).

79. S. Jordan and E. Farhi, *Phys. Rev. A* **77**, 062329 (2008).

80. M. Wolf, *Nat. Phys.*, **4**, 834–835 (2008).

81. J. Kempe, A. Kitaev, and O. Regev, *SIAM J. Comput.* **35**(5), 1070–1097 (2006).

82. A. Stern, *Ann. Phys.* **323**, 204–249 (2008).

83. J. M. Leinaas and J. Myrheim, *Nuovo Climento Soc. Ital. Fis. B* **37**, 1 (1977).

84. F. Wilczek, *Phys. Rev. Lett.* **48**, 1144 (1982).

85. F. Wilczek, *Fractional Statistics and Anyon Superconductivity*, World Scientific, Singapore, 1990.

86. L. S. Georgiev, *Phys. Rev. B* **74**, 235112 (2006).

87. C. Nayak, S. H. Simon, A. Stern, M. Freedman, and S. Das Sarma, *Rev. Mod. Phys.* **80**, 1083 (2008).
88. S. Bravyi, *Phys. Rev. A* **73**, 042313 (2006).
89. A. Einstein, B. Podolsky, and N. Rosen, *Phys. Rev.* **47**, 777–780 (1935).
90. L Amico, R. Fazio, A. Osterioh, and V. Vedral, *Rev. Mod. Phys.* **80**, 518 (2003).
91. W. K. Wootters, *Phys. Rev. Lett.* **80**, 2245–2248 (1998).
92. Z. Huang and S. Kais, *Chem. Phys. Lett.* **413**, 1–5 (2005).
93. Q. Wei, S. Kais, B. Fredrich, and D. Herschbach, *J. Chem. Phys.* **134**, 124107 (2011).
94. J. S. Bell, *Physics* **1**(3), 195–200 (1964).
95. W. Kutzelnigg, G. Del Re, and G. Berthier, *Phys. Rev.* **172**, 49 (1968).
96. Q. Shi and S. Kais, *J. Chem. Phys.* **121**, 5611 (2004).
97. H. Wang and S. Kais, *Israel J. Chem.* **47**, 59–65 (2007).
98. L. A. Wu, S. Bandyopadhyay, M. S. Sarandy, and D. A. Lidar, *Phys. Rev. A* **72**, 032309 (2005).
99. H. Wang and S. Kais, in *Handbook of Nanophysics*, K. Sattler, ed., Springer, 2011.
100. D. DeMille, *Phys. Rev. Lett.* **88**, 067901 (2002).
101. R. V. Krems, W. C. Stwalley, and B. Friedrich, *Cold Molecules: Theory, Experiment, Applications*, Taylor & Francis, 2009.
102. B. Friedrich and J. M. Doyle, *ChemPhysChem* **10**, 604 (2009).
103. S. F. Yelin, K. Kirby, and R. Cote, *Phys. Rev. A* **74**, 050301(R) (2006).
104. J. Z. Jones, *PhysChemComm* **11**, 1 (2001).
105. B. M. R. Schneider, C. Gollub, K. L. Kompa, and R. de Vivie-Riedle, *Chem. Phys.* **338**, 291 (2007).
106. B. M. R. Korff, U. Troppmann, K. L. Kompa, and R. de Vivie-Riedle, *J. Chem. Phys.* **123**, 244509 (2005).
107. U. Troppmann, C. M. Tesch, and R. de Vivie-Riedle, *Chem. Phys. Lett.* **378**, 273 (2003).
108. M. P. A. Branderhorst, et al., *Science* **638**, 320 (2008).
109. H. Rabitz, *Science* **314**, 264 (2006).
110. W. Warren, H. Rabitz, and M. Dahleh, *Science* **259**, 1581 (1993).
111. D. Babikov, *J. Chem. Phys.* **121**, 7577 (2004).
112. D. Sugny, L. Bomble, T. Ribeyre, O. Dulieu, and M. Desouter-Lecomte, *Phys. Rev. A* **80**, 042325 (2009).
113. D. L. Andrews and A. A. Demidov, *Resonance Energy Transfer*, Wiley, 1999.
114. G. D. Scholes, *J. Phys. Chem. Lett.* **1**(1), 2–8 (2010).
115. G. D. Scholes, *Nat. Phys.* **6**, 402–403 (2010).
116. R. E. Fenna and B. W. Matthews, *Nature* **258**, 573–577 (1975).
117. R. E. Fenna, B. W. Matthews, J. M. Olson, and E. K. Shaw, *J. Mol. Biol.* **84**, 231–234 (1974).
118. Y. F. Li, W. L. Zhou, R. E. Blankenship, and J. P. Allen, *J. Mol. Biol.* **271**, 456–471 (1997).
119. A. Camara-Artigas, R. E. Blankenship, and J. P. Allen, *Photosynth Res.* **75**, 49–55 (2003).
120. Y. C. Cheng and G. R. Fleming, *Annu. Rev. Phys. Chem.* **60**, 241–262 (2009).
121. G. S. Engel, T. R. Calhoun, E. L. Read, T. K. Ahn, T. Mancal, Y.-C. Cheng, R. E. Blankenship, and G. R. Fleming, *Nature* **446**, 782–786 (2007).
122. G. Panitchayangkoon, D. Hayes, K. A. Fransted, J. R. Caram, E. Harel, J. Wen, R. E. Blankenship, and G. S. Engel, *Proc. Natl. Acad. Sci. USA* **107**, 12766–12770 (2010).

123. M. Mohseni, P. Rebentrost, S. Lloyd, and A. Aspuru-Guzik, *J. Chem. Phys.* **129**, 174106 (2008).
124. P. Rebentrost, M. Mohseni, I. Kassal, S. Lloyd, and A. Aspuru-Guzik, *New J. Phys.* **11**, 033003 (2009).
125. P. Rebentrost, M. Mohseni, and A. Aspuru-Guzik, *J. Phys. Chem. B* **113**, 9942–9947 (2009).
126. A. Ishizaki and G. R. Fleming, *J. Chem. Phys.* **130**, 234111 (2009).
127. A. Ishizaki and G. R. Fleming, *Proc. Natl. Acad. Sci. USA* **106**, 17255–17260 (2009).
128. J. Zhu, S. Kais, P. Rebentrost, and A. Aspuru-Guzik, *J. Phys. Chem. B* **115**, 1531–1537 (2011).
129. Q. Shi, L. P. Chen, G. J. Nan, X. R. Xu, and Y. J. Yan, *J. Chem. Phys.* **130**, 084105 (2009).
130. T. Kramer, C. Kreisbeck, M. Rodriguez, and B. Hein, *American Physical Society March Meeting*, 2011.
131. T. C. Berkelbach, T. E. Markland, and D. R. Reichman, arXiv:1111.5026v1 (2011).
132. J. Prior, A. W. Chin, S. F. Huelga, and M. B. Plenio, *Phys. Rev. Lett.* **105**, 050404 (2010).
133. P. Huo and D. F. Coker, *J. Chem. Phys.* **133**, 184108 (2010).
134. J. Moix, J. Wu, P. Huo, D. Coker, and J. Cao, *J. Phys. Chem. Lett.* **2**, 3045–3052 (2011).
135. N. Skochdopole and D. A. Otti, *J. Phys. Chem. Lett.* **2**(23), 2989–2993 (2011).
136. D. A. Mazziotti, arXiv: 1112.5863v1 (2011).
137. P. Nalbach, D. Braun, and M. Thorwart, *Phys. Rev. E* **84**, 041926 (2011).
138. A. Shabani, M. Mohseni, H. Rabitz, and S. Lloyd, arXiv:1103.3823v3 (2011).
139. M. Mohseni, A. Shabani, S. Lloyd, and H. Rabitz, arXiv:1104.4812v1 (2011).
140. S. Lloyd, M. Mohseni, A. Shabani, and H. Rabitz, arXiv:1111.4982v1 (2011).
141. J. H. Kim and J. S. Cao, *J. Phys. Chem. B* **114**, 16189–16197 (2010).
142. J. L. Wu, F. Liu, Y. Shen, J. S. Cao, and R. J. Silbey, *New J. Phys.* **12**, 105012 (2010).
143. S. M. Vlaming and R. J. Silbey, arXiv:1111.3627v1 (2011).
144. N. Renaud, M. A. Ratner, and V. Mujica, *J. Chem. Phys.* **135**, 075102 (2011).
145. D. Abramavicius and S. Mukamel, *J. Chem. Phys.* **133**, 064510 (2010).
146. S. Yeh, J. Zhu, and S. Kais, *J. Chem. Phys.* **137**, 084110 (2012).
147. J. Zhu, S. Kais, A. Aspuru-Guzik, S. Rodriques, B. Brock, and P. Love, *J. Chem. Phys.* **137**, 074112 (2012).
148. S. Lloyd, *Nat. Phys.* **5**, 164–166 (2009).
149. E. M. Gauger, E. Rieper, J. L. Morton, S. C. Benjamin, and V. Vedral, *Phys. Rev. Lett.* **106**, 040503 (2011).
150. S. Johnsen and K. J. Lohmann, *Phys. Today* **61**(3), 29–35 (2008).
151. T. Ritz, S. Adem, and K. Schulten, *Biophys. J.* **78**, 707–718 (2000).
152. T. Ritz, P. Thalau, J. B. Phillips, R. Wiltschko, and W. Wiltschko, *Phys. Rev. Lett.* **106**, 040503 (2011).
153. C. R. Timmel and K. B. Henbest, *Philos. Trans. R. Soc. A* **362**, 2573 (2004).
154. T. Miura, K. Maeda, and T. Arai, *J. Phys. Chem.* **110**, 4151 (2006).
155. C. T. Rodgers, *Pure Appl. Chem.* **81**, 19–43 (2009).
156. C. T. Rodgers and P. J. Hore, *Proc. Natl. Acad. Sci. USA* **106**, 353–360 (2009).
157. K. Maeda, K. B. Henbest, F. Cintolesi, I. Kuprov, C. T. Rodgers, P. A. Liddell, D. Gust, C. R. Timmel, and P. J. Hore, *Nature* **453**, 387 (2008).
158. J. Cai, G. G. Guerreschi, and H. J. Briegel, *Phys. Rev. Lett.* **104**, 220502 (2010).
159. M. S. Byrd and D. A. Lidar, *J. Mod. Opt.* **50**, 1285 (2003).

160. P. W. Shor, *Phys. Rev. A* **52**, 2493–2496 (1995).
161. A. M. Steane, *Phys. Rev. Lett.* **77**, 793–797 (1996).
162. D. Gottesman, *Phys. Rev. A* **54**, 1862 (1996).
163. E. Knill and R. Laflamme, *Phys. Rev. A* **55**, 900–911 (1997).
164. A. M. Steane, *Introduction to Quantum Computation and Information*, World Scientific, Singapore, 1999.
165. P. Zanardi and M. Rasetti, *Phys. Rev. Lett.* **79**, 3306–3309 (1997).
166. L. Duan and G. C. Guo, *Phys. Rev. A* **57**, 737 (1998).
167. D. A. Lidar, I. L. Chuang, and K. B. Whaley, *Phys. Rev. Lett.* **81**, 2594 (1998).
168. E. Knill, R. Laflamme, and L. Viola, *Phys. Rev. Lett.* **84**, 2525 (2000).
169. J. Kempe, D. Bacon Dave, D. A. Lidar, and B. K. Whaley, *Phys. Rev. A* **63**, 042307 (2001).
170. D. A. Lidar, D. Bacon, J. Kempe, and K. B. Whaley, *Phys. Rev. A* **63**, 022306 (2001).
171. L. Viola and S. Lloyd, *Phys. Rev. A* **58**, 2733 (1998).
172. L.-M. Duan and G.-C. Guo, *Phys. Lett. A* **261**, 139 (1999).
173. L. Viola, E. Knill, and S. Lloyd, *Phys. Rev. Lett.* **82**, 2417 (1999).
174. P. Zanardi, *Phys. Lett. A* **77**, 258 (1999).
175. D. Vitali and P. Tombesi, *Phys. Lett. A* **59**, 4178–4186 (1999).
176. M. W. Johnson et al., *Nature* **473**, 194 (2011).
177. J. H. V. Nguyen, C. R. Viteri, E. G. Hohenstein, C. D. Sherrill, K. R Brown, and B. Odoml, *New J. Phys.* **13**, 063023 (2011).
178. L. M. K. Vandersypen and I. L. Chuang, *Rev. Mod. Phys.* **76**, 1037 (2005).
179. C. Negrevergne, T. S. Mahesh, C. A. Ryan, M. Ditty, F. Cyr-Racine, W. Power, N. Boulant, T. Havel, D. G. Cory, and R. Laflamme, *Phys. Rev. Lett.* **96**, 170501 (2006).
180. Y. Zhang, C. A. Ryan, R. Laflamme, and J. Baugh, *Phys. Rev. Lett.* **107**, 170503 (2011).
181. G. Brassad, I. Chuang, S. Lloyd, and C. Monroe, *Proc. Natl. Acad. Sci. USA* **95**, 11032 (1998).
182. D. Lu, et al., *Phys. Rev. Lett.* **107**, 020501 (2011).
183. E. S. Shuman, J. F. Barry, and D. DeMille, *Nature* **467**, 820 (2010).
184. J. Vala, Z. Amitay, B. Zhang, S. R. Leone, and R. Kosloff, *Phys. Rev. A* **66**, 062316 (2002).
185. D. R. Glenn, D. A. Lidar, and V. A. Apkarian, *Mol. Phys.* **104**, 1249 (2006).
186. B. P. Lanyon, J. D. Whitfield, G. G. Gillett, M. E. Goggin, M. P. Almeida, I. Kassal, J. D. Biamonte, et al., *Nat. Chem.* **2**, 106 (2010).
187. A. Aspuru-Guzik and P. Walther, *Nat. Phys.* **8**(4), 285–291 (2012).
188. D. Akoury, et al., *Science* **318**, 949 (2007).
189. J. P. Dowling and G. J. Milburn, *Philos. Trans. R. Soc. Lon. A* **361**, 1655–1674 (2003).

BACK TO THE FUTURE: A ROADMAP FOR QUANTUM SIMULATION FROM VINTAGE QUANTUM CHEMISTRY

PETER J. LOVE

Department of Physics, Haverford College, 370 Lancaster Avenue, Haverford, PA 19041, USA

I. INTRODUCTION

Quantum computation is the investigation of the properties of devices which use quantum mechanics to process information [1,2]. Such quantum computers exist in the laboratory now, but currently are only able to process a few bits of information. Many experimental proposals for quantum computing exist—all of which share the goal of controlled manipulation of the quantum state of the computer. These include superconducting systems [3], trapped atoms and ions [4,5], nuclear magnetic resonance (NMR) [6], optical implementations [7,8], Rydberg atoms [9], quantum dots [10,11], electrons on helium [12], and other solid-state approches [13,14].

Advances in Chemical Physics, Volume 154: Quantum Information and Computation for Chemistry, First Edition. Edited by Sabre Kais.

For certain specialized applications it is believed that quantum computers are more powerful than their classical counterparts [15–19].

One of the most promising areas of application for future quantum computers is quantum simulation: the use of a quantum computer to simulate or emulate another quantum system. Feynman's idea that quantum machines could be useful for simulating quantum systems was one of the founding ideas of quantum computing [20]. In the case of lattice Hamiltonians, specific techniques were first developed by Lloyd [21]. Zalka and Wiesner [22,23] developed methods for wave mechanics and Lidar and Wang used this to define an algorithm for the calculation of the thermal rate constant [24]. Work to date in the area of quantum simulation has considered both static [17] and dynamic properties of quantum systems [22,23]. Recently these ideas have been applied to the quantum statics and dynamics of molecular systems [25,26].

Accurate computation of the ground-state energy of molecules is a basic goal in chemistry, but an area where accurate results remain expensive and elusive in many areas. This is also an area where quantum computers are known to offer significant advantages over classical machines for which the cost of numerically exact calculation grows exponentially with the size of the problem [25]. In the context of physical chemistry, quantum computers are suited to full-CI electronic structure calculations and to the simulation of chemical reactions *without* the Born–Oppenheimer approximation [25,26]. In applicable cases, these quantum algorithms provide an exponential performance advantage over classical methods. Work to date indicates that a quantum computer with a few hundred qubits would be a revolutionary tool for quantum chemistry. This number of qubits is a realistic target for sustained experimental development. However, given the small number of qubits currently available any application of quantum computing faces the challenge: How do we get there from here?

Quantum computation is slowly moving out of its infancy and it is now possible to consider the implementation of the smallest realizations of many of the basic algorithms known to present an advantage over classical computation. Significant experimental progress has been made in NMR [27], in trapped atoms and ions [28–32], superconductors [33–35], optical implementations of the gate model [36–39], and other approaches [40]. While the field may be maturing beyond its earliest days, there remains a large amount of work to be done before quantum simulation can compete with existing classical computers. How should we think about the early efforts to implement quantum algorithms for chemistry, and how much progress do they represent toward the goal of real quantum chemical discovery on quantum devices? This chapter proposes a framework for evaluating partial progress toward useful quantum simulators for quantum chemistry. The basic idea is to use the more than 60 years of development of quantum chemical methods, implementations, and results on classical machines to guide and evaluate the development of methods for quantum computers.

A great deal of effort was focussed on quantum chemistry over the past 60 years, and it is possible to go back to the literature and find the calculations performed on early computers, some from the vacuum tube era [41]. These calculations are trivial by modern standards, but provide exemplars that could be implemented on early quantum computers (with, say, 5–20 quantum bits). By starting at this point, one can then trace the evolution of the classical calculations to the present day, and thereby lay out a roadmap leading from calculations that can be performed on quantum computers now to calculations at the research frontier.

For calculations in which 1–20 qubits are used to represent the wave function, we propose a historical approach to the development of experimental realizations of quantum chemistry on quantum computers. The first algorithm in this sequence has been implemented experimentally in linear optics quantum computing for a minimal basis set representation of the hydrogen molecule [42,43].

We first address the merit of such an approach: Why not simply plug in appropriate minimal bases from modern electronic structure methods for simple systems? One motivation is to reexamine the problems of quantum chemistry from a new viewpoint. In spite of the polynomial scaling of these algorithms with the problem size on quantum computers, gate and qubit estimates for simulation remain large. Theoretical improvements to these algorithms could significantly reduce the experimental requirements for simulation of technologically interesting molecules. Methods adapted for classical computers may not be the best for quantum computers, and so looking back at methods that were perhaps abandoned is worthwhile. Another motivation is to avoid the temptation to focus on those problems that are particularly suited to quantum simulation, and to instead evaluate the methods against a pre-selected set of benchmarks. As such, these historical benchmarks provide a complement to valuable investigations of a set of standard chemical examples, as is performed in Ref. [44]. Finally, it is interesting to measure progress in the quantum implementation against the historical development of the classical methods. This view gives a straightforward way of estimating the distance yet to be covered before quantum techniques can compete with classical methods.

This example-by-example approach should be contrasted with more general theoretical inquiries. Much theoretical work in quantum information is focussed on quantum complexity theory [45,46]. In particular, the local Hamiltonian problem has been extensively studied [1,47,48]. The local Hamiltonian problem formulates the task of finding the ground state of a Hamiltonian in the form of a decision problem. As such, the problem can be studied in terms of two solution concepts. First, one may consider versions of the problem where the solution may be found efficiently. Classically such examples would fall into the complexity classes P (polynomial time) or, if a probabilistic algorithm is required, they would fall into BPP (bounded-error probabilistic polynomial time). Second, one may consider those problems for which a given solution may be verified efficiently. Classically such problems would fall into the class NP (nondeterministic polynomial time), or again, if the verification procedure is polynomial, into MA (Merlin–Arthur; see in the following sections for an explanation of this nomenclature). Of course, the question of whether the two solution concepts—finding versus verifying a solution—are in fact the same is the question of whether P = NP [49].

In the classical world, it is known that the verification of the ground-state energy is as hard as any problem in NP [50]. At least one classical algorithm for fermionic quantum systems, quantum Monte Carlo, is also as difficult as any problem in NP in the worst case [51]. On the quantum side, it is known that the eigenvalue decision problem in general for a local Hamiltonian on qubits is as hard as any problem in QMA [1,47,48]. However, these complexity-theoretic results are mostly for worst-case complexity, and hence one may state the advantage of quantum algorithms for ground-state properties in complexity theoretic terms as dependent on the existence of a subset of physically interesting problems that are quantumly tractable but classically hard.

Here, we do not propose an approach to obtaining complexity-theoretical results on the efficiency of quantum algorithms for ground-state problems. Such a project might be described as Newtonian or Cartesian, the determination of grand theory covering all possible cases. Needless to say, such lofty goals are difficult to acheive in practice. Complexity theoretic proofs of the advantage of many widely used classical algorithms are few and far between. Indeed, in the case of density functional theory and quantum Monte Carlo the proofs are in the other direction, showing that the worst cases of these algorithms are likely to be hard, just as is the case for the local Hamiltonian problem [1,47,48,51,52].

We may compare the Cartesian approach of quantum complexity theory to the approach here. We propose instead a Baconian approach, where specific examples are investigated in detail one at a time. These two approaches were recently contrasted by Dyson [53], who characterized them in terms of Cartesian birds and Baconian frogs. The Cartesian birds fly over the landscape describing its broad features and general properties, whereas the Baconian frogs are happy to sit in the mud and investigate the details of what lies near them. Of course, we hope that these approaches are complementary. What we learn from the detailed investigation of a set of specific examples should be of use in correctly defining the right conditions under which quantum algorithms for ground-state problems can be rigorously proved to be efficient. Setting theoretical considerations to one side we also hope that the elucidation of a well-defined set of examples, and their use as a yardstick of progress, will help in the further development of experimental procedures and hardware for quantum simulation.

II. QUANTUM COMPUTING

In this section, we briefly review some salient techniques of quantum computation. Here, we shall focus on brute force methods aimed at producing the minimal number of qubits, suitable for the first experimental implementations. Such implementations are useful for several reasons, not least because as they determine whether the experiments have sufficient precision to obtain the eigenvalues to a reasonable accuracy. For further details on the fully scalable approach we refer the reader to the next chapter in this volume, and to Refs [25,54,55]. In those works the fully scalable approach to quantum simulation is reviewed in detail.

A. Phase Estimation

Phase estimation is a fundamental primitive of quantum computation [1,2,19]. Here, we review those elements of the procedure that are relevant for our subsequent discussion. The circuit we shall analyze is shown in Fig. 1. It is composed of two registers—the readout register—which we take to be one qubit, and the state register, which we consider to be an arbitrary number of qubits. The input state to the readout register is simply logical zero, $|0\rangle$, while the input state on the state register is an eigenstate of the Hamiltonian, and therefore of the time evolution operator:

$$U|\psi\rangle = e^{-i\phi}\,|\psi\rangle \tag{1}$$

where

$$\phi = \frac{E\tau}{\hbar} \tag{2}$$

for evolution time τ and eigenenergy E. A Hadamard gate (which is the Fourier transform over the group $Z/2Z$) is applied to the readout register, resulting in state:

$$\begin{aligned} |S1\rangle &= |+\rangle\,|\psi\rangle \\ &= \frac{1}{\sqrt{2}}\Big[|0\rangle\,|\psi\rangle + |1\rangle\,|\psi\rangle\Big] \end{aligned} \tag{3}$$

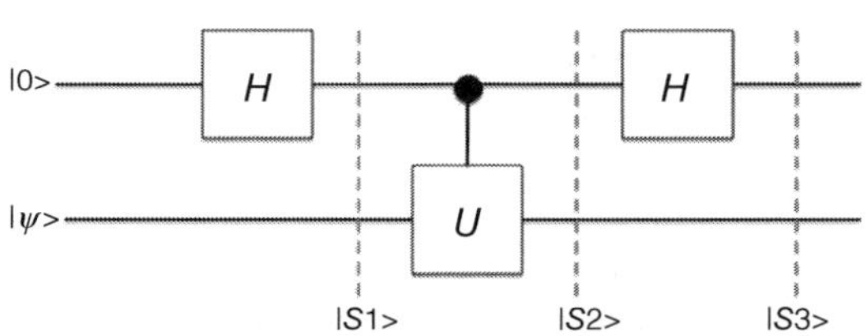

Figure 1. A schematic circuit for one step of the phase estimation algorithm. The upper line is the readout register (taken to be one qubit here) and the lower line is the state register, which carries an encoding of an eigenstate of the time evolution operator U.

The next stage is a controlled application of the unitary operator U corresponding to time evolution for a period τ, resulting in state:

$$|S2\rangle = \frac{1}{\sqrt{2}}\left[|0\rangle\,|\psi\rangle + e^{-i\phi}\,|1\rangle\,|\psi\rangle\right]$$
$$= \frac{1}{\sqrt{2}}\left[|0\rangle + e^{-i\phi}\,|1\rangle\right]|\psi\rangle \tag{4}$$

The final step before measurement is to again apply the Hadamard gate to the readout qubit, resulting in the state:

$$|S3\rangle = \frac{1}{2}\left[(1+e^{-i\phi})\,|0\rangle + (1-e^{-i\phi})\,|1\rangle\right]|\psi\rangle \tag{5}$$

It is worth noting that in the case where $|\psi\rangle$ is exactly equal to an eigenvector of U, the readout and state registers are not entangled at all in states $|S1\rangle$, $|S2\rangle$, $|S3\rangle$. A measurement performed on the readout qubit when the system is in state $|S3\rangle$ results in 0 or 1 with the following probabilities:

$$\text{Prob}(0) = \left|\frac{1+e^{-i\phi}}{2}\right|^2 = 1+\cos(\phi/2)$$
$$\text{Prob}(1) = \left|\frac{1-e^{-i\phi}}{2}\right|^2 = 1-\cos(\phi/2) \tag{6}$$

We can now proceed from our description of phase estimation to energy measurement, following Abrams and Lloyd [56]. Suppose that we can estimate $\arccos(\langle\sigma_z\rangle/2) = \phi$ to fixed precision. Let U be the time evolution operator of a system for time t:

$$U = \exp(-it\hat{H}/\hbar) \qquad \phi = tE/\hbar$$

suppose the energy scale is such that $E/\hbar < 1$ so that we can write $E/\hbar$ as a binary fraction:

$$E/\hbar = 0.E_2E_4E_8E_{16}\cdots = \frac{E_2}{2} + \frac{E_4}{4} + \frac{E_8}{8} + \frac{E_{16}}{16} + \ldots$$

then choose $t = 2\pi 2^n$ so that

$$\phi = N2\pi + \pi E_{2^n}/2 + \ldots \tag{7}$$

Using the recursive phase estimation procedure define in Refs [25,54,55] we can repeat this calculation to estimate E one bit at a time. This procedure has been implemented for calculations of the hydrogen molecule in Refs [42,43].

B. Time Evolution and the Cartan Decomposition

Evidently, the procedure outlined above requires the implementation of controlled operators corresponding to time evolution for a sequence of times $2^i\tau$ for $i = 1, 2, \ldots$. What is the difficulty of this problem if we know nothing about the form of U except its dimension (i.e., all that is known is the number of qubits in the state register)? The problem we wish to address is the decomposition of an arbitrary unitary operator into a sequence of one- and two-qubit gates. These one- and two-qubit gates are the basic primitive operations available to the experimentalist, and so we wish to proceed from an arbitrary unitary to a quantum circuit over a particular gate set. A typical choice is the controlled not (CNOT) operation combined with arbitrary operations on single qubits, and this choice is known to be *universal* (i.e., sufficient to implement any unitary operator) [57]. Much work in quantum computing has addressed this problem, including approaches that set bounds on the number of gates required [58,59], and approaches that can constructively obtain the full circuit, including all parameters for the one- and two-qubit gates [57,60–62]. For factoring a q qubit unitary over CNOT gates and arbitrary one qubit rotations the best-known performance is

$$\frac{23}{48}4^q - \frac{3}{2}2^q + \frac{4}{3} \tag{8}$$

CNOT gates and $4^q - 1$ one-qubit gates, where a one-qubit gate is either an X, Y, or Z rotation on a single qubit.

The number of gates for an arbitrary unitary grows exponentially with the number of qubits. This is expected for a method that pays no attention to any special structure that may be present in U beyond the number of qubits on which it acts and the tensor product structure of the Hilbert space of a set of qubits. Scalable implementations of time evolution operators must exploit the structure of the Hamiltonian in order to realize a polynomially scaling number of gates. Fortunately, for quantum chemical Hamiltonians in the second quantized formalism it is well understood how to do this, with a number of gates rising as q^5. For details of the scalable approach see next chapter in this volume, or Refs [54,55].

The bounds given in Ref. [8] arises from a technique referred to as the quantum Shannon decomposition [59]. The underlying mathematical tool used in much of the literature is the Cartan decomposition of a Lie algebra [63,64], and the corresponding decomposition for the Lie group, in this case the unitary group.

What is the Cartan decomposition for a Lie group? Let us restrict to unitary operators of dimension $U(2^n)$ with determinant one, meaning the Lie group $SU(2^q)$. Let us call this group G, with the corresponding algebra being the antihermitian matrices (which physically correspond to Hermitian observables, times i). The first step in the decomposition is to identify a subalgebra, and hence a subgroup of G. We call this subgroup K. For our purposes, we may consider two types of

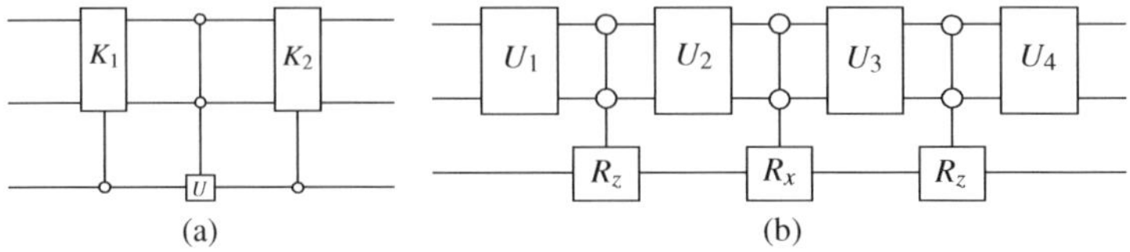

Figure 2. The two steps in the quantum Shannon decomposition treated in Ref. [62], illustrated for three qubits. (**a**) In the first stage of the decomposition the operator is broken into two operations controlled on the low qubit, separated by an operation on the low qubit controlled on the other qubits. (**b**) The two outer controlled operations are themselves decomposed, resulting in four operations on two (in general, $n - 1$) qubits. In the case of n qubits, these $n - 1$ qubit unitaries are factored again, until a circuit using only one- and two-qubit gates is obtained. This decomposition is the quantum Shannon decomposition [59], and using Cartan involutions one can implement the decomposition given an arbitrary unitary matrix and produce the factors [62].

subalgebra (and hence subgroup). First, we can choose the subgroup to be the real unitary matrices, in other words the orthogonal group $SO(2^q)$, with the corresponding subalgebra being real antisymmetric matrices. Second, we can consider block diagonal unitaries with blocks of size p and r such that $p + r = 2^n$. It is natural for systems composed of qubits to choose $p = r = 2^{q-1}$, so that the blocks are half the size of the original blocks.

Having selected our subalgebra we have the algebra g corresponding to the original group $G = SU(2^n)$, and the subalgebra k with corresponding subgroup K. The orthogonal complement of k in g is called m, and for a Cartan decomposition we find that the commutators obey the following relations. First, k is a subalgebra, so that

$$[k_1, k_2] \in k \text{ for } k_1 \text{ and } k_2 \text{ in } k \tag{9}$$

Second, we have

$$[k_1, m_1] \in m \text{ for } k_1 \text{ in } k \text{ and } m_1 \text{ in } m \tag{10}$$

Finally we find,

$$[m_1, m_2] \in k \text{ for } m_1 \text{ in and } m_2 \text{ in } m \tag{11}$$

In particular, we note that any subalgebra of m (m itself is not a subalgebra) must be abelian, as all commutators must be zero, or they would lie in k, violating the subalgebra property. Hence, abelian subalgebras of m may be found, and such a subalgebra of largest dimension will be called a. It is a fact, that we shall not prove here, that any element of m may be written

$$K_1^{\dagger} h K_1 = m_1 \tag{12}$$

where K_1 is an element of the subgroup K and h is an element of a maximal abelian subalgebra of m. Furthermore, every element of G can be written as

$$G_1 = K_2 M_1 = \exp(k_2) \exp m_1, \quad k_2 \in k \;\; m_1 \in m \tag{13}$$

and we may write

$$G_1 = \exp(k_2) \exp m_1 = \exp(k_2) \exp K_1^{\dagger} h K_1, \quad k_2, k_1 \in k \;\; m_1 \in m, \quad h \in a \tag{14}$$

using an elementary property of the exponential map we obtain

$$G_1 = \exp(k_2) \exp K_1^{\dagger} h K_1 = \exp(k_2) K_1^{\dagger} \exp h K_1, \quad k_2, k_1 \in k \;\; m_1 \in m, \quad h \in a \tag{15}$$

which we may write

$$G_1 = K_3 A K_1 \tag{16}$$

where $K_3 = \exp(k_2) K_1^{\dagger}$ and $A = \exp(h)$. To put this in somewhat more physical language, given a Cartan decomposition of a unitary group one can find a maximal commuting subalgebra h (i.e., a set of observables that may simultaneously take definite values) and a subalgebra K. Then the whole group can be written as a product of elements of K and of A, the abelian group corresponding to the maximal abelian subalgebra. For detailed proofs of the above assertions we refer the reader to the mathematics literature [63,64].

So far we have only sketched the existence of this decomposition. One further feature of these decompositions enables them to be obtained directly for a given element of G and specific K and a. A Cartan involution is a self-inverse map that fixes the subalgebra k and negates elements of the complement m

$$\begin{aligned} \theta(k) &= k \\ \theta(m) &= -m \\ \theta(\theta(x)) &= x \end{aligned} \tag{17}$$

The involution is commutator preserving

$$\theta([x, y]) = [\theta(x), \theta(y)] \tag{18}$$

and these maps also extend to the group, with the following useful properties

$$\Theta(K) = K, \quad \Theta(M) = M^{-1}, \quad \Theta(KM) = KM^{-1} \tag{19}$$

Perhaps the most straightforward example is the involution for the orthogonal subalgebra, where

$$\theta(x) = -x^T, \quad \Theta(X) = X^* \tag{20}$$

and the corresponding involution for the case where K is the subalgebra of block diagonal matrices, for example, where $p = q = 2^{n-1}$:

$$\theta(x) = \begin{pmatrix} I_{p\times p} & 0 \\ 0 & -I_{r\times r} \end{pmatrix} x \begin{pmatrix} I_{p\times p} & 0 \\ 0 & -I_{r\times r} \end{pmatrix} \tag{21}$$

where $I_{p\times p}$ is the $p \times p$ identity matrix. Similarly,

$$\Theta(X) = \begin{pmatrix} I_{p\times p} & 0 \\ 0 & -I_{r\times r} \end{pmatrix} X \begin{pmatrix} I_{p\times p} & 0 \\ 0 & -I_{r\times r} \end{pmatrix} \tag{22}$$

Given these involutions and their properties [19], it is straightforward to obtain M^2:

$$M^2 = (MK^{-1})KM = \Theta(M^\dagger K^\dagger)KM = \Theta(G^\dagger)G \tag{23}$$

From this point all other factors may be obtained, and recursive application of a sequence of alternating decompositions can obtain the quantum Shannon decomposition [60,62].

The utility of this decomposition is many-fold. First, for two-qubit circuits the accidental isomorphism $SU(2) \times SU(2) = SO(4)$ means that the Cartan decomposition over the orthogonal subalgebra can be used to obtain circuits in which the K factors are simply one-qubit rotations [65,66]. Second, it was recently shown that the quantum Shannon decomposition could be realized by using a pair of Cartan decompositions of the Lie algebra of the unitary group [62]. This insight provided a simple constructive method for obtaining circuits given a unitary operator. This method has also been implemented in the PYTHON programming language and extensively tested on random unitary matrices on up to eight qubits. The method is suprisingly fast, a fit to timing data gives a rule of thumb timing of $T = 2^{2.5(q-4.5)}$ s for a q-qubit unitary, corresponding to about a minute for a seven-qubit unitary operator (a 128×128 matrix).

III. QUANTUM CHEMISTRY: THE CI METHOD

The starting point for problems in full-CI calculations is the electronic Hamiltonian expressed in a basis. We briefly review the scalable second quantized approach, referred to as the direct mapping in Ref. [25], which is covered extensively in the next chapter, before moving on to describe the compact mapping of Ref. [25]. The compact mapping is based on a treatment of the CI matrix, and so we describe the construction of that matrix and the method for simulation arising from the use of the CI matrix. We close the section by comparing the direct and compact mappings for small numbers of qubits.

A. Second Quantization: Direct Mapping

The Hamiltonian for the full quantum treatment of a set of nuclei and electrons is

$$\hat{H}^{\text{mol}} = \hat{T}_e + \hat{T}_Z + \hat{V}_{ZZ}(L_{pq}) + \hat{V}_{ee}(r_{ij}) + \hat{V}_{eZ}(R_{pi}) \tag{24}$$

where $\hat{T}_e$ and $\hat{T}_Z$ are the electron and nuclear kinetic energies, respectively, $\hat{V}_{eZ}(R_{pi})$ and $\hat{V}_{ee}(r_{ij})$ are the electron–nuclear and electron–electron coulomb interactions, and $\hat{V}_{ZZ}(L_{pq})$ is the nuclear–nuclear coulomb interaction. If we invoke the Born–Oppenheimer approximation and neglect T_Z and treat V_{ZZ} classically, then we obtain

$$\hat{H}^{\text{elec}} = -\frac{1}{2}\sum_{i=1}^{N}\vec{\nabla}_i^2 - \sum_{i,L}\frac{Z_L}{r_{iL}} + \sum_{i>j}^{N}\frac{1}{r_{ij}} \tag{25}$$

which is the Hamiltonian for the electronic structure problem.

To attack this problem via quantum computation, one requires a mapping of the time evolution operator into a set of unitary operations performed on one or two qubits (a quantum circuit). In the direct mapping, this is accomplished by mapping the states of qubits into the occupation number basis. One therefore requires the same number of qubits as spin-orbitals, and a qubit in state one indicates that the corresponding spin-orbital is occupied. The Hamiltonian that acts on the fermionic states is

$$H = \sum_{p,q} h_{pq} a_p^\dagger a_q + \frac{1}{2}\sum_{p,q,r,s} h_{pqrs} a_p^\dagger a_q^\dagger a_r a_s \tag{26}$$

where $a_p^\dagger$ is the creation operator for spin-orbital p, and a_p is the annihilation operator for spin-orbital p.

The operator a_j acts on basis vectors as follows:

$$\begin{aligned} a_j \left| l_0, \ldots, l_{j-1}, 1, l_{j+1}, \ldots, l_{q-1} \right\rangle \\ = (-1)^{\sum_{s=0}^{j-1} s} \left| l_0, \ldots, l_{j-1}, 0, l_{j+1}, \ldots, l_{q-1} \right\rangle, \\ a_j \left| lp_0, \ldots, l_{j-1}, 0, l_{j+1}, \ldots, l_{q-1} \right\rangle = 0 \end{aligned} \tag{27}$$

where $a_j^\dagger$ is the Hermitian conjugate. In order to realize the Hamiltonian 26 as an operator on the qubits we must implement the action of the creation and annihilation operators on the qubit states. We do so using the Jordan–Wigner transformation following [67,68]. With the creation operator written in terms of Pauli matrices as $a_{\text{qubit}}^\dagger = \frac{1}{2}(\sigma^x - i\sigma^y) = \sigma^-$, we can completely express the fermionic annihilation and creation operators in terms of Pauli matrices:

$$a_j \equiv \sigma^{z\otimes j-1} \otimes \sigma^+ \otimes \mathbf{1}^{\otimes q-j} = \begin{bmatrix} 1 & 0 \\ 0 & -1 \end{bmatrix}^{\otimes j-1} \otimes \begin{bmatrix} 0 & 1 \\ 0 & 0 \end{bmatrix} \otimes \begin{bmatrix} 1 & 0 \\ 0 & 1 \end{bmatrix}^{\otimes q-j}$$
$$a_j^\dagger \equiv \sigma^{z\otimes j-1} \otimes \sigma^- \otimes \mathbf{1}^{\otimes q-j} = \begin{bmatrix} 1 & 0 \\ 0 & -1 \end{bmatrix}^{\otimes j-1} \otimes \begin{bmatrix} 0 & 0 \\ 1 & 0 \end{bmatrix} \otimes \begin{bmatrix} 1 & 0 \\ 0 & 1 \end{bmatrix}^{\otimes n-j} \tag{28}$$

This correspondence is the Jordan–Wigner transformation. The use of this method imposes a number of extra gate applications that scale as $O(q)$. As the general Hamiltonian 26 has order q^4 terms, the use of this method imposes a number of gates scaling at least as q^5, where q is the number of qubits (equivalently of spin-orbitals). Because the terms in the Hamiltonian 26 do not commute one may not simply separate the terms into gates. However, one may make use of the Trotter–Suzuki decompositions for small timesteps in order to obtain a circuit with a number of gates polynomial in q. The number of gates in such decompositions scales as the number of terms in the Hamiltonian, with a prefactor depending on the details of the method, which in turn set the accuracy with which the circuit reproduces the time evolution operator. The generic scaling with number of qubits is therefore q^5 for these scalable methods.

This may be improved to $q^4 \log q$ using the methods of [68] as explained in detail for chemical calculations in [69].

B. FCI: Compact Mapping

The alternative to second quantized is, of course, the first quantized approach in which all symmetry properties are retained in the wave function. We take as our starting point a set of $2K$ spin-orbitals (single electron basis functions), for a system of N electrons. From these basis functions we may form

$$\binom{2K}{N} \tag{29}$$

antisymmetrized N electron basis functions, or Slater determinants. It is convenient to classify these with reference to a Hartree–Fock basis state—the Slater determinant $|\Psi_0\rangle$ formed from the N lowest energy spin-orbitals. One may label states by their differences from the Hartree–Fock state. The single excitations are labeled $|\Psi_a^r\rangle$, where spin-orbital a in $|\Psi_0\rangle$ is replaced by spin-orbital r. Similarly, doubles and triples are labeled $|\Psi_{ab}^{rs}\rangle$, $|\Psi_{abc}^{rst}\rangle$, and so on. The number of n-tuple excitations is

$$\binom{N}{n}\binom{2K-N}{n} \tag{30}$$

summing over n gives the total number of Slater determinants:

$$\binom{2K}{N} = \sum_{n=0}^{N} \binom{N}{n}\binom{2K-N}{n} \tag{31}$$

a result obtained from the Vandermonde convolution and the symmetry property of binomial coefficients.

Given this basis of determinants one wishes to obtain the Hamiltonian matrix elements within this basis: the CI matrix. To obtain the matrix elements of the Hamiltonian one uses Slater's rules [70–72]. Given a matrix element $\langle A|\, H\, |B\rangle$, where A and B are configurations (lists of spin-orbital labels), we may compute the CI matrix elements as follows:

1. Diagonal elements. If $A = B$ so that the determinants are identical

$$\langle A|\, H\, |A\rangle = \sum_{p}^{n} \langle p|\, \hat{h}\, |p\rangle + \sum_{p<q}^{n} \Big(\langle pq|pq\rangle - \langle pq|qp\rangle \Big) \tag{32}$$

2. If the configurations A and B differ in one spin-orbital, p in A and q in B

$$\langle A|\, H\, |B\rangle = \langle p|\, \hat{h}\, |q\rangle + \sum_{l}^{n} \Big(\langle pl|ql\rangle - \langle pl|lq\rangle \Big) \tag{33}$$

3. If the configurations A and B differ in two spin-orbitals, pq in A and rs in B

$$\langle A|\, H\, |B\rangle = \langle pq|rs\rangle - \langle pq|sr\rangle \tag{34}$$

4. If the configurations A and B differ in more than two spin-orbitals

$$\langle A|\, H\, |B\rangle = 0 \tag{35}$$

where we have defined the one-electron integrals:

$$\langle p|\, \hat{h}\, |q\rangle = -\frac{\hbar^2}{2m} \int \phi_p^*(\mathbf{x}) \nabla^2 \phi_q(\mathbf{x}) d\mathbf{x} \tag{36}$$

and the two electron integrals:

$$\langle pq|rs\rangle = \int \phi_p^*(\mathbf{x}_1)\phi_q^*(\mathbf{x}_2) \frac{1}{|\mathbf{x}_1 - \mathbf{x}_2|} \phi_r(\mathbf{x}_1)\phi_s(\mathbf{x}_2) d\mathbf{x}_1 d\mathbf{x}_2 \tag{37}$$

At this stage, quantum simulation of FCI via the compact mapping comes into focus. The compact mapping is a mapping of the logical states of the qubits onto labels of Slater determinants. The matrix eigenvalue problem we wish to solve is defined by the CI matrix, and we implement time evolution according to this

Hamiltonian by, for example, using a Cartan decomposition of the relevant time evolution operator. Ironically, while FCI is typically the classical method that uses the largest number of states for a given problem, in the quantum case the compact mapping uses a smaller Hilbert space dimension than the direct mapping of the second quantized approach. The reason is that the second quantized approach maps the entire Fock space of the system into the Hilbert space of the qubits, whereas the compact mapping only maps the N electron sector into the quantum computer.

The procedure just outlined for the compact mapping is not scalable. Although the input data to the CI matrix is the same set of one- and two-electron integrals used to define the second quantized Hamiltonian, the compact mapping does not respect the tensor product structure of the qubit Hilbert space, and hence no known procedure is analogous to the Trotter–Suzuki decomposition that enables the efficient construction of circuits with a number of gates rising polynomially in the number of spin-orbitals. In addition, while for the second quantized formalism the one- and two-electron integrals appear directly in the circuit as rotation angles for quantum gates acting on a small number of qubits, at present no way is known for constructing the time evolution operator corresponding to the CI matrix on the quantum computer without computing it classically first.

The compact mapping remains of interest for a number of reasons, however. If we consider problems in which we have $2K$ spin-orbitals, and therefore the second quantized approach requires $2K$ qubits (for up to $2K$ electrons), the compact mapping requires:

$$c(k, N) = \left\lceil \log_2 \binom{2k}{N} \right\rceil \tag{38}$$

which is less than $2K$ for all $N < 2K$, being the nearest whole number of qubits into which the N particle sector of the full Fock space of up to $2K$ particles will fit. Hence, the compact mapping will always enable the mapping of larger problems than the direct mapping into the qubit Hilbert space.

This advantage is only of interest if one can in fact take advantage of the smaller qubit requirements by implementing the time evolution operator. Evidently, the nonscalable nature of the compact mapping makes it asymptotically uninteresting. However, the issue for small examples is not quite so clear cut. Suppose the number of gates required to implement the evolution operator that yields the U is $A(2K)^5$, where A is a constant. In order to obtain p bits of precision in the final answer one must run at least one phase estimation procedure with $2^p A(2K)^5$ gates. The corresponding numbers for performing the Cartan involution are much worse for the first stage of iterative phase estimation, where a Cartan involution requires

$$2^{2c(k,N)} = \binom{2k}{N}^2 \tag{39}$$

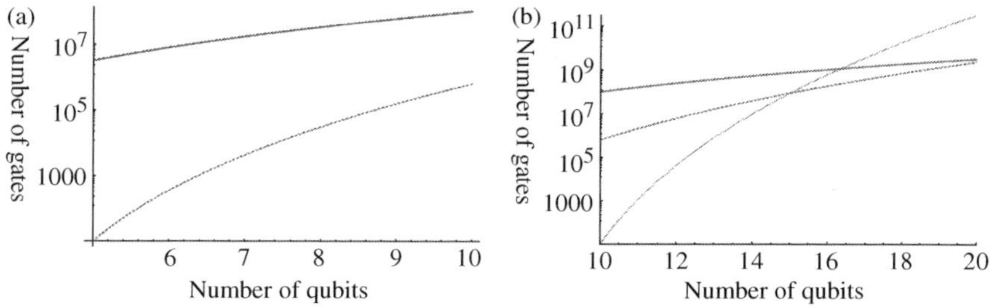

Figure 3. (**a**) A comparison of gate counts, as estimated by Eqs. (40) and (41) (with $A = 1$) for 5 electrons and numbers of qubits between 5 and 10. Solid line: G_{direct}. Dashed line: G_{compact}. This indicates that for fewer than 5 qubits the compact mapping is preferred over the scalable direct mapping. (**b**) A comparison of gate counts, as estimated by Eqs. (40) and (41) (with $A = 1$) for numbers of qubits between 5 and 10. Solid line: G_{direct}, 5 electrons. Dashed line: G_{compact}, 5 electrons. Dotted line: G_{compact}, 10 electrons. This indicates that the crossover between the compact mapping and the scalable direct mapping lies between 10 and 20 qubits. (*See the color version of this figure in the Color Plates section.*)

gates, as well as a comparable amount of classical computation to obtain the circuit. However, each stage of the iterative phase estimation requires only the same number of gates, as one simply computes new parameters in the Cartan decomposition for each evolution operator U^{2^p}. Hence, the total gate count comparison is roughly

$$G_{\text{compact}} \simeq p \binom{2k}{N}^2 \tag{40}$$

compared with

$$G_{\text{direct}} = 2^p A (2K)^5 \tag{41}$$

The crossover point at which the advantage in number of gates of the direct mapping outweighs the advantage in qubits of the compact mapping represents a significant milestone for quantum simulation—the point at which the advantage of scalable methods is realized in practice. We plot G_{compact} and G_{direct} in Fig. 3 for $A = 1$, for 5 and 10 electrons and qubit numbers between 5 and 20. These estimates indicate that the crossover point between compact and direct mappings is between 10 and 20 qubits.

IV. A SELECTION OF HISTORICAL CALCULATIONS IN QUANTUM CHEMISTRY

Here, we present a sequence of calculations spanning from the 1930s to the 1960s. Our intent here is not to give a historically faithful account of the development of quantum chemistry. We merely choose a few mileposts and evaluate the cost in qubits of the compact mapping already defined of those calculations on quantum hardware. The choice of these calculations was strongly influenced by the presentation of Ref. [73]. For translations of earlier foundational papers in quantum chemistry we refer the reader to Hinne Hettema's volume [74], which also includes a fascinating preface discussing the historical context of early work on quantum chemistry in both Germany and the United States.

A. The 1930s and 1940s

Perhaps the first exact *ab initio* calculation was performed by Coulson on a Brunsviga mechanical calculator for the molecular ion H_3^{+1}. Early calculations of the ground-state energy of the hydrogen molecule were conducted in the 1930s using the variational principle [75]. The first studies by Heitler and London were superceded by the methods of Hyelleras in which a variational Ansatz is constructed from symmetry considerations alone. The calculation of the variational energy in these calculations naturally involved integrals over the Ansatz functions.

B. The 1950s

In 1950, Boys introduced the use of Gaussian functions [76]. In the same year, Boys gave a presentation of the CI method and presented a CI calculation of the ground-state energy of beryllium [77]. Interestingly, Boys's first CI calculation used Slater-type LCAO orbitals rather than Gaussians because of the relatively simpler integrals that occur in this atomic problem. He used 3 s and 1 p single-electron functions and performed the calculation using 6 and 10 determinants formed from these functions. Here, we see perhaps the first application of CI, corresponding to a three- or four-qubit problem.

In 1952, Taylor and Parr applied the CI method to the helium atom, a system they note is "no longer an interesting problem *per se*" because of the accuracy of Hylleraas's variational solution obtained in 1930. However, they point out that in spite of the melancholy prospects for progress of the CI method helium is the first nontrivial example. Their calculation involved four configurations $1s1s'$, $2p^2$, $3d^2$, $4f^2$ and hence corresponds to a two-qubit calculation. Their paper reproduces 88% of the correlation energy of helium [78]. Three years later, Shull and Löwdin reconsidered the helium atom using natural orbitals [79]. They obtained the CI solution using six configurations from three orbitals (1s, 2s, 3s). The results were then analyzed in terms of natural orbitals, and they showed that similar

[1] For a biography of Coulson see the address given by Roy McWeeny, Professor of Theoretical Chemistry at University of Sheffield, which opened the Fifth Canadian Symposium on Theoretical Chemistry, being held at University of Ottawa in June, 1974, and was dedicated to the memory of Charles Coulson, F.R.S.

results could be obtained with only three configurations. By doing so they reduced a three-qubit (six-configuration) calculation to a two-qubit (three-configuration) calculation.

In 1956, CI calculations were reported on BH with 7 orbitals, 23 co-detors, and 3 internuclear distances, H_2O with 8 orbitals, 96 co-detors, at seven nuclear configurations. These results were obtained by Boys using the EDSAC computer at Cambridge [41]. The BH calculations are five-qubit problems, and the H_2O calculations are seven qubit problems. The following year Miller, Friedman, Hurst, and Matsen continue the extension of CI to molecular problems with calculations of LiH and BeH^+ [80]. Their calculations used 20 terms corresponding to all possible singlet states of $1s$, $2s$, $2p_\sigma$ Slater orbitals on the metal atoms and $1s$ on the hydrogen. They performed 6, 10, and 20 determinant calculations on IBM CPC and IBM 650 machines; these are three-, four-, and five-qubit problems, respectively. In the same year, Pekeris gave a new method for the determination of the ground-state energy of two-electron atoms [81].

In a largely philosophical paper published in 1959, Mulliken and Roothaan reflected on the status of exact quantum chemistry noting that "it so happens that the more difficult problems of chemistry and molecular physics are of a high order of complexity in terms of the mathematical analysis and the computational efforts required for quantitative results. However, and this is the point of our paper, those who are working in this field are now catching up, and we believe that the results of theoretical computations are going to compete more and more strongly with experiment from now on" [82].

C. The 1960s

In 1960 Watson published a CI study of Be involving 37 configurations constructed from a basis of STOs, a six-qubit problem [83]. Boys and Cook also laid out the progress made in developing the CI method over the previous decade in Ref. [96]. The following year, Sinanoglu published perhaps the first in-depth study of electron correlation, introducing the idea that the correlation energy would be well approximated by the sum of pair correlation energies [97].

In 1963, Barnett at MIT reports on the development of a FORTRAN code for the IBM 709/90 that implements SCF and CI calculations on basis sets of size <50 although no actual example calculations apart from integral evaluations are reported in this work [98]. In the same year, Hagstrom and Shull report on the efficacy of natural orbitals for H_2 and H_2^*. They report that calculations using 21 and 33 configurations may be replaced by natural orbital bases of size 8 and 15, respectively [85]. For our purposes, this finding is of particular interest as it converts five- and six-qubit calculations into three- and four-qubit calculations. Natural orbital calculations for HeH and Li_2 followed in 1966, providing seven-configuration (or three-qubit) calculations [86]. These calculations are of interest as

TABLE I
A Sequence of Quantum Chemical Problems and Their Minimum Qubit Requirements

Year	Calculation	Citation	Number of Qubits
1933	H_2	[75]	1
1950	Be	[77]	3, 4
1952	He	[78]	2
1955	He	[79]	2, 3
1956	BH	[41]	5
1956	H_2O	[41]	7
1957	LiH	[80]	3, 4, 5
1957	BeH^+	[80]	3, 4, 5
1960	Be	[83]	6
1960	CH_2	[84]	19
1963	H_2	[85]	3, 4, 5, 6
1966	HeH	[86]	3
1966	Li_2	[86]	3
1967	H_2O	[87]	10
1967	H_2O	[88]	24
1967	H_2O	[89,90]	38, 39
1968	H_2O	[91]	39, 46
1968	Be	[92]	11
1969	Li, Be^+, B^{++}	[93]	9, 10
1969	BH, FH	[94]	12, 14
1970	H_2O	[95]	23

Note: These calculations bridge the period of development of CI from the first variational calculations of the hydrogen molecule energy to the advent of the GAUSSIAN-70 program. The qubit costs are associated with the compact mapping of full-CI calculations on the system studied in the cited publication, although the methods used originally may not be full-CI.

the natural orbital wave functions provide a class of ansatz states whose complexity of preparation can be investigated (Table I).

In the same year (1968), Bunge published a new calculation of the ground-state energy of the Be atom, using a 180 term CI with 1492 Slater determinants, an 11-qubit calculation [92]. A study in 1969 of the $\pi\pi^*$ states of the ethylene molecule compared HF and CI calculations [99]. An improved CI algorithm was applied to three-electron problems (Li, Be^+, and B^{++}) using 7 to 9 Slater orbitals, (9- and 10-qubit problems, respectively) [93]. Calculations on H_3 were also performed with 9 Slater orbitals, another 10-qubit problem. The publication by Bender and Davidson in 1969 of a study of the first row diatomic hydrides using very large CI provides a rich set of examples [94]. Two example calculations are of BH, a 12-qubit calculation, and of FH, a 14-qubit calculation.

In 1970, a new study of water with a 14 function basis set, corresponding to a 23-qubit calculation, comparing STO and contracted Gaussian basis sets [95]. In the same year, one of the first calculations using the GAUSSIAN-70 code appeared, treating 30 species with STO-3G basis sets in which Slater-type functions are approximated by three Gaussians [100]. With this paper and the advent of codes such as GAUSSIAN, which achieved wide usage in the theoretical quantum chemistry community, full CI calculations became common. The Pople classification of basis sets makes the determination of a number of qubits for any large CI calculation straightforward, and the number of qubits is typically $20 - 50$ for calculations that historically lie between 1970 and the present. These basis sets and relevant citations are readily available from Ref. [101], for example.

In 1960, several papers applying variational and LCAO-SCF methods to molecules appeared. Mclean reports LCAO-SCF on C_2H_2 and CO_2 with molecular orbitals constructed from (for C_2H_2) 1s STOs on each H and $1s$, $2s$, $2p\sigma$, and $2p\pi$ STOs on each C. For CO_2 MOs are formed from $1s$, $2s$, $2p\sigma$, and $2p\pi$ STOs on each O and $1s$, $2s$, $2p\sigma$, and $2p\pi$ on the C [103]. In the same year, there was also a calculation of the ${}^1\Sigma_g^+$ excited state of H_2 by a variational method [104].

In 1960, Nesbet reports an LCAO Hartree–Fock calculation on HF using 9σ and 5π orbitals using the IBM 704 computer [105]. Foster and Boys report a study of the three lowest lying states of six different bond angles for the CH_2 radical using a basis of six exponential functions on the carbon and one on each hydrogen [84]. A CI calculation in this basis would represent a 19-qubit problem.

Much work in the previous century was devoted to approaching the Hartree–Fock limit using ever better SCF methods. For the water molecule, an example of this work is given by Ref. [91]. In this work, progress to date (prior to 1968) for the SCF treatment of water is summarized in Table IX, which we reproduce in Table II. These SCF calculations provide an interesting reference: repeating these calculations using CI yields more challenging, but still tractable problems. The data in Table II give us qubit requirements in the range 10–46.

TABLE II
SCF Calculations of the Water Molecule Ground-State Energy in 1967 and 1968

Calculation	Method	Size of Basis	Energy (a.u.)	Number of Qubits
Reeves and Boys [102]	MC STO	...	−75.776	...
McWeeny and Ohno [87]	MC STO	7	−75.761	10
Moccia	OC STO	28	−75.992	36
Harrison [88]	MC GLF	14	−76.002	24
Moskowitz and Harrison [90]	MC GTO	36	−76.034	39
Ritchie and King [89]	MC CGF	38	−76.034	38
[91]	MC GLF	32	−76.044	39
[91]	MC GTO	56	−76.002	46

Note: These calculations, aiming to approach the Hartree–Fock limit for water define large molecular orbital bases. The qubit numbers refer to the minimal requirements to perform a full-CI in the molecular orbital basis defined by the published SCF calculation. Unsuprisingly, SCF calculations that approach the Hartree–Fock limit define full-CI problems requiring several tens of qubits.

V. BOYS'S 1950 CALCULATION FOR Be

In this paper, Boys constructs a 12×12 CF Hamiltonian over a set of configuration state functions (CSFs), which are referred to as "co-detors," spin-adapted linear combinations of determinants. There are four sets of functions sA, sB, sC, and pA, where the s and p notation indicates that the functions in these sets are eigenvalues of the square of total angular momentum with quantum number $l = 0$ (s) and $l = 1$ (p). The letters A, B, C denote different radial dependence, and for each radial dependence and total orbital angular momentum there is one function for each possible combination of the eigenvalues of L_z and S_z. Because these are atomic orbitals (single-electron eigenfunctions) the total spin eigenvalue is $1/2$. Writing out the elements of each set we obtain

$$\begin{aligned}
sA &= \{sA(0, -1/2), sA(0, 1/2)\} \\
sB &= \{sB(0, -1/2), sB(0, 1/2)\} \\
sC &= \{sC(0, -1/2), sC(0, 1/2)\} \\
pA &= \{pA(-1, -1/2), pA(-1, 1/2), pA(0, -1/2), pA(0, 1/2), \\
&\quad pA(1, -1/2), pA(1, 1/2)\}
\end{aligned} \tag{42}$$

Hence, there are 12 atomic orbitals here. However, Boys uses linear combinations of functions in this basis, which are simultaneous zero eigenvalues of $\hat{L}^2$ and $\hat{S}^2$. He does so by taking the four electrons and placing them in two pairs, each pair being in a spin and orbital singlet.

The orbital coupling can use the special case of the Clebsch–Gordan coefficients relevant to the singlet state of total orbital angular momentum, $\langle l_1, l_2, m_1, m_2 | J, M \rangle$, where l_1, l_2, m_1, M_2 are the angular momentum quantum numbers labeling the orbitals and J and M are the total angular moment quantum numbers. By the recoupling rules for angular momentum, such a state can only be produced from a pair of angular momentum states with equal values of l:

$$\langle l, l, m_1, m_2 | 0, 0 \rangle = \delta_{m_1,-m_2} \frac{(-1)^{l-m_1}}{\sqrt{2l+1}} \tag{43}$$

This implies that the combination of orbital products is a uniform superposition of pairs with opposite z components with alternating signs.

$$\begin{aligned}
|0, 0\rangle &= \sum_{m_1=-l}^{l} \sum_{m_2=-l}^{l} \delta_{m_1,-m_2} \frac{(-1)^{l-m_1}}{\sqrt{2l+1}} |l, l, m_1, m_2\rangle \\
&= \sum_{m_1=-l}^{l} \frac{(-1)^{l-m_1}}{\sqrt{2l+1}} |l, l, m_1, -m_1\rangle
\end{aligned} \tag{44}$$

These are denoted

$$
\begin{aligned}
&(sAsA)^1 S(sBsB)^1 S \\
&(sAsA)^1 S(sBsC)^1 S \\
&(sAsA)^1 S(pApA)^1 S \\
&(sAsB)^1 S(pApA)^1 S
\end{aligned}
\tag{45}
$$

They are then antisymmetrized by applying the operator

$$
\mathcal{S} = \sum_{\pi \in S_4} \operatorname{sign}\pi \frac{\pi}{\sqrt{4!}} \tag{46}
$$

Resulting in the normalized wave functions given in Eq. (4) in Ref. [77].

The elementary integrals (overlaps and expectation values for the one-electron orbitals) whose radial parts are given by

$$
\langle r|A, l, a\rangle = r^{A+l} e^{-ar} \tag{47}
$$

These radial functions are not orthogonal. Their overlaps are given by

$$
\langle A, l, a|B, l, b\rangle = \int_0^\infty r^{A+B+2l} e^{-(a+b)r} r^2 dr = \frac{(A+B+2l+2)!}{(a+b)^{A+B+2l+3}} \tag{48}
$$

The values of these parameters for the functions sA, sB, sC, and pA are given in Table III.

The three p orbitals are orthogonal to the s orbitals and to each other by virtue of the orthogonality of the angular parts of the wave function (Table IV). The three s orbitals are not orthogonal, but an orthogonal basis may be constructed from them

TABLE III
Hydrogenic Orbitals for the Calculation of Ref. [77]

Orbital	$\|A, l, a\rangle$	$\langle r\|A, l, a\rangle$	A	l	a	$\langle \theta, \phi\|l, m\rangle$	m
sA	$\|0, 0, 4\rangle$	e^{-4r}	0	0	4	$\sqrt{\frac{1}{4\pi}}$	0
sB	$\|1, 0, 1\rangle$	re^{-r}	1	0	1	$\sqrt{\frac{1}{4\pi}}$	0
sC	$\|0, 0, 3\rangle$	e^{-3r}	0	0	3	$\sqrt{\frac{1}{4\pi}}$	0
pA	$\|0, 1, 1\rangle$	re^{-r}	0	1	1	$\sqrt{\frac{3}{8\pi}} \sin\theta e^{-i\phi}$	1
pA	$\|0, 1, 1\rangle$	re^{-r}	0	1	1	$\sqrt{\frac{3}{8\pi}} \cos\theta$	0
pA	$\|0, 1, 1\rangle$	re^{-r}	0	1	1	$\sqrt{\frac{3}{8\pi}} \sin\theta e^{i\phi}$	-1

TABLE IV
Overlap Integrals for the Calculation of Ref. [77]

Orbital	$\|A, l, a\rangle$	$\|0, 0, 4\rangle$	$\|1, 0, 1\rangle$	$\|0, 0, 3\rangle$
sA	$\langle 0, 0, 4\|$	$(1/16)^2$	$3!/5^4$	$2!/7^3$
sB	$\langle 1, 0, 1\|$	–	$4!/2^5$	$3!/4^4$
sC	$\langle 0, 0, 3\|$	–	–	$2!/6^3$

by the Gram–Schmidt procedure. The corresponding orthonormal orbitals are

$$
\begin{aligned}
|\phi_1\rangle &= \frac{1}{\sqrt{\langle 0,0,4|0,0,4\rangle}}\,|0,0,4\rangle = 16\,|0,0,4\rangle \\
|\phi_2\rangle &= |1,0,1\rangle - \frac{\langle 0,0,4|1,0,1\rangle}{\langle 0,0,4|0,0,4\rangle}\,|0,0,4\rangle \\
|\phi_3\rangle &= \frac{1}{\sqrt{\langle\psi_3|\psi_3\rangle}} 2^9 \left[\frac{3}{5^4}\frac{5^4 7^3 - 2^{17}}{2^5 5^4 7^3 - 3 2^{17} 7^3/5^4} - \frac{1}{\sqrt{\langle\psi_3|\psi_3\rangle}}\frac{1}{7^3}\right]|0,0,4\rangle \\
&\quad - \frac{1}{\sqrt{\langle\psi_3|\psi_3\rangle}}\frac{5^4 7^3 - 2^{17}}{2^5 5^4 7^3 - 3\times 2^{17} 7^3/5^4}\,|1,0,1\rangle + |0,0,3\rangle \\
|\phi_4\rangle &= \sqrt{\frac{4}{3}}\,|0,1,1\rangle
\end{aligned}
\tag{49}
$$

Where the coefficients in Eq. (18) in Ref. [77] can now be seen to arise as follows:

$$
\begin{aligned}
\frac{1}{\sqrt{\langle\psi_2|\psi_2\rangle}} &= 1.173302451 \\
\frac{1}{\sqrt{\langle\psi_2|\psi_2\rangle}}\frac{\langle 0,0,4|1,0,1\rangle}{\langle 0,0,4|0,0,4\rangle} = \frac{3!}{\sqrt{\langle\psi_2|\psi_2\rangle}}\left(\frac{4}{5}\right)^4 &= 2.883508103 \\
\frac{1}{\sqrt{\langle\psi_3|\psi_3\rangle}} 2^9 \left[\frac{3}{5^4}\frac{5^4 7^3 - 2^{17}}{2^5 5^4 7^3 - 3 2^{17} 7^3/5^4} - \frac{1}{\sqrt{\langle\psi_3|\psi_3\rangle}}\frac{1}{7^3}\right] &= -69.59716965 \\
\frac{1}{\sqrt{\langle\psi_3|\psi_3\rangle}}\frac{5^4 7^3 - 2^{17}}{2^5 5^4 7^3 - 3\times 2^{17} 7^3/5^4} &= 0.5968868832 \\
\frac{1}{\sqrt{\langle\psi_3|\psi_3\rangle}} &= 47.60738095 \\
\sqrt{\frac{4}{3}} &= 1.154700538
\end{aligned}
\tag{50}
$$

TABLE V
Compact Mapping of Boys's Basis for Beryllium

Boys Designation	Qubit State
$\mathcal{A}(sAsA)^1S(sBsB)^1S$	$\|0\rangle$
$\mathcal{A}(sBsB)^1S(sCsC)^1S$	$\|1\rangle$
$\mathcal{A}(sAsA)^1S(sBsC)^1S$	$\|2\rangle$
$\mathcal{A}(sAsC)^1S(sBsB)^1S$	$\|3\rangle$
$\mathcal{A}(sAsA)^1S(pApA)^1S$	$\|4\rangle$
$\mathcal{A}(sAsC)^1S(pApA)^1S$	$\|5\rangle$
$\mathcal{A}(sAsA)^1S(sCsC)^1S$	$\|6\rangle$
$\mathcal{A}(sAsB)^1S(sCsC)^1S$	$\|7\rangle$
$\mathcal{A}(sBsB)^1S(pApA)^1S$	$\|8\rangle$
$\mathcal{A}(sAsB)^1S(pApA)^1S$	$\|9\rangle$

TABLE VI
CI Matrix for Boys's Basis for Beryllium

	$\|0\rangle$	$\|1\rangle$	$\|2\rangle$	$\|3\rangle$	$\|4\rangle$	$\|5\rangle$	$\|6\rangle$	$\|7\rangle$	$\|8\rangle$	$\|9\rangle$
$\|0\rangle$	−14.4577	0.3967	0.2488	−0.9416	−0.1361	0	0.0541	0.0986	−0.1361	0.0276
$\|1\rangle$	0.3967	2.6051	−0.1257	−0.5601	0	0	0.0220	−0.1958	−0.0317	0
$\|2\rangle$	0.2488	−0.1257	−9.419	0.1499	−0.0099	0.0196	0.3534	1.0192	0	0.0070
$\|3\rangle$	−0.9416	−0.5601	0.1499	−5.7302	0	−0.1361	0.0269	−0.8085	0	−0.1780
$\|4\rangle$	−0.1361	0	−0.0099	0	−14.09	−0.9502	−0.0132	0	0.0220	−0.2816
$\|5\rangle$	0	0	0.0196	−0.1361	−0.9502	−5.3335	0.0215	0.0070	−0.0269	0.0407
$\|6\rangle$	0.0541	0.0220	0.3534	0.0269	−0.0132	0.0215	−3.7904	1.3259	0	0
$\|7\rangle$	0.0986	−0.1958	1.0192	−0.8085	0	0.0070	1.3259	1.5002	0	−0.0946
$\|8\rangle$	−0.1361	−0.0317	0	0	0.0220	−0.0269	0	0	−4.1937	0.0032
$\|9\rangle$	0.0276	0	0.0070	−0.1780	−0.2816	0.0407	0	0.0946	0.0032	−10.1121

After all that, the 6×6 CI matrix actually computed by Boys was

$$\begin{pmatrix} 0 & 0.3967 & 0.2488 & -0.9416 & -0.1361 & 0 \\ 0.3967 & 17.0628 & -0.1257 & -0.5601 & 0 & 0 \\ 0.2488 & -0.1257 & 5.0387 & 0.1499 & -0.0099 & 0.0196 \\ -0.9416 & -0.5601 & 0.1499 & 8.7275 & 0 & -0.1361 \\ -0.1361 & 0 & -0.0099 & 0 & 0.3677 & -0.9502 \\ 0 & 0 & 0.0196 & -0.1361 & -0.9502 & 9.1242 \end{pmatrix} \tag{51}$$

Given the CI matrix and established compact mapping for the co-detors, we can consider a nonscalable implementation of Boys's original calculation. Our starting point is the 6×6 version of the CI matrix. As noted previously, the task is to decompose powers of the evolution operator corresponding to this Hermitian matrix into a quantum circuit. This may be accomplished by a Cartan decomposition that treats the evolution operator as an action on a qubit and a qutrit. Although it is conventional to decompose unitaries over qubits, controlled quantum systems employing larger numbers of states than two have also been utilized in a number of experimental realizations [106]. A more conventional approach would utilize three-qubits in the state register with two three qubit states uninvolved in the calculation. The optimal three-qubit circuit requires 44 CNOT gates and 63 one-qubit gates [62] (Tables V and VI).

VI. CONCLUSIONS

Elucidation of small cases of quantum chemical calculations from the past literature provides an interesting experimental road map leading from calculations quantum computers can perform now, to calculations beyond the capabilities of comparable classical methods. Scalable methods for the calculation of ground-state properties using quantum computers have been studied extensively in the past few years. The known scalable techniques for implementing time evolution operators of fermionic systems, comprising the Jordan–Wigner transformation and Trotter–Suzuki decompositions, provide polynomial performance. However, they are most likely not yet optimized and further study of these scalable approaches is likely to be fruitful. At the other extreme, the approach detailed here considers the smallest calculations where nonscalable approaches that minimize the required number of qubits are used. These allow for early experimental implementation, the purpose of which is to demonstrate sufficient quantum control and coherence to obtain results of chemical accuracy for small systems. As we have argued here, a crossover point between scalable and nonscalable implementations will likely result in two experimental frontiers. At one frontier as much chemistry as possible will be crammed into the available qubit Hilbert space as possible, whether the approach used is scalable or not. At the other frontier we will be proceeding with the development of the scalable methods that we know will carry us beyond model systems and ultimately beyond the capabilities of any conceivable classical computing device.

Acknowledgments

The author would like to acknowledge productive discussions with Alan Aspuru-Guzik, James Whitfield, Ken Brown, and Jake Seeley. This work is supported by National Science Foundation awards CHE-1037992 and PHY-0955518.

REFERENCES

1. A. Y. Kitaev, A. H. Shen, and M. N. Vyalyi, *Classical and Quantum Computation*, Vol. 47 of *Graduate Studies in Mathematics*, American Mathematical Society, Providence, RI, 2000.
2. M. A. Nielsen and I. L. Chuang, *Quantum Computation and Quantum Information*, Cambridge University Press, Cambridge, UK, 2000.

3. Y. Makhlin, G. Schön, and A. Shnirman, *Rev. Mod. Phys.* **73**, 357–400 (2001).
4. C. E. Wieman, D. E. Pritchard, and D. J. Wineland, *Rev. Mod. Phys.* **71**, S253–S262 (1999).
5. D. J. Wineland, C. Monroe, W. M. Itano, D. Leibfried, B. E. King, and D. M. Meekhof, *J. Res. Natl. Inst. Stand. Technol.* **103**, 259 (1998).
6. L. M. K. Vandersypen and I. L. Chuang, *Rev. Mod. Phys.* **76**, 1037–1069 (2005).
7. P. Kok, W. J. Munro, K. Nemoto, T. C. Ralph, J. P. Dowling, and G. J. Milburn, *Rev. Mod. Phys.* **79**, 135–174 (2007).
8. E. Knill, R. Laflamme, and G. Milburn, *Nature* **409**, 46–52 (2001).
9. M. Saffman, T. Walker, and K. Mølmer, *Rev. Mod. Phys.* **82**(3), 2313–2363 (2010).
10. D. Loss and D. P. DiVincenzo, *Phys. Rev. A* **57**, 120–126 (1998).
11. D. D. Awschalom, D. Loss and N. Samarth, *Semiconductor Spintronics and Quantum Computation*, Springer-Verlag, Berlin, 2002.
12. P. M. Platzman and M. I. Dykman, *Science* **284**, 1967 (1999).
13. B. E. Kane, *Nature* **393**, 133 (1998).
14. B. E. Kane, *Fortschr. Phys.* **48**, 1023–1041 (2000).
15. P. Shor, *SIAM J. Comput.* **26**, 1484–1509 (1997).
16. L. K. Grover, *Phys. Rev. Lett.* **79**, 4709 (1997).
17. D. S. Abrams and S. Lloyd, *Phys. Rev. Lett.* **79**, 2586 (1997).
18. D. S. Abrams and S. Lloyd, *Phys. Rev. Lett.* **83**, 5162 (1999).
19. A. Y. Kitaev, Quantum measurements and the Abelian Stabilizer Problem. eprint quant-ph/9511026.
20. R. P. Feynman, *Intl. J. Theor. Phys.* **21**, 467 (1982).
21. S. Lloyd, *Science* **273**, 1073 (1996).
22. C. Zalka, *Proc. R. Soc. Lond. Ser. A* **454**, 313 (1998).
23. S. Wiesner, Simulations of Many-Body Quantum Systems by a Quantum Computer.
24. D. A. Lidar and H. Wang, *Phys. Rev. E* **59**, 2429 (1999).
25. A. Aspuru-Guzik, A. Dutoi, P. Love, and M. Head-Gordon, *Science* **309**, 1704–1707 (2005).
26. I. Kassal, S. Jordan, P. J. Love, M. Mohseni, and A. Aspuru-Guzik, *Proc. Nat. Acad. Sci.* **105**, (2008).
27. C. Negrevergne, T. S. Mahesh, C. A. Ryan, M. Ditty, F. Cyr-Racine, W. Power, N. Boulant, T. Havel, D. G. Cory, and R. Laflamme, *Phys. Rev. Lett.* **96**(17), 170501 (2006).
28. J. Benhelm, G. Kirchmair, C. F. Roos, and R. Blatt, *Nat. Phys.* **4**(6), 463–466 (2008).
29. B. P. Lanyon, C. Hempel, D. Nigg, M. Mueller, R. Gerritsma, F. Zaehringer, P. Schindler, J. T. Barreiro, M. Rambach, G. Kirchmair, M. Hennrich, P. Zoller, R. Blatt, and C. F. Roos, *Science* **333**(6052), 57–61 (2011).
30. M. Riebe, H. Häffner, C. F. Roos, W. Hänsel, J. Benhelm, G. P. T. Lancaster, T. W. Körber, C. Becher, F. Schmidt-Kaler, D. F. V. James, and R. Blatt, *Nature* **429**, 734 (2004).
31. J. T. Barreiro, M. Mueller, P. Schindler, D. Nigg, T. Monz, M. Chwalla, M. Hennrich, C. F. Roos, P. Zoller, and R. Blatt, *Nature* **470**(7335), 486–491 (2011).
32. T. Monz, P. Schindler, J. T. Barreiro, M. Chwalla, D. Nigg, W. A. Coish, M. Harlander, W. Haensel, M. Hennrich, and R. Blatt, *Phys. Rev. Lett.* **106**(13), 130506 (2011).
33. M. Mariantoni, H. Wang, T. Yamamoto, M. Neeley, R. C. Bialczak, Y. Chen, M. Lenander, E. Lucero, A. D. O'Connell, D. Sank, M. Weides, J. Wenner, Y. Yin, J. Zhao, A. N. Korotkov, A. N. Cleland, and J. M. Martinis, *Science* **333**(6052), 61–65 (2011).

34. T. Yamamoto, M. Neeley, E. Lucero, R. C. Bialczak, J. Kelly, M. Lenander, M. Mariantoni, A. D. O'Connell, D. Sank, H. Wang, M. Weides, J. Wenner, Y. Yin, A. N. Cleland, and J. M. Martinis, *Phys. Rev. B* **82**(18), 184515 (2010).
35. M. W. Johnson, M. H. S. Amin, S. Gildert, T. Lanting, F. Hamze, N. Dickson, R. Harris, A. J. Berkley, J. Johansson, P. Bunyk, E. M. Chapple, C. Enderud, J. P. Hilton, K. Karimi, E. Ladizinsky, N. Ladizinsky, T. Oh, I. Perminov, C. Rich, M. C. Thom, E. Tolkacheva, C. J. S. Truncik, S. Uchaikin, J. Wang, B. Wilson, and G. Rose, *Nature* **473**(7346), 194–198 (2011).
36. J. L. O'Brien, G. J. Pryde, A. Gilchrist, D. F. V. James, N. K. Langford, T. C. Ralph, and A. G. White, *Phys. Rev. Lett.* **93**(8), 080502 (2004).
37. A. Politi, J. C. F. Matthews, and J. L. O'Brien, *Science* **325**(5945), 1221–1221 (2009).
38. G. D. Marshall, A. Politi, J. C. F. Matthews, P. Dekker, M. Ams, M. J. Withford, and J. L. O'Brien, *Opt. Expr.* **17**(15), 12546–12554 (2009).
39. B. P. Lanyon, T. J. Weinhold, N. K. Langford, M. Barbieri, D. F. V. James, A. Gilchrist, and A. G. White, *Phys. Rev. Lett.* **99**(25), 250505 (2007).
40. P. Walther, K. Resch, T. Rudolph, and E. Schenck, *Nature* **434**, 169176 (2005).
41. S. F. Boys, G. B. Cook, C. M. Reeves, and I. Shavitt, *Nature* **178**, 1207 (1956).
42. B. P. Lanyon, J. D. Whitfield, G. G. Gillet, M. E. Goggin, M. P. Almeida, I. Kassal, J. D. Biamonte, B. J. Powell, M. Barbieri, A. Aspuru-Guzik, and A. G. White, *Molecular Energy Calculation on a Prototypical Optical Quantum Computer*, arXiv:0905.0887, 2009.
43. J. Du, N. Xu, X. Peng, P. Wang, S. Wu, and D. Lu, *Phys. Rev. Lett.* **104**(3), 1–4 (2010).
44. L. Veiss and J. Pittner, *J. Chem. Phys.* **133**, 194106 (2010).
45. R. Cleve, An Introduction to Quantum Complexity Theory, eprint quant-ph/9906111.
46. F. Bernstein and U. Vazirani, *SIAM* **26**, 1411–1473 (1997).
47. J. Kempe, A. Kitaev, and O. Regev, *SIAM J. Comput.* **35**, 1070 (2006).
48. T. Osborne, *Hamiltonian Complexity*, eprint arXiv:1106.5875.
49. M. R. Garey and D. S. Johnson, *Computers and Intractability: A Guide to the Theory of NP-Hardness*. W. H. Freeman, Gordonsville, 1979.
50. F. Barahona, *J. Phys. A* **15**, 3241 (1982).
51. M. Troyer and U.-J. Wiese, *Phys. Rev. Lett.* **94**, 170201 (2005).
52. N. Schuch and F. Verstraete, *Nat. Phys.* **5**(9), 1–4 (2009).
53. F. Dyson, *Notices Am. Math. Soc.* **56**(2), 212–223 (2009).
54. J. D. Whitfield, J. Biamonte, and A. Aspuru-Guzik, *Mol. Phys.* **109**, 735–750 (2011).
55. I. Kassal, J. D. Whitfield, A. Perdomo-Ortiz, M. H. Yung, and A. Aspuru-Guzik, *Ann. Rev. Phys. Chem.* **62**, 185–207 (2011).
56. D. Abrams and S. Lloyd, *Phys. Rev. Lett.* **83**, 5162–5165 (1999).
57. A. Barenco, C. H. Bennett, R. Cleve, D. P. DiVincenzo, N. Margolus, P. Shor, T. Sleator, J. A. Smolin, and H. Weinfurter, *Phys. Rev. A* **52**(5), 3457–3467 (1995).
58. N. Khaneja and S. J. Glaser, *Cartan Decomposition of SU(2N), Constructive Controllability of Spin Systems and Universal Quantum Computing. quant-ph/0502173*, 2000.
59. V. V. Shende, S. S. Bullock, and I. L. Markov, *Phys. Rev. A* **70**(1), 012310 (2004).
60. Y. Nakajima, Y. Kawano, and H. Sekigawa, *A New Algorithm for Producing Quantum Circuits Using kak Decompositions*, 2005.
61. H. Earp and J. K. Pachos, *A Constructive Algorithm for the Cartan Decomposition of* $su(2^n)$, arXiv.org:quant-ph/0505128, 2005.

62. B. Drury and P. J. Love, *J. Phys. A. Math. Theor.* **41**, 395305 (2008).
63. S. Helgason, *Differential Geometry, Lie Groups, and Symmetric Spaces*, Academic Press, New York, 1978.
64. G. Segal, R. Carter, and I. MacDonald, *Lectures on Lie Groups and Lie Algebras*, Cambridge University Press, Cambridge, UK, 1995.
65. Y. Makhlin, *Quant. Inf. Process.* **1**(4), 1–10 (2002).
66. J. Zhang, J. Vala, S. Sastry, and K. Whaley, *Phys. Rev. A* **67**(4), 042313 (2003).
67. G. Ortiz, J. E. Gubernatis, E. Knill, and R. Laflamme. *Phys. Rev. A* **64**(2), 022319 (2001).
68. S. Bravyi and A. Kitaev, *Ann. Phys.* **298**, 210–226 (2002).
69. J. T. Seeley, M. J. Richard, P. J. Love, *J. Chem. Phys.* **137**, 224109 (2012).
70. J. C. Slater, *Phys. Rev.* **38**, 1109 (1931).
71. J. C. Slater, *Phys. Rev.* **34**, 1293 (1929).
72. A. Szabo and N. Ostlund, *Modern Quantum Chemistry: Introduction to Advanced Electronic Structure Theory*, 1996.
73. H. F. Schaefer III, *Quantum Chemistry: The Development of* Ab Initio *Methods in Molecular Electronic Structure Theory*, Oxford University Press, New York, 1984.
74. H. Hettema. *Quantum Chemistry: Classic Scientific Papers*, World Scientific, Singapore, 2000.
75. H. M. James and A. S. Coolidge, *J. Chem. Phys.* **1**(12), 825–835 (1933).
76. S. F. Boys, *Proc. R. Soc. Lond. Ser. A* **200**, 542–554 (1950).
77. S. F. Boys, *Proc. R. Soc. Lond. Ser. A* **201**, 125–137 (1950).
78. G. R. Taylor and R. G. Parr, *Proc. Natl. Acad. Sci.* **38**, 154 (1952).
79. H. Shull and P.-O. Lowdin, *J. Chem. Phys.* **23**(8), 1565 (1955).
80. J. Miller, R. H. Friedman, R. P. Hurst, and F. A. Matsen, *J. Chem. Phys.* **27**(6), 1385–1387 (1957).
81. C. L. Pekeris, *Phys. Rev.* **112**(5), 1649–1658 (1958).
82. R. S. Mulliken and C. C. J. Roothaan, *Proc. Natl. Acad. Sci. USA* **45**(3), 394–398 (1959).
83. R. E. Watson, *Phys. Rev.* **119**(1), 170–177 (1960).
84. J. M. Foster and S. F. Boys, *Rev. Mod. Phys.* **32**(2), 305–307 (1960).
85. S. Hagstrom and H. Shull, *Rev. Mod. Phys.* **35**(3), 624–629 (1963).
86. C. F. Bender and E. R. Davidson, *J. Phys. Chem.* **70**(8), 2675–2685 (1966).
87. R. Mcweeny and K. A. Ohno, *Proc. R. Soc.* **A255**, 367 (1967).
88. J. F. Harrison, *J. Chem. Phys.* **47**, 2990 (1967).
89. C. D. Ritchie and H. F. King, *J. Chem. Phys.* **47**, 564 (1967).
90. J. W. Moskowitz and J. F. Harrison, *J. Chem. Phys.* **43**, 3550 (1967).
91. D. Neumann and J. W. Moskowitz, *J. Chem. Phys.* **49**(5), 2056–2070 (1968).
92. C. F. Bunge, *Phys. Rev.* **168**, 92 (1968).
93. R. C. Ladner and W. A. Goddard III, *J. Chem. Phys.* **51**(3), 1073–1087 (1969).
94. C. F. Bender and E. R. Davidson, *Phys. Rev.* **183**(1), 23–30 (1969).
95. R. P. Hosteny, R. R. Gilman, T. H. Dunning, A. Pipano, and I. Shavitt, *Chem. Phys. Lett.* **7**(3), 325–328 (1970).
96. S. F. Boys and G. B. Cook, *Rev. Mod. Phys.* **32**, 285 (1960).
97. O. Sinanoglu, *Proc. Natl. Acad. Sci. USA* **47**(8), 1217–1226 (1961).
98. M. P. Barnett, *Rev. Mod. Phys.* **35**(3), 571–572 (1963).

99. T. H. Dunning, W. J. Hunt, and W. A. Goddard, *Chem. Phys. Lett.* **4**(3), 147–150 (1969).
100. M. D. Newton, W. A. Lathan, W. J. Hehre, and J. A. Pople, *J. Chem. Phys.* **52**(8), 4064–4072 (1970).
101. K. L. Schuchardt, B. T. Didier, T. Elsethagen, L. Sun, V. Gurumoorthi, J. Chase, J. Li, and T. L. Windus, *J. Chem. Inf. Model.* **47**(3), 1045–1052 (2007).
102. C. M. REEVES, *Electronic Wavefunctions: A General Procedure for Automatic Computation Illustrated by an Application to the Water Molecule*. Ph.D. diss. Cambridge, 1957.
103. A. D. McLean, *J. Chem. Phys.* **32**(5), 1595–1597 (1960).
104. E. R. Davidson, *J. Chem. Phys.* **33**(5), 1577 (1960).
105. R. K. Nesbet, *Rev. Mod. Phys.* **32**(2), 272–273 (1960).
106. B. P. Lanyon, M. Barbieri, M. P. Almeida, T. Jennewein, T. C. Ralph, K. J. Resch, G. J. Pryde, J. L. O'Brien, A. Gilchrist, and A. G. White, *Nat. Phys.* **5**, 134–140 (2008).

INTRODUCTION TO QUANTUM ALGORITHMS FOR PHYSICS AND CHEMISTRY

MAN-HONG YUNG,[1,2] JAMES D. WHITFIELD,[2,3,4] SERGIO BOIXO,[2,5] DAVID G. TEMPEL,[2,6] and ALÁN ASPURU-GUZIK[2]

[1]*Center for Quantum Information, Institute for Interdisciplinary Information Sciences, Tsinghua University, Beijing 100084, P. R. China*
[2]*Department of Chemistry and Chemical Biology, Harvard University, 12 Oxford Street, Cambridge, MA 02138, USA*
[3]*NEC Laboratories America, 4 Independence Way, Princeton, NJ 08540, USA*
[4]*Department of Physics, Columbia University, 538 West 120th Street, New York, NY 10027, USA*
[5]*Google, 340 Main St, Venice, CA 90291, USA*
[6]*Department of Physics, Harvard University, 17 Oxford Street, Cambridge, MA 02138, USA*

Advances in Chemical Physics, Volume 154: Quantum Information and Computation for Chemistry, First Edition. Edited by Sabre Kais.

I. INTRODUCTION

Controllable quantum systems provide unique opportunities for solving problems in quantum chemistry and many-body physics that are intractable by classical computers. This approach is called "quantum simulation,"[1] and was pioneered by Feynman [1]. There are two different approaches for quantum simulation: analog or digital. In analog quantum simulation, dedicated physical systems are engineered to emulate the behavior of other quantum systems. A classic example is the use of atoms trapped in optical lattices to simulate the (Bose–)Hubbard model. Analog simulators are therefore special-purposed machines. On the other hand, digital simulation uses a universal quantum computer. Interestingly, a universal quantum computer is also capable, in principle, of factoring arbitrary long numbers [2], whereas a classical computer is not known to be able to perform the same task. For a recent review, see, for example, Refs [3] and [4].

One key advantage of simulations with quantum computers over classical computers is the huge Hilbert space available to faithfully represent quantum systems. Moreover, quantum simulation avoids many problems encountered in classical simulation. For example, many classical algorithms relying on Monte Carlo methods exhibit the so-called fermion sign problem that severely damages the performance of the algorithm. In quantum simulation, this problem can be avoided by either encoding the fully antisymmetrized wavefunction in the qubit states, or by performing Jordan–Wigner transformations in the second quantized Hamiltonian and turn it into a spin Hamiltonian as first suggested by Ortiz et al. [5]. The latter case will often result in nonlocal interaction terms, but it can still be simulated efficiently in a quantum computer.

The purpose of this chapter is to introduce the basic concepts of digital quantum simulation and several recent developments achieved in our group. We note that this is by no means a comprehensive review of the whole literature in quantum simulation. We will selectively cover materials that we find most useful to convey

[1] Unfortunately, the term "quantum simulation" in the community of computational physics refers to numerical simulation of quantum systems using classical computers.

an overall idea about the current status of quantum digital simulation. Several review articles [6–9] and book chapters [10–13] already present a different emphasis. This review also contains some new material, including new descriptions of the simulation in the first and second quantized representations. We also lay out a new point of view for the perturbative update of thermal states from smaller to bigger quantum systems, and a new bound for the change of a thermal state due to a perturbation.

A. Quantum Computational Complexity and Chemistry

1. An Exponential Wall for Many-Body Problems

The theory of computational complexity studies the scaling of the resources necessary to solve a given problem as a function of input size. Problems are considered to be "easy," or efficiently solvable, if the time (or number of steps) for solving the problem scales as a polynomial of the input size n. For example, sorting a list of n items will take at most $O(n^2)$ steps. On the other hand, problems are considered "hard" if the scaling is exponential in n. This exponential scaling is essentially true in the worst case for almost all many-body problems in physics and chemistry [14]. A concise discussion of this point is given by Kohn [15], where the exponential scaling of the Hilbert space of many-electron problems is referred to as the "Van Vleck catastrophe." The argument presented is as follows: if for each molecule, the accuracy to which one can approximate the state is $(1 - \epsilon)$ (under a suitable metric), then for n nonoverlapping (and nonidentical) molecules, the approximation worsens exponentially as $(1 - \epsilon)^n$. In the next section, we discuss the connection of many-body problems with computational complexity further. A more comprehensive review of quantum computational complexity appears in [16].

2. Computational Complexity of Quantum Simulation

The study of the computational complexity of problems in quantum simulation helps us better understand how quantum computers can surpass classical computers. It has also spurred new developments in computational complexity. For simplicity, computational complexity is often formulated using decision problems. A decision problem resolves if some condition is true or false (e.g., Is the ground-state energy of the system below a certain critical value?). Although the answer to decision problems is either "yes" or "no," one can keep asking questions in a binary search fashion. For instance, one could attempt to determine in this way the ground-state energy to an arbitrarily high accuracy.

A complexity class contains a set of computational problems that share some common properties about the computational resources required for solving them. We briefly summarize a few important examples of complexity classes of decision problems.

a. P *and* NP *Problems.* The complexity class P contains all decision problems that are solvable in a polynomial time with a classical computer (more precisely, a deterministic Turing machine). Roughly speaking, solving a problem in a polynomial time refers to the cases where the number of steps for solving the problem scale as a polynomial power instead of exponentially. This approach is considered "efficient" but, of course, exceptions are possible. For example, problems that scale as $O(n^{10000})$ may take very long time to finish, compared with ones that scale exponentially as $O(1.0001^n)$.

Nevertheless, from a theoretical perspective, this division allows for considerable progress to be made without considering the minutiae of the specific system or implementation. However from a practical standpoint, the order of the polynomial may be very important; especially in chemistry where an algorithm is applied to many molecules and many geometries. That said, the notion of polynomial versus exponential makes sense when considering Moore's "law": the density of transistors in classical computers doubles every 2 years.[2] If the algorithm runs in exponential time, one may be forced to wait several lifetimes in order for an instance to become soluble due to better classical hardware.

Practically, many hard problems typically fall into the complexity class NP, which contains decision problems whose "yes" instances can be efficiently verified to be true with a classical computer given an appropriate "solution" or witness. There is no doubt that P is a subclass of NP, that is,

$$\mathsf{P} \subset \mathsf{NP} \tag{1}$$

As an example, finding the prime factors of an integer belongs to an NP problem; once the factors are given, then it is easy to check the answer by performing a multiplication. Interestingly, finding the ground-state energy of the Ising model

$$\sum_{(i,j)\in E} \sigma_z^i \sigma_z^j + \sum_{i\in V} \sigma_z^i \tag{2}$$

where (V, E) is a planar graph, is an NP-complete [17]. It implies that if a polynomial algorithm for finding the ground-state energy is found, then all of the problems in NP could be solved in polynomial time. In other words, it will imply P = NP, a result considered highly unlikely. A rigorous proof or disproof of this statement would constitute a significant breakthrough.[3]

[2] The exponential growth in the computational density is expected to cease sometime this century, highlighting the importance of new methods of computation such as quantum computation. The growth in CPU clock speed has already ceased.

[3] P versus NP is one of the Millennium Problems of the Clay Mathematics Institute [http://www.claymath.org/millennium/P_vs_NP/].

It is believed, but not known with certainty, that quantum computers are not capable of solving all NP problems efficiently. Nevertheless, as already mentioned, they can solve the integer-factoring problem efficiently. It is believed that the complexity of integer factoring is intermediate between P and NP [2].

b. BQP and QMA Problems. The quantum analog of P and NP problems are, respectively, the BQP (bounded-error quantum polynomial time) and QMA (quantum Merlin Arthur) problems.[4] BQP is the class of (decision) problems that are solvable by a quantum computer in polynomial time. QMA is the class of (decision) problems that can be verified by a quantum computer in polynomial time. Like NP-problems, the QMA class covers many problems that are important to physics and chemistry [18–20]. For example, the ground-state problem of Hamiltonians involving local interaction terms is known to be QMA-complete [10,21]. For more discussion on topics of computational complexity and quantum simulation, readers may find the following references useful: [22–27].

The key point here is that so far it is not known whether quantum computers can solve NP and QMA problems efficiently. In fact, many attempts (see, e.g., Refs [28–34]) show that exponential resources are required to solve problems in these classes. Nevertheless, many problems in physics and chemistry do exhibit symmetries and structures that we could exploit to construct efficient quantum simulation algorithms. This is the main theme of the discussion in the rest of the chapter.

B. Basic Quantum Algorithms for Digital Quantum Simulation

Digital quantum simulation cannot be easily understood without a detour into the basics of quantum algorithms. Quantum algorithms are procedures for applying elementary quantum logic gates to complete certain unitary transformations of the input state. The quantum computer state is usually written in terms of qubits (two-level systems). In the two-dimensional Hilbert space of a single qubit, we label the upper and lower eigenstates of σ^z as $|0\rangle$ and $|1\rangle$. Note that the choice of σ^z as the computational basis is arbitrary. This is called the computational basis and the matrix representation of operators and states are written in this basis unless otherwise stated. The unitary transformations of the qubits may be visualized using quantum circuit diagrams introduced later to explain some of the more complex quantum algorithms.

It is known that any unitary gate can be decomposed into some sets of universal quantum logic gates that contain single- and two-qubit operations [12]. The

[4] More precisely, BQP is analogous to the classical complexity class BPP, which refers to problems that can be solved with randomized algorithms in a classical computer in polynomial time, subject to a bounded error probability.

first gate of interest is the single-qubit Hadamard transformation defined (in the computational basis) as

$$\mathsf{H} = \frac{1}{\sqrt{2}} \begin{bmatrix} 1 & 1 \\ 1 & -1 \end{bmatrix}$$

The Hadamard gate transforms between the σ^z basis and the σ^x basis ($|\pm\rangle = (|0\rangle \pm |1\rangle)/\sqrt{2}$) and will be used throughout the chapter. A second gate of interest is the CNOT (controlled not) gate, which is a nontrivial two-qubit gate. It leaves one input qubit unchanged and acts with $\sigma^x = |0\rangle\langle 1| + |1\rangle\langle 0|$ on the second qubit when the first qubit is in the state $|1\rangle$. The first qubit is called the control, and the NOT operation is applied coherently when the control qubit is in a superposition of computational basis states. Symbolically, the gate is written as CNOT$= |1\rangle\langle 1| \otimes \sigma^x + |0\rangle\langle 0| \otimes I$. The Hadamard and CNOT gates are not universal for quantum computation, and in fact quantum algorithms with only these gates can be simulated efficiently classically as shown by the Knill–Gottesman theorem [12]. Therefore, this gate set must be augmented by other single-qubit gates, which can always be expressed by single-qubit rotations, R_x, R_y, and R_z where R_x is defined as $\exp[-i\sigma^x\theta/2]$ for real angle θ.

There are two elementary algorithms, namely, quantum Fourier transform (QFT) and phase estimation algorithm (PEA), that play important roles in many applications in quantum simulation. We turn our attention to them now.

1. Quantum Fourier Transform

Given a vector with N elements $(x_0, x_1, \ldots, x_{N-1})$, in classical computation, the discrete Fourier transform outputs another vector of N numbers $(y_0, y_1, \ldots, y_{N-1})$ through the following relation:

$$y_k = \frac{1}{\sqrt{N}} \sum_{j=0}^{N-1} x_j e^{2\pi i jk/N} \tag{3}$$

In quantum computation, for any given quantum state, $|\phi\rangle = \sum_{x=0}^{N-1} \phi(x)\,|x\rangle$, the goal of the quantum Fourier transform U_{QFT} is to perform the following unitary transformation:

$$U_{\text{QFT}}\,|\phi\rangle = \sum_{k=0}^{N-1} \tilde{\phi}(k)\,|k\rangle \tag{4}$$

where $\tilde{\phi}(k) = (1/\sqrt{N}) \sum_{x=0}^{N-1} \phi(x) e^{2\pi i xk/N}$ is the Fourier-transform of the discrete function $\phi(x)$ [compare with Eq. (3)]. Due to the linearity of U_{QFT}, it is sufficient to consider the transformation of the basis vectors such that

$$U_{\mathrm{QFT}} |x\rangle = \frac{1}{\sqrt{N}} \sum_{k=0}^{N-1} e^{2\pi i x k/N} |k\rangle \tag{5}$$

For a system of n qubits, the number of gates required for such a transformation is $O(n^2)$ [12]. For the classical case [see Eq. (3)], one will require $O(n2^n)$ gates to complete the same transformation, for example, with fast Fourier transform (FFT). This may seem to suggest that quantum computers are exponentially more efficient in performing the task of discrete Fourier transformation. However, the caveat is that one cannot directly compare QFT with the classical FFT. The reason is that if we want to obtain a particular Fourier-transform coefficient, say $\tilde{\phi}(k)$, from the quantum state in Eq. (4), it would still require exponentially many steps to extract the information (phase and amplitude), for example, through quantum state tomography where many measurements are used to analyze the state [12].

Nevertheless, QFT is essential in many applications in digital quantum simulation. As we shall see in Section II.B.2, it allows us to simulate the time dynamics of particles efficiently by moving between the position and momentum representations. Another important application of the QFT is phase estimation, which is discussed next.

2. *Phase Estimation Algorithm*

The phase estimation algorithm U_{PEA} is an essential component for many quantum algorithms for quantum simulation, as well as the celebrated factoring algorithm [2]. Loosely speaking, the PEA can be considered as a realization of the von Neumann measurement scheme (without the last projective measurement) in the eigenvalue basis $|a_k\rangle$ of any Hermitian observable A (e.g., Hamiltonian H). More precisely, if we prepare a register of m ancilla qubits initialized in the state $|000...0\rangle$, then for any given state $|\phi\rangle = \sum_k c_k |a_k\rangle$, we have

$$U_{\mathrm{PEA}} |\phi\rangle |000...0\rangle \approx \sum_k c_k |a_k\rangle |A_k\rangle \tag{6}$$

where, for the moment, we assume that the A_ks are the m-integer-digit representations (i.e., $A_k \in \{0, 1, 2, \ldots, 2^m - 1\}$) of the eigenvalues of A. The projective measurement cannot be implemented perfectly in general (hence, the $\approx$ symbol). We will see where the errors come from as we go through the details of the algorithm.

Suppose that we are given an eigenstate $|a_k\rangle$ of the Hermitian observable A. The goal of PEA is to determine A_k, given that we are able to simulate a unitary operator W where

$$W |a_k\rangle = e^{2\pi i \phi_k} |a_k\rangle \tag{7}$$

and $\phi_k \equiv A_k/2^m$. The first step of the PEA is to apply Hadamard gates to each of the ancilla qubits. This results in an equal superposition of states

$$|S\rangle \equiv \frac{1}{\sqrt{2^m}} \sum_{x=0}^{2^m-1} |x\rangle \tag{8}$$

where x is a m-digit binary number. Then, taking each ancilla qubit j as a control qubit, we apply the controlled-W^{2^j-1} gate to the state $|S\rangle\,|a_k\rangle$; this effectively performs the following operation:

$$|x\rangle\,|a_k\rangle \rightarrow |x\rangle\, W^x\,|a_k\rangle \tag{9}$$

Of course, from Eq. (7), the right-hand side gives only a phase factor, namely $\exp(2\pi i x \phi_k)$. The resulting state is

$$\left(\frac{1}{\sqrt{2^m}} \sum_{x=0}^{2^m-1} e^{2\pi i x \phi_k}\,|x\rangle \right) |a_k\rangle \tag{10}$$

Comparing this state with that in Eq. (4), and assuming the special cases where the phase angle ϕ_x can be expressed exactly by m binary digits, the application of the inverse of the quantum Fourier transform U_{QFT} will convert the state in Eq. (10) into the following state

$$|A_k\rangle\,|a_k\rangle \tag{11}$$

Because the unitary operator U_{PEA} is linear, the procedure applies to any initial state. For this particular case, where A_ks are integers, we have shown that PEA is effectively a projective measurement as advertised in Eq. (6).

For the general case, where the A_ks are real numbers, the corresponding ϕ_ks will have precision beyond $1/2^m$, which is the source of the errors in the expression of Eq. (6). The overall error decreases when we increase the number of ancilla qubits and perform several QFTs in parallel (we refer to Ref. [35] for a detailed error analysis). More precisely, if we want to achieve a p-bit precision of ϕ_k with an error less than ϵ, one will need more than $m = p + \log(2 + 1/2\epsilon)$ ancilla qubits. In general, implementing the operator W^k requires k times as many resources as those needed for simulating W. Therefore, the scaling of the quantum gates of PEA grows exponentially when we increase the precision p of the phase measurement. This result is consistent with that of the general sampling theory in classical signal processing, where the precision of the Fourier spectrum $\delta\omega$ goes as the inverse of the total time T sampled, that is, $\delta\omega \sim O(1/T)$, because the cost of the quantum simulation is proportional to T, and T grows exponentially with the number of bits of precision.

II. DIGITAL QUANTUM SIMULATION

A. Overview

Broadly speaking, the steps involved in carrying out a digital quantum simulation consist of three parts: state preparation, time evolution, and measurement of observables. Measurement of Hermitian observables can be achieved via the phase estimation method [36–38] described before. Other applications [39–44] or quantities of physical interest such as the partition function [45,46], can be obtained through variants of the methods employed in state preparation and time evolution, and we will skip them in this review. Next, we give an overview of state preparation and simulation of time evolution. It turns out that many methods of state preparation also depend on the time evolution itself. Therefore, we will first cover the methods of time evolution before state preparation.

B. Simulation of Time Evolution

The simulation of the time evolution of quantum state $|\psi\rangle$ under Hamiltonian H according to the Schrödingers equation ($\hbar = 1$),

$$i\frac{\partial}{\partial t}|\psi\rangle = H(t)|\psi\rangle \tag{12}$$

is one of the key applications of quantum computation. If, for example, the time-evolution operator

$$U(t) = \exp(-iHt) \tag{13}$$

can be simulated efficiently, then the eigenvalues of H might be obtained through the phase estimation algorithm.[5] As mentioned in the introduction, Feynman [1] investigated the possibility of simulating quantum systems using another quantum system and conjectured that a class of universal quantum simulators evolved under a Hamiltonian with local interactions. This conjecture was justified by Lloyd [47], who argued that any Hamiltonian

$$H = \sum_{i=1}^{m} H_i \tag{14}$$

which can be decomposed into m local terms $\{H_i\}$ can be simulated efficiently by a universal quantum computer. Each H_i term acts on at most k qubits (or quantum subsystems). The key idea is based on the Trotter splitting or "trotterization" of

[5] Moreover, it can also be exploited for quantum cooling (see Section II.D).

all noncommuting operators,

$$e^{-iHt} \approx \left(e^{-iH_1t/n}e^{-iH_2t/n}\cdots e^{-iH_mt/n}\right)^n \tag{15}$$

where the approximation can be made arbitrarily tight by refining the time-slicing (i.e., increasing n).

There exist higher-order approximations (Suzuki–Trotter formulas) that reduce the error even further. For instance, the second-order approximation is given by

$$e^{-iHt} \approx \left(\left(e^{-ih_1\frac{\Delta t}{2}}\cdots e^{-ih_{N-1}\frac{\Delta t}{2}}\right)e^{-ih_N\Delta t}\left(e^{-ih_{N-1}\frac{\Delta t}{2}}\cdots e^{-ih_1\frac{\Delta t}{2}}\right)\right)^{\frac{t}{\Delta t}} + O(t(\Delta t)^2) \tag{16}$$

A quantum circuit on n qubits, which approximates $U(\tau)$, with error at most ϵ, is efficient if the number of one- and two-qubit gates involved is polynomial in the scaling of the problem (i.e., poly(n, τ, $1/\epsilon$) with $\tau = t/||H||$).

1. Suzuki–Trotter Formulas

We now briefly review the use of Suzuki–Trotter formulas in quantum simulation for time-independent sparse Hamiltonians, providing an introduction to the quantum simulation literature. Continuing the work of Lloyd [47], works by Aharonov and Ta-Shma [48] and Childs [49] show that black-box sparse Hamiltonians are too efficiently simulatable. Sparsity here means that the number of elements per row is bounded by some polynomial of n, while the dimension of the Hilbert space is $D = 2^n$. It is also required that each matrix element can be retrieved efficiently. Ref. [48] used a coloring scheme to decompose the Hamiltonian into a sum of 2×2 block diagonal matrices. This coloring scheme has been updated in several references [50–52]. The coloring scheme and black-box simulation will not be discussed further.

Berry et al. [50] were the first to approach the general problem of simulating noncommuting Hamiltonians by using higher order Suzuki–Trotter formulas. Papageorgiou and Zhang [53] returned to this issue and their contributions will be discussed later. The important results of Berry et al. [50] are the following:

1. The use of higher order Trotter-Suzuki decompositions to bound the number of noncommuting exponentials, N_{exp}, necessary to carry out a simulation for some amount of time t
2. A proof of a no-go theorem for sublinear black-box simulation
3. Improvements upon the coloring scheme of Refs [48] and [49] for black-box simulation of sparse Hamiltonians.

The simulations in this chapter are concerned with the first two results and they will be explained in more detail after describing the Suzuki–Trotter formulas.

Suzuki has studied and extended the Trotter formula essentially continuously since 1990 and this work was reviewed in Ref. [54]. The recursive formulas introduced by Suzuki define a fractal pattern where a combination of forward and backward propagation leads to an improved approximation of the desired exponential. Suzuki defines higher-order Trotter formulas in a recursive way. Beginning with the split operator formula, $e^{Ax/2}e^{Bx}e^{Ax/2}$, for m operators, the following series of equations were derived

$$S_2(x) = \left(\prod_{k=1}^{m} e^{h_k x}\right)\left(\prod_{k=m}^{1} e^{h_k x}\right) \tag{17}$$

$$S_4(x) = S_2(z_2 x)^2 S_2((1-4z_2)x) S_2(z_2 x)^2 \tag{18}$$

$$\vdots \quad \vdots$$

$$S_{2k}(x) = S_k(z_k x)^2 S_k((1-4z_k)x) S_k(z_k x)^2 \tag{19}$$

The values of the constants $\{z_j\}$ are selected so that S_{2j} is correct through $2j^{\text{th}}$ order and it can be shown [54] that $z_i = (4 - 4^{1/(2i-1)})^{-1}$. With m noncommuting Hamiltonians, the first-order approximation takes $m = N_{\text{exp}}$, and for the split operator formula, S_2, the number of exponentials is $2m - 1$. In general, $2(m-1)5^{k-1} + 1$ exponentials are used for the S_{2k} approximant.

For the kth order Suzuki–Trotter, with m Hamiltonians in the sum, and error tolerance given by ε, Berry et al. [50] give a bound on the number of exponentials by bounding each order of the Suzuki–Trotter formula. Papageorgiou and Zhang [53] presented an improvement by noting that the relative ratio of Hamiltonian norms is also important. The main idea is that if some of the Hamiltonians have small weight, then their exponentials can be effectively ignored.

The optimal order of Trotter decomposition, k^*, is determined by selecting the best compromise between time step length and a decomposition using more exponentials. In Berry et al. [50], this was worked out for unstructured sparse Hamiltonians $N_{exp} \geq ||H||t$. The lower bound on the generic cost of simulating an evolution was by contradiction, and relied on the lower bounds to quantum mechanical problems based on the polynomial method [55]. This bound could be violated given sublinear simulation time. In a departure from the methods discussed so far, Childs [56] used a quantum walk-based approach to push the scaling closer to linear in the reweighed time and Raesi et al. [57] looked at designing quantum circuits for quantum simulation.

For problems in chemistry, it is more natural to represent Hamiltonians in terms of first- and second-quantized forms. In the following, we will describe how to exploit the special structure of molecular Hamiltonians to simulate the time dynamics in quantum computers.

2. *First-Quantized Representation*

In the first-quantized form, the nonrelativistic molecular Hamiltonians H decomposes in a kinetic T and potential V terms, that is,

$$H = T + V \tag{20}$$

The kinetic term includes the contribution from the nuclei and electrons separately,

$$T = -\sum_i \frac{\hbar^2}{2M_i}\nabla_i^2 - \sum_j \frac{\hbar^2}{2m_e}\nabla_j^2 \tag{21}$$

where M_i is the mass of the nucleus i and m_e is the electron mass. The potential energy term comprises of the Coulomb interaction among the nuclei, among the electrons, and between the nuclei and electrons. Explicitly

$$V(\mathbf{r}, \mathbf{R}) = \frac{e^2}{4\pi\varepsilon_0}\sum_{i<j}\frac{Z_i Z_j}{|\mathbf{R}_i - \mathbf{R}_j|} + \frac{e^2}{4\pi\varepsilon_0}\sum_{i<j}\frac{1}{|\mathbf{r}_i - \mathbf{r}_j|} - \frac{e^2}{4\pi\varepsilon_0}\sum_{i,j}\frac{Z_i}{|\mathbf{R}_i - \mathbf{r}_j|} \tag{22}$$

where e is the electric charge and Z_i is the charge of nuclei i. The coordinates of nuclei i and electron j are denoted by $\mathbf{R}_i$ and $\mathbf{r}_j$. We will use the notation $\mathbf{r} = (\mathbf{r}_1, \mathbf{r}_2, \mathbf{r}_3, \ldots)$ (and similarly for $\mathbf{R}$). We also ignore the spin degrees of freedom, which can be incorporated easily.

The general wave function can be represented in the position basis as

$$|\Psi\rangle = \sum_{\mathbf{r},\mathbf{R}} \Psi(\mathbf{r}, \mathbf{R})\, |\mathbf{r}_1\mathbf{r}_2\mathbf{r}_3\cdots\rangle\, |\mathbf{R}_1\mathbf{R}_2\mathbf{R}_3\cdots\rangle \tag{23}$$

where each electronic or nuclear coordinate is represented on its own grid over m qubits resulting in a total of Bm qubits to represent the state of B particles. Note that the grid encoded in m qubits has 2^m points. The complex wave function $\Psi(\mathbf{r}, \mathbf{R})$ in addition to being properly normalized must also be antisymmetrized (or symmetrized for Bosons). Abrams and Lloyd [58] and Ward et al. [59] consider the necessary antisymmetrization process for fermions in first quantization.

To simulate the dynamics [6,60,61], we note that although the kinetic and potential terms do not commute with each other, both can be represented as diagonal operators in momentum and position basis, respectively. By using the quantum Fourier transform U_{QFT}, it is natural to decompose the time evolution as

$$e^{-iHt} \approx \left(U_{\mathrm{QFT}}^{\dagger} e^{-iTt/n} U_{\mathrm{QFT}} e^{-iVt/n}\right)^n \tag{24}$$

In fact, this method is known as the split-operator method [62,63]. Higher-order Suzuki–Trotter formulas can also be applied, as described earlier. This method was applied to quantum computing in a number of works [6,60,61,64–66].

In the context of quantum computing, it remains to find a method to induce a coordinate-dependent phase factor such that

$$|\mathbf{r}_1\mathbf{r}_2\mathbf{r}_3\cdots\rangle\,|\mathbf{R}_1\mathbf{R}_2\mathbf{R}_3\cdots\rangle \quad\rightarrow\quad e^{-iV(\mathbf{r},\mathbf{R})\delta t}\,|\mathbf{r}_1\mathbf{r}_2\mathbf{r}_3\cdots\rangle\,|\mathbf{R}_1\mathbf{R}_2\mathbf{R}_3\cdots\rangle \tag{25}$$

where $\delta t \equiv t/n$, and similarly for the kinetic term in the Fourier basis. An efficient method[6] is implicitly described in the book by Kitaev et al. [10] (pp. 131–135), which was further developed and adapted to the chemistry context by Kassal et al. [61]. We sketch the idea here for completeness. First, we will assume that the potential energy term is rescaled to become dimensionless, and projected into a range of integer values such that $0 \leq V(\mathbf{r}, \mathbf{R}) \leq 2^m - 1$, where m should be sufficiently large to allow appropriate resolution of $V(\mathbf{r}, \mathbf{R})$ in the integer representation. Next, we define a more compact notation $|\mathbf{r}, \mathbf{R}\rangle \equiv |\mathbf{r}_1\mathbf{r}_2\mathbf{r}_3...\rangle\,|\mathbf{R}_1\mathbf{R}_2\mathbf{R}_3...\rangle$, and an algorithmic operation $\mathcal{A}$ to be performed in the position basis

$$\mathcal{A}\,|\mathbf{r}, \mathbf{R}\rangle\,|\mathbf{s}\rangle \quad\rightarrow\quad |\mathbf{r}, \mathbf{R}\rangle\,|\mathbf{s} \oplus V(\mathbf{r}, \mathbf{R})\rangle \tag{26}$$

where $|\mathbf{s}\rangle, \mathbf{s} = \mathbf{1}, \mathbf{2}, \mathbf{3}, \ldots,$ is a quantum state of m ancilla qubits, and $\oplus$ is addition modulo 2^m. Suppose now that the ancilla qubits are initialized in the following state:

$$|\mathbf{q}\rangle \equiv \frac{1}{\sqrt{M}} \sum_{\mathrm{s}=0}^{M-1} e^{2\pi i \mathbf{s}/M}\,|\mathbf{s}\rangle \tag{27}$$

where $M \equiv 2^m$. This state is the Fourier transform of $|1\rangle$. Then the desired phase generating operation described in Eq. (25) can be achieved using controlled σ^z rotations after applying $\mathcal{A}$ to the state $|\mathbf{r}, \mathbf{R}\rangle\,|\mathbf{q}\rangle$. A similar procedure is applied to the kinetic term to complete the Trotter cycle.

An alternative approach to implement the controlled-phase operation described in Eq. (25) is the following: first include a register of qubits initialized as $|\mathbf{0}\rangle$. Then in a similar (but not identical) way as that described in Eq. (26), we define the operation

$$\tilde{\mathcal{A}}\,|\mathbf{r}, \mathbf{R}\rangle\,|\mathbf{0}\rangle \rightarrow |\mathbf{r}, \mathbf{R}\rangle\,|V(\mathbf{r}, \mathbf{R})\,\delta t\rangle \tag{28}$$

where we used $\mathbf{0} \oplus V(\mathbf{r}, \mathbf{R})\,\delta t = V(\mathbf{r}, \mathbf{R})\,\delta t$. The state $|V(\mathbf{r}, \mathbf{R})\,\delta t\rangle$ is the binary representation $\{x_1x_2x_3\ldots x_m\}$ defined through the following equality,

$$V(\mathbf{r}, \mathbf{R})\,\delta t \equiv 2\pi \times 0.x_1x_2x_3\ldots x_m = 2\pi \sum_{k=1}^{m} \frac{x_k}{2^k} \tag{29}$$

[6] An alternative method was proposed by Benenti and Strini [66], but it scales exponentially with the number qubits.

Now, we can decompose the overall phase as follows,

$$e^{-iV(\mathbf{r},\mathbf{R})\delta t} = e^{-i2\pi x_1/2} e^{-i2\pi x_2/2^2} \cdots e^{-i2\pi x_m/2^m} \tag{30}$$

This decomposition can be achieved through the application of m local phase gates $R_k \equiv |0\rangle\langle 0| + \exp\left(-2\pi i/2^k\right)|1\rangle\langle 1|$ for each ancilla qubit. This approach requires the ancilla to be uncomputed (i.e., the inverse of the operation in Eq. (28)) in the last step.

3. *Second-Quantized Representation*

The first-quantization method is universally applicable to any molecule. The shortcoming is that it does not take into account the physical symmetrization properties of the underlying quantum system. When a suitable set of basis functions is employed, the size of the problem can be significantly reduced. It is known as the second-quantization approach in quantum chemistry, which can be extended for quantum simulation.

Most studies on quantum simulation based on first quantization methods use grids to represent wave functions, while works employing second quantization methods generally use atomic or molecular orbitals as a basis set for the wave functions. We will take the latter approach here. Nevertheless, the choice of basis is not the key difference between the first and second quantization. Indeed, a basis set of delta functions (or approximations to delta functions) could be used to represent a grid within second quantization. On the other hand, the storage of the same wave function is different in second and first quantization. For example, a two-particle wave function with the first particle at site i and the second at site j, is represented as $|\text{coord}_i\rangle|\text{coord}_j\rangle$ in first quantization, and as $|0\cdots 1_i \cdots 1_j \cdots 00\rangle$ in second quantization.

The starting point of the second-quantization approach [67–70] is the Born-Oppenheimer approximation, where the nuclear coordinates $\mathbf{R}$ are taken to be classical variables. This allows us to focus on the electronic structure problem. Ignoring the nuclear kinetic and the nuclear–nuclear interaction terms, the molecular Hamiltonian in Eq. (20) can be expressed as

$$H = \sum_{pq} h_{pq} a_p^\dagger a_q + \frac{1}{2}\sum_{pqrs} h_{pqrs} a_p^\dagger a_q^\dagger a_r a_s \tag{31}$$

where the fermionic creation operator $a_p^\dagger$ creates an electron in the p mode from the vacuum, that is, $a_p^\dagger|\text{vac}\rangle = |p\rangle$. Denote $\chi_p(\mathbf{r})$ as the single-particle wave function corresponding to mode p.[7] Then, the explicit form for the single-electron integrals

[7] Here, **r** refers to the coordinates of one particular electron.

is given by

$$h_{pq} \equiv -\int d\mathbf{r}\chi_p^*(\mathbf{r})\left(\frac{\hbar^2}{2m_e}\nabla^2 + \frac{e^2}{4\pi\varepsilon_0}\sum_i \frac{Z_i}{|\mathbf{R}_i - \mathbf{r}|}\right)\chi_q(\mathbf{r}) \tag{32}$$

and the electron–electron Coulomb interaction term is,

$$h_{pqrs} \equiv \frac{e^2}{4\pi\varepsilon_0}\int d\mathbf{r}_1 d\mathbf{r}_2 \frac{\chi_p^*(\mathbf{r}_1)\chi_q^*(\mathbf{r}_2)\chi_r(\mathbf{r}_2)\chi_s(\mathbf{r}_1)}{|\mathbf{r}_1 - \mathbf{r}_2|} \tag{33}$$

These integrals have to be precalculated with classical computers before encoding them into the quantum algorithms. If we keep k single-particle orbitals, then there are $O(k^4)$ terms. More details of the formalism of second-quantized electronic structure theory in the Born-Oppenheimer approximation can be found in Ref. [71].

To simulate time dynamics in a quantum computer, we can apply the same Trotterization idea described earlier [see Eq. (15)], and simulate separately the terms

$$\exp(-ih_{pq}a_p^\dagger a_q \delta t) \quad \text{and} \quad \exp(-ih_{pqrs}a_p^\dagger a_q^\dagger a_r a_s \delta t) \tag{34}$$

Because the simulation of every single exponential term in a quantum computer is costly, due to error-correction overheads as discussed in Ref. [72], one simplification we can make is to group the terms of single-particle terms into two particle terms. This is possible for electronic problems with a fixed number N of electrons. Consider any N-fermionic state, then the identity operator I_N is equivalent to a summation of the following single-body number operators,

$$(1/N)\sum_s a_s^\dagger a_s \Leftrightarrow I_N \tag{35}$$

which means that we can write

$$a_p^\dagger a_q = \frac{1}{N-1}\sum_s a_p^\dagger a_s^\dagger a_s a_q \tag{36}$$

The last equation is a sum of two-electron terms, and can be absorbed into the precomputed values of h_{pqrs}. Now, denoting the new values as $\tilde{h}_{pqrs}$, the Hamiltonian H reduces to

$$H = \frac{1}{2}\sum_{pqrs}\tilde{h}_{pqrs}a_p^\dagger a_q^\dagger a_r a_s \tag{37}$$

Therefore, we are left only with simulating the two-body term in Eq. (34).

One challenge we need to overcome is the fermionic nature of the operators $a_p^\dagger$ and a_q, which comes from the antisymmetrization requirement of fermionic wave functions. A first step to overcome this challenge is to map the occupation representation to the qubit configuration. Explicitly, for each fermionic mode j, we represent the qubit state $|0\rangle_j \equiv |\downarrow\rangle_j$ as an unoccupied state, and similarly $|1\rangle_j \equiv |\uparrow\rangle_j$ as an occupied state. To enforce the exchange symmetry, we apply the Jordan-Wigner transformation [5,69]:

$$a_j^\dagger = \left(\prod_{m<j} \sigma_m^z\right) \sigma_j^- \quad \text{and} \quad a_j = \left(\prod_{m<j} \sigma_m^z\right) \sigma_j^+ \tag{38}$$

where

$$\sigma^\pm \equiv (\sigma^x \pm i\sigma^y)/2 \tag{39}$$

By using Eqs. (38) and (39), we can now write the fermionic Hamiltonian in Eq. (37) as a spin Hamiltonian involving products of Pauli matrices $\{\sigma^x, \sigma^y, \sigma^z\}$:

$$H_{\text{spin}} = \sum_{pqrs} \sum_{abcd} g_{pqrs}^{abcd} \theta_{pqrs} \sigma_p^a \sigma_q^b \sigma_r^c \sigma_s^d \tag{40}$$

where the set of indices $\{p, q, r, s\}$ is summed over the fermionic modes, and $\{a, b, c, d\}$ is either x or y. The operator θ_{pqrs} keeps track of the σ_zs; for example, if $p > q > r > s$, we then have

$$\theta_{pqrs} = \left(\prod_{p>i>q} \sigma_i^z\right) \times \left(\prod_{r>j>s} \sigma_j^z\right) \tag{41}$$

The punchline here is that the Hamiltonian becomes a polynomial sum of products of spin operators, and each operator is locally equivalent to σ_z. Therefore, the nontrivial part of simulating the time dynamics of the fermionic Hamiltonian is to simulate the nonlocal interaction terms of the following form:

$$\exp(-ig\sigma^z\sigma^z\sigma^z \cdots \sigma^z \delta t) \tag{42}$$

where g is some constant. It can be achieved by a series of controlled-NOT together with a local operation (see, e.g., Fig. 4.19 of Ref. [12]), or the phase-generating method similar to the one described in the previous section [cf. Eq. (25)]. The explicit circuits for simulating the time evolution operators can be found in Ref. [69].

4. *Open-System Dynamics*

In quantum mechanics, the time evolution dynamics of a closed system is always described by a unitary transformation of states, $U(t)\,\rho\,U^{\dagger}(t)$. However, nonunitary dynamics occur when the dynamics of the system of interest S is coupled to the environment B, as in

$$\rho_S(t) \equiv \mathrm{Tr}_B\left[U(t)\,\rho_{SB}\,U^{\dagger}(t)\right] \tag{43}$$

After some approximations this evolution can often be described by a (Markovian) quantum master equation in Lindblad form [73–75],

$$\frac{d}{dt}\rho_s(t) = -i\left[H_s, \rho_s\right] + \sum_{\alpha,\beta} m_{\alpha\beta}\left(\left[\Lambda_\alpha \rho_s, \Lambda_\beta^{\dagger}\right] + \left[\Lambda_\alpha, \rho_s \Lambda_\beta^{\dagger}\right]\right) \tag{44}$$

where H_s is the system Hamiltonian, $m_{\alpha\beta}$ is a positive matrix, and Λ_α is a linear basis of traceless operators. This quantum master equation is relevant in many physical, chemical, and biological processes at finite temperature [76,77]. Further, this equation has many applications in quantum information processing, including preparing entangled states (from arbitrary initial states) [78–82], quantum memories [83], and dissipative quantum computation [84]. It has been shown that the quantum master equation can be simulated by a unitary quantum circuit with polynomial resource scaling [85,86]. The basic idea is as follows: we first rewrite the master equation [Eq. (44)] in the form,

$$\frac{d}{dt}\rho_s(t) = \mathcal{L}(\rho_s) \tag{45}$$

where $\mathcal{L}$ is a superoperator. Similar to the unitary dynamics, we can define the superoperator version of the propagator $\mathcal{K}(t_1, t_0)$ through the relation

$$\rho_s(t_1) = \mathcal{K}(t_1, t_0)(\rho_s(t_0)) \tag{46}$$

for all values of time $t_1 \geq t_0$. Suppose we consider a finite time interval T, which can be divided into m small time intervals Δt (i.e., $T = m\Delta t$). Then similar arguments [86] based on Trotterization show that the following approximation,

$$\mathcal{K}(T) \approx \underbrace{\mathcal{K}(\Delta t)\,\mathcal{K}(\Delta t)\,\mathcal{K}(\Delta t)\ldots\mathcal{K}(\Delta t)}_{m\ \text{times}} \tag{47}$$

indeed converges when the division size goes to zero (i.e., $\Delta t \to 0$). The remaining part of the argument is to show that each of the small-time propagator terms $\mathcal{K}(\Delta t)$ can be simulated efficiently with a quantum circuit. This is generally true if the superoperator $\mathcal{L}$ is a finite (polynomial) sum of local terms [85].

C. State Preparation

We have discussed how quantum dynamics can be simulated efficiently with a quantum computer, but we have not yet discussed how quantum states of physical or chemical interest can be initialized on the quantum computer. In fact, both thermal and ground states of physical Hamiltonians can be prepared by incorporating the methods of simulating the time dynamics, as we shall explain later in this section.

We first consider a strategy to prepare quantum states that can be efficiently described by some integrable general function (e.g., a Gaussian wave packet). Before we provide a general description, it may be instructive to consider the case of creating a general (normalized) two-qubit state,

$$f_{00}\left|00\right\rangle + f_{01}\left|01\right\rangle + f_{10}\left|10\right\rangle + f_{11}\left|11\right\rangle \tag{48}$$

from the initial state $|00\rangle$. First of all, we will assume that all the coefficients f_{ij} are real numbers, as the phases can be generated by the method described in Eq. (25). Now, we can write the state in Eq. (48) as

$$g_0\left|0\right\rangle \otimes \left(\frac{f_{00}}{g_0}\left|0\right\rangle + \frac{f_{01}}{g_0}\left|1\right\rangle\right) + g_1\left|1\right\rangle \otimes \left(\frac{f_{10}}{g_1}\left|0\right\rangle + \frac{f_{11}}{g_1}\left|1\right\rangle\right) \tag{49}$$

where $g_0 \equiv \sqrt{f_{00}^2 + f_{01}^2}$ is the probability to find the first qubit in the state $|0\rangle$, and similarly for $g_1 \equiv \sqrt{f_{10}^2 + f_{11}^2}$. The form in Eq. (49) suggests that we can use the following method to generate the general state of Eq. (48) from $|00\rangle$.

1. Apply a rotation, such that $|0\rangle \to g_0\left|0\right\rangle + g_1\left|1\right\rangle$, to the first qubit. The resulting state becomes

$$\left(g_0\left|0\right\rangle + g_1\left|1\right\rangle\right)\left|0\right\rangle \tag{50}$$

2. Perform the following controlled operation:

$$\left|x\right\rangle\left|0\right\rangle \to \left|x\right\rangle\left(\frac{f_{x0}}{g_x}\left|0\right\rangle + \frac{f_{x1}}{g_x}\left|1\right\rangle\right) \tag{51}$$

   where $x = \{0, 1\}$.

The final state is exactly the same as that in Eq. (49) or Eq. (48).

Consider, more generally, the preparation of the following n-qubit quantum state [6,59,87,88]:

$$\sum_{x=0}^{2^n-1} f(x)\left|x\right\rangle \tag{52}$$

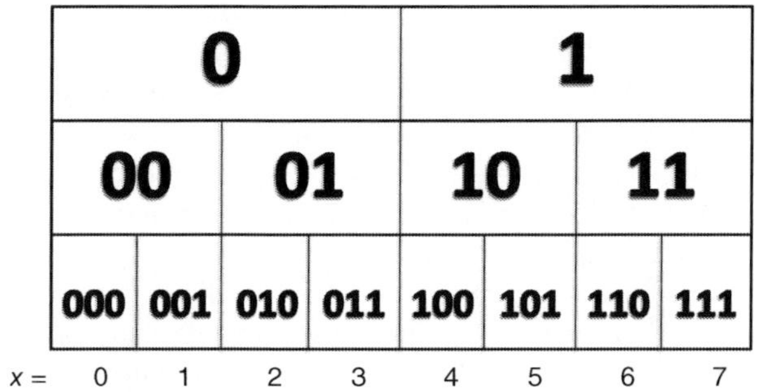

Figure 1. Example for the state preparation method. The space is divided in $L = 8$ divisions. The "0" division refers to the left half of the space, $(0 \leq x < L/2)$, and similarly for the "1" division. Finer resolution is achieved by increasing the number of labeling digits.

Here, again we will assume that $f(x)$ is real. We can image that this is the wave function of a particle in 1D. The first qubit describes whether the particle is located in the left half $|0\rangle$ or right half $|1\rangle$ of the line divided by $L \equiv 2^n$ divisions. The first step is therefore to rotate the first qubit as $\cos\theta_0 \, |0\rangle + \sin\theta_1 \, |1\rangle$, where

$$\cos^2\theta_0 = \sum_{0 \leq x < L/2} f(x)^2 \tag{53}$$

represents the probability of locating the particle at the left side (i.e., $0 \leq x < L/2$) (Fig. 1). The next step is to apply the following controlled rotation:

$$|x\rangle \, |0\rangle \rightarrow |x\rangle \left(\frac{\cos\theta_{x0}}{\cos\theta_x} \, |0\rangle + \frac{\cos\theta_{x1}}{\cos\theta_x} \, |1\rangle \right) \tag{54}$$

where

$$\cos^2\theta_{00} = \sum_{0 \leq x < L/4} f(x)^2 \quad \text{and} \quad \cos^2\theta_{01} = \sum_{L/4 \leq x < L/2} f(x)^2 \tag{55}$$

represents the probability for finding the particle in the "00" division $(0 < 0 < L/4)$ and the "01" division $(L/4 \leq 0 < L/2)$, respectively; analogous arguments apply for the "10" and "11" divisions. In the remaining steps, similar controlled operations described in Eq. (54) are applied, which depend on the division of the controlling qubits. The θ rotation angles have to be calculated explicitly. It is, therefore, necessary that the function $f(x)$ is efficiently integrable [87]. This is expected, as otherwise such a simple algorithm would be able to solve the random-field Ising spin models and other NP-complete problems. We will cover the creation of thermal states later. In the next section, we will consider methods for preparing ground states.

1. Preparing Ground States

a. Phase-Estimation Based Methods. Finding ground states of classical Hamiltonians (e.g., a random-field Ising model) is known to be NP-hard. Therefore, it is not expected that a quantum computer would be able to solve it efficiently in general. Furthermore, preparing the ground state of a general quantum Hamiltonian H is even more challenging as both eigenvalues and eigenvectors must be obtained, and this problem would solve the QMA complexity class, the quantum analog of NP. Fortunately, many problems in physics and chemistry exhibit structures and symmetries that allow us to arrive at solutions that are approximations of the exact solutions; for example, the BCS wave function related to superconductivity and superfluidity, and the Laughlin wave function related to the fractional quantum Hall effect (FQHE), both provide good predictions for the corresponding many-body problems. The quality of other approximated solutions, such as the mean-field or Hartree–Fock approximation, may vary from problem to problem.

The quality of the approximated solution (or trial solution) $|\psi_T\rangle$ can be quantified by the fidelity F defined by

$$F \equiv |\langle e_0 \, |\psi_T\rangle|^2 \tag{56}$$

where $|e_0\rangle$ is the target ground state (assumed unique) of the Hamiltonian H of interest. The physical meaning of F is that if one can implement a projective measurement $\{|e_k\rangle \langle e_k|\}$ in the eigenvector basis $\{|e_k\rangle\}$ of H to the trial state $|\psi_T\rangle$, then the probability of getting the ground state $|e_0\rangle$ is exactly equal to F, and can be implemented with the phase estimation algorithm [36]. A similar procedure can produce low energy eigenstates even if there is no gap [29].

With the methods of the previous paragraph, if the projection on the ground state fails, the initial approximation must be reconstructed again. Because the projection fails with probability $1 - F$, the approximate preparation must be done $1/(1 - F)$ times on average. Projection success can be improved using phase amplification (a trick similar to Grover's search) to $\sqrt{1/(1 - F)}$ "coherent" initial state preparations. A different method is possible if, as is often the case, we can evolve with a Hamiltonian $\tilde{H}$ for which the state approximation is a ground state. Assume that the approximated ground state has an energy gap bounded by Δ for $\tilde{H}$ and the exact ground state has a similar gap for H. Then we can transform a single preparation of the approximated state into the exact ground state using around $1/(1 - F)$ phase estimations, each implemented with a time evolution for a time of $1/\Delta$ [89].

Therefore, a quantum computer, even if it cannot solve all ground-state problems efficiently, is capable to leverage classical trial states and solve a broader class of problems than those efficiently solvable by classical computers.

b. Adiabatic State Preparation. The adiabatic method is an alternative way to prepare ground states [48,89–93]. The original idea is due to Farhi et al. [90]. We first must be able to efficiently prepare the ground state $|\psi(0)\rangle$ of a certain initial Hamiltonian $H(0) = H_i$. Then we change the Hamiltonian $H(t)$ slowly, for example,

$$H(t) = (1 - t/T)\, H_i + (t/T)\, H_f \tag{57}$$

Notice that for many reasonable choices of H_i and most physical Hamiltonians H_f the Hamiltonian $H(t)$ can be simulated using the methods of Section II.B. Nevertheless, common two-body Hamiltonians could be simulated directly.

If the change from H_i (when $t = 0$) to the target Hamiltonian H_f (when $t = T$) is slow enough, then the state $|\psi(t)\rangle$, satisfying the time-dependent Schrödinger equation

$$i\hbar \frac{d}{dt} |\psi(t)\rangle = H(t)\, |\psi(t)\rangle \tag{58}$$

follows the corresponding eigenstate of $H(t)$ adiabatically. Therefore, $|\psi(T)\rangle$ is close to the ground state of the target Hamiltonian H_f. A sufficient condition for the total time T to ensure the adiabaticity for a linear interpolation between two Hamiltonians is

$$T \gg \frac{\|\partial_s H(s)\|}{\Delta_{\min}^2} \tag{59}$$

where $s \equiv t/T$. Here,

$$\Delta_{\min} \equiv \min_{0 \leq s \leq 1} \left(E_1(s) - E_0(s)\right) \tag{60}$$

is the minimum gap between the instantaneous eigen-energies $E_1(s)$ and $E_0(s)$ of the first excited state and the ground state. The following bound has a better dependence on the minimum gap and it also holds for general (nonlinear) interpolations if the rate of change of the instantaneous eigenstate $|\partial_s \psi(s)\rangle$ is known [92]

$$T > \frac{\mathcal{L}^2}{\Delta_{\min}} \tag{61}$$

Here, $\mathcal{L}$ is the path length given by the equation

$$\mathcal{L} = \int \|\partial_s \psi(s)\rangle\| ds \tag{62}$$

Using the methods of Section II.B adiabatic evolutions can be simulated efficiently on a quantum circuit. That is, for cases where one may not be able to physically implement $H(t)$, it is still possible to turn the adiabatic state preparation into a quantum algorithm and simulate the adiabatic process in a digital quantum computer. Furthermore, in this case the total time of the adiabatic evolution can be improved to[8] [89]:

$$T > \frac{\mathcal{L}}{\Delta_{\min}} \tag{63}$$

The remaining question is, in terms of finding ground states: How good are adiabatic algorithms? As we have seen, the performance, or computational complexity, of adiabatic algorithms generically depends on the scaling of the minimal gap $\Delta_{\min}$. Even for classical target Hamiltonians H_f, whether adiabatic algorithms success in solving NP-problems is still a controversial issue [95,96]. Numerical results suggest that for the classical satisfiability (SAT) problems, the scaling of the gap would be exponential [28,31,34]. If the target Hamiltonian is quantum, the problem is QMA-complete. Nevertheless, we can in principle apply the adiabatic algorithm to the trial states to improve the ground-state fidelity [97], which gives us a higher probability of projecting into the exact ground state by the phase estimation algorithm discussed in the previous section.

2. *Preparing Thermal States Using Quantum Metropolis*

We now consider the preparation of density matrices for thermal states

$$\rho_{\text{th}} = e^{-\beta H}/\text{Tr}\left(e^{-\beta H}\right) \tag{64}$$

where H can be a quantum or classical Hamiltonian, and $\beta = 1/T$ is the inverse temperature. We simplify the notation by choosing our units so that the Boltzmann constant k_B is 1. Zalka [6] and Terhal and DiVincenzo [98] proposed to simulate the Markovian dynamics of the system by modeling the interaction with a heat-bath by some ancilla qubits. A similar idea has been recently investigated by Wang et al. [99]. Terhal and DiVincenzo [98] also attempted to prepare thermal states by generalizing classical Metropolis-type sampling [100]. This first quantum Metropolis algorithm was limited by the fact that it was not possible to control the update rule for the Metropolis algorithm, which would generally lead to a slow convergence rate of the underlying Markov chain. A significant improvement upon this work has been presented recently in Ref. [101] with the "quantum Metropolis sampling" algorithm. This algorithm also constructs a Markov-chain whose fixed point is a thermal state Eq. (64), but the transitions between states can be engineered to achieve faster convergence. The underlying time cost of this algorithm scales as $O(1/\Delta)$ [102], where Δ is the eigenvalue gap of the Markov matrix associated with the Metropolis algorithm.

[8] Boixo and Somma [94] have shown that this expression for the total evolution time is also optimal.

Szegedy [103] introduced a quantum algorithm to speed up classical Markov chains. Richter [104] extended this method to some quantum walks with decoherence. Szegedy's method has also been applied to achieve a quadratic speed up of classical simulated annealing algorithms [105,106]. Further, Yung and Aspuru-Guzik [107] achieved a similar speedup of the quantum Metropolis sampling algorithm for quantum Hamiltonians. This algorithm outputs a coherent encoding of the thermal state (CETS):

$$|\psi_{\text{th}}\rangle = \sum_k \sqrt{e^{-\beta E_k}/\mathcal{Z}}\,|e_k\rangle \tag{65}$$

where $\mathcal{Z}$ is the partition function and E_k is the eigenvalue associated with the eigenvector $|e_k\rangle$ of the Hamiltonian H.

Markov chain approaches are practical for many applications. However, for systems such as spin glasses, the eigenvalue gap Δ of the corresponding Markov matrices typically become exponentially small, making it inefficient. Several alternative approaches have been already introduced in the literature. Among them exploiting the transfer-matrix structure of spin systems [33,108]; mapping the CETS as the ground state of certain Hamiltonian [109,110]; and methods based on quantum phase estimation [30,32,111]. In the next section, we modified one of these algorithms to prepare thermal states, building up from smaller to bigger subsystems.

3. *Preparing Thermal States with Perturbative Updates*

The quantum Metropolis algorithms of the previous section extend the advantages of Markov chain Monte Carlo methods to the quantum case, even if we do not know how to diagonalize the quantum Hamiltonian. It is expected that, as in the classical case, they will exhibit good performance for most Hamiltonians. Nevertheless, for complex systems, such as strongly correlated molecules, it might be difficult to design rules to choose appropriate Markov chain update candidate states, or the convergence rate to the thermal state might still be too slow. In this section, we will show how, even in the worst case, quantum algorithms for preparing thermal states will exhibit a substantial speed up over classical algorithms, elaborating upon the method of Bilgin and Boixo [32].

The core of this algorithm is a perturbative update subroutine that builds the thermal state $\rho^{(\epsilon)} \propto e^{-\beta(H+\epsilon h)}$ from the state $\rho^{(0)} \propto e^{-\beta H}$. We can use this subroute to build thermal states of complex systems out of thermal states of their components (see Fig. 2). For this we start with the thermal states of subsystems with Hamiltonians H_1 and H_2, and use them to prepare the thermal state of the linked system with Hamiltonian $H_1 + H_2 + h$. The coupling Hamiltonian h is introduced perturbatively with a sequence of updates:

$$\rho^{(0)} \to \rho^{(\epsilon)} \to \rho^{(2\epsilon)} \to \cdots \to \rho^{(1)} \tag{66}$$

Figure 2. Pictorial representation of the perturbative update method. The top figure depicts two quantum systems with Hamiltonians H_1 and H_2 whose thermal states we can prepare (maybe through prior perturbative updates). The bottom figure depicts a single system where the two quantum systems of the top figure have been linked with Hamiltonian h. The perturbative update technique is a method to prepare the thermal state of the linked system from thermal states of the smaller subsystems.

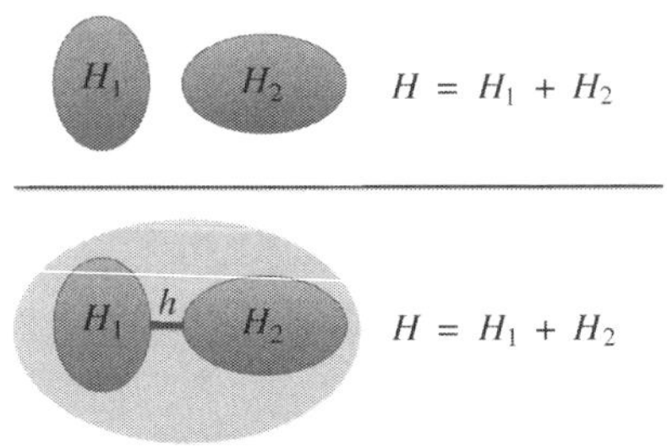

Quite possibly the thermal states of the smaller subsystems have themselves been prepared with a perturbative update over still smaller pieces. As in the quantum Metropolis case, it is not necessary to know how to diagonalize the corresponding Hamiltonian.

The perturbative update subroutine is probabilistic, and succeeds with probability $1 - \epsilon\beta \mathrm{Tr}\,\rho^{(0)} h$, which gives the dominant cost of the algorithm. If the perturbative update fails, we must reconstruct the state $\rho^{(0)}$. The probability of failure is given by the maximum change of a thermal state $\rho^{(0)} \propto e^{-\beta H}$ introduced by a perturbation ϵh, which we now bound. We denote with $Z = \mathrm{Tr}\,\rho^{(0)}$ the partition function of $\rho^{(0)}$. Using the Dyson series expansion in imaginary time we write

$$\frac{e^{-\beta(H+\epsilon h)\lambda}}{Z} = \frac{e^{-\beta H\lambda}}{Z} - \frac{\epsilon\beta}{Z}\int_0^{\lambda} d\lambda_1\, e^{-\beta H(\lambda-\lambda_1)} h e^{-\beta H\lambda_1} + \cdots \tag{67}$$

The appropriate measure of the difference between two density matrices is the trace norm $\|\cdot\|_{\mathrm{Tr}}$. The reason is that this norm bounds the difference of arbitrary measurement results for those two states. The trace norm for an arbitrary operator A is given by the sum of the eigenvalues of $\sqrt{A^\dagger A}$, and often scales with the dimension of the operator. We want to do better for the trace norm of the difference between a thermal state and its perturbation, because their dimension grows exponentially (in the number of subsystems). We give such a bound next.

We will use the following inequality, which applies to all unitarily invariant matrix norms [9]

$$\left|\left|\left| \int_0^1 A^t X B^{1-t} dt \right|\right|\right| \leq 1/2 |||AX + XB||| \tag{68}$$

[9] See, for instance, Theorem 5.4.7 in Ref. [112].

Applying this inequality to the trace norm of the Dyson series of a perturbed thermal state we obtain the bound

$$(1/Z)\left\|\epsilon\beta\int_0^1 d\lambda_1\, e^{-\beta H(1-\lambda_1)}he^{-\beta H\lambda_1}\right\|_{\mathrm{Tr}} \leq \epsilon\beta\|h\| \tag{69}$$

where $\|h\|$ is the operator norm of h. Notice that the operator norm $\|h\|$ is the highest eigenvalue of h, and does not scale with the dimension of H (or even h).[10]

The perturbative update subroutine is composed of two operations. The first operation implements the quantum map

$$(\rho^{(0)}) \rightarrow (1 - \epsilon\beta h/2)\rho^{(0)}(1 - \epsilon\beta h/2)\mathcal{N} \tag{70}$$

where $\mathcal{N}$ is just a normalization factor. Similar to the algorithms of the previous section, this map is implemented with phase estimation and a conditional rotation on an ancillary system. The ancillary system is then measured. This measurement can fail, which corresponds to implementing the wrong transformation in the thermal state. The success rate is $1 - \epsilon\beta\mathrm{Tr}\,\rho^{(0)}h$. When the measurement of the ancilla system fails, the thermal state can not be recovered, and we must start again from the beginning. The cost of the phase estimation is $\epsilon^{-1}\beta^{-1}\|h\|^{-2}$. This operation can be understood as an update of the Gibbs probabilities of $\rho^{(0)}$ to those of $\rho^{(\epsilon)}$. The second operation of the perturbative update is a transformation to the eigenbasis of $\rho^{(\epsilon)}$. It is implemented by "dephasing" in that basis, which is achieved by evolving for a random amount of time (with expectation time $\epsilon\|h\|$) using the Hamiltonian $H + \epsilon h$. This completes the perturbative update subroutine.

D. Algorithmic Quantum Cooling

Yung et al. [113] presented an algorithmic quantum cooling approach that transforms any input state ρ_{in} into an output state ρ_{out} that is guaranteed to have lower energy with respect to a given Hamiltonian H. Explicitly,

$$\mathrm{Tr}\,(H\rho_{out}) < \mathrm{Tr}\,(H\rho_{in}) \tag{71}$$

In principle, this algorithm can cool the resulting quantum state to a state arbitrarily close to the ground state of H. Nevertheless, like the ground-state algorithms of Section II.C.1, the efficiency is related to the energy gap Δ between the ground state and the excited state(s). Depending on how the algorithm is implemented, this dependence can scale like $O(1/\Delta^2)$ or $O(1/\Delta)$.

[10] Although strictly speaking we have derived the bound on the change of the thermal state here only to the second order, it can be shown to be valid to all orders. For that, we use the exact formula for the perturbation $\epsilon\beta\int_0^1 d\lambda_1\, e^{-\beta H(1-\lambda_1)}he^{-\beta(H+h)\lambda_1}$, and the same matrix norm inequality. In addition, we need to use the bound for the change on the partition from Poulin and Wocjan [30].

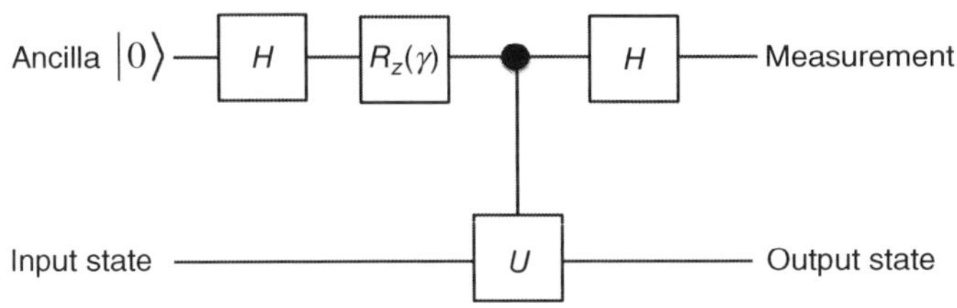

Figure 3. Quantum circuit diagram for the quantum cooling algorithm. Here, $\mathsf{H} = \frac{1}{\sqrt{2}}(|0\rangle + |1\rangle)\langle 0| + \frac{1}{\sqrt{2}}(|0\rangle - |1\rangle)\langle 1|$ is the Hadamard gate, $\mathsf{R_Z}(\gamma) = |0\rangle\langle 0| - ie^{i\gamma}|1\rangle\langle 1|$ is a local phase gate, and $U(t) = e^{-iH_s t}$, is the time t evolution operator simulating the dynamics of the system under the Hamiltonian H_s.

Algorithmic quantum cooling first entangles an ancilla qubit with the system state. When the ancilla qubit is measured, a result of $|0\rangle$ correlates with a cooler system state. On average, however, no gain or loss of energy occurs. This measurement is used to gain information, just like a Maxwell's demon. The measurement outcome of the ancilla qubit in algorithmic quantum cooling can be mapped into a 1D random walk. The walker starts at $x = 0$. For the cooling outcome, the walker makes a step toward the positive side $x > 0$, and toward the negative side $x < 0$ for the heating outcome. If the walker moves too far to the negative side, the procedure is restarted. For some range of parameters, whenever the walker goes to the negative side $x < 0$, the quantum state is guaranteed to be hotter than the original state. Therefore, removing these hot walkers will reduce the average energy over an ensemble of walkers, just like in evaporative (or "coffee") cooling of gas molecules. The procedure stops once the walker has moved sufficiently to the positive side.

1. Basic Idea of the Quantum Cooling Method

We now sketch the basic working mechanism of algorithmic quantum cooling. The core component of this cooling algorithm consists of four quantum gates (see Fig. 3).[11] The first gate is a so-called Hadamard gate

$$\mathsf{H} \equiv \frac{1}{\sqrt{2}}(|0\rangle + |1\rangle)\langle 0| + \frac{1}{\sqrt{2}}(|0\rangle - |1\rangle)\langle 1| \tag{72}$$

It is followed by a local phase gate

$$\mathsf{R_Z}(\gamma) \equiv |0\rangle\langle 0| - ie^{i\gamma}|1\rangle\langle 1| \tag{73}$$

where the parameter γ plays a role in determining the overall efficiency of the cooling performance of the algorithm. The interaction with the Hamiltonian H,

[11] Similar quantum circuits are used in DQC1 and phase estimation, for instance.

which can be either quantum or classical, is encoded in the time evolution operator

$$U(t) = e^{-iH_s t} \tag{74}$$

As already explained, time evolution can be implemented efficiently in a quantum computer.

The operation of the circuit in Fig. 3 on input state $|\psi_{in}\rangle$ is as follows.

Step 1. State initialization,

$$|\psi_{in}\rangle\,|0\rangle \tag{75}$$

with the ancilla state in $|0\rangle$.

Step 2. Apply the Hadamard gate, $\mathsf{H} = \frac{1}{\sqrt{2}}(|0\rangle + |1\rangle)\langle 0| + \frac{1}{\sqrt{2}}(|0\rangle - |1\rangle)\langle 1|$, and the local phase gate $\mathsf{R_Z}(\gamma) = |0\rangle\langle 0| - ie^{i\gamma}|1\rangle\langle 1|$ to the ancilla qubit

$$|\psi_{in}\rangle\left(|0\rangle - ie^{i\gamma}|1\rangle\right)/\sqrt{2} \tag{76}$$

Step 3. Apply the controlled-$U(t)$ to the system state,

$$\left(|\psi_{in}\rangle\,|0\rangle - ie^{i\gamma}U(t)\,|\psi_{in}\rangle\,|1\rangle\right)/\sqrt{2} \tag{77}$$

Step 4. Apply the Hadamard to the ancilla qubit again, which produces the following output state:

$$\Lambda_0\,|\psi_{in}\rangle\,|0\rangle + \Lambda_1\,|\psi_{in}\rangle\,|1\rangle \tag{78}$$

where $\Lambda_j \equiv \left(I + (-1)^{j+1} ie^{i\gamma}U\right)/2$ for $j = \{0, 1\}$.

A projective measurement on the ancilla qubit in the computational basis $\{|0\rangle, |1\rangle\}$ yields one of the two (unnormalized) states

$$\left(I \pm ie^{i\gamma}U\right)|\psi_{in}\rangle \tag{79}$$

Their mean energy is either higher (for outcome $|1\rangle$, x is decreased by 1) or lower (for outcome $|0\rangle$, x is increased by 1) than that of the initial state $|\psi_{in}\rangle$.

To justify this assertion, let us expand the input state,

$$|\psi_{in}\rangle = \sum\nolimits_k c_k\,|e_k\rangle \tag{80}$$

in the eigenvector basis $\{|e_k\rangle\}$ of the Hamiltonian H. Note that

$$\left\|\left(1 \pm ie^{i\gamma}U\right)|e_k\rangle\right\|^2 = 2(1 \pm \sin\phi_k) \tag{81}$$

where $\phi_k \equiv E_k t - \gamma$ depends on the eigen-energy E_k of H. For simplicity, we will assume that one can always adjust the two parameters, γ and t, such that

$$-\frac{\pi}{2} \leq \phi_k < \frac{\pi}{2} \tag{82}$$

for all nonnegative integers k. Then, the factors $(1 - \sin \phi_k)$ are in descending order of the eigen-energies, and the opposite is true for the factors $(1 + \sin \phi_k)$. Therefore, apart from an overall normalization constant, the action of the operator $(I \pm ie^{i\gamma} U)$ is to scale each of the probability weights $|c_k|^2$ by an eigen-energy dependent factor $(1 \pm \sin \phi_k)$, that is,

$$|c_k|^2 \rightarrow |c_k|^2 (1 \pm \sin \phi_k) \tag{83}$$

The probability weights scale to larger values, that is,

$$(1 - \sin \phi_k) / \left(1 - \sin \phi_j\right) > 1 \tag{84}$$

for the eigen-energy $E_k < E_j$ in the cooling case (i.e., for outcome $|0\rangle$), and vice versa for the heating case (i.e., for outcome $|1\rangle$). Further cooling can be achieved by applying the quantum circuit repeatedly and reject/recycle the random walker when $x < 0$.

2. *Connection with Heat-Bath Algorithmic Cooling*

The algorithmic quantum cooling approach is related to the well-known heat bath algorithmic cooling (HBAC) [114–116]. HBAC aims to polarize groups of spins as much as possible, that is, to prepare the state

$$|\uparrow\uparrow\uparrow \ldots \uparrow\rangle \tag{85}$$

This state is important for providing fresh ancilla qubits for quantum error correction as well as for NMR quantum computation. In HBAC, some reversible operations are first performed to redistribute the entropy among a group of spins. Some of the spins will become more polarized. For a closed system, a so-called Shannon bound [116] limits the compression of the entropy. In order to decrease the entropy of the whole system, the depolarized spins interact with a physical heat bath that acts as an entropy sink. We note that from an algorithm point of view, the existence of a physical heat bath can be replaced by the (imperfect) preparation of polarized spins by other methods. The method of algorithmic quantum cooling from Ref. [113] may be considered as a generalization of the HBAC, because it is applicable to cool *any* physical system that is simulable by a quantum computer, not just noninteracting spins.

III. SPECIAL TOPICS

A. Adiabatic Nondestructive Measurements

In Section II.C, we reviewed several methods to prepare ground states and thermal states of quantum systems of interest in physics and chemistry. In particular, in Section II.C.1 we gave an overview of the adiabatic method for preparing ground states. The adiabatic model may be naturally more robust against noise, offering a method to perform small to medium-size simulations without using sophisticated error correction schemes. Because of this and other reasons, adiabatic-based quantum computation is possibly easier to realize physically than quantum computation based on the circuit model. In this section, we review a method to effect nondestructive measurements of constants of the motion within the adiabatic model.

As explained in Section II.C.1, it is in principle possible to adiabatically prepare the ground state of a physical or chemical system with Hamiltonian H_f. There we said that this can be done by interpolating slowly enough between a simple initial Hamiltonian H_i and the final Hamiltonian H_f. Following [93], we now add an ancillary qubit subsystem with orthonormal basis $\{|p_0\rangle, |p_1\rangle\}$. This auxiliary system will be use for the adiabatic nondestructive measurements. During the adiabatic ground state preparation, this subsystem is acted upon with Hamiltonian $\delta|p_1\rangle\langle p_1|$, and therefore it remains in the state $|p_0\rangle$. The choice of $\delta > 0$ will be explained shortly.

The measurement procedure begins by bringing the ancillary qubit and the system being simulated into interaction, adiabatically.[12] We choose the interaction Hamiltonian $H_{\text{int}} = A \otimes |p_1\rangle\langle p_1|$. Here, A is any observable corresponding to a constant of the motion, that is, $[A, H] = 0$. In particular, the Hamiltonian H_f itself can be used to obtain the ground-state energy. The total Hamiltonian becomes

$$H_f + \delta|p_1\rangle\langle p_1| + \underbrace{A \otimes |p_1\rangle\langle p_1|}_{H_{SP}} \tag{86}$$

If the energy bias δ is bigger than the expectation value of the observable A, the state does not change during this initial interaction [93].

After the initial interaction, we apply a Hadamard gate to the ancillary qubit. We denote the time at which we apply this gate as $t = 0$. Let $|s_0\rangle$ be the ground state of H_f. After a further time t the system plus ancilla qubit evolves to

$$|\psi(t)\rangle = \frac{1}{\sqrt{2}}|s_0\rangle \otimes (|p_0\rangle + e^{-i\omega t}|p_1\rangle) \tag{87}$$

[12] The interaction Hamiltonian is typically a three-body Hamiltonian, which makes direct simulations more difficult. This difficulty can be overcome using gadgets [21,93,117,118] or the average Hamiltonian method [119].

where $\omega = (a_0 + \delta)/\hbar$, and $a_0 = \langle s_0|A|s_0\rangle$ is the expectation value we wish to measure. Finally, we again apply a Hadamard gate to the probe. The resulting state is

$$|\psi(t)\rangle = |s_0\rangle \otimes (\cos(\omega t/2)\,|p_0\rangle + i\sin(\omega t/2)\,|p_1\rangle) \tag{88}$$

yielding probability

$$P_0(t) = \frac{1}{2}(1 + \cos(\omega t)) = \cos^2(\omega t/2) \tag{89}$$

Measuring the probe does not disturb the state of the simulator, which can be reused for another measurement. This measurement can be repeated until sufficient statistics have been accumulated to reconstruct ω. We refer to Ref. [93] for details on numerical simulations and considerations of the influence of noise.

B. TDDFT and Quantum Simulation

Density functional theory (DFT) and its time-dependent extension (TDDFT) have become arguably the most widely used methods in computational chemistry and physics. In DFT and TDDFT, the properties of a many-body system can be obtained as functionals of the simple one-electron density rather than the correlated many-electron wave function. This represents a great conceptual leap from usual wave function-based methods such as Hartree-Fock, configuration interaction, and coupled cluster methods, and therefore the connections between DFT/TDDFT and quantum computation have just begun to be explored. Because TDDFT is a time-dependent theory, it is more readily applicable to quantum simulation than DFT, which is strictly a ground state theory. For recent developments in the connections between DFT and quantum complexity[13] (see Ref. [19]), while for applications of DFT to adiabatic quantum computation (see Ref. [120]). In this section, we provide a brief overview of the fundamental theorems of TDDFT, which establish its use as a tool for simulating quantum many-electron atomic, molecular, and solid-state systems, and we mention recent extensions of TDDFT to quantum computation [121].

In its usual formulation, TDDFT is applied to a system of N-electrons described by the Hamiltonian

$$\hat{H}(t) = \sum_{i=1}^{N} \frac{\hat{p}_i^2}{2m} + \sum_{i<j}^{N} w(|\hat{\mathbf{r}}_i - \hat{\mathbf{r}}_j|) + \int v(\mathbf{r}, t)\hat{n}(\mathbf{r})d^3\mathbf{r} \tag{90}$$

[13] It turns out that finding a universal funcional for DFT is QMA hard. Liu [18] proved a related result: N-representability is also QMA hard.

where $\hat{\mathbf{p}}_i$ and $\hat{\mathbf{r}}_i$ are, respectively, the position and momentum operators of the ith electron, $w(|\hat{\mathbf{r}}_i - \hat{\mathbf{r}}_j|)$ is the electron–electron repulsion and $v(\mathbf{r}, t)$ is a time-dependent one-body scalar potential, which includes the potential due to nuclear charges as well as any external fields. The electron-electron repulsion, $w(|\hat{\mathbf{r}}_i - \hat{\mathbf{r}}_j|)$, leads to an exponential scaling of the Hilbert space with system-size and makes simulation of the many-electron Schrödinger equation on a classical computer intractable. $\hat{n}(\mathbf{r}) = \sum_i^N \delta(\mathbf{r} - \hat{\mathbf{r}}_i)$ is the electron density operator, whose expectation value yields the one-electron probability density $\langle \hat{n}(\mathbf{r}) \rangle \equiv n(\mathbf{r}, t)$. The basic theorems of TDDFT prove that, in principle, one can simulate the evolution of the Hamiltonian in Eq. (90) using $n(\mathbf{r}, t)$ *directly* and thereby avoid calculating and storing the exponential amount of information in the many-electron wave function.

The first basic theorem of TDDFT, known as the "Runge–Gross (RG) theorem" [122], establishes the existence of a one-to-one mapping between the expectation value of $\hat{n}(\mathbf{r})$ and the scalar potential $v(\mathbf{r}, t)$. That is,

$$n(\mathbf{r}, t) \leftrightarrow v(\mathbf{r}, t) \tag{91}$$

However, $v(\mathbf{r}, t)$ is the only part of the Hamiltonian in Eq. (90) that is nonuniversal. That is, $\sum_{i=1}^N \frac{\hat{p}_i^2}{2m} + \sum_{i<j}^N w(|\hat{\mathbf{r}}_i - \hat{\mathbf{r}}_j|)$ is the same operator for each electronic system. Therefore, due to the uniqueness of the solution to the time-dependent Schrödinger equation, the RG theorem establishes a one-to-one mapping between the density and the wave function, which implies that the wave function is in fact a unique functional of the density

$$\psi(\mathbf{r}_1, \ldots, \mathbf{r}_N; t) \equiv \psi[n](\mathbf{r}_1, \ldots, \mathbf{r}_N; t) \tag{92}$$

as is any observable of the system. The RG theorem implies the remarkable fact that the one-electron density contains the same quantum information as the many-electron wave function. In principle, then, if one had a means of directly simulating $n(\mathbf{r}, t)$, one could extract all observables of the system without ever needing to simulate the many-body wave function.

The second basic TDDFT theorem is known as the "van Leeuwen (VL) theorem" [123]. It gives an analytic expression for a time-dependent one-body scalar potential that applied to another system with a different, and possibly simpler, electron-electron repulsion $w'(|\hat{\mathbf{r}}_i - \hat{\mathbf{r}}_j|)$ gives the same density evolution as the original Hamiltonian of Eq. (90). When $w'(|\hat{\mathbf{r}}_i - \hat{\mathbf{r}}_j|) = 0$, this auxiliary system is referred to as the "Kohn–Sham system" [124]. Due to its simplicity and accuracy, the Kohn-Sham system is in practice used in most DFT and TDDFT calculations. Because the Kohn–Sham system is noninteracting, its wavefunction is simply described by a Slater determinant of single-electron orbitals, which satisfy the time-dependent Kohn–Sham equations

$$\imath \frac{\partial}{\partial t} \phi_i(\mathbf{r}, t) = \left[-\frac{1}{2}\nabla^2 + v_{ks}[n](\mathbf{r}, t) \right] \phi_i(\mathbf{r}, t) \tag{93}$$

The true interacting density is obtained from the orbitals by square-summing; that is, $n(\mathbf{r}, t) = \sum_i |\phi_i(\mathbf{r}, t)|^2$. Naturally, the set of single-particle equations in Eq. (93) are far easier to solve than evolution under the Hamiltonian in Eq. (90). In practice, the Kohn–Sham potential, $v_{ks}[n](\mathbf{r}, t)$, must be approximated as a density functional, but the VL theorem rigorously guarantees its existence and uniqueness.

Tempel and Aspuru-Guzik [121] recently extended the RG and VL theorems to systems of interacting qubits described by the class of universal 2-local Hamiltonians

$$\hat{H}(t) = \sum_{i=1}^{N-1} J^{\perp}_{i,i+1} \left(\hat{\sigma}^x_i \hat{\sigma}^x_{i+1} + \hat{\sigma}^y_i \hat{\sigma}^y_{i+1} \right) + \sum_{i=1}^{N-1} J^{\parallel}_{i,i+1} \hat{\sigma}^z_i \hat{\sigma}^z_{i+1} + \sum_{i=1}^{N} h_i(t) \hat{\sigma}^z_i \tag{94}$$

These Hamiltonians apply to a variety of different systems, particularly in solid state quantum computing. Benjamin and Bose [125] and [126] have shown that any set of universal two- and single-qubit gates can be implemented with the Hamiltonian of Eq. (94), and therefore it can be used to perform universal quantum computation. The RG theorem applied to such universal Hamiltonians establishes a one-to-one mapping between the set of local fields $\{h_1, h_2, \ldots h_N\}$ used to implement a given computation, and the set of single-qubit expectation values $\{\sigma^z_1, \sigma^z_2, \ldots \sigma^z_N\}$ (see Fig. 4). This implies that one can use single-qubit expectation values as the basic variables in quantum computations rather than wave functions and *directly* extract all observables of interest with only knowledge of the spin densities. Naturally, certain properties such as entanglement will be difficult to extract as functionals of the set of spin densities $\{\sigma^z_1, \sigma^z_2, \ldots \sigma^z_N\}$. Nevertheless, Tempel and Aspuru-Guzik [121] give an explicit entanglement functional.

In addition to the RG theorem, one can derive a VL theorem for qubits. The VL theorem for qubits provides an exact prescription for simulating universal Hamiltonians with other universal Hamiltonians that have different, and possibly easier-to-realize, two-qubit interactions. In analogy to the Kohn–Sham system in electronic TDDFT, one can consider an auxiliary Hamiltonian

$$\begin{aligned} \hat{H}'(t) = & \sum_{i=1}^{N-1} J'^{\perp}_{i,i+1} (\hat{\sigma}^x_i \hat{\sigma}^x_{i+1} + \hat{\sigma}^y_i \hat{\sigma}^y_{i+1}) + \sum_{i=1}^{N-1} J'^{\parallel}_{i,i+1} \hat{\sigma}^z_i \hat{\sigma}^z_{i+1} \\ & + \sum_{i=1}^{N} h'_i[\sigma^z_1, \sigma^z_2, \ldots \sigma^z_N](t) \hat{\sigma}^z_i \end{aligned} \tag{95}$$

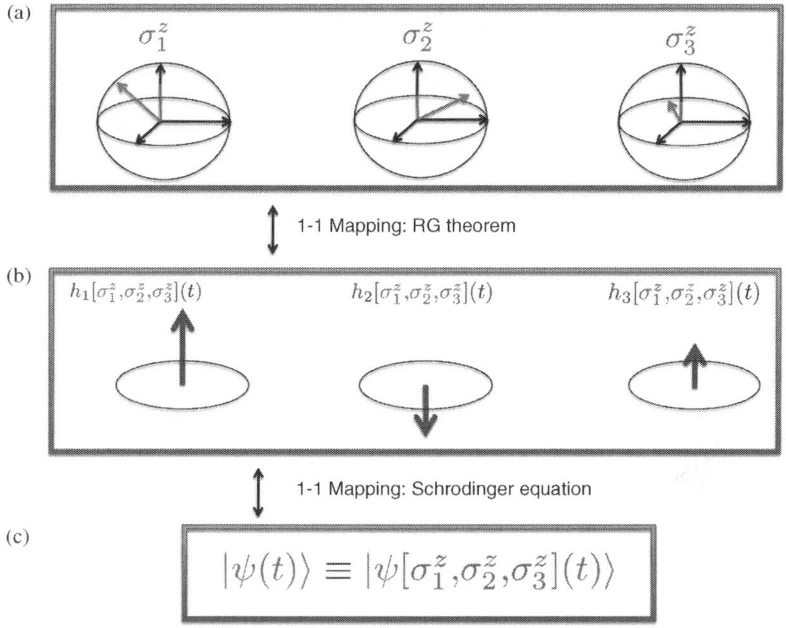

Figure 4. *Runge–Gross theorem for a three-qubit example.* The set of expectation values $\{\sigma_1^z, \sigma_2^z, \ldots, \sigma_N^z\}$, defined by the the Bloch vector components of each qubit along the z-axis in (a), is uniquely mapped onto the set of local fields $\{h_1, h_2, \ldots, h_N\}$ in (b) through the RG theorem. Then, through the Schrödinger equation, the set of fields is uniquely mapped onto the wave function. These two mappings together imply that the N-qubit wave function in (c) is in fact a unique functional of the set of expectation values $\{\sigma_1^z, \sigma_2^z, \ldots, \sigma_N^z\}$. *(See the color version of this figure in the Color Plates section.)*

with simpler two-qubit couplings $\{J'^{\perp}, J'^{\parallel}\}$. The VL theorem guarantees the existence of the auxiliary fields $\{h'_1, h'_2, \ldots h'_N\}$ as functionals of the spin densities that reproduce any given set $\{\sigma_1^z, \sigma_2^z, \ldots \sigma_N^z\}$ that one might wish to simulate. In this way, one can construct an entire class of density functionals for quantum computing that map between different universal Hamiltonians as illustrated in Fig. 5.

TDDFT applied to universal-qubit Hamiltonians provides a potentially powerful tool for simulating quantum computations on classical computers, similar to how it has been applied in computational chemistry for simulating electronic systems. By choosing the auxiliary Kohn–Sham system to be less entangled than the original system, one can hope to simplify simulations of quantum algorithms by finding simple approximations to the auxiliary local fields $\{h'_1, h'_2, \ldots h'_N\}$ as functionals of the spin density. As in electronic TDDFT, the development of approximate density functionals for qubit systems will be a necessary next step, which is discussed in Ref. [121].

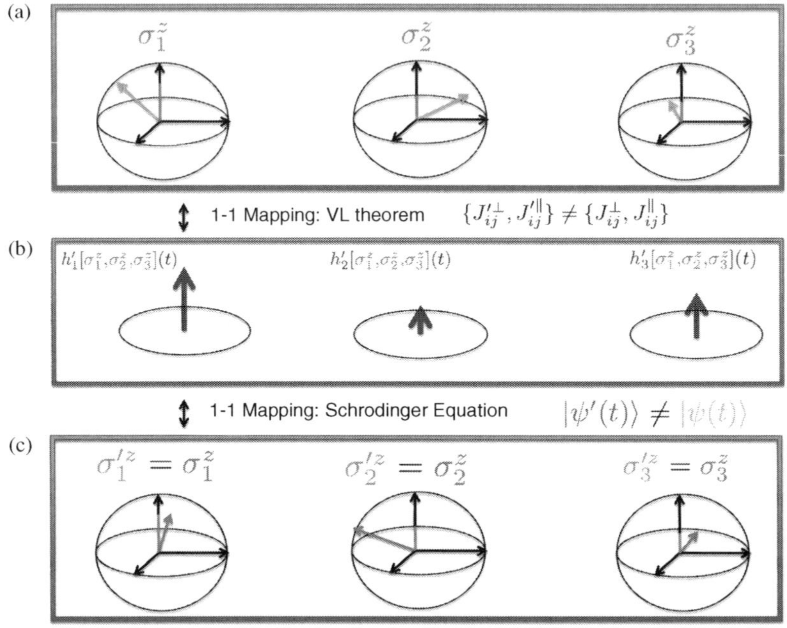

Figure 5. *Van Leeuwen theorem for a three-qubit example.* The set $\{\sigma_1^z, \sigma_2^z, \dots \sigma_N^z\}$ (a) obtained from evolution under Eq. (94), is uniquely mapped to a new set of fields $\{h_1', h_2', \dots h_N'\}$ (b) for a Hamiltonian with different two-qubit interactions. Evolution under this new Hamiltonian returns the same expectation values $\{\sigma_1^z, \sigma_2^z, \dots \sigma_N^z\}$, although the wave function is different and hence projections of the Bloch vectors along other axes are in general different (c). (*See the color version of this figure in the Color Plates section.*)

IV. CONCLUSION AND OUTLOOK

To the best of our knowledge, the first quantum simulation experiment was performed by Somaroo et al. [136] in a two-qubit NMR system in 1999, where a truncated quantum harmonic or anharmonic oscillator (four levels) was simulated. Strictly speaking, this belongs to an "analog" simulation; because the Hamiltonian of the quantum oscillator was directly simulated by the Hamiltonian of the spins, instead of applying quantum algorithms. The progress of quantum simulation is still gaining momentum. In Table I, we list several recent experiments on digital quantum simulation. Earlier references may be found from them and from Refs [7,9,137].

So far, none of the simulations implement active error correction. An important aspect of digital quantum simulation is the resource estimation when fault-tolerant

TABLE I
A Brief Survey of Recent Experiments on Digital Quantum Simulation

Physical Implementations	What Is Simulated?	Scale
Nuclear magnetic resonance (NMR)	• Thermal states of a frustrated magnet [127]	4 qubits
	• Ground state of a pair of interacting Heisenberg spins subject to simulated external fields [128]	3 qubits
	• Isomerization reaction dynamics [129]	3 qubits
	• Ground state of hydrogen molecule H_2 [130]	2 qubits
Trapped ions	• Time dynamics of spin systems [131]	6 qubits
	• Dissipative open-system dynamics [132]	5 qubits
Quantum optics	• 1D quantum walk of a topological system [133]	4 steps
	• 1D quantum walk with tunable decoherence [134]	6 steps
	• Ground state of hydrogen molecule H_2 [135]	2 qubits

structures are considered. We refer to Refs [138] and [72] for further exploring this area. In short, to achieve large-scale quantum simulation, many technological challenges must yet be overcome. For example, in many currently available setups, the performances of the two-qubit gates are still too noisy for fault-tolerant simulation. However, by experimenting with small-scale quantum simulation experiments, we believe that valuable lessons can be learned to optimize the performance of the currently available technology.

A related question is "What is the best way to implement quantum simulation?" For classical computers, silicon-based semiconductors work successfully. For quantum computers, a general feature of the currently proposed technologies, such as quantum dots, quantum optics, trapped ions, nuclear and electron spins, impurity, superconducting devices, and so on, is a trade-off between controllability and reliability. Usually, systems that can be controlled easily suffer more from decoherence from the environment. Two approaches to tackle this problem include looking for new systems that are good for both control and can be isolated from the environment, or developing hybrid structures that combine the advantages from both sides. For example, progress has been made in coupling superconducting devices with spin ensembles [139]. The former provides the controllability and the latter provides reliability. In short, the future of quantum computation and quantum simulation is still full of challenges and opportunities. We hope this chapter can stimulate more ideas that can help move this field forward.

Acknowledgments

The authors would like to acknowledge NSF CCI grant number: CHE-1037992. JDW would like to acknowledge NSF grant no. 1017244. We thank L. Veis for the careful reading of the manuscript. MHY acknowledges funding support from the National Basic Research Program of China Grant 2011CBA00300, 2011CBA00301, the National Natural Science Foundation of China Grant 61033001, 61061130540.

REFERENCES

1. R. P. Feynman, *Int. J. Theor. Phys.* **21**(6-7), 467–488 (1982).
2. P. W. Shor, *SIAM J. Comput.* **26**(5), 1484 (1997).
3. T. D. Ladd, F. Jelezko, R. Laflamme, Y. Nakamura, C. Monroe, and J. L. O'Brien, *Nature* **464**(7285), 45–53 (2010).
4. P. Hauke, F. M. Cucchietti, L. Tagliacozzo, I. Deutsch, and M. Lewenstein, *Rep. Prog. Phys.* **75**, 082401 (2012).
5. G. Ortiz, J. Gubernatis, E. Knill, and R. Laflamme, *Phys. Rev. A* **64**(2), 1–14 (2001).
6. C. Zalka, *Proc. R. Soc. A* **454**(1969), 313–322 (1998).
7. I. Buluta and F. Nori, *Science* **326**(5949), 108–11 (2009).
8. K. L. Brown, W. J. Munro, and V. M. Kendon, *Entropy* **12**(11), 2268–2307 (2010).
9. I. Kassal, J. D. Whitfield, A. Perdomo-Ortiz, M.-H. Yung, and A. Aspuru-Guzik, *Ann. Rev. Phys. Chem.* **62**, 185–207 (2011).
10. A. Kitaev, A. Shen, and M. Vyalyi, *Classical and Quantum Computation, Graduate studies in mathematics*, American Mathematical Society, Providence, RI, 2002.
11. J. Stolze and D. Suter, *Quantum Computing: A Short Course from Theory to Experiment, Physics Textbook*, Wiley-VCH, Weinheim, 2008.
12. M. A. Nielsen and I. L. Chuang, *Quantum Computation and Quantum Information*, 10th Anniversary Edition, Cambridge University Press, Cambridge, 2011.
13. C. P. Williams, *Explorations in Quantum Computing*, Springer, London, 2010.
14. J. Pople, *Rev. Modern Phys.* **71**(5), 1267–1274 (1999).
15. W. Kohn, *Rev. Modern Phys.* **71**(5), 1253–1266 (1999).
16. J. D. Whitfield, P. J. Love, and A. Aspuru-Guzik, *PCCP* **15**, 397–411 (2013).
17. F. Barahona, *J. Phys. A* **15**(10), 3241–3253 (1982).
18. Y.-K. Liu, M. Christandl, and F. Verstraete, *Phys. Rev. Lett.* **98**(11), 110503 (2007).
19. N. Schuch and F. Verstraete, *Nat. Phys.* **5**(10), 732–735 (2009).
20. T.-C. Wei, M. Mosca, and A. Nayak, *Phys. Rev. Lett.* **104**(4), 1–4 (2010).
21. J. Kempe, A. Kitaev, and O. Regev, *SIAM J. Comput.* **35**(5), 1070–1097 (2006).
22. D. Aharonov and T. Naveh, *Quantum NP: A Survey, arXiv:quant-ph/0210077*, 2002, p. 23.
23. V. A. Rassolov and S. Garashchuk, *Chem. Phys. Lett.* **464**(4–6), 262–264 (2008).
24. S. Aaronson, *Nat. Phys.* **5**(10), 707–708 (2009).
25. T. J. Osborne, *Hamiltonian Complexity. arXiv:1106.5875*, 2011, p. 14.

26. S. Aaronson, *Why Philosophers Should Care About Computational Complexity. arXiv:1108.1791*, 2011, p. 58.
27. J. D. Whitfield, M.-H. Yung, D. G. Tempel, S. Boixo, and A. Aspuru-Guzik, *Computational Complexity of Time-Dependent Density Functional Theory. arXiv:1310.1428.*
28. A. Young, S. Knysh, and V. Smelyanskiy, *Phys. Rev. Lett.* **101**(17), 1–4 (2008).
29. D. Poulin and P. Wocjan, *Phys. Rev. Lett.* **102**(13), 1–4 (2009).
30. D. Poulin and P. Wocjan, *Phys. Rev. Lett.* **103**(22), 220502 (2009).
31. A. P. Young and V. N. Smelyanskiy, *Phys. Rev. Lett.* **104**(2), 1–4 (2010).
32. E. Bilgin and S. Boixo, *Phys. Rev. Lett.* **105**(17) (2010).
33. M.-H. Yung, D. Nagaj, J. Whitfield, and A. Aspuru-Guzik, *Phys. Rev. A* **82**(6), 5 (2010).
34. I. Hen and A. P. Young, *Phys. Rev. E* **84**(6), 061152 (2011).
35. P. Kaye, R. Laflamme, and M. Mosca, *An Introduction to Quantum Computing*, 1 ed., Oxford University Press, USA, 2007.
36. D. Abrams and S. Lloyd, *Phys. Rev. Lett.* **83**(24), 5162–5165 (1999).
37. P. Jaksch and A. Papageorgiou, *Phys. Rev. Lett.* **91**(25), 1–4 (2003).
38. E. Knill, G. Ortiz, and R. D. Somma, *Phys. Rev. A* **75**(1), 012328–13 (2007).
39. D. Lidar and H. Wang, *Phys. Rev. E* **59**(2), 2429–2438 (1999).
40. L.-A. Wu, M. Byrd, and D. Lidar, *Phys. Rev. Lett.* **89**(5), 1–4 (2002).
41. R. Somma, G. Ortiz, J. E. Gubernatis, E. Knill, and R. Laflamme, *Phys. Rev. A* **65**(4), 42323 (2002).
42. R. Somma, G. Ortiz, E. Knill, and J. Gubernatis, *Proc. SPIE* **5105**, 12 (2003).
43. T. Byrnes and Y. Yamamoto, *Phys. Rev. A* **73**(2), 1–16 (2006).
44. I. Kassal and A. Aspuru-Guzik, *J. Chem. Phys.* **131**(22), 224102 (2009).
45. C. Master, F. Yamaguchi, and Y. Yamamoto, *Phys. Rev. A* **67**(3), 3 (2003).
46. P. Wocjan, C.-F. Chiang, D. Nagaj, and A. Abeyesinghe, *Phys. Rev. A* **80**(2), 022340 (2009).
47. S. Lloyd, *Science* **273**(5278), 1073–1078 (1996).
48. D. Aharonov and A. Ta-Shma, *Proceedings of the Thirty-Fifth Annual ACM Symposium on Theory of Computing*, ACM, (2003), pp. 20–29.
49. A. M. Childs, *Quantum Information Processing in Continuous Time*, MIT, Cambridge, MA, 2004.
50. D. W. Berry, G. Ahokas, R. Cleve, and B. C. Sanders, *Commun. Math. Phys.* **270**(2), 359–371 (2006).
51. D. W. Berry and A. M. Childs, *Quant. Inform. Comput.* **12**, 29 (2012).
52. A. M. Childs and R. Kothari, *Theory of Quantum Computation Communication and Cryptography (TQC 2010), Lecture Notes in Computer Science*, Vol. 6519, 2011, pp. 94–103.
53. A. Papageorgiou and C. Zhang, *Quant. Inf. Process.* 1–21 (2010).
54. N. Hatano and M. Suzuki, *in Quantum Annealing and Other Optimization Methods, Lectures Notes in Physics*, A. Das and B. Chakrabarti, eds., Springer-Verlag, Berlin, 2005, pp. 37–68.
55. R. Beals, H. Buhrman, R. Cleve, M. Mosca, and R. de Wolf, *Proc. FOCS 98*, 352–361 (1998).
56. A. M. Childs, *Commun. Math. Phys.* **294**, 581 (2010).
57. S. Raesi, N. Wiebe, and B. C. Sanders, *New J. Phys.* **14**, 103017 (2012).
58. D. Abrams and S. Lloyd, *Phys. Rev. Lett.* **79**, 2586 (1997).

59. N. J. Ward, I. Kassal, and A. Aspuru-Guzik, *J. Chem. Phys.* **130**(19), 194105 (2009).
60. S. Wiesner, *Simulations of Many-Body Quantum Systems by a Quantum Computer. quant-ph/9603028*, 1996.
61. I. Kassal, S. P. Jordan, P. J. Love, M. Mohseni, and A. Aspuru-Guzik, *Proc. Natl. Acad. Sci. USA* **105**(48), 18681–6 (2008).
62. M. Feit, J. Fleck Jr., and A. Steiger, *J. Comput. Phys.* **47**(3), 412–433 (1982).
63. R. Kosloff, *J. Phys. Chem.* **92**(8), 2087–2100 (1988).
64. C. Zalka, *Fortschr. der Phys.* **46**(6–8), 877–879 (1998).
65. G. Strini, *Fortschr. Phys.* **50**(2), 171 (2002).
66. G. Benenti and G. Strini, *Am. J. Phys.* **76**(7), 657 (2008).
67. A. Aspuru-Guzik, A. D. Dutoi, P. J. Love, and M. Head-Gordon, *Science* **309**(5741), 1704–7 (2005).
68. H. Wang, S. Kais, A. Aspuru-Guzik, and M. R. Hoffmann, *Phys. Chem.* **10**(35), 5388–5393 (2008).
69. J. D. Whitfield, J. Biamonte, and A. Aspuru-Guzik, *Molec. Phys.* **109**(5), 735–750 (2011).
70. L. Veis, J. Višňák, T. Fleig, S. Knecht, T. Saue, L. Visscher, and J.c.v. Pittner, *Phys. Rev. A* **85**, 030304 (2012).
71. T. Helgaker, P. Jorgensen, and J. Olsen, *Molecular Electronic-Structure Theory*, John Wiley and Sons, Chichester, 2000.
72. C. Clark, T. Metodi, S. Gasster, and K. Brown, *Phys. Rev. A* **79**(6), 1–9 (2009).
73. H.-P. Breuer and F. Petruccione, *The theory of open quantum systems*, Oxford University Press, New York, 2002.
74. G. Lindblad, *Commun. Math. Phys.* **48**, 119 (1975).
75. V. Gorini, A. Kossakokowski, and E. C. G. Sudarshan, *J. Math. Phys.* **17**, 821 (1976).
76. M. Mohseni, P. Rebentrost, S. Lloyd, and A. Aspuru-Guzik, *J. Chem. Phys.* **129**, 174106 (2008).
77. P. Rebentrost, M. Mohseni, I. Kassal, S. Lloyd, and A. Aspuru-Guzik, *New J. Phys.* **11**, 033003 (2009).
78. B. Kraus, H. Büchler, S. Diehl, A. Kantian, A., Micheli, and P. Zoller, *Phys. Rev. A* **78**(4), 1–9 (2008).
79. H. Krauter, C. Muschik, K. Jensen, W. Wasilewski, J. Petersen, J. Cirac, and E. Polzik, *Phys. Rev. Lett.* **107**(8), 1–5 (2011).
80. C. Muschik, E. Polzik, and J. Cirac, *Phys. Rev. A* **83**(5), 1–19 (2011).
81. J. Cho, S. Bose, and M. Kim, *Phys. Rev. Lett.* **106**(2), 1–4 (2011).
82. M. Müller, K. Hammerer, Y. L. Zhou, C. F. Roos, and P. Zoller, *New J. Phys.* **13**(8), 085007 (2011).
83. F. Pastawski, L. Clemente, and J. Cirac, *Phys. Rev. A* **83**(1), 1–12 (2011).
84. F. Verstraete, M. M. Wolf, and J. Ignacio Cirac, *Nat. Phys.* **5**(9), 633–636 (2009).
85. D. Bacon, A. Childs, I. Chuang, J. Kempe, D. Leung, and X. Zhou, *Phys. Rev. A* **64**(6), (2001).
86. M. Kliesch, T. Barthel, C. Gogolin, M. Kastoryano, and J. Eisert, *Phys. Rev. Lett.* **107**(12), 1–5 (2011).
87. L. Grover and T. Rudolph, *Creating Superpositions that Correspond to Efficiently Integrable Probability Distributions. quant-ph/0208112*, 2002.
88. P. Kaye and M. Mosca, *Proceedings, International Conference on Quantum Information (ICQI)*. Rochester, New York, USA, 2001.

89. S. Boixo, E. Knill, and R. D. Somma, *Fast Quantum Algorithms for Traversing Paths of Eigenstates. arXiv:1005.3034*, 2010.
90. E. Farhi, J. Goldstone, S. Gutmann, and M. Sipser, *Quantum Computation by Adiabatic Evolution. arXiv:quant-ph/0001106v1*, 2000.
91. A. Perdomo, C. Truncik, I. Tubert-Brohman, G. Rose, and A. Aspuru-Guzik, *Phys. Rev. A* **78**(1), 1–15 (2008).
92. S. Boixo, E. Knill, and R. D. Somma, *Quant. Inform. Comput.* **9**, 833–855 (2009).
93. J. D. Biamonte, V. Bergholm, J. D. Whitfield, J. Fitzsimons, and A. Aspuru-Guzik, *Adiabatic Quantum Simulators*, University Computing, 2010, p. 9.
94. S. Boixo and R. D. Somma, *Phys. Rev. A* **81**(3), 032308 (2010).
95. B. Altshuler, H. Krovi, and J. Roland, *Proc. Natl. Acad. Sci. USA* **107**(28), 12446–50 (2010).
96. S. Knysh and V. Smelyanskiy, *On the Relevance of Avoided Crossings Away from Quantum Critical Point to the Complexity of Quantum Adiabatic Algorithm. arXiv:1005.3011*, 2010, p. 8.
97. S. Oh, *Phys. Rev. A* **77**(1), 012326 (2008).
98. B. M. Terhal and D. P. DiVincenzo, *Phys. Rev. A,* **61**(2), 040301 (2000).
99. H. Wang, S. Ashhab, and F. Nori, *Phys. Rev. A* **83**(6), 1–11 (2011).
100. H. Gould, J. Tobochnik, and W. Christian, *An introduction to computer simulation methods: Applications to physical systems*, Pearson Addison Wesley, New York, 2007.
101. K. Temme, T. J. Osborne, K. G. Vollbrecht, D. Poulin, and F. Verstraete, *Nature* **471**(7336), 87–90 (2011).
102. D. J. Aldous, *J. Lond. Math. Soci.* **s2-25**(3), 564–576 (1982).
103. M. Szegedy, *45th Annual IEEE Symposium on Foundations of Computer Science*, IEEE (2004), pp. 32–41.
104. P. Richter, *Phys. Rev. A* **76**(4), 042306 (2007).
105. R. Somma, S. Boixo, H. Barnum, and E. Knill, *Phys. Rev. Lett.* **101**(13), 130504 (2008).
106. P. Wocjan and A. Abeyesinghe, *Phys. Rev. A* **78**(4), 042336 (2008).
107. M.-H. Yung and A. Aspuru-Guzik, *Proc. Natl. Acad. Sci. USA* **109**, 754–759 (2012).
108. D. Lidar and O. Biham, *Phys. Rev. E* **56**(3), 3661–3681 (1997).
109. R. Somma, C. Batista, and G. Ortiz, *Phys. Rev. Lett.* **99**(3), 1–4 (2007).
110. M. Ohzeki and H. Nishimori, *J. Comput. Theor. Nanosci.* **8**(6), 963–971 (2011).
111. A. Riera, C. Gogolin, and J. Eisert, *Phys. Rev. Lett.* **108**, 080402 (2012).
112. R. Bhatia, *Positive Definite Matrices*, Princeton University Press, Princeton, 2007.
113. M.-H. Yung, S. Boixo, and A. Aspuru-Guzik, *in preparation* (2011).
114. P. O. Boykin, T. Mor, V. Roychowdhury, F. Vatan, and R. Vrijen, *Proc. Natl. Acad. Sci. USA* **99**, 3388–3393 (2002).
115. J. Baugh, O. Moussa, C. a. Ryan, A. Nayak, and R. Laflamme, *Nature* **438**(7067), 470–3 (2005).
116. L. Schulman, T. Mor, and Y. Weinstein, *Phys. Rev. Lett.* **94**(12), 1–4 (2005).
117. R. Oliveira and B. M. Terhal, *Quant. Inform. Comput.* **8**(10), 0900–0924 (2008).
118. J. D. Biamonte and P. J. Love, *Phys. Rev. A* **78**(1), 012352–7 (2008).
119. J. Waugh, L. Huber, and U. Haeberlen, *Phys. Rev. Lett.* **20**(5), 180–182 (1968).
120. F. Gaitan and F. Nori, *Phys. Rev. B* **79**, 205117 (2009).
121. D. G. Tempel and A. Aspuru-Guzik, *Sci. Rep.* **2**, 391 (2012).
122. E. Runge and E. K. U. Gross, *Phys. Rev. Lett.* **52**, 997 (1984).

123. R. van Leeuwen, *Phys. Rev. Lett.* **82**(19), 3863–3866 (1999).
124. W. Kohn and L. J. Sham, *Phys. Rev.* **140**, 1133 (1965).
125. S. C. Benjamin and S. Bose, *Phys. Rev. Lett.* **90**, 247901 (2003).
126. S. C. Benjamin and S. Bose, *Phys. Rev. A* **70**, 032314 (2004).
127. J. Zhang, M.-H. Yung, R. Laflamme, A. Aspuru-Guzik, and J. Baugh, *Nat. Commun.* **3**, 880 (2012).
128. Z. Li, M-H. Yung, H. Chen, D. Lu, J. D. Whitfield, X. Peng, A. Aspuru-Guzik, and J. Du, *Scientific Reports* **1**, 88 (2011).
129. D. Lu, N. Xu, R. Xu, H. Chen, J. Gong, X. Peng, and J. Du, *Phys. Rev. Lett.* **107**(2), 8–11 (2011).
130. J. Du, N. Xu, X. Peng, P. Wang, S. Wu, and D. Lu, *Phys. Rev. Lett.* **104**(3), 1–4 (2010).
131. B. P. Lanyon, C. Hempel, D. Nigg, M. Muller, R. Gerritsma, F. Zahringer, P. Schindler, J. T. Barreiro, M. Rambach, G. Kirchmair, M. Hennrich, P. Zoller, R. Blatt, and C. F. Roos, *Science* **334**, 57–61 (2011).
132. J. T. Barreiro, M. Müller, P. Schindler, D. Nigg, T. Monz, M. Chwalla, M. Hennrich, C. F. Roos, P. Zoller, and R. Blatt, *Nature* **470**(7335), 486–491 (2011).
133. T. Kitagawa, M. A. Broome, A. Fedrizzi, M. S. Rudner, E. Berg, I. Kassal, A. Aspuru-Guzik, E. Demler, and A. G. White, *Nat. Commun.* **3**, 882 (2012).
134. M. A. Broome, A. Fedrizzi, B. P. Lanyon, I. Kassal, A. Aspuru-Guzik, and A. G. White, *Phys. Rev. Lett.* **104**(15), 1–4 (2010).
135. B. P. Lanyon, J. D. Whitfield, G. G. Gillett, M. E. Goggin, M. P. Almeida, I. Kassal, J. D. Biamonte, M. Mohseni, B. J. Powell, M. Barbieri, A. Aspuru-Guzik, and A. G. White, *Nat. Chem.* **2**(2), 106–111 (2010).
136. S. Somaroo, C. Tseng, T. Havel, R. Laflamme, and D. Cory, *Phys. Rev. Lett.* **82**(26), 5381–5384 (1999).
137. J. A. Jones, *Prog. Nucl. Magn. Reson. Spectrosc.* **59**(2), 91–120 (2011).
138. K. R. Brown, R. J. Clark, and I. L. Chuang, *Phys. Rev. Lett.* **97**(5), 050504 (2006).
139. T. Duty, *Physics* **3**, 80 (2010).

QUANTUM COMPUTING APPROACH TO NONRELATIVISTIC AND RELATIVISTIC MOLECULAR ENERGY CALCULATIONS

LIBOR VEIS[1,2] and JIŘÍ PITTNER[1]

[1]*J. Heyrovský Institute of Physical Chemistry, Academy of Sciences of the Czech Republic, v.v.i., Dolejškova 3, 18223 Prague 8, Czech Republic*
[2]*Department of Physical and Macromolecular Chemistry, Faculty of Science, Charles University in Prague, Hlavova 8, 12840 Prague 2, Czech Republic*

Advances in Chemical Physics, Volume 154: Quantum Information and Computation for Chemistry, First Edition. Edited by Sabre Kais.

I. OVERVIEW

An *exact* simulation of quantum systems on a classical computer is computationally hard. The problem lies in the dimensionality of the Hilbert space needed for the description of our system that in fact grows exponentially with the size of this system. No matter whether we simulate the dynamics or calculate some static property such as the energy, this limitation is always present. Richard Feynman came up with an alternative to the classical simulation [1]. His idea was to convert the aforementioned drawback of quantum systems into their benefit. He suggested mapping the Hilbert space of a studied quantum system on another one (both of them being exponentially large) and thus to *efficiently* simulate a quantum system on another one (i.e., on a quantum computer).

This was the original idea of quantum computers. Today, quantum computing is undoubtedly a well-established discipline of computer science. Apart from the *efficient* simulation of quantum systems [2–6], other interesting problems where quantum computers could beat their classical counterparts have been discovered. The most famous examples are integer factorization for which quantum computers supply an exponential speedup [7,8] or database search [9]. However, for our purposes the *efficient* (polynomially scaling) quantum algorithm of Abrams and Lloyd for obtaining eigenvalues of local Hamiltonians [10] is particularly important.

The first paper connecting quantum computation and chemistry was published by Lidar and Wang [11] and concerned the *efficient* calculations of thermal rate constants of chemical reactions. This work in fact founded the new field of computational chemistry that is the subject of this book, namely the "computational chemistry on quantum computers." Aspuru-Guzik et al. in their seminal article [12] reduced the number of quantum bits (qubits) needed by the Abrams and Lloyd's algorithm [10] and applied it to molecular ground state energy calculations. Since these two pioneering works, other papers involving energy calculations of excited states [13], quantum chemical dynamics [14], calculations of molecular properties and geometry optimizations [15], state preparations [16,17], or global minima search [18] were published. The list of all chemical applications for quantum computers is quite rich and an interested reader can find a comprehensive review in Ref. [19].

Aspuru-Guzik et al. [12] also proposed that quantum computers with tens of (noise free) qubits would already exceed the limits of classical full configuration interaction (FCI) calculations, which is in contrast to other quantum algorithms, such as the Shor's algorithm [7,8] for integer factorization that would for practical tasks in cryptography require thousands of qubits. For this reason, calculations and simulations of quantum systems will belong to the first practical applications of quantum computers. Recent proof-of-principle few-qubit experiments covering energy calculations of the hydrogen molecule [20,21] or Heisenberg spin model [22] and the simulation of a chemical reaction dynamics [23] confirm that interesting applications might be just beyond the door.

The aim of this chapter is to review quantum algorithms for exact molecular energy calculations (within a finite one-particle basis set), that is, quantum analogs of the FCI calculations. On a classical computer, a computational cost of the FCI method scales exponentially with the size of the system. This fact stems from the dimension of the Hilbert space in which we diagonalize the Hamiltonian matrix, and it is the reason why this method is limited only to the smallest systems (diatomics, triatomics). For example, in the nonrelativistic case, the number of Slater determinants that build up the FCI wave function for a closed-shell system with n electrons in m orbitals is equal to

$$N_{\text{non-rel.}} = \binom{m}{n/2}^2 \tag{1}$$

It is evident that this number quickly grows into huge values with increasing m. On a quantum computer on the other hand, it has been shown [20,24] and will be discussed later in the chapter that the FCI cost has a polynomial scaling [$\mathcal{O}(m^5)$], therefore, it is *exponentially* faster.

The structure of this chapter is as follows: First we briefly give the necessary basics of quantum computing including the quantum Fourier transform (QFT), the phase estimation algorithm (PEA), and their semiclassical variants (Section II). Despite being textbook topics [25], we believe that it may be convenient for readers from the quantum chemistry community to make this chapter more self-contained. Section III presents details of the quantum algorithm for the FCI method (qFCI). In Section IV, we show its applications to nonrelativistic computations, namely the computations of the ground and excited state energies of the methylene molecule (CH_2). We then generalize this approach to the relativistic four-component (4c) calculations and apply it to the problem of spin-orbit splitting in the SbH molecule in Section V.

II. QUANTUM COMPUTING BACKGROUND

In this section, we briefly mention the necessary quantum computing background. For a more detailed description, we refer the reader, for example, to the excellent book by Nielsen and Chuang [25].

A. Quantum Fourier Transform

The classical discrete Fourier transform takes as an input a vector of complex numbers $(x_0, \ldots, x_{N-1})$ and outputs the elements of another vector $(y_0, \ldots, y_{N-1})$

according to the equation

$$y_k = \frac{1}{\sqrt{N}} \sum_{j=0}^{N-1} x_j e^{2\pi i jk/N} \tag{2}$$

Similarly, the quantum Fourier transform operates on an orthonormal basis of n qubits: $|0\rangle \ldots |2^n - 1\rangle$ and is defined as an operator $\hat{U}_{\mathrm{QFT}}$

$$\hat{U}_{\mathrm{QFT}} |k\rangle = \frac{1}{\sqrt{N}} \sum_{j=0}^{N-1} e^{2\pi i jk/N} |j\rangle, \qquad N = 2^n \tag{3}$$

where the kets are numbered by a binary representation of integers

$$|j\rangle = |j_n \ldots j_1\rangle, \qquad j_i \in \{0, 1\}, \\ j = \sum_{i=1}^{n} j_i \cdot 2^{i-1} \tag{4}$$

It can be shown that the QFT is an unitary operator [25].

The advantage of the QFT is that it can be performed with just $\mathcal{O}(n^2)$ operations (quantum gates). This is in sharp contrast to the classical fast Fourier transform (FFT) with the scaling $\mathcal{O}(N\log N = n2^n)$. Without the derivation, which can be found, for example, in Ref. [25], the quantum circuit for the QFT is shown in Fig. 1. H (Hadamard) and R_j gates are represented by the following matrices

$$H = \frac{1}{\sqrt{2}} \begin{pmatrix} 1 & 1 \\ 1 & -1 \end{pmatrix}, \qquad R_j = \begin{pmatrix} 1 & 0 \\ 0 & e^{2\pi i/2^j} \end{pmatrix} \tag{5}$$

It should be noted that even though the QFT can be done *exponentially* faster than the FFT, it cannot be used as an *efficient* straightforward replacement of the Fourier transform itself. It would indeed require preparing an arbitrary state of n qubits and also measure all of the complex amplitudes at the end, which cannot

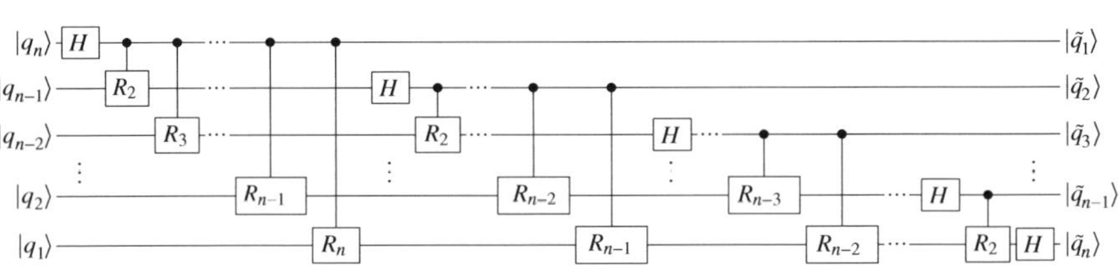

Figure 1. The quantum Fourier transform circuit. Note that qubits of the result are in a reversed order after the application of this circuit.

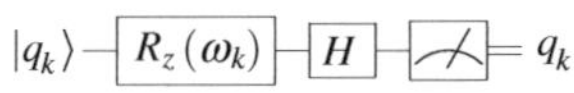

Figure 2. Simplified measurement-based circuit for the kth qubit of the QFT.

be done *efficiently*. Nevertheless the QFT is a key part of the phase estimation algorithm [25] (contained also in the Shor's factoring algorithm [7,8]) as will be shown in Section II.C.

B. Semiclassical Approach to Quantum Fourier Transform

The QFT circuit from Fig. 1 can in fact be greatly simplified using the semiclassical (measurement-based) approach [26]. Notice that gates acting on each qubit [except the first (top most) and the last ones where the corresponding parts are missing] obey the general structure. First, R_j gates controlled by previous qubits are applied, then the Hadamard gate is applied, and finally they serve as control qubits for the subsequent ones. Because the state of each qubit does not change after the application of the Hadamard gate, we can in fact do the measurement just after this gate. Instead of employing controlled R_j gates, we then apply only the corresponding one-qubit gates depending on the results of individual measurements. Moreover, all R_j gates acting on a kth qubit can be merged into a single rotation gate

$$R_z(\omega_k) = \begin{pmatrix} 1 & 0 \\ 0 & e^{2\pi i \omega_k} \end{pmatrix} \tag{6}$$

whose angle ω_k depends on the results of previously measured qubits (q_i) according to the formula

$$\omega_k = \sum_{l=2}^{n-k+1} \frac{q_{k+i-1}}{2^i}, \qquad k : n \longrightarrow 1 \tag{7}$$

Figure 2 shows the semiclassical QFT circuit pattern, which is the same for all qubits. The big advantage of this approach is that we have actually replaced two-qubit gates by one-qubit ones (controlled by a classical signal). This technique is especially useful in connection with the phase estimation algorithm where it leads to the formulation of its iterative version (IPEA, Section II.D).

C. Phase Estimation Algorithm

The phase estimation algorithm [25] is a quantum algorithm for obtaining the eigenvalue of an unitary operator $\hat{U}$, based on a given initial guess of the corresponding eigenvector. Since an unitary $\hat{U}$ can be written as $\hat{U} = e^{i\hat{H}}$, with $\hat{H}$ Hermitian, the PEA can be viewed as a quantum substitute of the classical diagonalization.

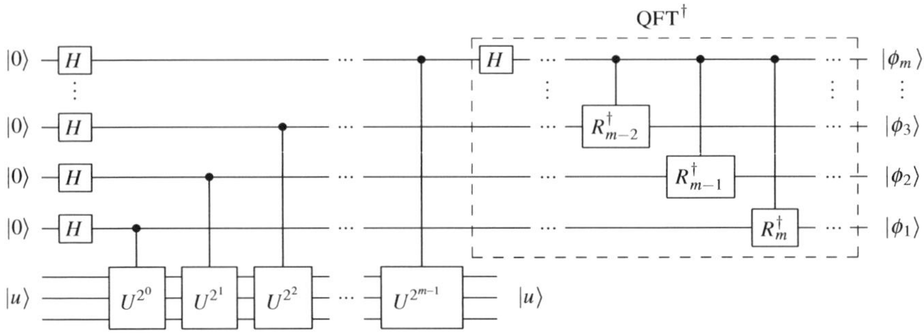

Figure 3. The PEA circuit with the highlighted part corresponding to the inverse QFT.

Suppose that $|u\rangle$ is an eigenvector of $\hat{U}$ and that it holds

$$\hat{U}|u\rangle = e^{2\pi i\phi}|u\rangle, \qquad \phi \in \langle 0, 1) \tag{8}$$

where ϕ is the phase that is estimated by the algorithm. Quantum register is divided into two parts. The first one, called the read-out part, is composed of m qubits on which the binary representation of ϕ is measured at the end and is initialized to the state $|0\rangle^{\otimes m}$. The second part contains the corresponding eigenvector $|u\rangle$.

The PEA quantum circuit is shown in Fig. 3. The application of Hadamard gates on all qubits in the first part of the register gives

$$|\text{reg}\rangle = \frac{1}{\sqrt{2^m}} \sum_{j=0}^{2^m-1} |j\rangle|u\rangle \tag{9}$$

Next, after the application of a sequence of controlled powers of $\hat{U}$, the register is transformed into

$$|\text{reg}\rangle = \frac{1}{\sqrt{2^m}} \sum_{j=0}^{2^m-1} \hat{U}^j|j\rangle|u\rangle = \frac{1}{\sqrt{2^m}} \sum_{j=0}^{2^m-1} e^{2\pi i j\phi}|j\rangle|u\rangle \tag{10}$$

The heart of the PEA is the inverse quantum Fourier transform (QFT†, highlighted in Fig. 3) performed on the read-out part of the register, which is transformed to $|2^m\phi\rangle|u\rangle$.

If the phase can be expressed exactly in m bits

$$\phi = 0.\phi_1\phi_2\ldots\phi_m = \frac{\phi_1}{2} + \frac{\phi_2}{2^2} + \ldots + \frac{\phi_m}{2^m}, \qquad \phi_i \in \{0, 1\} \tag{11}$$

it (and consequently the eigenvalue) is recovered with unity probability by a measurement on the first part of the quantum register.

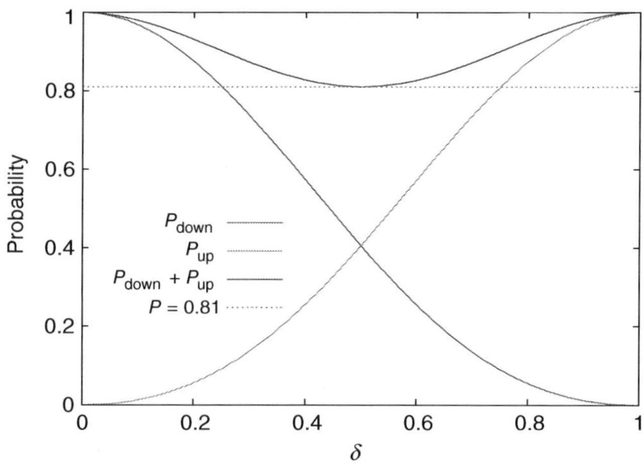

Figure 4. The dependence of success probabilities of the PEA on δ for $m = 20$. P_{down} and P_{up} denote the success probabilities corresponding to rounding the exact phase up/down to m binary digits.

The situation is more complicated when ϕ cannot be expressed exactly in m bits. Then we can write

$$\phi = \tilde{\phi} + \delta 2^{-m} \tag{12}$$

where $\tilde{\phi} = \phi_1\phi_2 \ldots \phi_m$ denotes the first m bits of the binary expansion and $\delta : 0 \leq \delta < 1$ is a remainder. The closest m-bit estimators of ϕ correspond to either $\tilde{\phi}$ (rounding down) or $\tilde{\phi} + 2^{-m}$ (rounding up). When we label the probabilities of measuring these two estimators by P_{down} and P_{up}, it can be shown (e.g., [27]) that the sum $P_{\text{down}} + P_{\text{up}}$ decreases monotonically with increasing m. The dependence of P_{down} and P_{up} on δ for $m = 20$ is presented in Fig. 4. In the limit $m \to \infty$, the lower bound reads [27]

$$P_{\text{down}}(\delta = 1/2) + P_{\text{up}}(\delta = 1/2) = \frac{4}{\pi^2} + \frac{4}{\pi^2} > 0.81 \tag{13}$$

If the desired eigenvector is not known explicitly (as is typically the case in quantum chemistry), we can start the algorithm with an arbitrary initial guess vector $|\psi\rangle$, which can be expanded in terms of eigenvectors of $\hat{U}$

$$|\psi\rangle = \sum_i c_i |u_i\rangle \tag{14}$$

The probability of obtaining the *exact* m-bit ϕ_i is due to linearity of the algorithm $|c_i|^2$. It is important to note that the initial guess does not influence the accuracy

Figure 5. The kth iteration of the IPEA. The feedback angle depends on the previously measured bits; k is iterated backwards from m to 1.

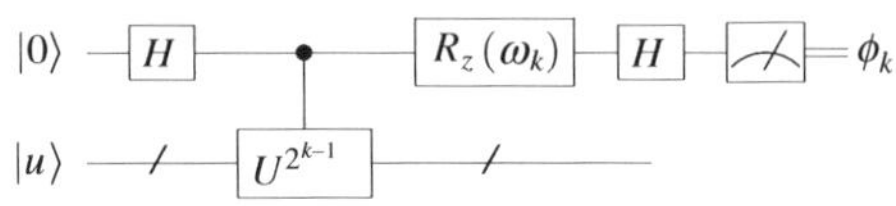

of the phase, only the probability with which the phase of a particular eigenstate is measured. When ϕ_i cannot be expressed in m bits, the lower bound for $P_{\text{down}} + P_{\text{up}}$ corresponding to ϕ_i is equal to $0.81 \cdot |c_i|^2$.

D. Iterative Phase Estimation Algorithm

Using the semiclassical QFT [26] (Section II.B), the PEA circuit can be simplified, having only one ancillary qubit in the read-out part of the quantum register. The algorithm then proceeds in an iterative manner [iterative phase estimation algorithm (IPEA)]. The kth iteration of this scheme is presented in Fig. 5. Note that as the PEA uses the inverse QFT, the angle ω_k in Eq. (7) must be negative now. The algorithm is iterated backward from the least significant bits of ϕ to the most significant ones, for k going from m to 1. For our purposes, the presented IPEA, which is the *unitary* matrix eigenvalue algorithm, is completely adequate, but we would like to note that Wang et al. recently presented a modified version of the IPEA capable of finding eigenvalues of *nonunitary* matrices [28].

The IPEA is in fact completely equivalent to the original (multiqubit) PEA [27]. It exhibits the same behavior, decreasing the probability of success when the phase cannot be expressed exactly in a particular number of bits. One possibility of amplifying the probability of success is performing more iteration steps (more than the desired accuracy of ϕ): when extracting $m' = m + \log(2 + 1/2\epsilon)$ bits, the phase is accurate to m binary digits with probability at least $1 - \epsilon$ [25]. This method is, however, not very useful, because implementing $\hat{U}^{2^{k-1}}$ for large k is the algorithm's bottleneck in a realistic noisy environment [29].

Another alternative [29] is to repeat the measurement for the least important bits of the phase binary expansion. Using the majority voting (for bit value 0 or 1), the effective error probability decreases exponentially with the number of repetitions according to the binomial distribution. This measurement repetition only for the few least important bits of ϕ is unfortunately possible only if the exact eigenstates of $\hat{U}$ are available.

When working with general initial states (as in a practical application to quantum chemistry), two scenarios are possible [30], as shown in Fig. 6. Maintaining the second part of the quantum register during all iterations and amplification of the success probabilities by repeating the whole process a number of times is the first possibility. This version was denoted as **A** version of IPEA. The biggest advantage of this approach is that one always ends up with one of the eigenstates of $\hat{U}$ in the second part of the quantum register as was also the case of the PEA. It happens through successive collapses of the system state into the

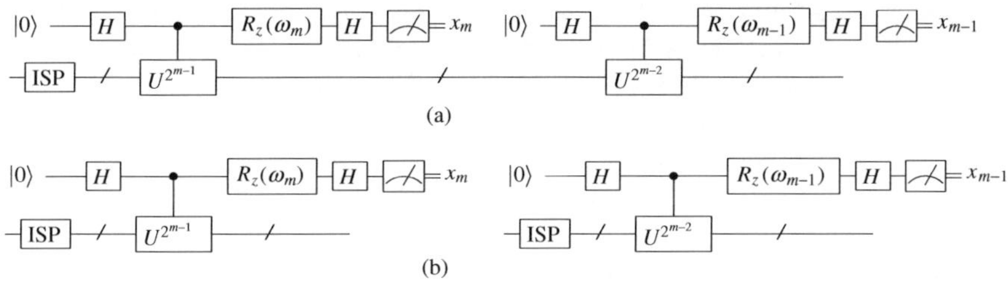

Figure 6. Comparison of the two versions of IPEA. ISP denotes the initial state preparation. (**a**) Maintaining the second part of the quantum register during all iterations (version **A**). (**b**) Repeated initial state preparation in each iteration (version **B**).

corresponding eigensubspace and is demonstrated on the hydrogen molecule with random initial states in Fig. 7. The biggest disadvantage that complicates the potential physical realization of this scheme is the requirement for a long coherence time of the quantum register. We would like to note here that when amplifying the success probability by repeating the whole process, it must be higher than 0.5 to be sure that we get the energy of the right state. This, however, is not necessary for the ground-state energy, which can always be identified by the lowest eigenvalue [22,30].

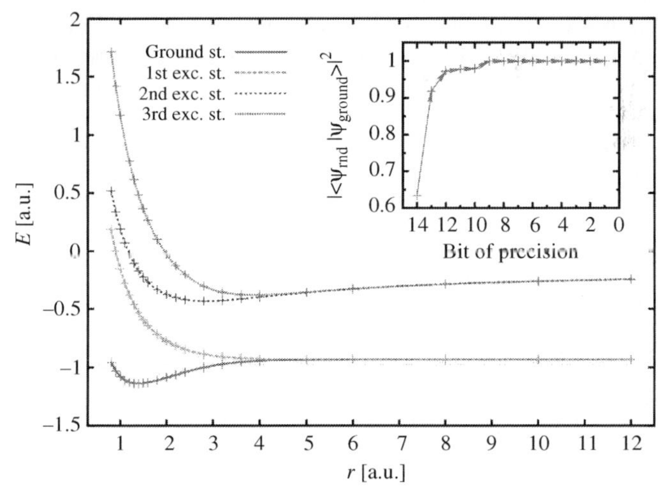

Figure 7. Energies of the four electronic states of H_2 in STO-3G basis that were obtained by the qFCI method (IPEA version **A**) with randomly generated initial guess states. Inset presents the increasing overlap between the actual state of the second part of the quantum register and the exact wave function for one of random runs of the algorithm leading to the ground state. Reprinted with permission from Ref. [30]. Copyright 2010, American Institute of Physics.

Another possibility is to initialize the second part of the quantum register at every iteration step (**B** version of IPEA). Every iteration step (not only the least important bits of ϕ as in Ref. [29]) must be repeated and measurement statistics performed. One could otherwise possibly mix bits belonging to different eigenvalues in different iterations and obtain an unphysical result. The biggest advantage of this approach is avoidance of the long coherence times and therefore potentially easier physical implementation. On the other hand, the biggest disadvantage is that no improving of the overlap between the actual state of the quantum register and the exact wave function occurs during the iterations and one must "fight" the overlap problem at every iteration step. But as will be shown in Section IV in the example of the methylene molecule [30], a small number of repetitions of each iteration is sufficient for amplification of the success probability to unity, when a suitable initial state of the quantum register is used.

At the end of this section, we would like to mention a different way of reducing the number of read-out qubits required by the PEA, which was suggested in the seminal work by Aspuru–Guzik et al. [12]. Their recursive variant of the PEA uses four qubits in the read-out part of the quantum register on which the phase (and therefore also the energy) is successively improved. It starts with measuring the first four bits of the phase ϕ. The Hamiltonian is then shifted by this reference value and a four-bit estimate of the deviation of the phase from the reference one on the half of the interval computed. The procedure is iteratively repeated and the overall effect is a gain of one additional bit of ϕ at each iteration step (thus the same as in the IPEA). In spite of the fact that this method uses four read-out qubits instead of one, which is used by the IPEA, it is worth mentioning. First of all, it was the first iterative version of the PEA applied to the Hamiltonian eigenvalue problem [12]. Second, it recovers the energy starting from the most important bits toward the least important ones (in contrast to the IPEA), which can be advantageous in certain situations.

III. QUANTUM FULL CONFIGURATION INTERACTION METHOD

The PEA/IPEA can in fact be used for efficient computations of eigenvalues of local Hamiltonians [10]. If we take $\hat{U}$ in the form

$$\hat{U} = e^{i\tau\hat{H}} \tag{15}$$

where $\hat{H}$ is a local Hamiltonian and τ a suitable parameter assuring ϕ of being in the interval $\langle 0, 1)$, then the algorithm reveals the energy spectrum of $\hat{H}$. The whole procedure can be simply viewed as a time propagation of a trial wave function followed by the QFT switching from the time to energy domain and a measurement projecting out a certain eigenstate.

In this section, we discuss the application to molecular Born–Oppenheimer electronic Hamiltonians. We will start with a mapping of quantum chemical wave functions onto a quantum register (Section III.A). Section III.B briefly mentions the initial state preparation and Section III.C deals with the crucial part of the algorithm, namely the implementation of controlled powers of the exponential of molecular Hamiltonians (controlled "time propagation").

A. Mapping of Quantum Chemical Wave Functions onto Quantum Register

Several possible mappings of a quantum chemical wave function onto a register of qubits have been proposed. The simplest, but the least economical one in terms of the number of employed qubits, is so-called direct mapping [12]. In this approach, individual spin orbitals are directly assigned to qubits, because each spin orbital can be either occupied or unoccupied, corresponding to $|1\rangle$ or $|0\rangle$ states. The inefficiency lies in the fact that it actually maps the whole Fock space of the system (states with different number of electrons) on the Hilbert space of the quantum register. Relativistic generalization of this approach [31] assigns one qubit to one Kramers pair bispinor (A or B, analogy to α and β spin in nonrelativistic treatment). The advantage of the direct mapping stems from the fact that a general factorization scheme, that is, an algorithm to systematically generate a quantum circuit implementing the exponential of a Hamiltonian, has been discovered (see Section III.C.1).

Compact mappings from a subspace of fixed-electron-number wave functions, spin-adapted [12], or even symmetry-adapted wave functions employing the point group [13] or double group symmetry [31] to the register of qubits have also been proposed. However, general factorization schemes are unfortunately not known for these mappings. The factorization to elementary quantum gates can be for small circuits performed either with numerical optimization techniques (e.g., with genetic algorithms [32]) or analytically [33], but neither *efficiently*. Its use is motivated by the need to employ as few qubits as possible in today's experimental realizations [20,21].

B. Initial States for the Algorithm

The quantum full configuration interaction (qFCI) algorithm must be started with some initial guess state. Generally, it holds that the closer is the initial guess to the exact wave function corresponding to the calculated energy, the higher is the success probability of measuring the energy. As was shown in Refs [17,30], the simplest one-determinantal Hartree–Fock guess may not be successful in situations where correlation (particularly the static one) plays an important role. In these situations, initial guess states from more sophisticated *polynomially* scaling methods can be used (e.g., complete active space self-consistent field (CASSCF) method in a limited orbital CAS).

Preparing a general initial state (an arbitrarily chosen vector from the Hilbert space of n qubits) is a hard task when this vector can contain up to 2^n nonzero components in general and it cannot be performed *efficiently*. Fortunately, initial guesses including only a few determinants in a superposition are sufficient for most purposes of quantum chemistry [30]. These states can be prepared, for example, with the procedure described by Ortiz et al. [4], which scales as $\mathcal{O}(N^2)$ in the number of determinants N. Preparation of general molecular-like states from the combinatorial space of dimension $\binom{n}{m}$ corresponding to distributing m electrons among n spin orbitals was presented in Ref. [17]. Preparation of many-particle states in a superposition on a lattice, which can be then propagated by a quantum chemical dynamics algorithm [14], was studied in Ref. [16].

1. *Adiabatic State Preparation*

A different approach of the initial state preparation is the adiabatic state preparation (ASP) method of Aspuru-Guzik et al. [12]. In the ASP method, one slowly varies the Hamiltonian of the quantum register, starting with a trivial one and the register in its (exactly known) ground state and ending with the final exact one in the following simple way

$$\hat{H} = (1-s)\hat{H}_{\text{init}} + s\hat{H}_{\text{exact}} \qquad s : 0 \longrightarrow 1 \tag{16}$$

If the change is slow enough (depending on the gap between the ground and the first excited state), the register remains in its ground state according to the adiabatic theorem [34]. In the compact mapping, $\hat{H}_{\text{init}}$ can be defined to have all matrix elements equal to zero, except H_{11}, which is equal to the (Dirac–)Hartree–Fock energy [12,31].

Figure 8a demonstrates in the example of the SbH molecule the improvement of the IPEA (version **A**) ground-state success probability during the ASP procedure. The dependence of the energy gap (ΔE) between the ground and the first excited state on the adiabatic transition parameter s is shown in Fig. 8b. Although ΔE is getting close to 0 for $r = 8.0\ a_0$ and s going to 1, the ground state is becoming degenerate at this internuclear distance and this fact thus does not influence the IPEA success probability. We will discuss the relativistic qFCI method and its application to the SbH spin-orbit splitting in more detail in Section V.

Recently, the ASP of the hydrogen molecule ground state has been realized experimentally on a NMR quantum simulator [21].

C. Controlled "Time Propagation"

To study the overall scaling of the qFCI algorithm, one must decompose the only multi-qubit gate from Fig. 5 (i.e., controlled powers of $\hat{U} = e^{i\tau\hat{H}}$ to elementary one- and two-qubit gates).

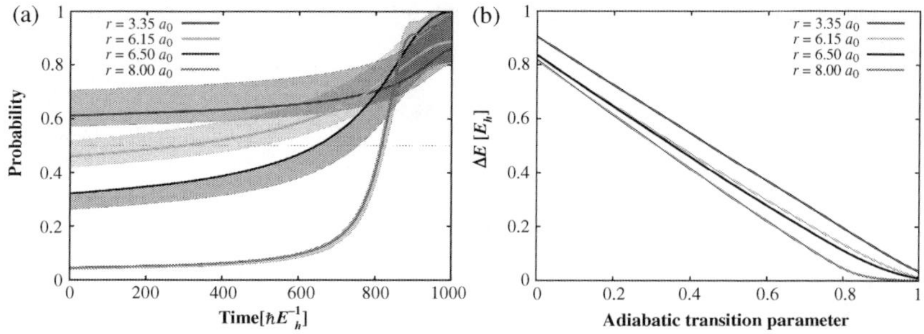

Figure 8. Adiabatic state preparation (ASP) of the SbH ground state for different internuclear distances. (**a**) Dependence of the IPEA success probability on time during the ASP (1000 $\hbar E_h^{-1} \approx 10^{-14}$ s). Solid lines correspond to the success probabilities, $|\langle\psi_{\text{ASP}}|\psi_{\text{exact}}\rangle|^2 \cdot (0.81, 1)$ interval is coloured. Reprinted with permission from Ref. [31]. Copyright (2010) by the American Physical Society. (**b**) Dependence of the energy gap between the ground and the first excited state on the adiabatic transition parameter s. (*See the color version of this figure in the Color Plates section.*)

For this purpose, it is convenient to express the electronic Hamiltonian in the second quantized form [35]

$$\hat{H} = \sum_{pq} h_{pq}\hat{a}_p^\dagger\hat{a}_q + \frac{1}{2}\sum_{pqrs} g_{pqrs}\hat{a}_p^\dagger\hat{a}_q^\dagger\hat{a}_s\hat{a}_r = \sum_{X=1}^{L} \hat{h}_X \tag{17}$$

where h_{pq} and g_{pqrs} are one- and two-electron integrals and $\hat{a}_p^\dagger$ and $\hat{a}_p$ are fermionic creation and annihilation operators. The whole summation is formally expressed as a sum of individual terms $\hat{h}_X$. The molecular integrals h_{pq} and g_{pqrs} can be *efficiently* precalculated on a conventional computer [36] and represent a classical input to the quantum algorithm. In nonrelativistic case, they are real-valued, while in relativistic case, they are in general complex.

Because the creation and annihilation operators generally do not commute, the exponential of a Hamiltonian cannot be written as a product of the exponentials of individual $\hat{h}_X$, but a numerical approximation must be used [2]. The first-order Trotter approximation [37] can be expressed as

$$e^{i\tau\hat{H}} = e^{i\tau\sum_{X=1}^{L}\hat{h}_X} = \left(\prod_{X=1}^{L} e^{i\hat{h}_X\tau/N}\right)^N + \mathcal{O}(\tau^2/N) \tag{18}$$

By choosing $N \geq (\tau^2/\epsilon)$, we can implement $\hat{U}$ within an error tolerance of $\mathcal{O}(\epsilon)$ using $\mathcal{O}(L(\tau^2/\epsilon))$ particular terms $e^{i\hat{h}_X\tau/N}$.

Before discussing the factorization of these terms to elementary quantum gates in Section III.C.1, we would like to mention the implementation modification we use [30]. Two more external inputs are necessary in our case. They are maximum

($E_{\max}$) and minimum ($E_{\min}$) energies expected in the studied system and we use $\hat{U}$ in the form

$$\hat{U} = e^{i\tau(E_{\max} - \hat{H})} \tag{19}$$

For τ, it holds

$$\tau = \frac{2\pi}{E_{\max} - E_{\min}} \tag{20}$$

and the final energy is obtained according to the formula

$$E = E_{\max} - \frac{2\pi\phi}{\tau} \tag{21}$$

The modification assures ϕ to be in the interval $\langle 0, 1)$.

$E_{\min}$ and $E_{\max}$ can in fact be chosen arbitrarily, but one must be sure that the calculated energy is within this interval, otherwise one would end up with a nonphysical energy, due to the periodicity of $e^{2\pi i\phi}$. The maximum energy can be, for example, the upper bound provided by any classical variational (*polynomially* scaling) method, techniques for calculation of lower bounds [38–40] can on the other hand give the minimum energy. The smaller the interval between them is, the fewer iterations of IPEA are necessary for desired precision of E.

Taking $\hat{U}$ in the form of Eq. (19) does not pose any difficulties, and as the following circuit equality shows, just one more one-qubit rotation is needed.

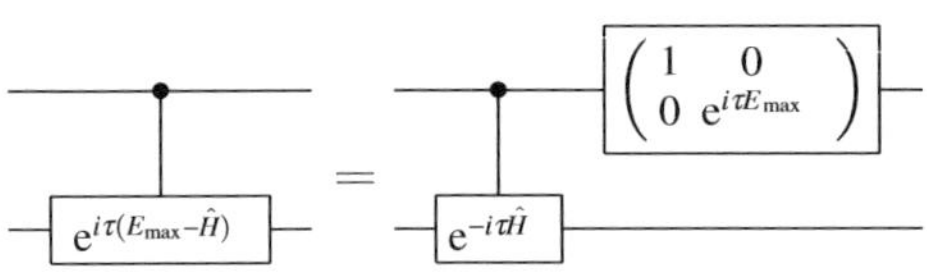

1. *Decomposition of Unitary Propagator to Elementary Quantum Gates*

The decomposition of the unitary propagator $e^{i\tau\hat{H}}$ to elementary quantum gates [4,24,41] proceeds in the following manner. First, the Jordan–Wigner transform [42] is used to express the fermionic second quantized operators in terms of Pauli σ matrices. The Jordan–Wigner transform has the form

$$\hat{a}_n^\dagger = \left(\bigotimes_{j=1}^{n-1} \sigma_z^j \right) \otimes \sigma_-^n, \quad \hat{a}_n = \left(\bigotimes_{j=1}^{n-1} \sigma_z^j \right) \otimes \sigma_+^n \tag{22}$$

where $\sigma_\pm = 1/2(\sigma_x \pm i\sigma_y)$ and the superscript denotes the qubit on which the matrix operates. The Hamiltonian Eq. (17) then can be rewritten using strings of

σ matrices. Finally, the exponentials of these strings are built up from single-qubit gates and controlled NOT operations (CNOTs) [25].

We will demonstrate this approach on the one-electron part of the Hamiltonian (with complex-valued molecular integrals)

$$\hat{H}_1 = \sum_{pq} h_{pq}\hat{a}_p^\dagger \hat{a}_q = \sum_{pp} h_{pp}\hat{a}_p^\dagger \hat{a}_p + \sum_{p>q} \left(h_{pq}\hat{a}_p^\dagger \hat{a}_q + h_{qp}\hat{a}_q^\dagger \hat{a}_p\right) \tag{23}$$

Employing the Jordan–Wigner transform, the diagonal terms can be written as

$$h_{pp} a_p^\dagger a_p = \frac{h_{pp}^{\mathrm{R}}}{2}(\mathbf{1} - \sigma_z^p) \tag{24}$$

where h_{pp}^{R} is the real part of h_{pp} [h_{pp}^{I} (the imaginary part) is equal to zero due to the Hermicity of $\hat{H}$]. For the exponentials holds

$$e^{i\hat{h}_X \tau/N} = e^{ih_{pp}a_p^\dagger a_p \tau/N} = \begin{pmatrix} 1 & 0 \\ 0 & e^{ih_{pp}\tau/N} \end{pmatrix}^{(p)} \tag{25}$$

The superscript (p) at the matrix denotes the qubit on which the one-qubit gate operates.

Similarly, the off-diagonal terms read

$$\begin{aligned} h_{pq}a_p^\dagger a_q + h_{qp}a_q^\dagger a_p = \\ = \frac{h_{pq}^{\mathrm{R}}}{2}\left[\sigma_x^p \otimes \left(\sigma_z^{p\to q}\right) \otimes \sigma_x^q + \sigma_y^p \otimes \left(\sigma_z^{p\to q}\right) \otimes \sigma_y^q\right] + \\ + \frac{h_{pq}^{\mathrm{I}}}{2}\left[\sigma_y^p \otimes \left(\sigma_z^{p\to q}\right) \otimes \sigma_x^q - \sigma_x^p \otimes \left(\sigma_z^{p\to q}\right) \otimes \sigma_y^q\right] \end{aligned} \tag{26}$$

where $\sigma_z^{p\to q}$ represents the direct product

$$\sigma_z^{p\to q} \equiv \sigma_z^{p+1} \otimes \sigma_z^{p+2} \otimes \ldots \otimes \sigma_z^{q-2} \otimes \sigma_z^{q-1} \tag{27}$$

Note that Eq. (26) contains the four aforementioned strings of σ matrices.

The exponential of the string of σ_z matrices $\exp[i\tau(\sigma_z \otimes \ldots \otimes \sigma_z)]$ is in fact diagonal in the computational basis with the phase shift $e^{\pm i\tau}$ on the diagonal. The sign of this phase shift depends on the parity of the corresponding basis state ("+" if the number of ones in the binary representation is even, "−" otherwise). The exponential can be implemented with the following circuit [25]

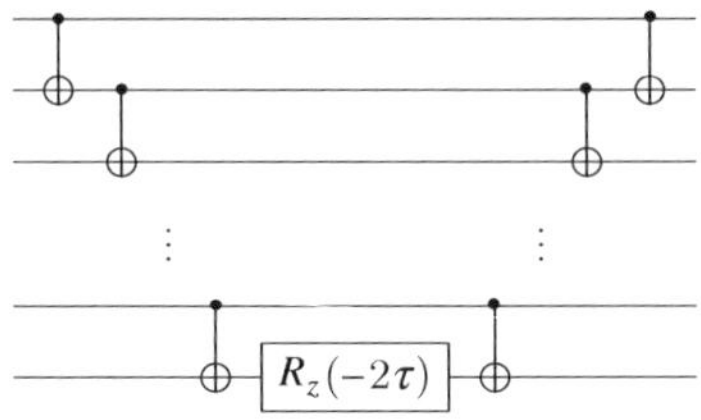

(28)

where for the z-rotations holds

$$R_z(\theta) = \begin{pmatrix} e^{-i\theta/2} & 0 \\ 0 & e^{i\theta/2} \end{pmatrix} \tag{29}$$

and CNOTs assure the correct sign of the phase shift according to the parity of the state.

Due to the following change-of-basis identities [25]

$$\sigma_x = H\sigma_z H^{\dagger} \tag{30}$$

$$\sigma_y = Y\sigma_z Y^{\dagger} \tag{31}$$

where

$$Y = R_x(-\pi/2) = \frac{1}{\sqrt{2}} \begin{pmatrix} 1 & i \\ i & 1 \end{pmatrix} \tag{32}$$

the exponentials

$$\begin{aligned}
&\exp\left[\frac{ih^R_{pq}\tau}{2N}\sigma^p_x \otimes \left(\sigma^{p\rightarrow q}_z\right) \otimes \sigma^q_x\right] \\
&\exp\left[\frac{ih^R_{pq}\tau}{2N}\sigma^p_y \otimes \left(\sigma^{p\rightarrow q}_z\right) \otimes \sigma^q_y\right] \\
&\exp\left[\frac{ih^{\mathrm{I}}_{pq}\tau}{2N}\sigma^p_y \otimes \left(\sigma^{p\rightarrow q}_z\right) \otimes \sigma^q_x\right] \\
&\exp\left[\frac{-ih^{\mathrm{I}}_{pq}\tau}{2N}\sigma^p_x \otimes \left(\sigma^{p\rightarrow q}_z\right) \otimes \sigma^q_y\right]
\end{aligned} \tag{33}$$

can be implemented with the following circuit pattern,
where A and B are for the individual exponentials of Eq. (33) equal to $\{H, H\}$, $\{Y, Y\}$, $\{Y, H\}$, and $\{H, Y\}$, respectively, and θ to $-h^R_{pq}\tau/N$, $-h^R_{pq}\tau/N$, $-h^{\mathrm{I}}_{pq}\tau/N$,

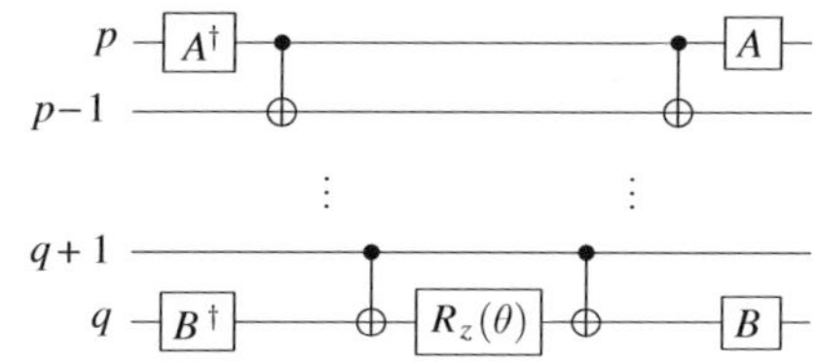

(34)

and $h^{\mathrm{I}}_{pq}\tau/N$, respectively. Note that although the two strings of σ matrices in the first parenthesis in Eq. (26) commute as do the two strings in the second parenthesis, they do not commute mutually. This, however, is not a complication because the Trotter approximation in Eq. (18) must be employed anyway.

We have demonstrated the decomposition technique for the *direct mapping* approach on the one-electron part of the Hamiltonian. The procedure for the two-electron part is more elaborate, but completely analogous and we refer an interested reader to Ref. [24], where all the cases are presented in a systematic way.

The overall scaling of the algorithm is given by the scaling of a single controlled action of the unitary propagator without repetitions enforced by the Trotter approximation of Eq. (18). These repetitions increase only the prefactor to the polynomial scaling, not the scaling itself. Also the required precision is limited, about 20 binary digits of ϕ are sufficient to achieve the chemical accuracy [30].

The single controlled action of the exponential of a one-body Hamiltonian in Eq. (23) results in $\mathcal{O}(n^3)$ scaling: the $\mathcal{O}(n^2)$ different h_{pq} terms each requires $\mathcal{O}(n)$ elementary quantum gates [see the circuit of Eq. (34)]. Because the same decomposition technique applied to the two-electron part of the Hamiltonian gives rise to similar circuit patterns [24], each term g_{pqrs} requires $\mathcal{O}(n)$ elementary quantum gates as well (this in fact holds for general m-body Hamiltonians [41]). The total scaling is thus $\mathcal{O}(n^5)$ [20,24], where n is the number of molecular spin orbitals (or bispinors in the relativistic case [31]) and the qFCI achieves an *exponential* speedup over the conventional FCI. This speedup is demonstrated in Fig. 9.

At this point, we would like to make few remarks. First, we assumed that the initial state preparation is an *efficient* step, as was already mentioned. Second, when a quantum chemical method with a scaling worse than $\mathcal{O}(n^5)$ is used for calculation of the initial guess state on a conventional computer, then this classical step becomes a rate-determining one. In addition, the classical computation of the integrals in the molecular orbital basis scales as $\mathcal{O}(n^5)$ (due to the integral transformation). We also assumed noise-free qubits and thus did not take into account any quantum error correction [43]. Clark et al. studied the resource requirements for a similar, but fault-tolerant computation of the ground state of a one-dimensional transverse Ising model [44] on a proposed scalable quantum computing architecture [45]. They showed that due to the exponential scaling of the resource requirements with the desired energy precision as well as due to the Trotter approximation

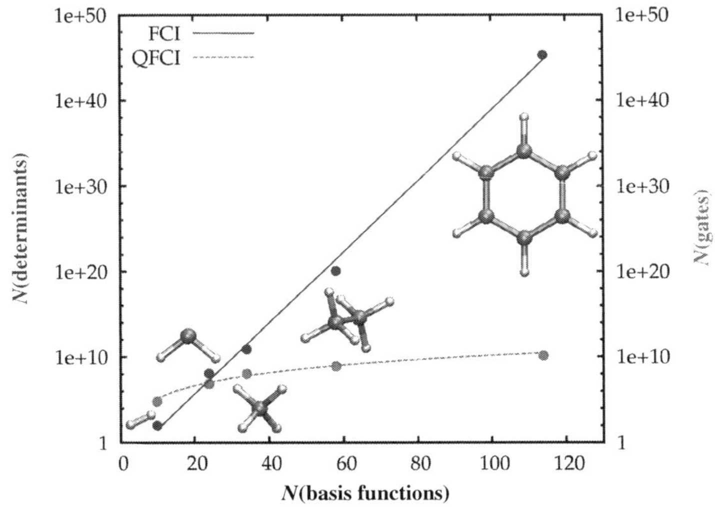

Figure 9. The *exponential* speedup of the qFCI over the FCI. In case of the FCI (solid line), dependence of the number of Slater determinants in the FCI expansion on the number of basis functions is shown. In case of the qFCI (dashed line), dependence of the number of one- and two-qubit gates needed for a single controlled action of the unitary operator on the number of basis functions is presented. Points in the graph correspond to the depicted molecules (hydrogen, methylene, methane, ethane, and benzene) in the cc-pVDZ basis set. Reprinted with permission from Ref. [30]. Copyright 2010, American Institute of Physics.

employed, an elaborate error correction is required, which leads to a huge increase of a computational time. They also gave the values of the experimental parameters (e.g., the physical gate time) needed for acceptable computational times. However, the question of reducing the resource requirements needed for fault tolerant qFCI computations is still open and undergoes an active research.

IV. APPLICATION TO NONRELATIVISTIC MOLECULAR HAMILTONIANS

We will demonstrate an application of the qFCI method to non-relativistic ground and excited state energy calculations on the example of the methylene molecule [30].

A. Example of CH_2 Molecule

Methylene molecule (CH_2) in a minimal basis set (STO-3G) is a simple, yet computationally interesting system suitable for simulations and performance testing of the qFCI method. CH_2 is well known for the multireference character of its lowest-lying singlet electronic state ($\tilde{a}\ ^1A_1$) and is often used as a benchmark system for

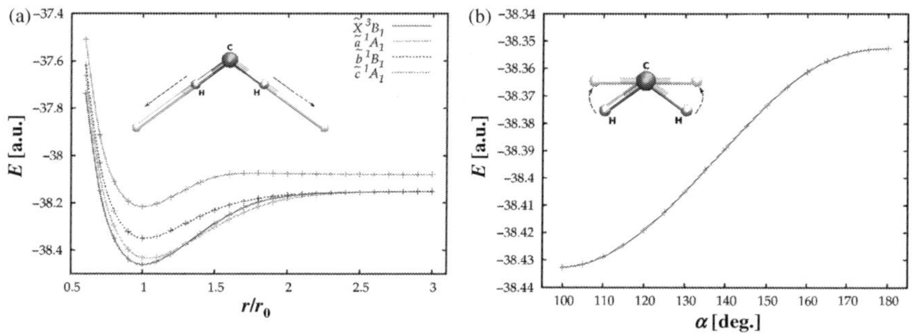

Figure 10. (**a**) Energies of the four simulated states of CH_2 for the C–H bond stretching, r_0 denotes the equilibrium bond distance. (**b**) Energy of $\tilde{a}\,^1A_1$ state of CH_2 for the H–C–H angle bending, α denotes the H–C–H angle. Reprinted with permission from Ref. [30]. Copyright 2010, American Institute of Physics.

testing of newly developed computational methods (see, e.g., Refs [46–49]). In the STO-3G basis set, the total number of molecular (spin)orbitals is 7(14), which means 15 qubits in the direct mapping approach (one qubit is needed in the read-out part of the register). Because classical simulations of 15-qubit computations are feasible, the CH_2 molecule serves as an excellent candidate for the first benchmark simulations.

We simulated the qFCI energy calculations of the four lowest-lying electronic states of CH_2: $\tilde{X}\,^3B_1$, $\tilde{a}\,^1A_1$, $\tilde{b}\,^1B_1$, and $\tilde{c}\,^1A_1$. Two processes shown in Fig. 10 were studied: C–H bond stretching (both C–H bonds were stretched, Fig. 10b), and H–C–H angle bending for $\tilde{a}\,^1A_1$ state (Fig. 10b). These processes were chosen purposely because description of bond breaking is a hard task for many of computational methods and H–C–H angle bending since the $\tilde{a}\,^1A_1$ state exhibits a strong multireference character at linear geometries.

Our work followed up the work by Wang et al. [13] who studied the influence of initial guesses on the performance of the quantum FCI method on two singlet states of water molecule across the bond-dissociation regime. They found out that the Hartree–Fock (HF) initial guess is not sufficient for bond dissociation and suggested the use of CASSCF method. Few configuration state functions added to the initial guess in fact improved the success probability dramatically.

We, therefore, also tested different initial guesses for the qFCI calculations. Those denoted as HF guess were composed only from spin-adapted configurations that qualitatively describe certain states. Here, for simplicity, we will present the results just for the $\tilde{X}\,^3B_1$ ground state described by the configuration $(1a_1)^2(2a_1)^2(1b_2)^2(3a_1)(1b_1)$ and $\tilde{a}\,^1A_1$ state characterized by $(1a_1)^2(2a_1)^2(1b_2)^2(3a_1)^2$ configuration. Initial guesses denoted as CAS(x,y) were based on complete active space configuration interaction (CASCI) calculations with small active spaces, which contained x electrons in y orbitals. Initial guesses

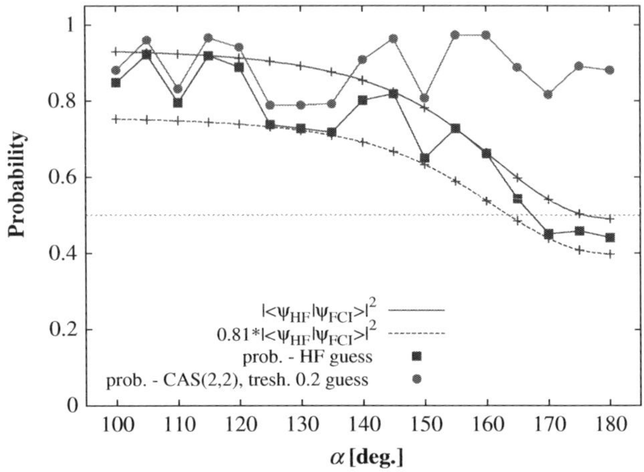

Figure 11. Success probabilities of the **A** version of IPEA for the $\tilde{a}\ {}^1A_1$ state with HF and CAS(2,2) initial guesses, tresh. 0.2 means that only configurations with absolute values of amplitudes higher than 0.2 were involved in the initial guess, α denotes the H–C–H angle. Reprinted with permission from Ref. [30]. Copyright 2010, American Institute of Physics.

were constructed only from the configurations whose absolute values of amplitudes were higher than 0.1. Those constructed from the configurations whose absolute values of amplitudes were higher than 0.2 are denoted as CAS(x,y), tresh. 0.2 guess. All the initial guesses were normalized before the simulations.

In our simulations, similarly as in Ref. [12], the exponential of a Hamiltonian operator was implemented as an n-qubit gate, which is motivated by the fact that once the decoherence is not involved in the model, the exponential of a Hamiltonian can be obtained with an arbitrary precision simply by a proper number of repetitions in Eq. (18). But we would like to note that Whitfield et al. [24] among others also numerically studied the length of a Trotter time step needed for a required energy precision on the example of the helium atom. We ran 20 iterations of the IPEA, which correspond to the final energy precision $\approx 10^{-6}\ E_h$. All the other computational details including the exact definition of the CA spaces can be found in Ref. [30].

Figure 11 presents the results of the **A** version of IPEA for the angle-bending process. The overlap and scaled overlap of the initial HF guess wave function and the exact FCI wave function are shown as well as the success probabilities for the HF and CAS(2,2) initial guesses and dotted line bounding the safe region. The results numerically confirm that the success probabilities always lie in the interval $\left|\langle\psi_{\text{init}}|\psi_{\text{exact}}\rangle\right|^2 \cdot \left(0.81, 1\right)$, depending on the value of the remainder δ in Eq. (12). This algorithm can be *safely* used when the resulting success probability is higher than 0.5 (as it can then be amplified by repeating the whole process). When going to the linear geometry, where the $\tilde{a}\ {}^1A_1$ state exhibits a strong multireference character and the restricted HF (RHF) description is no more adequate, the CAS(2,2) initial guess improves the success probabilities dramatically. Moreover, these initial states correspond to only two configurations and are thus easy to prepare [4].

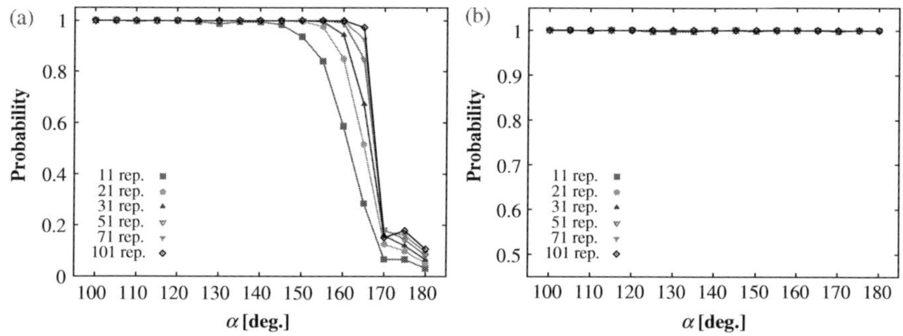

Figure 12. Success probabilities of the **B** version of IPEA with (**a**) HF and (**b**) CAS(2,2), tresh. 0.2 initial guesses and different number of repetitions of individual bit measurements for the $\tilde{a}\ {}^1A_1$ state, α denotes the H–C–H angle. Reprinted with permission from Ref. [30]. Copyright 2010, American Institute of Physics.

Performance of the **B** version of IPEA with the HF and CAS(2,2) initial guesses is illustrated in Fig. 12. As can be seen, in the situations where the particular initial guess state can be used, few repetitions are enough to amplify the success probability to unity. The HF initial guess is again not sufficient for the linear and close to linear geometries.

Results corresponding to the $\tilde{a}\ {}^1A_1$ and $\tilde{X}\ {}^3B_1$ states for the C–H bond stretching are summarized in Figs. 13 and 14. Figure 13 presents the performance of the **A** version of IPEA. When going to more stretched C–H bonds, the RHF initial guess again fails. The CAS(2,2) initial guess improves the success probabilities in case of $\tilde{a}\ {}^1A_1$ state near the equilibrium geometry but in the region of more stretched C–H bonds it also fails. In this region, CAS(4,4) initial guess must be

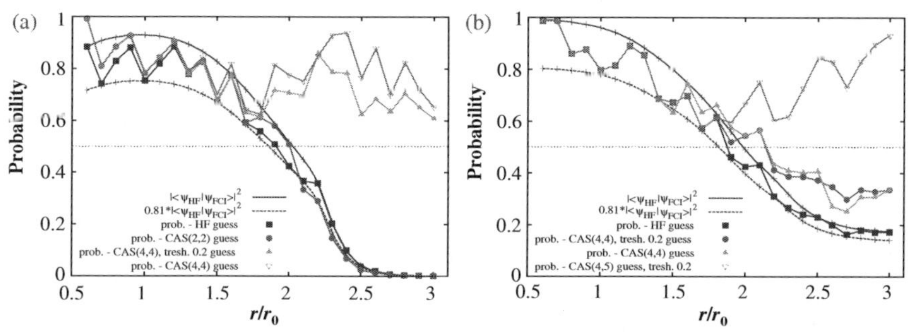

Figure 13. Success probabilities of the **A** version of IPEA for (**a**) $\tilde{a}\ {}^1A_1$ state, (**b**) $\tilde{X}\ {}^3B_1$ state of CH_2. Different initial guesses were used, tresh. 0.2 means that only configurations with absolute values of amplitudes higher than 0.2 were involved in the initial guess, r_0 denotes the equilibrium bond distance. Reprinted with permission from Ref. [30]. Copyright 2010, American Institute of Physics.

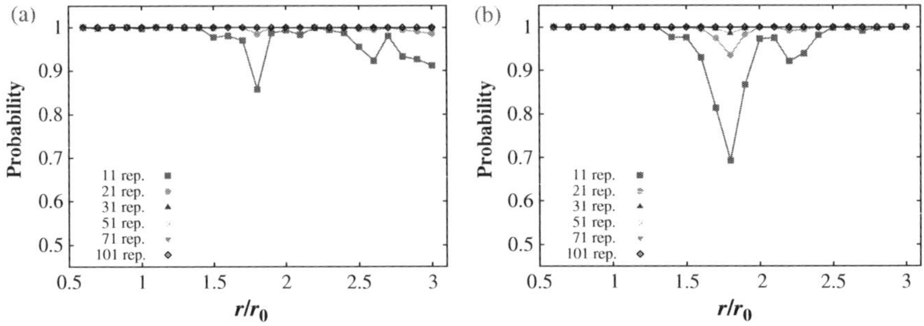

Figure 14. Success probabilities of the **B** version of IPEA with "best" initial guesses and different number of repetitions of individual bit measurements, r_0 denotes the equilibrium bond distance. (**a**) $\tilde{a}\ ^1A_1$ state, CAS(4,4), tresh. 0.2 guess, (**b**) $\tilde{X}\ ^3B_1$ state, CAS(4,5), tresh. 0.2 guess. Reprinted with permission from Ref. [30]. Copyright 2010, American Institute of Physics.

used [CAS(4,4), tresh. 0.2 guess is sufficient]. The situation is even more difficult for the $\tilde{X}\ ^3B_1$ state when the C–H bonds are stretched. Here, even the CAS(4,4) initial guess fails and bigger active space—CAS(4,5)—must be used for initial guess state calculations. However, apart from the CAS size, initial guess states always contained at most 12 configurations, but usually 8 or even fewer for nearly dissociated molecules. This observation is in agreement with the results of Ref. [13], where few configuration state functions added to the initial guess improved the success probability dramatically.

Figure 14 presents the performance of the **B** version of IPEA for the "best" initial guesses in terms of price/performance ratio. The evident result is again that relatively small number of repetitions ($\approx$31) amplifies the success probability nearly to unity. This is important, because the **B** version of IPEA seems to be a better candidate for the first real larger-scale qFCI calculations, because it does not require a long coherence time.

V. EXTENSION TO RELATIVISTIC MOLECULAR HAMILTONIANS

So far, we addressed nonrelativistic computations only. But it is well known that relativistic effects can be very important in chemistry. In fact, accurate description of systems with heavy elements requires adequate treatment of relativistic effects [50]. The most rigorous approach (besides the quantum electrodynamics (QED), which is not feasible for quantum chemical purposes) is the four-component (4c) formalism [51]. Recently, we therefore developed the qFCI method for 4c relativistic computations [31] and the details of this method are the subject of this section.

The 4c formalism brings three major computational difficulties: (1) working with 4c orbitals (bispinors), (2) complex algebra when molecular symmetry is low, and (3) rather large Hamiltonian matrix eigenvalue problems (due to larger mixing of states than in the nonrelativistic case). All of them can nevertheless be solved on a quantum computer. Before discussing how this is done, we would like to note that in our work, we restricted ourselves to the 4c Dirac–Coulomb Hamiltonian (with the rest mass term mc^2 subtracted):

$$\hat{H} = \sum_{i=1}^{N} \left[c(\boldsymbol{\alpha}_i \cdot \mathbf{p}_i) + \beta_i' mc^2 + \sum_A \frac{1}{r_{iA}} \right] + \sum_{i<j} \frac{1}{r_{ij}} \tag{35}$$

$$\alpha_k = \begin{pmatrix} 0 & \sigma_k \\ \sigma_k & 0 \end{pmatrix}, \quad \beta' = \beta - I_4, \quad \beta = \begin{pmatrix} I_2 & 0 \\ 0 & -I_2 \end{pmatrix}$$

where σ_k ($k = x, y, z$) are Pauli matrices and I_2 the 2×2 unit matrix. This type of Hamiltonian covers the major part of the spin–orbit interaction (including two-electron spin–own orbit) and also scalar relativistic effects. It is in fact without loss of generality sufficient for our purposes because going to Dirac–Coulomb–Breit Hamiltonian [51] correct to $\mathcal{O}(c^{-2})$ is conceptually straightforward as the inclusion of the corresponding integrals requires a classically polynomial effort.

We will start the description of the algorithm with a mapping of a *relativistic* quantum chemical wave function onto a quantum register. We have already briefly mentioned this topic in Section III. We, in fact, do not need any qubits to represent positronic bispinors in the direct mapping due to the no-pair approximation [51], which is widely used in relativistic quantum chemistry. In this approach, one actually builds up an N-electron wave function only from Slater determinants containing positive-energy bispinors. This procedure neglects all QED effects, but it is justifiable at the energy scale relevant to chemistry. Moreover, because of the time-reversal symmetry of the Dirac equation, bispinors occur in degenerate Kramers pairs [51] denoted A and B (analogy to α and β spin in nonrelativistic treatment). The direct mapping thus employs one qubit for bispinor A and one for B. The 4c character of molecular bispinors does not complicate the approach substantially, because the Hartree–Fock part of a calculation is done on a classical computer and only the exponentially scaling FCI part on a quantum one.

Assuming the no-pair approximation and the empty Dirac picture, the relativistic Hamiltonian has the same second quantized structure in Eq. (17) as the nonrelativistic one. The only difference is that the molecular integrals h_{pq} and g_{pqrs} are now in general complex (and have thus lower index permutation symmetry). It poses no difficulty for a quantum computer, because our working environment is a complex vector space of qubits anyway and we do the exponential of a complex matrix even if the Hamiltonian is real. Moreover, on the example of the one-electron part of the Hamiltonian [see Eq. (26) and the circuit in Eq. (34)], one can see that the decomposition of the unitary propagator $e^{i\tau\hat{H}}$ with complex molecular integrals requires twice as many gates compared to real-valued Hamiltonians [24], while complex arithmetic on a classical computer requires four times more operations.

The last of the difficulties of the 4c formalism is the size of a Hamiltonian matrix eigenvalue problem. When we put the double group symmetry aside and employ Kramers restricted (KR) approach, the relativistic Hamiltonian, unlike the nonrelativistic one, mixes determinants with different values of the pseudo-quantum number M_K

$$M_K = 1/2(N_A - N_B) \tag{36}$$

where N_A and N_B denote the number of electrons in A and B Kramers pair bispinors (in the nonrelativistic case, M_K corresponds to M_S and Hamiltonian is block diagonal in M_S). Therefore, in the case of a closed shell system with n electrons in m molecular orbitals (bispinors), the size of the relativistic Hamiltonian matrix (number of Slater determinants in the FCI expansion) reads

$$N_{\text{rel.}} = \sum_{x=0}^{n} \binom{m}{x} \binom{m}{n-x} = \binom{2m}{n} \tag{37}$$

The last equality is due to the Vandermonde's convolution [52]. When compared to the nonrelativistic case of Eq. (1) and using the Stirling's approximation in the form

$$\ln m! \approx \frac{1}{2}\ln(2\pi m) + m\ln m - m \qquad \text{for } m \to \infty \tag{38}$$

the ratio between the relativistic and nonrelativistic Hamiltonian matrix sizes is given by the expression

$$k_{\text{rel./non-rel.}} = \frac{N_{\text{rel.}}}{N_{\text{non-rel.}}} = \left(\frac{\sqrt{\pi}(2k-1)}{2k}\right) \cdot m^{1/2} \tag{39}$$

where we set $m = k \cdot n$.

On a quantum computer on the other hand, this problem does not occur. When employing the *direct mapping*, which maps the whole Fock space of the system on the Hilbert space of qubits, the Hamiltonian Eq. (17) in fact implicitly works with all possible values of M_K. The scaling of the relativistic qFCI method is therefore the same as the nonrelativistic one, namely $\mathcal{O}(m^5)$, where m is the number of molecular orbitals/bispinors (see Section III.C.1).

It is quite remarkable and deserves an emphasis that the relativistic qFCI method not only achieves an *exponential* speedup over its classical counterpart, but, as we

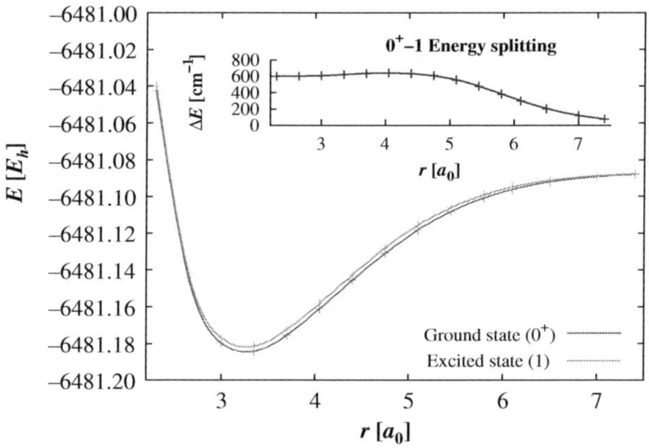

Figure 15. Simulated potential energy curves of ground (0^+) and excited (1) states of SbH, and spin-orbit energy splitting. Reprinted with permission from Ref. [31]. Copyright (2010) by the American Physical Society.

have just discussed, also has the same cost (in terms of scaling) as its nonrelativistic analogue.

A. Example of SbH Molecule

We will demonstrate the performance of the relativistic qFCI method on the example of the SbH molecule. This molecule is of particular interest to us, because its nonrelativistic ground state $^3\Sigma^-$ splits due to spin-orbit effects into $X\ 0^+$ and $A\ 1$ states. In the approximate $\lambda\omega$-projection, these states are dominated by $\sigma^2_{1/2}\pi^2_{1/2}\pi^0_{3/2}$ and $\sigma^2_{1/2}\pi^1_{1/2}\pi^1_{3/2}$ configurations. The splitting is truly of "molecular nature" as it disappears for dissociated atoms and its experimental value is $\Delta E_{\mathrm{SO}} = 654.97\ \mathrm{cm}^{-1}$ [53].

For this reason, we simulated the SbH bond dissociation process. Simulated potential energy curves of both states are shown in Fig. 15. Because we employed rather large basis sets of triple-ζ quality, we could not manage to simulate the FCI calculations with all electrons. We instead simulated general active space (GAS) KRCI computations [54] with the occupation constraints that give rise to CI spaces of approximately 30,000 determinants. Definition of the GA spaces together with all the computational details can be found in Ref. [31].

We worked solely with a compact mapping employing the double-group symmetry (C^*_{2v}) and exponential of a Hamiltonian was again simulated as an n-qubit gate. We used the DIRAC program[1] for calculations of Hamiltonian matrices

[1] DIRAC, a relativistic *ab initio* electronic structure program, Release DIRAC08 (2008), written by L. Visscher, H. J. Aa. Jensen, and T. Saue, with new contributions from R. Bast, S. Dubillard, K. G. Dyall, U. Ekström, E. Eliav, T. Fleig, A. S. P. Gomes, T. U. Helgaker, J. Henriksson, M. Iliaš, Ch. R. Jacob, S. Knecht, P. Norman, J. Olsen, M. Pernpointner, K. Ruud, P. Sa lek, and J. Sikkema (see http://dirac.chem.sdu.dk).

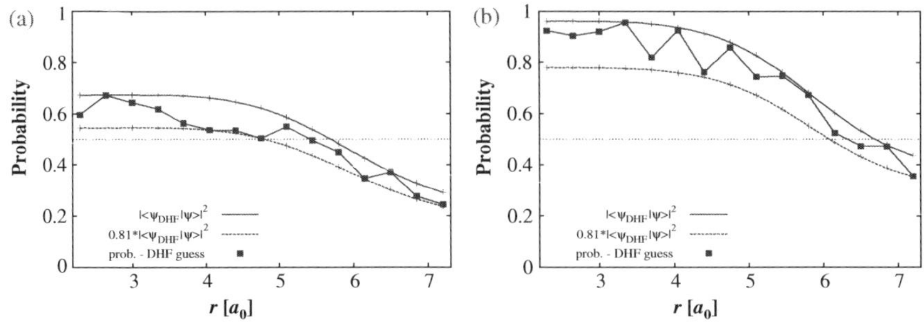

Figure 16. SbH (**a**) ground (0^+) and (**b**) excited (1) state qFCI (IPEA version **A**) success probabilities corresponding to the HF initial guesses.

and ran 17 iterations of the IPEA, which correspond to the final energy precision $\approx 3.81 \times 10^{-6}\ E_h$.

Based on our KRCI setup we obtain a vertical ΔE_{SO} of 617 cm^{-1}. Success probabilities of the algorithm with the HF initial guesses (represented by the dominant configurations of both states) are presented in Fig. 16. They correspond to the **A** version of IPEA.

Ground state success probabilities confirm that relativistic states have, due to near degeneracies caused by the spin-orbit coupling, often a stronger multireference character than nonrelativistic ones. The upper bound of the success probability is less than 0.7 even for the equilibrium geometry and the lower bound is higher than 0.5 only up to 4.8 a_0. The success probabilities of the $A1$ state are higher and the HF initial guesses can be in a noise-free environment used up to 6 a_0. For longer distances, initial guesses from more accurate *polynomially* scaling relativistic methods must be used (analogously to the nonrelativistic example, e.g., the relativstic 4c CASSCF method with a small orbital CAS). The low ground state success probabilities can be also improved by the ASP procedure as is shown in Fig. 8.

VI. CONCLUSIONS

In this chapter, we have reviewed quantum algorithms for the FCI energy calculations. Starting with the seminal paper by Aspuru-Guzik et al. [12] and ending with our most recent work on relativistic generalization of the algorithm [31], we gave a detailed description of the qFCI method, which is applicable to the ground as well as excited state energy calculations in the nonrelativistic and also relativistic regimes. Both variants share the scaling in the number of molecular orbitals/bispinors m, namely $\mathcal{O}(m^5)$ and exhibit thus an *exponential* speedup over their classical counterparts. We have demonstrated their performance by numerical simulations of the CH_2 and SbH energy calculations. The results indicate that the ground and excited state energies at equilibrium geometries are accessible with HF initial guesses, which are easy to prepare. In the example of the CH_2 molecule, we have shown that CASCI initial guess states with small complete active spaces composed of relatively few configurations ($\approx$10) are sufficient even for a nearly dissociated molecule to achieve the probability amplification regime of the IPEA algorithm. The first proof-of-principle experimental realizations recently achieved by several groups [20–22] are very promising and confirm the real applicability of the method.

Acknowledgments

The authors thank Prof. Timo Fleig for fruitful discussions about the relativistic quantum chemistry and gratefully acknowledge the financial support from the Grant Agency of the Czech Republic-GAČR (203/08/0626) and the Grant Agency of the Charles University-GAUK (114310). Libor Veis also acknowledges support from the Hlávka Foundation.

REFERENCES

1. R. P. Feynman, *Int. J. Theor. Phys.* **21**, 467 (1982).
2. S. Lloyd, *Science* **273**, 1073 (1996).
3. C. Zalka, *Proc. R. Soc. Lond. Ser. A* **454**, 313 (1998).
4. G. Ortiz, J. E. Gubernatis, E. Knill, and R. Laflamme, *Phys. Rev. A* **64**, 022319 (2001).
5. R. Somma, G. Ortiz, J. E. Gubernatis, E. Knill, and R. Laflamme, *Phys. Rev. A* **65**, 042323 (2002).
6. D. S. Abrams and S. Lloyd, *Phys. Rev. Lett.* **79**, 2586 (1997).
7. P. W. Shor, in *Proceedings of 35th IEEE Symposium on Foundations of Computer Science*, IEEE Press, 1994, pp. 124–134.
8. P. W. Shor, *SIAM J. Comput.* **25**, 1484 (1997).
9. L. K. Grover, *Phys. Rev. Lett.* **79**, 325 (1997).
10. D. S. Abrams and S. Lloyd, *Phys. Rev, Lett.* **83**, 5162 (1999).
11. D. A. Lidar and H. Wang, *Phys. Rev. E* **59**, 2429 (1999).
12. A. Aspuru-Guzik, A. D. Dutoi, P. J. Love, and M. Head-Gordon, *Science* **309**, 1704 (2005).
13. H. Wang, S. Kais, A. Aspuru-Guzik, and M. R. Hoffmann, *Phys. Chem. Chem. Phys.* **10**, 5388 (2008).
14. I. Kassal, S. P. Jordan, P. J. Love, M. Mohseni, and A. Aspuru-Guzik, *Proc. Natl. Acad. Sci.* **105**, 18681 (2008).
15. I. Kassal and A. Aspuru-Guzik, *J. Chem. Phys.* **131**, 224102 (2009).
16. N. J. Ward, I. Kassal, and A. Aspuru-Guzik, *J. Chem. Phys.* **130**, 194105 (2009).
17. H. Wang, S. Ashhab, and F. Nori, *Phys. Rev. A* **79**, 042335 (2009).
18. J. Zhu, Z. Huang, and S. Kais, *Mol. Phys.* **107**, 2015 (2009).
19. I. Kassal, J. D. Whitfield, A. Perdomo-Ortiz, M. H. Yung, and A. Aspuru-Guzik, *Annu. Rev. Phys. Chem.* **62**, 185 (2011).

20. B. P. Lanyon, J. D. Whitfield, G. G. Gillett, M. E. Goggin, M. P. Almeida, I. Kassal, J. D. Biamonte, M. Mohseni, B. J. Powell, M. Barbieri, A. Aspuru-Guzik, and A. G. White, *Nat. Chem.* **2**, 106 (2010).
21. J. Du, N. Xu, X. Peng, P. Wang, S. Wu, and D. Lu, *Phys. Rev. Lett.* **104**, 030502 (2010).
22. Z. Li, M.-H. Yung, H. Chen, D. Lu, J. D. Whitfield, S. Peng, A. Aspuru-Guzik, and J. Du, *Nat. Sci. Rep.* **1**, 88 (2011).
23. D. Lu, N. Xu, R. Xu, H. Chen, J. Gong, X. Peng, and J. Du, *Phys. Rev. Lett.* **107**, 020501 (2011).
24. J. D. Whitfield, J. Biamonte, and A. Aspuru-Guzik, *Mol. Phys.* **109**, 735 (2011).
25. M. A. Nielsen and I. L. Chuang, *Quantum Computation and Quantum Information*, Cambridge University Press, Cambridge, 2000.
26. R. B. Griffiths and C.-S. Niu, *Phys. Rev. Lett.* **76**, 3228 (1996).
27. M. Dobšíček, *Quantum Computing, Phase Estimation and Applications*, Ph.D. thesis, Faculty of Electrical Engineering, Czech Technical University in Prague (2008).
28. H. Wang, L.-A. Wu, Y. xi Liu, and F. Nori, *Phys. Rev. A* **82**, 062303 (2010).
29. M. Dobšíček, G. Johansson, V. Shumeiko, and G. Wendin, *Phys. Rev. A* **76**, 030306 (2007).
30. L. Veis and J. Pittner, *J. Chem. Phys.* **133**, 194106 (2010).
31. L. Veis, J. Višňák, T. Fleig, S. Knecht, T. Saue, L. Visscher, and J. Pittner, *arXiv:quant-ph/1111.3490v1*, 2011.
32. A. Daskin and S. Kais, *J. Chem. Phys.* **134**, 144112 (2011).
33. V. V. Shende, S. S. Bullock, and I. L. Markov, *IEEE Trans. Comput. Aided Design* **25**, 1000 (2006).
34. E. Fahri, J. Goldstone, S. Gutmann, J. Lapan, A. Lundgren, and D. Preda, *Science* **292**, 472 (2001).
35. A. Szabo and N. Ostlund, *Modern Quantum Chemistry: Introduction to Advanced Electronic Structure Theory*, Dover Publications, 1996.
36. T. Helgaker , P. Jorgensen, and J. Olsen, *Molecular Electronic-Structure Theory*, Wiley, 2000.
37. N. Hatano and M. Suzuki, Finding Exponential Product Formulas of Higher Orders, in *Lecture Notes in Physics*, Springer, Heidelberg, 2005.
38. N. W. Bazley, *Phys. Rev.* **120**, 144 (1960).
39. P.-O. Löwdin, *J. Math. Phys.* **3**, 969 (1962).
40. P.-O. Löwdin, *Phys. Rev. A* **139**, 357 (1965).
41. E. Ovrum and M. Hjorth-Jensen, *arXiv:quant-ph/0705.1928v1*, 2007.
42. P. Jordan and E. Wigner, *Z. Phys. A* **47**, 631 (1928).
43. F. Gaitan, *Quantum Error Correction and Fault Tolerant Quantum Computing*, CRC Press, 2008.
44. C. R. Clark, T. S. Metodi, S. D. Gasster, and K. R. Brown, *Phys. Rev. A* **79**, 062314 (2009).
45. T. S. Metodi, D. D. Thaker, A. W. Cross, F. T. Chong, and I. L. Chuang, in *Proceeding of the 38th Annual IEEE/ACM International Symposium on Microarchitechture*, 2005, p. 305.
46. J. Pittner, P. Nachtigall, P. Čársky, J. Mášik, and I. Hubač, *J. Chem. Phys.* **110**, 10275 (1999).
47. F. A. Evangelista, W. D. Allen, and H. F. Schaefer III, *J. Chem. Phys.* **125**, 154113 (2006).
48. K. Bhaskaran-Nair, O. Demel, and J. Pittner, *J. Chem. Phys.* **132**, 154105 (2010).
49. O. Demel and J. Pittner, *J. Chem. Phys.* **128**, 104108 (2008).
50. B. A. Hess and C. M. Marian, in *Computational Molecular Spectroscopy*, P. Jensen and P. R. Bunker, eds., Wiley, Sussex, 2000, pp. 169–219.

51. K. G. Dyall and K. Faegri, *Introduction to Relativistic Quantum Chemistry*, Oxford University Press, New York, 2007.

52. R. L. Graham, D. E. Knuth, and O. Patashnik, *Concrete Mathematics*, Addison-Wesley Publishing Company, 1990.

53. K. Balasubramanian, *Chem. Rev.* **89**, 1801 (1989).

54. T. Fleig, J. Olsen, and L. Visscher, *J. Chem. Phys.* **119**, 2963 (2003).

DENSITY FUNCTIONAL THEORY AND QUANTUM COMPUTATION

FRANK GAITAN[1] and FRANCO NORI[2,3]

[1] *Laboratory for Physical Sciences, 8050 Greenmead Drive, College Park, MD 20740-4004, USA*
[2] *CEMS, RIKEN, Saitama 351-0198, Japan*
[3] *Physics Department, University of Michigan, Ann Arbor, MI 48109-1040, USA*

I. INTRODUCTION

The inability of a classical computer to efficiently simulate the dynamics of a quantum system is well known. The problem is that the dimension of the Hilbert space grows exponentially with the number of degrees of freedom of the quantum system, which in turn causes an exponential growth in the amount of memory and CPU time required to carry out the simulation. This inefficiency is a major stumbling block for numerical studies aiming to determine the asymptotic performance of quantum algorithms. For example, numerical simulation of the dynamics of the quantum adiabatic evolution (QAE) algorithm applied to the NP-Complete problem Exact Cover 3 has been limited to systems containing $N \leq 20$ qubits [1].

Advances in Chemical Physics, Volume 154: Quantum Information and Computation for Chemistry, First Edition. Edited by Sabre Kais.

Because the algorithm requires adiabatic dynamics, its performance is highly sensitive to the minimum energy gap $\Delta(N)$ separating the instantaneous ground- and first-excited states, with the gap's asymptotic limit, $(N \gg 1)$, largely determining the algorithm's computational efficiency. Attempts to evaluate $\Delta(N)$ using exact diagonalization have been limited to $N \leq 20$ qubits [2]. Recently, however, the minimum gap $\Delta(N)$ for QAE applied to Exact Cover 3 has been determined for $N \leq 128$ qubits using quantum Monte Carlo methods [3]. This substantial technical advance has stirred great interest in finding other computational approaches that might allow quantum algorithm performance to be determined for still larger qubit systems.

Quantum computation is not the only research area struggling with the difficulties of simulating quantum systems. Condensed-matter physicists and quantum chemists have been working under the shadow of this problem for decades. A number of computational approaches have been developed that, together with increasingly more powerful computers, have allowed much progress to be made, despite the ultimately unavoidable difficulties involved. Among these approaches, density functional theory (DFT) has proven to be one of the most successful [4]. DFT applies to *interacting* fermion systems. It captures all many-body effects through the exchange-correlation energy functional. Ground state density functional theory (GS-DFT) has been used to determine a wide range of ground state properties of atomic, molecular, and solid-state systems [5,6]; while time-dependent density functional theory (TD-DFT) has been used to determine excited-state properties, as well as the linear and nonlinear response of interacting many-electron systems to electromagnetic fields [7,8]. For our purposes, it is especially significant that DFT has been successfully applied to quantum systems containing $N \sim 10^3$ interacting degrees of freedom [9–11].

Here, we show how DFT can be applied to quantum computing systems [12]. Although our analysis can be extended to the case of N qudits (d-level systems) residing on a D-dimensional lattice, we restrict the presentation to N qubits residing on a 2D lattice because it corresponds to the experimentally interesting cases of qubits placed in a 2D ion trap, or restricted to a planar superconducting qubit circuit. We begin by showing how the N-qubit system can be transformed into a system of N lattice fermions. We illustrate this transformation by using it to rewrite the dynamics of the QAE algorithm applied to the NP-Complete problem MAXCUT [13]. For the resulting interacting fermion system, we establish the Hohenberg-Kohn (HK) [5] and Runge-Gross [7] theorems, and setup the auxiliary Kohn–Sham (KS) system of noninteracting fermions [6]. This justifies the application of GS- and TD-DFT to the dynamics of the interacting fermion system. The linear response of the interacting fermions is determined using TD-DFT, which is then used to obtain the energy gap $\Delta(N)$ for the MAXCUT dynamics. In light of earlier remarks, this approach would allow evaluation of the minimum gap $\Delta(N)$ for $N \sim 10^3$ qubits.

II. QUBIT/FERMION TRANSFORMATION

Consider N qubits residing on an N-site 2D lattice with basis vectors $\hat{\mathbf{e}}_k$ $(k = 1, 2)$, and sites specified by the position vector $\mathbf{r}$. Let $\boldsymbol{\sigma}(\mathbf{r})$ denote the Pauli matrices associated with the qubit at $\mathbf{r}$. We now show how the qubits can be converted into lattice fermions via the 2D Jordan–Wigner (JW) transformation [14]. Note that the following analysis can be extended to N qudits on a D-dimensional lattice using the generalized JW transformation [15] that fermionizes a spin s system in D spatial dimensions $(d = 2s + 1)$.

For a 2D system of qubits, the JW transformation is

$$\begin{aligned}\sigma^+(\mathbf{r}) &= 2a_{\mathbf{r}}^{\dagger}Q_{\mathbf{r}} \\ \sigma^-(\mathbf{r}) &= 2Q_{\mathbf{r}}^{\dagger}a_{\mathbf{r}} \\ \sigma_z(\mathbf{r}) &= 2\hat{n}_{\mathbf{r}} - 1\end{aligned} \tag{1}$$

Here, $\sigma^{\pm}(\mathbf{r}) = \sigma_x(\mathbf{r}) \pm i\sigma_y(\mathbf{r})$; $a_{\mathbf{r}}^{\dagger}$ $(a_{\mathbf{r}})$ creates (annihilates) a lattice fermion at $\mathbf{r}$; $\hat{n}_{\mathbf{r}} = a_{\mathbf{r}}^{\dagger}a_{\mathbf{r}}$; and

$$Q_{\mathbf{r}} = \exp\left[-i\phi_{\mathbf{r}}\right]; \qquad \phi_{\mathbf{r}} = (1/2\pi\theta)\sum_{\mathbf{r}'} \Phi(\mathbf{r}, \mathbf{r}')\,\hat{n}_{\mathbf{r}'} \tag{2}$$

In Eq. (2), $\Phi(\mathbf{r}, \mathbf{r}')$ is the angle made by $(\mathbf{r} - \mathbf{r}')$ with respect to some reference direction, say $\hat{\mathbf{e}}_1$. Thus, (i) $\Phi(\mathbf{r}, \mathbf{r}')$ changes by 2π when $(\mathbf{r} - \mathbf{r}')$ traces out a closed loop around $\mathbf{r}'$; and (ii) $\Phi(\mathbf{r}, \mathbf{r}) \equiv 0$. The requirement that the Pauli operators $\boldsymbol{\sigma}(\mathbf{r})$ commute at different lattice sites forces θ to satisfy

$$\theta = \frac{1}{(2m + 1)2\pi} \qquad (m = 0, \pm 1, \pm 2, \ldots)$$

in Eq. (2).

As shown in Ref. [14], the lattice fermions are spinless and minimally coupled to a gauge field $A_k(\mathbf{r}) = \Delta_k\phi_{\mathbf{r}} = \phi_{\mathbf{r}+\hat{\mathbf{e}}_k} - \phi_{\mathbf{r}}$. The action for the gauge field $A_\mu(\mathbf{r})$ is given by the Chern–Simons term [16]

$$\mathcal{A}_{CS} = -\frac{\theta}{4}\int dt \sum_{\mathbf{r}} \epsilon^{\mu\nu\lambda} A_\mu(\mathbf{r}) F_{\nu\lambda}(\mathbf{r})$$

Maxwell's equations for this system take the form

$$j^{\mu} = \epsilon^{\mu\nu\lambda} F_{\nu\lambda} \tag{3}$$

where j^{μ} is the fermion current, $F_{\nu\lambda}$ is the gauge field tensor, $\epsilon^{\mu\nu\lambda}$ is the totally antisymmetric Levi-Civita tensor, and $\mu, \nu, \lambda = 0, 1, 2$. Fermion current conservation, $\partial_\mu j^{\mu} = 0$, follows immediately from Maxwell's equations.

III. NP-COMPLETE PROBLEM MAXCUT

In this problem, one considers an N-node undirected graph with nodes specified by $\mathbf{r}$. The nodes (edges) are assigned weights $w_{\mathbf{r}}$ ($w_{\mathbf{r},\mathbf{r}'}$) and a binary variable $s_{\mathbf{r}}$ is associated with each node $\mathbf{r}$. A cut of the graph is a partition of the nodes into two sets $\mathcal{S}_0$ and $\mathcal{S}_1$. For the nodes belonging to $\mathcal{S}_0$ ($\mathcal{S}_1$), $s_{\mathbf{r}}$ is assigned the value 0 (1). The problem is to find the cut that maximizes the payoff function

$$P(s) = \sum_{\mathbf{r}} w_{\mathbf{r}} s_{\mathbf{r}} + \sum_{\mathbf{r},\mathbf{r}'} s_{\mathbf{r}}(1 - s_{\mathbf{r}'}) w_{\mathbf{r},\mathbf{r}'}$$

The QAE algorithm was applied to MAXCUT in Ref. [13], where the dynamics is driven by the time-dependent Hamiltonian

$$H(t) = (1 - t/T)H_0 + (t/T)H_P \tag{4}$$

Here, T is the algorithm runtime; the initial Hamiltonian H_0 is

$$H_0 = \sum_{\mathbf{r}} \frac{1}{2}\left[1 - \sigma_x(\mathbf{r})\right] \tag{5}$$

and the final/problem Hamiltonian H_P is

$$H_P = \sum_{\mathbf{r}} w_{\mathbf{r}}(1 - \sigma_z(\mathbf{r}))/2 + \sum_{\mathbf{r},\mathbf{r}'} w_{\mathbf{r},\mathbf{r}'}(1 - \sigma_z(\mathbf{r})\sigma_z(\mathbf{r}'))/2 \tag{6}$$

Using Eqs. (1) in $H(t)$ gives the fermionized QAE Hamiltonian for MAXCUT:

$$\begin{aligned} H(t) = {} & (1 - t/T) \sum_{\mathbf{r}} [a_{\mathbf{r}}^{\dagger} Q_{\mathbf{r}} + Q_{\mathbf{r}}^{\dagger} a_{\mathbf{r}}] \\ & + (t/T) \sum_{\mathbf{r}} v_{\mathbf{r}} \hat{n}_{\mathbf{r}} + (t/T) \sum_{\mathbf{r}} \sum_{\mathbf{r}' \neq \mathbf{r}} w_{\mathbf{r},\mathbf{r}'} \hat{n}_{\mathbf{r}} \hat{n}_{\mathbf{r}'} \end{aligned} \tag{7}$$

where $w_{\mathbf{r}} \equiv v_{\mathbf{r}} + W_{\mathbf{r}}$; $W_{\mathbf{r}} = \sum_{\mathbf{r}' \neq \mathbf{r}} w_{\mathbf{r},\mathbf{r}'}$; and a term proportional to the identity has been suppressed. We now show that GS-DFT can be applied to the dynamics of the instantaneous MAXCUT Hamiltonian $H(t)$ for each $t \in (0, T)$.

Note: Throughout the remainder of this chapter, the instantaneous ground-state is assumed to be nondegenerate as would be appropriate for the case of a nonvanishing gap Δ.

IV. GS-DFT

The QAE algorithm has an adiabatic dynamic that is driven by a slowly varying Hamiltonian $H(t)$. In our discussion of GS-DFT, we focus on the MAXCUT Hamiltonian $H(t)$ at a *fixed instant of time* $t = t_*$. By fixing the time, we obtain a well-defined *static* Hermitian operator $H_* \equiv H(t = t_*)$. We aim to show that the

Hohenberg–Kohn theorem applies to H_*. With this theorem in place, GS-DFT can be used to study the ground state properties of $H_* = H(t = t_*)$ for any specific intermediate time $0 < t_* < T$. We stress that even though the QAE algorithm works with a slowly varying Hamiltonian, the following GS-DFT discussion is restricted to the *static* Hermitian operator H_* that is the value of $H(t)$ at $t = t_*$.

Our starting point is the energy functional for the instantaneous MAXCUT Hamiltonian $H_* = H(t = t_*)$,

$$E[n] = \min_{|\psi\rangle \to n} \langle \psi | H_* | \psi \rangle \tag{8}$$

The domain of $E[n]$ is the set of all N-representable site occupation functions (SOF) $n_{\mathbf{r}}$ that can be obtained from an N-fermion wave function. The minimization in Eq. (8) is over all $|\psi\rangle$ for which

$$n \equiv n_{\mathbf{r}} = \langle \psi | \hat{n}_{\mathbf{r}} | \psi \rangle$$

and the minimizing state $|\psi_{\min}[n]\rangle$ is thus a functional of $n_{\mathbf{r}}$. Let $|\psi^g\rangle$ denote the ground state of H_*; E^g the ground state energy; and $n_{\mathbf{r}}^g$ the ground state SOF. Inserting Eq. (7) into Eq. (8) gives

$$E[n] = (t_*/T) \sum_{\mathbf{r}} v_{\mathbf{r}} n_{\mathbf{r}} + Q[n]$$

where

$$Q[n] \equiv \min_{|\psi\rangle \to n} \langle \psi | (T_{t_*} + U_{t_*}) | \psi \rangle$$

and T_{t_*} and U_{t_*} are the first and third terms, respectively, on the right-hand side of Eq. (7) at $t = t_*$.

To establish the HK theorem for H_* we must show [17]: (i) $E[n^g] = E^g$; (ii) for $n_{\mathbf{r}} \neq n_{\mathbf{r}}^g$, $E[n] > E^g$; and (iii) the ground state expectation value of any observable is a unique functional of the ground state SOF $n_{\mathbf{r}}^g$. By the variational principle, $\langle \psi | H_* | \psi \rangle \geq E^g$, with equality when $|\psi\rangle = |\psi^g\rangle$. Thus, for $n = n^g$, the search in Eq. (8) returns the ground state $|\psi^g\rangle$ as the state $|\psi_{\min}[n^g]\rangle$ that minimizes $E[n^g]$. It follows that

$$E[n^g] = \langle \psi^g | H(t) | \psi^g \rangle = E^g$$

This establishes condition (i). For $n \neq n^g$, the minimizing state $|\psi_{\min}[n]\rangle \neq |\psi^g[n^g]\rangle$, and so by the variational principle,

$$E[n] = \langle \psi_{\min}[n] | H(t) | \psi_{\min}[n] \rangle > E^g$$

This establishes condition (ii). Finally, because the ground state $|\psi^g\rangle = |\psi_{\min}[n^g]\rangle$, it is a unique functional of n^g and, consequently, so are all ground state expectation values,

$$\langle \hat{O} \rangle_{gs} = \langle \psi^g | \hat{O} | \psi^g \rangle = \langle \psi_{\min}[n^g] | \hat{O} | \psi_{\min}[n^g] \rangle = \mathcal{O}[n^g]$$

Condition (iii) is thus established, completing the proof of the HK theorem for $H_* = H(t_*)$.

To obtain a practical calculational scheme, an auxiliary system of noninteracting KS fermions is introduced [6], and it is assumed that the ground state SOF $n_{\mathbf{r}}^g$ can be obtained from the ground state density of the KS fermions moving in an external potential $v_{\mathbf{r}}^{ks}$. For $H_* = H(t_*)$, the KS Hamiltonian $H_{ks} = T_t' + V^{ks}$ is defined to be

$$H_{ks} = \sum_{\mathbf{r}} (1 - t_*/T)\{q_{\mathbf{r}} a_{\mathbf{r}}^{\dagger} + q_{\mathbf{r}}^* a_{\mathbf{r}}\} + \sum_{\mathbf{r}} (t_*/T) v_{\mathbf{r}}^{ks} \hat{n}_{\mathbf{r}}$$

where $q_{\mathbf{r}} = \langle Q_{\mathbf{r}} \rangle$ is the ground state expectation value of $Q_{\mathbf{r}}$. The effects of $Q_{\mathbf{r}}$ are thus incorporated into the KS dynamics through the mean-field $q_{\mathbf{r}}$. The KS energy functional $\epsilon_{ks}[n]$ is

$$\epsilon_{ks}[n] = \min_{|\psi\rangle \to n} \langle \psi | H_{ks} | \psi \rangle = T_{t_*}'[n] + \sum_{\mathbf{r}} (t_*/T) v_{\mathbf{r}}^{ks} n_{\mathbf{r}} \tag{9}$$

To determine the KS external potential $v_{\mathbf{r}}^{ks}$, we rewrite Eq. (8) as

$$E[n] = T_{t_*}'[n] + \sum_{\mathbf{r}} (t_*/T) v_{\mathbf{r}} n_{\mathbf{r}} + \xi_{xc}[n] \tag{10}$$

where

$$\xi_{xc}[n] \equiv Q[n] - T_{t_*}'[n]$$

is the exchange-correlation energy functional. Recall that it is through the exchange-correlation energy functional $\xi_{xc}[n]$ that DFT accounts for *all many-body effects*. Because $n_{\mathbf{r}}^g$ minimizes $\epsilon_{ks}[n]$ and $E[n]$, Eqs. (9) and (10) are stationary about $n = n^g$. Taking their functional derivatives with respect to n, evaluating the result at $n = n^g$, and eliminating $\delta T'/\delta n|_{n=n^g}$ gives

$$v_{\mathbf{r}}^{ks} = v_{\mathbf{r}} + (T/t_*) v_{xc}[n^g] \tag{11}$$

for $t_* \neq 0$. Here, $v_{xc}[n^g](\mathbf{r})$ is the exchange-correlation potential, which is the functional derivative of the exchange-correlation energy functional $\xi_{xc}[n^g]$,

$$v_{xc}[n^g](\mathbf{r}) = \frac{\delta \xi_{xc}[n^g]}{\delta n_{\mathbf{r}}^g}$$

This sets in place the formulas for a self-consistent calculation of the ground state properties of $H_* = H(t_*)$ using GS-DFT. Entanglement [18] and its links to quantum phase transitions [19] have been studied using GS-DFT.

V. TD-DFT

Here, we establish the Runge-Gross theorem [7] for the instantaneous MAXCUT dynamics. Thus, we focus on the instantaneous Hamiltonian $H_* = H(t_*)$ for a fixed t_* $(0 < t_* < T)$. Now, however, we suppose that the external potential $v_{\mathbf{r}}$ in $H(t_*)$ begins to vary at a moment we call $t = 0$. For $t \leq 0$, $v_{\mathbf{r}}(t) = v_{\mathbf{r}}$, and the fermions are in the ground-state $|\psi_0\rangle$ of $H(t_*)$. The Runge-Gross theorem states that the SOFs $n_{\mathbf{r}}(t)$ and $n'_{\mathbf{r}}(t)$ evolving from a common initial state $|\psi(0)\rangle = |\psi_0\rangle$ under the influence of the respective potentials $V_{\mathbf{r}}(t)$ and $V'_{\mathbf{r}}(t)$ (both Taylor-series expandable about $t = 0$) will be different provided that $[V_{\mathbf{r}}(t) - V'_{\mathbf{r}}(t)] \neq C(t)$. For us,

$$
\begin{aligned}
V_{\mathbf{r}}(t) &= (t_*/T)(1 - t_*/T)v_{\mathbf{r}}(t) \\
V'_{\mathbf{r}}(t) &= (t_*/T)(1 - t_*/T)v'_{\mathbf{r}}(t)
\end{aligned}
$$

and

$$
\begin{aligned}
V_{\mathbf{r}}(t) &= \sum_{k=0}^{\infty} a_k(\mathbf{r})t^k/k! \\
V'_{\mathbf{r}}(t) &= \sum_{k=0}^{\infty} a'_k(\mathbf{r})t^k/k!
\end{aligned}
$$

Let $C_k(\mathbf{r}) \equiv a_k(\mathbf{r}) - a'_k(\mathbf{r})$. The condition that $[V_{\mathbf{r}}(t) - V'_{\mathbf{r}}(t)] \neq C(t)$ means a smallest integer K exists such that $C_k(\mathbf{r})$ is a nontrivial function of $\mathbf{r}$ for all $k \geq K$, while for $k < K$, it is a constant C_k, which can be set to zero without loss of generality.

It follows from Eq. (3) that the conserved fermion current $j_{\mathbf{r},\mu}$ has components

$$
\begin{aligned}
\hat{j}_{\mathbf{r},0}(t) &= n_{\mathbf{r}}(t) \\
\hat{j}_{\mathbf{r},k}(t) &= (1/2\pi)\sum_{\mathbf{y}}(\Delta_k G_{\mathbf{r},\mathbf{y}})\partial_t n_{\mathbf{y}}(t)
\end{aligned}
$$

with $k = 1, 2$ [14]. Here, $G_{\mathbf{r},\mathbf{y}}$ is the Green's function for the lattice Laplacian,

$$
\sum_{k=1,2} \Delta_k \Delta_k G_{\mathbf{r},\mathbf{y}} = -2\pi\delta_{\mathbf{r},\mathbf{y}}
$$

Defining $j_{\mathbf{r},k}(t) = \langle\psi_0|\hat{j}_{\mathbf{r},k}(t)|\psi_0\rangle$, it follows that

$$
\partial_t\{j_{\mathbf{r},k}(t) - j'_{\mathbf{r},k}(t)\} = \langle\psi_0|[\hat{j}_{\mathbf{r},k}(t), H(t) - H'(t)]|\psi_0\rangle \tag{12}
$$

Here, $j_{\mathbf{r},k}(t)$ $[j'_{\mathbf{r},k}(t)]$ and $H(t)$ $[H'(t)]$ are the expected fermion current and the Hamiltonian, respectively, when the external potential is $v_{\mathbf{r}}(t)$ $[v'_{\mathbf{r}}(t)]$. The Hamiltonians $H(t)$ and $H'(t)$ differ only in the external potential. Defining

$$\delta j_{\mathbf{r},k}(t) = j_{\mathbf{r},k}(t) - j'_{\mathbf{r},k}(t)$$

and

$$\delta V_{\mathbf{y}}(t) = V_{\mathbf{y}}(t) - V'_{\mathbf{y}}(t)$$

evaluation of the commutator in Eq. (12) eventually gives

$$\partial_t\{\delta j_{\mathbf{r},k}(t)\} = -(1/2\pi)\sum_{\mathbf{y}}(\Delta_k G_{\mathbf{r},\mathbf{y}})\,\delta V_{\mathbf{y}}(t)\,\mathcal{M}_{\mathbf{y}}(t) \tag{13}$$

where

$$\mathcal{M}_{\mathbf{y}}(t) = \langle\psi_0|(a^\dagger_{\mathbf{y}}Q_{\mathbf{y}} + Q^\dagger_{\mathbf{y}}a_{\mathbf{y}})|\psi_0\rangle$$

With K defined, taking K time-derivatives of Eq. (13) and evaluating the result at $t = 0$ gives

$$\left.\frac{\partial^{K+1}}{\partial t^{K+1}}(\delta j_{\mathbf{r},k}(t))\right|_0 = \frac{-1}{2\pi}\sum_{\mathbf{y}}(\Delta_k G_{\mathbf{r},\mathbf{y}})\mathcal{M}_{\mathbf{y}}(0)C_K(\mathbf{r}) \tag{14}$$

where we have used that

$$\partial^k/\partial t^k(\delta V_{\mathbf{y}}(t))|_{t=0} = C_k(\mathbf{y}) = 0$$

for $k < K$. It is important to note that $\mathcal{M}_{\mathbf{y}}(0) \neq 0$. This follows because

$$[H(t_*), n_{\mathbf{r}}(t_*)] \neq 0$$

for $t_* \neq T$, and so the eigenstates of $H(t_*)$ (specifically, its ground state $|\psi_0\rangle$) cannot be fermion number eigenstates, which ensures the ground state expectation value

$$\mathcal{M}_{\mathbf{y}}(0) = \langle\psi_0|(a^\dagger_{\mathbf{y}}Q_{\mathbf{y}} + Q^\dagger_{\mathbf{y}}a_{\mathbf{y}})|\psi_0\rangle \neq 0$$

for $t_* \neq T$. It follows from the continuity equation for the fermion current that

$$\partial_t(n_{\mathbf{r}}(t) - n'_{\mathbf{r}}(t)) = -\sum_{k=1,2}\Delta_k(\delta j_{\mathbf{r},k}(t))$$

Taking K time-derivatives of this equation, evaluating the result at $t = 0$, and using Eq. (14) gives

$$\frac{\partial^{K+2}}{\partial t^{K+2}}(n_{\mathbf{r}}(t) - n'_{\mathbf{r}}(t))|_{t=0} = -C_K(\mathbf{r})\mathcal{M}_{\mathbf{r}}(0) \neq 0 \tag{15}$$

where we have used the equation of motion for $G_{\mathbf{r},\mathbf{y}}$. Eq. (15) indicates that $n_{\mathbf{r}}(t)$ cannot equal $n'_{\mathbf{r}}(t)$ because it ensures that they will be different at $t = 0^+$, and so cannot be the same function, which proves the Runge-Gross theorem for the instantaneous MAXCUT dynamics.

We have just seen that when potentials $V_{\mathbf{r}}(t)$ and $V'_{\mathbf{r}}(t)$ differ by a time-dependent function $C(t)$, they give rise to the same SOF $n_{\mathbf{r}}(t)$. However, the wave functions produced by these potentials from the same initial state will differ by a time-dependent phase factor. For our purposes, it is important to note that this extra phase factor cancels out when calculating the expectation value of an operator. In particular, it will cancel out when calculating the instantaneous energy eigenvalues $E_n(t) = \langle E_n(t)|H(t)|E_n(t)\rangle$. As a result, this phase factor will not affect our calculation of the minimum energy gap to be described next. Having said this, it is worth noting that this subtlety is not expected to cause difficulties in practice because the probe potential $V_{\mathbf{r}}(t)$ is assumed to be under the direct control of the experimenter, and so the precise form of $V_{\mathbf{r}}(t)$ is *known*. When an experimentalist says a sinusoidal probe potential has been applied, this means $V_{\mathbf{r}}(t) = V_{\mathbf{r}} \sin \omega t$; it does *not* mean $V_{\mathbf{r}}(t) = V_{\mathbf{r}} \sin \omega t + C(t)$. Thus, in a well-defined experiment $C(t) = 0$.

The KS system of noninteracting fermions can also be introduced in TD-DFT [7]. We must still assume that the interacting SOF $n_{\mathbf{r}}(t)$ can be obtained from the SOF of the noninteracting KS fermions moving in the external potential $v_{\mathbf{r}}^{ks}(t)$. The potentials $v_{\mathbf{r}}^{ks}(t)$ and $v_{\mathbf{r}}(t)$ are related via ($t_* \neq 0$)

$$v_{\mathbf{r}}^{ks}(t) = v_{\mathbf{r}}(t) + (T/t_*)\, v_{xc}[n_{\mathbf{r}}(t)] \tag{16}$$

though Eq. (16) is to be thought of as defining the exchange correlation potential $v_{xc}[n_{\mathbf{r}}(t)]$.

VI. MINIMUM GAP

A problem of long-standing treachery in GS-DFT is the calculation of the excitation energies of a fermion system. TD-DFT was able to find these energies by determining the system's frequency-dependent linear response, and relating the excitation energies to poles appearing in that response [20]. The arguments used are quite general and can be easily adapted to determine the energy gap for the instantaneous MAXCUT dynamics.

Previously, we considered an external potential that becomes time-varying for $t \geq 0$. Our interest is the *interacting* fermion linear response, and so we assume that the total potential has the form

$$v^{\text{tot}}_{\mathbf{r}}(t) = v_{\mathbf{r}} + v^{1}_{\mathbf{r}}(t)$$

with $v^{1}_{\mathbf{r}}(t)$ a suitably small time-varying perturbation. The probe potential $v^{1}_{\mathbf{r}}(t)$ generates a first-order response $n^{1}_{\mathbf{r}}(t)$ in the SOF:

$$n^{\text{tot}}_{\mathbf{r}}(t) = n^{g}_{\mathbf{r}} + n^{1}_{\mathbf{r}}(t)$$

The susceptibility $\chi_{\mathbf{r},\mathbf{r}'}(t-t')$ connects the first-order probe potential to the SOF response. The total potential $v^{\text{tot}}_{\mathbf{r}}(t)$ is related to the KS potential $v^{ks}_{\mathbf{r}}(t)$ through Eq. (16), and by assumption, the SOF for both the interacting and KS fermions is the same. The time-Fourier transform of the SOF response $n^{1}_{\mathbf{r}}(\omega)$ can then be determined from the time-Fourier transforms of the KS susceptibility $\chi^{ks}_{\mathbf{r},\mathbf{r}'}(\omega)$, the exchange-correlation kernel $f_{xc}[n^{g}](\mathbf{r},\mathbf{r}';\omega)$, and the probe potential $v^{1}_{\mathbf{r}}(\omega)$:

$$\sum_{\mathbf{y}'}\{\delta_{\mathbf{r},\mathbf{y}'} - \sum_{\mathbf{r}'}\chi^{ks}_{\mathbf{r},\mathbf{r}'}(\omega) f_{xc}[n^{g}](\mathbf{r}',\mathbf{y}';\omega)\} n^{1}_{\mathbf{y}'}(\omega) = \sum_{\mathbf{r}'}\chi^{ks}_{\mathbf{r},\mathbf{r}'}(\omega) v^{1}_{\mathbf{r}'}(\omega) \qquad (17)$$

The KS susceptibility [8] depends on the KS static unperturbed orbitals $\phi_j(\mathbf{r})$; and the corresponding energy eigenvalues ε_j and orbital occupation numbers f_j,

$$\chi^{ks}_{\mathbf{r},\mathbf{r}'}(\omega) = \sum_{j,k}\left(f_k - f_j\right)\frac{\phi_j(\mathbf{r})\overline{\phi}_k(\mathbf{r})\overline{\phi}_j(\mathbf{r}')\phi_k(\mathbf{r}')}{\omega - (\varepsilon_j - \varepsilon_k) + i\eta} \qquad (18)$$

The exchange-correlation kernel $f_{xc}[n^{g}]$ incorporates all many-body effects into the linear response dynamics and is related to the exchange-correlation potential $v_{xc}[n^{g}]$ through a functional derivative,

$$f_{xc}[n^{g}] = \frac{\delta v_{xc}[n^{g}]}{\delta n^{g}}$$

In general, the *interacting* fermion excitation energies

$$\Omega_{jk} = E_j - E_k$$

differ from the KS excitation energies

$$\omega_{jk} = \varepsilon_j - \varepsilon_k$$

The right-hand side of Eq. (17) remains finite as $\omega \to \Omega_{jk}$, while the first-order SOF response $n^{1}_{\mathbf{y}'}(\omega)$ has a pole at each Ω_{jk}. Thus, the operator on the left-hand side acting on $n^{1}_{\mathbf{y}'}(\omega)$ cannot be invertible. Otherwise, its inverse could be applied to both sides of Eq. (17) with the result that the right-hand side would remain finite as $\omega \to \Omega_{jk}$, while the left-hand side would diverge. To avoid this inconsistency, the operator must have a zero eigenvalue as $\omega \to \Omega_{jk}$. Following Ref. [20], one is led to the following eigenvalue problem:

$$\sum_{k',j'} \frac{M_{kj;k'j'}(\omega)}{\omega - \omega_{j'k'} + i\eta} \xi_{k'j'}(\omega) = \lambda(\omega)\xi_{kj}(\omega) \tag{19}$$

where, writing

$$\alpha_{k'j'} = f_{k'} - f_{j'}$$

and

$$\Phi_{\mathbf{r}}^{kj} = \overline{\phi}_{\mathbf{r}}^{k} \phi_{\mathbf{r}}^{j}$$

we have

$$M_{k'j';kj}(\omega) = \alpha_{k'j'} \sum_{\mathbf{r}',\mathbf{y}'} \overline{\Phi}_{\mathbf{r}'}^{kj} f_{xc}[n^g](\mathbf{r}', \mathbf{y}'; \omega) \Phi_{\mathbf{y}'}^{k'j'}(\omega)$$

It can be shown that $\lambda(\Omega_{jk}) = 1$.

So far, the argument is exact. To proceed further, some form of approximation must be introduced. In the single-pole approximation [20], the KS poles are assumed to be well-separated so that we can focus on a particular KS excitation energy $\omega_{jk} = \omega_*$. The eigenvectors $\xi_{k'j'}(\omega)$ and the matrix operator $M_{kj;k'j'}(\omega)$ are finite at ω_*, while the eigenvalue $\lambda(\omega)$ must have a pole there to match the pole on the left-hand side of Eq. (19):

$$\lambda(\omega) = \frac{A(\omega_*)}{\omega - \omega_*} + \mathcal{O}(1)$$

Let ω_* be d-fold degenerate: $\omega_{k_1 j_1}, \ldots, \omega_{k_d j_d} = \omega_*$. Matching singularities in Eq. (19) gives

$$\sum_{l=1}^{d} M_{k_i j_i; k'_l j'_l}(\omega_*) \xi^n_{k'_l j'_l} = A^n(\omega_*) \xi^n_{k_i j_i}(\omega_*) \tag{20}$$

where $i, n = 1, \ldots, d$. For our purposes, the eigenvalues $A^n(\omega_*)$ are of primary interest and are found from Eq. (20). From each $A^n(\omega_*)$, we find

$$\lambda^n(\omega) = \frac{A^n(\omega_*)}{\omega - \omega_*}$$

Since $\lambda^n(\Omega_{jk}) = 1$, it follows that the sum of $\lambda^n(\Omega_{jk})$ and its complex conjugate is 2. Plugging into this sum the singular expressions for $\lambda^n(\Omega_{jk})$ and that of its complex conjugate, and solving for Ω^n_{jk} gives

$$\Omega^n_{jk} = \omega_* + Re[A^n(\omega_*)]$$

Interactions will thus generally *split* the ω_*-degeneracy. Now let

$$\delta E = \min_n Re[A^n(\omega_*)]$$

and

$$\overline{\Omega}_{jk} = \min_n \Omega^n_{jk}$$

Our expression for Ω^n_{jk} then gives

$$\overline{\Omega}_{jk} = \omega_* + \delta E$$

In the context of the QAE algorithm, our interest is the energy gap

$$\Delta(t_*) = E_1(t_*) - E_0(t_*)$$

separating the instantaneous ground- and first-excited states. In this case, our expression for $\overline{\Omega}_{jk}$ gives

$$\Delta(t_*) = [\varepsilon_1(t_*) - \varepsilon_0(t_*)] + \delta E(t_*) \tag{21}$$

To obtain the minimum gap Δ for QAE numerically, one picks a sufficiently large number of $t_* \in (0, T)$; solves for $\Delta(t_*)$ using the KS system associated with $H(t_*)$ to evaluate the right-hand side of Eq. (21); and then uses the minimum of the resulting set of $\Delta(t_*)$ to upper bound Δ. Because the KS dynamics is noninteracting, it has been possible to treat KS systems with $N \sim 10^3$ KS fermions [9–11], which allows evaluation of the minimum gap $\Delta(N)$ for the QAE algorithm for $N \sim 10^3$.

VII. DISCUSSION

As with all KS calculations, the minimum gap calculation requires an approximation for the exchange-correlation energy functional $\xi_{xc}[n]$. Note that, because the qubits in a quantum register must be located at fixed positions for the register to function properly, the associated JW fermions are distinguishable because they are each pinned to a specific lattice site. Consequently, antisymmetrization of the fermion wave function is not required, with the result that the exchange energy vanishes in the MAXCUT dynamics. The exchange-correlation energy functional $\xi_{xc}[n]$ is then determined solely by the correlation energy, which can be calculated using the methods of Ref. [21]. Parametrization of these results yields analytical expressions for the correlation energy per particle that upon differentiation, give $v_{xc}[n]$ and $f_{xc}[n]$. Replacing $n \rightarrow n_{\mathbf{r}}$ in $\xi_{xc}[n]$ gives the local density approximation (LDA) for GS-DFT; while $n \rightarrow n_{\mathbf{r}}(t)$ gives the adiabatic local density approximation (ALDA) for TD-DFT. These simple approximations have proven to be remarkably successful and provide a good starting point for the minimum gap calculation. Self-interaction corrections to $\xi_{xc}[n]$ are not necessary since the two-fermion interaction [see Eq. (7)] has no self-interaction terms. Finally, because the fermions are pinned, it will be necessary to test the gap for sensitivity to derivative discontinuities [22] in $\xi_c[n]$.

After our initial work appeared [12], others followed this line of research [23].

Acknowledgments

We thank A. Satanin for stimulating conversations. Franco Nori is partially supported by the ARO, RIKEN iTHES Project, MURI Center for Dynamic Magneto-Optics, JSPS-RFBR contract no. 12-02-92100, Grant-in-Aid for Scientific Research(s), MEXT Kakenhi on Quantum Cybernetics and the JSPS via its FIRST program; and F. Gaitan thanks T. Howell III for continued support.

REFERENCES

1. E. Farhi et al., *Science* **292**, 472 (2001).
2. T. Hogg, *Phys. Rev. A* **67**, 022314 (2003).
3. A. Young et al., *Phys. Rev. Lett.* **101**, 170503 (2008).
4. R. Dreizler and E. Gross, *Density Functional Theory*, Springer-Verlag, New York, 1990.
5. P. Hohenberg and W. Kohn, *Phys. Rev.* **136**, B864 (1964).
6. W. Kohn and L. J. Sham, *Phys. Rev.* **140**, A1133 (1965).
7. E. Runge and E. Gross, *Phys. Rev. Lett.* **52**, 997 (1984).
8. M. Marques and E. Gross, *Annu. Rev. Phys. Chem.* **55**, 427 (2004).
9. F. Shimojo et al., *Phys. Rev. B* **77**, 095103 (2008).
10. H. Jiang et al., *Phys. Rev. B* **68**, 165337 (2003).
11. D. Sanchez-Portal et al., *Int. J. Quantum Chem.* **65**, 453 (1997).
12. F. Gaitan and F. Nori, *Phys. Rev. B* **79**, 205117 (2009).
13. M. Steffen et al., *Phys. Rev. Lett.* **90**, 067903 (2003).
14. E. Fradkin, *Phys. Rev. Lett.* **63**, 322 (1989).
15. C. D. Batista and G. Ortiz, *Phys. Rev. Lett.* **86**, 1082 (2001).
16. The component $A_0(\mathbf{r})$ remains classical because its canonically conjugate momentum vanished: $\Pi_0 = 0$. Its equation of motion (Gauss's law) becomes $\partial L/\partial A_0 = 0$, which is imposed as a constraint on physical states. See, for example, R. Jackiw, in *Current Algebras and Anomalies*, S. B. Treiman et al., eds., Princeton University Press, Princeton, NJ, 1985.
17. M. Levy, *Proc. Natl. Acad. Sci. USA* **76**, 6062 (1979).
18. V. V. Franca and K. Capelle, *Phys. Rev. Lett.* **100**, 070403 (2008); *Phys. Rev. A* **77**, 062324 (2008).
19. L.-A. Wu et al., *Phys. Rev. A* **74**, 052335 (2006).
20. M. Petersilka et al., *Phys. Rev. Lett.* **76**, 1212 (1996).
21. D. Ceperley and B. Adler, *Phys. Rev. Lett.* **45**, 566 (1980).
22. J. Perdew and M. Levy, *Phys. Rev. Lett.* **51**, 1884 (1983); L. Sham and M. Schlüter, *Phys. Rev. Lett.* **51**, 1888 (1983).
23. D. G. Tempel and A. Aspuru-Guzik, *Sci. Rep.* **2**, 391 (2012).

QUANTUM ALGORITHMS FOR CONTINUOUS PROBLEMS AND THEIR APPLICATIONS

A. PAPAGEORGIOU and J. F. TRAUB

Department of Computer Science, Columbia University, New York, NY 10027, USA

I. INTRODUCTION

Problems in science and engineering are frequently formulated using continuous models. In most cases, they can only be solved numerically and, therefore, approximately to within a given accuracy ε. Examples include multivariate integration, path integration, function approximation, the solution of ordinary and partial differential equations, optimization, and eigenvalue problems. Numerous applications require the solution of these problems, ranging from physics and chemistry to economics and finance.

Advances in Chemical Physics, Volume 154: Quantum Information and Computation for Chemistry, First Edition. Edited by Sabre Kais.

Typically, the solution to a problem depends on an underlying function, often of many variables. The algorithm solving the problem must obtain information about the function, for example, by sampling it at a number of points, and then it must combine the information to produce the result. The computational complexity (for brevity, the complexity) of a problem is the least number of resources required to solve a problem with accuracy ε. In this chapter, the resources will be the information operations, the combinatorial operations, and the space (e.g., the number of qubits).

Decades of research on the classical complexity of continuous problems include, for example, the monographs [1–9]. Over the past decade a significant amount of work has been done on algorithms and complexity of continuous problems in the quantum setting. This research was motivated by the results of Shor [10] for integer factorization, and Grover [11] for searching an unstructured database.

The challenges of quantum computing are

- To find quantum algorithms that are better than any known classical algorithm for solving certain continuous problems.
- To determine for which continuous problems quantum computers are provably more powerful than classical computers.

Similar challenges apply to discrete problems, however, we do not deal with them in this chapter. An extensive review of quantum algorithms for discrete problems can be found in Ref. [12]. The first challenge, which is weaker than the second, allows us to consider problems for which we do not know optimal classical algorithms or we do not have sharp bounds for the classical complexity. Important problems, such as eigenvalue estimation for multiparticle systems, fall in this category. On the other hand, in problems, such as path integration, we know the quantum complexity and that the optimal quantum algorithm is faster than any classical algorithm with the same accuracy.

Certain ideas and techniques are broadly applicable and have led to a significant number of results for continuous problems in the quantum setting. We discuss them briefly.

The amplitude amplification and estimation algorithm of Brassard et al. [13] has been applied in the study of integration, path integration, the solution of ordinary differential equations, and other problems. Using it we obtain a quantum algorithm that approximates the mean of a Boolean function. This can be extended to an algorithm approximating the weighted average of N numbers. Hence, it can be used to approximate integrals. As a result, using the amplitude amplification and estimation algorithm we can convert classical algorithms for integration, as well as modules of algorithms for other problems that need to compute integrals, to quantum algorithms more or less directly.

The quantum lower bounds of Nayak and Wu [14] establishing the optimality of the algorithm in Ref. [13] for computing the Boolean mean also lead to lower bounds for the cost of quantum algorithms and the complexity not only of integration but other continuous problems as well.

For the simulation of quantum systems, splitting formulas [15,16] have been used to derive quantum algorithms [17–19]. They are used to implement efficiently approximations of matrix exponentials that are required in a number of quantum algorithms. An example is the algorithm solving linear systems in Ref. [20]. Phase estimation [17] has had an impact. It is used in eigenvalue estimation problems such as those of differential operators. The solution of the time-independent Schrödinger equation is an example.

II. THE MODEL OF COMPUTATION

We begin by summarizing the model of computation that is used for classical algorithms solving continuous problems before we discuss the quantum model of computation. This will motivate the approach taken in the analysis of quantum algorithms, allowing one to draw an analogy between the classical and quantum models of computation.

In the study of the classical complexity of continuous problems, the *real number model* with *oracles* is often used. In this model, one can compute function evaluations or linear functionals as information operations, see for example, [4]. The information operations are represented as *black-box* or *oracle* calls. One can also perform arithmetic operations, make comparisons, and evaluate elementary functions. This model of computation is an abstraction of fixed precision floating point arithmetic used in science and engineering. It has also been used in the study of the complexity of algebraic problems such as matrix multiplication [21]. A comparison of the real number model and the Turing machine model of computation can be found in Ref. [22].

Continuous problems are typically defined for classes of functions of $d \geq 1$ variables and are to be solved with accuracy ε. As a result the cost of the algorithms and the problem complexity are studied with respect to the parameters d and ε.

For some continuous problems, such as certain zero finding problems, convex optimization, and the solution of linear systems with well-conditioned matrices, the complexity depends logarithmically on ε^{-1}. However, for the majority of continuous problems, the complexity grows much faster. A continuous problem is considered to be tractable if its complexity is proportional to

$$d^{p_1}\varepsilon^{-p_2} \quad \text{for some } p_1, p_2 \in \mathbb{R} \tag{1}$$

We stress that a problem's complexity depends on the setting in which it is studied. Indeed, for multivariate integration of smooth functions, the cost of any classical algorithm with worst-case accuracy ε can grow at least as $\varepsilon^{-\alpha d}$, for some $\alpha > 0$. Then the problem suffers from the curse of dimensionality and is intractable in the worst case. In some cases, classical randomized algorithms can break intractability because they exhibit a polynomial dependence on ε^{-1} and, as we will see, quantum algorithms provide an additional speedup.

A. Quantum Queries

Inputs to quantum algorithms are often given using quantum queries. They correspond to black-box or oracle calls returning evaluations of some function f. The model of Beals et al. [23] has been used in the study of discrete problems and, with a slight change in the definition of the queries, it has also been used in the study of continuous problems.

In the case of a Boolean function $f : \{0, \ldots, 2^m - 1\} \to \{0, 1\}$ the quantum query providing information about f is defined by the unitary operator

$$Q_f|j\rangle|k\rangle = |j\rangle|k \oplus f(i)\rangle \tag{2}$$

where $|j\rangle$ is an m-qubit computational basis state, $|k\rangle$ is single-qubit computational basis state, and $\oplus$ denotes addition modulo 2. This type of query is used in Grover's search algorithm.

In the case of real-valued bounded functions, different quantum queries have been studied in the literature. Without loss of generality assume that $f : \{0, \ldots, 2^m - 1\} \to [0, 1]$. Abrams and Williams [24] in their study of integration used the query

$$\begin{aligned} Q_f|j\rangle|0\rangle &= \sqrt{1 - f(j)^2}|j\rangle|0\rangle + f(j)|j\rangle|1\rangle \\ Q_f|j\rangle|1\rangle &= -f(j)|j\rangle|0\rangle + \sqrt{1 - f(j)^2}|j\rangle|1\rangle \end{aligned} \tag{3}$$

We point out that this query is defined using the real number $f(j)$. Therefore, depending on f it may be hard to implement the query exactly using elementary quantum gates. A truncation of the value $f(j)$ to a finite number of most significant bits can be used to overcome this difficulty. Often, truncations of the function evaluations to a number of significant bits proportional to ε^{-1} can be used without loss of generality.

Novak [25] in his paper studying the complexity of integration on Hölder[1] classes used the query

$$\begin{aligned} Q_f|j\rangle|0\rangle &= \sqrt{f(j)}|j\rangle|0\rangle + \sqrt{1 - f(j)}|j\rangle|1\rangle \\ Q_f|j\rangle|1\rangle &= -\sqrt{1 - f(j)}|j\rangle|0\rangle + \sqrt{f(j)}|j\rangle|1\rangle \end{aligned} \tag{4}$$

[1] A function belongs to a Hölder class or a Sobolev space if it satisfies certain smoothness expressed by conditions on its partial derivatives. A reader not familiar with these concepts may think of a function with bounded partial derivatives up to a given order.

The considerations about truncating the function evaluations apply to this case too.

A third kind of query was introduced by Heinrich in Ref. [26]. Namely

$$Q_f|j\rangle|k\rangle = |j\rangle|k \oplus \hat{f}(j)\rangle \tag{5}$$

where $|j\rangle$ and $|k\rangle$ are m- and ν-qubit computational basis states, respectively, $\hat{f}(j)$ is obtained from $f(j)$ using a ν bit discretization of the range of f and $\oplus$ denotes addition modulo 2^ν. Moreover, in the case of real functions of real variables $g : [0, 1]^d \to [0, 1]$, $d \geq 1$, it suffices to consider a discretization $\tau : \{0, 1, \ldots, 2^m - 1\} \to [0, 1]^d$ of the domain of g and to use the query definition with the function $f(j) = g(\tau(j))$, $j = 0, \ldots, 2^m - 1$.

The three queries are not equivalent, in general. Query (5) can be used to efficiently simulate queries (3) and (4), but the converse is not true [27].

B. Quantum Algorithms

Consider a problem defined using a linear or nonlinear operator S such that

$$S : \mathcal{F} \to \mathcal{G} \tag{6}$$

Typically, $\mathcal{F}$ is a linear space of real functions of several variables, and $\mathcal{G}$ is a normed linear space. We wish to approximate $S(f)$ to within ε for $f \in \mathcal{F}$. We approximate $S(f)$ using n function evaluations $f(t_1), \ldots, f(t_n)$ at deterministically and a priori chosen sample points. The quantum query Q_f encodes this information and provides it to the algorithm.

A quantum algorithm consists of a sequence of unitary transformations applied to an initial state. The result of the algorithm is obtained by measuring its final state. The quantum model of computation is discussed in detail in Refs [17,23,26,28,29]. We summarize it here as it applies to continuous problems.

The initial state $|\psi_0\rangle$ of the algorithm is a unit vector of the Hilbert space $\mathcal{H}_\nu = \mathbb{C}^2 \otimes \cdots \otimes \mathbb{C}^2$, ν times, for some appropriately chosen integer ν, where $\mathbb{C}^2$ is the two dimensional space of complex numbers. The dimension of $\mathcal{H}_\nu$ is 2^ν. The number ν denotes the number of qubits used by the quantum algorithm.

The final state $|\psi\rangle$ is also a unit vector of $\mathcal{H}_\nu$ and is obtained from the initial state $|\psi_0\rangle$ through a sequence of unitary $2^\nu \times 2^\nu$ matrices, that is,

$$|\psi\rangle_f := U_T Q_f U_{T-1} Q_f \cdots U_1 Q_f U_0 |\psi_0\rangle \tag{7}$$

The unitary matrix Q_f is a quantum query and as we already mentioned, it is used to provide information about a function f. Q_f depends on n function evaluations $f(t_1), \ldots, f(t_n)$, at deterministically chosen points, $n \leq 2^\nu$. The selection of the query among the types (2), (3), (4), or (5) is often a matter of convenience.

The matrices $U_0, U_1, \ldots, U_T$ are unitary and do not depend on f. The integer T denotes the number of quantum queries.

At the end of the quantum algorithm, the final state $|\psi_f\rangle$ is measured. The measurement produces one of M outcomes, where $M \leq 2^\nu$. Outcome $j \in \{0, 1, \ldots, M-1\}$ occurs with probability $p_f(j)$, which depends on j and the input f. Knowing the outcome j, we classically compute the final result $\phi_f(j)$ of the algorithm.

In principle, quantum algorithms may have measurements applied between sequences of unitary transformations of the form presented previously. However, any algorithm with multiple measurements can be simulated by a quantum algorithm with only one measurement [30].

We consider algorithms that approximate $S(f)$ with probability $p \geq \frac{2}{3}$. We can boost the success probability of an algorithm to become arbitrarily close to 1 by repeating the algorithm a number of times. The success probability becomes at least $1 - \delta$ with a number of repetitions proportional to $\log \delta^{-1}$.

The local error of the quantum algorithm (7) that computes the approximation $\phi_f(j)$, for $f \in \mathcal{F}$ and the outcome $j \in \{0, 1, \ldots, M-1\}$, is defined by

$$e(\phi_f, S) = \min \left\{ \alpha : \sum_{j:\ \|S(f)-\phi_f(j)\| \leq \alpha} p_f(j) \geq \tfrac{2}{3} \right\}$$

where $p_f(j)$ denotes the probability of obtaining outcome j for the function f. The worst-case error of a quantum algorithm ϕ is defined by

$$e^{\text{quant}}(\phi, S) = \sup_{f \in \mathcal{F}} e(\phi_f, S)$$

The query complexity $\text{comp}^{\text{query}}(\varepsilon, S)$ of the problem S is the minimal number of queries necessary for approximating the solution with accuracy ε, that is,

$$\text{comp}^{\text{query}}(\varepsilon) = \min\{T : \exists\, \phi \text{ such that } e^{\text{quant}}(\phi, S) \leq \varepsilon\}$$

The query complexity gives a sense of the depth of the quantum circuit realizing the algorithm and provides a complexity lower bound. It allows one to study algorithms and obtain complexity results for classes of functions in a way that is unobscured by the cost of a query, which varies with f. This is how the cost is measured in Grover's search algorithm [11,17].

The cost for combining the queries to produce the result (i.e., the implementation cost of the unitary operators $U_0, \ldots, U_t$) must also be taken into account. The complexity of the problem is the minimal cost, including the queries and other quantum operations, of an algorithm solving the problem with accuracy ε. An algorithm with cost equal to the complexity, modulo an absolute constant, is considered to be optimal.

Some papers in the literature consider only the query complexity. In some cases, this is a simplification. In other cases, the query complexity, modulo polylog factors, reflects the total cost of the optimal algorithm as well. On the other hand, not all papers we review consider the query complexity alone. A number of them give estimates of the query complexity as well as the cost of other quantum operations.

Finally, the qubit complexity of the problem S is the minimal number of qubits necessary for approximating the solution with accuracy ε

$$\mathrm{comp}^{\mathrm{qubit}}(\varepsilon) = \min\{\nu : \exists\, \phi \text{ such that } e^{\mathrm{quant}}(\phi, S) \leq \varepsilon\} \tag{8}$$

III. APPLICATIONS

A. Integration

For numerous applications, one seeks the expected value of a quantity and, therefore, is confronted with an integral. Often, in these integrals of multivariate functions, the number of variables can be huge, say, in the hundreds or thousands. In most cases, the integrals cannot be computed analytically and their values are approximated numerically.

Classical algorithms for integration have been extensively studied in the literature and optimal algorithms are known for numerous classes of functions (e.g., see Ref. [4] and the references therein).

In many cases, the optimal algorithms are linear. Thus, the optimal algorithm approximating

$$S(f) = \int_{I_d} f(x)\, dx \tag{9}$$

where $I_d \subset \mathbb{R}^d$, has the form

$$A(f) = \sum_{j=0}^{N-1} a_j f(x_j) \tag{10}$$

where a_j, $j = 0, \ldots, N-1$, are independent of f and x_j, $j = 0, \ldots, N-1$, are deterministic or random sample points. An example of particular interest is the algorithm

$$A(f) = \frac{1}{N} \sum_{j=0}^{N-1} f(x_j) \tag{11}$$

For $I_d = [0, 1]^d$ this is the midpoint rule in d dimensions that samples f on a uniform grid. The Monte Carlo algorithm that samples f at random points also has this form.

Quantum algorithms can compute an approximation to $A(f)$ fast. This gives them an advantage over classical algorithms for integration as we will see. This idea was used by Abrams and Williams [24] who were the first to derive a quantum algorithm for integration. They use query (3) to provide the necessary function evaluations to their algorithm. Novak [25] was the first to study the quantum complexity of integration in Hölder classes of functions. His algorithm uses query (4). Soon after, Heinrich studied the quantum complexity of integration in Sobolev spaces using the query (5). He has obtained a large number of results [26,31–34].

We will use [25] to illustrate the key ideas in the derivation of quantum algorithms for integration. We will start with an algorithm approximating the average of N real numbers and then use it to obtain an algorithm for integration in Hölder classes. Consider a function $f : \{0, \ldots, N-1\} \to [0, 1]$ and $N = 2^n$ and the average of Eq. (11). To approximate the more general sums, Eq. (10) suffices to express them as an average using a suitable transformation.

The approximation of the Boolean mean is a special simple case. For a function $f : \{0, \ldots, N-1\} \to \{0, 1\}$, the amplitude amplification and estimation algorithm of Brassard et al. [13] computes an approximation of the mean (11) with error ε using $O(\varepsilon^{-1})$ queries. Moreover, the lower bounds of Nayak and Wu [14] show that this algorithm is optimal modulo constants.

Considering the mean (11) of a real-valued function, using the query (4) we have

$$
\begin{aligned}
Q_f(H^{\otimes n} \otimes I)|0\rangle^{\otimes n}|0\rangle &= \frac{1}{\sqrt{N}} \sum_{j=0}^{N-1} \left[\sqrt{f(j)}|j\rangle|0\rangle + \sqrt{1-f(j)}|j\rangle|1\rangle\right] \\
&= a_0|\psi_0\rangle + a_1|\psi_1\rangle
\end{aligned}
$$

where H is the Hadamard gate, $N = 2^n$, $a_0^2 = \frac{1}{N}\sum_{j=0}^{N-1} f(j)$, $a_0^2 + a_1^2 = 1$, $\langle\psi_0|\psi_1\rangle = 0$ and

$$
\begin{aligned}
|\psi_0\rangle &= \frac{1}{a_0\sqrt{N}} \sum_{j=0}^{N-1} \sqrt{f(j)}|j\rangle|0\rangle \\
|\psi_1\rangle &= \frac{1}{a_1\sqrt{N}} \sum_{j=0}^{N-1} \sqrt{1-f(j)}|j\rangle|1\rangle
\end{aligned}
$$

have unit length.

Then $A(f) = a_0^2$ and the modification of the amplitude amplification and estimation algorithm that uses T queries of the form (4) (instead of queries of the form (2)) approximates $A(f) = \alpha_0^2$ with error [13, Th. 12]

$$
2\pi k \frac{\sqrt{a_0^2(1-a_0^2)}}{T} + k^2 \frac{\pi^2}{T^2},
$$

and its success probability is at least

$$\begin{cases} 1 - \dfrac{1}{2(k-1)} & k > 1 \\ \dfrac{8}{\pi^2} & k = 1 \end{cases}$$

For functions of real variables, the smoothness of the functions determines lower bounds on the cost of the algorithms and the complexity of integration. There are different ways to define classes of smooth functions using a condition on their (partial) derivatives. Recall that in the Hölder classes of functions the growth of all partial derivatives up to a given order is restricted in a certain way, as we see in the following definition. These classes have been considered extensively in the study of classical algorithms for integration. Similarly, Sobolev spaces are also classes of smooth functions and we discuss them briefly at the end of this section. We define Hölder classes before stating Novak's integration results.

Definition 1 *The Hölder class $F_d^{k,\alpha}$ is the class of all k-times continuously differentiable functions $f : [0,1]^d \to \mathbb{R}$ with $\|f\|_\infty \leq 1$ that satisfy*

$$|\partial^l f(x) - \partial^l f(y)| \leq \|x - y\|^\alpha$$

for all partial derivatives $\partial^l = \partial_1^{l_1} \cdots \partial_d^{l_d}$ of order $|l| = l_1 + \cdots + l_d = k$.

The integration algorithm of Novak [25] proceeds as follows. A function $f \in F_d^{k,\alpha}$ is approximated using a piecewise polynomial $P_n(f)$ that interpolates f at n points. This way an evaluation of $f - P_n(f)$ has cost independent of n, although the cost may depend on d and k. Also $|f - P_n(f)| = \Theta(n^{-\gamma})$, $\gamma = (k+\alpha)/d$.

Then

$$\int_{[0,1]^d} f(x)\,dx = \int_{[0,1]^d} P_n(f)(x)\,dx + \int_{[0,1]^d} (f - P_n(f))(x)\,dx$$

Because the integral of $P_n(f)$ can be computed exactly it suffices to approximate the third integral. Using the notation of Eq. (9) we approximate $I(f - P_n(f))$ by the average $A(f - P_n(f))$ as in Eq. (11). All the steps are classical except the approximation of $A(f - P_n(f))$, which is done by the quantum algorithm. Appropriately selecting the values of n and N in a way that the algorithm approximates the integral with accuracy ε and with probability at least $\frac{3}{4}$, Novak shows that, modulo polylog factors, the query complexity of integration in Hölder classes satisfies

$$\text{comp}^{\text{query}}(\varepsilon) \asymp \varepsilon^{-1/(1+\gamma)} \qquad \gamma = (k+\alpha)/d$$

For other classes of functions, such as Sobolev spaces $W^r_{p,d}$, variations of this approach lead to algorithms with optimal query complexity. Table I summarizes

TABLE I
Complexity of Integration in Sobolev Spaces and Hölder Classes

	Worst Case	Randomized	Quantum
$F_d^{k,\alpha}$	$\varepsilon^{-d/(k+\alpha)}$	$\varepsilon^{-2d/(2(k+\alpha)+d)}$	$\varepsilon^{-d/(k+\alpha+d)}$
$W^r_{p,d}$, $2 \leq p \leq \infty$	$\varepsilon^{-d/r}$	$\varepsilon^{-2d/(2r+d)}$	$\varepsilon^{-d/(r+d)}$
$W^r_{p,d}$, $1 \leq p \leq 2$	$\varepsilon^{-d/r}$	$\varepsilon^{-pd/(rp+pd-d)}$	$\varepsilon^{-d/(r+d)}$
$W^r_{1,d}$	$\varepsilon^{-d/r}$	$\varepsilon^{-d/r}$	$\varepsilon^{-d/(r+d)}$

the query complexity results (up to polylog factors) for multivariate integration in the worst case, randomized and quantum setting for functions belonging to Hölder classes $F_d^{k,\alpha}$ and Sobolev spaces $W^r_{p,d}$. An example of a classical randomized algorithm for integration is the well-known Monte Carlo algorithm. Observe that the integration problem suffers from the curse of dimensionality in the classical worst case, see for example, Ref. [4]. Quantum algorithms offer an exponential speedup over classical algorithms in the worst case and a polynomial speedup over classical randomized algorithms. Heinrich who obtained most of the quantum query complexity results in a series of papers, which we cited earlier, summarized his results in Ref. [31] where a corresponding table showing error bounds can be found.

Quantum algorithms for integration have been used to derive optimal quantum algorithms for other continuous problems, such as path integration, certain approximation problems, and the solution of ordinary differential equations.

B. Path Integration

Traub and Woźniakowski [35] study quantum algorithms and the complexity of path integration. Path integrals can be viewed as infinite dimensional integrals. They are defined by

$$I(f) = \int_X f(x)\,\mu(dx)$$

where μ is a probability measure on X, an infinite dimensional space in general, $f : X \to \mathbb{R}$ belongs to a class F, of μ-integrable functions. In particular, they consider a Gaussian measure μ where the eigenvalues of its covariance operator are of order j^{-k}, $k > 1$. The Wiener measure is an example with $k = 2$. They also assume that F is the class of functions with rth Frechet derivative ($r < \infty$) continuous and uniformly bounded by one.

We describe the idea leading to the algorithm in Ref. [35]. First approximate $I(f)$, with error ε, by a d-dimensional integral

$$I_d(f) = \int_{\mathbb{R}^d} f_d(t)\,\mu_d(df)$$

where μ_d is a zero mean Gaussian measure. This implies that $d = d(\varepsilon)$ is a polynomial in ε^{-1} and its degree depends on k. Then approximate $I_d(f)$, with error ε, using an algorithm $A(f)$ of the form (10). The number of terms in the sum is $N = m^d$ and m is a polynomial in ε^{-1}. Thus, N is an exponential function of ε^{-1}. Finally, use a quantum algorithm for integration to approximate the value $A(f)$ with error ε.

The resulting algorithm has error proportional to ε. It uses a number of queries proportional to ε^{-1}. The number of additional quantum operations is polynomial in ε^{-1}. The number of qubits is also polynomial in ε^{-1}.

On the other hand, the classical complexity of path integration has been considered in Refs [36,37]. In the worst case it is of order $\varepsilon^{-\varepsilon^{-\beta}}$, where β is a positive number that depends on r. Hence, the problem is intractable in the worst case. Approximating the finite dimensional integral $I_d(f)$ using Monte Carlo leads to a classical randomized algorithm with cost proportional to ε^{-2}, which is an optimal classical algorithm.

In summary we have

- Path integration on a quantum computer is tractable.
- Path integration on a quantum computer can be solved roughly ε^{-1} times faster than on a classical computer using randomization, and exponentially faster than on a classical computer with a worst-case assurance.

The Feynman–Kac path integral is a special case of a path integral and occurs in many applications [38]. In this case $X = C$ is the space of continuous functions, and the measure is the Wiener measure $\mu = w$. For example, the diffusion equation

$$\frac{\partial z}{\partial t}(u,t) = \frac{1}{2}\Delta z(u,t) + V(u)z(u,t)$$
$$z(u,0) = v(u)$$

with $u \in \mathbb{R}^d$, $t > 0$, and $V, v : \mathbb{R}^d \to \mathbb{R}$, are the potential and the initial value functions, respectively, and Δ denotes the Laplacian. The solution is given by the Feynman-Kac path integral

$$z(u,t) = \int_C v(x(t)+u)e^{\int_0^t V(x(s)+u)\,ds}\, w(dx) \tag{12}$$

where C is the set of continuous functions $x : \mathbb{R}_+ \to \mathbb{R}^d$ such that $x(0) = 0$.

Note the two kinds of dimension here. A Feynman-Kac path integral is infinite dimensional because we are integrating over continuous functions and u is a function of d variables.

Kwas [39], following an approach similar to that of Traub and Woźniakowski, derived a quantum algorithm for Feynman-Kac path integration that uses a number of queries proportional to ε^{-1}. He also showed as optimal a slightly more complicated algorithm that uses a number of queries proportional

$$\varepsilon^{-1/(1+r/d)}$$

For comparison, we briefly discuss the classical complexity of Feynman-Kac path integration. For $d = 1$ when u is a scalar, a number of papers deal with the solution of integral (12), see for example, Ref. [40]. In particular, for $u = 1$ and V four times continuously differentiable, Chorin's well-known algorithm has cost proportional to $\varepsilon^{-2.5}$. Plaskota et al. [41] were the first to study the complexity in the worst case. They construct an algorithm with cost $\varepsilon^{-0.25}$ and show it is optimal. We remark that the algorithm depends on a numerically difficult precomputation. Multivariate Feynman-Kac integration is studied in Ref. [42] in the worst case and in Ref. [39]. In the worst case the complexity is $\varepsilon^{-d/r}$, for v and V that are $r < \infty$ times continuously differentiable. In the randomized case the curse of dimensionality is broken. An algorithm based on Monte Carlo has cost of order ε^{-2}. We remark that the quantum algorithm of Kwas uses a quantum algorithm for integration as a module instead of Monte Carlo. Finally, as in the quantum case, a more complicated randomized algorithm with cost proportional to $\varepsilon^{-2/(1+2r/d)}$ is optimal.

C. Approximation

Classically, approximation of functions of d variables has been studied for functions in Sobolev spaces $W^r_{p,d}$ with error measured using the norm of L_q. Its complexity depends on the values of the parameters [1,6] and the accuracy ε. Here, r is a smoothness parameter, d is the number of variables, and $1 \leq p \leq \infty$ indicates the norm is the L_p norm. For $p = \infty$ the problem suffers the curse of dimensionality in the classical worst and randomized cases (i.e., the cost of any classical deterministic or randomized algorithm grows exponentially with the number of variables d). Recently, Heinrich [43] showed that quantum algorithms do not provide an advantage.

Table II summarizes the classical and quantum complexity of approximation (modulo polylog factors) in Sobolev spaces for the various values of the parameters p, q, r, d.

For other approximation problems, quantum algorithms have an advantage over classical algorithms. Novak et al. [44] study such problem. They consider a space of functions of d variables, where certain variables are more important than others. Weights are used to define the relative importance of the variables. They show a quantum algorithm that is exponentially faster than any classical algorithm in the worst case, and is roughly $\varepsilon^{-(1+r)}$ times faster than any classical randomized algorithm. The parameter r depends on the weights and can be large. Moreover, the quantum algorithm uses about $d + \log \varepsilon^{-1}$ qubits.

TABLE II
Complexity of Approximation in Sobolev Spaces

	Worst Case	Randomized	Quantum
$1 \le p < q \le \infty$, $r/d \ge 2/p - 2/q$	$\varepsilon^{-dpq/(rpq-d(q-p))}$	$\varepsilon^{-dpq/(rpq-d(q-p))}$	$\varepsilon^{-d/r}$
$1 \le p < q \le \infty$, $r/d < 2/p - 2/q$	$\varepsilon^{-dpq/(rpq-d(q-p))}$	$\varepsilon^{-dpq/(rpq-d(q-p))}$	$\varepsilon^{-dpq/(2rpq-2d(q-p))}$
$1 \le q \le p \le \infty$	$\varepsilon^{-d/r}$	$\varepsilon^{-d/r}$	$\varepsilon^{-d/r}$

D. Ordinary Differential Equations

Quantum algorithms for initial value problems for systems of first-order equations and scalar equations of higher order have been studied in the literature. In both cases the algorithms are derived from classical algorithms by taking the modules that compute integrals classically and replacing them via a quantum algorithm for integration.

Kacewicz [45] studied the problem

$$z'(t) = f(z(t)), \quad t \in [a, b], \quad z(a) = \eta$$

where $f : \mathbb{R}^d \to \mathbb{R}^d$, $z : [a, b] \to \mathbb{R}^d$, and $\eta \in \mathbb{R}^d$ with $f(\eta) \neq 0$. For the right-hand-side function $f = [f_1, \ldots, f_d]$, where $f_j : \mathbb{R}^d \to \mathbb{R}$, he assumed that the f_j belong to the Hölder class $F_d^{k,\alpha}$, $k + \alpha \ge 1$. He wanted to compute a bounded function on the interval $[a, b]$ that approximates the solution z.

He derived a quantum algorithm with cost that differs from the lower bound by only an arbitrarily small parameter in the exponent. In particular, its cost (modulo polylog factors) is

$$O(\varepsilon^{-1/(k+\alpha+1-\gamma)})$$

where $\gamma \in (0, 1)$ is arbitrarily small, while the quantum complexity satisfies

$$\Omega(\varepsilon^{-1/(k+\alpha+1)})$$

The complexity of classical randomized algorithms is also studied by Kacewicz in the same paper. He derived a randomized algorithm with cost

$$O(\varepsilon^{-1/(k+\alpha+1/2-\gamma)})$$

modulo polylog factors, where $\gamma \in (0, 1)$ is arbitrarily small, and showed the complexity lower bound

$$\Omega(\varepsilon^{-1/(k+\alpha+1/2)})$$

Much earlier he had studied the classical worst-case complexity of ordinary differential equations [46].

Goćwin and Szczesny [47] considered quantum and classical randomized algorithms for the solution of

$$\begin{cases} u^{(k)}(x) = g(x, u(x), u'(x), \dots, u^{(q)}), & x \in [a, b], \\ u^{(j)}(a) = u_a^j, & j = 0, 1, \dots, k-1 \end{cases}$$

where $0 \leq q < k$, $g : [a.b] \times \mathbb{R}^{q+1} \to \mathbb{R}$, $u : [a, b] \to \mathbb{R}$ $(a < b)$. They showed the same complexity upper and lower bounds as the ones already shown hold for any k.

Finally we mention that Heinrich and Milla [48] recently showed that the randomized complexity lower bound of Kacewicz holds with $\gamma = 0$.

E. Partial Differential Equations

The numerical solution of partial differential equations is a vast subject. Here we confine ourselves to elliptic equations. They have many applications and classical algorithms for solving them have been extensively studied in the literature, see Ref. [7] and the references within. A simple example is the Poisson equation, for which we want to find a function $u : \bar{\Omega} \to \mathbb{R}$, that satisfies

$$\begin{aligned} -\Delta u(x) &= f(x), & x \in \Omega \\ u(x) &= 0, & x \in \partial\Omega \end{aligned}$$

where Δ denotes the Laplacian and $\Omega \subset \mathbb{R}^d$.

Heinrich [49] studied the quantum query complexity of elliptic partial differential equations of order $2m$ on a smooth bounded domain $\Omega \subset \mathbb{R}^d$ with smooth coefficients and homogeneous boundary conditions with the right-hand-side function belonging to $C^r(\Omega)$ and the error measured in the L_∞ norm, and (modulo polylog factors) found it proportional to

$$\varepsilon^{-\max\{d/(r+2m),\, d/(r+d)\}}$$

We note that classical randomized algorithms have cost at least proportional to

$$\varepsilon^{-\max\{d/(r+2m),\, 2d/(2r+d)\}}$$

and this lower bound is sharp [50]. In the worst case, the problem has complexity proportional to $\varepsilon^{-d/r}$ and is intractable.

Hence, quantum algorithms may have a polynomial advantage over classical algorithms, but not always. For fixed m and r and for $d > 4m$ the problem is intractable in the quantum and classical settings.

F. Optimization

The query complexity of finding the maximum of a multivariate function belonging to the Hölder class $F_d^{r,\alpha}$ is studied in Ref. [51]. Lower bounds are derived using the results of Ref. [14]. An optimal quantum algorithm is also shown. The main idea is to discretize the function and then use an algorithm that finds the maximum of a finite sequence. The latter algorithm is based on that in Ref. [52]. In particular, the query complexity is

$$\text{comp}^{\text{query}}(\varepsilon) = \Theta(\varepsilon^{-d/(2(r+\alpha))})$$

The classical worst-case and randomized complexity of this problem is well known [1]. In both cases it is $\Theta(\varepsilon^{-d(k+\alpha)})$. Thus, quantum algorithms provide a quadratic speedup relative to classical algorithms.

G. Gradient Estimation

Approximating the gradient of a function $f : \mathbb{R}^d \to \mathbb{R}$ with accuracy ε requires a minimum of $d + 1$ function evaluations on a classical computer. Jordan [53] shows how this can be done using a single query on a quantum computer.

We present Jordan's algorithm for the special case where the function is a plane passing through the origin, that is, $f(x_1, \ldots, x_d) = \sum_{j=1}^{d} a_j x_j$, and is uniformly bounded by 1. Then $\nabla f = (a_1, \ldots, a_d)^T$. Using a single query and *phase kickback* we obtain the state

$$\frac{1}{\sqrt{N^d}} \sum_{j_1=0}^{N-1} \cdots \sum_{j_d=0}^{N-1} e^{2\pi i f(j_1,\ldots,j_d)} |j_1\rangle \cdots |j_d\rangle$$

where N is a power of 2. Equivalently, we have

$$\frac{1}{\sqrt{N^d}} \sum_{j_1=0}^{N-1} \cdots \sum_{j_d=0}^{N-1} e^{2\pi i (a_1 j_1 + \cdots + a_d j_d)} |j_1\rangle \cdots |j_d\rangle$$

This is equal to the state

$$\frac{1}{\sqrt{N}} \sum_{j_1=0}^{N-1} e^{2\pi i a_1 j_1} |j_1\rangle \ldots \frac{1}{\sqrt{N}} \sum_{j_d=0}^{N-1} e^{2\pi i a_d j_d} |j_d\rangle$$

We apply the Fourier transform to each of the d registers and then measure each register in the computational basis to obtain $m_1, \ldots, m_d$. If a_j can be represented with finitely many bits and N is sufficiently large, then $m_j/N = a_j$, $j = 1, \ldots, d$.

For functions with second-order partial derivatives not identically equal to zero, the analysis is more complicated. We refer the reader to Ref. [53] for the details.

H. Simulation

In 1982, Richard Feynman [18] observed that simulating quantum systems would be difficult or impossible on a classical computer. The number of parameters describing the quantum states grows exponentially with the system size and so does the computational cost of the best classical deterministic algorithms known. In some cases classical randomized algorithms have been used to overcome these difficulties, however, randomized algorithms also have limitations. As an alternative to simulation with a classical computer, Feynman proposed simulation with a quantum computer. He conjectured that quantum computers might be able to carryout the simulation more efficiently than classical computers. For an overview of quantum simulation see, for example, [18,19,54,55].

In the Hamiltonian simulation problem one is given a Hamiltonian H, $t \in \mathbb{R}$, and an accuracy demand ε and the goal is to derive an algorithm approximating the unitary operator e^{-iHt} with error at most ε. The size of the quantum circuit realizing the algorithm is its cost. Assuming that H is a matrix of size $2^q \times 2^q$ the algorithm is efficient if its cost is a polynomial in q, t, and ε^{-1}.

Lloyd [19] showed that local Hamiltonians can be simulated efficiently on a quantum computer. About the same time, Zalka [56,57] showed that many-particle systems can be simulated efficiently on a quantum computer. Later, Aharonov and Ta-Shma [58] generalized Lloyd's results to sparse Hamiltonians. We note that Hamiltonian simulation is also related to adiabatic evolution and quantum walks [59–63].

Berry et al. [64] extended the complexity results of [58] for sparse Hamiltonians. They assume that the Hamiltonian H is given by an *oracle* (a *black-box*) and that H can be decomposed efficiently, by a quantum algorithm using oracle calls, into a sum of Hamiltonians H_j, $j = 1, \ldots, m$, that individually can be simulated efficiently. They approximate e^{-iHt} with error ε by a sequence of N unitary operators of the form $e^{-iH_{j_\ell} t_{j_\ell}}$, $\ell = 1, \ldots, N$. The cost of the simulation is the total number of oracle calls. All the unitary operators in the sequence have to be considered in the simulation, one after the other. The algorithm has to make oracle calls to each Hamiltonian appearing in the sequence and to simulate it. Each oracle call to any H_j is simulated by making oracle calls to H; see Ref. [64, Sec. 5] for details. Thus, the total number of oracle calls is proportional to N, although it is not equivalent because of the potential overhead in implementing each $e^{-iH_{j_\ell} t_{j_\ell}}$, $\ell = 1, \ldots, N$.

In particular, let $H = \sum_{j=1}^{m} H_j$, where e^{-iH_jt}, $t \in \mathbb{R}$, can be implemented efficiently, and the H_j do not commute, $j = 1, \ldots, m$. Consider algorithms approximating e^{-iHt}, $t \in \mathbb{R}$, that are obtained using Suzuki's high order splitting formulas [15,16]. These algorithms have the form

$$\prod_{l=1}^{N} e^{-iH_{j_l}t_{j_l}} \tag{13}$$

for suitable $t_{j_l} \in \mathbb{R}$, where $j_l \in \{1, \ldots, m\}$. The cost of the simulation of H is proportional to the number of exponentials, N, so that

$$\left\| e^{-iHt} - \prod_{l=1}^{N} e^{-iH_{j_l}t_{j_l}} \right\| \leq \varepsilon$$

Berry et al. [64] show that

$$N \leq N_{\text{prev}} := m5^{2k}(m\|H_1\|t)^{1+\frac{1}{2k}}\varepsilon^{-1/(2k)} \tag{14}$$

where the splitting formula is of order $2k+1$ and $\|H_1\| \geq \|H_1\| \geq \cdots \geq \|H_m\|$. They also derive the value of k that minimizes the upper bound. Papageorgiou and Zhang [65] improve this estimate for N by showing

$$N \leq N_{\text{new}} := 2(2m-1)\,5^{k-1}\|H_1\|t\left(\frac{4emt\|H_2\|}{\varepsilon}\right)^{1/(2k)}\frac{4me}{3}\left(\frac{5}{3}\right)^{k-1}$$

From this they also derive an improved estimate for the k that minimizes the upper bound.

Among the many applications of these estimates, in Ref. [64] they are used along with the decomposition cost of H and the simulation cost of the individual H_j, $j = 1, \ldots, m$, to derive the overall simulation cost. Recently, Childs and Kothari [66] used the estimates in Ref. [64] in the simulation of sparse Hamiltonians with star decompositions.

A more general Hamiltonian simulation problem is studied by Wiebe et al. [67] who derive an estimate similar to Eq. (14).

Besides the papers already mentioned, a large and varied literature covers Hamiltonian simulation. Many papers deal with particular algorithms. There are no tight complexity bounds. The following list of papers, which is by no means complete, is of interest. Abrams and Lloyd [68] show algorithms for the simulation of many-body Fermi systems. Brown et al. [69] establish limits of quantum simulation. Boghosian and Taylor [70] present efficient algorithms simulating quantum mechanical systems. Bravyi et al. [71] show an efficient algorithm for the simulation of weakly interacting quantum spin systems. Buluta and Nori [55] provide an overview of quantum simulators. Chen et al. [72] study the simulation of the

Burgers equation. Kassal et al. [54,73] deal with quantum simulation in chemistry. Ortiz et al. [74] study algorithms for fermionic simulations. Paredes et al. [75] present an algorithm that exploits quantum parallelism to simulate randomness. Somma et al. [76,77] study the quantum simulation of physics problems. Whitfield et al. [78] study the quantum simulation of electronic Hamiltonians. Wiesner [79] studies the quantum simulation of many-body systems. Wu et al. [80] study the simulation of pairing models on a quantum computer. Yepez [81] presents an efficient algorithm for the many-body three-dimensional Dirac equation.

I. Eigenvalue Estimation

The estimation of the ground-state eigenvalue of a time-independent Hamiltonian corresponding to a multiparticle system is an important problem in physics and chemistry. Decades of calculating ground-state eigenvalues of systems with a large number of particles have suggested that such problems are hard on a classical computer. That is why researchers have been experimenting with quantum computers to solve eigenvalue problems in quantum chemistry with encouraging results [82,83].

In fact, a fair amount of work deals with eigenvalue problems; see, for example, [71,84–90]. See also Refs [54,73] and the references therein.

Abrams and Lloyd [91] were the first to observe that the ground-state eigenvalue of the Born–Oppenheimer electronic Hamiltonian [92, p. 43] can be approximated on a quantum computer using the phase estimation algorithm [17, Fig. 5.2]. Phase estimation is not limited to this particular Hamiltonian eigenvalue problem but is broadly applicable, provided its requirements are met with reasonable cost for the problem at hand.

One requirement is that the second register of its initial state should contain an approximation of the eigenvector corresponding to the eigenvalue of interest. For instance, for the estimation of the ground-state eigenvalue one needs an approximate ground-state eigenvector. This approximation does not need to be very precise. It suffices that the magnitude of its projection on the actual eigenvector is not exponentially small. The success probability of the algorithm depends on the quality of the approximate eigenvector. In some cases such approximations can be computed efficiently by quantum algorithms [93], in other cases quantum algorithms designed to prepare general quantum states [89] are used to prepare the approximate eigenvector, or it is empirically or randomly chosen [84,91].

The second requirement is the implementation of powers of the unitary matrix U that phase estimation uses. In the case of the time-independent Schrödinger equation $U = e^{i\gamma H}$, where H denotes the system Hamiltonian and γ is a suitable constant that aims to ensure the phase corresponding to the eigenvalue of interest belongs to $[0, 1)$. Then one needs to derive the cost simulating the U^{2^t},

$t = 0, \ldots, b-1$, so that the algorithm has accuracy $2^{-b} \le \varepsilon$, with high probability. We remark that the powers of U do not have to be simulated very accurately, because the simulation error only affects the success probability of phase estimation. The algorithms for Hamiltonian simulation we discussed previously are used to approximate the powers of U. Thus, the total cost of phase estimation includes the simulation cost.

For the approximation of the ground-state eigenvalue (ground-state energy) of the time-independent Schrödinger equation the form of the Hamiltonian H used in phase estimation depends on the way the eigenvalue problem is approached. One possibility is to obtain H by spatially discretizing the time-independent Schrödinger equation. An advantage of this is that one solves the problem for a class of potentials. Another possibility is to use the Born–Oppenheimer electronic Hamiltonian in the second quantized form [92, p. 89] in phase estimation. We discuss both alternatives next.

Papageorgiou et al. [94], using a spatial discretization of the Schrödinger equation, provide rigorous estimates of the cost and the success probability of the phase estimation. Their algorithm prepares the initial state and simulates all the U^{2^t}, $t = 0, \ldots, b-1$. We remark that some similarities are found between their approach and the approach already used by Lidar and Wang [84] for the calculation of the thermal rate constant.

In particular, Papageorgiou et al. consider the approximation of the smallest eigenvalue E_1 of the equation

$$\begin{aligned} (-\tfrac{1}{2}\Delta + V)\Psi_1(x) &= E_1\Psi_1(x) \quad \text{for all } \ x \in I_d := (0, 1)^d \\ \Psi_1(x) &= 0 \quad \text{for all } \ x \in \partial I_d \end{aligned}$$

where ∂I_d denotes the boundary of the unit cube, x is the position variable, and Ψ_1 is a normalized eigenfunction. For simplicity, they assume that all masses and the normalized Planck constant are 1. The boundary conditions are for particles in a box. Multiparticle systems on bounded domains with the wave function equal to zero on the boundary have been studied in the literature; see, for example, [95, p. 621].

Here, Δ is the d-dimensional Laplacian and $V \ge 0$ is a function of d variables. The dimension is proportional to the number of particles p, such as $d = 3p$. For many applications the number of particles p and hence d is huge. Moreover, it is assumed that V and its first-order partial derivatives $\partial V/\partial x_j$, $j = 1, \ldots, d$, are continuous and uniformly bounded by 1.

To approximate E_1 with relative error proportional to ε observe that the finite difference discretization of the operator $-\frac{1}{2}\Delta + V$ on a regular grid with mesh size $h = \varepsilon$ yields a matrix $H = -\frac{1}{2}\Delta_h + V_h$, whose smallest eigenvalue $E_{h,1}$ approximates E_1 with relative error $O(\varepsilon)$. The matrix size is $\varepsilon^{-d} \times \varepsilon^{-d}$.

Phase estimation approximates $E_{h,1}$ using

1. The state

$$|\psi_1\rangle^{\otimes d}$$

in the second register of the initial state. This is an estimate of the eigenvector corresponding to $E_{h,1}$, where $|\psi_1\rangle^{\otimes d}$ is the ground-state eigenvector of $-\Delta_h$, which are implemented efficiently using the quantum Fourier transform with a number of quantum operations proportional to $d(\log \varepsilon^{-1})^2$.

2. Suzuki's [15,16] high-order splitting formulas to simulate the unitaries U^{2^t}, $t = 0, \ldots, b-1$, where $U = e^{iH/(2d)}$. These splitting formulas use exponentials involving $-\frac{1}{2}\Delta_h$ and V_h, respectively. The former are implemented using the quantum Fourier transform with cost proportional to $d(\log \varepsilon^{-1})^2$. The latter, involving the evaluations of the potential, are implemented using quantum queries.

   The overall simulation error is at most $1/20$ using

$$O\left(\varepsilon^{-3} e^{\sqrt{\ln \frac{1}{d\varepsilon}}}\right) \quad \text{as } d\varepsilon \to 0$$

matrix exponentials.

The errors due to the approximation of the ground-state eigenvector and the simulation of the exponentials affect the success probability of phase estimation. However, it remains at least $2/3$. Its total cost, including the number of queries and the number of all other quantum operations, is

$$Cd\varepsilon^{-(3+\delta)}$$

where $\delta > 0$ is arbitrarily small and C is a constant. The number of qubits is

$$cd \log \varepsilon^{-1}$$

where c is a constant.

Tight quantum complexity bounds for the ground-state eigenvalue problem are not known. On the other hand, the cost of any classical algorithm in the worst case with respect to V grows exponentially with d. Indeed, consider a potential function V, and let $\bar{V}$ be a perturbation of V. Then the eigenvalue $E_1(V)$ corresponding to V and the eigenvalue $E_1(\bar{V})$ corresponding to $\bar{V}$ are related according to the formula

$$E_1(V) = E_1(\bar{V}) + \int_{I_d} (V(x) - \bar{V}(x))\Psi_1^2(x; \bar{V})\, dx + O(\|V - \bar{V}\|_\infty^2)$$

where $\Psi_1(\cdot;\bar{V})$ denotes the eigenfunction corresponding to $E_1(\bar{V})$. This implies that approximating E_1 is at least as hard as approximating the multivariate integral involving V in the worst case. As a result, any classical deterministic algorithm for the eigenvalue problem with accuracy ε must use a number of function evaluations of V that grows as ε^{-d}; see Ref. [96] for details.

As we indicated, for multiparticle systems many papers consider the Born–Oppenheimer electronic Hamiltonian in the second quantized form, see, for example, [73,78,90,97–99]. It is given by

$$H = \sum_{p,q} h_{pq} a_p^\dagger a_q + \frac{1}{2} \sum_{p,q,r,s} h_{pqrs} a_p^\dagger a_q^\dagger a_r a_s$$

where h_{pq} and h_{pqrs} are one- and two-electron integrals, respectively, in a molecular spin orbital basis, $p, q, r, s = 1 \ldots, M$; see Refs [78, Eqs. 3 and 4], and M is the number of basis functions. So we have $M^2 + M^4$ integrals. The values of these integrals are considered known since they are computed classically once the basis functions are chosen. The $a_j, a_j^\dagger$ are fermionic annihilation and creation operators, respectively, $j = 1, \ldots, M$. We have the anticommutation relations

$$a_j a_i + a_i a_j = 0 \quad \text{and} \quad a_j a_i^\dagger + a_i^\dagger a_j = \delta_{ij} I, \quad i, j = 1, \ldots, M \tag{15}$$

where δ_{ij} is the Kronecker delta and I is the identity operator.

Whitfield et al. [78], Ovrum and Hjorth-Jensen [98], and Veis and Pittner [99] take similar approaches for estimating the ground-state eigenvalue of H. Our discussion is based on Ref. [78]. The Jordan–Wigner transformation is used to map the creation and annihilation operators $a_j^\dagger$ and a_j to products of Pauli matrices. This transformation is defined by

$$\begin{aligned} a_j &\to \sigma_j^+ \left(\prod_{k=j+1}^{M} \sigma_k^z \right) \\ a_j^\dagger &\to \sigma_j^- \left(\prod_{k=j+1}^{M} \sigma_k^z \right) \end{aligned} \tag{16}$$

where $\sigma_j^s = 1 \otimes \cdots \otimes \sigma^s \otimes 1 \cdots \otimes 1$, σ^s is applied to the jth qubit, $s \in \{z, +, -\}$, and

$$\sigma^z = \begin{pmatrix} 1 & 0 \\ 0 & -1 \end{pmatrix}, \quad \sigma^+ = \begin{pmatrix} 0 & 1 \\ 0 & 0 \end{pmatrix}, \quad \sigma^- = \begin{pmatrix} 0 & 0 \\ 1 & 0 \end{pmatrix}$$

See Ref. [78, Eqs. 5a and 5b] for details.

This yields the Pauli representation of the Hamiltonian

$$H = \sum_{j=1}^{K} H_j$$

with $K = O(M^4)$. For simplicity, we assume here that each of the H_j corresponds to a term of the original Hamiltonian. Phases estimation is used to approximate the ground-state eigenvalue of this Hamiltonian. The exponentials e^{-iH_jt}, $t \in \mathbb{R}$, $j = 1, \ldots, K$, can be implemented using $O(M)$ elementary quantum gates. We remark that in Ref. [78] extra care is taken to group the terms in a way that each of the resulting ones can be implemented using quantum circuit primitives with cost $O(M)$, which is an important feature of their algorithm. In either case, the cost of implementing the exponentials of all the terms in H is $O(M^5)$.

In phase estimation, the Trotter formula is used to simulate the exponential of the Hamiltonian as follows

$$e^{i\gamma H 2^t} \approx \left(\prod_{j=1}^{K} e^{i\gamma H_j 2^t / N_t} \right)^{N_t} \qquad t = 0, \ldots, b-1$$

where b is the number of qubits in the first register of phase estimation upon which the accuracy depends. The error of this approximation is bounded by $c\gamma 2^t \sum_{j=1}^{K} \|H_j\| N_t^{-1}$, see Ref. [100, Th. 3], where c is constant independent of K and the H_j, $j = 1, \ldots, K$. Often, the values of the N_t, $t = 0, \ldots, b-1$, are chosen empirically in practice. Similarly, the value of b is determined by considerations, such as reasonable chemical accuracy, and is relatively small.

J. Linear Systems

Many applications require the solution of systems of linear equations. An extensive literature about classical algorithms for this problem includes Refs [101,102] and the references therein. Recently Harrow et al. [20] derived a quantum algorithm for this problem. We sketch this algorithm.

Consider the linear system $Ax = b$, where A is an $N \times N$ Hermitian matrix. Harrow et al. derive an quantum algorithm that computes the solution of $A|x\rangle = |b\rangle$. They assume that the singular values of A belong to $[\kappa^{-1}, 1]$ (so the condition number $K(A)$ of A satisfies $K(A) \leq \kappa$) and that b is a unit vector that has been quantum mechanically implemented and is given as state $|b\rangle$. The algorithm does not output $|x\rangle$ classically. The solution is available as a quantum state so that one can compute functionals involving it, for instance, an expectation $\langle x|M|x\rangle$, for a given M.

Let λ_j be the eigenvalues and $|u_j\rangle$, $j = 1, \ldots, N$ be the normalized eigenvectors of A. Then $|b\rangle = \sum_{j=1}^{N} \beta_j |u_j\rangle$. Consider phase estimation as in Ref. [17, Fig. 5.2] with the state $|b\rangle$ in the second register, and the conditional Hamiltonian evolution

$$\sum_{k=0}^{T-1} |k\rangle\langle k| \otimes e^{iAkt_0/T}$$

where $t_0 = O(\kappa\varepsilon^{-1})$, T is a sufficiently large number, and ε is the desired accuracy in the solution of the system. After the inverse Fourier transform is applied at the top register the state is

$$\sum_{j=1}^{N} \beta_j \sum_{k=0}^{T-1} \alpha(j,k)|k\rangle|u_j\rangle$$

Then $|\alpha(j,k)|$ is large for the indices $k_j \in \{0,\ldots,T-1\}$ that lead to good approximations $\lambda_j \approx \frac{2\pi k_j}{t_0} =: \tilde{\lambda}_j$, and are small for the remaining indices, $j = 1,\ldots,N$.

Neglecting the terms that do not lead to good approximations of the eigenvalues of A we have

$$\sum_{j=1}^{N} \beta_j \alpha(j,k_j)|k_j\rangle|u_j\rangle$$

Adding a qubit and performing a conditional rotation we get

$$\sum_{j=1}^{N} \beta_j \alpha(j,k_j)|k_j\rangle|u_j\rangle \left(\sqrt{1 - C^2/\tilde{\lambda}_j^2}|0\rangle + C/\tilde{\lambda}_j|1\rangle\right)$$

where $C = O(\kappa^{-1})$. At this point we do not need the $|k_j\rangle$ any longer and we undo phase estimation. To further simplify the analysis we assume that $\alpha(j,k_j) = 1$, $j = 1,\ldots,N$. So we have the state

$$\sum_{j=1}^{N} \beta_j |u_j\rangle \left(\sqrt{1 - C^2/\tilde{\lambda}_j^2}|0\rangle + C/\tilde{\lambda}_j|1\rangle\right)$$

If we measure the last qubit and the outcome is 1, the system collapses to the state

$$\gamma \sum_{j=1}^{N} \frac{\beta_j}{\tilde{\lambda}_j}|u_j\rangle$$

which is an approximation to the solution $|x\rangle$ of the linear system, modulo the factor $\gamma = (\sum_{j=1}^{N} C^2|\beta_j|^2/|\tilde{\lambda}_j|^2)^{-1/2}$, which is the reciprocal of the square root of the probability to obtain outcome 1. This probability is $\Omega(\kappa^{-2})$ because $C = O(\kappa^{-1})$. Thus, $O(\kappa)$ steps of amplitude amplification [13] are sufficient to boost this probability.

Assuming that A is s-sparse and taking into account the cost of simulating e^{iAt}, as well as the error of phase estimation, the total cost of the quantum algorithm is proportional to

$$s \log N \, \kappa^2 \varepsilon^{-1}$$

The best general purpose classical algorithm for the solution of linear systems is the conjugate gradient algorithm [102] and has cost proportional to

$$sN\sqrt{\kappa} \log \varepsilon^{-1}$$

for a positive definite matrix A, and

$$sN\kappa \log \varepsilon^{-1}$$

otherwise.

This quantum algorithm is especially efficient for systems involving matrices of huge size with condition number polynomial in $\log N$, assuming ε is not arbitrarily small. However, in many applications the matrix size N depends on the desired accuracy ε, and grows as $\varepsilon \to 0$, which can be true for the condition number $K(A)$ as well. Thus, the dependence of N and $K(A)$ on ε determines whether there is an advantage in the quantum algorithm.

Two recent papers apply this quantum algorithm to the solution of first-order differential equations [103,104]. Because neither addresses the relationship between N and ε for solving the differential equations with error ε it is hard to draw a conclusion about the efficiency of the quantum algorithm in these cases.

However, a detailed analysis of the performance of the quantum algorithm for the linear systems involved in the solution of differential equations with error ε may reveal a significant advantage. Consider, for example, the systems obtained from the discretization of second-order elliptic partial differential equations in d dimensions. Then we can have $N = \varepsilon^{-d}$ and $K(A) = \Theta(\varepsilon^{-2})$. Observe that a classical algorithm that solves the system must have a cost at least proportional to the matrix size $N = \varepsilon^{-d}$ (i.e., exponential in d). On the other hand, the quantum algorithm whose cost depends on the logarithm of N, and the condition number $K(A)$, is exponentially faster than the classical algorithm when d is large.

Acknowledgments

We thank Iasonas Petras for his comments and suggestions. This work has been supported in part by the National Science Foundation/Division of Mathematical Sciences.

REFERENCES

1. E. Novak, *Deterministic and Stochastic Error Bounds in Numerical Analysis*, Lecture Notes in Mathematics 1349, Springer-Verlag, Berlin, 1988.
2. L. Plaskota, *Noisy Information and Computational Complexity*, Cambridge University Press, Cambridge, UK, 1996.
3. K. Ritter, *Average-Case Analysis of Numerical Problems*, Lecture Notes in Mathematics, Springer, Berlin, 2000, p. 1733.
4. J. F. Traub and A. G. Werschulz, *Complexity and Information*, Cambridge University Press, Cambridge, UK, 1998.
5. J. F. Traub and H. Woźniakowski, *A General Theory of Optimal Algorithms*, ACM Monograph Series, Academic Press, New York, 1980.
6. J. F. Traub, G. Wasilkowski, and H. Woźniakowski, *Information-Based Complexity*, Academic Press, New York, 1988.
7. A. G. Werschulz, *The Computational Complexity of Differential and Integral Equations*, Oxford University Press, Oxford, UK, 1991.
8. N. Novak and H. Woźniakowski, *Tractability of Multivariate Problems, Volume I: Linear Information*, European Mathematical Society, Zurich, 2008.
9. N. Novak and H. Woźniakowski, *Tractability of Multivariate Problems, Volume II: Standard Information for Functionals*, European Mathematical Society, Zurich, 2010.
10. P. W. Shor, *SIAM J. Comput.* **26**(5), 1484–1509 (1997).
11. L. Grover, *Proceedings of the 28th Annual Symposium on the Theory of Computing*, ACM Press, New York, 1996, pp. 212–219.
12. M. Mosca, in *Encyclopedia of Complexity and Systems Science*, Vol. 8, Springer, New York, 2009, p. 7088.
13. G. Brassard, P. Hoyer, M. Mosca, and A. Tapp, in *Contemporary Mathematics, Quantum Computation and Information*, S. J. Lomonaco, Jr. and H. E. Brandt, eds. AMS, Providence, RI, 2002, p. 53. Also http://arXiv.org/abs//quant-ph/0005055.
14. A. Nayak and F. Wu, *Proc. 31st Annual ACM Symposium on Theory of Computing (STOC)*, 1999, pp. 384–393.
15. M. Suzuki, *Phys. Lett. A* **146**, 319–323 (1990).
16. M. Suzuki, *J. Math. Phys.* **32**, 400–407 (1991).
17. M. A. Nielsen and I. L. Chuang, *Quantum Computation and Quantum Information*, Cambridge University Press, Cambridge, UK, 2000.
18. R. P. Feynman, *Int. J. Theoret. Phys.* **21**, 467–488 (1982).
19. S. Lloyd, *Science* **273**, 1073–1078 (1996).
20. A. W. Harrow, A. Hassidim, and S. Lloyd, *Phys. Rev. Lett.* **103**, 150502 (2009).
21. V. Strassen, *Num. Math.* **13**, 354–356 (1969).

22. J. F. Traub, *Phys. Today* **1999**, 39–44 (1999).
23. R. Beals, H. Buhrman, R. Cleve, M. Mosca, and R. de Wolf, *Proc. FOCS* (1998), p. 352–361. Also http://arxiv.org/abs/quant-ph/9802049.
24. D. S. Abrams and C. P. Williams, *Report NASA Jet Propulsion Laboratory*, 1999. Also http://arXiv.org/abs/quant-ph/9908083.
25. E. Novak, *J. Complexity* **19**(1), 19–42 (2001).
26. S. Heinrich, *J. Complexity* **18**(1), 1–50 (2002).
27. A. Bessen, *J. Complexity* **20**(5), 699–712 (2004).
28. C. H. Bennet, E. Bernstein, G. Brassard, and U. Vazirani, *SIAM J. Comput.* **26**(5), 1510–1523 (1997).
29. R. Cleve, A. Ekert, C. Macchiavello, and M. Mosca, *Proc. R. Soc. Lond. A.* **454**, 339–354 (1996).
30. E. Bernstein and U. Vazirani, *SIAM J. Comput.* **26**(5), 1411–1473 (1997).
31. S. Heinrich, From Monte Carlo to quantum computation. *Proceedings of the 3rd IMACS Seminar on Monte Carlo Methods MCM2001, Salzburg, Special Issue of Mathematics and Computers in Simulation*, K. Entacher, W. Ch. Schmid, and A. Uhl, eds., Vol. 62, pp. 219–230 (2003).
32. S. Heinrich, *J. Complexity* **19**, 19–42 (2003).
33. S. Heinrich and E. Novak, in *Monte Carlo and Quasi-Monte Carlo Methods 2000*, K.-T. Fang, F. J. Hickernell, and H. Niederreiter eds., Springer-Verlag, Berlin, 2002.
34. S. Heinrich, M. Kwas, and H. Woźniakowski, in *Monte Carlo and Quasi-Monte Carlo Methods 2002*, H. Niederreiter, ed., 2004, pp. 27–49.
35. J. F. Traub and W. Woźniakowski, *Quantum Inf. Process.* **1**(5), 365–388 (2002).
36. G. W. Wasilkowski and H. Woźniakowski, *J. Math. Phys.* **37**(4), 2071–2088 (1996).
37. F. Curbera, *J. Complexity* **16**(2), 474–506 (2000).
38. A. D. Egorov, P. I. Sobolevsky, and L. A. Yanovich, *Functional Integrals: Approximate Evaluation and Applications*, Kluwer Academic Publishers, Dordrecht, 1993.
39. M. Kwas, Quantum algorithms and complexity for certain continuous and related discrete problems, Ph.D. Thesis, Department of Computer Science, Columbia University, 2005.
40. R. M. Cameron, *Duke Math. J.* **8**, 111–130 (1951).
41. L. Plaskota, G. W. Wasilkowski, and H. Woźniakowski, *J. Comp. Phys.* **164**(2), 335–353 (2000).
42. M. Kwas and Y. Li, *J. Complexity* **19**, 730–743 (2003).
43. S. Heinrich, *J. Complexity* **20**, 27–45 (2004).
44. E. Novak, I. H. Sloan, and H. Woźniakowski, *J. Foundations Comput. Math.* **4**, 121–156 (2004).
45. B. Z. Kacewicz, *J. Complexity* **22**(5), 676–690 (2006).
46. B. Z. Kacewicz, *Numer. Math.* **45**, 93–104 (1984).
47. M. Goćwin and M. Szczesny, *Opuscula Math.* **28**(3), 247–277 (2008).
48. S. Heirich and B. Milla, *J. Complexity* **24**(2), 77–88 (2008).
49. S. Heinrich, *J. Complexity* **22**(5), 691–725 (2006).
50. S. Heinrich, *J. Complexity* **22**(2), 220–249 (2006).
51. M. Goćwin, *Quant. Inf. Process.* **5**(1), 31–41 (2006).
52. C. Dür and P. Hoyer, in *Proc. of the 30th Ann. ACM Symposium on Theory of Computing*, 1998, p. 1516–1524.
53. S. P. Jordan, *Phys. Rev. Lett.* **95**, 050501 (2005).

54. I. Kassal, S. P. Jordan, P. J. Love, M. Mohseni, and A. Aspuru-Guzik, *Proc. Nat. Acad. Sci.* **105**, 18681 (2008).
55. I. Buluta and F. Nori, *Science* **326**, 108 (2009).
56. C. Zalka, *Proc. R. Soc. Lond. A* **454**, 313 (1998).
57. C. Zalka, *Fortschr. Phys.* **46**, 877 (1998).
58. D. Aharonov and A. Ta-Shma, *Proc. 35th Annual ACM Symp. on Theory of Computing* (2003), p. 359.
59. E. Farhi, J. Goldstone, and M. Sipser, *Quantum Computation by Adiabatic Evolution*, http://arXiv.org/abs/quant-ph/0001106, 2000.
60. A. Childs, E. Farhi, and S. Gutmann, *J. Quant. Inf. Proc.* **1**, 35–43 (2002).
61. E. Farhi, J. Goldstone, and S. Gutmann, *A Quantum Algorithm for the Hamiltonian NAND Tree*, http://arXiv.org/abs/quant-ph/0702144, 2007.
62. A. Childs, *Phys. Rev. Lett.* **102**, 180501 (2009).
63. A. Childs, *Commun. Math. Phys.* **294**, 581–603 (2010).
64. D. W. Berry, G. Ahokas, R. Cleve, and B. C. Sanders, *Commun. Math. Phys.* **270**, 359 (2007).
65. A. Papageorgiou and C. Zhang, *Quantum Inf. Process* **11**, 541–561 (2012).
66. A. Childs and R. Kothari, in *Proceedings of the 5th Conference on Theory of Quantum Computation, Communication, and Cryptography*, TQC'10, Springer-Verlag, Berlin, Heidelberg, 2011, pp. 94–103. Also http://arXiv.org/abs/1003.3683.
67. N. Wiebe, D. Berry, P. Hoyer, and B. C. Sanders, *J. Phys. A: Math. Theor.* **43**, 065203 (2010).
68. D. S. Abrams and S. Lloyd, *Phys. Rev. Lett.* **79**(13), 2586–2589 (1997).
69. K. R. Brown, R. J. Clark, and I. L. Chuang, *Phys. Rev. Lett.* **97**(5), 050504 (2006).
70. B. M. Boghosianm and W. Taylor, *Proc. 4th Workshop on Physics and Computation*, Boston, MA, 1998, p. 30–42. Also http://arXiv.org/abs/quant-ph/9701019.
71. S. Bravyi, D. DiVincenzo, and D. Loss, *Comm. Math. Phys.* **284**, 481–507 (2008).
72. Z. Chen, J. Yepez, and D. G. Cory, *Phys. Rev. A* **74**, 042321 (2006).
73. I. Kassal, J. D. Whitfield, A. Perdomo-Ortiz, M.-H. Yung, and A. Aspuru-Guzik, *Ann. Rev. Phys. Chem.* **62**, 185–207 (2011).
74. G. Ortiz, J. E. Gubernatis, E. Knill, and R. Laflamme, *Phys. Rev. A* **64**(2), 022319 (2001).
75. B. Paredes, F. Verstraete, and J. I. Cirac, *Phys. Rev. Lett.* **95**, 140501 (2005).
76. R. Somma, G. Ortiz, J. E. Gubernatis, E. Knill, and R. Laflamme, *Phys. Rev. A* **65**, 042323 (2002).
77. R. Somma, G. Ortiz, E. Knill, and J. E. Gubernatis, *Proc. SPIE 2003 Quantum Information and Computation*, Andrew R. Pirich and Howard E. Brant eds., Vol. 5105, pp. 96–103 (2003).
78. J. Whitfield, J. Biamonte, and A. Aspuru-Guzik, *Mol. Phys.* **109**(5), 735–750 (2011).
79. S. Wiesner, *Simulations of Many-Body Quantum Systems by a Quantum Computer*, http://arXiv.org/abs/quant-ph/9603028, 1996.
80. L.-A. Wu, M. S. Byrd, and D. A. Lidar, *Phys. Rev Lett.* **89**(5), 057904 (2002).
81. J. Yepez, *An Efficient and Accurate Quantum Algorithm for the Dirac Equation*, http://arXiv.org/abs/quant-ph/0210093, 2002.
82. J. Du, N. Xu, X. Peng, P. Wang, S. Wu, and D. Lu, *Phys. Rev. Lett.* **104**, 030502 (2010).
83. B. P. Lanyon, J. D. Whitfield, G. G. Gillett, M. E. Goggin, M. P. Almeida, I. Kassal, J. D. Biamonte, M. Mohseni, B. J. Powell, M. Barbieri, A. Aspuru-Guzik, and A. G. White, *Nat. Chem.* **2**, 106–111 (2010).

84. D. A. Lidar and H. Wang, *Phys. Rev. E* **59**, 2429–2438 (1999).
85. A. Hams and H. DeRaedt, *Phys. Rev. E* **62**(3), 4365–4377 (2000).
86. S. Oh, *Phys. Rev. A* **77**(1), 012326 (2008).
87. T. Szkopek, V. Roychowdhury, E. Yablonovitch, and D. S. Abrams, *Phys. Rev. A* **72**(6), 062318 (2005).
88. P. Varga and B. Apagyi, *Phys. Rev. A* **78**(2), 022337 (2008).
89. H. Wang, S. Ashhab, and F. Nori, *Phys. Rev. A* **79**, 042335 (2009).
90. H. Wang, S. Kais, A. Aspuru-Guzik, and M. R. Hoffmann, *Phys. Chem. Chem. Phys.* **10**, 5388–5393 (2008).
91. D. S. Abrams and S. Lloyd, *Phys. Rev. Lett.* **83**, 5162 (1999).
92. A. Szabo and N. S. Ostlund, *Modern Quantum Chemistry*, Dover Publications Inc., New York, 1996.
93. P. Jaksch and A. Papageorgiou, *Phys. Rev. Lett.* **91**, 257902 (2003).
94. A. Papageorgiou, I. Petras, J. F. Traub, and C. Zhang, *Math. Comp.* **82**(284), 2293–2304 (2013).
95. P. G. Ciarlet and C. Le Bris, *Handbook of Numerical Analysis, Special Volume Computational Chemistry*, Vol. X, North Holland, Amsterdam, 2003.
96. A. Papageorgiou, *J. Complexity* **23**(4-6), 802–827 (2007).
97. A. Aspuru-Guzik, A. D. Dutoi, P. J. Love, and M. Head-Gordon, *Science* **309**(5741), 1704–1707 (2005).
98. E. Ovrum and M. Hjorth-Jensen, *Quantum Computation Algorithm for Many-Body Studies, arXiv:quant-ph/0705.1928*, 2007.
99. L Veis and J. Pittner, *J. Chem. Phys.* **133**, 194106 (2010).
100. M. Suzuki, *Comm. Math. Phys.* **51**(2), 183–190 (1976).
101. J. W. Demmel, *Applied Numerical Linear Algebra*. SIAM, Philadelphia, PA, 1997.
102. Y. Saad, *Iterative Methods for Sparse Linear Systems*, SIAM, Philadelphia, PA, 2003.
103. S. K. Leyton and T. Osborne, A Quantum Algorithm to Solve Nonlinear Differential Equations, http://arxiv.org/abs/0812.4423, 2008.
104. D. W. Berry, *Quantum Algorithms for Solving Linear Differential Equations*, http://arxiv.org/abs/1010.2745, 2010.

ANALYTIC TIME EVOLUTION, RANDOM PHASE APPROXIMATION, AND GREEN FUNCTIONS FOR MATRIX PRODUCT STATES

JESSE M. KINDER, CLAIRE C. RALPH, and GARNET KIN-LIC CHAN

Department of Chemistry and Chemical Biology, Cornell University, Ithaca, NY 14850, USA

I. INTRODUCTION

Matrix product states (MPS) are a powerful class of quantum states for analyzing strongly correlated one-dimensional quantum systems [1]. They underpin the density matrix renormalization group (DMRG) algorithm, which has yielded unprecedented accuracy in many systems [2–4]. A matrix product state encodes a quantum state as a contraction of independent tensors, each associated with a site on the lattice. Because of the contracted product structure, the theory of matrix product states can be viewed as a site-based mean field theory.

Another class of mean field theories is based on independent particles. One example is Hartree–Fock (HF) theory, in which the quantum state of a collection of fermions is approximated by a Slater determinant of independent orbitals. HF theory is the foundation on which many other approximations are built. For example, time-dependent Hartree–Fock theory (TDHF) approximates the true evolution of a quantum state with a single Slater determinant. Linearizing the equations of

Advances in Chemical Physics, Volume 154: Quantum Information and Computation for Chemistry, First Edition. Edited by Sabre Kais.

motion yields the random phase approximation (RPA), which gives the excitation spectrum and dynamical properties [5]. The random phase approximation is also the simplest treatment of correlation beyond the Hartree–Fock mean field, which can be demonstrated via the fluctuation–dissipation theorem [6]. Finally, the independent particle picture described by HF theory is the starting point for a particle-based Green function approach to many-body quantum systems.

Much of the recent work in MPS has focused on extending the ground-state formalism to study time evolution [1,7–12], Green functions, and dynamical (response) properties [13–16] and to extend the method from one dimension to higher-dimensional systems [17–23]. In this chapter, we describe how developments in these directions can proceed in analogy with Hartree–Fock theory. Despite the different physical content of the independent-site mean field of the MPS and independent-particle mean field of HF, the two theories share a similar mathematical structure. This suggests that many familiar concepts from HF theory can be directly applied to MPS. We concentrate on two familiar concepts from Hartree–Fock theory—the Dirac–Frenkel variational principle and the random phase approximation—and explore how they can be adapted to MPS. The resulting formulations for time evolution and dynamical properties differ from existing approaches and offer potential advantages. For example, the analog of time dependent Hartree–Fock theory yields equations of motion for MPS where evolution occurs entirely within the space of MPS with fixed auxiliary dimension. This eliminates the need for compression steps to control entanglement, and is optimal in the limit of small time steps. Similarly, the MPS RPA yields dynamical quantities and Green functions from the linear response of the entire MPS basis, in contrast to a single, averaged MPS basis as used in the dynamical DMRG approach [15]. As a result, the MPS RPA results in more accurate calculations for MPS with small auxiliary dimension. Finally, the MPS RPA can introduce additional correlation to stationary states, which can be demonstrated via the fluctuation–dissipation theorem. This additional correlation provides a way to improve MPS stationary-state properties and opens up the possibility of adapting other Hartree–Fock theories of correlation to the MPS language.

The chapter is organized as follows. First, we briefly summarize the HF and MPS approaches to stationary states to establish notation and illustrate the parallel structure of the theories. We then derive analytic equations of motion for MPS time evolution using the Dirac–Frenkel variational principle. We show that the resulting evolution is optimal for MPS of fixed auxiliary dimension, and discuss the relationship of this approach to time evolution to schemes currently in use. Next, we derive an MPS analog of the RPA by linearizing the equations of motion and show how excitation energies and dynamical properties can be obtained from a linear eigenvalue problem. We also discuss the relationship of this MPS RPA to other dynamical approaches for matrix product states. Finally, we explore the site-based Green functions that emerge naturally within the theory of MPS and use the fluctuation–dissipation theory to analyze the stationary-state correlations introduced at the level of the MPS RPA.

II. STATIONARY STATES

We begin with a review of stationary states in the Hartree–Fock and MPS theories to establish notation and illustrate the parallel mathematical structure in a familiar context.

In HF theory, an N-particle wave function is approximated by a single Slater determinant of orbitals $\phi_i(\mathbf{r})$:

$$\psi(\mathbf{r}_1, \mathbf{r}_2, \ldots, \mathbf{r}_k) = \hat{\mathcal{A}} \prod_{i=1}^{k} \phi_i(\mathbf{r}_i) \tag{1}$$

$\hat{\mathcal{A}}$ is an operator that transforms the orbital product into a determinant.

The variational objects of the theory are the orbitals, and the best mean field approximation to the ground state is the set of orbitals that minimizes the expectation value of the Hamiltonian:

$$E = \frac{\langle \psi | \mathcal{H} | \psi \rangle}{\langle \psi | \psi \rangle} \tag{2}$$

This leads to the Hartree–Fock equations:

$$\hat{\mathbf{f}}\, \phi_i(\mathbf{r}) = \varepsilon_i \phi_i(\mathbf{r}) \tag{3}$$

where $\hat{\mathbf{f}}$ is the Fock operator. Thus, each orbital is an eigenstate of a one-particle Hamiltonian that depends on all of the orbitals. In this sense, HF theory is a mean field theory for independent particles. If the orbitals are expanded in a finite basis, then the Hartree–Fock equations may be expressed in matrix form:

$$\mathsf{F} \cdot \mathsf{C} = \mathsf{S} \cdot \mathsf{C} \boldsymbol{\varepsilon} \tag{4}$$

where F is the Fock matrix, C is the coefficient matrix of the orbitals in the finite basis, S is the overlap matrix, and $\boldsymbol{\varepsilon}$ is the diagonal matrix of orbital energies. Because F is a function of C, this equation must be solved self-consistently [24].

The defining equations of an optimal MPS wave function can be derived in an analogous way. In an MPS wave function for k sites, the amplitude of a Fock state is given by the trace of a product of matrices:

$$|\psi\rangle = \sum_{\mathbf{n}} \mathrm{Tr} \left(\prod_{i=1}^{k} \mathsf{A}_i^{n_i} \right) |\mathbf{n}\rangle \tag{5}$$

where $|\mathbf{n}\rangle = |n_1, n_2, \ldots\rangle$ labels a state in Fock space. A matrix $\mathsf{A}_i^{n_i}$ is associated with each site i and orbital occupancy n_i. The size of the matrices defines the auxiliary dimension M, and the maximum occupation number of a site defines the physical dimension d. The d-independent $M \times M$ matrices associated with each site can be collected into a single tensor $(\mathsf{A}_i)_{\alpha\beta}^{n_i}$, where α and β are the auxiliary indices. This tensor can be flattened into a vector labeled by a single compound index $I = (n_i, \alpha, \beta)$. Derivatives with respect to the components of the tensor define a (nonorthogonal) basis for the wave function:

$$|\psi_I\rangle = \frac{\partial|\psi\rangle}{\partial \mathsf{A}_i^I} \tag{6}$$

From the linear dependence of $|\psi\rangle$ on each matrix, we have $|\psi\rangle = \sum_i \mathsf{A}_i^I |\psi_I\rangle$.

Requiring the variational energy to be stationary with respect to variations in a single tensor defines an eigenvalue problem:

$$\mathsf{H}_i \cdot \mathsf{A}_i = E\,\mathsf{S}_i \cdot \mathsf{A}_i \tag{7}$$

The matrices H_i and S_i are the Hamiltonian and overlap matrices in the local basis defined by A_i. Because the basis defined in Eq. (6) depends on the other tensors, A_i is the eigenvector of an effective Hamiltonian defined by the other tensors, analogous to Eq. (3) for the HF orbitals.

The entire set of equations, one for each tensor, can be combined into a single eigenvalue problem by defining the compound index $\mu = (i, n_i, \alpha, \beta)$ and collecting all the elements of all the tensors into a single vector $\mathbf{A}$. This vector contains kdM^2 elements. The wave function can be expanded as

$$|\psi\rangle = \frac{1}{k}\sum_{\mu} A_\mu |\psi_\mu\rangle \tag{8}$$

and the optimal MPS wave function is a self-consistent solution of

$$\mathsf{H} \cdot \mathbf{A} = E\,\mathsf{S} \cdot \mathbf{A} \tag{9}$$

The Hamiltonian and overlap matrices are defined as

$$\mathsf{H}_{\mu\nu} = \frac{1}{k}\langle\psi_\mu|\mathcal{H}|\psi_\nu\rangle \tag{10}$$

$$\mathsf{S}_{\mu\nu} = \frac{1}{k}\langle\psi_\mu|\psi_\nu\rangle \tag{11}$$

The power of MPS wave functions comes from their numerical efficiency. For MPS with open boundary conditions (as used in the DMRG algorithm), expectation values of local operators and Hamiltonians can be calculated with $\mathcal{O}(kM^3)$ complexity. The action of local operators and Hamiltonians, such as $\mathsf{H} \cdot \mathbf{A}$ and

$\mathsf{S} \cdot \mathbf{A}$, is also obtained with $\mathcal{O}(kM^3)$ complexity. Solving the eigenvalue problem Eq. (9) iteratively requires a cost proportional to the operations $\mathsf{H} \cdot \mathbf{A}$ and $\mathsf{S} \cdot \mathbf{A}$ and is thus also of $\mathcal{O}(kM^3)$ complexity. The prefactor depends on preconditioning for both H and S. The DMRG preconditions S by solving Eq. (7) for a single tensor A_i at a time. In this case, the overlap S_i can be exactly removed by canonicalization.

III. TIME EVOLUTION AND EQUATIONS OF MOTION

Time-dependent Hartree–Fock theory can be derived from the Dirac–Frenkel variational principle [5]. Here we derive an analogous time-dependent theory for MPS. (A time-dependent variational principle has also been considered in Ref. [25].) We show that the resulting equations of motion yield efficient time evolution algorithms for MPS of fixed auxiliary dimension without compression. In fact, the equations of motion give the *optimal* compression of the evolving state in the limit of an infinitesimal time-step.

The time-dependent Schrödinger equation can be obtained by minimizing the Dirac–Frenkel action

$$S = \int \mathrm{d}t \Big(i\hbar \langle \psi | \dot{\psi} \rangle - \langle \psi | \mathcal{H} | \psi \rangle \Big) \tag{12}$$

with respect to arbitrary variations $\langle \delta\psi |$. When the wave function is constrained to a particular form, such as a Slater determinant or an MPS, variations may only be taken with respect to the parameters $\{\lambda_i\}$:

$$\langle \delta\psi | = \sum_i \delta\lambda_i \cdot \langle \partial_i \psi | \tag{13}$$

The time derivative of $|\psi\rangle$ is also constrained in this way. Minimizing the action then gives the best approximation to the true evolution of the wave function *within the space of variational wave functions*.

When applied to Hartree–Fock theory, the resulting equations of motion are the time-dependent Hartree–Fock equations. The time-dependent version of Eq. (4) is

$$i\hbar \mathsf{S} \cdot \frac{\mathrm{d}\mathsf{C}(t)}{\mathrm{d}t} = \mathsf{F}(t) \cdot \mathsf{C}(t) \tag{14}$$

The Fock matrix depends on t through the time-dependence of the orbitals. It may also depend explicitly on t through a time-dependent external potential.

When we apply the same variational principle to the action for a matrix product state, we find the following equations of motion:

$$i\hbar \mathsf{S}(t) \cdot \frac{\mathrm{d}\mathbf{A}(t)}{\mathrm{d}t} = \mathsf{H}(t) \cdot \mathbf{A}(t) \tag{15}$$

Here, H and S are the time-dependent versions of Eqs. (10) and (11). The solution of Eq. (15) is a time-ordered exponential. While it is impractical to simulate time evolution using the formal solution, the equations of motion can be efficiently propagated if the time interval is discretized into units of duration Δt. In that case, eq. (15) can be used to obtain the MPS at t_{n+1} from the MPS at t_n:

$$\mathsf{S}_n \cdot \Delta \mathbf{A}_n = \epsilon \, \mathbf{B}_n \tag{16}$$

where $\epsilon = \Delta t / i\hbar$ and $\mathbf{B}_n = \mathsf{H}_n \cdot \mathbf{A}_n$. This is a linear equation that can be solved iteratively with complexity $\mathcal{O}(kM^3)$. In practice, more sophisticated time propagation schemes can be used, such as the norm- and energy-conserving propagators developed for time-dependent Hartree–Fock theory.

We now discuss the merits of the equation of motion approach and the connection to existing time evolution algorithms, including time evolution by block decimation [8], time-dependent density matrix renormalization group [9–11], and time-dependent matrix product states (tMPS) [12]. The computational cost of the MPS equation of motion (per time step) is the same as existing algorithms, so the relevant question is which method provides a better approximation to the true wave function.

The basic difficulty in simulating the evolution of a matrix product state is that the matrix dimension required to faithfully represent the exact wave function grows exponentially in time. The equation of motion approach restricts the evolution of the wave function to the space of MPS with fixed M; thus, the auxiliary dimension *is fixed at M throughout the simulation.* In contrast, existing methods for the time evolution of MPS first allow the dimension of the matrix to grow at each time-step and then project back onto an MPS of auxiliary dimension M. This projection is called the compression step. Methods differ in how this compression is carried out.

In time evolution by block decimation (TEBD) [8] and the time-dependent DMRG (tDMRG) [9–11], the evolution operator $\exp(\epsilon\mathcal{H})$ is factored into a product of local operators using a Trotter decomposition. Each local evolution operator is then applied in sequence. A two-site evolution operator $\mathcal{U}_{i,i+1}$ between neighboring sites joins two tensors of the MPS, A_i and A_{i+1}, into a single object, $\mathsf{T}_{i,i+1}$. This tensor is then approximately factored into a product of two tensors of the same size as A_i and A_{i+1} using singular value decomposition. For a single time-step (sweeping over all evolution operators for a local Hamiltonian), this update is efficient, with complexity of $\mathcal{O}(kM^3)$. However, the approximate projection is only performed locally, and does not involve all the tensors. Thus, the compressed matrix product state is not the optimal representation of the evolved wave function in the space of MPS with a given auxiliary dimension.

In time-dependent matrix product states [12], the full evolution operator is applied to the current state before compression. In the projection, tMPS attempts to minimize the cost function

$$\Delta[|\psi_{n+1}\rangle] = \left\| |\psi_{n+1}\rangle - e^{\epsilon\mathcal{H}}|\psi_n\rangle \right\|^2 \tag{17}$$

$|\psi_{n+1}\rangle$ and $|\psi_n\rangle$ are the new and old MPS, respectively. The minimization is of complexity $\mathcal{O}(kM^3)$ and yields, in principle, the optimal projected MPS wave function. However, because $|\psi_{n+1}\rangle$ depends nonlinearly on A_i, the minimization is not guaranteed to find the optimal solution in practice, and may instead converge to a local minimum of the cost function.

For a vanishingly small time-step, the projection can be done *without* any nonlinear minimization. We see this by recognizing that

$$i\hbar\epsilon\frac{\partial|\psi_n\rangle}{\partial t} = |\psi_{n+1}\rangle - |\psi_n\rangle + O(\mathrm{d}t^2) \tag{18}$$

Substituting this expression into Eq. (17) and minimizing with respect to changes in **A** yields precisely the discretized linear equation of motion obtained in Eq. (16) to leading order in ϵ. Therefore, propagation of the MPS equation of motion *exactly determines the optimal projection in a tMPS algorithm in the limit* $\Delta t \to 0$. In this sense, propagating the equations of motion yields the best possible time evolution of an MPS with a given auxiliary dimension, *without the need for explicit compression*.

IV. RANDOM PHASE APPROXIMATION

We now turn to the random phase approximation for time-dependent Hartree–Fock theory and develop an analogous theory of excited states and dynamical quantities for MPS. Unlike the correction vector and dynamical density matrix renormalization group formalisms [4,14,15], which use a single MPS at each frequency to evaluate dynamical quantities, the MPS RPA expresses dynamical quantities using a linear combination of many MPS, while still being of the same computational cost. This has the potential to yield more accurate results, as demonstrated by recent analytic response calculations using the density matrix renormalization group [16].

The Hartree–Fock RPA is based on the idea that excited states and dynamical properties such as the spectrum and other expectation values can be obtained without studying the full time evolution of the system. Only the linearized time evolution, or linear response, is considered. Linearizing the equations of motion around a HF stationary state yields the random phase approximation. This is achieved in Eq. (14) by taking $\mathsf{C}(t) = \mathsf{C}_0 + \mathsf{D}(t)$ and expanding all quantities to linear order in $\mathsf{D}(t)$.

A similar approach to the MPS equations of motion in Eq. (15) yields an MPS analog of the RPA. We take the zeroth-order MPS to be a stationary state defined by $\mathbf{A}$, with energy E_0. The time-dependent MPS is defined by

$$\mathbf{A}(t) = \mathbf{A} + \mathbf{b}(t). \tag{19}$$

The wave function and its derivatives are given by

$$|\psi\rangle = \frac{1}{k}\sum_{\mu}\left[\mathbf{A}_\mu + \mathbf{b}_\mu(t)\right]\left|\psi^{(0)}_\mu\right\rangle \tag{20}$$

$$|\psi_\mu\rangle = \frac{1}{k}\sum_{\nu}\left[\mathbf{A}_\nu + \mathbf{b}_\nu(t)\right]\left|\psi^{(0)}_{\mu\nu}\right\rangle \tag{21}$$

where $|\psi^{(0)}\rangle$ and its derivatives are evaluated with $\mathbf{b} = 0$.

Expanding Eq. (15) to linear order in $\mathbf{b}$ gives

$$i\hbar \mathsf{S}\frac{\mathrm{d}\mathbf{b}}{\mathrm{d}t} = E_0\, \mathsf{S}\cdot\mathbf{A} + \mathsf{H}\cdot\mathbf{b} + \mathsf{W}\cdot\mathbf{b}^* \tag{22}$$

The matrices S and H are the zeroth-order overlap and Hamiltonian matrices. The matrix W couples $\mathbf{b}$ to its complex conjugate and is defined by

$$\mathsf{W}_{\mu\nu} = \frac{1}{k}\left\langle \psi^{(0)}_{\mu\nu}|\mathcal{H}|\psi^{(0)}\right\rangle \tag{23}$$

W is symmetric, but not Hermitian.

The first term on the right-hand-side of Eq. (22) can be eliminated by multiplying the entire wave function $|\psi\rangle$ by the phase factor $\mathrm{e}^{-iE_0t/\hbar}$. The equations of motion for $\mathbf{b}(t)$ are harmonic and may be solved by taking

$$\mathbf{b}(t) = \mathbf{X}\,\mathrm{e}^{-i\omega t} + \mathbf{Y}^*\,\mathrm{e}^{i\omega t} \tag{24}$$

Equation (22) then gives

$$\hbar\omega\begin{bmatrix}\mathsf{S} & 0\\ 0 & -\mathsf{S}^*\end{bmatrix}\begin{pmatrix}\mathbf{X}\\ \mathbf{Y}\end{pmatrix} = \begin{bmatrix}\mathsf{H} & \mathsf{W}\\ \mathsf{W}^* & \mathsf{H}^*\end{bmatrix}\begin{pmatrix}\mathbf{X}\\ \mathbf{Y}\end{pmatrix} \tag{25}$$

The eigenvectors of this system of equations define the normal modes of the system, and the (positive) eigenvalues approximate its excitation spectrum. They both are efficiently obtained with $\mathcal{O}(kM^3)$ complexity. The normal modes define the response matrix $\Pi(\omega)$, which determines all the dynamical properties of the system:

$$\Pi(\omega) = \sum_q \begin{pmatrix}\mathbf{X}_q\\ \mathbf{Y}_q\end{pmatrix}\frac{1}{\omega-\omega_q}\begin{pmatrix}\mathbf{X}^*_q & \mathbf{Y}^*_q\end{pmatrix} \tag{26}$$

For example, consider an arbitrary harmonic perturbation $\mathcal{Q}(t)$. This defines a source $\mathbf{q}(t)$ for $\mathbf{b}$ in Eq. (22):

$$\langle\psi|\mathcal{Q}(t)|\psi\rangle = \mathbf{q}\cdot\mathbf{b}^* + \mathbf{b}\cdot\mathbf{q}^* \tag{27}$$

The elements of $\mathbf{q}$ are given by

$$\mathbf{q}_\mu = \left\langle \psi_\mu^{(0)}|\mathcal{Q}(t)|\psi^{(0)}\right\rangle \tag{28}$$

(We assume that the expectation value of $\mathcal{Q}(t)$ in the stationary state vanishes.) The time-dependent variation in the expectation value of an observable $\langle\delta\mathcal{P}\rangle$ due to the perturbation $\lambda\mathcal{Q}$ is

$$\langle\delta\mathcal{P}\rangle = \mathbf{p}\cdot\mathbf{b}^* + \mathbf{b}\cdot\mathbf{p}^* \tag{29}$$

This may be expressed using the response matrix $\Pi(\omega)$:

$$\langle\delta\mathcal{P}(\omega)\rangle = \begin{bmatrix}\mathbf{p}^*(\omega) & \mathbf{p}(\omega)\end{bmatrix}\cdot\Pi(\omega)\cdot\begin{bmatrix}\mathbf{q}(\omega)\\ \mathbf{q}^*(\omega)\end{bmatrix} \tag{30}$$

We now contrast the MPS RPA formulation for dynamical quantities with that used in the dynamical DMRG. From Eq. (21), the response (correction) vector in the MPS RPA is written as $\sum_\mu \mathbf{b}_\mu|\psi_\mu^{(0)}\rangle$. This is a linear superposition of k different matrix product states:

$$\begin{aligned}&\mathsf{b}_1\cdot\mathsf{A}_2\cdots\mathsf{A}_k + \mathsf{A}_1\cdot\mathsf{b}_2\cdots\mathsf{A}_k + \cdots\\ &+\mathsf{A}_1\cdot\mathsf{A}_2\cdots\mathsf{b}_i\cdots\mathsf{A}_k + \cdots + \mathsf{A}_1\cdot\mathsf{A}_2\cdots\mathsf{b}_k\end{aligned}$$

In contrast, in the dynamical DMRG, the ground-state wave function, perturbation $\mathcal{Q}$, and response vector are all expressed using a common MPS basis. For example, the response vector at the ith step of the DMRG sweep can be written as a single MPS:

$$\mathsf{A}_1\cdot\mathsf{A}_2\cdots\mathsf{b}_i\cdots\mathsf{A}_k$$

The tensors A are obtained from a density matrix that averages contributions from the ground state, response, and perturbation [15]. Because a single MPS basis is used for several quantities, we expect dynamical quantities to be less accurate than in the MPS RPA, at least for small auxiliary dimension. An algorithm for the MPS RPA in which each response tensor b_i is determined sequentially gives the DMRG algorithm for analytic response recently described by us in Ref. [16]. As shown there, the MPS RPA is more accurate than the dynamical DMRG for MPS with small auxiliary dimension.

V. GREEN FUNCTIONS AND CORRELATIONS

The MPS RPA provides a framework for studying response functions and correlation functions. One particular class of correlation functions, Green functions, is an extremely useful tool in the study of many-body physics. In this section, we explore the natural Green functions for the theory of MPS: site-based Green functions. We then use the fluctuation–dissipation theorem to analyze the nature of the additional correlations introduced at the mean field level.

HF theory is an independent particle mean field theory and the natural Green functions that emerge are labeled by particle indices (e.g., the single-particle Green function). Such Green functions can be obtained within an MPS response formalism through the dynamical DMRG [13–15,26] or the MPS RPA approach described earlier. However, because MPS is a site-based mean field theory, the more natural Green functions are labeled by site indices: instead of describing correlations between particles, these Green functions describe correlations between the quantum states of individual sites of the lattice. Site-based Green functions are linear combinations of the standard Green functions, but the operators involved are more natural for the MPS formalism.

We first define a one-site density matrix, $\Gamma^{(i)}_{nn'}$, which is the site-based analog of the one-particle density matrix, $\rho(\mathbf{r}, \mathbf{r}')$. $\Gamma^{(i)}_{nn'}$ is obtained from $|\psi\rangle\langle\psi|$ by tracing out all sites of the lattice except site i. The expectation value of a one-site operator $\mathcal{P}^{(i)}$ is then given by the trace

$$\langle \mathcal{P}^{(i)} \rangle = \sum_{nn'} \Gamma^{(i)}_{nn'} \mathsf{P}^{(i)}_{nn'} \tag{31}$$

Each element $\Gamma^{(i)}_{nn'}$ is an expectation value of an operator $\hat{\gamma}_{nn'}$ defined by

$$\langle \bar{n} | \hat{\gamma}_{nn'} | \bar{n}' \rangle = \delta_{n\bar{n}} \delta_{n'\bar{n}'} \tag{32}$$

For a system with two physical degrees of freedom per site, the operators a_i and $a_i^\dagger$ correspond to $\hat{\gamma}_{01}$ and $\hat{\gamma}_{10}$, and the number operator is $\hat{\gamma}_{11}$. Creation, annihilation, and number operators of systems with more degrees of freedom per site can be constructed from $\hat{\gamma}$ as well. Thus, the one-site density matrix contains components of one-, two-, and mixed-particle density matrices.

We now define a site–site (retarded) Green function $\mathbb{G}(ij; t)$ by considering the response of the site density matrix at site i and time $t > 0$, $\Gamma^{(i)}(t)$, to a perturbation at site j at time $t = 0$,

$$\mathcal{Q}^{(j)} = \delta(t) \sum_{nn'} V_{nn'} \hat{\gamma}^{(j)}_{nn'} \tag{33}$$

We then have

$$\mathbb{G}^{nn'}_{\bar{n}\bar{n}'}(ij;t) = \frac{\partial \Gamma^{(i)}_{nn'}(t)}{\partial V_{\bar{n}\bar{n}'}} \tag{34}$$

Note that by choosing appropriate combinations of $\gamma^{(i)}_{nn'}$ and $\gamma^{(j)}_{\bar{n}\bar{n}'}$, it is possible to construct spectral functions, one-particle Green functions, two-particle Green functions, and other types of conventional Green functions.

Within the RPA framework, the elements of $\mathbb{G}(ij;\omega)$ are determined by the response matrix of Eq. (30) with

$$\mathbf{p}_\mu(\omega) = \left\langle \psi_\mu | \gamma^{(i)}_{nn'} | \psi \right\rangle \tag{35}$$

$$\mathbf{q}_\mu(\omega) = \left\langle \psi_\mu | \gamma^{(j)}_{\bar{n}\bar{n}'} | \psi \right\rangle \tag{36}$$

The response function and the correlation functions of the ground state are related through the fluctuation–dissipation theorem [6]. Using this theorem, we can deduce explicit expressions for the correlations introduced by site-based Green functions at the level of the MPS RPA. For example, two-site correlation functions are given by

$$\left\langle \gamma^{(i)}_{nn'} \gamma^{(j)}_{\bar{n}\bar{n}'} \right\rangle = -\frac{1}{2\pi i} \int_{-\infty}^{+\infty} \mathrm{d}\omega \, \mathbb{G}^{nn'}_{\bar{n}\bar{n}'}(ij;\omega + i\eta) \tag{37}$$

From the expression for the response matrix $\Pi(\omega)$ in Eq. (26), we see that the correlation function is a sum of four terms, each of the form

$$\begin{aligned} &\sum_{q,\mu\nu} \left\langle \psi | \gamma^{(i)}_{nn'} | \psi_\mu \right\rangle X_{q\mu} X^*_{q\nu} \left\langle \psi_\nu | \gamma^{(j)}_{\bar{n}\bar{n}'} | \psi \right\rangle \\ &= \left\langle \psi | \gamma^{(i)}_{\bar{n}\bar{n}'} \cdot \Lambda_{XX} \cdot \gamma^{(j)}_{nn'} | \psi \right\rangle \end{aligned} \tag{38}$$

Λ_{XX}, Λ_{XY}, Λ_{YX}, and Λ_{YY} are matrix product operators (MPOs) responsible for additional correlation introduced by the RPA relative to the ground state (which is obtained by setting all $\Lambda = 1$). By regrouping terms in the sum, we obtain

$$\Lambda_{XX} = \sum_q \left(\sum_i \mathsf{X}^I_{qi} | \psi_I \rangle \right) \left(\sum_j \langle \psi_J | \mathsf{X}^{J*}_{qj} \right) \tag{39}$$

Each term in parentheses is a sum of k MPS of dimension M, and their outer product is formally an MPO of dimension $2kM$. An MPO for each normal mode makes Λ_{XX} an MPO of very large auxiliary dimension. Thus, the MPS RPA introduces entanglement and correlation beyond that of an MPS with dimension M.

The Hartree–Fock wave function is an uncorrelated many-body state in the sense that all many-body correlation functions factor into products of one-body expectation values. The random phase approximation is the first rung in a series of approximations that introduce correlation [24]. These include both time-independent correlation methods, such as coupled cluster theory [27,28], and correlation methods based on Green functions, such as the GW approximation and its extensions [29,30].

In the same way as the RPA provides a starting point for treating correlations beyond the Hartree–Fock mean field, the MPS RPA is a starting point for introducing correlations beyond the MPS mean field description. The MPS RPA is based on a state with auxiliary dimension M, but describes correlations that are only possible with matrices of much larger dimension. It is possible that other methods based on Hartree–Fock theory also have analogs in the theory of matrix product states. These would provide an alternative to the traditional method for increasing the accuracy of MPS calculations, which is to increase the auxiliary dimension.

VI. CONCLUSION

In this work, we have drawn on parallels between the product structure of HF theory and MPS to develop analytic equations of motion for time evolution, an MPS RPA for response and dynamical properties, and site-based Green functions that introduce correlations not described by the MPS ground state. In each case, the MPS analog of the Hartree–Fock theory offers potential advantages over standard MPS approaches to the problems while retaining the favorable $\mathcal{O}(kM^3)$ complexity that makes MPS so attractive for numerical calculations. For instance, the MPS equations of motion are formally optimal in the limit of small time-steps and avoid the need for compression. Likewise, the MPS RPA describes a linear combination of matrix product states adapted to the response, unlike the dynamical density matrix renormalization group that works with only a single, averaged state. In addition, the site-based Green functions of the MPS RPA introduce correlations that cannot be encoded in the MPS ground state. The RPA forms the first step on a ladder of increasingly sophisticated methods for introducing correlation in Hartree–Fock theory, and we plan to explore analogous methods for matrix product states and tensor networks in the future.

Acknowledgments

This work was supported by the Cornell Center for Materials Research, the Center for Molecular Interfacing, NSF CAREER, the Camille and Henry Dreyfus Foundation, the David and Lucile Packard Foundation, and the Alfred P. Sloan Foundation. Claire C. Ralph would like to acknowledge the DOE CSGF program for support.

REFERENCES

1. U. Schollwöck, *Ann. Phys.* **326**, 96 (2010).
2. S. White, *Phys. Rev. B* **48**, 10345 (1993).
3. S. Rommer and S. Östlund, *Phys. Rev. B* **55**, 2164 (1997).
4. U. Schollwöck, *Rev. Mod. Phys.* **77**, 259 (2005).
5. A. McLachlan and M. Ball, *Rev. Mod. Phys.* **36**, 844 (1964).
6. D. Pines and P. Nozieres, *Theory of Quantum Liquids: Normal Fermi Liquids*, Westview Press, 1994.
7. H. G. Luo, T. Xiang, and X. Q. Wang, *Phys. Rev. Lett.* **91**, 049701 (2003).
8. G. Vidal, *Phys. Rev. Lett.* **93**, 40502 (2004).
9. S. White and A. Feiguin, *Phys. Rev. Lett.* **93**, 76401 (2004).
10. A. Daley, C. Kollath, U. Schollwöck, and G. Vidal, *J. Stat. Mech.* **2004**, P04005 (2004).
11. P. Schmitteckert, *Phys. Rev. B* **70**, 121302(R) (2004).
12. F. Verstraete, J. Garcia-Ripoll, and J. Cirac, *Phys. Rev. Lett.* **93**, 207204 (2004).
13. K. Hallberg, *Phys. Rev. B* **52**, 9827 (1995).
14. T. Kühner and S. White, *Phys. Rev. B* **60**, 335 (1999).
15. E. Jeckelmann, *Phys. Rev. B* **66**, 45114 (2002).
16. J. Dorando, J. Hachmann, and G. Chan, *J. Chem. Phys.* **130**, 184111 (2009).
17. A. Gendiar, N. Maeshima, and T. Nishino, *Prog. Theor. Phys.* **110**, 691 (2003).
18. F. Verstraete and J. Cirac, arXiv:cond-mat/0407066 (2004).
19. F. Verstraete, M. Wolf, D. Perez-Garcia, and J. Cirac, *Phys. Rev. Lett.* **96**, 220601 (2006).
20. Y.-Y. Shi, L.-M. Duan, and G. Vidal, *Phys. Rev. A* **74**, 022320 (2006).
21. G. Vidal, *Phys. Rev. Lett.* **101**, 110501 (2008).
22. G. Evenbly and G. Vidal, *Phys. Rev. B* **79**, 144108 (2009).
23. E. Stoudenmire and S. White, arXiv:1105.1374 (2011).
24. A. Szabo and N. Ostlund, *Modern Quantum Chemistry*, Dover Publications, 1996.
25. K. Ueda, C. Jin, N. Shibata, Y. Hieida, and T. Nishino, arXiv:cond-mat/0612480 (2006).
26. S. Ramasesha, S. K. Pati, H. Krishnamurthy, Z. Shuai, and J. Brédas, *Synth. Met.* **85**, 1019 (1997).
27. G. E. Scuseria, T. M. Henderson, and D. C. Sorensen, *J. Chem. Phys.* **129**, 231101 (2008).
28. A. Grüneis, M. Marsman, J. Harl, L. Schimka, and G. Kresse, *J. Chem. Phys.* **131**, 154115 (2009).
29. G. Onida, L. Reining, and A. Rubio, *Rev. Mod. Phys.* **74**, 601 (2002).
30. P. Romaniello, S. Guyot, and L. Reining, *J. Chem. Phys.* **131**, 154111 (2009).

FEW-QUBIT MAGNETIC RESONANCE QUANTUM INFORMATION PROCESSORS: SIMULATING CHEMISTRY AND PHYSICS

BEN CRIGER,[1,2] DANIEL PARK,[1,2] and JONATHAN BAUGH[1,2,3]

[1]*Institute for Quantum Computing, University of Waterloo, Waterloo, Ontario, N2L 3G1, Canada*
[2]*Departments of Physics and Astronomy, University of Waterloo, Waterloo, Ontario, N2L 3G1, Canada*
[3]*Department of Chemistry, University of Waterloo, Waterloo, Ontario, N2L 3G1, Canada*

Advances in Chemical Physics, Volume 154: Quantum Information and Computation for Chemistry, First Edition. Edited by Sabre Kais.

I. NUCLEAR MAGNETIC RESONANCE QIP

A. Introduction

Quantum information science is an interdisciplinary area of study combining computer science, physics, mathematics, and engineering, the main aim of which is to perform exponentially faster computation using systems governed by the laws of quantum mechanics. The task of performing quantum algorithms or simulations is referred to as quantum information processing (QIP). More fundamentally, quantum information science provides a new lens through which to view quantum physics. An interesting consequence of this new perspective is that any system possessing an accurate description in terms of quantum mechanics may be simulated by any other quantum system of similar size [1]. Such simulations provide a link between chemistry and quantum information science, through the discipline of quantum chemistry.

In order to accurately predict the results of an atomic or molecular interaction, it is sometimes necessary to formulate a Hamiltonian model for the molecules in question. The Hamiltonian assigns energy values to certain states of the physical system, called basis states. This model is then incorporated into an appropriate equation of motion; either the Schrödinger equation (for the non-relativistic case) or the Dirac equation (for the relativistic case). The number of basis states grows exponentially with the physical size of the system, making the solution of the equation of motion a difficult task for a classical computer.

These interactions can be efficiently simulated on a quantum computer, however, because a quantum operation is effectively performed in parallel across all basis states taken into a superposition. Quantum simulations of the type described in this chapter are thought to comprise a set of attainable milestones for quantum information processing, given that there exist relatively simple quantum systems whose dynamics are not easily simulated using classical computers (e.g., the 10-body Schrödinger equation for electrons in a water molecule) [2].

The remainder of this chapter is organized as follows: We first detail a series of quantum algorithms that are useful in simulation of chemical phenomena, and then describe a class of QIP implementations using nuclear magnetic resonance (NMR) and electron spin resonance (ESR). In conclusion, we discuss some recent experiments and progress toward scalable spin-based implementations.

B. Quantum Algorithms for Chemistry

The ostensible goal of a chemical simulation is to extract a small amount of data about a given process: a rate constant, ground state energy, or other quantity of interest. However, it is often necessary to manipulate a large data structure, such as a full molecular electronic configuration, in order to obtain the desired output. It is the exact class of mathematical problem for which a quantum computer is thought to be superior [1,3], often qualified with the notion that using a quantum system to simulate another quantum system provides a physically elegant intuition. In this section, we describe how quantum information processing can provide the desired data, without an exponential growth in the required computational resources. The algorithms used include straightforward digital simulation of Hamiltonian mechanics using the Trotter expansion, as well as implementations of the adiabatic algorithm to simulate ground state properties of a large class of physical Hamiltonians. These comprise the vast majority of proposed simulation techniques; although they are far from a complete list of QIP architectures.

1. Digital Quantum Simulation

In order to simulate continuous degrees of freedom, such as position and momentum, it is often necessary to *discretize* these degrees of freedom onto a finite space, to ensure that the amount of memory required to store their values is bounded, while also ensuring that the resolution remains high enough to mimic the dynamics of the continuous-variable system under consideration. This method was pioneered by Zalka [4] and Wiesner [5] to simulate Hamiltonians of the form

$$\hat{H} = \frac{\hat{p}^2}{2m} + V(\hat{x}) \tag{1}$$

To accomplish this, the position is discretized as detailed previously, and encoded into a quantum register. In this way, an arbitrary superposition of 2^q $\hat{x}$-eigenstates can be stored in q qubits.

To simulate the evolution under this Hamiltonian, it is useful to perform a Trotter decomposition to first order [6]:

$$\hat{U}_{\text{evol}} = \exp\left[-i\left(\frac{\hat{p}^2}{2m} + V(\hat{x})\right)t\right] \approx \exp\left[-i\frac{\hat{p}^2}{2m}t\right]\exp[-iV(\hat{x})t] \tag{2}$$

This approximation, valid for small t, expresses the evolution under the Hamiltonian to be simulated in terms of operators that are diagonalized in known bases (momentum and position, respectively). Since $\hat{V}$ depends only on $\hat{x}$, it is diagonal in the basis selected. If diagonal operators can be implemented quickly on a quantum computer, what remains is to find an efficient quantum circuit that will transform operators that are diagonal in the $\hat{x}$-basis into operators that are diagonal in the $\hat{p}$-basis. The quantum Fourier transform fulfills this requirement and has been widely studied [7,8] and implemented [9–11]. The effective evolution operator, for a single particle in one dimension, is then

$$\mathrm{QFT}^{\dagger} \exp\left[-i\frac{\hat{D}_{p^2}}{2m}t\right] \mathrm{QFT} \exp\left[-iV(\hat{x})t\right] \tag{3}$$

where $\hat{D}_{p^2}$ is the diagonal operator whose entries are the eigenvalues of $\hat{p}^2$.

Using n q-qubit registers, one can simulate a system of n interacting degrees of freedom. This simulation involves the class of Hamiltonians

$$\hat{H} = \sum_{j=1}^{n} \frac{\hat{p}_j^2}{2m_j} + V(\hat{x}_1, \hat{x}_2, \ldots \hat{x}_n) \tag{4}$$

Here, the index j lists the degrees of freedom in the Hamiltonian, which may correspond to distinct particles or dimensions. The Trotter expansion of the evolution under this Hamiltonian requires one quantum Fourier transform per degree of freedom.

In order to benefit from the efficiency offered by quantum simulation, it is also necessary to ensure that measurement can be achieved in polynomial time. Kassal et al. [12] demonstrated that reaction probabilities, state-to-state transition probabilities, and rate constants can be extracted from a quantum simulation of a chemical process in polynomial time. To accomplish these measurements, the classical algorithm for generating a transition state dividing surface is invoked, subdividing the simulation space into regions corresponding to products and reactants. Measurement of the single bit corresponding to the presence of the wave function in the reactant or product regions can be used in conjunction with phase estimation [13], so that the precision of the transition probability scales as N^{-1}, where N is the number of single-bit measurements.

Using a limited capacity for quantum control, it is possible to simulate quantum dynamics, with immediate applications to small numbers of reacting atoms. Note that no Born–Oppenheimer approximation [14] has been used. Indeed, the quantum algorithm is more efficient without this approximation, because the Born–Oppenheimer approximation requires the calculation of potential energy surfaces at many points throughout a given simulation. The realization of chemical simulations by the preceding algorithm, while scalable, requires hundreds of qubits in order to simulate a few particles with sufficient spatial precision, and $\sim 10^{12}$ elementary quantum gates to obtain sufficient precision in time.

To reduce the computational requirements for quantum simulations, second-quantized Hamiltonians (those expressed in terms of mode occupation numbers, instead of position/momentum operators for individual particles) are used to represent systems of interest. Second quantization presents four immediate advantages for the design of simulation algorithms. First, the number of bits required to represent an occupation number for a given mode is usually far smaller than that required to accurately represent a numerical value of position or momentum. This is due to the integer nature of occupation numbers. This reduction in overhead is especially evident in fermionic systems, where the Pauli exclusion principle results in occupation numbers that can be stored in single bits. Second, second-quantized Hamiltonians allow for the elimination of modes which are irrelevant to the quantity being measured. Also, simplifying approximations in second-quantized Hamiltonians are easy to formulate. Take, for example, a restriction to local interactions for electrons. In a second-quantized system, this can be accomplished by adding an on-site interaction term of the form $\hat{H}_{\text{int}} = U \sum_j n_{j,\uparrow} n_{j,\downarrow}$ to the Hamiltonian, where j enumerates the modes and n is an occupancy number. Since the modes $\{j, \uparrow\}$ and $\{j, \downarrow\}$ can each be represented on one qubit, evolution under this interaction Hamiltonian can be expressed as a simple one-parameter, two-qubit gate

$$\begin{bmatrix} 1 & 0 & 0 & 0 \\ 0 & 1 & 0 & 0 \\ 0 & 0 & 1 & 0 \\ 0 & 0 & 0 & e^{-iUt} \end{bmatrix} \tag{5}$$

One can compare this to the addition of a contact interaction to a first-quantized simulation. The interaction term in the Hamiltonian can be expressed simply, as $U \sum_{j<k} \delta(\vec{r}_j - \vec{r}_k)$. However, evolution under this term requires comparison of the coordinates $\vec{r}_j$ and $\vec{r}_k$, each of which are represented on an n-bit register. Operations which perform this comparison are far more complicated than the two-qubit gate above. Finally, second-quantized Hamiltonians are often sparse in canonical bases, which permits them to be simulated more efficiently by the algorithms presented in next section.

One of the first applications of second-quantized simulations was to the problem of graph traversal [15], which is the minimization of the number of steps taken in an abstract discrete space to locate a state with a special property, given limited information about the global structure of the space. The authors solve this problem by defining a Hamiltonian proportional to the adjacency matrix of the graph, and developing an algorithm for approximating the evolution of a Hamiltonian H acting on at most 2^m local systems, each of which is coupled to at most k others, acting over time t with error ϵ using $O(k^2t^2/\epsilon)$ queries to a function which returns partial data about the Hamiltonian, and $O(k^2t^2m/\epsilon)$ elementary quantum gates.

The computational cost of a simulation in terms of the number of queries is important, but not a complete description of the difficulty involved in performing the simulation. This is because there is an unknown amount of computational work required to calculate the response to a black-box query, which is unaccounted for. In 2007, Berry et al. [16] considered a different Hamiltonian simulation scheme in which Suzuki–Trotter formulae [17] of order k are used instead of the first-order formula used in Eq. (2). The authors were able to reduce the query complexity to $O\left(\log^*(n)t^{1+\frac{1}{2k}}\right)$, where n is the number of qubits, for Hamiltonians with at most a constant number of non-zero entries in each column. Here, $\log^*(n) \equiv \min\left\{r|\log^{(r)}(n) < 2\right\}$, where $\log^{(r)}$ is the iterated logarithm. The authors show that this scaling is near-optimal, as no generic simulation can scale sublinearly with t.

This scaling law makes the study and implementation of second-quantized simulations an attractive proposition, and much recent work has been done to improve this reuslt. Quantum walks can result in strictly linear scaling of the query complexity in time [18,19], higher-order Suzuki–Trotter formulae can be applied to the simulation of time-dependent Hamiltonians [20], and adaptive time steps can be used to increase accuracy per time step [21]. Simulations of the type discussed above are likely candidates for medium-to-long-term practical applications to chemistry and other disciplines.

2. *Adiabatic Quantum Simulation*

Given that the number of quantum gates used in the preceding simulation is prohibitive for implementation in the near future, it is preferable to search for alternatives to gate-based quantum computation for the purpose of simulating certain chemical processes. One such alternative is adiabatic quantum computation [22]. This method of quantum information processing is naturally suited to problems that relate to ground states of complicated Hamiltonians, for example, the problem of protein folding [23].

The central mechanism of adiabatic quantum computation is the transformation of a prepared state to a final desired state, by the gradual alteration of the system Hamiltonian. The algorithm requires a Hamiltonian $\left(\hat{H}_A\right)$ whose ground state is easily found, and a Hamiltonian $\left(\hat{H}_P\right)$ whose ground state encodes the solution to an appropriate mathematical problem. The adiabatic algorithm consists of three steps: prepare the ground state of $\hat{H}_A$; vary the Hamiltonian according to

$$\hat{H}(t) = \left(1 - \frac{t}{\tau}\right)\hat{H}_A + \frac{t}{\tau}\hat{H}_P \tag{6}$$

with τ sufficiently large so that the process is approximately adiabatic; and perform an appropriate measurement on the final state, which encodes the solution to the problem.

This method can be straightforwardly applied to energy minimization problems, such as protein folding. Perdomo et al. [23] produced an instance of the adiabatic algorithm that is designed to find low-energy configurations for lattice-based hydrophobic-polar protein models. Although it is unknown whether this model can be solved for arbitrary cases by any computer, quantum, or classical, it is reasonable to conclude that a quantum algorithm would be of use, as the process of protein folding is inherently quantum-mechanical [24,25].

The hydrophobic-polar model of protein structure [26] is a simplified model of protein folding, taking the individual peptides to be either hydrophobic (H) or polar (P). These peptides are distributed in a chain on a 2D or 3D lattice, and an energy is assigned to the configuration of the chain by counting the number of adjacencies between P peptides. This energy assignment based on spatial coordinates is mapped to a Hamiltonian consisting of three terms: an on-site term that penalizes conformations that have multiple peptides on the same site, a pairwise interaction term that rewards assignments with adjacent hydrophobic peptides, and a structure constraint Hamiltonian that penalizes spatial conformations that do not form chains. Because the positions of the peptides are the quantities of interest, a measurement in the computational basis is sufficient; no advanced measurement technique is required.

We have seen that quantum information processing methods can be of great benefit to theoretical chemistry, providing properties of ground states, as well as reaction parameters. In the next section, we describe in further detail the requirements for a quantum computer. This content is followed by a report on the current state of QIP implementations using nuclear and electronic spins in an external magnetic field. These NMR and ESR experimental approaches are well suited to digital QIP, but not adiabatic methods, so we will focus on digital methods in the remaining sections.

C. NMR and the DiVincenzo Criteria

In order to perform the simulation algorithms mentioned earlier, there must first be a system that displays the unique properties of quantum mechanics, while allowing experimental control and measurement. This idea has been formalized into five well-accepted requirements known as the DiVincenzo criteria [27]: (1) a scalable physical system with well-characterized qubits, (2) the ability to initialize the register to a simple fiducial state, such as $|0\rangle^{\otimes n}$, (3) a universal set of quantum gates, (4) a qubit-specific measurement capability, and (5) decoherence times much longer than the average gate operation time. These criteria are partially satisfied by NMR QIP, where algorithms are implemented on nuclear spins in molecules, subjected to a large, constant external magnetic field and radio-frequency (RF) pulses [28]. This system has the advantage of being a good testbed for the development of quantum control techniques in the few-qubit regime (up to $\sim$12 qubits as of this writing [29]). Next, we describe the relationship between the desired properties of a quantum information processor and the particular features of the NMR system.

1. Scalability with Well-Characterized Qubits: Spin-1/2 Nuclei

In order for a quantum algorithm or processor to be considered scalable, the resources it requires must grow only polynomially with the size of the system.

Any implementation of QIP using k two-level subsystems (referred to as qubits) is scalable in the total energy of the system, or equivalently, the precision with which the energy is measured. NMR QIP employs such an architecture, as each spin-1/2 nucleus has two well-defined energy levels in a magnetic field, described by a two-dimensional Hamiltonian

$$\hat{H} = \frac{1}{2}(\hbar\gamma B_0 + \delta)\hat{Z} = \frac{1}{2}(\hbar\gamma B_0 + \delta)\begin{bmatrix} 1 & 0 \\ 0 & -1 \end{bmatrix} \tag{7}$$

where B_0 is the external magnetic field, γ is the nuclear gyromagnetic ratio, and δ is the chemical shift term, imposed on each nucleus by its local molecular environment. When nuclei have distinct δ, they provide qubits that can easily be individually addressed through RF pulses. If this is not the case, more advanced techniques are required to address the qubits.

NMR spectroscopy has been widely used in analytic chemistry and other disciplines for decades before quantum information research began, resulting in a large body of literature dedicated to measuring and modeling the NMR spectra for a given molecule [30,31]. The techniques developed in conventional NMR can be used to determine the spin Hamiltonian, and thus characterize qubits, to a sufficient level of precision for quantum information. However, the number of qubits in NMR QIP systems is constrained in a practical sense, because the chemical shifts do not grow with the number of qubits, limiting the available frequency space for qubit addressing. Using solid-state NMR techniques, it is possible to use a magnetic field gradient to render the qubits distinguishable without the need for distinct δ. However, it is difficult in practice to achieve gradients large enough to provide frequency resolution for nearby spins without inducing decoherence through gradient non-uniformity.

Liquid-state NMR experiments have successfully demonstrated universal control over NMR systems ranging from a few qubits in the late 1990s and early 2000s [32,33] to up to 12 qubits more recently [29]. Although this number is not expected to appreciate greatly, NMR QIP experiments with up to $\sim$20 qubits are likely achievable. The few-qubit experiments we have today are sufficient for testing small quantum simulations and general concepts of QIP. Perhaps more importantly, the techniques developed in liquid- and solid-state NMR QIP are applicable to other, more scalable quantum information architectures that surpass the capability of the best classical computers could be performed.

2. *Initialization: The Pseudopure State*

To perform a quantum algorithm, we need the register to begin in a known state such as $|0\rangle^{\otimes n}$. In NMR, the Boltzmann distribution for a single spin has a bias toward the ground state of $\epsilon \sim \tanh(\hbar\gamma B_0/k_B T)$. In order to emulate initialization of the register into a pure state, it is possible to produce a *pseudo-pure* initial state,

$$\rho_{\text{pseudopure}} = \frac{1-\alpha}{2^n}\hat{I} + \alpha\,(|0\rangle\langle 0|)^{\otimes n} \tag{8}$$

where α is determined by ϵ and the experimental procedure used for preparation. For typical values of γ and B_0, ϵ and α are on the order $10^{-(5-6)}$. Unitary operations due to applied pulses will alter the $(|0\rangle\langle 0|)^{\otimes n}$ term *just as they would a pure state*, while leaving the identity term unchanged (note, the identity term is not measurable in NMR QIP, as it does not contribute to the total spin magnetization). As long as ϵ and α are large enough to produce a measurable NMR signal, the final pseudo-pure state can be measured.

As the size of the system increases, the methods that have been used to produce pseudo-pure initial states will result in values of α that decay exponentially in the system size, precluding scalability in liquid state NMR QIP. However, experimental procedures such as dynamic nuclear polarization and algorithmic cooling, to be discussed in Section II.C, could increase nuclear polarizations to values near unity in suitable solid-state systems that include electron spins.

3. *A Universal Set of Quantum Gates: RF Pulses and Spin Coupling*

It is possible to implement an arbitrary n-qubit unitary gate using only one- and two-qubit gates [34]. Single-qubit gates are realized in NMR QIP by exerting a time-dependent, radio frequency magnetic field over the sample in the $\hat{X}-\hat{Y}$ plane (the $\hat{Z}$ direction being determined by the background static magnetic field), which implements the rotation operators $\hat{U}_x(\alpha)$ and $\hat{U}_y(\beta)$ (or a rotation along any axis in the $\hat{X}-\hat{Y}$ plane of the Bloch sphere), where α and β are arbitrary angles determined by the strength and duration of the RF pulses. Placing unitaries of this type in sequence can produce any possible single-qubit operation. Conditional logic (also known as *if-then* logic) in NMR QIP is achieved by allowing the nuclear magnetic state to evolve under coupling mechanisms present in the molecule. These coupling mechanisms are direct dipolar coupling and electron-mediated dipolar coupling (i.e., indirect dipolar coupling), the latter also known as J-coupling.

One can also express the universality of an implementation of QIP directly in terms of the Hamiltonian, in contrast to examining the unitary operators generated by the Hamiltonian. This is more useful for quantum simulations, because the class of Hamiltonians that can be simulated by a physical system is easily obtained from the form of the natural and control Hamiltonians [35]. Universal control over a system, in this sense, is the ability to simulate any Hamiltonian with the same number of degrees of freedom. Importantly, any coupled Hamiltonian, together with universal single-qubit control, can be used to simulate any other coupled Hamiltonian that has the same degree of connectivity [35].

Because all spin-1/2 nuclei in a molecule possess magnetic moments, they will possess a pairwise direct dipolar coupling term in the NMR Hamiltonian:

$$\hat{H}_{jk}^{DD} = -\frac{\mu_0}{4\pi}\frac{\hbar\gamma_j\gamma_k}{r_{jk}^3}(3(\vec{\sigma}_j \cdot \vec{e}_{jk})(\vec{\sigma}_k \cdot \vec{e}_{jk}) - \vec{\sigma}_j \cdot \vec{\sigma}_k) \tag{9}$$

where r_{jk} is the distance between the two nuclei, $\vec{\sigma} = [\hat{X},\ \hat{Y},\ \hat{Z}]$ and $\vec{e}_{jk}$ is a unit vector along the line connecting the two nuclei. In the liquid state, Θ_{jk} (the angle between $\vec{e}_{jk}$ and the external $\vec{B}$-field axis) is, to a good approximation, uniformly distributed by fast molecular tumbling. This results in zero average interaction strength. Therefore, direct dipolar coupling is only useful in systems in which the Θ_{jk} are not uniformly distributed, such as solid-state NMR (where all of the Θ_{jk} are fixed by crystal orientation) and liquid crystal NMR (where the Θ_{jk} are distributed non-uniformly). In addition to the direct dipolar coupling, there exists an indirect coupling, known as the J-coupling. It is a relatively weak coupling mediated by the electron cloud in the molecule, according to

$$\hat{H}_{jk}^{J} = 2\pi\vec{\sigma}_j \cdot \hat{J}_{jk} \cdot \vec{\sigma}_k^{\mathsf{T}} \tag{10}$$

where $\hat{J}_{jk}$ is a 3×3 real matrix. This coupling term does not vanish under molecular motion, since the electrons and nuclei are co-moving. In isotropic liquids, it has the Heisenberg form $\vec{\sigma}_j \cdot \vec{\sigma}_k$, which reduces to $\hat{Z}_j\hat{Z}_k$ when the two spins have a chemical shift difference much larger than the value of J; this is the typical generator of a two-qubit gate in liquid-state NMR. The variety of natural coupling terms renders QIP possible in many different molecules, in the liquid, solid, or liquid crystalline states.

4. *Measurement: Free Induction Decay*

The apparatus (an RF coil) used to implement rotations about axes in the $\hat{X} - \hat{Y}$ plane can also be used to detect ensemble magnetization signals from the sample in the $\hat{X} - \hat{Y}$ plane. In order to translate the logical state of an NMR QIP register along the $\hat{Z}$-axis into an $\hat{X}$ observable value, a readout pulse of angle $\frac{\pi}{2}$ is applied about the $\hat{Y}$-axis at each qubit resonance frequency. This results in a signal, proportional to the pseudopure parameter α (see Section I.C.2), from which the logical state can be extracted. This process does not result in projective measurement, but still allows universal computation including full state tomography [36]. In addition, the expectation value of any Hermitian operator can be obtained in a single shot, allowing the measurement of arbitrary operators by quadrature detection [31].

5. *Noise and Decoherence: T_1/T_2 versus Coupling Strength*

In order to perform an algorithm, the time used to implement the appropriate pulses must be much shorter than the characteristic timescale of decoherence. In addition to the noise related to imperfect implementation of the gates (these are typically

TABLE I
Characteristic Timescales for Control in Solid-State NMR QIP

	C_1(Hz)	C_2(Hz)	C_m(Hz)	T_2^*(ms)	T_1(s)
C_1	5693	237	828	2.4	162
C_2		1748	1020	2.6	326
C_m			−3358	3.1	314

Notes: These timescales, using the ^{13}C-labeled malonic acid molecule, can be deduced from the dipolar coupling strengths (off-diagonal elements in second and third columns) and chemical shifts (diagonal elements), and compared with relaxation and dephasing times (rightmost columns) [37]. Note the description of decoherence in terms of T_2^*, which combines decoherence from inhomogeneous and homogeneous sources. Much longer $T_2 \approx 100$ ms was observed with suitable dipolar refocusing/dynamical decoupling sequences applied [38].

"coherent" errors), the unavoidable interaction with the surrounding environment leads to true decoherence. In NMR, the two main mechanisms for this decoherence are thermal equilibration (called *relaxation*) and dephasing, characterized by timescales T_1 and T_2^*, respectively. T_2^* represents the ensemble, inhomogeneous dephasing time of the system, whereas T_2 is the intrinsic decoherence time of each single quantum subsystem. T_2^* can often be increased to a larger effective timescale T_2 by applying decoupling pulses as described in Section II.B. In NMR QIP, the coupling terms between qubits are usually the smallest terms in the Hamiltonian, leading to relatively slow two-qubit gates. The solid-state malonic acid system provides an example, shown in Table I; the carbon–carbon coupling ranges from 200 Hz to 1 kHz and indicates that the associated controlled-$\hat{Z}$ gate requires a time of about 0.5–2.5 ms, allowing the implementation of several two-qubit gates before a single decay period T_2^* has elapsed. This suffices for simple algorithms, but a larger ratio of decoherence time to gate time is required in order to implement more complex algorithms; in particular, to implement error correction. If the noise is below a threshold [39], it is possible to use fault-tolerant methods to counteract its effects. If the noise exceeds this threshold, control must be improved until the threshold is attained. It is possible to assess the noise level using benchmarking [40,41]. In an NMR system where control was optimized, it has been possible to reach an error per gate rate of 10^{-4} for single-qubit gates and 10^{-3} for two-qubit gates [42].

Ongoing improvements to the implementation of one- and two-qubit gates in magnetic resonance QIP have arisen, in part, due to the following:

- Optimization of RF pulses to produce high-fidelity unitary operations
- Isolation of the system of interest from unwanted degrees of freedom
- Exploitation of the unique properties of the electron spin.

In the following sections, we describe recent progress in these directions.

II. QUANTUM CONTROL IN MAGNETIC RESONANCE QIP

A. Advances in Pulse Engineering

In NMR QIP, pulse engineering is the practice of developing control techniques for either generating coherent transfer from an initial spin state to a desired spin state (state-to-state transfer) or producing an effective Hamiltonian that implements a desired unitary gate $\hat{U}$ (unitary propagator) by manipulating the external RF field. This can be achieved by augmenting the internal Hamiltonian with a control Hamiltonian (in the laboratory frame):

$$\hat{H}_c(t) = \sum_{k,l} \frac{\omega_k(t)}{2}[\cos(\Omega_k t + \phi_k(t))\hat{X}_l + \sin(\Omega_k t + \phi_k(t))\hat{Y}_l]$$

where $\omega_k(t)$ and $\phi_k(t)$ denote the amplitude and the phase applied to spin l at a transmitter frequency Ω_k.The time evolution is typically divided into a set of N steps. At each step j, the evolution is given by the unitary propagator $\hat{U}_j = \exp\{-i\Delta t\hat{H}_j(t)\}$, where $\hat{H}_j(t)$ is the total Hamiltonian with a control set $\{\omega_k^j,\ \phi_k^j\}$. The final density operator is given by $\rho(T) = \hat{U}_N \cdots \hat{U}_1\rho_0\hat{U}_1^\dagger \cdots \hat{U}_N^\dagger$. The task of pulse engineering is to find a set of $\{\omega_k^j,\ \phi_k^j\}$ so that the resulting dynamics are sufficiently close to those desired. The quality of the pulse is evaluated by a fidelity function Φ that is proportional to overlap (inner product) between the obtained operator and the target operator.

Pulse engineering techniques must address experimental limitations for practical applications: qubit selectivity due to finite frequency bandwidth, instrumental errors (such as static and RF field inhomogeneity), miscalibration of pulse power or duration, and frequency offset.

Pulse design must take these artifacts into account in order to achieve robust pulses with high fidelity on the ensemble. This causes a substantial increase in the complexity of the pulse design problem. For example, errors caused by chemical shift dispersion, RF inhomogeneity, and RF miscalibration can be suppressed if the parameters are sampled over a range of discrete values determined by the uncertainty. Then the total fidelity function to be optimized can be measured as $\Phi_{\text{tot}} = \sum_\alpha \Phi(x_\alpha)$, where x_α is a particular value of some parameter 'seen' by a fraction of spins (e.g., RF field amplitude). In this case, the optimization problem becomes significantly more difficult, because the effective parameter space is much larger.

Over the past 30 years, numerous techniques have been developed for control in NMR QIP. Traditionally, average Hamiltonian theory (AHT) has been a powerful tool that provides intuitive guidelines for constructing pulse sequences in relatively simple cases [43]. In particular, composite pulses [44], adiabatic pulses [45–48], and shaped pulses [49] were introduced during the earlier development of NMR QIP to better compensate for static and RF field inhomogeneity by increasing the number of degrees of freedom in the pulse shape. The long pulse times produced by these techniques lead to greater decoherence and relaxation effects, and interference of selective pulses simultaneously applied to different spins. Also, not all errors can be corrected using these techniques [50]. Strongly modulated pulses [50] average out unwanted evolution by using strong control fields to drive the spins, so that a desired unitary operator can be synthesized directly. This method uses high-power pulses that decrease the required pulse duration, and hence reduce the effect of decoherence and relaxation. Penalty functions on power, frequency, and total pulse time can be applied to obtain more experimentally acceptable solutions. Moreover, it allows for the incorporation of robustness against errors caused by static or RF field inhomogeneities. Strongly modulated pulses with ideal fidelities on the order of 99% are performed for universal control of liquid-state NMR system with up to six qubits [50]. The method was also studied in a three-qubit solid-state NMR system, single crystal malonic acid. Universal control with ideal fidelity $\sim$98% was achieved in the strong-coupling regime of the direct nuclear dipole–dipole couplings, in which evolution under the internal Hamiltonian is more complex than those of the typical liquid-state NMR system [38]. In addition, the transformations generated by the strongly modulated pulses were shown to be robust against the ensemble inhomogeneities that are present in solid-state NMR systems [38]. The cost of this method (as well as the optimal control methods described next) is that the full system dynamics must be simulated in the course of finding pulses, so that even using clever methods such as gradient-based approaches to improve efficiency does not render the method scalable to arbitrary numbers of qubits.

Recently, optimal control theory (OCT), originally developed for problems in engineering [51–53], has been used for systematic optimization of pulse designs in NMR QIP. Analytical solutions to the problem of time-optimal realization of unitary operations can be obtained by formulating a variational principle to reduce the problem to a set of first order ordinary differential equations [54,55]. For more than two qubits, the analytical approach is intractable, and numerical methods are required.

For larger systems, the majority of control methods for optimizing the efficiency of coherent transfer are based on a gradient approach, such as the Gradient Ascent Pulse Engineering (GRAPE) algorithm [56]. The GRAPE algorithm can be summarized as follows: It begins with an initial guess for a set of control parameters $u_k(j)$ for all time-steps $j \leq N$. Then in each iteration the algorithm evaluates $\delta\Phi_{\text{tot}}/\delta u_k(j)$ for $j \leq N$ and updates $u_k(j)$:

$$u_k(j) \rightarrow u_k(j) + \epsilon \frac{\delta\Phi_{\text{tot}}}{\delta u_k(j)}$$

where ϵ is a step parameter.

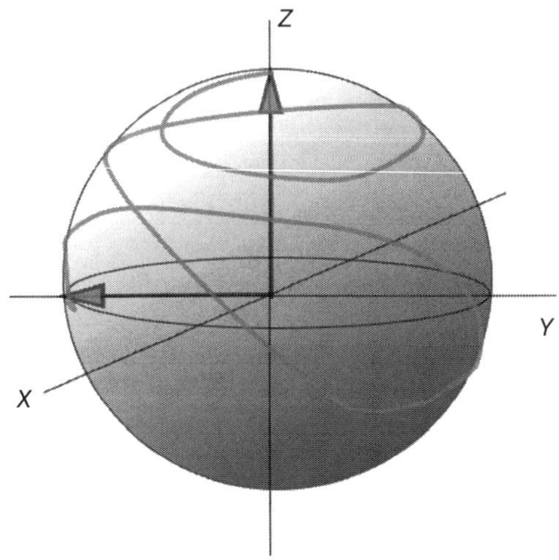

Figure 1. Trajectory of the spin magnetization vector on the Bloch sphere for a pulse found by the GRAPE method [56]. The pulse is designed for a $\frac{\pi}{2}$ rotation around the $\hat{X}$-axis.

The iteration continues until the improvement in the performance index Φ_{tot} is smaller than a chosen threshold value. GRAPE also evaluates penalty functions to reflect realistic experimental constraints. The gradient approach enables the GRAPE algorithm to calculate full time evolutions much faster than conventional (brute-force) numerical difference methods. Consequently, the number of pulse parameters to be optimized can be orders of magnitude larger than in conventional approaches. A number of recent NMR QIP experiments [57–63] have used GRAPE pulses to achieve high-control fidelities. However, just as with any optimal control method, the GRAPE algorithm cannot deterministically locate global minima but instead finds local minima in the search space. Therefore, the ultimate fidelities that can be achieved are limited by the initial guess, even if unit fidelity pulses are possible in principle. In addition, pulse shapes determined by gradient ascent are often convoluted, rendering little information to the experimentalist seeking to further improve fidelity. Figure 1 illustrates an example of a path taken by the spin magnetization vector on the Bloch sphere during a GRAPE pulse designed for a $\frac{\pi}{2}$ rotation around the $\hat{X}$-axis.

Another well-developed numerical method for quantum control is that of Krotov [64–67]. The Krotov-based method allows large changes in control parameters from one iteration to the next, and immediately exploits all available information at each time-step, monotonically improving the objective fidelity functional at each iteration. Maximov et al. [53] analyzed the Krotov-based algorithm in the context of NMR spectroscopy and compared it to GRAPE, concluding that the Krotov-based algorithm consumes less computational resources per iteration and is much better than GRAPE for making an initial guess toward a global maximum.

However, the larger step size in the Krotov-based method limits efficiency as the solution gets closer to the desired fidelity. An open question centers on whether high efficiency can be obtained by combining the two methods, using the Krotov algorithm to quickly prepare a good initial pulse sequence to load into GRAPE for final refinement [53].

As mentioned earlier, a major drawback with these methods is that computation cost grows quickly with increased system size. As quantum processors become larger, new techniques will need to be developed to render high-fidelity control algorithms scalable, perhaps by allowing these OCT methods to be applied within blocks of nearby qubits (in frequency space), using traditional pulse design techniques between blocks [68].

B. Advances in Dynamical Decoupling

In order to achieve high-fidelity control in NMR QIP, one must address system–environment interaction, as well as unintended evolution under internal couplings between spins. The coupling interaction is "always on" and therefore must be suppressed to perform high-fidelity single qubit gates. This process is called refocusing and is relatively simple in liquid-state NMR, because the dominant coupling term is of the form $\hat{Z}_j\hat{Z}_k$. Solid-state NMR Hamiltonians have more complex coupling terms and are therefore more difficult to refocus. Techniques based on average Hamiltonian theory, such as magic angle spinning [69], Lee and Goldburg decoupling [70], and multiple-pulse techniques [43], have proven useful in refocusing these terms. Furthermore, recent developments in optimal control theory (see Section II.A) allow the refocusing of unwanted internal interactions for arbitrary coupling forms, as long as the Hamiltonian is well-known.

Hereafter, we narrow the discussion to system–environment, or dynamical decoupling (DD). A DD scheme is a sequence of control fields applied to a system for some time with the objective of increasing coherence time by attenuating the system–environment interaction. Starting from Hahn's discovery of spin-echo in 1950 [71], many methods to accomplish this have been actively studied. One seminal example is the Carr–Purcell–Meiboom–Gill (CPMG) sequence [72], which is successful at suppressing single-axis, low-frequency noise. The performance of realistic DD sequences is limited by experimental imperfections (see Section II.A), so it is important to design a DD scheme that is more robust to the most relevant errors.

In traditional DD, a pulse sequence is periodically applied to reduce undesired terms of the system–bath interaction Hamiltonian, known as periodic dynamical decoupling (PDD). These sequences were improved in 2005 by Khodjasteh and Lidar [73], who introduced concatenated DD (CDD) pulse sequences, such as $p_{n+1} = p_n\hat{X}p_n\hat{Z}p_n\hat{X}p_n\hat{Z}$, where $p_0 = \tau_0$ is a delay between pulses and n is a concatenation level. CDD is significantly more efficient at decoupling than PDD with equal pulse numbers and is more robust to both random and systematic control errors. CDD can increase fidelity to $\sim (J\tau_n)(\beta\tau_n/2^n)^n$ where J and β are system–environment coupling strengths, τ_n is the duration of the sequence p_n, and n is the concatenation level. The caveat of this method is that due to the nature of concatenation the number of necessary pulses grows exponentially (4^n) in the concatenation level.

In 2007, Uhrig introduced an optimized dynamical decoupling sequence based on gradient moment nulling in NMR [74], called Uhrig dynamical decoupling (UDD) [75–77]. UDD was first introduced for a specific spin-boson model, and was later discovered [76,77] to be system independent. What distinguishes UDD from conventional DD schemes is that the time delay blocks between π pulses are not equal. Uhrig showed that splitting the total time interval t into smaller intervals $0, \delta_1 t, \ldots, \delta_n t$ (where $\delta_j = \sin^2[\pi j/(2n+2)]$), can suppress decoherence up to order t^{n+1} without exponential cost. Moreover, Uhrig's simulation showed that the performance of decoupling from the environment becomes independent of system–bath coupling strengths for long sequences. One drawback of UDD is that it works for only single-axis noise, and cannot, for example, simultaneously suppress T_1 and T_2 processes.

In an attempt to refocus the effects of noise about multiple axes, Uhrig concatenated the UDD sequence (CUDD) [78]. The total number of pulses to suppress decoherence and relaxation is proportional to 2^n, decreasing the resources by 2^n from CDD sequences. West et al. further improved on this by creating a new UDD-based sequence that suppresses decoherence and relaxation to order n using $(n+1)^2$ pulse intervals [79]. Their scheme integrates two sequences, T_1-correcting UDD and T_2-correcting UDD, and is known as quadratic UDD (QDD).

UDD-based methods are optimal when the noise has a sharp high-frequency cutoff [76,80–83]. However, in a low-frequency-dominated noise environment, conventional CP (or CPMG) is preferred [81–83]. Borneman et al. [84] pointed out that, although the CPMG sequence is inherently tolerant of field inhomogeneities and pulse calibration errors, the robustness of the sequence is limited by the quality of the RF pulses used. They adapted the GRAPE algorithm (see Section II.A) and applied an iterative optimization method to design general refocusing pulses (universal π rotations around the $\hat{Y}$-axis) with high fidelity over a wide range of resonance offset frequencies and RF amplitudes. Borneman et al. were able to find a pulse that refocuses 99% of the initial magnetization over a range of frequency offsets of ± 10 kHz (four times greater than the maximum RF amplitude) for uniform RF fields. For RF inhomogeneity of $\pm 10\%$, they were able to find a pulse that refocuses 98% of the initial magnetization over a frequency range of ± 8 kHz (3.2 times the maximum RF amplitude), an improvement over any previously published refocusing pulses of similar duration and maximum RF amplitude when applied in a CPMG sequence.

Souza et al. [85] compared the performance of a number of pulse sequences in order to find a decoupling scheme that is robust against pulse imperfections or control errors. They considered two possible approaches to compensate the imperfections: use composite pulses that the error correction is done "inside" the pulse (robust pulse), or create the sequence where noise from one pulse is averaged out in subsequent pulses (self-correcting sequence). In the experiment, a standard pulse (non-robust) and the Knill pulse [86] $\pi_{\pi/6+\phi} - \pi_{\phi} - \pi_{\pi/2+\phi} - \pi_{\phi} - \pi_{\pi/6+\phi}$, which is robust against flip-angle and off-resonance errors, are inserted into the CPMG, periodic XY-4 [87], and CDD sequences. Then decay times of the magnetization as a function of the duty cycle (the total pulsing time in the sequence divided by total duration of the sequence) are analyzed and compared for each case. A new sequence called Knill dynamical decoupling (KDD), which is achieved by inserting delays between the π pulses of the Knill pulse, is also examined. For small duty cycles, sequences with non-robust pulses are superior, due to the shorter cycle time of the nonrobust pulse sequence when constant duty cycles are compared. However, the performance of these sequences saturates or decreases with increasing duty cycle. Comparatively, robust pulses continue to increase in performance for large duty cycles. KDD is found to have the best performance for large duty cycles, and comparable to sequences without robust pulses for small duty cycles.

C. Control in the Electron-Nuclear System

Many of the techniques used to control spin-$1/2$ nuclei in NMR can also be applied to electrons, comprising electron spin resonance (ESR). When the γ of an electron is $\approx$660 times greater than that of a proton, it not only leads to a much larger Larmor frequency and higher polarization, but also faster decoherence and relaxation than nuclear states. The coupling between electron and nuclear spins is governed by the hyperfine interaction

$$\hat{H}_{\mathrm{hf}} = \vec{\sigma}_E \cdot \mathcal{A} \cdot \vec{\sigma}_N^{\mathsf{T}} \tag{11}$$

where E denotes the electron, N denotes the nucleus, and $\mathcal{A}$ denotes the hyperfine coupling tensor. The combined electron-nuclear solid-state spin system is appealing for QIP because it is possible to exploit the properties of each type of spin: electron spins possess higher polarization for initialization and readout, and can be manipulated on faster timescales, while nuclear spins are ideal for long-term storage of coherent states [88].

Although ESR provides much better signal-to-noise ratio than NMR, it usually suffers from broader line widths, as a result of faster decoherence. For an electron-nuclear spin system, one is required to characterize the internal Hamiltonian as precisely as possible in order to achieve high-fidelity control. However, the nuclear Zeeman term in the Hamiltonian is small relative to terms involving the electron. Also, it is often not possible to measure directly in an ESR experiment due to spin selection rules [89]. Moreover, the ESR line width can be comparable to the hyperfine splitting, which in turn makes it difficult to study the hyperfine interaction term. Fortunately, the well-established magnetic resonance techniques of electron spin echo envelope modulation (ESEEM) [89,90] and electron nuclear double resonance (ENDOR) spectroscopy [89,91,92] can precisely quantify nuclear transition frequencies and hyperfine couplings. In ENDOR, both NMR and ESR transitions are driven directly. Thus, ENDOR spectroscopy is a promising technique for controlling electron-nuclear systems for QIP. However, it should be noted that ENDOR experiments require additional RF hardware, and nuclear spin flip times are limited by the small nuclear gyromagnetic ratio and the strength of the RF field that can be achieved.

The anisotropy of hyperfine interaction allows controlling nuclear spins solely by irradiating electron spin transitions [93]. The indirect control technique is advantageous: it simplifies the instrument design because an additional RF interface is not needed, and faster control of a nuclear spin is achievable when the hyperfine coupling strength exceeds the Larmor frequency of the nucleus in a given external field [94]. The following section reviews this approach and recent experiments demonstrating universal control in electron-nuclear systems.

1. *Indirect Control via the Anisotropic Hyperfine Interaction*

If the coupling frequency due to hyperfine interaction is larger than the nuclear Larmor frequency, efficient nuclear control can be obtained by manipulating the electron [94], producing an effective magnetic field seen by the nucleus that depends on the electron spin state. With the anisotropic coupling, the nucleus sees an effective field

$$\vec{B}_{\text{eff}} = \left(B_0 \pm \frac{A}{2\gamma_N}\right)\hat{z} \pm \frac{B}{2\gamma_N}\hat{x} \tag{12}$$

where A is the zz-component of the hyperfine tensor [89] and B is $\sqrt{\mathcal{A}_{zx}^2 + \mathcal{A}_{zy}^2}$, the component of the hyperfine tensor in the $\hat{x}$-direction on the nucleus, in a specially chosen frame. B_0 is the static magnetic field and γ_N is the nuclear gyromagnetic ratio. The $\pm$ sign takes the value $+$ when the electron spin is parallel to the external field, and $-$ when it is antiparallel. Any rotation of the Bloch sphere can be generated by repeated rotations about these two distinct axes. Universal single-qubit control of the nuclear spin can then be achieved indirectly, because the free precession of the nucleus about these axes does not require RF pulses resonant with nuclear transitions. Hodges et al. implemented the GRAPE algorithm (see Section II.A) and showed universal control of a 1e-1n system (malonic acid radical) [93]. In [88], Zhang and others also used the GRAPE method to perform an entangling gate between two nuclear spins in a $1e{-}2n$ system (a singly ^{13}C-labeled malonic acid radical). Schematics of the malonic acid radical, an energy level diagram showing creation and detection of double-nuclear coherence,

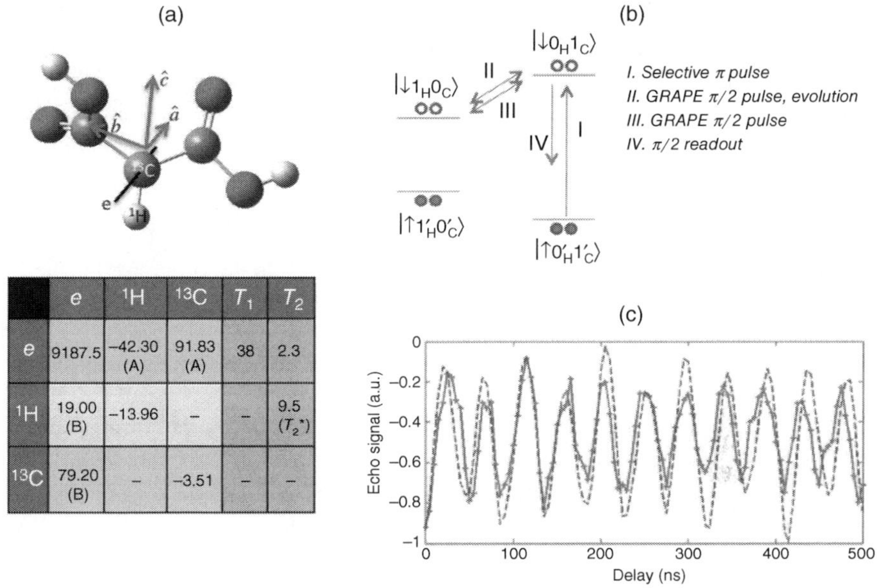

	e	^{1}H	^{13}C	T_1	T_2
e	9187.5	−42.30 (A)	91.83 (A)	38	2.3
^{1}H	19.00 (B)	−13.96	–	–	9.5 (T_2^*)
^{13}C	79.20 (B)	–	−3.51	–	–

Figure 2. (**a**) Schematic of the singly ^{13}C-labeled malonic acid radical and a table of Hamiltonian parameters for the crystal orientation used in the experiment. Larmor frequencies are along the diagonal and hyperfine coupling coefficients are off-diagonal with all frequencies in MHz. T_1 and T_2 are shown in the two right-most columns in microseconds. (**b**) Schematic of the experimental sequence for creating and detecting a double-nuclear coherence in the electron spin down manifold. (**c**) The double-nuclear coherence echo signal versus delay time in experiment (solid line with crosses) and simulation (dashed line) for the transition between $|\downarrow 1_H 0_C\rangle$ and $|\downarrow 0_H 1_C\rangle$ in (b) (electron spin down manifold). Reprinted with permission from Ref. [88]. Copyright (2011) by the American Physical Society.

and the double-nuclear coherence echo signal from [88] are shown in Fig. 2. These experiments represented the first steps toward reaching high-fidelity universal control of a one-electron, N-nuclear spin hyperfine coupled system using the electron as an actuator.

In the following sections, we discuss additional quantum information techniques that exploit the properties of combined NMR/ESR systems:

- Dynamic nuclear polarization and algorithmic cooling, which increase the polarizations of nuclear spins
- Spin-bus implementations and parallel information transfer, which produce two-qubit gates between nuclear spins mediated by electrons

These two techniques, described next, reinforce the potential of combining nuclear and electron spin control to better satisfy the DiVincenzo criteria in NMR QIP.

2. *Dynamic Nuclear Polarization and Algorithmic Cooling*

A nuclear spin in a magnetic field at room temperature has an equilibrium density operator described by a Boltzmann distribution, with a typical bias of $\sim 10^{-(5-6)}$ as discussed in Section I.C.2. Although this is sufficient to demonstrate the principles of QIP, it does not allow for resetting ancilla qubits as needed, for example, in quantum error correction. Because the nuclear magnetic moment is so weak, this problem cannot be simply resolved by brute-force cooling or increasing the magnetic field. For example, reducing the temperature to 1 K and using the largest available magnetic fields increase the polarization only to the order 10^{-3}. Therefore, it is necessary to develop novel techniques to provide initial polarization that approaches unity.

The much larger electron magnetic moment can be used as a polarization source. A swap gate between a nuclear spin and an electron, each at thermal equilibrium, acts as a polarization swap and gives the nuclear spin a large nonequilibrium polarization. This type of process is called *dynamic nuclear polarization*, or DNP. In the context of QIP, such polarization transfers have been implemented in $^{15}N@C_{60}$ [95], achieving a single-spin nuclear polarization of 62% at 4.2 K, in an 8.6 T field. Also, DNP has been used in conjunction with spin diffusion in silicon microparticles to produce bulk spin polarizations up to 5% at 1.5 K and 2.35 T [96]. Nuclear T_1 for these systems was shown to be dependent on the particle size, making the polarization decay controllable in principle. The ^{15}N nucleus within a nitrogen vacancy (NV) center has been initialized to a polarization of 98% by optically pumping the NV defect [97], and similar polarizations can be reached for proximate ^{13}C nuclei. These examples show that control of a single electron spin coupled to an NMR QIP register can be used to great effect.

It is possible to obtain nuclear qubits with still greater polarization by an implementation-independent method, known as algorithmic cooling. This technique combines reversible entropy compression (a unitary process) and thermal equilibration (a nonunitary process). For an initial register which is in thermal equilibrium with an external bath, a unitary operation is used to produce a state on the register which has high entropy on a qubit that is strongly coupled to the bath, and lower entropy on the remaining $n - 1$ bits. Thermal equilibration quickly restores the state of the strongly-coupled qubit without greatly affecting the remainder, resulting in an overall decrease in entropy (corresponding to an increase in polarization) on the register.

This two-step process is applied iteratively until no further cooling is possible. Given n qubits and a heat bath polarization of $\epsilon_b > 2^{-n}$, it is possible to cool the system very close to its ground state with only polynomially-increasing resources [98]. However, for $\epsilon_b < 2^{-n}$, the maximum polarization attainable by a single spin is $\epsilon_b 2^{n-2}$ [37]. Algorithmic cooling has been implemented in liquid and solid-state NMR [58,99,100]. Because electron spins possess higher polarization and much have shorter T_1 than nuclear spins at a given magnetic field strength and temperature, having nuclear qubits able to interact with an electron spin "bath" is a promising path to nuclear qubit initialization. Single quantum systems, such as NV centers with a sufficient number of coupled ^{13}C spins, would be ideal for demonstrating nuclear qubit purification by algorithmic cooling.

3. Spin Buses and Parallel Information Transfer

Because a single electron can couple to multiple nuclei, it can be used to transfer information between them, creating an effective coupling. This indirect coupling is the basis of the S-bus [101], useful for performing multi-qubit gates when the electron–nuclear and electron–electron coupling are much stronger than the nuclear–nuclear coupling. This concept was used to perform Deutsch's algorithm [101], using a system of two nuclear qubits coupled by an electron in $CaF_2:Ce^{3+}$.

The indirect coupling of nuclei using information transfer of the type already described has two flaws that must be overcome to confirm its utility as a nuclear control method. The first is that the state being transferred is subject to strong decoherence and relaxation while it is stored on the electron. The second is that only two nuclei can be coupled through the bus at the same time; two-qubit gates cannot be performed in parallel. These problems were recently resolved [102], where the assumed architecture consists of two local nodes (taken to be sets of n nuclei, each coupled to an electron via anisotropic hyperfine interaction), with the only internode coupling (dipolar or exchange interaction) being between electrons. By using the interaction frame, Borneman, Granade, and Cory showed that states of the multi-nucleus nodes can be swapped in parallel, effectively performing n two-qubit gates simultaneously. Also, the effect of decoherence on the electron is mitigated by ensuring that no computational state is stored on the electron. Figure 3 illustrates the idea presented in Ref. [102].

The electron spin, when used as a component of the NMR QIP system, possesses properties that complement those of the nuclear spins. While the nuclear spin has a longer coherence time and low initial polarization, the electron has a short coherence time and high initial polarization. The electron–nuclear coupling also permits an array of techniques, making the electron a valuable asset to NMR QIP.

An example of electron–nuclear system application for QIP that has been extensively investigated recently is entanglement generation [103–108]. Preparing entangled states in traditional NMR QIP is impossible because the required spin polarization (according to the positive partial transpose (PPT) criterion [109,110]) for entanglement is far above what is reachable. Electron–nuclear QIP at high field and low temperature, in conjunction with optimal control techniques, can potentially provide a highly entangled quantum state, which is a key ingredient in many quantum algorithms and devices.

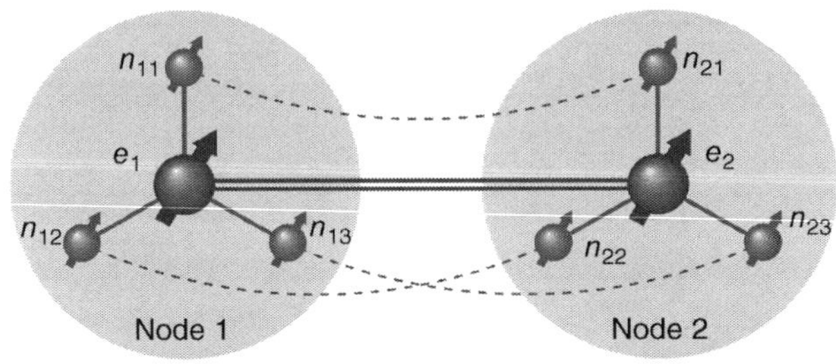

Figure 3. Schematic of $2 \times (1e - 3n)$ node: The nodes are taken to be identical, with resolved anisotropic hyperfine interactions (solid single-stroke lines) between electron actuator spins and nuclear processor spins. The local processors are initially disjointed, but may be effectively coupled (dashed lines) by modulating an isotropic actuator exchange or dipolar interaction (double-stroke line) and moving into an appropriate microwave Hamiltonian interaction frame. The spin labeling is e_i for electron actuator spins and n_{ij} for nuclear processor spins, where i and j label the nodes and the qubits, respectively. Reprinted with permission from Ref. [102]. Copyright (2012) by the American Physical Society.

III. NMR QIP FOR CHEMISTRY

A. Recent Experiments in NMR Quantum Simulation

The advances described Section II can enable the development of larger, more precise processors for quantum information. We focus now on three recent experiments that display the current capabilities of the NMR QIP system, and prove its utility as a testbed for future implementations. These experiments are motivated by the study of physical systems rather than chemical processes; though the methods used are easily adapted to problems of chemical interest.

1. Simulation of Burgers' Equation

One challenge that presents itself to scientists of many disciplines is the solution of nonlinear differential equations. Numerical solution of these equations is costly, this cost having motivated the development of numerous simulation methods. One such method, which takes advantage of quantum resources, is the quantum lattice gas algorithm on a type-II quantum computer.

Type-II quantum computers [112] are networks of small quantum processors that communicate classically to form a large analog computer. Operations that can be carried out in parallel are well-suited to implementation on such a processor, because the individual QIP nodes are distinct and can therefore be addressed simultaneously. The quantum lattice gas algorithm consists of such operations; it is divided into three steps that are iterated:

1. A state is prepared on the individual QIP nodes, corresponding to initial (or transient) conditions of the dynamical model being studied. The state on a

given node is

$$(\sqrt{1 - f_a(x_l, t_0)}|0\rangle + \sqrt{f_a(x_l, t_0)}|1\rangle)^{\otimes 2} \tag{13}$$

denoting two copies of the state encoding the value of the function to be propagated.

2. A *collision operator* $\hat{U}$ is applied in parallel across all of the nodes, which simulates a single finite time-step for the dynamics being modeled.
3. A measurement is made on the nodes and the resulting transient state is fed back to step 1. Classical communication is used, in this step, to transfer the results of step 2 between QIP nodes.

The resulting class of finite difference equations, which can be numerically solved by this method, is that of the form

$$f_a(x_{l+e_a}, t_{n+1}) = f_a(x_l, t_n) + \langle \psi(x_l, t_n)|\hat{U}^\dagger \hat{n}_a \hat{U} - \hat{n}_a|\psi(x_l, t_n)\rangle \tag{14}$$

where $e_a = \pm 1$, $\hat{U}$ is the collision operator and f_a is the function whose evolution is being simulated.

Discretization of Burgers' equation yields a finite difference equation of the form described, it is a nonlinear differential equation that describes shock formation in fluid dynamics:

$$\partial_t u(z, t) + u(z, t)\partial_z u(z, t) = \nu \partial_{zz} u(z, t) \tag{15}$$

Due to the simplicity of this equation, it can be solved analytically. From Ref. [111], the analytical solution to Burgers' equation is compared to the simulation on a 16-node type-II NMR quantum processor, shown in Figure 4.

2. *Simulation of the Fano–Anderson Model*

Another challenge with which physicists are often presented is that of simulating the evolution under a Hamiltonian representing the energy landscape of a large number of identical particles. The Fano–Anderson model possesses such a Hamiltonian, acting on n spinless fermions constrained to a ring, surrounding an impurity:

$$\hat{H} = \sum_{l=0}^{n-1} \epsilon_{k_l} c_{k_l}^\dagger c_{k_l} + \epsilon b^\dagger b + V(c_{k_0}^\dagger b + b^\dagger c_{k_0}) \tag{16}$$

where c_{k_l} is a fermionic annihilation operator on the conduction mode k_l, b is a fermionic annihilation operator on the impurity in the center, $c_{k_l}^\dagger$ and $b^\dagger$ are the respective creation operators, and ϵ_{k_l}, ϵ, and V denote the strengths of the terms in the Hamiltonian corresponding to occupation of a conduction mode, occupation of the impurity and tunnelling between modes and the impurity, respectively. This Hamiltonian is exactly diagonalizable; however, its simulation is an important

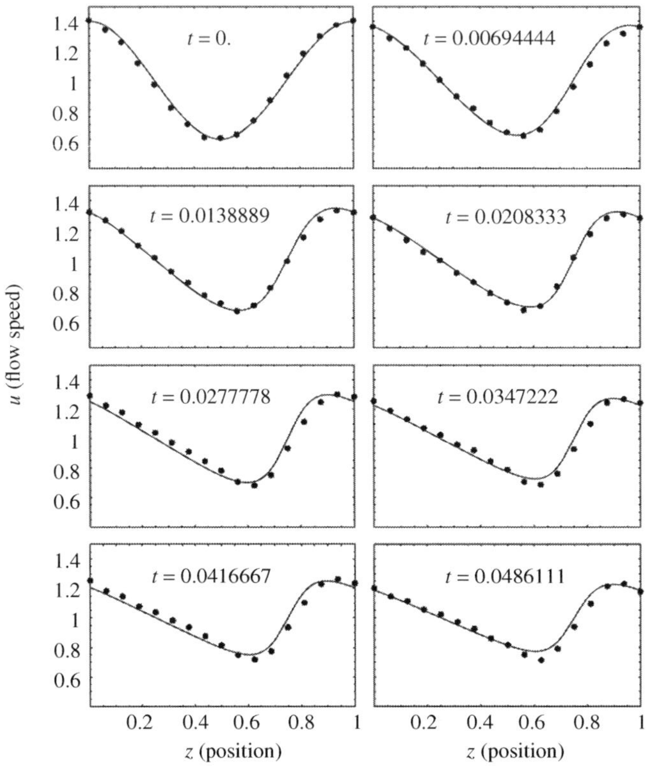

Figure 4. Comparison of analytic (solid line) and numerical solutions (dots) to Burgers' equation [111]. Numerical data have been obtained on a 16-node type-II quantum computer. Reprinted with permission from Ref. [111]. Copyright (2006) by the American Physical Society.

step in validating QIP methods for application to simulations in condensed matter theory.

The simulation performed by Negrevergne et al. [113] concerns the smallest possible Fano-Anderson model, one site interacting with an impurity. This algorithm requires three qubits, one for the site, one for the impurity, and one ancilla qubit, required for indirect measurement. It is implemented in a liquid-state NMR system, using transcrotonic acid, whose Hamiltonian parameters are shown in Fig. 5.

Again, the simulation is divided into three broad steps. However, iteration is unnecessary in this experiment, because the Hamiltonian to be simulated can be exponentiated continuously. First, the system is prepared in an initial state, related to the pseudopure state. Next, in order to perform an indirect measurement, a two-qubit gate couples the ancilla to the simulator, which is then evolved according

	C_1	C_2	C_3	C_4	M	H_1	H_2
C_1	−1914.06	40.5	1.5	7	127	3.9	6.3
C_2		−18115.10	69.9	1.3	−7.1	155.1	−0.6
C_3			−15157.41	73.2	6.6	−1.8	163
C_4				−21148.90	−0.9	6.5	3.6
M					230.43	6.9	−1.7
H_1						−2370.80	15.5
H_2							−1774.47

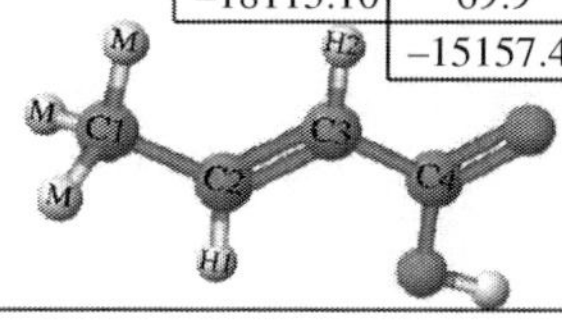

Figure 5. Spin Hamiltonian parameters and molecular diagram for transcrotonic acid. Reprinted with permission from Ref. [113]. Copyright (2005) by the American Physical Society.

to the Fano–Anderson Hamiltonian, and decoupled using another two-qubit gate, shown in Fig. 6.

Two quantities of interest were obtained in this experiment. $G(t)$, the correlation between the states $b^\dagger(t)|FS\rangle$ and $b^\dagger(0)|FS\rangle$, where $|FS\rangle$ is the state corresponding to the filled Fermi sea. In the NMR QIP system, this correlation function is $\langle 10|\exp\left(i\bar{H}t\right)\sigma_x^1\exp\left(-i\bar{H}t\right)\sigma_x^1|10\rangle$. The results of the measurement of this function in the simulation are shown in Fig. 7. Another quantity, of more general interest, is the spectrum of the Hamiltonian. This can be measured with a simpler quantum circuit, and results in excellent agreement with theory were also obtained in Ref. [113].

3. *Simulation of Frustrated Magnetism*

Any spin system with interaction Hamiltonians that cannot be simultaneously minimized is frustrated. Frustration is a fundamental problem in the study of magnetism; its presence means that the Hamiltonian cannot be divided into small subsystems that can then be studied individually to obtain information regarding a global property. This exacerbates the central problem of spin system simulation

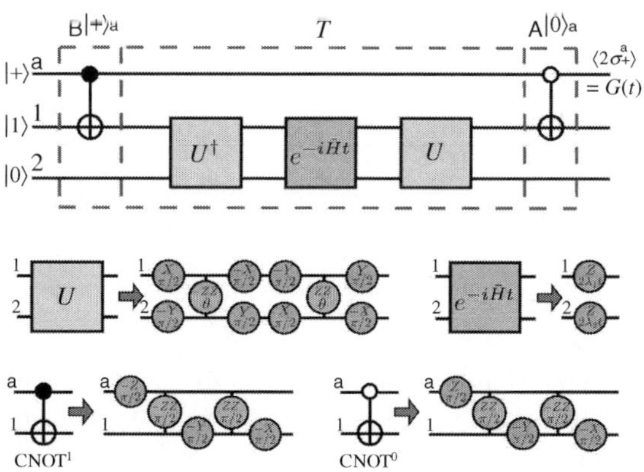

Figure 6. The kernel of the quantum simulation is the implementation of a unitary operator that evolves the qubit states according to the desired Hamiltonian [113]. Given the capability to produce controlled operations between an ancilla qubit and the simulation register, correlation functions of interest can be directly measured. Reprinted with permission from Ref. [113]. Copyright (2005) by the American Physical Society.

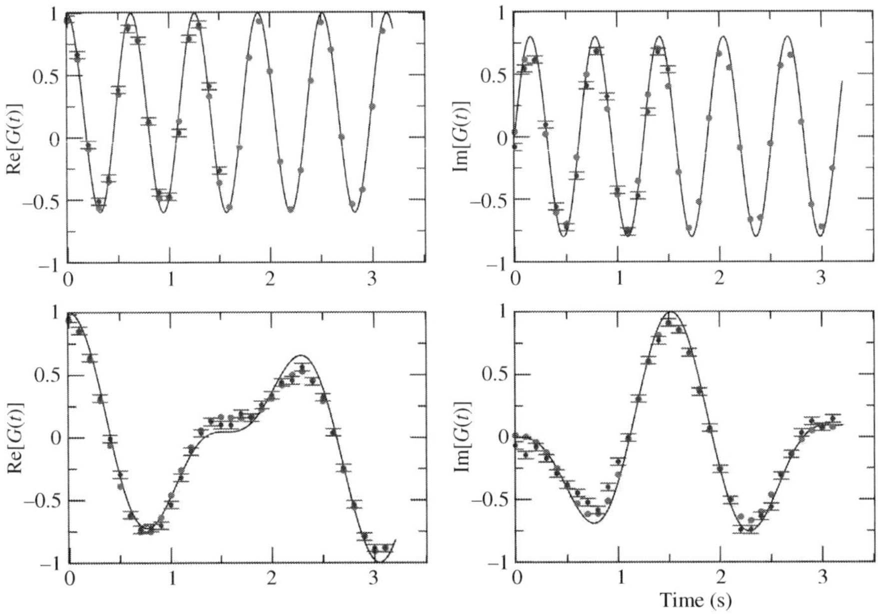

Figure 7. Analytic (solid line) and numerical solutions (dots) for the correlation function $G(t)$, whose evaluation is given diagrammatically in Fig. 6. The top panel is obtained for $\epsilon_{k_0} = -2, \epsilon = -8$, $V = 4$, and the bottom panel is obtained for $\epsilon_{k_0} = -2, \epsilon = 0, V = 4$. Reprinted with permission from Ref. [113]. Copyright (2005) by the American Physical Society.

(i.e., the exponential number of degrees of freedom required to simulate such a system). Zhang et al. [114] have performed a digital quantum simulation of the fundamental building block of such a frustrated system: the three-spin antiferromagnetic Ising model (Fig. 8):

$$\hat{H} = J(\hat{Z}_1\hat{Z}_2 + \hat{Z}_2\hat{Z}_3 + \hat{Z}_1\hat{Z}_3) + h(\hat{Z}_1 + \hat{Z}_2 + \hat{Z}_3) \tag{17}$$

where $J > 0$. This simple Hamiltonian can be solved analytically, due to its small size. The central technique in this work is to simulate an arbitrary *thermal* state using pseudopure state preparation. In this manner, the phase diagram in h, J, and T can be explored, deriving multiple properties of the resulting states. The pseudopure portion of the prepared state is

$$|\psi_\beta\rangle = \sum_k \sqrt{\exp(-\beta E_k)/\mathcal{Z}}|\phi_k\rangle \tag{18}$$

where the $|\phi_k\rangle$ are the eigenstates of the Hamiltonian with energy E_k, $\mathcal{Z}$ is the partition function and $\beta = 1/T$. The two physical quantities that Zhang et al. focus on are the total magnetization of the system $\hat{Z}_1 + \hat{Z}_2 + \hat{Z}_3$, and the entropy of the

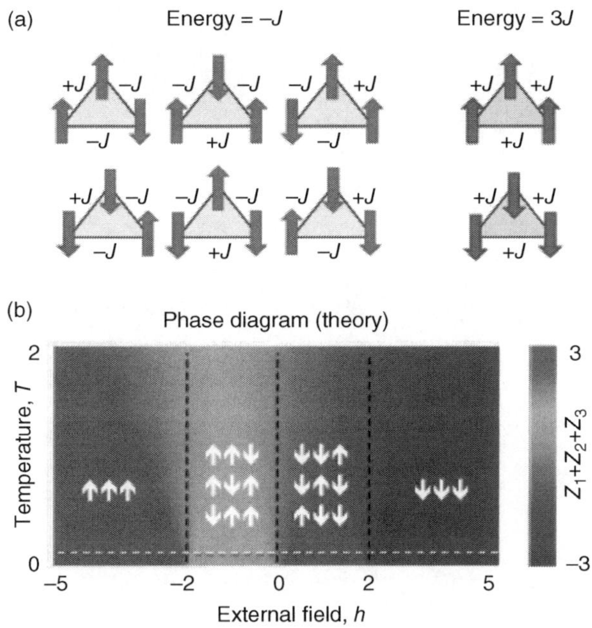

Figure 8. (**a**) Spin configurations of the Ising Hamiltonian at zero temperature and field h. There is a six-fold degeneracy in the ground state, leading to a nonzero ground state entropy at zero external field. (**b**) Thermal phase diagram for the same Hamiltonian showing total magnetization versus field and temperature [114]. Reprinted with permission from Ref. [114]. Copyright (2012) by the Macmillan Publishers Ltd: Nature Communications. (*See the color version of this figure in the Color Plates section.*)

resulting state $S = \mathrm{Tr}(\rho_\beta \log \rho_\beta)$, where ρ_β is the thermal density matrix of the Ising system (see Fig. 9). A peak in entropy is observed at $h = 0$ indicating frustration of the magnet, in agreement with theory. These three experiments serve to demonstrate the validity of the principles of quantum simulation, and demonstrate the necessary control to achieve scientific results on small registers. The greatest challenge to implementation of these algorithms on a larger scale is the inability to exert control in very large Hilbert spaces. The advances presented in Section II.C, for example, may be helpful in obtaining this control.

IV. PROSPECTS FOR ENGINEERED SPIN-BASED QIPs

Nuclear and electron magnetic resonance on bulk materials containing natural spin systems has been an excellent testing ground for quantum information processing techniques in small registers. The progression to larger numbers of qubits will most likely be accomplished by transitioning to engineered single spin systems, and much progress has taken place in this direction recently. Promising candidates include quantum dot electron spins [115], nitrogen-vacancy centers in diamond [116], donor electrons/nuclei in Si [117], and nitrogen atoms trapped in C_{60} [118], among others. A common theme among many of these approaches is to use the electron spin for initialization, fast gate operations and readout, and to use nuclear spins for long-time qubit storage or, in the case of quantum dots, as a controllable

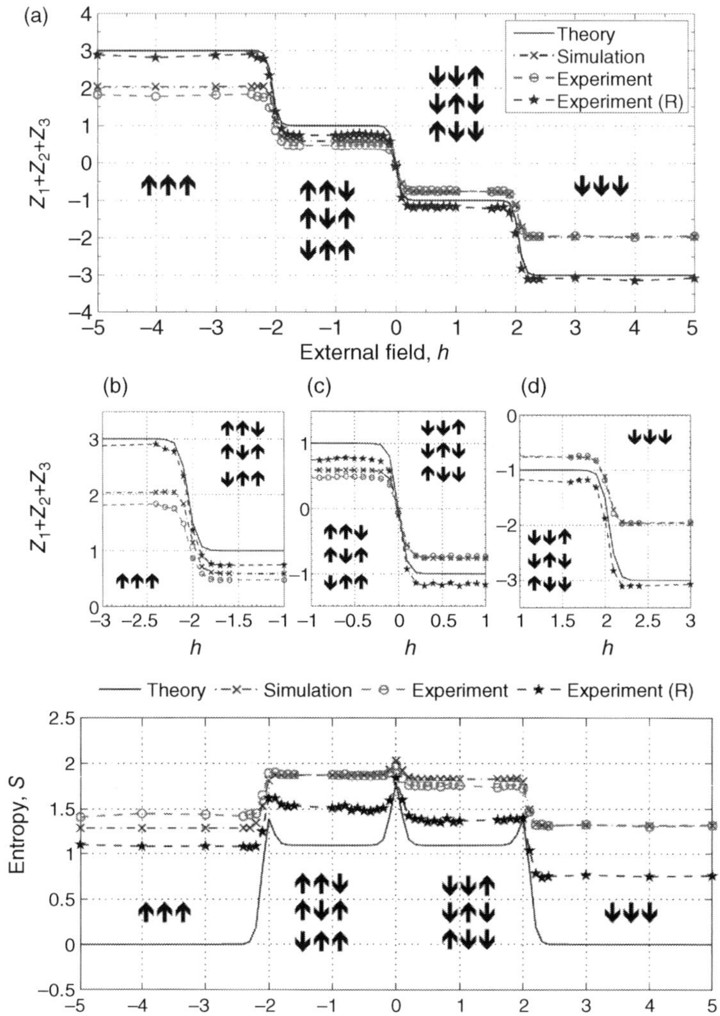

Figure 9. (**a–d**) The average magnetization of the three-spin antiferromagnetic Ising system versus simulated field h at low temperature, showing theoretical predication, numerical simulation of the NMR experiment including decoherence, and experimental results [114]. Below, entropy (S) versus external magnetic field (h), also at low temperature. The entropy peak at $h = 0$ indicates frustration. Reprinted with permission from Ref. [114]. Copyright (2012) by the Macmillan Publishers Ltd: Nature Communications.

local effective magnetic field [119] (the nuclear spin in III–V quantum dots present an unfortunate decoherence problem when left uncontrolled [120]). Single-qubit control is realized by some form of magnetic resonance in each of these approaches (with the exception of the singlet–triplet qubit in quantum dots [121]), whereas two-qubit coupling strategies vary widely. Because these are single quantum systems rather than ensembles, spatial addressing or a mix of spatial and frequency addressing becomes possible, an important advantage for scalability over frequency-only addressing. These systems are, in principle, more readily scalable than bulk magnetic resonance of molecules; however, developing reliable qubits with single quantum systems typically carries a host of engineering challenges. Although none of the approaches listed here has yet moved beyond a small handful of qubits, increasingly higher-quality single- and few-qubit systems are being realized at a rapid pace; some examples include demonstration of dynamical decoupling [122] and multi-qubit control [123] in NV centers, single-spin readout in Si [124], and high-fidelity single-qubit control and refocussing in quantum dots [125,126].

A significant challenge is to achieve fast, high-fidelity readout of a single spin (note that projective readout of the spin state is quite different from simply detecting the presence of a magnetic moment). So far, the most promising methods are optical (e.g., spin-dependent optical transitions in NV centres) or via electron transport (e.g., using the Pauli spin blockade in a double quantum dot [121,127]). Atomic force [128] and nano-magnetometry methods [129] are able to detect single spins, but require orders of magnitude improvement in sensitivity/$\sqrt{\mathrm{Hz}}$ before single-shot spin readout becomes feasible. These magnetometry experiments have nonetheless opened up a wide range of possible applications in quantum sensing (i.e., exploiting quantum coherence to surpass classical limits on measurement sensitivity) [129], and this technology may play an eventual role in spatial readout of spin-based quantum processors. See Ref. [130] for a seminal demonstration of quantum sensing of magnetic fields using liquid-state NMR.

Another common feature of these approaches is the solid-state environment surrounding the qubits, which typically leads to shorter decoherence times than more isolated systems such as ion traps or liquid-state NMR. Although spin-1/2 particles are immune to direct coupling with electric fields (the dominant noise source in solids), the spin-orbit coupling together with phonons provides a pathway for spin relaxation of electrons, which can then in turn act as magnetic noise sources for nuclei. Here, dynamical decoupling cannot typically improve the situation, because the correlation time of the electron spin relaxation process is usually much shorter than the timescale of control. Low temperature is thus required in order to reduce the density of phonons and suppress relaxation, typically leading to electron spin relaxation times of milliseconds for defect centers in dielectric crystals and for spins in quantum dots [131], and up to minutes for electrons at shallow donors in high purity Si at 1.2 K [132]. The latter work in high-purity Si demonstrated electron spin coherence times exceeding one second, a groundbreaking result for a solid-state system [132].

A set of nuclear spins, evolving under nuclear–nuclear dipole couplings, can also act as a magnetic noise source for an electron coupled to one or more of them.

This is the case for III–V quantum dots, where an electron spin is coupled to $\sim 10^6$ nuclear spins at once due to the contact hyperfine interaction [133]. The electron spin is dephased on a timescale ~ 10 ns due to statistical fluctuations of nuclear polarization, but orders of magnitude longer dephasing times have been achieved with dynamical decoupling [125]. Similar spin dynamics take place for the electron spin in NV centers, which is coupled to a small number of proximate natural abundance ^{13}C nuclei, or for donor electrons in Si coupled to ^{29}Si nuclei; again dynamical decoupling is seen to improve coherence times significantly [122,134] due to the slow correlation time of the nuclear bath. In the case of electronic defects in insulating or semiconducting crystals, the nearest "shell" of nuclei may have resolvable hyperfine couplings and can therefore be used as qubits, whereas more distant nuclei have hyperfine couplings which cannot be resolved, and simply generate (dephasing) magnetic noise.

Hybridizing spin and other quantum degrees of freedom, such as photons, may solve some of these challenges and is an active area of research. The aforementioned NV centre is already a type of hybrid system in which electron spin can be read out and initialized optically, and furthermore, coherent quantum information stored in the spin state can be converted to a photonic "flying qubit" [135]. The strong coupling cavity quantum electrodynamics regime has been demonstrated with NV center optical dipole transitions coupled to high finesse optical cavities [136,137], opening the door to photon-mediated coupling of spatially distant spin qubits. Similar efforts are underway to achieve strong coupling between quantum dot spins and microwave photon modes in on-chip superconducting resonators, for example, using strong spin-orbit coupling to couple the spin qubit to the cavity electric field [138]. Besides applications in quantum communication, this kind of approach could allow for distributing processing tasks optimally across different qubit realizations, and for minimizing the number of gates needed in a computation by increasing the effective dimensionality of the coupled network.

Chemistry has played a role in the early development of quantum control in the context of bulk magnetic resonance of molecular ensembles, and one can envision "bottom-up" architectures emerging for scalable QIP based on patterned molecular arrays or monolayers (e.g., those investigated in molecular electronics research) [139,140]. A viable approach would likely use some form of stable radical to provide electron spins for initialization, fast manipulation, and readout, and nuclear spins for ancillae and quantum memory. The principal challenges in such a system would be addressing and readout of spins on individual molecules. Addressing could be achieved using suitably strong magnetic field gradients to encode spatial information in the frequency domain, and with suitable improvements in sensitivity it might be possible to use a nano-diamond NV center scanning probe "read head" to achieve single-shot spatial readout of spin qubits [129]. The techniques for electron–nuclear hyperfine control discussed in Section II.C would then be invaluable tools for implementing a universal set of high-fidelity quantum gates in such a system.

Acknowledgments

We thank G. Passante for helpful discussions, and Y. Zhang for help in compiling references. The Natural Science and Engineering Research Council of Canada provided support for the writing of this document.

REFERENCES

1. R. P. Feynman, *Int. J. Theor. Phys.* **21**(6-7), 467–488 (1982).
2. G. Chan and M. Head-Gordon, *J. Chem. Phys.* **116**(11), 4462–4476 (2002).
3. S. Lloyd, *Science* **273**(5278), 1073–1078 (1996).
4. C. Zalka, *Proc. R. Soc. A* **454**(1969), 313–322 (1998).
5. S. Wiesner. *Simulations of Many Body Quantum Systems by a Quantum Computer, arXiv:quant-ph/9603028*, 1996.
6. H. F. Trotter, *Proc. Amer. Math. Soc.* **10**(4), 545–551 (1959).
7. A. Ekert and R. Jozsa, *Rev. Mod. Phys.* **68**, 733–753 (1996).
8. R. Jozsa, *Proc. R. Soc. Lond.* **454**(1969), 323–337 (1998).
9. Y. S. Weinstein, M. A. Pravia, E. M. Fortunato, S. Lloyd, and D. G. Cory, *Phys. Rev. Lett.* **86**, 1889–1891 (2001).
10. J. Chiaverini, J. Britton, D. Leibfried, E. Knill, M. D. Barrett, R. B. Blakestad, W. M. Itano, J. D. Jost, C. Langer, R. Ozeri, T. Schaetz, and D. J. Wineland. *Science* **308**(5724), 997–1000 (2005).
11. M. O. Scully and M. S. Zubairy, *Phys. Rev. A* **65**, 052324 (2002).
12. I. Kassal, S. P. Jordan, P. J. Love, M. Mohseni, and A. Aspuru-Guzik, *Proc. Natl. Acad. Sci.* **105**, 18681 (2008).
13. E. Knill, G. Ortiz, and R. D. Somma, *Phys. Rev. A* **75**, 012328 (2007).
14. C. Eckart, *Phys. Rev.* **46**, 383–387 (1934).
15. A. M. Childs, R. Cleve, E. Deotto, E. Farhi, S. Gutmann, and D. A. Spielman, in *Proceedings of the Thirty-Fifth Annual ACM Symposium on Theory of Computing*, STOC '03, ACM, New York, NY, USA, 2003, pp. 59–68.
16. D. W. Berry, G. Ahokas, R. Cleve, and Barry C, *Commun. Math. Phys.* **270**, 359–371 (2007).
17. M. Suzuki, *Phys. Lett. A* **146**(6), 319–323 (1990).
18. A. M. Childs, *Commun. Math. Phys.* **294**(2), 581–603 (2010).
19. A. M. Childs and D. W. Berry, *Quantum Inf. Commun.* **12**(1), 29–62 (2012).
20. D. Poulin, A. Qarry, R. Somma, and F. Verstraete, *Phys. Rev. Lett.* **106**, 170501 (2011).
21. N. Wiebe, D. W. Berry, P. Høyer, and B. C. Sanders, *J. Phys. A: Math. Theor.* **44**(44), 445308 (2011).
22. E. Farhi, J. Goldstone, S. Gutmann, and M. Sipser, *Quantum Computation by Adiabatic Evolution, eprint arXiv:quant-ph/0001106*, 2000.
23. A. Perdomo, C. Truncik, I. Tubert-Brohman, G. Rose, and A. Aspuru-Guzik, *Phys. Rev. A* **78**, 012320 (2008).

24. L. Luo and J. Lu, *Temperature Dependence of Protein Folding Deduced from Quantum Transition. ArXiv e-prints*, February 2011.
25. L. Luo, *Front. Phys. China* **6**(1), 133–140 (2011).
26. K. A. Dill, *Biochemistry* **24**(6), 1501–1509 (1985).
27. D. P. DiVincenzo, *Fortschr. Phys.* **48**(9-11), 771–783 (2000).
28. L. M. K. Vandersypen and I. L. Chuang, *Rev. Mod. Phys.* **76**, 1037–1069 (2005).
29. C. Negrevergne, T. S. Mahesh, C. A. Ryan, M. Ditty, F. Cyr-Racine, W. Power, N. Boulant, T. Havel, D. G. Cory, and R. Laflamme, *Phys. Rev. Lett.* **96**, 170501 (2006).
30. A. Abragam, *Principles of Nuclear Magnetism (International Series of Monographs on Physics)*, Oxford University Press, 1983.
31. M. H. Levitt, *Spin Dynamics: Basics of Nuclear Magnetic Resonance*, 2nd ed., John Wiley & Sons, Ltd., 2008.
32. D. G. Cory, M. D. Price, W. Maas, E. Knill, R. Laflamme, W. H. Zurek, T. F. Havel, and S. S. Somaroo, *Phys. Rev. Lett.* **81**, 2152–2155 (1998).
33. L. M. K. Vandersypen, M. Steffen, G. Breyta, C. S. Yannoni, M. H. Sherwood, and I. L. Chuang, *Nature* **414**(6866), 883–887 (2001).
34. D. P. DiVincenzo, *Phys. Rev. A* **51**, 1015–1022 (1995).
35. D. D'Alessandro, *Introduction to Quantum Control and Dynamics*, CRC Press, Wiley, 2008.
36. C. Ramanathan, N. Boulant, Z. Chen, D. G. Cory, I. Chuang, and M. Steffen, *Quantum Inf. Process.* **3**, 15 (2004).
37. O. Moussa, *On Heat-Bath Algorithmic Cooling and its Implementation in Solid-State NMR*. Master's thesis, University of Waterloo, Waterloo, Ontario, 2005.
38. J. Baugh, O. Moussa, C. A. Ryan, R. Laflamme, C. Ramanathan, T. F. Havel, and D. G. Cory, *Phys. Rev. A* **73**, 022305 (2006).
39. M. A. Nielsen and I. L. Chuang, *Quantum Computation and Quantum Information*, Cambridge University Press, Cambridge, UK, 2000.
40. E. Knill, D. Leibfried, R. Reichle, J. Britton, R. B. Blakestad, J. D. Jost, C. Langer, R. Ozeri, S. Seidelin, and D. J. Wineland, *Phys. Rev. A* **77**, 012307 (2008).
41. J. Emerson, *Can. J. Phys.* **86**(4), 557–561 (2008).
42. C. A. Ryan, M. Laforest, and R. Laflamme, *New J. Phys.* **11**(1), 013034 (2009).
43. J. S. Waugh, *Average Hamiltonian Theory*, John Wiley & Sons, Ltd, 2007.
44. M. H. Levitt, *Prog. in Nucl. Magn. Reson. Spect.* **18**, 61–122 (1986).
45. J. Baum, R. Tycko, and A. Pines, *Phys. Rev. A* **32**, 3435–3447 (1985).
46. E. Kupce and R. Freeman, *J. Magn. Reson. A* **115**, 273–276 (1995).
47. R. A. Degraaf, Y. Luo, M. Terpstra, H. Merkle, and M. Garwood, *J. Magn. Reson. B* **106**, 245–252 (1995).
48. M. Garwood and L. DelaBarre, *J. Magn. Reson.* **153**, 155–177 (2001).
49. R. Freeman, *Prog. Nucl. Magn. Reson. Spect.* **32**, 59–106 (1998).
50. E. M. Fortunato, M. A. Pravia, N. Boulant, G. Teklemariam, T. F. Havel, and D. G. Cory, *J. Chem. Phys.* **116**, 7599–7606 (2002).
51. L. Pontryagin, B. Boltyanskii, R. Gamkrelidze, and E. Mishchenko, *The Mathematical Theory of Optimal Processes*, Wiley Interscience, New York, 1972.
52. A. Bryson, Jr. and Y. C. Ho, *Applied Optimal Control*, Hemisphere, Washington, DC, 1975.
53. I. I. Maximov, Z. Tosner, and N. C. Nielsen, *J. Chem. Phys.* **128**, 184505 (2008).

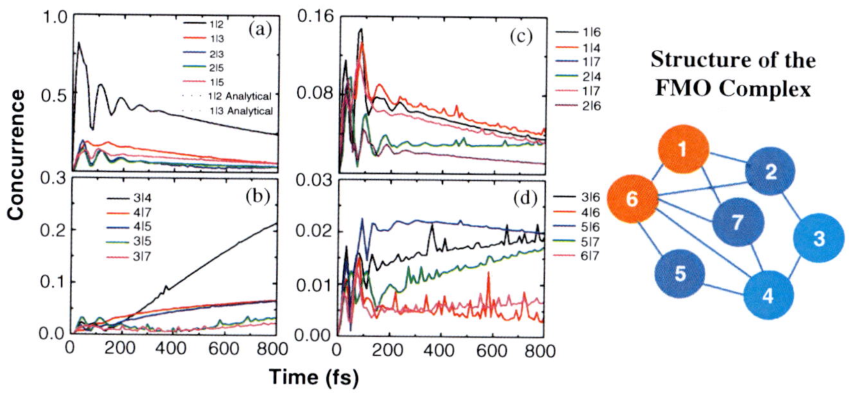

Figure 8, Page 27. The quantum entanglement evolution for the pairwise entanglement in the FMO complex with the site 1 initially excited. The left panel shows the entanglement with all pairs. Based the amplitude of the concurrence, all pairs had been divided into four groups, from the largest pairs (**a**) to the smallest pairs (**d**). The solid line indicates the entanglement computed via the convex roof approach [147], while the dotted line shows the evolution calculated through the concurrence method. The right panel is the geometry structure of the FMO complex.

Advances in Chemical Physics, Volume 154: Quantum Information and Computation for Chemistry, First Edition. Edited by Sabre Kais.

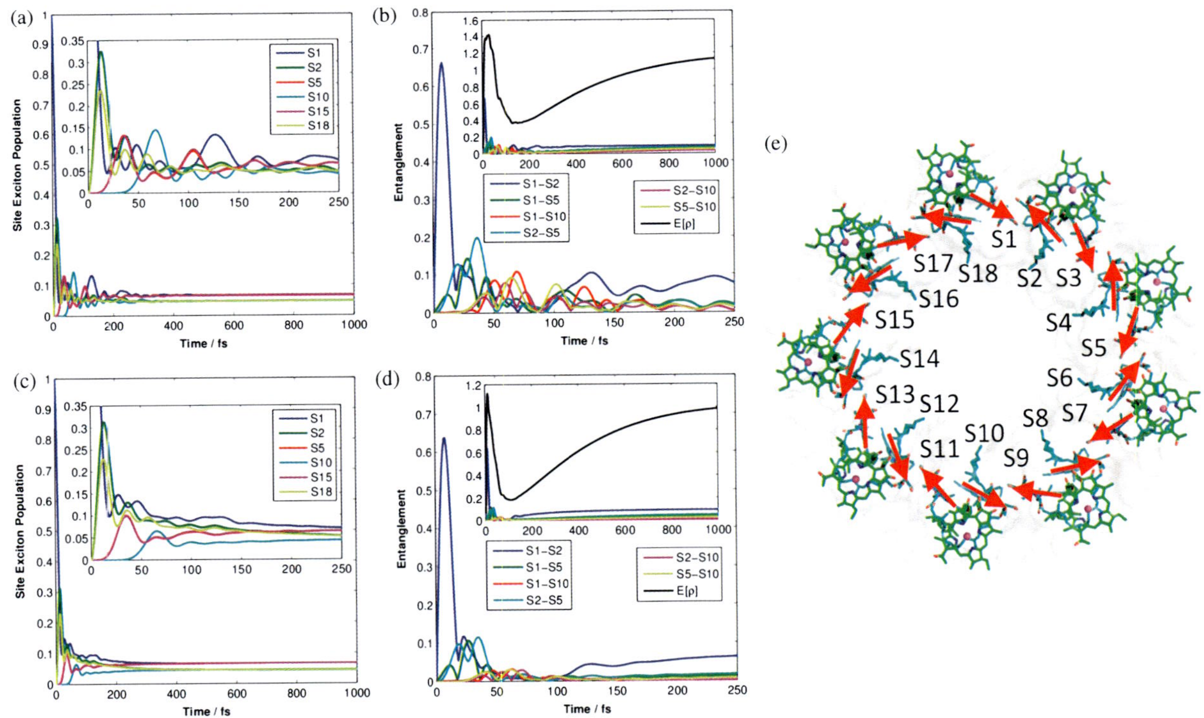

Figure 9, Page 29. (**a**) The excitation population dynamics of LH2 B850 18 sites at 77 K. (**b**) The concurrence and global entanglement (black line in the inset) in the 18 sites at 77 K. (**c**) The excitation population dynamics of LH2 B850 18 sites at 300 K. (**d**) The concurrence and global entanglement (black line in the inset) in the 18 sites at 300 K. (**e**) Top view (from cytoplasmic to periplasmic side) of 2.0 Å resolution LH2 structure of purple bacteria (*Rhodopseudomonas acidophila*). The 9 B800 and 18 B850 bacteriochlorophylls (BChls) are shown in green and cyan, respectively. The red arrows represent the transition dipoles of B850 BCHls Q_Y excitation. Top view (from cytoplasmic to periplasmic side) of 2.0 Å resolution LH2 structure of purple bacteria. The 9 B800 and 18 B850 bacteriochlorophylls (BChls) are shown in green and cyan, respectively. The red arrows represent the transition dipoles of B850 BCHls Q_Y excitation [146].

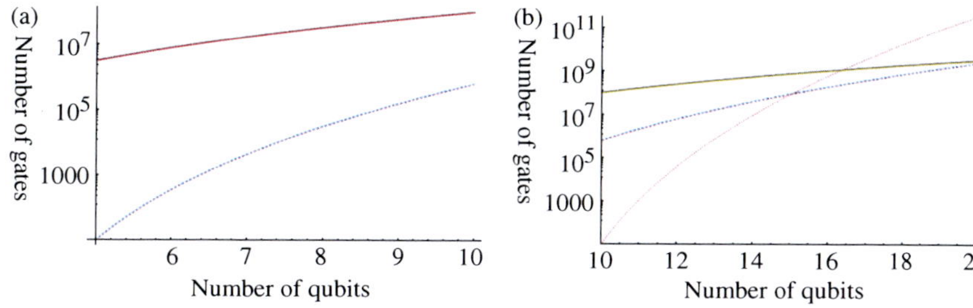

Figure 3, Page 53. (**a**) A comparison of gate counts, as estimated by Eqs. (40) and (41) (with $A = 1$) for 5 electrons and numbers of qubits between 5 and 10. Solid line: G_{direct}. Dashed line: G_{compact}. This indicates that for fewer than 5 qubits the compact mapping is preferred over the scalable direct mapping. (**b**) A comparison of gate counts, as estimated by Eqs. (40) and (41) (with $A = 1$) for numbers of qubits between 5 and 10. Solid line: G_{direct}, 5 electrons. Dashed line: G_{compact}, 5 electrons. Dotted line: G_{compact}, 10 electrons. This indicates that the crossover between the compact mapping and the scalable direct mapping lies between 10 and 20 qubits.

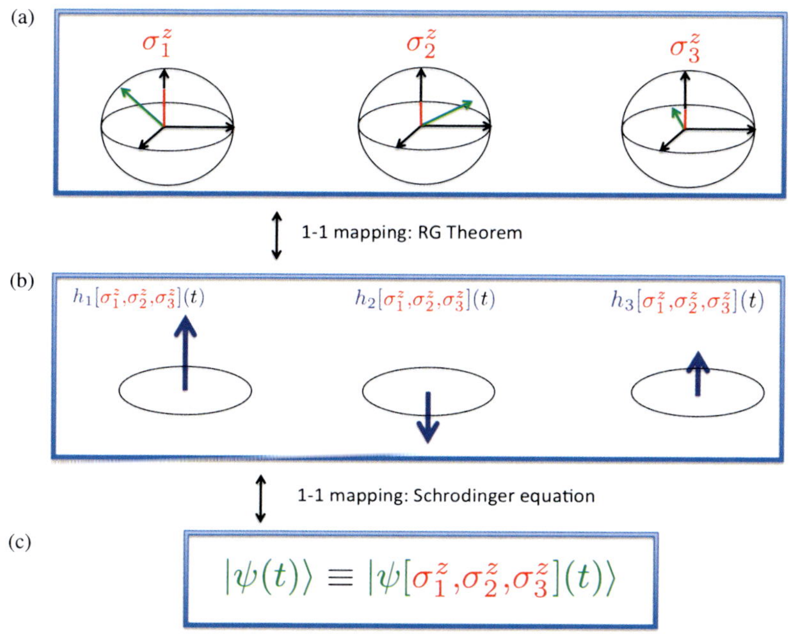

Figure 4, Page 99. *Runge–Gross theorem for a three-qubit example.* The set of expectation values $\{\sigma_1^z, \sigma_2^z, \ldots, \sigma_N^z\}$, defined by the the Bloch vector components of each qubit along the z-axis in (a), is uniquely mapped onto the set of local fields $\{h_1, h_2, \ldots, h_N\}$ in (b) through the RG theorem. Then, through the Schrödinger equation, the set of fields is uniquely mapped onto the wave function. These two mappings together imply that the N-qubit wave function in (c) is in fact a unique functional of the set of expectation values $\{\sigma_1^z, \sigma_2^z, \ldots, \sigma_N^z\}$.

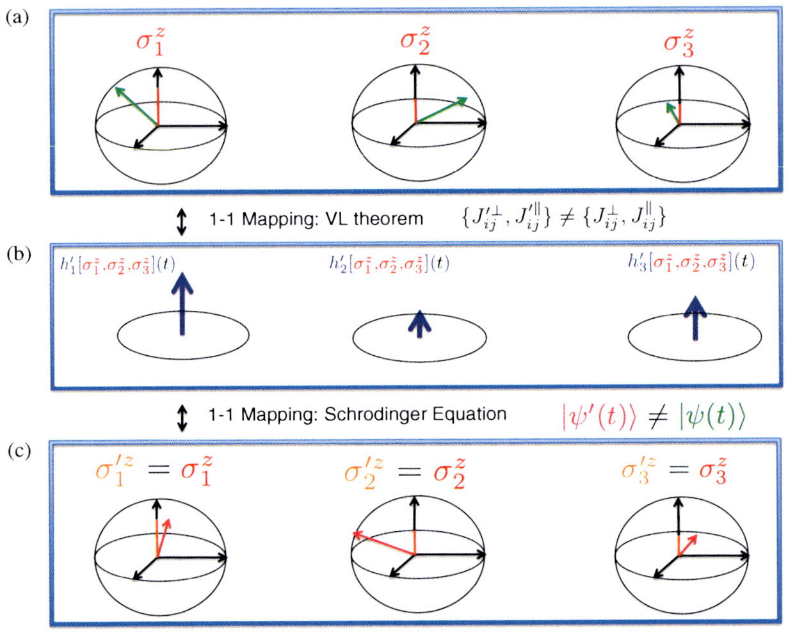

Figure 5, Page 100. *Van Leeuwen theorem for a three-qubit example*. The set $\{\sigma_1^z, \sigma_2^z, \ldots \sigma_N^z\}$ (**a**) obtained from evolution under Eq. (94), is uniquely mapped to a new set of fields $\{h'_1, h'_2, \ldots h'_N\}$ (**b**) for a Hamiltonian with different two-qubit interactions. Evolution under this new Hamiltonian returns the same expectation values $\{\sigma_1^z, \sigma_2^z, \ldots \sigma_N^z\}$, although the wave function is different and hence projections of the Bloch vectors along other axes are in general different (**c**).

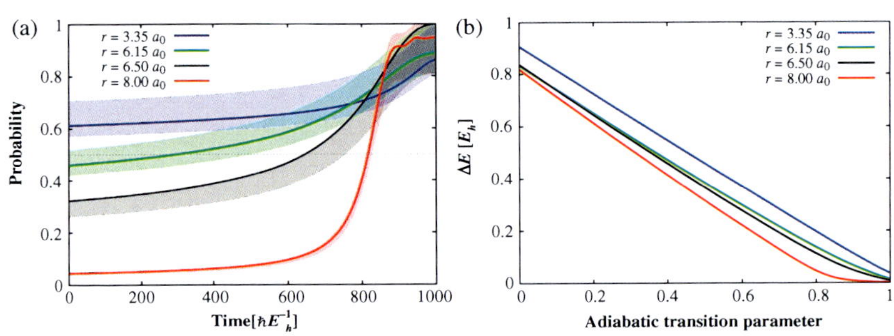

Figure 8, Page 119. Adiabatic state preparation (ASP) of the SbH ground state for different internuclear distances. (**a**) Dependence of the IPEA success probability on time during the ASP (1000 $\hbar E_h^{-1} \approx 10^{-14}$ s). Solid lines correspond to the success probabilities, $|\langle\psi_{\text{ASP}}|\psi_{\text{exact}}\rangle|^2 \cdot (0.81, 1\rangle$ interval is coloured. Reprinted with permission from Ref. [31]. Copyright (2010) by the American Physical Society. (**b**) Dependence of the energy gap between the ground and the first excited state on the adiabatic transition parameter *s*.

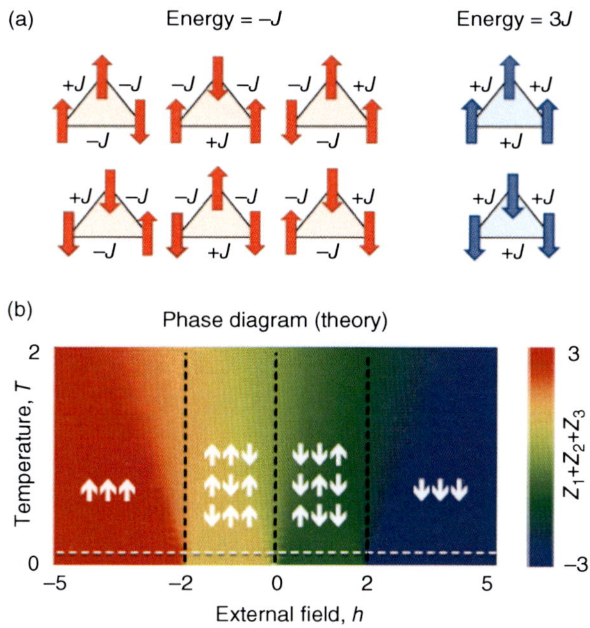

Figure 8, Page 219. (**a**) Spin configurations of the Ising Hamiltonian at zero temperature and field h. There is a six-fold degeneracy in the ground state, leading to a nonzero ground state entropy at zero external field. (**b**) Thermal phase diagram for the same Hamiltonian showing total magnetization versus field and temperature [114]. Reprinted with permission from Ref. [114]. Copyright (2012) by the Macmillan Publishers Ltd: Nature Communications.

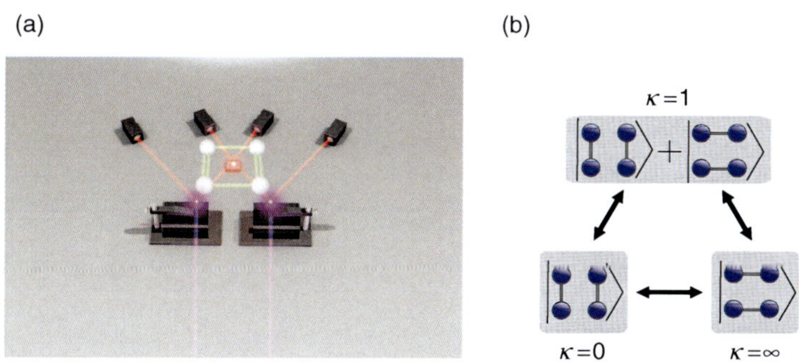

Figure 4, Page 237. (**a**) Artistic view of the photonic quantum simulation of valence-bond state. Two entangled photon pairs are generated via SPDC. Superimposing one single photon from each pair at a tunable beam splitter results in quantum interference such that the measured four-photon coincidences correspond to the ground state, for example, of a Heisenberg-interacting spin tetramer. Dependent on the beam splitter reflectivity, frustration in valence bond states or so-called spin liquid states could be investigated. (**b**) Dependent on the neighboring spin–spin interaction strength (κ) controlled by the splitting ratio of the beam splitter, various spin-paring configurations can be simulated. Figure 4a is taken from Ref. [55].

Figure 5, Page 237. Future experiments using more entangled photon pairs may allow the study of ground-state properties of molecular ground states, such as the delocalized bonds in benzene. This figure is taken from Ref. [55].

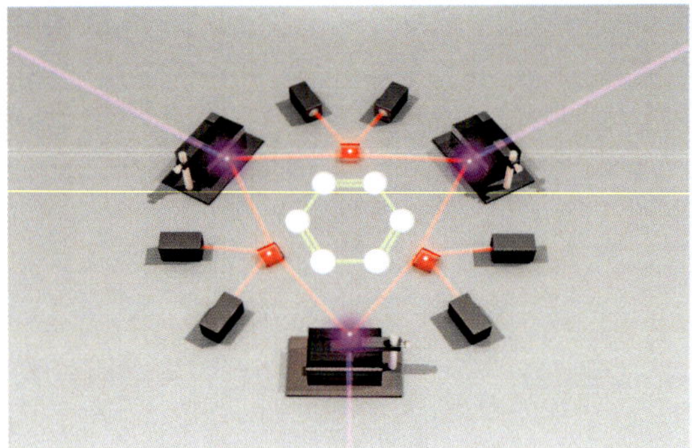

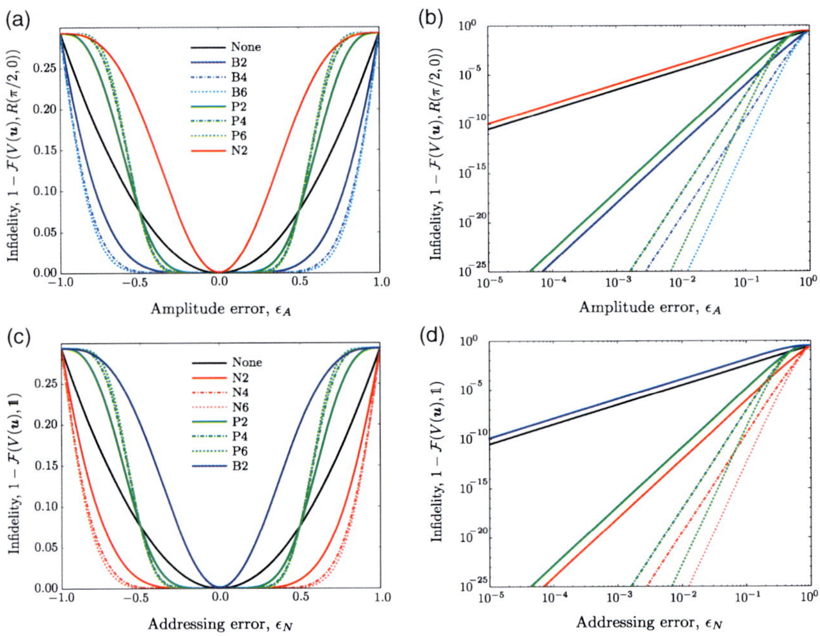

Figure 9, Page 275. Infidelity of the Trotter–Suzuki sequences B2j, P2j, and N2j. (**a, b**) In the amplitude error model, $M(\theta, \phi) = M(\theta(1 + \epsilon_A), \phi)$ and $U_T = R(\pi/2, 0)$. (**c, d**) In the addressing error model on the unaddressed qubits $M(\theta, \phi) = M(\theta\epsilon_N, \phi)$ and $U_T = \mathbb{1}$, while on the addressed spins $R(\pi/2, 0)$ is applied. Each error model establishes a separate preferred interaction frame; when transformed into the appropriate pictures, the B2j and N2j sequences are homologous. The passband sequences P2j can correct both amplitude and addressing errors at the cost of reduced efficacy.

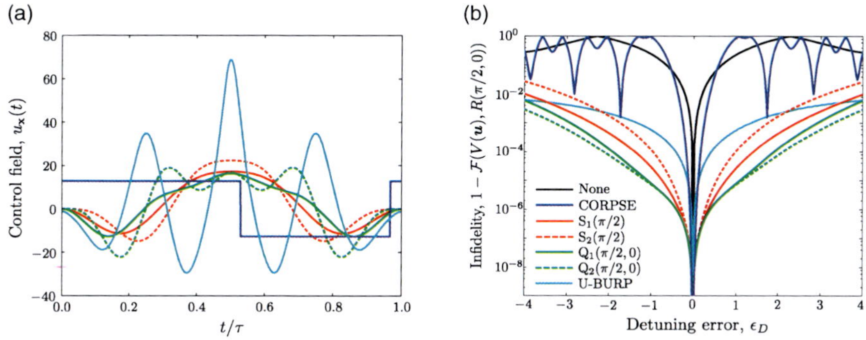

Figure 12, Page 284. (**a**) Control function $u_x(t)$ for various pulse sequences designed to compensate systematic detuning errors. (**b**) Performance of several shaped pulses over a wide detuning range.

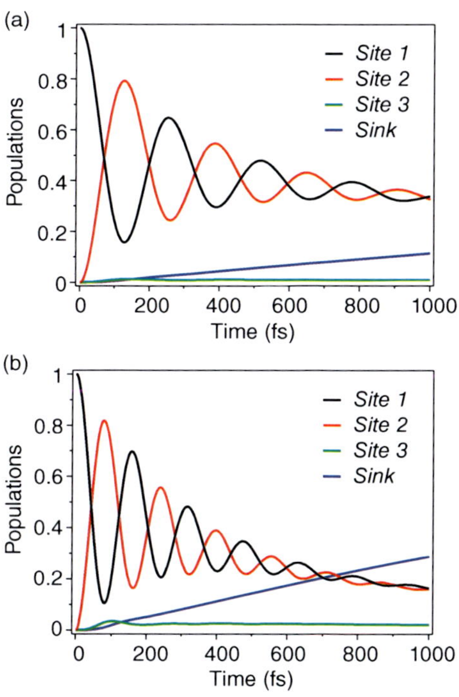

Figure 3, Page 361. Populations of chromophores 1–3 and sink with (**b**) and without (**a**) electron correlation per site. The exciton populations in chromophores 1, 2, and 3 as well as the sink population are shown as functions of time in femtoseconds (fs) for $N = 4$ with (**a**) $V = 0.0$ and (**b**) $V = 0.8$. Correlating the four electrons on each chromophore significantly accelerates the increase in the sink population with time. Reprinted with permission from Ref. [32]. Copyright 2012 American Institute for Physics.

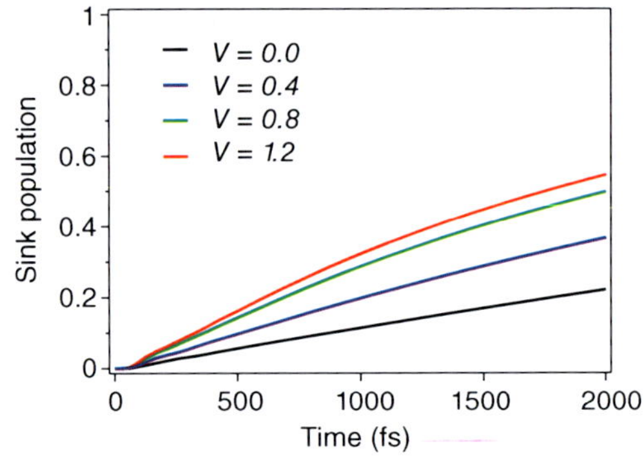

Figure 4, Page 361. Correlation-enhanced transfer to the reaction center. The reaction center (sink) population as a function of time (fs) is shown in for a range of V with $N = 4$. Correlating the electrons within the chromophores significantly increases the efficiency of energy transfer to the reaction center (sink). Reprinted with permission from Ref. [32]. Copyright 2012 American Institute for Physics.

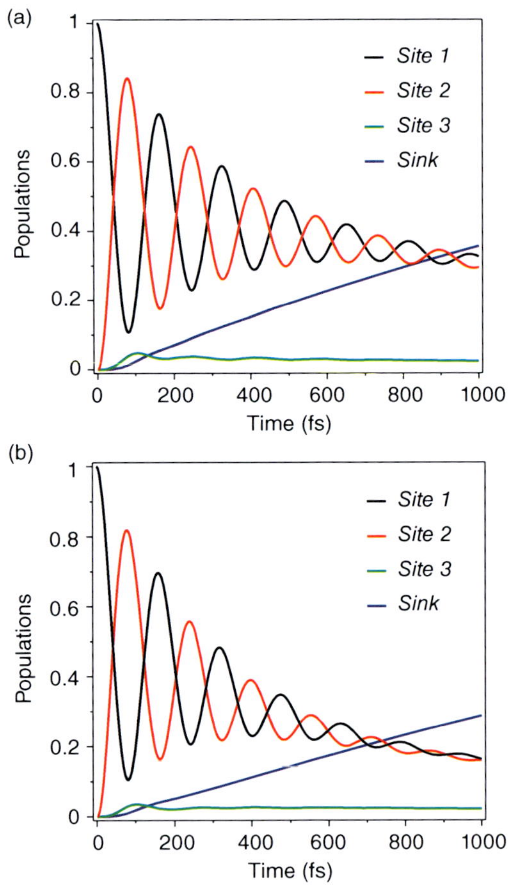

Figure 6, Page 365. Populations of chromophores 1–3 and the reaction center for (**a**) the subsystem 123 and (**b**) the full FMO system 1234567. Subsystem 123 of the FMO complex transfers energy more efficiently from chromophore 1 to the reaction center (sink) than the full FMO system. Reprinted with permission from Ref. [4]. Copyright 2011 American Chemical Society.

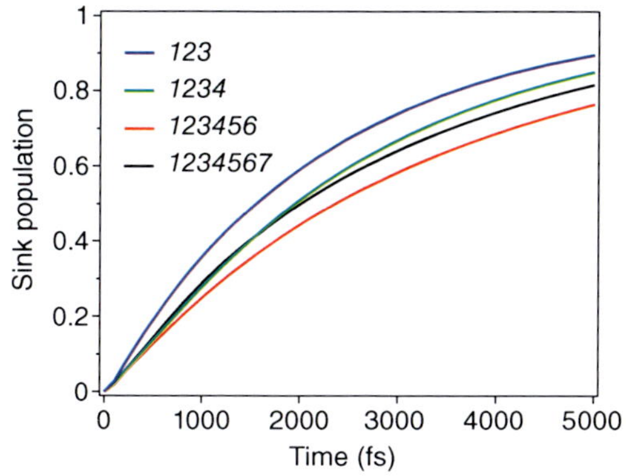

Figure 7, Page 365. Population in the reaction center for several chromophore subsystems. The subsystems 123, 1234, and 12345 (not shown) are more efficient than the full FMO system at transferring energy from chromophore 1 to the reaction center (sink). Reprinted with permission from Ref. [4]. Copyright 2011 American Chemical Society.

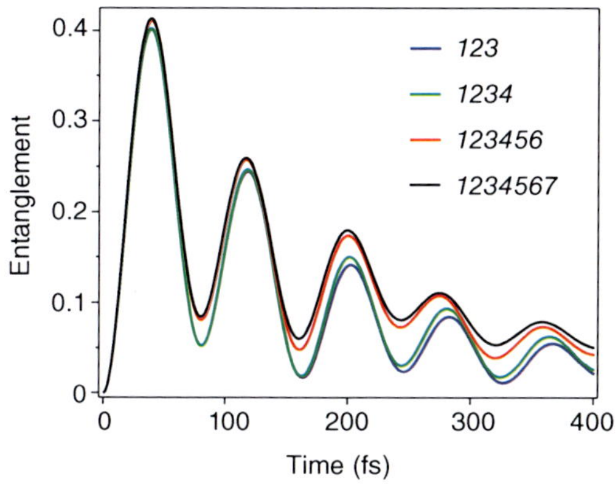

Figure 8, Page 366. Entanglement of excitons for several chromophore subsystems as a function of time. A measure of global entanglement is shown as a function of time (fs) for several chromophore subsystems, 123, 1234, and 123456, as well as the full system 1234567. Like the full system, the subsystems exhibit entanglement of excitons between the chromophores. Reprinted with permission from Ref. [4]. Copyright 2011 American Chemical Society.

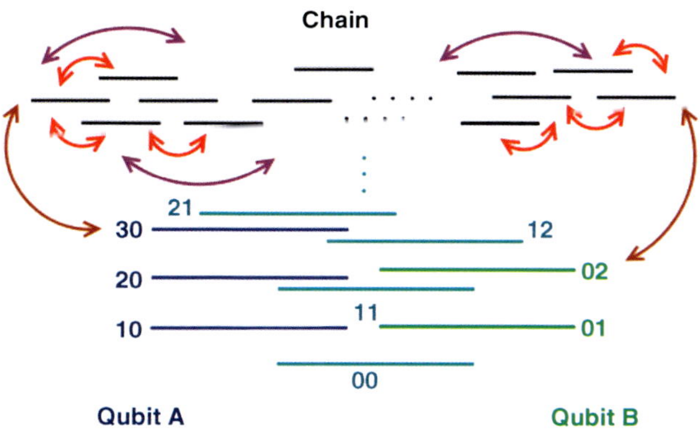

Figure 2, Page 391. Model system for two qubits connected by local mode states of a molecular chain (black). As indicated by the green and blue qubit levels, the fundamental frequencies of the qubit modes differ in this setup. The red and purple arrows indicate the kinetic and potential couplings between the local mode chain states, and additionally a coupling between a qubit overtone state and one chain state at each site is assumed (brown arrows).

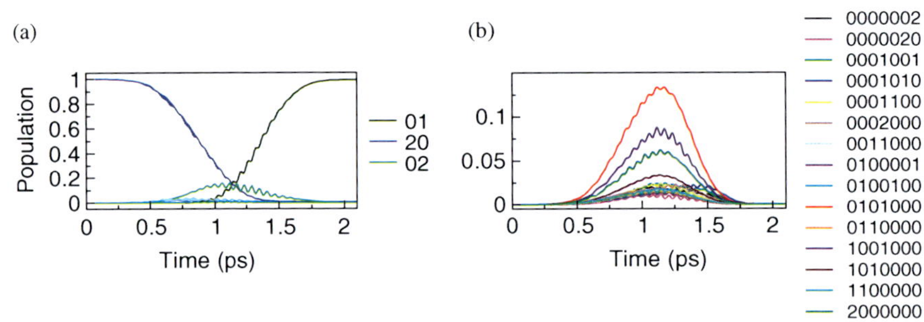

Figure 5, Page 393. Mechanism induced by the laser field depicted in Fig. 4a. (**a**) The population in the qubit mode states are depicted, where the vibrational quantum numbers of the local chain states are 0. The blue line refers to the initial state 20 and the dark-green line to the target state 01 of the transfer process. (**b**) The intermediate population of the local chain states is shown. Here, only the vibrational quantum number of the local chain states are given.

Figure 6, Page 393. From the change of the local mode to the normal mode basis, different transition pathways from the initial and target states into the normal mode states can be found. They are indicated by the blue lines for the initial state and by the green lines for the target state. The varying width of the lines indicates that the transition dipole strengths differ.

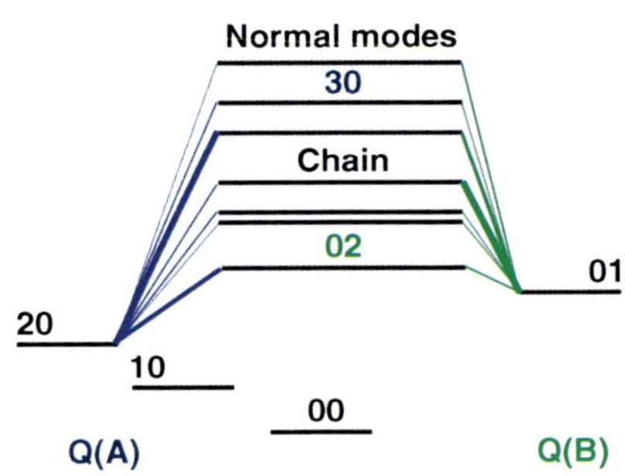

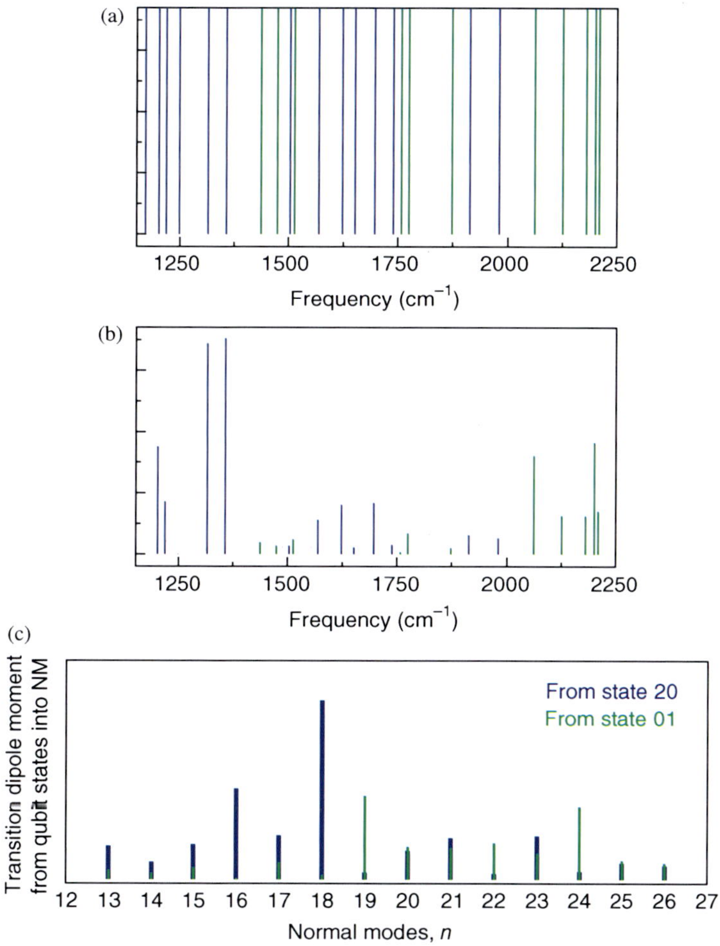

Figure 7, Page 394. (**a**) Transition pathways from the initial (20, blue) and target (01, green) states into the normal mode states, plotted as a function of the transition frequency. The spectrum of the blue transitions is plotted for $n = 13 - 26$, for the green transitions for $n = 14 - 23$. (**b**) Transition pathways scaled by the size of the respective dipole matrix elements. (**c**) Pathways from both qubit sites, plotted against the normal modes n.

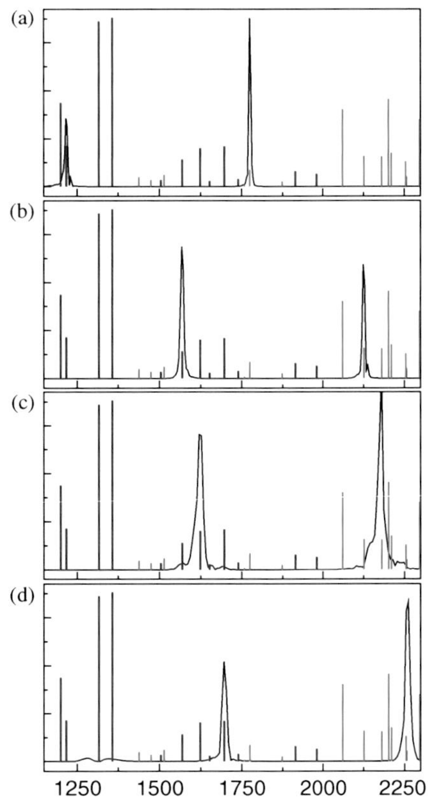

Figure 8, Page 396. Transition pathways with spectra of the laser pulses driving the state transfer across the chain molecule through different quantum channels. (**a**) $n = 17$, (**b**) $n = 20$, (**c**) $n = 21$, and (**d**) $n = 23$ (Fig. 7c).

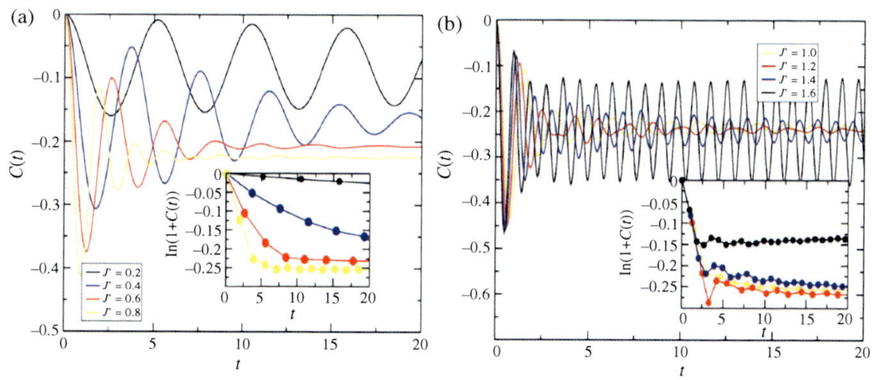

Figure 3, Page 462. The spin correlation function $C(t)$ of centered spin for $N = 501$, $h = 0.5$, and $\gamma = 1.0$ versus time t for different values of the coupling (**a**) $J' \leq J$; (**b**) $J' \geq J$ at zero temperature. The decay profile for each case is shown in the inner panel.

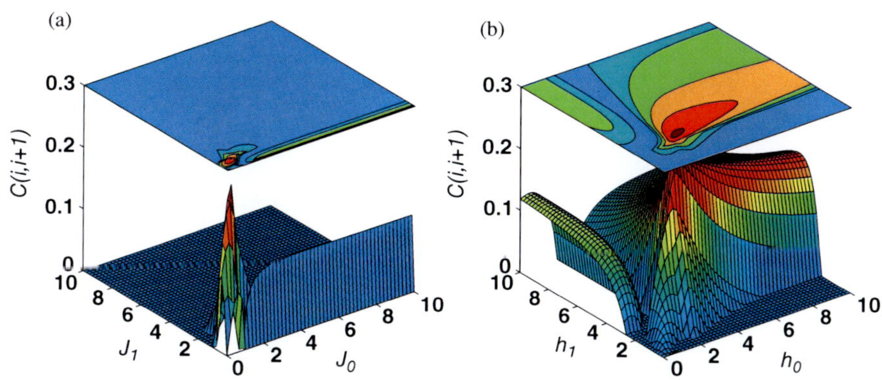

Figure 6, Page 468. The asymptotic behavior of $C(i, i+1)$ as a function of (**a**) J_0 and J_1 with $h_0 = h_1 = 1$; (**b**) h_0 and h_1 with $J_0 = J_1 = 2$, where h_0, h_1, and J_0 are in units of J_1, $kT = 0$ and $\gamma = 1$.

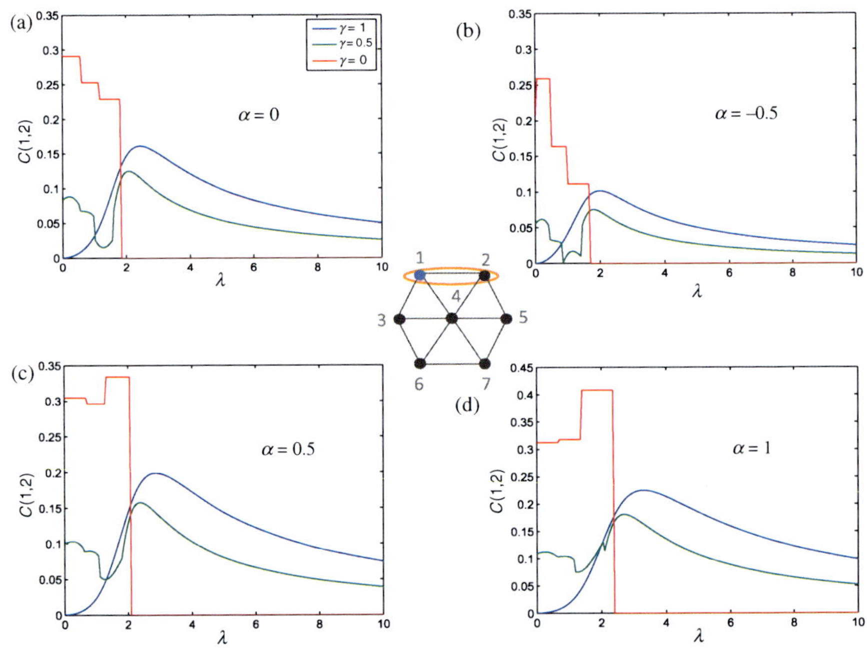

Figure 28, Page 494. The concurrence $C(1, 2)$ versus the parameter λ with a single impurity at the border site 1 with different impurity coupling strengths $\alpha = -0.5, 0, 0.5, 1$ for different degrees of anisotropy $\gamma = 1, 0.5, 0$ as shown. The legend for all subfigures is as shown in part (a).

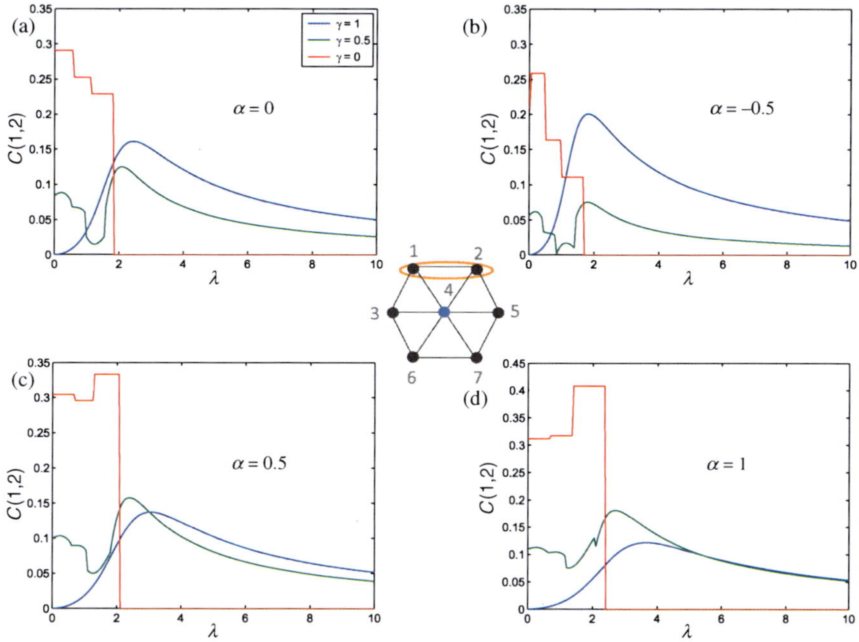

Figure 29, Page 495. The concurrence $C(1, 2)$ versus the parameter λ with a single impurity at the central site 4 with different impurity coupling strengths $\alpha = -0.5, 0, 0.5, 1$ for different degrees of anisotropy $\gamma = 1, 0.5, 0$ as shown. The legend for all parts is as shown in part (a).

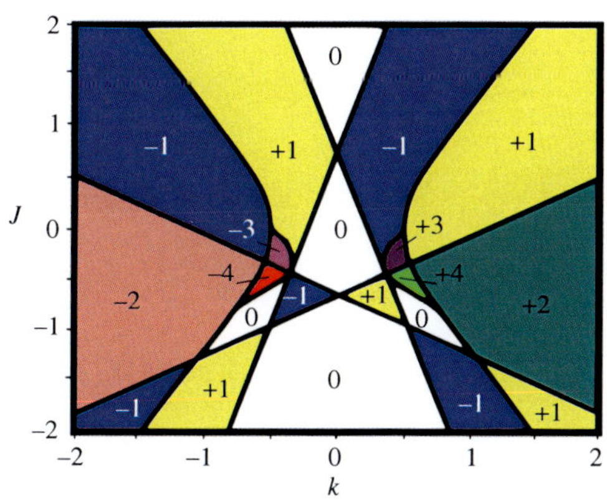

Figure 13, Page 548. The extended phase diagram of the square–octagon model. Here, the parameters of the model are $J_z = 1$, $J = J_x = J_y$, and $\kappa_1 = \kappa_2$. The phases with odd spectral Chern number are non-abelian topological phases: $\nu = \pm 1$ characterizes the Ising topological phase and $\nu = \pm 3$ is the $SU(2)_2$ Chern–Simons phase.

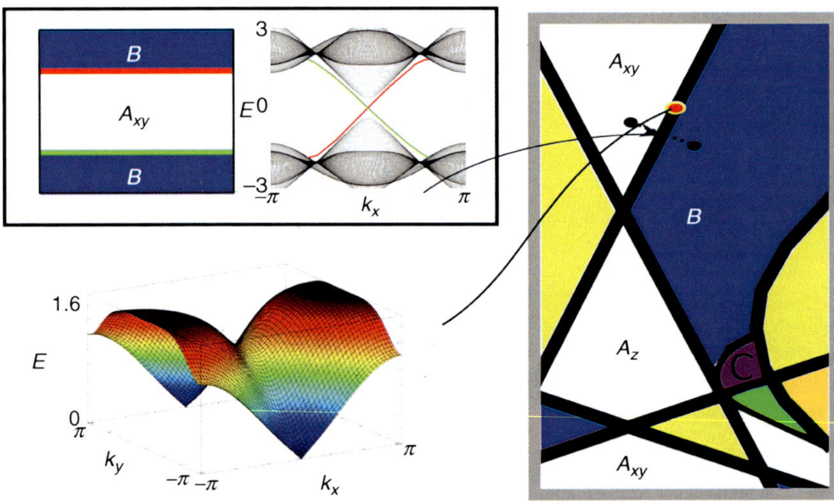

Figure 17, Page 552. Edge modes of the square–octagon model on a cylinder. Plots of the energies associated with the edge modes at a phase boundary.

54. A. Carlini, A. Hosoya, T. Koike, and Y. Okudaira, *Phys. Rev. Lett.* **96**, 060503 (2006).
55. A. Carlini, A. Hosoya, T. Koike, and Yosuke Okudaira, *Phys. Rev. A* **75**, 042308 (2007).
56. N. Khaneja, T. Reiss, C. Kehlet, and T. Schulte-Herbrüggen, *J. Magn. Reson.* **172**, 296–305 (2005).
57. J. Zhang, D. Gangloff, O. Moussa, and R. Laflamme, *Phys. Rev. A* **84**, 034303 (2011).
58. C. A. Ryan, O. Moussa, J. Baugh, and R. Laflamme, *Phys. Rev. Lett.* **100**(14), 140501 (2008).
59. G. Passante, O. Moussa, C. A. Ryan, and R. Laflamme, *Phys. Rev. Lett.* **103**, 250501 (2009).
60. G. Passante, O. Moussa, D. A. Trottier, and R. Laflamme, *Phys. Rev. A* **84**, 044302 (2011).
61. O. Moussa, C. A. Ryan, D. G. Cory, and R. Laflamme, *Phys. Rev. Lett.* **104**, 160501 (2010).
62. A. M. Souza, J. Zhang, C. A. Ryan, and R. Laflamme, *Nat. Commun.* **2**(169) (2011).
63. O. Moussa, J. Baugh, C. A. Ryan, and R. Laflamme, *Phys. Rev. Lett.* **107**, 160501 (2011).
64. A. I. Konnov and V. F. Krotov, *Automat. Rem. Control* **60**, 1427 (1999).
65. D. J. Tannor, V. Kazakov, and V. Orlov, *Time-Dependent Quantum Molecular Dynamics*, Plenum, New York, 1992.
66. W. Zhu and H. Rabitz, *J. Chem. Phys.* **109**, 385 (1995).
67. Y. Maday and G. Turinici, *J. Chem. Phys.* **118**, 8191 (2003).
68. C. A. Ryan, C. Negrevergne, M. Laforest, E. Knill, and R. Laflamme, *Phys. Rev. A* **78**, 012328 (2008).
69. E. R. Andrew, A. Bradbury, and R. G. Eades, *Nature* **182**(4650), 1659–1659, (1958).
70. M. Lee and W. I. Goldburg, *Phys. Rev.* **140**, A1261–A1271 (1965).
71. E. L. Hahn, *Phys. Rev.* **80**, 580–594 (1950).
72. H. Y. Carr and E. M. Purcell, *Phys. Rev.* **94**, 630–638 (1954).
73. K. Khodjasteh and D. A. Lidar, *Phys. Rev. Lett.* **95**, 180501 (2005).
74. P. J. Keller and F. W. Wehrli, *J. Magn. Reson.* **78**(1), 145–149 (1988).
75. G. S. Uhrig, *Phys. Rev. Lett.* **98**, 100504 (2007).
76. B. Lee, W. M. Witzel, and S. Das Sarma, *Phys. Rev. Lett.* **100**, 160505 (2008).
77. W. Yang and R.-B. Liu, *Phys. Rev. Lett.* **101**, 180403 (2008).
78. G. S. Uhrig, *Phys. Rev. Lett.* **102**, 120502 (2009).
79. J. R. West, B. H. Fong, and D. A. Lidar, *Phys. Rev. Lett.* **104**, 130501 (2010).
80. G. S. Uhrig, *New J. Phys.* **10**, 083024 (2008).
81. S. Pasini and G. S. Uhrig, *Phys. Rev. A* **81**, 012309 (2010).
82. H. Uys, M. J. Biercuk, and J. J. Bollinge, *Phys. Rev. Lett.* **103**, 040501 (2009).
83. M. J. Biercuk, H. Uys, A. P. VanDevender, N. Shiga, W. M. Itano, and J. J. Bollinge, *Nature* **458**, 996–1000 (2009).
84. T. W. Borneman, M. D. Hürlimann, and D. G. Cory, *J. Magn. Reson.* **207**, 220–233 (2010).
85. A. M. Souza, G. A. Álvarez, and D. Suter, *Phys. Rev. Lett.* **106**, 240501 (2011).
86. C. A. Ryan, J. S. Hodges, and D. G. Cory, *Phys. Rev. Lett.* **105**, 200402 (2010).
87. T. Gullion, D. B. Baker, and M. S. Conradi, *J. Magn. Reson.* **89**, 479–484 (1990).
88. Y. Zhang, C. A. Ryan, R. Laflamme, and J. Baugh, *Phys. Rev. Lett.* **107**, 170503 (2011).
89. A. Schweiger and G. Jeschke, *Principles of Pulse Electron Paramagnetic Resonance*, Oxford University Press, New York, 2001.
90. W. B. Mims, *Phys. Rev. B* **5**, 2409–2419 (1972).
91. G. Feher, *Phys. Rev.* **103**, 834–835 (1956).

92. W. B. Mims, *Proc. R. Soc. Lond. A* **283**(1395), 452–457 (1965).

93. J. S. Hodges, J. C. Yang, C. Ramanathan, and D. G. Cory, *Phys. Rev. A* **78**, 010303 (2008).

94. N. Khaneja, *Phys. Rev. A* **76**(3), 032326 (2007).

95. G. W. Morley, J. van Tol, A. Ardavan, K. Porfyrakis, J. Zhang, G. Andrew, and D. Briggs, *Phys. Rev. Lett.* **98**(22), 220501 (2007).

96. A. E. Dementyev, D. G. Cory, and C. Ramanathan, *Phys. Rev. Lett.* **100**(12), 127601 (2008).

97. V. Jacques, P. Neumann, J. Beck, M. Markham, D. Twitchen, J. Meijer, F. Kaiser, G. Balasubramanian, F. Jelezko, and J. Wrachtrup, *Phys. Rev. Lett.* **102**, 057403 (2009).

98. L. J. Schulman, T. Mor, and Y. Weinstein, *Phys. Rev. Lett.* **94**, 120501 (2005).

99. G. Brassard, Y. Elias, J. M. Fernandez, H. Gilboa, J. A. Jones, T. Mor, Y. Weinstein, and L. Xiao, *Experimental Heat-Bath Cooling of Spins*, *arXiv:quant-ph/0511156*, 2005.

100. Y. Elias, H. Gilboa, T. Mor, and Y. Weinstein, *Chem. Phys. Lett.* **517**, 126–131 (2011).

101. M. Mehring and J. Mende, *Phys. Rev. A* **73**(5), 052303 (2006).

102. T. W. Borneman, C. E. Granade, and D. G. Cory, *Phys. Rev. Lett.* **108**, 140502 (2012).

103. M. Mehring, J. Mende, and W. Scherer, *Phys. Rev. Lett.* **90**, 153001 (2003).

104. M. Mehring, W. Scherer, and A. Weidinger, *Phys. Rev. Lett.* **93**, 206603 (2004).

105. W. Scherer and M. Mehring, *J. Chem. Phys.* (2008).

106. R. Rahimi, *Studies on Entanglement in Nuclear and Electron Spin Systems for Quantum Computing*, Ph.D. thesis, Department of Systems Innovation, Graudate School of Engineering Science, Osaka University, 2006.

107. K. Sato, S. Nakazawa, R. Rahimi, T. Ise, S. Nishida, T. Yoshino, N. Mori, K. Toyota, D. Shiomi, Y. Yakiyama, Y. Morita, M. Kitagawa, K. Nakasuji, M. Nakahara, H. Hara, P. Carl, P. Hofer, and T. Takui, *J. Mater. Chem.* **19**, 3739–3754 (2009).

108. S. Simmons, R. M. Brown, H. Riemann, N. V. Abrosimov, P. Becker, H.-J. Pohl, M. L. W. Thewalt, K. M. Itoh, and J. J. L. Morton, *Nature* **470**(7332), 69–72 (2011).

109. A. Peres, *Phys. Rev. Lett.* **77**, 1413–1415 (1996).

110. M. Horodecki, P. Horodecki, and R. Horodecki, *Phys. Rev. Lett.* **78**, 574–577 (1997).

111. Z. Chen, J. Yepez, and D. G. Cory, *Phys. Rev. A* **74**, 042321 (2006).

112. G. P. Berman, A. A. Ezhov, D. I. Kamenev, and J. Yepez, *Phys. Rev. A* **66**, 012310 (2002).

113. C. Negrevergne, R. Somma, G. Ortiz, E. Knill, and R. Laflamme, *Phys. Rev. A* **71**, 032344 (2005).

114. J. Zhang, M.-H. Yung, R. Laflamme, A. Aspuru-Guzik, and J. Baugh, *Nat. Commun.* **3**, 880 (2012).

115. M. Ciorga, A. S. Sachrajda, P. Hawrylak, C. Gould, P. Zawadzki, Y. Feng, and Z. Wasilewski, *Phys. E* **11**(35-40) (2001).

116. M. Steiner, P. Neumann, J. Beck, F. Jelezko, and J. Wrachtrup, *Phys. Rev. B* **81**(3), 035205 (2010).

117. B. E. Kane, *Nature* **393**, 133–137 (1998).

118. J. J. L. Morton, A. M. Tyryshkin, A. Ardavan, K. Porfyrakis, S. A. Lyon, G. Andrew, and D. Briggs, *J. Chem. Phys.* **124**(1), 014508 (2006).

119. C. W. Lai, P. Maletinsky, A. Badolato, and A. Imamoglu, *Phys. Rev. Lett.* **96**, 167403 (2006).

120. A. V. Khaetskii, D. Loss, and L. Glazman, *Phys. Rev. Lett.* **88**, 186802 (2002).

121. J. R. Petta, A. C. Johnson, J. M. Taylor, E. A. Laird, A. Yacoby, M. D. Lukin, C. M. Marcus, M. P. Hanson, and A. C. Gossard, *Science* **309**(2180) (2005).

122. C. A. Ryan, J. S. Hodges, and D. G. Cory, *Phys. Rev. Lett.* **105**, 200402 (2010).

123. F. Jelezko, T. Gaebel, I. Popa, M. Domhan, A. Gruber, and J. Wrachtrup, *Phys. Rev. Lett.* **93**, 130501 (2004).

124. A. R. Stegner, C. Boehme, H. Huebl, M. Stutzmann, K. Lips, and M. S. Brandt, *Nat. Phys.* **2**, 835–838 (2006).

125. H. Bluhm, S. Foletti, I. Neder, M. Rudner, D. Mahalu, V. Umansky, and A. Yacoby, *Nat. Phys.* **7**, 109–113 (2011).

126. S. Nadj-Perge, S. M. Frolov, E. P. A. M. Bakkers, and L. P. Kouwenhoven, *Nature* **468**, 1084–1087 (2011).

127. N. Shaji, C. B. Simmons, M. Thalakulam, L. J. Klein, H. Qin, H. Luo, D. E. Savage, M. G. Lagally, A. J. Rimberg, R. Joynt, M. Friesen, R. H. Blick, S. N. Coppersmith, and M. A. Eriksson, *Nat. Phys.* **4**, 540–544 (2008).

128. D. Rugar, R. Budakian, H. J. Mamin, and B. W. Chui, *Nature* **430**, 329–332 (2004).

129. M. S. Grinolds, P. Maletinsky, S. Hong, M. D. Lukin, R. L. Walsworth, and A. Yacoby, *Nat. Phys.* **7**, 687–692 (2011).

130. J. A. Jones, S. D. Karlen, J. Fitzsimons, A. Ardavan, S. C. Benjamin, G. Andrew, D. Briggs, and J. J. L. Morton, *Science* **324**(5931), 1166–1168 (2009).

131. J. M. Elzerman, R. Hanson, L. H. Willems van Beveren, B. Witkamp, L. M. K. Vandersypen, and L. P. Kouwenhoven, *Nature* **430**, 431–435 (2004).

132. A. M. Tyryshkin, S. Tojo, J. J. L. Morton, H. Riemann, N. V. Abrosimov, P. Becker, H.-J. Pohl, T. Schenkel, M. L. W. Thewalt, K. M. Itoh, and S. A. Lyon, *Nat. Mater.* **33** (2011).

133. W. A. Coish and J. Baugh, *Phys. Status Solidi (b)* **246**(10), 2203–2215 (2009).

134. J. J. L. Morton, A. M. Tyryshkin, R. M. Brown, S. Shankar, B. W. Lovett, A. Ardavan, T. Schenkel, E. E. Haller, J. W. Ager, and S. A. Lyon, *Nat. Phys.* **455**, 1085–1088 (2008).

135. K.-M. C. Fu, C. Santori, P. E. Barclay, I. Aharonovich, S. Prawer, N. Meyer, A. M. Holm, and R. G. Beausoleil, *Appl. Phys. Lett.* **93**(23), 234107 (2008).

136. Y.-S. Park, A. K. Cook, and H. Wang, *Nano Lett.* **6**(9), 2075–2079 (2006).

137. D. Englund, B. Shields, K. Rivoire, F. Hatami, J. Vuckovic, H. Park, and M. D. Lukin, *Nano Lett.* **10**(10), 3922–3926 (2010).

138. M. Trif, V. N. Golovach, and D. Loss, *Phys. Rev. B* **77**, 045434 (2008).

139. J. E. Green, J. W. Choi, A. Boukai, Y. Bunimovich, E. Johnston-Halperin, E. DeIonno, Y. Luo, B. A. Sheriff, K. Xu, Y. S. Shin, H.-R. Tseng, J. F. Stoddart, and J. R. Heath, *Nature* **445**, 414–417 (2007).

140. J. M. Tour. *Acc. Chem. Res.* **33**(11), 791–804 (2000).

PHOTONIC TOOLBOX FOR QUANTUM SIMULATION

XIAO-SONG MA,[1,2] BORIVOJE DAKIĆ,[1] and PHILIP WALTHER[1,2]

[1] *Faculty of Physics, University of Vienna, Boltzmanngasse 5, A-1090 Vienna, Austria*
[2] *Institute for Quantum Optics and Quantum Information (IQOQI), Austrian Academy of Sciences, Boltzmanngasse 3, A-1090 Vienna, Austria*

I. INTRODUCTION

The functionalities of new exotic materials, such as high-temperature superconductors, carbon nanotubes, graphene, and magneto-resistance materials, stem from complicated quantum phenomena. In order to study the properties of them, we need powerful tools to model these materials in microscopic scale with high accuracy. Unfortunately, modeling such materials is inhibited by our limited knowledge of strongly correlated many-body quantum physics and intrinsic incapability of classical computers. The number of equations required to describe the quantum system and its dynamics governed by many-body entanglement and the superposition principle of quantum mechanics grow exponentially with the number of particles involved, which is intractable for computer. About three decades ago, Feynman proposed the ingenious idea that one could use a controllable quantum system to simulate the dynamics and reproduce the quantum state of other system of interest [1]. Recently, various quantum simulators based on different

Advances in Chemical Physics, Volume 154: Quantum Information and Computation for Chemistry, First Edition. Edited by Sabre Kais.

quantum architectures are being successfully constructed, such as atoms [2–7], trapped ions [8–13], NMR [14,15], superconducting circuits [16], as well as single photons [17–24].

For understanding molecular systems, quantum simulators are required to have very precise quantum control of the individual particles. In principle, the two main types of such quantum simulators are the analog and the digital quantum simulators [25]. Analog quantum simulators utilize the adiabatic theorem [26], where a simple Hamiltonian can be adiabatically mapped to the Hamiltonian of interest [27–29]. An adiabatic quantum simulator can be built by engineering interactions among particles using tunable external parameters, where the system will remain in its ground state if the system parameters change gradually enough. The other method of realizing a quantum simulator is based on discrete gate operations and the phase estimation algorithm [19,30] referred to as a digital quantum simulator [25]. Typically, such quantum simulators are built out of single- and two-qubit quantum gates [31–35].

Photons have the feature that they do not interact with anything easily, which provides the advantages and disadvantages for using them to realize quantum simulators. The advantageous side is a natural decoherence-free system that allows the implementations without complicated cryogenic experimental setups. With commercially available components one can individually address and manipulate single photons with high precision easily. The second advantage of photonic simulators is that photons are easily moved either in free space or waveguides and thus are not restricted to nearest–neighbor interaction. Therefore, the mobility of photons, ideally on a single chip, will allow almost arbitrary interconnections and consequently facilitates the simulation of complex and nonlocal many-body interactions. The disadvantages are the complications of generation of many-body entanglement and scalability. It has been proven that by using measurement-induced interaction, one can circumvent the difficulty of photon–photon interaction and generate genuine many-body entangled photonic states [36].

II. TOOLBOX OF PHOTONIC QUANTUM SIMULATOR

A. Entangled Photon-Pair Source

Spontaneous parametric down-conversion (SPDC) is commonly used to generate entangled photons in which the $\chi^{(2)}$ optical nonlinearity of a birefringent crystal is used. In the photon picture, SPDC can be viewed as a decay of a pump photon into two photons: signal photon and idler photon. SPDC photons can only be emitted if the so-called phase matching conditions are fulfilled. In this conversion process, energy and momentum are conserved. Energy conservation requires that

$$h\nu_{\text{pump}} = h\nu_s + h\nu_i \tag{1}$$

where ν_{pump}, ν_s, and ν_i are the frequencies of the pump, signal, and idler photons, respectively. And momentum conservation requires that

$$\hbar\vec{k}_{\text{pump}} \simeq \hbar\vec{k}_s + \hbar\vec{k}_i \tag{2}$$

where $\vec{k}_{\text{pump}}$, $\vec{k}_s$, and $\vec{k}_i$ are the wave numbers of the pump, signal, and idler photons, respectively, and $|\vec{k}| = \frac{n2\pi}{\lambda}$ (the vacuum wavelength λ and the refractive index n of the media). Given a well-defined pump, these two conditions govern the wavelengths and the directions of emission of the down-converted photons. Due to energy and momentum conservation the down-converted photons are strongly correlated in these two degrees of freedom (DoF): frequency and direction.

In a birefringent crystal, the refractive indices of the extraordinary and ordinary polarization mode are different. When a birefringent crystal is pumped by an extraordinarily polarized beam, there are two different types of phase-matching conditions. Type-I: The signal and idler photons have the same polarization and are ordinarily polarized. Type-II: Signal and idler photons are orthogonally polarized, that is, one ordinarily (o-photon) and the other extraordinarily (e-photon). Most of the experiments described in this work are based on type-II SPDC.

By employing the SPDC process, one can also generate polarization-entangled photon pairs. As shown in Fig. 1, an incident pump photon can spontaneously decay into two photons that are entangled in momentum and energy. Each photon can be emitted along a cone in such a way that two photons of a pair are found opposite each other on the respective cones. The two photons are orthogonally polarized. Along the directions where the two cones overlap, one obtains polarization-entangled pairs. Moreover, in order to obtain high-quality entangled

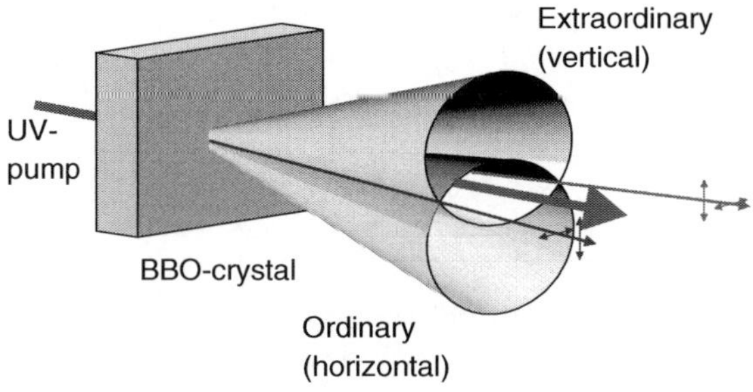

Figure 1. A schematic overview of the noncollinear type-II spontaneous parametric down-conversion (SPDC). An incident pump photon can spontaneously decay into two photons. Each photon can be emitted along a cone in such a way that two photons of a pair are found opposite to each other on the respective cones. The two photons are orthogonally polarized. Along the directions where the two cones overlap, one obtains polarization-entangled pairs. The figure is taken and adapted from Ref. [37].

photon pairs from the noncollinear configuration, it is necessary to erase timing information about the photons stemming from the walk-off effect. Because of the birefringent properties of BBO, the e-photon and the o-photon travel with different speeds. Therefore, this time information is correlated with the polarization of the photon in the same spatial mode and provides distinguishability. One uses a combination of a half-wave plate (HWP) and compensation BBO crystals (Comp BBO) in each arm to erase this information. The polarization of the down-converted photon is rotated by 90° via the HWP, oriented along 45° such that the e-photon and o-photon are exchanged. The Comp BBO is oriented the same as the down-conversion crystal, but is half the thickness of the down-conversion crystal. This reverses the walk-off effects by half, which consequently erases the time information and reduces the effect of the transverse walk-off as well [38].

The ideal polarization-entangled state produced by down-conversion has the form

$$|\Psi\rangle = \frac{1}{\sqrt{2}}(|H_1 V_2\rangle - e^{i\varphi}\,|V_1 H_2\rangle) \tag{3}$$

where φ is the phase. By adjusting this phase φ, we can change the state $|\Psi\rangle$ between $|\Psi^-\rangle$ or $|\Psi^+\rangle$. Experimentally, this can be realized by tilting one of the compensation BBOs.

B. Encoding and Processing Quantum Information on Single Photon

The qubit, $|Q\rangle$, can be written as

$$|Q\rangle = \alpha\,|0\rangle + \beta\,|1\rangle \tag{4}$$

where α, β are complex numbers and $|\alpha|^2 + |\beta|^2 = 1$. It has been realized physically utilizing different DoF of many different particles and systems [39]. For instance, qubits can be encoded in the internal energy states of a single atom, or the spin directions of an electron in a quantum dot. In this review, we mainly focus on qubits encoded in the polarization and path DoF of photons. For a polarization qubit, we assign

$$\begin{aligned} |0\rangle &= |H\rangle \\ |1\rangle &= |V\rangle \end{aligned} \tag{5}$$

where $|H\rangle$ and $|V\rangle$ are quantum states of photons with horizontal and vertical polarizations, respectively. Therefore, a polarization-encoded qubit $|Q\rangle$ can be represented as

$$|Q\rangle = \alpha\,|0\rangle + \beta\,|1\rangle = \alpha\,|H\rangle + \beta\,|V\rangle \tag{6}$$

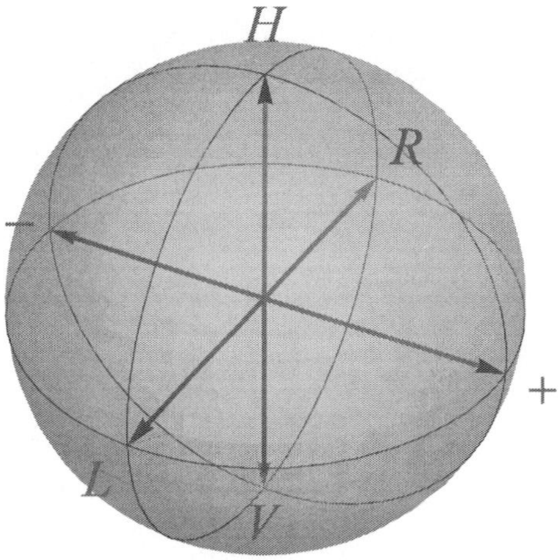

Figure 2. Bloch sphere of a polarization qubit. A pure qubit state, as in Eq. (7), is a point on the surface of the Bloch sphere. $|H\rangle$ and $|V\rangle$ lie on the poles of the sphere while $|+\rangle$ and $|-\rangle$, and $|L\rangle$ and $|R\rangle$ (refer to Table I) are located on the equatorial plane. These states lie on the mutually orthogonal axes. θ and φ uniquely define a point on the surface of the Bloch sphere.

The Hilbert space of one qubit can be represented graphically by the so-called Bloch sphere. A one-to-one mapping between the Bloch sphere and the Poincaré sphere is widely used in polarimetry. A pure qubit state may be rewritten as

$$|Q\rangle = \cos\frac{\theta}{2}\,|H\rangle + e^{i\varphi}\sin\frac{\theta}{2}\,|V\rangle \tag{7}$$

where θ and φ are the angles of the three-dimensional polar coordinate system shown in Fig. 2. In Table I, we list the mapping from the qubit states to polarization states and their Jones matrix representations.

To manipulate and measure polarization-encoded qubits, one can use a HWP, a quarter-wave plate (QWP), and a polarizing beam splitter (PBS). The unitary operation of a HWP oriented along θ can be represented as

$$\mathrm{U_{HWP}}(\theta) = \begin{pmatrix} -\cos 2\theta & -\sin 2\theta \\ -\sin 2\theta & \cos 2\theta \end{pmatrix} \tag{8}$$

The operation of a QWP oriented along θ can be represented as

$$\mathrm{U_{QWP}}(\theta) = \begin{pmatrix} 1-(1+i)\cos^2\theta & -(1+i)\sin\theta\cos\theta \\ -(1+i)\sin\theta\cos\theta & 1-(1+i)\sin^2\theta \end{pmatrix} \tag{9}$$

A PBS is a device that transmits horizontally polarized light and reflects vertically polarized light. The combination of a PBS, an HWP, and a QWP is able to measure the polarization state of a photon in an arbitrary basis.

Another remarkable feature of single photons is the capability of using various DoFs at the same time. For example, in addition to the internal spin or polarization,

TABLE I
Some Qubit States That Correspond to Polarization States of Photons and Their Jones Matrix Representation

Qubit State	Polarization State	Jones Matrix	Polarization Name
$\|0\rangle$	$\|H\rangle$	$\begin{pmatrix}1\\0\end{pmatrix}$	Horizontal linear
$\|1\rangle$	$\|V\rangle$	$\begin{pmatrix}0\\1\end{pmatrix}$	Vertical linear
$\frac{1}{\sqrt{2}}(\|0\rangle + \|1\rangle)$	$\frac{1}{\sqrt{2}}(\|H\rangle + \|V\rangle) = \|+\rangle$	$\frac{1}{\sqrt{2}}\begin{pmatrix}1\\1\end{pmatrix}$	+45° Linear
$\frac{1}{\sqrt{2}}(\|0\rangle - \|1\rangle)$	$\frac{1}{\sqrt{2}}(\|H\rangle - \|V\rangle) = \|-\rangle$	$\frac{1}{\sqrt{2}}\begin{pmatrix}1\\-1\end{pmatrix}$	−45° Linear
$\frac{1}{\sqrt{2}}(\|0\rangle + i\|1\rangle)$	$\frac{1}{\sqrt{2}}(\|H\rangle + i\|V\rangle) = \|L\rangle$	$\frac{1}{\sqrt{2}}\begin{pmatrix}1\\i\end{pmatrix}$	Left-handed circular
$\frac{1}{\sqrt{2}}(\|0\rangle - i\|1\rangle)$	$\frac{1}{\sqrt{2}}(\|H\rangle - i\|V\rangle) = \|R\rangle$	$\frac{1}{\sqrt{2}}\begin{pmatrix}1\\-i\end{pmatrix}$	Right-handed circular

also the external orbital angular momentum [40,41] as well as path encoding can be used. In particular, the later is of interest due to its feasible manipulation. A general path qubit can be represented as

$$|Q\rangle = \alpha\,|a\rangle + \beta\,|b\rangle \tag{10}$$

where $|a\rangle$ and $|b\rangle$ are the orthogonal path states. It can be controlled by interferometric alignments using beam splitters and phase shifters. A 50:50 beam splitter (BS) is equivalent to a Hadamard gate for a path qubit. In general, the operation of a 50:50 BS can be represented as

$$\begin{aligned} |a\rangle &\underset{\rightarrow}{\text{BS}} \frac{|a\rangle + e^{i\chi}\,|b\rangle}{\sqrt{2}} \\ |b\rangle &\underset{\rightarrow}{\text{BS}} \frac{|b\rangle + e^{i(\pi-\chi)}\,|a\rangle}{\sqrt{2}} \end{aligned} \tag{11}$$

where $\chi = \frac{\pi}{2}$ for a symmetric beam splitter. A phase shifter (PS) on spatial mode $|a\rangle$ can be represented as

$$\begin{aligned} |a\rangle &\underset{\rightarrow}{\text{PS}}\, e^{i\phi}\,|a\rangle \\ |b\rangle &\underset{\rightarrow}{\text{PS}}\, |b\rangle \end{aligned} \tag{12}$$

where ϕ is the amount of the phase shift. Note that this section has been reviewed in Ref. [42] in greater details.

C. Photon–Photon Interaction via Measurement

One of the major challenges for processing quantum information with photons is the photonic two-qubit gate due to the extremely weak photon–photon interactions. In 2001, Knill, Laflamme, and Milburn (KLM) showed that efficient linear optical quantum computation was possible using only single-photon sources, linear optical interferometers, and single photon detectors [36]. The essential ingredients of KLM scheme were measurement-induced nonlinearities relying on nonclassical two-photon interference. This two-photon interference is also called Hong–Ou–Mandel (HOM) effect [43], which is crucial for many quantum information-processing protocols, including quantum computation [31–35,44], quantum communication [45–49], and photonic quantum simulation [19,22]. HOM interference originates from the statistics of bosons and the unitary transformation of a beam splitter. As shown in Fig. 3, when two indistinguishable single photons impinge on a balanced beam splitter, they bunch together and leave together from either one of the two possible outputs of the beam splitter. If one measures the coincidence counts between two outputs, no coincidence counts will be obtained.

Note that when we use entangled photons as inputs for the beam splitter, HOM interference allows us to stay in spin-zero singlet subspace. In the context of quantum simulation, this feature has significant importance, because many strongly correlated spin systems exhibit spin-zero ground states. This originates from Marshall's theorem [50] and its extension [51], which state that the absolute ground state of Heisenberg-type Hamiltonian has total spin zero ($S^2 = 0$) for N spins

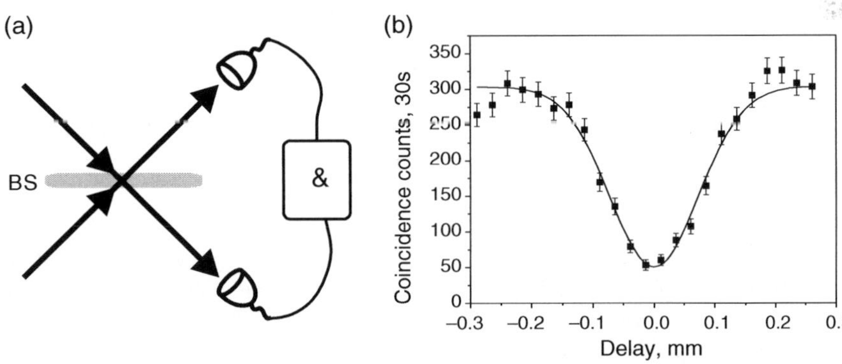

Figure 3. Hong–Ou–Mandel interference. (a) Two indistinguishable single photons impinge on a balanced beam splitter (BS). No coincidences will be obtained between two detectors if these two photons arrive at the BS at the same time. This interference effect stems from the statistics of bosons. (b) When we tune the relative spatial delay between two photons, the maximal drop in coincidence counts occurs for the optimal temporal overlap of the two photons.

on a bipartite lattice with nearest-neighbor antiferromagnetic Heisenberg-type interactions.

This constraint leads to the fact that the ground state can be built as a superposition of pairs of spins forming singlets or equally so-called valence bonds. If all the spins are covered by valence bonds, which are maximally entangled states, then the ground state's total spin is zero and nonmagnetic. This is established by valence bonds that are either static and localized or fluctuating as a superposition of different partitionings of spins. In general, the equally weighted superposition of all different localized valence-bond states corresponds to a quantum spin liquid, the so-called resonating valence-bond state [52,53].

III. QUANTUM CHEMISTRY ON A PHOTONIC QUANTUM COMPUTER

Photonic quantum systems are ideally suited to promise potential returns in understanding detailed quantum phenomenon of small-sized quantum chemical systems. By using two entangled photons and taking advantage of additional photonic DoF to implement an arbitrary controlled-unitary evolution, Lanyon et al. simulated the hydrogen molecule. This experiment opened the field of photonic chemical quantum simulations [19]. The hydrogen molecule Hamiltonian in the smallest atom-centered chemistry basis results in a 6×6 matrix that has block diagonal form with two 2×2 blocks and two 1×1 blocks. The 2×2 blocks can be diagonalized by carrying out the iterative phase estimation algorithm [30].

Ma et al. [22] have extended the simulation of molecular systems to a spin-1/2 tetramer and have shown that frustration in such Heisenberg-interacting spin systems can be investigated by using a photonic quantum simulator. The pairing of quantum correlations of spin systems is of crucial importance for understanding high-temperature superconductivity in cuprates [54] and quantum magnetism in solid-state physics, as well as for chemical reactions. The so-called valence-bond state, where two electrons from different atoms share an anti-correlated spin state, is a useful tool for investigating such problems. The same quantum correlation of valence-bond states can be simulated by a pair of photons that is maximally entangled in polarization (i.e., that the two photons are always orthogonally polarized). Ma's experiment used two entangled photon pairs in a singlet state to simulate the spin of a Heisenberg-interacting spin tetramer, where the singlet state is analogous to the antiferromagnetic coupling of two spin-1/2 particles or valence-bond state. Then an analog quantum simulation was performed by superimposing photons from each pair at a beam splitter with a tunable splitting ratio, followed by a measurement of the photons in the output ports (Fig. 4a). Depending on the interaction strength, the transition from local to resonant valence bonds ground state could be observed (Fig. 4b). The precise addressability of individual photons provided

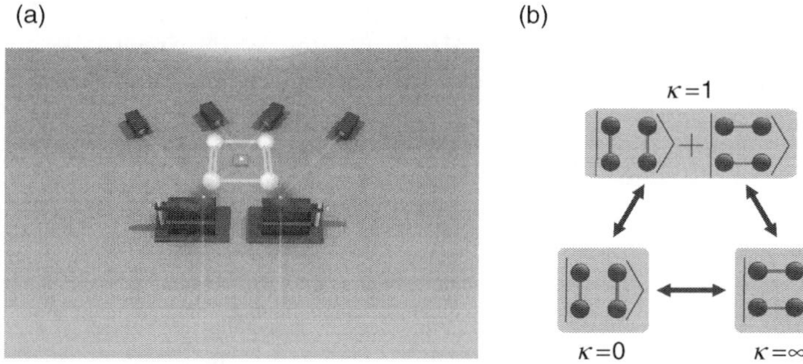

Figure 4. (a) Artistic view of the photonic quantum simulation of valence-bond state. Two entangled photon pairs are generated via SPDC. Superimposing one single photon from each pair at a tunable beam splitter results in quantum interference such that the measured four-photon coincidences correspond to the ground state, for example, of a Heisenberg-interacting spin tetramer. Dependent on the beam splitter reflectivity, frustration in valence bond states or so-called spin liquid states could be investigated. (b) Dependent on the neighboring spin–spin interaction strength (κ) controlled by the splitting ratio of the beam splitter, various spin-paring configurations can be simulated. Figure 4a is taken from Ref. [55]. (*See the color version of this figure in the Color Plates section.*)

insight into the pairwise quantum correlations and allowed the observation that the energy distributions are restricted by the role of quantum monogamy [56,57]. Such quantum simulation experiments will be of interest for quantum chemistry with small numbers of particles and might allow in the near future the simulation of aromatic systems such as benzene (Fig. 5).

However, in the case of photonic quantum systems, arbitrary unitary matrices can be also implemented by interferometric beam splitter arrays or so-called multiport arrays [58]. Such interferometric setups are promising for complex quantum walk experiments that enable, for example, the simulation of excitation transfer in

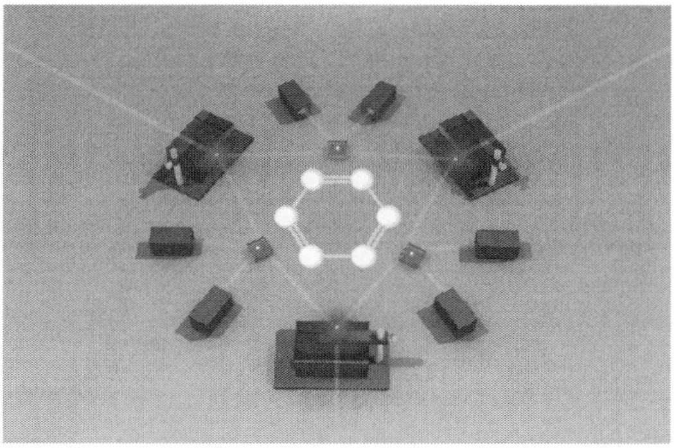

Figure 5. Future experiments using more entangled photon pairs may allow the study of ground-state properties of molecular ground states, such as the delocalized bonds in benzene. This figure is taken from Ref. [55]. (*See the color version of this figure in the Color Plates section.*)

biochemical systems. Quantum walks have been realized using bulk optics [20,59] and waveguides [21,23,24]. Single-particle quantum walks can also be carried out with coherent classical light, but truly novel effects happen when more than one photon is employed.

IV. CONCLUSIONS

The ability to monitor the full dynamics of individual particles and bonds provides a fascinating perspective for the quantum simulation of small molecules, even for the case that such simulators would not be immediately competitive with traditional classical computers. In the future, quantum simulators based on optics are expected to tackle grand challenges such as accurate simulation of molecules and materials, including aromaticity and reaction centers.

Acknowledgments

This work was supported by the European Commission, Q-ESSENCE (No. 248095), QUILMI (No. 295293), EQUAM (No. 323714), PICQUE (No. 608062), and the ERA-Net CHISTERA project QUASAR, the John Templeton Foundation, the Vienna Center for Quantum Science and Technology (VCQ), the Austrian Nanoinitiative NAP Platon, the Austrian Science Fund (FWF) through the SFB FoQuS (No. F4006-N16 and SFB F40-FoQus F4012-N16), START (No. Y585-N20), grant (No. P24273-N16) and the doctoral programme CoQuS, the Vienna Science and Technology Fund (WWTF) under grant ICT12-041, and the Air Force Office of Scientific Research, Air Force Material Command, United States Air Force, under grant number FA8655-11-1-3004.

REFERENCES

1. R. P. Feynman, *Int. J. Theor. Phys.* **21**, 467–488 (1982).
2. M. Lewenstein, et al., *Adv. Phys.* **56**, 243–379 (2007).
3. W. S. Bakr, J. I. Gillen, A. Peng, S. Folling, and M. Greiner, *Nature* **462**, 74–77 (2009).
4. W. S. Bakr, et al., *Science* **329**, 547–550 (2010).
5. S. Trotzky, et al., *Nat. Phys.*, **6**, 988–1004, (2010).
6. J. F. Sherson, et al., *Nature* **467**, 68–72 (2010).
7. C. Weitenberg, et al., *Nature* **471**, 319–324 (2011).
8. A. Friedenauer, H. Schmitz, J. T. Glueckert, D. Porras, and T. Schaetz, *Nat. Phys.* **4**, 757–761 (2008).
9. R. Gerritsma, et al., *Nature* **463**, 68–71 (2010).
10. K. Kim, et al., *Nature* **465**, 590–593 (2010).
11. J. T. Barreiro, et al., *Nature* **470**, 486–491 (2011).

12. R. Islam, et al., *Nat. Comm.* **2**, 377 (2011).
13. B. P. Lanyon, et al., *Science* **334**, 57–61 (2011).
14. X. Peng, J. Zhang, J. Du, and D. Suter, *Phys. Rev. Lett.* **103**, 140501 (2009).
15. J. Du, et al., *Phys. Rev. Lett.* **104**, 030502 (2010).
16. M. Neeley, et al., *Science* **325**, 722–725 (2009).
17. C.-Y. Lu, et al., *Phys. Rev. Lett.* **102**, 030502 (2009).
18. J. K. Pachos, et al., *New J. Phys.* **11**, 083010 (2009).
19. B. P. Lanyon, et al., *Nat. Chem.* **2**, 106–111 (2010).
20. M. A. Broome, et al., *Phys. Rev. Lett.* **104**, 153602 (2010).
21. A. Peruzzo, et al., *Science* **329**, 1500–1503 (2010).
22. X.-s. Ma, B. Dakić, W. Naylor, A. Zeilinger, and P. Walther, *Nat. Phys.* **7**, 399–405 (2011).
23. J. C. F. Matthews, et al., *Sci. Rep.* **3**, 1539 (2013).
24. L. Sansoni, et al., *Phys. Rev. Lett.* **108**, 010502 (2012).
25. I. Buluta and F. Nori, *Science* **326**, 108–111 (2009).
26. M. Born and V. Fock, *Z. Phys. A* **51**, 165–180 (1928).
27. S. Lloyd, *Science* **273**, 1073–1078 (1996).
28. E. Farhi, J. Goldstone, S. Gutmann, and M. Sipser, *Quantum Computation by Adiabatic Evolution. arXiv:quant-ph/0001106v1* (2000).
29. J. Biamonte, V. Bergholm, J. Whitfield, J. Fitzsimons, and A. Aspuru-Guzik, *AIP Adv.* **1**, 022126 (2011).
30. A. Aspuru-Guzik, A. D. Dutoi, P. J. Love, and M. Head-Gordon, *Science* **309**, 1704–1707 (2005).
31. J. L. O'Brien, G. J. Pryde, A. G. White, T. C. Ralph, and D. Branning, *Nature* **426**, 264–267 (2003).
32. S. Gasparoni, J. W. Pan, P. Walther, T. Rudolph, and A. Zeilinger, *Phys. Rev. Lett.* **93**, 020504 (2004).
33. N. K. Langford, et al., *Phys. Rev. Lett.* **95**, 210504 (2005).
34. N. Kiesel, C. Schmid, U. Weber, R. Ursin, and H. Weinfurter, *Phys. Rev. Lett.* **95**, 210505 (2005).
35. R. Okamoto, H. F. Hofmann, S. Takeuchi, and K. Sasaki, *Phys. Rev. Lett.* **95**, 210506 (2005).
36. E. Knill, R. Laflamme, and G. Milburn, *Nature* **409**, 46–52 (2001).
37. G. Weihs, Ph.D. thesis, University of Vienna, 1999.
38. P. G. Kwiat, et al., *Phys. Rev. Lett.* **75**, 4337–4341 (1995).
39. T. D. Ladd, et al., *Nature* **464**, 45–53 (2010).
40. A. Mair, A. Vaziri, G. Weihs, and A. Zeilinger, *Nature* **412**, 313–316 (2001).
41. G. Molina-Terriza, J. P. Torres, and L. Torner, *Nat. Phys.* **3**, 305–310 (2007).
42. X. s. Ma, Ph.D. thesis, University of Vienna, 2010.
43. C. K. Hong, Z. Y. Ou, and L. Mandel, *Phys. Rev. Lett.* **59**, 2044–2046 (1987).
44. P. Kok, et al., *Rev. Mod. Phys.* **79**, 135 (2007).
45. D. Bouwmeester, et al., *Nature* **80**, 575–579 (1997).
46. J.-W. Pan, D. Bouwmeester, H. Weinfurter, and A. Zeilinger, *Phys. Rev. Lett.* **80**, 3891–3894 (1998).
47. T. Jennewein, G. Weihs, J.-W. Pan, and A. Zeilinger, *Phys. Rev. Lett.* **88**, 017903 (2001).
48. J.-W. Pan, S. Gasparoni, R. Ursin, G. Weihs, and A. Zeilinger, *Nature* **423**, 417–422 (2003).

49. P. Walther, et al., *Phys. Rev. Lett.* **94**, 040504 (2005).
50. W. Marshall, *Proc. R. Soc. A* **232**, 48–68 (1955).
51. E. Lieb and D. Mattis, *J. Math. Phys.* **3**, 749–751 (1962).
52. I. Affleck, T. Kennedy, E. H. Lieb, and H. Tasaki, *Phys. Rev. Lett.* **59**, 799–802 (1987).
53. L. Balents, *Nature* **464**, 199–208 (2010).
54. P. W. Anderson, *Science* **235**, 1196 (1987).
55. A. Aspuru-Guzik and P. Walther, *Nat. Phys.* **8**, 285–291 (2012).
56. V. Coffman, J. Kundu, and W. K. Wootters, *Phys. Rev. A* **61**, 052306 (2000).
57. T. J. Osborne and F. Verstraete, *Phys. Rev. Lett.* **96**, 220503 (2006).
58. M. Reck, A. Zeilinger, H. J. Bernstein, and P. Bertani, *Phys. Rev. Lett.* **73**, 58–61 (1994).
59. T. Kitagawa et al., *Nat. Commun.* **3**, 882 (2012).

PROGRESS IN COMPENSATING PULSE SEQUENCES FOR QUANTUM COMPUTATION

J. TRUE MERRILL and KENNETH R. BROWN

School of Chemistry and Biochemistry, School of Computational Science and Engineering, School of Physics, Georgia Institute of Technology, Ford Environmental Science and Technology Building, 311 Ferst Dr, Atlanta, GA 30332-0400, USA

Advances in Chemical Physics, Volume 154: Quantum Information and Computation for Chemistry, First Edition. Edited by Sabre Kais.

I. INTRODUCTION

In any experiment, external noise sources and control errors limit the accuracy of the preparation and manipulation of quantum states. In quantum computing, these effects place an important fundamental limit on the size and accuracy of quantum processors. These restrictions may be reduced by quantum error correction. Although sophisticated quantum error-correcting codes are robust against any general error, these codes require large-scale multipartite entanglement and are especially challenging to implement in practice [1]. Therefore, it is of great interest to investigate schemes that reduce errors with a smaller resource overhead. One alternative strategy involves replacing an error-prone operation by a pulse sequence that is robust against the error.

The basis of our strategy is that all noises and errors can be treated as an unwanted dynamic generated by an error Hamiltonian. This error Hamiltonian can arise either through interactions with the environment or by the misapplication of control fields. This view unifies the pulse sequences developed for combating unwanted interactions (e.g., dynamic decoupling and dynamically corrected gates [2–4]), with those for overcoming systematic control errors, compensating composite pulse sequences [5]. In each case, the methods are limited by the rate at which control occurs relative to the time scale over which the error Hamiltonian fluctuates. However, many experiments are limited by control errors and external fields that vary slowly relative to the time scale of a single experimental run but vary substantially over the number of experiments required to obtain precise results.

In this chapter, we examine a number of techniques for handling unwanted control errors, including amplitude errors, timing errors, and frequency errors in the control field. We emphasize the common principles used to develop compensating pulse sequences and provide a framework in which to develop new sequences, which will be of use to both quantum computation and coherent atomic and molecular spectroscopy.

II. COHERENT CONTROL OVER SPIN SYSTEMS

The accurate control of quantum systems is an important prerequisite for many applications in precision spectroscopy and in quantum computation. In complex experiments, the task reduces to applying a desired unitary evolution using a finite set of controls, which may be constrained by the physical limitations of the experimental apparatus. Stimulated by practical utility, quantum control theory has become an active and diverse area of research [6–10]. Although originally developed using nuclear-magnetic resonance (NMR) formalism, compensating pulse sequences can be approached from the perspective of quantum control theory with unknown systematic errors in the controls [11]. Here, we review several fundamental concepts in quantum control and apply these ideas using NMR as an instructive example. Although we restrict the discussion to control in NMR spectroscopy, the following analysis is quite general as several other coherent systems (e.g., semiconductor quantum dots [12], superconducting qubits [13–15], and trapped ions [16,17]) may be considered by minor modifications to the Hamiltonian.

In practice, a desired evolution is prepared by carefully manipulating the coupling of the system to a control apparatus, such as a spectrometer. In the absence of relaxation, the coherent dynamics are governed by the quantum propagator $U(t)$, which in nonrelativistic quantum mechanics must satisfy an operational Schrödinger equation:

$$\dot{U}(t) = -\mathrm{i}\left(\sum_{\mu} u_\mu(t) H_\mu\right) U(t), \qquad U(0) = \mathbb{1} \tag{1}$$

In this model, the unitless Hamiltonians $H_\mu \in \{H_1, H_2, \ldots, H_n\}$ are modulated by real-valued control functions $u_\mu(t) \in \{u_1(t), u_2(t), \ldots, u_n(t)\}$ and represent the n available degrees of control for a particular experimental apparatus. In analogy with linear vector spaces, we interpret the vector $\boldsymbol{u}(t) = (u_1(t), u_2(t), \ldots, u_n(t))$ as a vector function over the manifold of control parameters, with components $u_\mu(t)$ representing the magnitudes of the control Hamiltonians with units of angular frequency. It is convenient to introduce a second vector of control Hamiltonians $\boldsymbol{H} = (H_1, H_1, \ldots, H_n)$ and the short-hand notation $H(t) = \boldsymbol{u}(t) \cdot \boldsymbol{H}$. We omit a term that represents the portion of the total Hamiltonian outside direct control (i.e., a drift Hamiltonian); in principle it is always possible to work in an interaction picture where this term is removed. Alternatively, one may assign a Hamiltonian H_0 to represent this interaction, with the understanding that $u_0(t) = 1$ for all t.

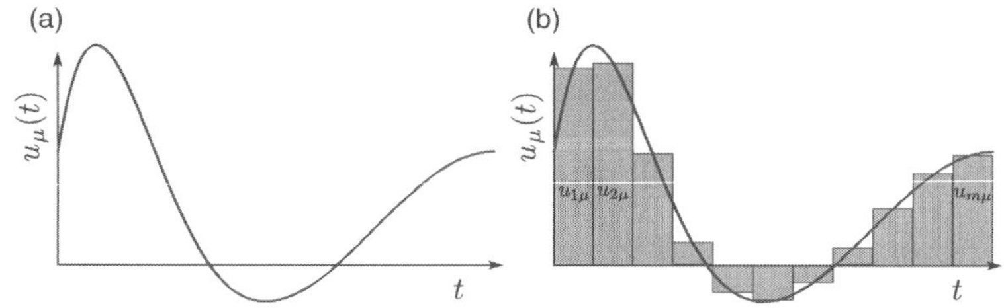

Figure 1. The experimental controls available to manipulate a quantum system are modeled using a set of real-valued control functions $\{u_\mu(t)\}$ that modify a set of dimensionless Hamiltonians $\{H_\mu\}$. (**a**) An example control function $u_\mu(t)$. (**b**) A discrete approximation for $u_\mu(t)$ composed of square pulses.

For a given control system, a natural question concerns the optimal approximation of a desired unitary propagation using a set of constrained control functions. Constraints may include limitations on the total operation length, control amplitudes, or derivatives. The study of this question requires the solution of Eq. (1) for a particular set of controls; such solutions may be obtained using several methods, including the Dyson series [18], and the Magnus [19,20], Fer [21], and Wilcox [22] expansions. We label particular solutions to the control equation over the interval $t_i \leq t \leq t_f$ as $U(\boldsymbol{u}(t); t_f, t_i)$. If the set of all possible solutions to Eq. (1) is the set of all unitary gates on the Hilbert space (i.e., the solutions form a representation of $\mathsf{U}(n)$ or $\mathsf{SU}(n)$, to be discussed in Section III.A) then the system is *operator controllable* [10,23].

For some applications, it is convenient to assume that the operation time τ is discretized into m-many time intervals, over which the control functions u_μ are constant; that is, during the kth time interval Δt_k the Hamiltonian is time-independent and the resulting evolution operator is $U_k = U(\boldsymbol{u}_k; t_k + \Delta t_k, t_k) = \exp(-\mathrm{i}\Delta t_k \boldsymbol{u}_k \cdot \boldsymbol{H})$. Figure 1 illustrates an example control function and a possible discretization scheme. Over each time interval, the applied unitary operation is a square pulse. In many experiments, the application of gates using sequences of square pulses is preferred for simplicity. The discretization of the control functions into square pulses allows one to solve Eq. (1) in a piecewise fashion. The total propagator for a sequence of m-many time-steps is given by the time-ordered product

$$U(\boldsymbol{u}(t), \tau, 0) = U_m U_{m-1} \dots U_k \dots U_2 U_1 = \prod_{k=1}^{m} U_k \qquad (2)$$

where the differentiable multiplication of each successive operator is understood to be taken on the left; this is in agreement with standard quantum mechanics conventions, where operations are ordered right-to-left, but at odds with some

NMR literature where successive operations are ordered left-to-right. Frequently, we will consider propagators over the entire duration of a sequence $0 \leq t \leq \tau$; as a matter of notational convenience we drop the time interval labels whenever there is no risk of confusion. If a pulse sequence $U(\boldsymbol{u}(t))$ is equivalent to a target operation U_T in the sense that during an experiment, $U(\boldsymbol{u}(t))$ may be substituted for U_T, then $U(\boldsymbol{u}(t))$ is a *composite pulse sequence* [24,25].

The usefulness of composite pulse sequences lies in that in many cases, one may simulate a target unitary transformation U_T, which may be difficult to directly implement, by instead implementing a sequence of simpler pulses that under some set of conditions is equivalent to U_T. Composite pulse sequences may be designed to have several important advantages over a directly applied unitary, such as improved resilience to errors. The properties of pulses or pulse sequences may be considered either by the transformations produced on a particular initial state ρ, or by comparing the sequence to an ideal operation, which contains information on how the sequence transforms all initial states. Following Levitt [5], we assign composite pulses into two classes: the fully compensating class A, and the partially compensating class B. The properties of these classes are briefly reviewed.

Class A: All composite pulse sequences in class A may be written in the form

$$e^{i\phi} U(\boldsymbol{u}(t)) = U_T \tag{3}$$

where it is assumed that the individual pulses in the sequence are error-free. The global phase ϕ is irrelevant to the dynamics; for any initial state ρ, both the pulse sequence and the target operation apply the same transformation $(U(\boldsymbol{u}(t))\rho U^\dagger(\boldsymbol{u}(t)) = U_T \rho U_T^\dagger)$. Sequences in this class are suited for use in quantum computation, because the transformation is independent of the initial quantum state. The study of these sequences will be the primary topic of this article.

Class B: Composite pulse sequences in class B transform one particular initial condition to a set of final conditions, which for the purposes of the experiment, are equivalent. For example, consider an NMR experiment on a spin $I = 1/2$ nucleus where the nuclear magnetization, initially oriented "spin-up" (i.e., $\rho = \mathbb{1}/2 + H_z$, where H_x, H_y, H_z are the nuclear angular momentum operators) is transferred to the H_x–H_y plane by a pulse sequence. In this sense, all sequences that apply the transformations are equivalent up to a similarity transform $\exp(-i\beta H_z)$, which applies a H_z phase to the spin. Other sequences in class B may satisfy $U(\boldsymbol{u}(t))\rho U^\dagger(\boldsymbol{u}(t)) = U_T \rho U_T^\dagger$ for a particular initial state ρ, but fail to satisfy Eq. (3). All sequences in class B are generally not well suited for use in quantum computation, because implementation requires specific knowledge of the initial and final states of the qubit register. Class B sequences are however useful in other applications, including NMR [24–26], MRI [27], control over nitrogen vacancy centers [28], and in ion trapping experiments [29]. We close our discussion by noting that class B sequences may be converted into a fully compensating class A sequence by a certain symmetrical construction [30].

A. Errors in Quantum Control

We now consider the effects of unknown errors in the control functions. Recall that a pulse sequence may be specified by a set of control functions $\{u_\mu(t)\}$, which we group into a control vector $\boldsymbol{u}(t)$. Suppose, however, during an experiment an unknown systematic error deforms each of the applied controls from $u_\mu(t)$ to $v_\mu(t)$. In the presence of unknown errors, the perfect propagators $U(\boldsymbol{u}(t); t_f, t_i)$ are replaced with their imperfect counterparts $V(\boldsymbol{u}(t); t_f, t_i) = U(\boldsymbol{v}(t); t_f, t_i)$, which may be regarded as an image of the perfect propagator under the deformation of the controls.

In practice, systematic errors typically arise from a miscalibration of the experimental system, for example, an imprecise measurement of the intensity or frequency of a controlling field. In these cases, it is appropriate to introduce a deterministic model for the control deformation. Specifically, $\boldsymbol{v}(t)$ must be a function of the controls $\boldsymbol{u}(t)$, and for each component of the imperfect control vector we may write

$$v_\mu(t) = f_\mu[\boldsymbol{u}(t); \epsilon] \tag{4}$$

where the differentiable functional f_μ is the *error model* for the control and the variable ϵ is an unknown real parameter that parametrizes the magnitude of the error. This construction may be generalized to the case of multiple systematic errors by considering error models of the form $f_\mu[\boldsymbol{u}(t); \epsilon_i, \epsilon_j, \ldots, \epsilon_k]$. The model f_μ is set by the physics of the problem, and is chosen to produce the correct evolution under the imperfect controls. Formally, we may perform the expansion

$$f_\mu[\boldsymbol{u}(t); \epsilon] = f_\mu[\boldsymbol{u}(t); 0] + \epsilon \frac{d}{d\epsilon} f_\mu[\boldsymbol{u}(t); 0] + \frac{\epsilon^2}{2!} \frac{d^2}{d\epsilon^2} f_\mu[\boldsymbol{u}(t); 0] + \mathcal{O}(\epsilon^3) \tag{5}$$

then by the condition that when $\epsilon = 0$ the control must be error free, it is trivial to identify $u_\mu(t) = f_\mu[\boldsymbol{u}(t); 0]$. Frequently, it is sufficient to consider models that are linear in the parameter ϵ. In this case, we introduce the shorthand notation $\delta u_\mu(t) = \frac{d}{d\epsilon} f_\mu[\boldsymbol{u}(t); 0]$ and the corresponding vector $\boldsymbol{\delta u}(t)$ to represent the first-order deformation of the controls, so that the imperfect controls take the form $v_\mu(t) = u_\mu(t) + \epsilon \delta u_\mu(t)$. Figure 2 illustrates two common error models: constant offsets from the ideal control value ($f_\mu[\boldsymbol{u}(t); \epsilon] = u_\mu(t) + \epsilon$), and errors in the control amplitude ($f_\mu[\boldsymbol{u}(t); \epsilon] = (1 + \epsilon) u_\mu(t)$).

A natural question to ask is what effect unknown errors have on the evolution of the system. It is obvious that imperfect pulses make accurate manipulation of a quantum state difficult. One may be surprised to find that for some cases, the effects

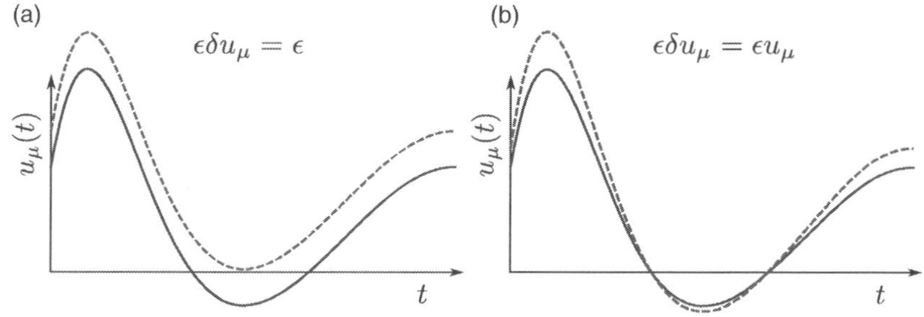

Figure 2. Systematic errors induce deformations of the ideal control functions $u_\mu(t)$ (solid curves) to the imperfect controls $v_\mu(t) = u_\mu(t) + \epsilon\delta u_\mu(t)$ (dashed curves). Common error models include (**a**) a constant unknown offset in the control and (**b**) an error in the amplitude of the control function.

of errors on the controls may be systematically removed, without knowledge of the amplitude ϵ. The method we describe involves implementing a compensating composite pulse sequence that is robust against distortion of the controls by a particular error model. As an example, consider a case where an experimentalist would like to approximate a target unitary $U_T = U(\boldsymbol{u}(t); \tau, 0)$, where at least one of the controls is influenced by a systematic error. The target operation may be simulated up to $\mathcal{O}(\epsilon^n)$ if a set of control functions $\boldsymbol{u}(t)$ exists such that

$$V(\boldsymbol{u}(t); \tau, 0) = U(\boldsymbol{u}(t) + \epsilon\boldsymbol{\delta u}(t); \tau, 0) = U(\boldsymbol{u}(t); \tau, 0) + \mathcal{O}(\epsilon^{n+1}) \tag{6}$$

Many (infinite in most cases) sets of control functions $\boldsymbol{u}(t)$ implement a target unitary transformation, however, only a small subset of possible control functions are robust to distortion by a particular systematic error. If robust controls can be found, then pulse sequence $V(\boldsymbol{u}(t); \tau, 0)$ can be applied in the place of U_T, and the leading order terms of the offset $\epsilon\boldsymbol{\delta u}(t)$ are suppressed. A sequence with these properties is called a *compensating pulse sequence*, and may be thought of as a set of control functions that are optimized to remove the effect of leading-order terms of unknown systematic errors in the controls. By construction, sequences of this form are fully compensating (Class A).

When an experimentalist implements a compensating pulse sequence they attempt to apply the ideal operations $U(\boldsymbol{u}(t); \tau, 0)$, ignorant of the amplitude ϵ of a systematic error. However, the operations are not ideal and to emphasize this we introduce $V(\boldsymbol{u}(t); \tau, 0)$ to represent the *imperfect propagators* that are actually implemented when $U(\boldsymbol{u}(t); \tau, 0)$ is attempted. The functional dependence of the error in $V(\boldsymbol{u}(t); \tau, 0)$ is not explicitly written, allowing us to study different error models with the same pulse sequence.

Let us consider the dynamics of the system under the interaction frame Hamiltonian $H^I(t) = \sum_\mu \epsilon\delta u_\mu(t) H^I_\mu(t)$, where $H^I_\mu(t) = U^\dagger(\boldsymbol{u}(t'); t, 0) H_\mu U(\boldsymbol{u}(t'); t, 0)$

are the control Hamiltonians in the interaction frame. In this picture, $H^I(t)$ is regarded as a perturbation, and we associate the propagator $U^I(\boldsymbol{\delta u}(t); \tau, 0)$ as the particular solution to the interaction picture Schrödinger equation over the interval $0 \leq t \leq \tau$. Hence,

$$V(\boldsymbol{u}(t); \tau, 0) = U(\boldsymbol{u}(t); \tau, 0)U^I(\epsilon\boldsymbol{\delta u}(t); \tau, 0) \tag{7}$$

and from Eq. (6)

$$U^I(\epsilon\boldsymbol{\delta u}(t); \tau, 0) = \mathbb{1} + \mathcal{O}(\epsilon^{n+1}) \tag{8}$$

Quite generally, when a fully compensating pulse sequence is transformed into the interaction frame the resulting propagator must approximate the identity operation [31]. The techniques for constructing compensating pulse sequences discussed in the present chapter rely on performing a series expansion by powers of ϵ for the interaction frame propagator $U^I(\epsilon\boldsymbol{\delta u}(t); \tau, 0)$, then choosing a set of controls that remove the leading terms of the distortion, and finally transforming back out of the interaction frame.

B. NMR Spectroscopy as a Model Control System

In this section, we will apply the ideas developed thus far to a model one-qubit NMR quantum computer [32,33], which serves as a relevant example of a system where coherent control is possible. In Section V.A, multi qubit operations are considered. The many possible physical implementations for a qubit are often based on a two-level subsystem of a larger Hilbert space. In this case, the qubit is defined on the angular momentum states of a spin-1/2 nucleus.

Consider an ensemble of spin $I = 1/2$ nuclei undergoing Larmor precession under a static magnetic field $\boldsymbol{B}_0$ oriented along the $\hat{\mathbf{z}}$ axis. The static field $\boldsymbol{B}_0$ induces a net magnetization among the nuclei. A transverse radiofrequency (rf) field near nuclear resonance $\boldsymbol{B}_1(t) = \boldsymbol{B}_1 \cos(\omega_1 t - \phi)$ is applied in the $\hat{\mathbf{x}}$–$\hat{\mathbf{y}}$ plane. A simplified diagram of an NMR spectrometer is provided in Fig. 3. The analysis is simplified by assuming that individual nuclei in the ensemble are decoupled by

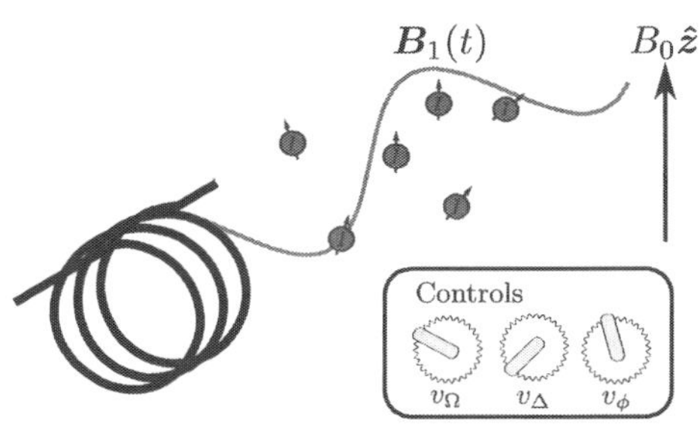

Figure 3. Simplified diagram of an NMR spectrometer. Control over an ensemble of nuclear spins is applied by a radiofrequency (rf) field $\boldsymbol{B}_1(t)$, where the Rabi frequency $\Omega(t)$, field detuning $\Delta(t)$, and phase $\phi(t)$ are control parameters. Unknown systematic control errors (e.g., poor intensity control, detuning errors) may be present.

the rapid tumbling of spins in the sample [25]. After transforming into the rotating frame, the Hamiltonian for a single spin may be written as

$$H(t) = \Delta(t)H_{\mathrm{z}} + \Omega(t)\big(\cos(\phi(t))H_{\mathrm{x}} + \sin(\phi(t))H_{\mathrm{y}}\big) \tag{9}$$

where

$$H_{\mathrm{x}} = \frac{1}{2}\begin{pmatrix} 0 & 1 \\ 1 & 0 \end{pmatrix}, \qquad H_{\mathrm{y}} = \frac{1}{2}\begin{pmatrix} 0 & -\mathrm{i} \\ \mathrm{i} & 0 \end{pmatrix}, \qquad H_{\mathrm{z}} = \frac{1}{2}\begin{pmatrix} 1 & 0 \\ 0 & -1 \end{pmatrix}$$

and the counter-rotating terms have been neglected under the rotating-wave approximation. Manipulation of the rf field provides a convenient set of controls to guide the evolution of the spins; it is assumed that the rf field detuning $\Delta(t)$, phase $\phi(t)$, and the Rabi frequency $\Omega(t)$ are each independently controllable and may suffer from independent systematic errors. Let the controls $\boldsymbol{u}(t)$ denote the vector $(\Omega(t)\cos(\phi(t)), \Omega(t)\sin(\phi(t)), \Delta(t))$. Then the Hamiltonian may be written as $H(t) = \boldsymbol{u}(t) \cdot \boldsymbol{H}$, and the resulting Schrödinger equation for the propagator is of the form of Eq. (1). Thus, the coherent spin dynamics in magnetic resonance spectroscopy can be reformulated in terms of a problem in quantum control.

1. *Error Models*

In practice, systematic errors in the controls caused by instrumental limitations prohibit the application of perfect pulse propagators. We consider several models for errors in the controls of a one-qubit NMR quantum computer. Often compensating pulse sequences are well optimized for one type of error, but provide no advantage against a different error model.

a. Amplitude Errors. An amplitude error arises from slow systematic variation in the amplitude of the rf field, resulting in a small offset in the applied Rabi frequency. Let Ω represent the ideal Rabi frequency and Ω' represent the offset. From the form of the control vector $\boldsymbol{u}(t)$ it follows that in the presence of the error, the $u_{\mathrm{x}}(t)$ and $u_{\mathrm{y}}(t)$ controls are distorted; that is, they are replaced by their imperfect counterparts $v_{\mathrm{x/y}}(t) = u_{\mathrm{x/y}}(t) + \epsilon_A \delta u_{\mathrm{x/y}}(t)$, where $\delta u_{\mathrm{x/y}}(t) = u_{\mathrm{x/y}}(t)$ is proportional to the ideal control value, and the error parameter $\epsilon_A = \Omega'/\Omega < 1$ is the relative amplitude of the field offset. The imperfect pulses take the form $V(\boldsymbol{u}(t)) = U(\boldsymbol{u}(t)(1 + \epsilon_A))$.

b. Pulse Length Errors. A pulse length error is a systematic error in the duration of individual pulses, perhaps due to an offset in the reference oscillator frequency. The imperfect propagator takes the form $V(\boldsymbol{u}(t); t_0 + \Delta t, t_0) = U(\boldsymbol{u}(t); t_0 + \Delta t + \delta t, t_0)$, where Δt is the ideal pulse length and δt is the unknown timing error. In some cases, errors on the clock may be rewritten in terms of equivalent errors on the control functions. For simplicity, we restrict ourselves to square pulses, where the controls $\boldsymbol{u}(t)$ are constant and the imperfect propagator may be rewritten as $V(\boldsymbol{u}; t_f, t_i) = U(\boldsymbol{u} + \epsilon_T \boldsymbol{\delta u}; t_0 + \Delta t, t_0)$, where again $\boldsymbol{\delta u} = \boldsymbol{u}$ is proportional to the ideal control and now $\epsilon_T = \delta t/\Delta t$, similar to the result for an amplitude error. Amplitude and pulse length errors in square pulses are similar in some senses; although they arise from distinct physical processes, they both act as errors in the angle of the applied rotation.

c. Addressing Errors. Individual spins among the ensemble may experience slightly different Rabi couplings owing to the spatial variation in the strength of the control field. In some applications this variation is exploited to yield spatially localized coherent operations, such as in magnetic resonance imaging, and in addressing single atoms in optical lattices or single ions in ion trap experiments [17,34]. In fact, many proposed scalable architectures for quantum processors rely on this effect to discriminate between qubits. In these cases, it is important to distinguish between the evolution applied to the addressed spins and the evolution of spins outside of the addressed region, where ideally no operation is applied.

Consider an experiment where the ensemble of spins is divided into a high field, Ω, and low field, $\Omega' < \Omega$, region. We assume that the control of the addressed (high-field) spins is perfect (i.e., for addressed spins $V(\boldsymbol{u}(t)) = U(\boldsymbol{u}(t))$); however, the spins in the low-field region experience an undesired correlated rotation $V(\boldsymbol{u}(t)) = U(\epsilon_N \boldsymbol{u}(t))$, where $\epsilon_N = \Omega'/\Omega$ (the subscript N denotes neighboring spins). In many cases, the imperfect pulses on the unaddressed qubits may be regarded as very small rotations. From a mathematical point of view, sequences composed of these rotations are more easily attacked using the expansion techniques that will be developed in Section III.B.

d. Detuning Errors. Systematic errors may also arise in the control of the frequency of the rf field. Consider the case where an experimentalist attempts to perform an operation at a particular field tuning $\Delta(t)$, however, a slow unknown frequency drift $\delta(t)$ is present. From the form of the NMR controls, it follows that the error distorts the H_z component to $v_z(t) = u_z(t) + \epsilon_D \delta u_z$, where $u_z = \Delta(t)$, and $\epsilon_D \delta u_z(t) = \delta(t)$. Unlike the other models discussed thus far, the detuning error applies a shift in the effective rotation axis, rotating it in the direction of the H_z axis. Therefore, the ideal propagator and the imperfect counterpart do not commute in general.

C. Binary Operations on Unitary Operators

In our discussion of the properties of compensating pulse sequences, it will be necessary to study the structure of the control Hamiltonians $\{H_\mu\}$, and to calculate the accuracy of gates. Here, we discuss the Hilbert–Schmidt product and the fidelity measure.

1. *Hilbert–Schmidt Inner Product*

The Hilbert–Schmidt inner product (also known as the Frobenius product) is a natural extension of the vector inner product over the field of complex numbers to matrices with complex coefficients. Let U and V be $n \times n$ matrices with entries U_{ij} and V_{ij}, respectively. The Hilbert–Schmidt inner product $\langle U, V\rangle$ is defined as

$$\langle U, V\rangle = \sum_{i=1}^{n}\sum_{j=1}^{n} U_{ji}^{*} V_{ij} = \mathrm{tr}(U^{\dagger} V) \tag{10}$$

This inner product has several properties that closely mirror the inner product for complex vectors, namely $\langle U, V\rangle = \langle V, U\rangle^{*}$, and $|\langle U, V\rangle| \leq ||U||_{\mathrm{HS}}\, ||V||_{\mathrm{HS}}$, where the Hilbert–Schmidt norm $||U||_{\mathrm{HS}} = \sqrt{\langle U, U\rangle}$ also satisfies a corresponding triangle inequality $||U + V||_{\mathrm{HS}} \leq ||U||_{\mathrm{HS}} + ||V||_{\mathrm{HS}}$. This strong correspondence to Euclidean vector spaces will be exploited in the next section, where Lie-algebraic techniques are employed to construct compensating pulse sequences.

2. *Fidelity*

A useful measure for evaluating the effects of a systematic control error is the operational fidelity, defined as

$$\mathcal{F}(U, V) = \min_{\psi} \sqrt{\langle\psi|U^{\dagger}V|\psi\rangle\langle\psi|V^{\dagger}U|\psi\rangle} \tag{11}$$

For any two unitary matrices in the group $\mathsf{SU}(2)$, the fidelity may be written as $\mathcal{F}(U, V) = |\langle U, V\rangle|/2$. In this chapter, we show error improvement graphically by plotting the infidelity, $1 - \mathcal{F}(U, V)$.

III. GROUP THEORETIC TECHNIQUES FOR SEQUENCE DESIGN

In the present work, we briefly discuss several aspects of the theory of Lie groups, which are useful for constructing compensating pulse sequences, with emphasis on conceptual clarity over mathematical rigor. The interested reader is referred to Refs [10] and [35] for additional details and a rigorous treatment of the subject. Quantum control theory and the related study of continuous transformation groups are rich and active subjects, with applications in chemistry [36–38], and other applications requiring the precise manipulation of quantum states [39,40]. We turn our attention to continuous groups of quantum transformations, specifically transformations that induce qubit rotations in quantum computing.

A. Lie Groups and Algebras

Consider the family of unitary operators that are solutions to the control equation

$$\dot{U}(t) = \sum_{\mu} u_{\mu}(t)\tilde{H}_{\mu}U(t), \qquad U(0) = \mathbb{1} \tag{12}$$

where $\tilde{H}_{\mu} = -\mathrm{i}H_{\mu}$ are skew-symmetrized Hamiltonians with the added condition that the (possibly infinite) set of control Hamiltonians $\{\tilde{H}\}$ is closed under the commutation operation (in the sense that for all $\tilde{H}_{\mu}$, $\tilde{H}_{\nu} \in \{\tilde{H}\}$, $\alpha[\tilde{H}_{\mu}, \tilde{H}_{\nu}] \in \{\tilde{H}\}$, where $\alpha \in \mathbb{R}$). This extra condition does not exclude any of the "physical" solutions that are generated by a subset of $\{\tilde{H}\}$, because these operators correspond to solutions where the remaining control functions $u_{\mu}(t)$ have been set to zero. Observe that these solutions form a representation of a group, here denoted as G, because the following properties are satisfied: for all solutions U_1 and U_2, the product U_1U_2 is also a solution (see Section III.B); the associative property is preserved (i.e., $U_1(U_2U_3) = (U_1U_2)U_3$); the identity $\mathbb{1}$ is a valid solution; and for all solutions U_1, the inverse $U_1^{\dagger}$ is also a valid solution. Moreover, the group forms a continuous differentiable manifold, parametrized by the control functions. A continuous group that is also a differentiable manifold with analytic group multiplication and group inverse operations is called a *Lie group*. We identify the group G of solutions as a Lie group, and note that differentiation of the elements $U \in \mathsf{G}$ is well defined by nature of Eq. (12).

We now turn our attention to elements of G in the neighborhood of the identity element, that is, infinitesimal unitary operations. In analogy with differentiation on Euclidean spaces, note that for any $U(t) \in \mathsf{G}$ one may find a family of tangent curves at $U(0) = \mathbb{1}$,

$$\left.\frac{dU(t)}{dt}\right|_{t=0} = \sum_{\mu} u_{\mu}(0)\tilde{H}_{\mu} \tag{13}$$

The set of skew-symmetrized Hamiltonians $\{\tilde{H}\}$ and the field of real numbers $\mathbb{R}$ (corresponding to the allowed values for the components $u_{\mu}(0)$) form a linear vector space under matrix addition and the Hilbert–Schmidt inner product [35]. Here, the Hamiltonians $\tilde{H}$ take the place of Euclidean vectors and span the tangent space of G at the identity, denoted by $T_{\mathbb{1}}\mathsf{G}$. A homomorphism exists between the control functions $\boldsymbol{u}(0)$ and vectors in $T_{\mathbb{1}}\mathsf{G}$. On this space is defined the binary Lie bracket operation between two vectors $[\tilde{H}_{\mu}, \tilde{H}_{\nu}]$, which for our purposes is synonymous with the operator commutator. A vector space that is closed under the Lie bracket is an example of a *Lie algebra*. By construction, the set $\{\tilde{H}\}$ is closed under commutation and therefore forms a Lie algebra, here denoted as $\mathfrak{g}$, corresponding to the Lie group G. We note that in general the Hamiltonians may be linearly dependent; however, an orthogonal basis under the Hilbert–Schmidt product may be generated using an orthogonalization algorithm.

The power of most Lie algebraic techniques relies on the mapping between group elements that act on a manifold (such as the manifold of rotations on a Bloch sphere) to elements in a Lie algebra, which are members of a vector space. In the groups we study here, the mapping is provided by the exponential function $\mathsf{G} = e^{\mathfrak{g}}$ (i.e., every element $U \in \mathsf{G}$ may be written as $U = e^{g}$ where $g \in \mathfrak{g}$). The object of this method is to study the properties of composite pulse sequences, which are products of members of a Lie group, in terms of vector operations on the associated Lie algebra.

1. The Spinor Rotation Group SU(2)

As a relevant example, consider the group of single-qubit operations generated by the NMR Hamiltonian Eq. (9). This is a representation of the special unitary group $\mathsf{SU}(2)$. The skew-symmetrized control Hamiltonians are closed under the Lie bracket and thus form a representation of the Lie algebra $\mathfrak{su}(2) = \text{span}\{-\mathrm{i}H_x, -\mathrm{i}H_y, -\mathrm{i}H_z\}$.[1] Therefore, any element in $U \in \mathsf{SU}(2)$ may be written as $U = e^{-\mathrm{i}t\boldsymbol{u}\cdot\boldsymbol{H}}$, where $-\mathrm{i}t\boldsymbol{u}\cdot\boldsymbol{H} \in \mathfrak{su}(2)$ may now be interpreted as a vector on the the Lie algebra. Furthermore, since $\langle -\mathrm{i}H_\mu, -\mathrm{i}H_\nu\rangle = \delta_{\mu,\nu}/2$, the spin operators form an orthogonal basis for the algebra. Topologically $\mathsf{SU}(2)$ is compact and is homomorphic to rotations of the 2-sphere, the group $\mathsf{SO}(3)$ [35]; exactly two group elements, U and $-U \in \mathsf{SU}(2)$, map to the same rotation.

B. Baker–Campbell–Hausdorff and Magnus Formulas

A composite pulse sequence may be studied from the perspective of successive products between elements of a Lie group. In the following analysis, it will be useful to relate the product of two members of a Lie group to vector operations on the Lie algebra. The relationship allows us to map pulse sequences to effective Hamiltonians. This correspondence is provided by the Baker–Campbell–Hausdorff (BCH) formula [35,41], which relates group products to a series expansion in the Lie algebra. For rapid convergence, it is most convenient to consider products of infinitesimal unitary operations, that is, operations of the form $e^{\epsilon g}$, where $g \in \mathfrak{g}$ and $\epsilon < 1$ is a real expansion parameter. We assume ϵ is sufficiently small to guarantee that the group product of propagators always lies within the radius of convergence for the expansion [20]. Let $U_1 = e^{\epsilon\tilde{H}_1}$ and $U_2 = e^{\epsilon\tilde{H}_2}$ be members of a Lie group G, where the Hamiltonians $\tilde{H}_1, \tilde{H}_2 \in \mathfrak{g}$ are members of the

[1] The operation span denotes all linear combinations with real coefficients

associated algebra. The BCH representation for the product $U_1U_2 = U_3$ involves the calculation of an effective Hamiltonian $\tilde{H}_3 \in \mathfrak{g}$ by the expansion

$$U_3 = \exp(\tilde{H}_3) = \exp\left(\sum_n^{\infty} \epsilon^n F_n\right) \tag{14}$$

where the terms

$$\begin{aligned}
\epsilon F_1 &= \epsilon(\tilde{H}_1 + \tilde{H}_2) \\
\epsilon^2 F_2 &= \frac{\epsilon^2}{2}[\tilde{H}_1, \tilde{H}_2] \\
\epsilon^3 F_3 &= \frac{\epsilon^3}{12}\Big([\tilde{H}_1, [\tilde{H}_1, \tilde{H}_2]] + [\tilde{H}_2, [\tilde{H}_2, \tilde{H}_1]]\Big)
\end{aligned}$$

are calculated from $\tilde{H}_1$, $\tilde{H}_2$, and nested commutators of elements of the Lie algebra. A combinatoric formula found by Dykin [42] exists to calculate F_n for arbitrary n. The expansion may be truncated once a desired level of accuracy is reached. In principle, group products of any length may be approximated to arbitrary accuracy using BCH formulas; however, these formulas rapidly become unwieldy and difficult to use without the aid of a computer. The BCH expansion is most useful for sequences of square pulses, where for each pulse the Hamiltonian is time independent.

1. The Magnus Expansion

A related expansion developed by Magnus [19] may be used to compute the propagator generated by a general time-dependent Hamiltonian. The solution to a control equation (e.g., $\dot{U}(t) = \epsilon\tilde{H}(t)U(t)$, where $\epsilon\tilde{H}(t) = \sum_\mu \epsilon\delta u_\mu(t)\tilde{H}_\mu$) over the interval $t_i \leq t \leq t_f$ may be written as the power series

$$U(\epsilon\boldsymbol{\delta u}; t_f, t_i) = \exp\left(\sum_n^{\infty} \epsilon^n \Omega_n(t_f, t_i)\right) \tag{15}$$

where the first few expansion terms are

$$\begin{aligned}
\epsilon\Omega_1(t_f, t_i) &= \epsilon\int_{t_i}^{t_f} dt\tilde{H}(t) \\
\epsilon^2\Omega_2(t_f, t_i) &= \frac{\epsilon^2}{2}\int_{t_i}^{t_f} dt\int_{t_i}^{t} dt'[\tilde{H}(t), \tilde{H}(t')] \\
\epsilon^3\Omega_3(t_f, t_i) &= \frac{\epsilon^3}{6}\int_{t_i}^{t_f} dt\int_{t_i}^{t} dt'\int_{t_i}^{t'} dt''([\tilde{H}(t), [\tilde{H}(t'), \tilde{H}(t'')]] \\
&\quad + [\tilde{H}(t''), [\tilde{H}(t'), \tilde{H}(t)]])
\end{aligned}$$

Again as a matter of notational convenience, we drop the time interval labels (t_f, t_i) on the expansion terms when there is no risk of confusion. Formulas for higher-order terms may be found in Ref. [43]. The BCH and Magnus expansions are in fact intimately related; when considering piecewise-constant controls the techniques are equivalent. We refer the interested reader to Ref. [20] for further details regarding both the Magnus expansion and BCH formulas.

The BCH and Magnus expansions are well known in composite pulse literature, and techniques that utilize these expansions are collectively referred to as average Hamiltonian theory. A variant of this technique, pioneered by Waugh [44,45], has been a mainstay of composite pulse design in the NMR community for decades. In some formalisms, the BCH expansion of the product of two propagators is interpreted as a power series in the rotation axis and angle. For the study of composite single-qubit rotations, this picture is extremely useful as it allows rotations on the sphere to guide the mathematics. However, this picture of composite rotations cannot be generalized to more complex groups, such as the group of n-qubit operations $\mathsf{SU}(2^n)$.

In this work, we emphasize a Lie algebraic interpretation of these methods, which also leads to a second geometric picture for the terms of the expansions. Observe that the first-order terms F_1 and Ω_1 may be regarded as simple vector sums on the Lie algebra (i.e., the sum of $\tilde{H}_1$ and $\tilde{H}_2$ in the BCH expansion and the sum of each of the infinitesimal vectors $\tilde{H}(t)dt$ in the Magnus expansion). In an analogous way, one may interpret the higher-order terms as the addition of successively smaller vectors on $\mathfrak{g}$. From this insight, one may construct composite sequences that simulate a target unitary from geometric considerations on the Lie algebra.

2. *A Method for Studying Compensation Sequences*

At this point, it is useful to introduce the general method that will be used to study compensating pulse sequences. Recall from Section II.A that in the presence of an unknown systematic control error, the ideal control functions are deformed into imperfect analogues. The imperfect propagator may be decomposed as $V(\boldsymbol{u}(t); \tau, 0) = U(\boldsymbol{u}(t); \tau, 0)U^I(\epsilon\boldsymbol{\delta u}(t); \tau, 0)$, where $U^I(\epsilon\boldsymbol{\delta u}(t); \tau, 0)$ represents the portion of the evolution produced by the systematic distortion of the ideal controls. Provided that the displacements $\epsilon\boldsymbol{\delta u}(t)$ are sufficiently small relative to the ideal controls, we may perform a Magnus expansion for the interaction frame propagator

$$U^I(\epsilon\boldsymbol{\delta u}(t); \tau, 0) = \exp\left(\sum_{n=1}^{\infty} \epsilon^n \Omega_n(\tau, 0)\right) \tag{16}$$

where the integrations in the expansion terms are performed in the appropriate frame. For example, the first-order term is

$$\epsilon\Omega_1(\tau, 0) = -i\epsilon \int_0^{\tau} dt\, \boldsymbol{\delta u}(t) \cdot \boldsymbol{H}^I(t) \tag{17}$$

where the components $H^I_\mu(t) = U^\dagger(\boldsymbol{u}(t'); t, 0) H_\mu U(\boldsymbol{u}(t'); t, 0)$ are the interaction frame control Hamiltonians. The reader should recall from Eq. (8) that if the controls $\boldsymbol{u}(t)$ form an nth order compensating pulse sequence, then the interaction frame propagator $U^I(\epsilon \boldsymbol{\delta u}(t); \tau, 0) = \mathbb{1} + \mathcal{O}(\epsilon^{n+1})$ must approximate the identity to sufficient accuracy. It immediately follows that this condition is satisfied for any ϵ when the leading n-many Magnus expansion terms $\Omega_1(\tau, 0), \Omega_2(\tau, 0), \ldots, \Omega_n(\tau, 0)$ over the pulse interval $0 \leq t \leq \tau$ simultaneously equal zero.

This condition may also be understood in terms of geometric properties of vector paths on the Lie algebra. For example, consider the first-order Magnus expansion term for the interaction frame propagator and observe that $-i\epsilon \boldsymbol{\delta u}(t) \cdot \boldsymbol{H}^I(t)$ may be regarded as a vector path on the dynamical Lie algebra. From the condition $\Omega_1(\tau, 0) = 0$ and Eq. (17), we see that the path must form a closed cycle in order for the first-order term to be eliminated (see Fig. 4). The elimination of higher-order expansion terms will place additional geometric constraints on the path, which will depend on the structure of the Lie algebra (i.e., the commutators between paths on the algebra). In the case of piecewise constant control functions, the resulting propagator may be understood as a sequence of square pulses. It is clear that in this case the Magnus series reduces to a BCH expansion for the total pulse propagator, and on the Lie algebra the corresponding path forms a closed polygon.

This Lie theoretic method is a useful tool in determining whether a control function $\boldsymbol{u}(t)$ is also a compensating sequence; we may directly calculate the interaction frame Magnus expansion terms in a given error model and show that they equal zero. The inverse problem (i.e., solving for control functions) is typically much more difficult. In general, the interaction Hamiltonians $H^I_\mu(t)$ are highly nonlinear functions of the ideal controls $\boldsymbol{u}(t)$, which impedes several analytical

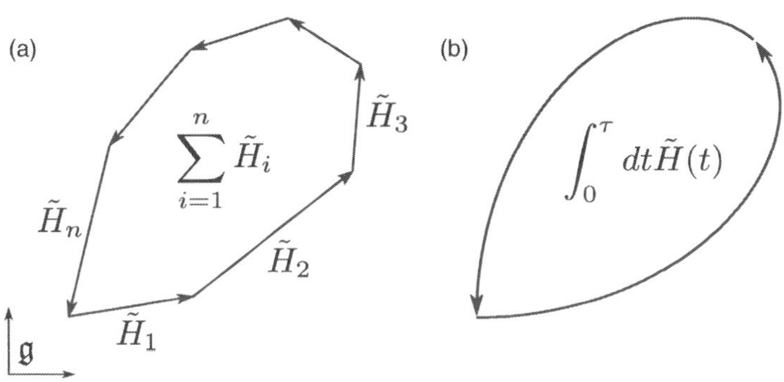

Figure 4. Vector paths on a Lie algebra $\mathfrak{g}$ may be used to represent a pulse sequence. (**a**) The BCH expansion relates group multiplication of several square pulses to vector addition on $\mathfrak{g}$. (**b**) A shaped pulse is represented by a vector curve on $\mathfrak{g}$, parametrically defined by the control functions. Both pulse sequences form a closed loop on $\mathfrak{g}$ and therefore the first-order terms equal zero.

solution methods. However, we note that in several special cases the problem is considerably simplified, such as when the ideal operation is to perform the identity, and the interaction and Schrödinger pictures are equivalent.

C. Decompositions and Approximation Methods

Several useful techniques in sequence design involve decompositions that may be understood in terms of the structure of a Lie group and its corresponding algebra. In this section, we discuss several important methods and identities.

1. *Basic Building Operations*

Given a limited set of controls $\{\tilde{H}_1, \tilde{H}_2\}$ that generate the algebra $\mathfrak{g}$, one may produce any unitary operation in the corresponding Lie group $\mathsf{G} = \mathrm{e}^{\mathfrak{g}}$ using only two identities. The first identity, the Lie–Trotter formula [46], describes how to produce a unitary generated by the sum of two noncommuting control operators. Using the BCH formula one may compute that $\mathrm{e}^{\tilde{H}_1/n}\mathrm{e}^{\tilde{H}_2/n} = \mathrm{e}^{(\tilde{H}_1+\tilde{H}_2)/n} + \mathcal{O}([\tilde{H}_1, \tilde{H}_2]/n^2)$. In terms of physical pulses, this corresponds to dividing the propagators $\mathrm{e}^{\tilde{H}_1}$ and $\mathrm{e}^{\tilde{H}_2}$ into n equal intervals to produce the propagators $\mathrm{e}^{\tilde{H}_1/n}$ and $\mathrm{e}^{\tilde{H}_2/n}$. Suppose we perform n such successive products, so that the resulting propagator is

$$\left(\mathrm{e}^{\tilde{H}_1/n}\mathrm{e}^{\tilde{H}_2/n}\right)^n = \mathrm{e}^{\tilde{H}_1+\tilde{H}_2} + \mathcal{O}([\tilde{H}_1, \tilde{H}_2]/n) \tag{18}$$

Although the Hamiltonians $\tilde{H}_1$ and $\tilde{H}_2$ do not commute in general, we may approximate $U = e^{\tilde{H}_1+\tilde{H}_2}$ to arbitrary accuracy by dividing the evolution into n-many time intervals and using the construction Eq. (18). By extension, it follows that any unitary generated by a Hamiltonian in the Lie algebra subspace $\mathrm{span}\{\tilde{H}_1, \tilde{H}_2\}$ may be approximated to arbitrary accuracy using a Trotter sequence. A number of improved sequences were developed by Suzuki [47] that remove errors to higher commutators and scale more strongly with n. The Trotter-Suzuki formulas may be used to eliminate successively higher-order errors at the cost of increased operation time [48].

The second identity, which we refer to as the balanced group commutator, enables the synthesis of a unitary generated by the Lie bracket $[\tilde{H}_1, \tilde{H}_2]$. Again the BCH formula may be used to show $\mathrm{e}^{\tilde{H}_1/n}\mathrm{e}^{\tilde{H}_2/n}\mathrm{e}^{-\tilde{H}_1/n}\mathrm{e}^{-\tilde{H}_2/n} = \mathrm{e}^{[\tilde{H}_1,\tilde{H}_2]/n^2} + \mathcal{O}([\tilde{H}_1 + \tilde{H}_2, [\tilde{H}_1, \tilde{H}_2]]/n^3)$. If we now consider n^2-many successive balanced group commutator constructions, the resulting propagator is

$$\left(\mathrm{e}^{\tilde{H}_1/n}\mathrm{e}^{\tilde{H}_2/n}\mathrm{e}^{-\tilde{H}_1/n}\mathrm{e}^{-\tilde{H}_2/n}\right)^{n^2} = \mathrm{e}^{[\tilde{H}_1,\tilde{H}_2]} + \mathcal{O}([\tilde{H}_1 + \tilde{H}_2, [\tilde{H}_1, \tilde{H}_2]]/n) \tag{19}$$

Then, as in the case of the Trotter formula, we may approximate $U = \mathrm{e}^{[\tilde{H}_1,\tilde{H}_2]}$ to arbitrary accuracy by increasing the number of intervals n. By assumption the entire

Lie algebra may be generated by nested Lie brackets between the Hamiltonians $\tilde{H}_1$ and $\tilde{H}_2$, which implies that any $U \in \mathsf{G}$ may be produced by a combination of balanced group commutator and Trotter formulas. However, we emphasize that in almost all cases much more efficient constructions exist. The balanced group commutator construction also forms the basis of the Solovay-Kitaev theorem [49], an important result regarding the universality of a finite gate set in quantum computation. Later, we will study the Solovay-Kitaev method (see Section IV.A.1), which produces compensating sequences of arbitrary accuracy by using a balanced group commutator. Specifically we will use the formula

$$\exp(\tilde{H}_1\epsilon^k)\exp(\tilde{H}_2\epsilon^l)\exp(-\tilde{H}_1\epsilon^k)\exp(-\tilde{H}_2\epsilon^l) = \exp([\tilde{H}_1, \tilde{H}_2]\epsilon^{k+l}) + \mathcal{O}(\epsilon^{k+l+1}) \tag{20}$$

where the parameter $\epsilon < 1$ will represent the strength of a systematic error.

2. *Euler Decomposition*

In Section III.A.1, it was shown that any one-qubit operation $U \in \mathsf{SU}(2)$ may be written in the form $U = \exp(-\mathrm{i}t\boldsymbol{u} \cdot \boldsymbol{H})$. It is well known that an alternative representation exists, namely the Euler decomposition

$$U = \exp(-\mathrm{i}\alpha_3 H_\mathrm{x})\exp(-\mathrm{i}\alpha_2 H_\mathrm{y})\exp(-\mathrm{i}\alpha_1 H_\mathrm{x}) \tag{21}$$

which is given by sequential rotations by the angles $\{\alpha_1, \alpha_2, \alpha_3\}$ about the H_x, H_y, and H_x axes of the Bloch sphere. The Euler decomposition gives the form of a pulse sequence that produces any arbitrary one-qubit gate. The Euler decomposition also gives a method of producing rotations generated by a Hamiltonian outside of direct control. For example, at perfect resonance $U = \exp(-\mathrm{i}\theta H_\mathrm{z})$ cannot be directly produced; however, the Euler decomposition for the same operation $U = \exp(-\mathrm{i}\frac{\pi}{2}H_\mathrm{x})\exp(-\mathrm{i}\theta H_\mathrm{y})\exp(\mathrm{i}\frac{\pi}{2}H_\mathrm{x})$ may be implemented.

3. *Cartan Decomposition*

Let $\mathfrak{g}$ be a semi-simple Lie algebra that may be decomposed into two subspaces $\mathfrak{g} = \mathfrak{k} \oplus \mathfrak{m}$, $\mathfrak{m} = \mathfrak{k}^\perp$ satisfying the commutation relations

$$[\mathfrak{k}, \mathfrak{k}] \subseteq \mathfrak{k}, \qquad [\mathfrak{m}, \mathfrak{k}] \subseteq \mathfrak{m}, \qquad [\mathfrak{m}, \mathfrak{m}] \subseteq \mathfrak{k} \tag{22}$$

Such a decomposition is called a Cartan decomposition of $\mathfrak{g}$ [10]. Suppose for the moment a subalgebra $\mathfrak{a}$ of $\mathfrak{g}$ is in a subspace of $\mathfrak{m}$. Because $\mathfrak{a}$ is an algebra of its own right, it is closed under the Lie bracket $[\mathfrak{a}, \mathfrak{a}] \subseteq \mathfrak{a}$. However, note $\mathfrak{a} \subseteq \mathfrak{m}$ implies that $[\mathfrak{a}, \mathfrak{a}] \subseteq [\mathfrak{m}, \mathfrak{m}] \subseteq \mathfrak{k}$. Because the subspaces $\mathfrak{k}$ and $\mathfrak{m}$ are mutually orthogonal, then $[\mathfrak{a}, \mathfrak{a}] = \{0\}$ and the subalgebra $\mathfrak{a}$ must be abelian. A maximal abelian subalgebra $\mathfrak{a} \subseteq \mathfrak{m}$ for a Cartan decomposition pair $(\mathfrak{k}, \mathfrak{m})$ is called a Cartan subalgebra [10].

For brevity, we state without proof an important theorem regarding the decomposition of an operator in a group G with a Lie algebra admitting a Cartan decomposition. Consider a Lie algebra $\mathfrak{g}$ with a Cartan subalgebra $\mathfrak{a}$ corresponding to the decomposition pair $(\mathfrak{k}, \mathfrak{m})$. Every U in the group $\mathsf{G} = e^{\mathfrak{g}}$ may be written in the form

$$U = K_2 A K_1 \tag{23}$$

where $K_1, K_2 \in e^{\mathfrak{k}}$, and $A \in e^{\mathfrak{a}}$. This is called the KAK Cartan decomposition for the group G. The interested reader is referred to Ref. [35] for additional details regarding the KAK decomposition.

As a relevant example, here we show how the Euler decomposition for a propagator $U \in \mathsf{SU}(2)$ is a special case of a KAK decomposition. The algebra is spanned by the orthogonal basis matrices $\mathfrak{su}(2) = \text{span}\{-iH_x, -iH_y, -iH_z\}$. Observe that $\mathfrak{k} = \text{span}\{-iH_x\}$ and $\mathfrak{m} = \text{span}\{-iH_y, -iH_z\}$ form a Cartan decomposition for $\mathfrak{su}(2)$. The maximal abelian subalgebra of $\mathfrak{m}$ is one-dimensional; we choose $\mathfrak{a} = \text{span}\{-iH_y\}$ although the choice $\mathfrak{a}' = \text{span}\{-iH_z\}$ would serve just as well (i.e., the different axis conventions of the Euler decomposition differ in the choice of a maximal abelian subalgebra). Then by Eq. (23), every element $U \in \mathsf{SU}(2)$ may be expressed in the form $U = \exp(-i\alpha_3 H_x)\exp(-i\alpha_2 H_y)\exp(-i\alpha_1 H_x)$, where the parameters α_j are real. This restatement of Eq. (21) completes the proof. The KAK decomposition is an existence theorem, and does not provide a direct method for the calculation of the required rotation angles α_j.

The Cartan decomposition has important implications for universality. For instance, if one may generate any unitary operation over the subgroups $e^{\mathfrak{k}}$ and $e^{\mathfrak{a}}$, then any gate in the larger group $e^{\mathfrak{g}}$ may be produced. Similarly, if compensation sequences exist for operations in these subgroups, then they may be combined in the KAK form to yield a compensating pulse sequence for an operation in the larger group. Another important application is the decomposition of large Lie groups into products of more simple ones. Of special interest to quantum computing is the inductive decomposition of n-qubit $\mathsf{SU}(2^n)$ gates into products of one-qubit $\mathsf{SU}(2)$ and two-qubit $\mathsf{SU}(4)$ rotations [50].

IV. COMPOSITE PULSE SEQUENCES ON SU(2)

In this section, we study pulse sequences that compensate single-qubit operations, which form a representation of the group $\mathsf{SU}(2)$. Our approach is to use Lie theoretic methods to study the effects of composite rotations. Topologically, sequences of infinitesimal rotations may be interpreted as paths in the neighborhood of the group identity. It is frequently easier to construct sequences of infinitesimal paths on the Lie algebra, and then map these sequences to the manifold of group operations. Several techniques will be used to construct compensating pulse sequences of arbitrary accuracy, including techniques that use Solovay–Kitaev methods and Trotter formulas.

A. Solovay–Kitaev Sequences

The Solovay–Kitaev sequences, so named because the construction of the higher-order sequences involves the identity Eq. (20) used by Kitaev in his proof of universal control from finite gate sets [49], are among the simplest families of fully compensating composite pulse sequences. The SK sequences were first introduced by Brown et al. [51], and are designed to compensate pulse length, amplitude, and addressing errors using resonant pulses. Here, we show that the SK family of sequences may be derived using the method outlined in Section III.B.1.

1. *Narrowband Behavior*

Narrowband composite pulse sequences apply a spin rotation over only a narrow range of strengths of the control field [52]. Therefore, they are most suited for correcting addressing errors [34] (i.e., situations where the spatial variation in the field strength is used to discriminate between spins in an ensemble). The operations on the addressed qubits are assumed to be error-free, whereas on the unaddressed qubits the imperfect pulses take the form $V(\boldsymbol{u}(t)) = U(\epsilon_N \boldsymbol{u}(t))$, where $\epsilon_N < 1$ is the systematic addressing error amplitude (see Section II.B.1). Narrowband sequences are a means of applying a target operation U_T on the addressed spins while removing the leading effects of the operation on the unaddressed spins. For a composite pulse sequence to exhibit nth-order narrowband behavior, we require that two conditions must be satisfied: first that for the unaddressed qubits the sequence $V(\boldsymbol{u}(t))$ approximates the identity up to $\mathcal{O}(\epsilon_N^{n+1})$, and second, that on the targeted spins the desired operation is applied without error.

From a mathematical perspective, systematic addressing errors are among the easiest to consider. Note that for the unaddressed qubits the ideal values for the controls $\boldsymbol{u}(t)$ is zero (ideally no operation takes place). This implies $H_\mu^I(t) = H_\mu$, and that the Schrödinger and interaction frames in Eq. (8) are identical. We may develop a sequence that corrects the control distortion without the added complication of the passage into an interaction picture. For this reason, we first study narrowband sequences before considering other error models.

Sequences with first-order narrowband properties may be constructed using the method described in Section III.B.1. We consider sequences composed of three square pulses (i.e., piecewise constant control functions), which produce the ideal propagation:

$$U(\boldsymbol{u}(t)) = \prod_{k=1}^{3} U_k, \qquad U_k = \exp(-\mathrm{i}t_k \boldsymbol{u}_k \cdot \boldsymbol{H}) \tag{24}$$

On an addressed qubit, $U(\boldsymbol{u}(t))$ is implemented, whereas on the unaddressed qubits $V(\boldsymbol{u}(t)) = U(\epsilon_N \boldsymbol{u}(t))$ is applied. We may use either a BCH or Magnus expansion to compute the applied operation on the unaddressed qubit. From Eq. (17) and the error model $v_{\mathrm{x/y}}(t) = \epsilon_N u_{\mathrm{x/y}}(t)$, the first-order term is

$$\epsilon_N \Omega_1 = -\mathrm{i}\epsilon_N (t_3 \boldsymbol{u}_3 + t_2 \boldsymbol{u}_2 + t_1 \boldsymbol{u}_1) \cdot \boldsymbol{H} \tag{25}$$

The spin operators $-\mathrm{i}H_\mu \in \{-\mathrm{i}H_\mathrm{x}, -\mathrm{i}H_\mathrm{y}, -\mathrm{i}H_\mathrm{z}\}$ form an orthogonal basis for $\mathfrak{su}(2)$, and the terms $-\mathrm{i}\epsilon_N t_k \boldsymbol{u}_k \cdot \boldsymbol{H}$ may be regarded as vectors on the dynamical Lie algebra. In order to eliminate the first-order Magnus term $\epsilon_N \Omega_1$, the sum of the components must equal zero, that is, the vectors must form a closed triangular path. Such paths may be found using elementary geometric methods.

It is important at this point to allow experimental considerations to place constraints on the sequences under study. For instance, in many cases it is desirable to perform coherent operations at resonance, a condition that forces the vectors to lie in the H_x–H_y plane. One possible choice is

$$\begin{aligned}
-\mathrm{i}\epsilon_N t_1 \boldsymbol{u}_1 \cdot \boldsymbol{H} &= -\mathrm{i}\epsilon_N \theta H_\mathrm{x} \\
-\mathrm{i}\epsilon_N t_2 \boldsymbol{u}_2 \cdot \boldsymbol{H} &= -\mathrm{i}\epsilon_N 2\pi(\cos\phi_{\mathrm{SK1}} H_\mathrm{x} + \sin\phi_{\mathrm{SK1}} H_\mathrm{y}) \\
-\mathrm{i}\epsilon_N t_3 \boldsymbol{u}_3 \cdot \boldsymbol{H} &= -\mathrm{i}\epsilon_N 2\pi(\cos\phi_{\mathrm{SK1}} H_\mathrm{x} - \sin\phi_{\mathrm{SK1}} H_\mathrm{y})
\end{aligned} \tag{26}$$

where the phase $\phi_{\mathrm{SK1}} = \arccos(-\theta/4\pi)$ is selected so that $\epsilon_N \sum_k t_k \boldsymbol{u}_k = 0$, and therefore the first-order expansion term $\epsilon_N \Omega_1 = 0$ is eliminated. Figure 5a is a diagram of these vectors on $\mathfrak{su}(2)$, where the sequence may be represented as a closed isosceles triangle with one segment aligned on the $-\mathrm{i}H_\mathrm{x}$ axis.

For clarity, we use the notation $R(\theta, \phi) = \exp(-\mathrm{i}\theta(\cos\phi H_\mathrm{x} + \sin\phi H_\mathrm{y}))$ to represent propagators that induce rotations about an axis in the H_x–H_y plane. Observe that for resonant square-pulse operations in $\mathsf{SU}(2)$, $R(\theta, \phi) = U(\boldsymbol{u}_k; t_k, 0)$, where $\boldsymbol{u}_k = \Omega(\cos\phi, \sin\phi, 0)$ and $\theta = t_k|\boldsymbol{u}_k|$. We also define the corresponding imperfect propagator $M(\theta, \phi) = V(\boldsymbol{u}; t_k, 0)$, and recall that the imperfect propagators on the unaddressed qubits $M(\theta, \phi) = R(\theta\epsilon_N, \phi)$ and addressed qubits $M(\theta, \phi) = R(\theta, \phi)$ have different implied dependencies on the systematic error. Then combining Eqs. (24) and (26), the propagator may be written as

$$\begin{aligned}
U(\epsilon_N \boldsymbol{u}(t)) &= R(2\pi\epsilon_N, -\phi_{\mathrm{SK1}}) R(2\pi\epsilon_N, \phi_{\mathrm{SK1}}) R(\theta\epsilon_N, 0) = \mathbb{1} + \mathcal{O}(\epsilon_N^2) \\
M_{\mathrm{SK1}}(\theta, 0) &= M(2\pi, -\phi_{\mathrm{SK1}}) M(2\pi, \phi_{\mathrm{SK1}}) M(\theta, 0)
\end{aligned} \tag{27}$$

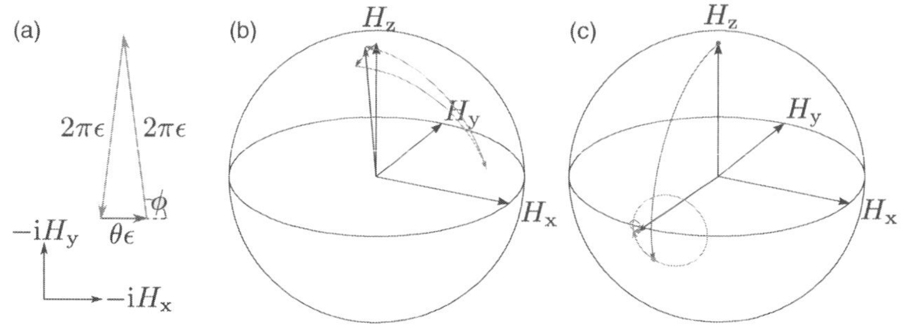

Figure 5. (**a**) Vector path followed by SK1 on the Lie algebra. (**b**) Trajectory of an unaddressed spin during an SK1 sequence, using imperfect rotations of the form $M(\theta, \phi) = R(\theta\epsilon_N, \phi)$. (**c**) SK1 correcting an amplitude error, using imperfect $M(\theta, \phi) = R(\theta(1 + \epsilon_A), \phi)$. In these plots $\epsilon_N = \epsilon_A = 0.2$.

for the unaddressed qubits, and as

$$\begin{aligned} U(\boldsymbol{u}(t)) &= R(2\pi, -\phi_{\mathrm{SK1}})R(2\pi, \phi_{\mathrm{SK1}})R(\theta, 0) \\ M_{\mathrm{SK1}}(\theta, 0) &= M(2\pi, -\phi_{\mathrm{SK1}})M(2\pi, \phi_{\mathrm{SK1}})M(\theta, 0) \end{aligned} \tag{28}$$

for addressed qubits. This is the first-order Solovay–Kitaev sequence, here denoted as SK1. On the addressed qubits the rotations $R(2\pi, \phi_{\mathrm{SK1}}) = R(2\pi, -\phi_{\mathrm{SK1}}) = -\mathbb{1}$ are resolutions of the identity and the effect of the sequence is to apply the target unitary $U_T = R(\theta, 0)$, whereas on the unaddressed spins the sequence applies $\mathbb{1} + \mathcal{O}(\epsilon_N^2)$ and the leading first-order rotation of the unaddressed spins is eliminated. Therefore, SK1 satisfies the conditions for narrowband behavior. When the sequence SK1 is used in the place of the simple rotation $R(\theta, 0)$, the discrimination between addressed and unaddressed spins is enhanced.

2. *Broadband Behavior*

Broadband composite pulses apply a spin rotation over a large range of strengths of the control field and are best suited for correcting systematic amplitude and pulse-length errors that correspond to systematic over- and underrotations during qubit manipulations. Broadband sequences are a means of applying a target operation in the presence of inaccurate field strengths or pulse durations. When we require for a sequence to exhibit nth-order broadband behavior, the effect of the sequence is to approximate the target operation U_T up to $\mathcal{O}(\epsilon_A^n)$ in the case of amplitude errors, or up to $\mathcal{O}(\epsilon_T^n)$ in the case of pulse-length errors. As before, it is convenient to consider sequences comprised of resonant square pulses. In this case the amplitude and pulse length error models are in some sense equivalent because they both apply a proportional distortion to the $u_x(t)$ and $u_y(t)$ controls. In the following discussion we will explicitly use the amplitude error model, where imperfect rotations take the form $M(\theta, \phi) = R(\theta(1 + \epsilon_A), \phi)$.

We now show how a Solovay–Kitaev sequence with first-order broadband properties may be derived. Although the method presented in Section III.A.1 may be applied, here a more direct technique is used. Our strategy is to construct a pulse sequence entirely out of imperfect rotations $M(\theta, \phi) = R(\theta(1 + \epsilon_A), \phi) = R(\theta\epsilon_A, \phi)R(\theta, \phi)$ using Eq. (27) as a template. Explicitly, this is achieved with the matrix manipulations

$$\begin{aligned} R(\theta, 0) + \mathcal{O}(\epsilon_A^2) &= R(2\pi\epsilon_A, -\phi_{\mathrm{SK1}})R(2\pi\epsilon_A, \phi_{\mathrm{SK1}})R(\theta\epsilon_A, 0)R(\theta, 0) \\ &= R(2\pi\epsilon_A, -\phi_{\mathrm{SK1}})R(2\pi, -\phi_{\mathrm{SK1}})R(2\pi\epsilon_A, \phi_{\mathrm{SK1}}) \\ &\quad \times R(2\pi, \phi_{\mathrm{SK1}})R(\theta\epsilon_A, 0)R(\theta, 0) \end{aligned}$$

where we have right-multiplied Eq. (27) by $R(\theta, 0)$ (after substituting ϵ_A for ϵ_N) and inserted the identity $R(2\pi, \phi_{\mathrm{SK1}}) = R(2\pi, -\phi_{\mathrm{SK1}}) = -\mathbb{1}$. Then, by combining rotations about the same axis, one obtains the result,

$$\begin{aligned} U((1 + \epsilon_A)\boldsymbol{u}(t)) &= R(2\pi(1 + \epsilon_A), -\phi_{\mathrm{SK1}})R(2\pi(1 + \epsilon_A), \phi_{\mathrm{SK1}})R(\theta(1 + \epsilon_A), 0) \\ M_{\mathrm{SK1}}(\theta, 0) &= M(2\pi, -\phi_{\mathrm{SK1}})M(2\pi, \phi_{\mathrm{SK1}})M(\theta, 0) = R(\theta, 0) + \mathcal{O}(\epsilon_A^2) \end{aligned} \tag{29}$$

where again $\phi_{\mathrm{SK1}} = \arccos(-\theta/4\pi)$. In the presence of unknown amplitude errors, the sequence reduces the effect of the error while applying the effective rotation $R(\theta, 0)$. Observe that in terms of imperfect rotations this is the same SK1 sequence derived earlier. SK1 has both narrowband and broadband behavior, and the sequence may correct both addressing and amplitude errors simultaneously. Such a sequence is called a *passband* sequence.

3. *Generalization to Arbitrary Gates in* SU(2)

The sequence SK1 is designed to compensate single-qubit rotations about the H_x axis. If this sequence is to be useful in quantum computation, it must be generalized so that any single-qubit rotation may be corrected. One method involves transforming the sequence by a similarity transformation of the pulse propagators. As an example, suppose we require an SK1 sequence that performs the rotation $R(\theta, \phi) = \exp(-i\phi H_z)R(\theta, 0)\exp(i\phi H_z)$ on the addressed spins. Similarity transformation of pulses under $\exp(-i\phi H_z)$ represent a phase advance in the rotating frame. It is evident that the transformed sequence

$$M_{\mathrm{SK1}}(\theta, \phi) = M(2\pi, \phi - \phi_{\mathrm{SK1}})M(2\pi, \phi + \phi_{\mathrm{SK1}})M(\theta, \phi) \tag{30}$$

performs the desired compensated rotation. In this manner, a compensating pulse sequence for any target operation $U_T \in \mathrm{SU}(2)$ may be constructed. One may solve for the operation Υ that performs the planar rotation $\Upsilon U_T \Upsilon^\dagger = R(\theta, 0)$, where

$\theta = ||\log U_T||_{\mathrm{HS}}/||H_{\mathrm{x}}||_{\mathrm{HS}}$. Then the transformed sequence $\Upsilon^\dagger M_{\mathrm{SK1}}(\theta, 0)\Upsilon$ performs a first-order compensated U_T operation. Similarity transformations of pulse sequence propagators can be extremely useful and are frequently applied in composite pulse sequences.

Alternative methods exist for generating an arbitrary compensated rotations. Recall from Section III.C.2 that any operation $U_T \in \mathsf{SU}(2)$ may be expressed in terms of a Euler decomposition $U_T = R(\alpha_3, 0)R(\alpha_2, \pi/2)R(\alpha_1, 0)$. An experimentalist may apply U_T by implementing each of the Euler rotations in sequence, however, in the presence of an unknown systematic error each of the applied rotations is imperfect and the fidelity of the applied gate is reduced. The error may be compensated by replacing each imperfect pulse with a compensating pulse sequence. For example, the sequence

$$M_{\mathrm{SK1}}(\alpha_3, 0)M_{\mathrm{SK1}}(\alpha_2, \pi/2)M_{\mathrm{SK1}}(\alpha_1, 0) = U_T + \mathcal{O}(\epsilon_A^2) \tag{31}$$

compensates amplitude errors to first-order by implementing SK1 sequences for each of the rotations in an Euler decomposition for U_T. This construction is an example of pulse sequence concatenation. By concatenating two independent pulse sequences, it is sometimes possible to produce a sequence with properties inherited from each parent sequence.

4. *Arbitrarily Accurate SK Sequences*

In this section, we discuss the Solovay–Kitaev method for constructing arbitrarily accurate composite pulse sequences, which may be used to systematically improve the performance of an initial seed sequence. The Lie algebraic picture is particularly helpful in the description of the algorithm. The method is quite general, and can be used on sequences other than SK1.

Suppose that we have an nth-order compensating pulse sequence, here denoted as W_n. The problem we consider is the identification of a unitary operator A_{n+1} such that $W_{n+1} = A_{n+1}W_n = U_T + \mathcal{O}(\epsilon^{n+2})$, where W_{n+1} is an $(n+1)$th-order sequence. Assume for now that such an operator exists, and consider that in the presence of systematic errors, the application of the correction gate A_{n+1} is imperfect. However, if it is possible to implement a compensating pulse sequence B_{n+1}, which is an $\mathcal{O}(\epsilon^{n+2})$ approximation of A_{n+1}, then it is still possible to construct an $(n+1)$th-order sequence

$$W_{n+1} = B_{n+1}W_n = U_T + \mathcal{O}(\epsilon^{n+2}) \tag{32}$$

We may then continue constructing pulse sequences of increasing accuracy in this fashion if there exists a family of operators $\{A_{n+1}, A_{n+2}, A_{n+3}, \cdots A_m\}$ and a corresponding family of pulse sequences $\{B_{n+1}, B_{n+2}, B_{n+3}, \cdots B_m\}$ that implement the operators to the required accuracy. This immediately suggests an inductive construction for the sequence W_m

$$W_m = B_m\, B_{m-1} \cdots B_{n+3}\, B_{n+2}\, B_{n+1}\, W_n = U_T + \mathcal{O}(\epsilon^{m+1}) \tag{33}$$

This is the basis of the Solovay–Kitaev method [51]. To apply the method, we must first have a means of calculating the correction A_{n+1}, and second we must find a compensating sequence B_{n+1} robust to the systematic error model considered.

We turn our attention to the calculation of the correction terms A_{n+1}. It is convenient to decompose the pulse sequence propagator using an interaction frame as $W_n = U_T U^I(\epsilon\boldsymbol{\delta u}(t))$, where $U_T = U(\boldsymbol{u}(t))$ is the target gate. In the spirit of Eq. (16), a Magnus expansion for $U^I(\epsilon\boldsymbol{\delta u}(t))$ may be used. Observe that $W_n = U_T \exp(\epsilon^{n+1}\Omega_{n+1}) + \mathcal{O}(\epsilon^{n+2})$. Then letting $A_{n+1} = U_T \exp(-\epsilon^{n+1}\Omega_{n+1})U_T^\dagger$ it may be verified using the BCH formula that $A_{n+1} W_n = U_T + \mathcal{O}(\epsilon^{n+2})$. This result may also be interpreted in terms of vector displacements on the Lie algebra. In the interaction frame, $\epsilon^{n+1}\Omega_{n+1}$ may be interpreted as a vector in $\mathfrak{su}(2)$. Similarly, the infinitesimal rotation A_{n+1} corresponds to the vector $-\epsilon^{n+1}\Omega_{n+1}$, equal in magnitude and opposite in orientation. The first-order term of the BCH series corresponds to vector addition on the Lie algebra, and the $\mathcal{O}(\epsilon^{n+1})$ terms cancel.

What remains is to develop a compensating pulse sequence that implements A_{n+1} to the required accuracy under a given error model. Let us define $P_{jz}(\alpha) = \exp(-\mathrm{i}\alpha\epsilon^j H_z) + \mathcal{O}(\epsilon^{j+1})$ and also the rotated analogues $P_{jx}(\alpha) = \exp(-\mathrm{i}\alpha\epsilon^j H_x) + \mathcal{O}(\epsilon^{j+1})$ and $P_{jy}(\alpha) = \exp(-\mathrm{i}\alpha\epsilon^j H_y) + \mathcal{O}(\epsilon^{j+1})$. Frequently if one such P_j may be produced, then often the remaining two may be produced by similarity transformation of the propagators or by using an Euler decomposition. At this point we use Eq. (20), used in the proof for the Solovay–Kitaev theorem, to construct relation

$$P_{kx}(-\alpha)P_{\ell y}(-\beta)P_{kx}(\alpha)P_{\ell y}(\beta) = P_{jz}(\alpha\beta), \qquad k + \ell = j \tag{34}$$

Continuing in this manner, each of the P_js may be recursively decomposed into a product of first-order propagators $P_{1k}(\alpha) = \exp(-\mathrm{i}\alpha\epsilon H_k) + \mathcal{O}(\epsilon^2)$. Our strategy is to use Eq. (34) to implement $P_{(n+1)z}(\xi)$ where $\xi = ||\Omega_{n+1}||_{\mathrm{HS}}/||H_z||_{\mathrm{HS}}$. On the Lie algebra, this operation is represented by vector of length $\epsilon^{n+1}||\Omega_{n+1}||_{\mathrm{HS}}$ oriented along the $-\mathrm{i}H_z$ axis. Let Υ be the rotation that performs $\Upsilon\Omega_{n+1}\Upsilon^\dagger = \mathrm{i}\xi H_z$ (i.e., the operator that rotates Ω_{n+1} onto the $\mathrm{i}H_z$ axis). Then, by similarity transformation under $U_T\Upsilon^\dagger$

$$U_T\Upsilon^\dagger P_{(n+1)z}(\xi)\Upsilon U_T^\dagger = U_T \exp(-\epsilon^{n+1}\Omega_{n+1})U_T^\dagger + \mathcal{O}(\epsilon^{n+2}) = B_{n+1} \tag{35}$$

the sequence $P_{(n+1)z}(\xi)$ may be transformed into precisely the required correction sequence needed for the Solovay–Kitaev method. There are two approaches for applying the transformation by $U_T\Upsilon^\dagger$: we may either calculate the transformed analogues of each of the pulses in $P_{(n+1)z}(\xi)$ and then apply the transformed pulses, or we may, when possible, directly include the transformation pulses (or an estimate for them) as physically applied pulses in the sequence. The second approach is only viable when it is possible to generate accurate inverse operations $\Upsilon U_T^\dagger$ [53].

Our construction will be complete once we have a method for generating the simple "pure-error" propagator $P_{1x}(\alpha)$. In many cases it is sufficient to only consider $P_{1x}(\alpha)$ because the other propagators $P_{1y}(\alpha)$ and $P_{1z}(\alpha)$ are related by a similarity transformation. In general, the method will depend on the error model under consideration; here we explicitly show how to construct this term in the amplitude and pulse-length error models and also in the addressing error model.

a. Addressing Errors. We wish to perform the evolution $P_{1x}(\alpha)$ using a product of imperfect square-pulse propagators $M(\theta, \phi)$. Recall that in this model rotations on the addressed qubit are error free $M(\theta, \phi) = R(\theta, \phi)$, whereas on the unaddressed qubit the applied unitary depends on the systematic addressing error $M(\theta, \phi) = R(\theta\epsilon_N, \phi)$. Similarly, the sequence implementing $P_{1x}(\alpha)$ must resolve to the identity on the addressed qubit, while on the unaddressed qubit $\exp(-\mathrm{i}\alpha\epsilon_N H_x) + \mathcal{O}(\epsilon_N^2)$ is applied. This behavior may be achieved by using a pulse sequence to implement P_{1x}. Let

$$S_x(\alpha) = M(2\pi a, -\phi_\alpha)M(2\pi a, \phi_\alpha) \tag{36}$$

where $a = \lceil |\alpha|/4\pi \rceil$ is a integer number of 2π rotations and $\phi_\alpha = \arccos(\alpha/4\pi a)$. On the addressed qubit, $S_x(\alpha)$ resolves to the identity, whereas for the unaddressed spins,

$$S_x(\alpha) = M(2\pi a, -\phi_\alpha)M(2\pi a, \phi_\alpha) = R(2\pi a\epsilon_N, -\phi_\alpha)R(2\pi a\epsilon_N, \phi_\alpha) = P_{1x}(\alpha)$$

is applied. In the Lie algebraic picture, $S_x(\alpha)$ is composed of two vectors constructed so that their vector sum is $-\mathrm{i}\alpha\epsilon_N H_x$. Similarly, $S_y(\beta) = M(2\pi b, \pi/2 - \phi_\beta)M(2\pi b, \pi/2 + \phi_\beta) = P_{1y}(\beta)$. From these basic sequences, we may construct $P_{2z}(\alpha\beta)$ by using the balanced group commutator Eq. (34). In Fig. 6a, we plot the

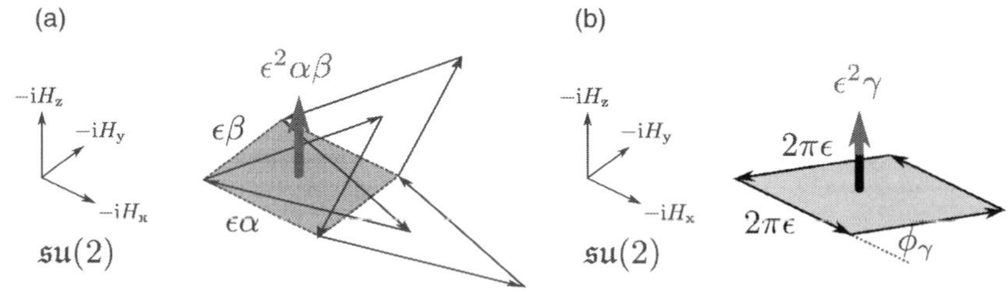

Figure 6. Generation of the pure error term in the Solovay–Kitaev method. (**a**) From Eq. (34), $P_{2z}(\alpha\beta)$ may be produced by the sequence $S_x(-\alpha)S_y(-\beta)S_x(\alpha)S_y(\beta)$. On the Lie algebra the sequence corresponds to a closed rectangular path with an enclosed area of $\epsilon^2\alpha\beta$. (**b**) Alternatively, the rhombus construction may be used to generate $P_{2z}(\alpha)$ using four pulses.

sequence $P_{2z}(\alpha\beta)$ as a vector path on the Lie algebra. The sequence encloses a signed area $\epsilon_N^2\alpha\beta$, which is denoted by the shaded rectangular figure. By tuning the rotation angles α and β, one may generate a term that encloses any desired area, thus allowing the synthesis of an arbitrary pure-error term.

At this point, we discuss a subtle feature of the addressing error model, which at first appears to complicate the application of the SK method. Observe that on the unaddressed spin, the imperfect propagators may only apply small rotations (i.e., rotations by angles $\theta\epsilon_N$). If we restrict ourselves to sequences composed of resonant square pulses, then the term proportional to $P_{1z}(\alpha)$ may not be produced; we may not prepare such a term by similarity transformation (e.g., $R(\pi/2, 0)S_y(\alpha)R^\dagger(\pi/2, 0)$) because such an operation would either require a large rotation or if instead the transformation was carried out on the individual sequence propagators, the rotation axes would be lifted out of the H_x–H_y plane. Similar arguments show that the Euler decomposition is also unavailable.

Fortunately, this restriction is not as serious as it first appears; the SK method may be used provided that the sequence terms are chosen with care. We are ultimately saved by the orientation of the error terms in the Lie algebra. Using the BCH formula, it is straightforward to show that for sequences composed of resonant pulses, the even-order error terms are always aligned along the $-\mathrm{i}H_z$ axis, whereas the odd-order terms are confined to $H_x - H_y$ plane. Likewise, using only $S_x(\alpha)$ and $S_y(\beta)$, it is possible to generate correction terms that follow the same pattern. As a consequence, in this case it is possible to generate the correction terms $U_T \exp(-\epsilon_N^{n+1}\Omega_{n+1})U_T^\dagger$ by carefully choosing the rotation angles and phases in the correction sequence B_{n+1}.

As an instructive example, we shall derive a second-order passband sequence using the Solovay–Kitaev method. We begin by calculating the Magnus expansion for the seed sequence $M_{\mathrm{SK1}}(\theta, 0) = U_T \exp(\epsilon_N^2\Omega_2 + \epsilon_N^3\Omega_3 + \cdots)$ where the target operation $U_T = \mathbb{1}$ for the unaddressed qubit. To cancel the second-order term, we simply need to apply the inverse of $\exp(\epsilon_N^2\Omega_2) = \exp(-\mathrm{i}2\pi^2\epsilon_N^2 \sin(2\phi_{\mathrm{SK1}})H_z)$. The planar rotation $\Upsilon = \mathbb{1}$, since $B_2 = P_{2z}(-2\pi^2 \sin(2\phi_{\mathrm{SK1}}))$ is already oriented in the correct direction. One possible choice for B_2 is the sequence

$$B_2 = S_x(-2\pi\cos\phi_{\mathrm{SK1}})S_y(2\pi\sin\phi_{\mathrm{SK1}})S_x(2\pi\cos\phi_{\mathrm{SK1}})S_y(-2\pi\sin\phi_{\mathrm{SK1}}) \tag{37}$$

The sequence $M_{\mathrm{SK2}}(\theta, 0) = B_2 M_{\mathrm{SK1}}(\theta, 0) = U_T + \mathcal{O}(\epsilon_N^3)$ corrects addressing errors to second order. We denote an nth order compensating sequence produced by the Solovay–Kitaev method using SK1 as an initial seed as SKn; here we have produced an SK2 sequence.

More efficient constructions for the correction sequence B_2 exist. Observe that we may directly create the pure error term $P_{2z}(\gamma)$ by using four pulses in the balanced group commutator arrangement $P_{2z}(\gamma) = M^\dagger(2\pi c, \phi_\gamma)M^\dagger(2\pi c, \phi'_\gamma)$ $M(2\pi c, \phi_\gamma)M(2\pi c, \phi'_\gamma)$, where $c = \lceil|\gamma|/4\pi\rceil$ and the phases are chosen to be $\phi_\gamma = \arcsin(\gamma/4\pi^2c^2)$ and $\phi'_\gamma = (1 - \mathrm{sign}\gamma)\pi/2$. The phase ϕ'_γ is only necessary to ensure that the construction also works for negative γ. We call this arrangement the rhombus construction. In Fig. 6b, we plot this sequence as a vector path on $\mathfrak{su}(2)$. In this construction the magnitude of the error term is tuned by adjusting the phase ϕ_γ (i.e., adjusting the area enclosed by the rhomboidal path of the sequence in the Lie algebra). The rhombus construction has the advantage of requiring half as many pulses as the standard method. We will now use this construction to produce an alternative form for SK2. Let

$$B_2' = M(2\pi, \phi_\gamma + \pi)M(2\pi, 0)M(2\pi, \phi_\gamma)M(2\pi, \pi) \tag{38}$$

where $\phi_\gamma = \arcsin(\sin(2\phi_{\text{SK1}})/2)$. Then $M'_{\text{SK2}}(\theta, 0) = B_2' M_{\text{SK1}}(\theta, 0) = U_T + \mathcal{O}(\epsilon_N^3)$.

b. Amplitude/Pulse-Length Errors. We now turn our attention to the compensation of amplitude and pulse-length errors. When considering these error models, the imperfect propagators of the form $M(\theta, \phi) = R(\theta(1 + \epsilon_A), \phi)$. In this case, we may also construct P_{1x} using Eq. (36) since the imperfect 2π rotations reduce to $M(2\pi a, \phi) = R(2\pi a\epsilon_A, \phi)$ and $S_x(\alpha) = P_{1x}(\alpha)$. As a consequence, if the initial seed sequence W_n is a passband sequence then W_{n+1} is also a passband sequence. We note however, that in this error model, one has more flexibility in the synthesis of the correction sequences B_{n+1} because now the imperfect propagators apply large rotations, then we may perform similarity transformations of sequences by directly implementing the required pulses. Notably, we may use imperfect propagators of order $\mathcal{O}(\epsilon_A^n)$ to apply a desired transformation at a cost of an error of order $\mathcal{O}(\epsilon_A^{n+1})$, for example, $M(\theta, \phi)P_{jz}M^\dagger(\theta, \phi) = R(\theta, \phi)P_{jz}R^\dagger(\theta, \phi) + \mathcal{O}(\epsilon_A^2)$.

The SK method may be used to calculate higher-order compensation sequences for the amplitude error model. We begin by calculating the relevant Magnus expansion for the seed sequence $M_{\text{SK1}}(\theta, 0) = M(2\pi, -\phi_{\text{SK1}})M(2\pi, \phi_{\text{SK1}})M(\theta, 0) = U_T \exp(\epsilon_A^2 \Omega_2 + \epsilon_A^3 \Omega_3 + \cdots)$ where now in the amplitude error model the target operation $U_T = R(\theta, 0)$. To cancel the second-order term we must apply the inverse of $U_T \exp(\epsilon_A^2 \Omega_2) U_T^\dagger = \exp(-\mathrm{i}2\pi^2 \epsilon_A^2 \sin(2\phi_{\text{SK1}}) H_z)$. This is precisely the same term that arose previously for addressing errors and the error may be compensated the same way. In fact every nth-order SK sequence is passband and works for both error models.

B. Wimperis/Trotter-Suzuki Sequences

In this section, we study a second family of fully compensating pulse sequences, first discovered and applied by Wimperis [52], which may be used to correct pulse length, amplitude, and addressing errors to second order. These sequences have been remarkably successful and have found extensive use in NMR and quantum information [54–56]. Furthermore, the Wimperis sequences may be generalized to a family of arbitrarily high-order sequences by connecting them to Trotter-Suzuki formulas [51]. An analogous composite sequence composed of rotations in SU(4) may be used to correct two-qubit operations [57,58]. We study the two-qubit case in Section V.A.

1. Narrowband Behavior

We begin with the problem of identifying narrowband sequences that correct addressing errors. In the addressing error model, operations on addressed qubits are error free, whereas on the unaddressed qubits the imperfect propagators takes the form $V(\boldsymbol{u}(t)) = U(\epsilon_N \boldsymbol{u}(t))$ and $U_T = \mathbb{1}$. Further, we shall constrain ourselves to sequences composed of resonant square-pulse propagators. Specifically, we search for arrangements of four pulses that eliminate both the first- and second-order Magnus expansion term for the imperfect propagator.

Before explicitly describing the construction of the Wimperis sequences, we digress shortly to point out a certain symmetry property that may be used to ensure that $\epsilon_N^2 \Omega_2 = 0$. Consider the group product of propagators of the form,

$$U(\epsilon_N \boldsymbol{u}(t)) = U_2 U_3 U_2 U_1, \qquad U_k = \exp(-\mathrm{i}\epsilon_N t_k \boldsymbol{u}_k \cdot \boldsymbol{H}) \tag{39}$$

with the added condition that the first-order expansion term for $U(\epsilon_N \boldsymbol{u}(t))$ has been already eliminated (i.e. $\epsilon_N \Omega_1 = -\mathrm{i}\epsilon_N (t_3 \boldsymbol{u}_3 + 2t_2 \boldsymbol{u}_2 + t_1 \boldsymbol{u}_1) \cdot \boldsymbol{H} = 0$). In this arrangement, the second and fourth propagators are identical; this pulse symmetry along with $t_1 \boldsymbol{u}_1 + 2t_2 \boldsymbol{u}_2 + t_3 \boldsymbol{u}_3 = 0$ eliminates the second-order term

$$\epsilon_N^2 \Omega_2 = \frac{\epsilon_N^2}{2} \sum_{i=1}^{4} \sum_{j=1}^{i} [-\mathrm{i}t_i \boldsymbol{u}_i \cdot \boldsymbol{H}, -\mathrm{i}t_j \boldsymbol{u}_j \cdot \boldsymbol{H}] = 0 \tag{40}$$

As a consequence, $U(\epsilon_N \boldsymbol{u}(t)) = \mathbb{1} + \mathcal{O}(\epsilon_N^3)$. Alternatively, one may regard the product $T = U_2 U_3 U_2$ as a second-order symmetric Trotter-Suzuki formula for the inverse operation $U_1^\dagger$ [47], which approximately cancels the undesired rotation U_1. Considering the control fields applied during the application of the corrector sequence T, the applied control Hamiltonian is symmetric with respect to time inversion. By a well-known theorem, all even-order expansion terms produced by a time-symmetric Hamiltonian cancel [43,59] (i.e., $\epsilon_N^{2j} \Omega_{2j} = 0$, for all positive integers j). Thus by implementing symmetric corrector sequences T_{2j}, it is sufficient to only consider the cancellation of the remaining odd-order error terms. In Section IV.B.1, we will inductively develop a series of corrector sequences of increasing accuracy based on symmetric Trotter-Suzuki formulas [47,51].

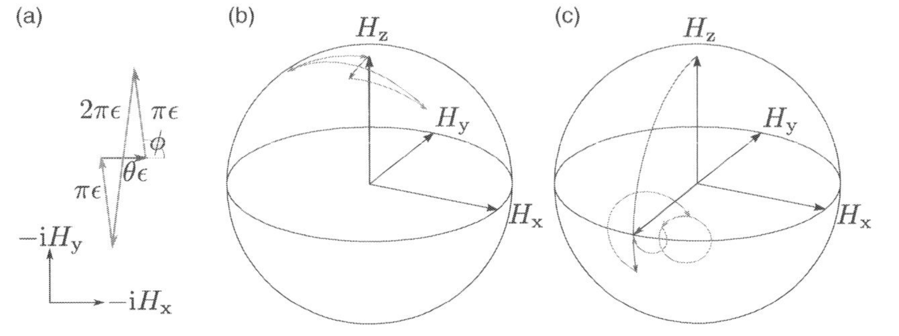

Figure 7. (**a**) Vector path followed by N2 on the Lie algebra. (**b**) Trajectory of an unaddressed spin during an N2 sequence, using imperfect rotations of the form $M(\theta, \phi) = R(\theta\epsilon_A, \phi)$. (**c**) B2 correcting an amplitude error, using imperfect rotations of the form $M(\theta, \phi) = R(\theta(1 + \epsilon_A), \phi)$. In these plots $\epsilon_N = \epsilon_A = 0.2$.

The cancellation of the second-order term may also be inferred from geometric considerations on the Lie algebra. To be concrete, consider a sequence of the form Eq. (39) where

$$\begin{aligned}
\epsilon_N t_1 \boldsymbol{u}_1 \cdot \boldsymbol{H} &= \theta\epsilon_N H_x \\
\epsilon_N t_2 \boldsymbol{u}_2 \cdot \boldsymbol{H} &= \pi\epsilon_N(\cos\phi_{N2} H_x + \sin\phi_{N2} H_y) \\
\epsilon_N t_3 \boldsymbol{u}_3 \cdot \boldsymbol{H} &= 2\pi\epsilon_N(\cos\phi_{N2} H_x - \sin\phi_{N2} H_y)
\end{aligned} \tag{41}$$

and the phase is $\phi_{N2} = \arccos(-\theta/4\pi)$ is chosen so that $t_1\boldsymbol{u}_1 + 2t_2\boldsymbol{u}_2 + t_3\boldsymbol{u}_3 = 0$ (i.e., the vectors form a closed path on the dynamical Lie algebra). Figure 7a is a diagram of these vectors on $\mathfrak{su}(2)$. In $\mathfrak{su}(2)$, the Lie bracket is equivalent to the vector cross-product [35], therefore, we may interpret $||\epsilon_N^2\Omega_2||_{\mathrm{HS}}$ as the signed area enclosed by the vector path on the Lie algebra. We note that the sequence under study encloses two regions of equal area and opposite sign, ensuring that the second-order term is eliminated.

With this insight, the construction of a second-order narrowband sequence is straightforward. The vectors Eq. (41) correspond to the sequence

$$M_{N2}(\theta, 0) = M(\pi, \phi_{N2})M(2\pi, -\phi_{N2})M(\pi, \phi_{N2})M(\theta, 0) \tag{42}$$

Wimperis refers to this sequence as NB1, which is the established name in the literature. In the current chapter, we label this sequence N2 in anticipation of the generalization of this form to N2j, which compensates addressing errors to $\mathcal{O}(2j)$. We use this language to avoid confusion with other established sequences, namely NB2, NB3, and so on [52]. The N2 sequence may be used to compensate addressing errors. For unaddressed qubits $M(\theta, \phi) = R(\theta\epsilon_N, \phi)$ and thus from Eqs. (40) and (41) it follows that $M_{N2}(\theta, 0) = \mathbb{1} + \mathcal{O}(\epsilon_N^3)$. Thus, on unaddressed qubits the sequence performs the identity operation up to second order. Furthermore, for addressed qubits $M(\theta, \phi) = R(\theta, \phi)$ and therefore $M_{\text{N2}}(\theta, 0) = R(\theta, 0)$. As a result, when the sequence $M_{\text{N2}}(\theta, 0)$ is used in the place of the imperfect operation $M(\theta, 0)$, the discrimination between addressed and unaddressed spins is enhanced. In Fig. 7b, we plot the magnetization trajectory for an unaddressed qubit under an N2 sequence.

2. *Broadband Behavior*

As previously discussed, broadband sequences are best suited for correcting amplitude or pulse-length errors. In the following, we will explicitly consider the amplitude error model, where $M(\theta, \phi) = R(\theta(1 + \epsilon_A), \phi) = R(\theta, \phi)R(\theta\epsilon_A, \phi)$. Although a pulse sequence may be studied by considering the interaction frame propagator, the method originally used by Wimperis is simpler. Wimperis's insight was that for lowest orders the toggled frame could be derived geometrically, using the relation $R(\theta, -\phi) = R(\pi, \phi)R(\theta, 3\phi)R(\pi, \phi + \pi)$. Consider the application of the target gate $U_T = R(\theta, 0)$ using the sequence

$$M_{\text{B2}}(\theta, 0) = M(\pi, \phi_{\text{B2}})M(2\pi, 3\phi_{\text{B2}})M(\pi, \phi_{\text{B2}})M(\theta, 0) \tag{43}$$

where $\phi_{\text{B2}} = \arccos(-\theta/4\pi)$. This Wimperis broadband sequence is traditionally called BB1, but here it is denoted as B2. When rewritten in terms of proper rotations one obtains

$$\begin{aligned} M_{\text{B2}}(\theta, 0) &= R(\pi\epsilon_A, \phi_{\text{B2}})\Big(R(\pi, \phi_{\text{B2}})R(2\pi\epsilon_A, \phi_{\text{B2}})R^\dagger(\pi, \phi_{\text{B2}})\Big) \\ &\quad \times R(\pi\epsilon_A, \phi_{\text{B2}})R(\theta\epsilon_A, 0)R(\theta, 0) \\ &= \Big[R(\pi\epsilon_A, \phi_{\text{B2}})R(2\pi\epsilon_A, -\phi_{\text{B2}})R(\pi\epsilon_A, \phi_{\text{B2}})R(\theta\epsilon_A, 0)\Big] R(\theta, 0) \end{aligned}$$

where the identity $R(2\pi, 3\phi_{\text{B2}}) = -\mathbb{1} = R(\pi, \phi_{\text{B2}} + \pi)R(\pi, \phi_{\text{B2}} + \pi)$ was used. Let Q denote the quantity enclosed in square brackets, so that we may write $M_{\text{B2}}(\theta, 0) = QU_T$. Observe that Q is precisely the form considered previously in N2. From our previous result, we may conclude $Q = \mathbb{1} + \mathcal{O}(\epsilon_A^3)$ and $M_{\text{B2}}(\theta, 0) = R(\theta, 0) + \mathcal{O}(\epsilon_A^3)$. As a result, when $M_{\text{B2}}(\theta, 0)$ is used in the place of the imperfect operation $M(\theta, 0)$ the effect of the systematic amplitude error is reduced. Note that errors for the B2 and N2 sequences follow equivalent paths on the Lie algebra in their respective interaction frames. In Fig. 7c, we plot the magnetization trajectory for a qubit under a B2 sequence with amplitude errors.

3. *Passband Behavior*

In some cases, it is convenient to have a passband pulse sequence that corrects for both addressing errors and for amplitude errors, as we saw with the Solovay–Kitaev sequences. The passband Wimperis sequence P2 is simply two

Solovay–Kitaev correction sequences in a row where the order of pulses is switched, $M_{P2}(\theta, 0) = M(2\pi, \phi_{P2})M(2\pi, -\phi_{P2})M(2\pi, -\phi_{P2})M(2\pi, \phi_{P2}) M(\theta, 0)$, with $\phi_{P2} = \arccos(-\theta/8\pi)$. The switching of pulse order naturally removes the second-order error term. One may verify that this sequence works for both addressing and amplitude errors.

4. *Arbitrarily Accurate Trotter-Suzuki Sequences*

The Wimperis sequences rely on a certain symmetrical ordering of pulse sequence propagators to ensure that even-order Magnus expansion terms are eliminated. We may further improve the performance of these sequences by taking advantage of additional symmetries that cancel higher-order terms. In the following, we shall show how by using symmetric Trotter-Suzuki formulas [46,47,60] a family of arbitrarily accurate composite pulse sequences may be constructed [51,53]. The Lie algebraic picture along with the Magnus and BCH series will be important tools in this process.

a. Symmetrized Suzuki Formulas. Before discussing the particular form of these sequences, it is helpful to briefly mention a few important results regarding symmetric products of time-independent propagators [47]. Given a series of skew-Hermitian time-independent Hamiltonians $\{\tilde{H}_1, \tilde{H}_2, \ldots, \tilde{H}_m\}$ such that $\sum_i^m \tilde{H}_i = \tilde{H}_T$, the BCH expansion tells us

$$\prod_{i=1}^{m} \exp(\lambda \tilde{H}_i) = \exp\left(\lambda \tilde{H}_T + \sum_{n=2}^{\infty} \lambda^n \Omega_n\right) \tag{44}$$

where λ is a real parameter and the expansion terms Ω_n depend on the specific ordering of the sequence. If we choose to apply the operators in a time-symmetric manner such that $\exp(\lambda \tilde{H}_i) = \exp(\lambda \tilde{H}_{m+1-i})$, then the symmetry of the pulse removes all even-order terms. For symmetric products, we have

$$\prod_{\text{symmetric}} \exp(\lambda \tilde{H}_i) = \exp\left(\lambda \tilde{H}_T + \sum_{j=2}^{\infty} \lambda^{2j-1} \Omega_{2j-1}\right) \tag{45}$$

An important observation concerning the elimination of the remaining odd terms was made by Suzuki [47]. Provided that $\tilde{H}_i = p_i \tilde{H}_T + (p_i)^{2j-1} \tilde{H}_B$, where the coefficients p_i are real numbers, there exist certain choices of coefficients such that $\sum_i^m \tilde{H}_i = \tilde{H}_T$. This requires that $\sum_i^m p_i = 1$ and $\sum_i^m (p_i)^{2j-1} = 0$. The situation is considerably simplified if we restrict ourselves to sequences composed of just two kinds of propagators, $U_1 = \exp(\tilde{H}_1)$ and $U_2 = \exp(\tilde{H}_2)$. In this case the previous expression simplifies to $n_1 p_1 + n_2 p_2 = 1$ and $n_1 p_1^{2j-1} + n_2 p_2^{2j-1} = 0$, where the integers n_1 and n_2 are the number of U_1 and U_2 pulses required to produce a sequence that is independent of H_B up to $\mathcal{O}(p^{2j+1})$. We solve for a set of coefficients by setting $p_2 = -2p_1$ and $n_2 = 1$, thus yielding $n_1 = 2^{2j-1}$ and $p_1 = 1/(2^{2j-1} - 2)$.

Combining these observations, if $W_{2j-2}(p)$ is a $(2j-2)$th approximation of $\exp(p\tilde{H}_T)$ and $\Omega_{2j-1} = p^{2j-1}\tilde{H}_B$ where $\tilde{H}_B$ is independent of p, then we can construct $W_{2j}(1)$ from the lower approximations, $W_{2j}(1) = (W_{2j-2}(p))^{2^{2j-2}} W_{2j-2}(-2p)(W_{2j-2}(p))^{2^{2j-2}}$. As a result, Suzuki formulas provide a path of producing higher-order sequences from a symmetric combination of lower-order sequences. Notice that the symmetric decomposition is used to keep the even-order terms zero.

b. Passband Behavior. In the following, we will seek to generalize the second-order passband sequence P2 to an arbitrarily accurate passband sequence P2j. Our goal is to develop a correction sequence $T_{2j}(k, \phi) = \exp(\mathrm{i}\theta\epsilon_A H_x) + \mathcal{O}(\epsilon_A^{2j+1})$ that cancels the unwanted rotation of a $M(\theta, 0)$ operation.

When considering passband sequences, it is convenient to use rotation angles that are integer multiples of 2π. Let us define the triangular motif,

$$\begin{aligned} T_1(k, \phi) &= M(2k\pi, -\phi)M(2k\pi, \phi) \\ &= \exp(-\mathrm{i}4k\pi\epsilon_A \cos\phi H_x - \mathrm{i}(2\pi k\epsilon_A)^2 \cos\phi \sin\phi H_z) + O(\epsilon_A^3) \end{aligned} \tag{46}$$

Observe the passband sequence SK1 may now be written as $M_{\mathrm{SK1}}(\theta, 0) = T_1(1, \phi_{\mathrm{SK1}})M(\theta, 0)$ and where the phase ϕ_{SK1} is as defined previously. The remaining second-order term is odd with respect to ϕ, and is related to the vector cross-product on the Lie algebra. A second-order sequence may be constructed by combining two $T_1(k, \phi)$ terms so that the correction sequence is symmetric and the cross-product cancels. Let us define the symmetrized product

$$\begin{aligned} T_2(k, \phi) &= T_1(k, -\phi)T_1(k, \phi) \\ &= \exp(p\tilde{H}_T + p^3\tilde{H}_B) + \mathcal{O}(p^5) \end{aligned} \tag{47}$$

where the length $p = -(8k\pi/\theta)\cos\phi$, the target Hamiltonian $\tilde{H}_T = \mathrm{i}\theta\epsilon_A H_x$, and $\tilde{H}_B = \epsilon_A^3\Omega_3/p^3$ is the remaining term we wish to cancel. Ω_3 is a function that depends on k and ϕ such that $\tilde{H}_B$ is a function of ϕ but not of the length scale k. For fixed ϕ and variable k, this makes H_B independent of p. Observe that the passband sequence P2 may now be written as $M_{\mathrm{P2}}(\theta, 0) = T_2(1, \phi_{\mathrm{P2}})M(\theta, 0)$, where again $\phi_{\mathrm{P2}} = \arccos(-\theta/8\pi)$.

Our strategy is to use a Suzuki formula to construct higher-order $T_4(k, \phi)$ (i.e., $T_{2j}(k, \phi)$ for $j = 2$), using a symmetric combination of $(n_1 = 8)$-many $\exp(\tilde{H}_1) = T_2(k, \phi)$ sequences and a single $\exp(\tilde{H}_2) = T_2(-2k, \phi)$ sequence, yielding

$$T_4(k, \phi) = \Big(T_2(k, \phi)\Big)^4 T_2(-2k, \phi)\Big(T_2(k, \phi)\Big)^4 \tag{48}$$

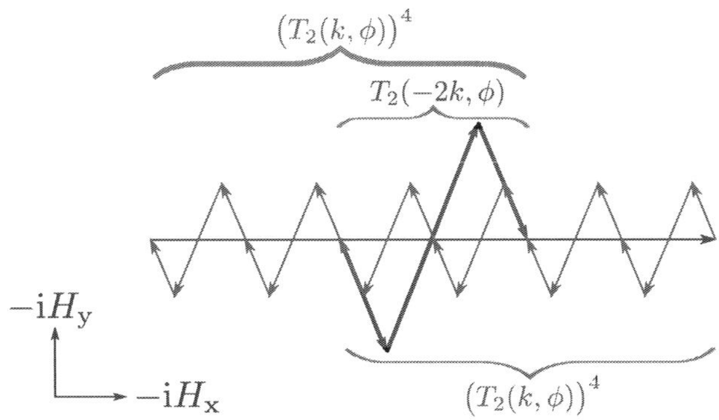

Figure 8. Vector path followed by P4 on the Lie algebra.

In order to produce required correction term, the parameters (k, ϕ) must be chosen such that $(n_1 - 2)p = 1$. The fourth-order passband sequence P4 is $M_{\text{P4}}(\theta, 0) = T_4(1, \phi_{\text{P4}})M(\theta, 0)$, where $\phi_{\text{P4}} = \arccos(-\theta/48\pi)$. In Fig. 8, we plot the vector path followed by P4 on the Lie algebra. This result may be further generalized. To produce $T_{2j}(k, \phi)$ requires $(n_1 = 2^{2j-1})$-many $T_{2j-2}(k, \phi)$ sequences and a single $T_{2j-2}(-2k, \phi)$ sequence in the symmetric ordering

$$T_{2j}(k, \phi) = \Big(T_{2j-2}(k, \phi)\Big)^{2^{2j-2}} T_{2j-2}(-2k, \phi)\Big(T_{2j-2}(k, \phi)\Big)^{2^{2j-2}} \tag{49}$$

We then fix ϕ so that the first-order term cancels the unwanted rotation, yielding

$$\phi_{\text{P}2j} = \arccos\left(-\frac{\theta}{2\pi f_j}\right) \tag{50}$$

where $f_j = (2^{2j-1} - 1)f_{j-1}$ and for the sequence P2j, $f_1 = 4$. Then the $2j$th-order passband sequence P2j is $M_{\text{P}2j}(\theta, 0) = T_{2j}(1, \phi_{\text{P}2j})M(\theta, 0)$.

The same method can be used to develop exclusively narrowband or broadband sequences called N2j and B2j, respectively [51]. This requires redefining the bottom recursion layer $T_2(k, \phi)$ to have either narrowband or broadband properties. For N2j, $f_1 = 2$ and $T_2(k, \phi) = T_1(k/2, \phi)T_1(k/2, -\phi)$. For B2$j$ $f_1 - 2$, but T_2 is slightly more complicated; when k is even $T_2(k, \phi) = T_1(k/2, -\phi)T_1(k/2, \phi)$ just like N2j, however, when k is odd and $T_2(k, \phi) = M(k\pi, \phi)M(k\pi, 3\phi)M(k\pi, 3\phi)$. In Fig. 9, we compare several of the generalized Trotter–Suzuki sequences to the ideal unitaries $U_T = R(\pi/2, 0)$ in the case of amplitude errors (top row) and $U_T = \mathbb{1}$ in the case of addressing errors on unaddressed qubits (bottom row).

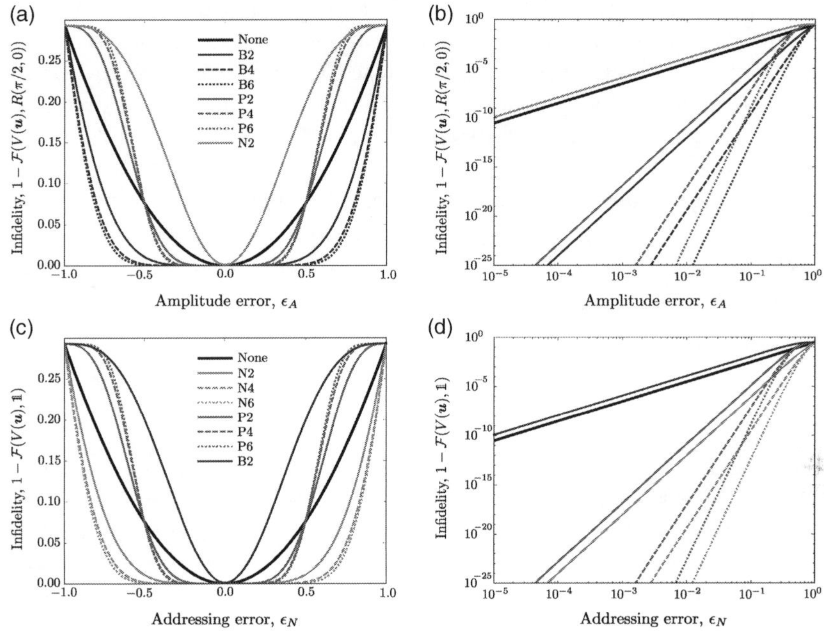

Figure 9. Infidelity of the Trotter–Suzuki sequences B2j, P2j, and N2j. (**a, b**) In the amplitude error model, $M(\theta, \phi) = M(\theta(1 + \epsilon_A), \phi)$ and $U_T = R(\pi/2, 0)$. (**c, d**) In the addressing error model on the unaddressed qubits $M(\theta, \phi) = M(\theta\epsilon_N, \phi)$ and $U_T = \mathbb{1}$, while on the addressed spins $R(\pi/2, 0)$ is applied. Each error model establishes a separate preferred interaction frame; when transformed into the appropriate pictures, the B2j and N2j sequences are homologous. The passband sequences P2j can correct both amplitude and addressing errors at the cost of reduced efficacy. (*See the color version of this figure in the Color Plates section.*)

C. CORPSE

So far, the sequences considered here have been designed to correct systematic amplitude and addressing errors. The correction of errors arising from an inaccurate tuning of the control field are also of practical interest. The treatment of detuning errors is similar in principle to the error models already considered, however, in practice the construction of compensating sequences is complicated by the noncommutivity of the ideal Hamiltonian and the erroneous Hamiltonian generated by the control distortion $[\boldsymbol{u}(t) \cdot \boldsymbol{H}, \boldsymbol{\delta u}(t) \cdot \boldsymbol{H}] \neq 0$.

Fully compensating pulse sequences for detuning errors were originally studied by Tycko [61] and later generalized by Cummins and Jones into the humorously named ROTTEN (resonance offset tailoring to enhance nuition) [62] and CORPSE (compensating for off-resonance with a pulse sequence) [63,64] family of sequences. Cummins and Jones initially derived the sequence using quaternion algebra to represent simple rotations, and optimized the angles to eliminate the first-order effects of the detuning error. The CORPSE family of sequences has found application in NMR [63] and SQUID [65] experiments. Here, we reexamine CORPSE using the techniques outlined in Section III.B.2.

CORPSE is a sequence that performs a compensated rotation about the H_x axis. Following Cummins and Jones, the sequence is constructed from three square pulses, which in the case of perfect resonance, induce rotations about the H_x, $-H_x$, and H_x axes, sequentially. The pulse sequence is parametrized by the piecewise constant control function

$$u_x(t) = \begin{cases} u_x & 0 \le t < t_1 \\ -u_x & t_1 \le t < t_2 \\ u_x & t_2 \le t \le \tau \end{cases} \tag{51}$$

where u_x is a constant H_x control amplitude and the t_j are times at which the field direction is switched. This construction is particularly amenable to analytic methods, as the ideal control Hamiltonian $H(t) = u_x(t)H_x$ commutes with itself at all times. In the absence of systematic detuning errors the control Hamiltonian produces the following ideal unitary evolution,

$$U(u_x(t')\mathbf{x}; t, 0) = \exp(-i\vartheta(t)H_x) = R(\vartheta(t), 0), \qquad \vartheta(t) = \int_0^t dt' u_x(t') \tag{52}$$

Over the entire time interval $0 \le t \le \tau$, the pulse sequence produces the gate $U_T = R(\theta, 0)$, where $\theta = \vartheta(\tau)$.

In the presence of an unknown detuning error, $\boldsymbol{\delta u} = \epsilon_D \mathbf{z}$, the rotation axis is lifted in the direction of the H_z axis on the Bloch sphere. A Magnus expansion may be used for the interaction frame propagator $U^I(\epsilon_D \mathbf{z}; \tau, 0)$, which produces the evolution generated by the systematic detuning error. Combining Eqs. (17) and (52) the first-order term is

$$\epsilon_D \Omega_1(\tau, 0) = -i\epsilon_D \int_0^\tau dt \cos(\vartheta(t))H_z + \sin(\vartheta(t))H_y \tag{53}$$

Direct integration yields:

$$\begin{aligned} \epsilon_D \Omega_1(\tau, 0) = &-\frac{i\epsilon_D}{u_x}[\sin(\theta_1)H_z + (1 - \cos(\theta_1))H_y] + \cdots \\ &+ \frac{i\epsilon_D}{u_x}[(\sin(\theta_1 - \theta_2) - \sin(\theta_1))H_z \\ &+ (\cos(\theta_1) - \cos(\theta_1 - \theta_2))H_y] + \cdots \\ &- \frac{i\epsilon_D}{u_x}[(\sin(\theta_1 - \theta_2 + \theta_3) - \sin(\theta_1 - \theta_2))H_z \\ &+ (\cos(\theta_1 - \theta_2) - \cos(\theta_1 - \theta_2 + \theta_3))H_y] \end{aligned} \tag{54}$$

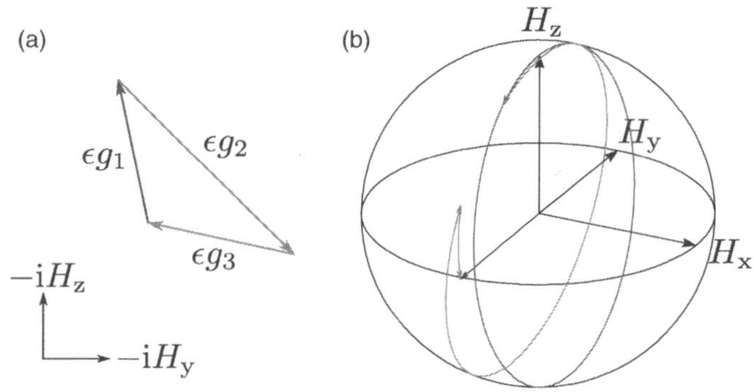

Figure 10. (**a**) Vector path followed by CORPSE on the Lie algebra, with the choice of parameters $n_1 = n_2 = n_3 = 0$. Each vector g_k corresponds to a term in Eq. (54). (**b**) Trajectory of a spin under a CORPSE sequence for $U_T = R(\pi/2, 0)$, with $n_1 = n_2 = 1$ and $n_3 = 0$ chosen to produce positive angles. The sequence is constructed of imperfect rotations of the form $M(\theta_k, 0) = \exp(-\mathrm{i}(\theta_k H_\mathrm{x} + \epsilon_D \theta_k H_\mathrm{z}))$, with $\epsilon_D = 0.2$.

where $\theta_k = \vartheta(t_k) - \vartheta(t_{k-1})$ are the effective rotation angles applied during the kth square pulse. At this point, we may interpret each of the terms of Eq. (54) as vectors on the dynamical Lie algebra. Figure 10a is a diagram of these vectors on $\mathfrak{su}(2)$. In order to eliminate the first-order expansion term, the rotation angles θ_k must be chosen so that the vectors must form a closed path.

Then by choosing the rotation angles to be

$$\begin{aligned} \theta_1 &= 2\pi n_1 + \theta/2 - \arcsin(\sin(\theta/2)/2) \\ \theta_2 &= 2\pi n_2 - 2\arcsin(\sin(\theta/2)/2) \\ \theta_3 &= 2\pi n_3 + \theta/2 - \arcsin(\sin(\theta/2)/2) \end{aligned} \tag{55}$$

where n_1, n_2, and n_3 are integers, we find that both $\epsilon_D \Omega_1(\tau, 0) = 0$ and $\theta = (\theta_1 - \theta_2 + \theta_3 \mod 2\pi)$. The extra factors of 2π are added so that the individual pulse rotation angles may be made positive. In principle, any choice of integers is sufficient to compose a first-order sequence, however, the choice $n_1 = n_2 = 1$ and $n_3 = 0$ minimizes the remaining second-order term while still producing a positive set of rotation angles [64]. The CORPSE family of sequences

$$\begin{aligned} V(\boldsymbol{u}(t)) &= \exp(-\mathrm{i}(\theta_3 H_\mathrm{x} + \epsilon_D \theta_3 H_\mathrm{z}))\exp(-\mathrm{i}(-\theta_2 H_\mathrm{x} + \epsilon_D \theta_2 H_\mathrm{z})) \\ &\quad \times \exp(-\mathrm{i}(\theta_1 H_\mathrm{x} + \epsilon_D \theta_1 H_\mathrm{z})) \\ M_{\mathrm{C1}}(\theta, 0) &= M(\theta_3, 0)M(\theta_2, \pi)M(\theta_1, 0) = R(\theta, 0) + \mathcal{O}(\epsilon_D^2) \end{aligned} \tag{56}$$

are fully compensating first-order sequences. In the presence of an unknown detuning error, a CORPSE sequence, $M_{\mathrm{C1}}(\theta, 0)$, may be implemented in the place of the simple rotation $R(\theta, 0)$ and the erroneous evolution is suppressed. In Fig. 10b, we plot the magnetization trajectory for a qubit under a CORPSE sequence with detuning error $\epsilon_D = 0.2$.

1. Arbitrarily Accurate CORPSE

We turn our attention to sequences that compensate detuning errors to arbitrarily high order. Once again, the Solovay–Kitaev method may be used to construct higher-order sequences, now using CORPSE as the seed sequence. This problem was first studied by Alway and Jones [53]. Recall that in the Solovay–Kitaev method, one synthesizes a correction sequence B_{n+1} in two steps. First, one generates a propagator $P_{(n+1)z}(\xi)$ with an amplitude proportional to the leading-order error term. Second, the similarity transformation $U_T\Upsilon^\dagger$ is used so that $B_{n+1} = U_T\Upsilon^\dagger P_{(n+1)z}(\xi)\Upsilon U_T^\dagger$ cancels the leading error term of the seed sequence (see Eq. (35)). Equation (34) shows that the term $P_{(n+1)z}(\xi)$ may be constructed recursively using a product of first-order P_1s. Therefore, the Solvay-Kitaev method may be extended to correct detuning errors if (1) procedures for generating P_{1x}, P_{1y}, and P_{1z} using imperfect pulses have been found, and (2) when a method for applying the similarity transformation $U_T\Upsilon^\dagger$ using imperfect pulses has been identified. This task is complicated by the difficulty of generating inverse operations for imperfect pulses affected by detuning errors.

Earlier, the first-order P_1s were applied using a composite pulse sequence (see Eq. (37)). We employ a similar strategy here. Let

$$S_z(\alpha) = M(\alpha/2, \phi + \pi)M(\alpha/2, \phi) = \exp(-\mathrm{i}\alpha\epsilon H_z) = P_{1z}(\alpha) \tag{57}$$

Then we may apply $P_{1z}(\alpha)$ by implementing the sequence $S_z(\alpha)$ instead. What remains is to develop sequences that apply $P_{1x}(\alpha)$ and $P_{1y}(\alpha)$. Observe that if it were possible to perform ideal rotations (where $R^\dagger(\theta, \phi) = R(\theta, \phi + \pi)$), then one could easily implement $S_x(\alpha) = R(\pi/2, \pi/2)S_z(\alpha)R^\dagger(\pi/2, \pi/2)$ and $S_y(\alpha) = R^\dagger(\pi/2, 0)S_z(\alpha)R(\pi/2, 0)$ in the place of $P_{1x}(\alpha)$ and $P_{1y}(\alpha)$. However, in the presence of systematic detuning errors, we may not apply this transformation using a simple imperfect pulse since $M^\dagger(\theta, \phi) \neq M(\theta, \phi + \pi)$. We may avoid this complication by using a CORPSE sequence to approximate the transformation to sufficient accuracy. We use the first-order CORPSE sequence, $M_{C1}(\theta, \phi)$ to implement the target operation $U_T = R(\theta, \phi)$. Then we may write

$$\begin{aligned} S_x(\alpha) &= M_{C1}(\pi/2, \pi/2)S_z(\alpha)M_{C1}(\pi/2, 3\pi/2) \\ &= \exp(-\mathrm{i}\alpha\epsilon H_x) + \mathcal{O}(\epsilon^2) = P_{1x}(\alpha) \end{aligned} \tag{58}$$

and

$$S_y(\alpha) = M_{C1}(\pi/2, \pi)S_x(\alpha)M_{C1}(\pi/2, 0) = \exp(-\mathrm{i}\alpha\epsilon H_y) + \mathcal{O}(\epsilon^2) = P_{1y}(\alpha) \tag{59}$$

Furthermore, by careful choice of θ and ϕ, we can implement $P_{1\eta}(\alpha)$ for any axis η about any angle α. Following the Solovay–Kitaev method, we can then construct $P_{jz}(\xi)$ to any order j.

What remains is to implement the similarity transformation by $U_T\Upsilon^\dagger$. However, we note that the available approximate rotations $M_{C1}(\theta,\phi) = R(\theta,\phi) + \mathcal{O}(\epsilon_D)$ are only accurate to first order. Hence, if we were to attempt to apply a similarity transformation on a higher-order $P_{jz}(\xi)$ term using CORPSE sequences, additional second-order errors would be introduced. This difficulty may be avoided by applying the appropriate rotation at the level of the first-order P_1 operations. As an instructive example, we work through the process for the second-order correction. The correction sequences may be constructed in a rotated coordinate system, determined by the basis transformation $H'_\mu = U_T\Upsilon^\dagger H_\mu \Upsilon U_T^\dagger$ for $\mu \in \{\mathrm{x}, \mathrm{y}, \mathrm{z}\}$. One may calculate four rotations $R(\theta_{\pm\mu'}, \phi_{\pm\mu'})$, that map the H_z axis to the $H_{\pm\mathrm{x}'}$ and $H_{\pm\mathrm{y}'}$ axes $H_{\pm\mu'} = R(\theta_{\pm\mu'}, \phi_{\pm\mu'}) H_\mathrm{z} R^\dagger(\theta_{\pm\mu'}, \phi_{\pm\mu'})$ and also $P_{1\pm\mu'}(\alpha) = R(\theta_{\pm\mu'}, \phi_{\pm\mu'}) P_{1\mathrm{z}}(\alpha) R^\dagger(\theta_{\pm\mu'}, \phi_{\pm\mu'})$. This transformation may also be applied using a first-order CORPSE sequence, because the resulting error is absorbed into the second-order error term in $P_{1\pm\mu'}(\alpha)$. Once transformed terms have been obtained, the correction sequence may be constructed in the usual manner

$$
\begin{aligned}
B_2 &= P_{2z'}(\xi) = P_{1x'}(-\sqrt{\xi}) P_{1y'}(-\sqrt{\xi}) P_{1x'}(\sqrt{\xi}) P_{1y'}(\sqrt{\xi}) \\
&= M_{C1}(\theta_{-x'}, \phi_{-x'}) P_{1z}(\sqrt{\xi}) M_{C1}(\theta_{-x'}, \phi_{-x'} + \pi) M_{C1}(\theta_{-y'}, \phi_{-y'}) P_{1z}(\sqrt{\xi}) \\
&\quad \times M_{C1}(\theta_{-y'}, \phi_{-y'} + \pi) M_{C1}(\theta_{x'}, \phi_{x'}) P_{1z}(\sqrt{\xi}) M_{C1}(\theta_{x'}, \phi_{x'} + \pi) \\
&\quad \times M_{C1}(\theta_{y'}, \phi_{y'}) P_{1z}(\sqrt{\xi}) M_{C1}(\theta_{y'}, \phi_{y'} + \pi)
\end{aligned}
$$

We can then define a second-order CORPSE sequence as $M_{C2}(\theta,0) = B_2 M_{C1}(\theta,0) = R(\theta,0) + \mathcal{O}(\epsilon_D^3)$. One can then continue using Solvay-Kitaev techniques to remove the detuning error to all orders.

2. *Concatenated CORPSE: Correcting Simultaneous Errors*

So far, we have considered the problem of quantum control in the presence of a single systematic error, while in a real experiment several independent systematic errors may affect the controls. In many situations one error dominates the imperfect evolution; it is appropriate in these cases to use a compensating sequence to suppress the dominant error. However, it is also important to consider whether a sequence reduces the sensitivity to one type of error at the cost of increased sensitivity to other types of errors [64]. Such a situation may occur if an error model couples two error sources. Consider a set of control functions $\{u_\mu(t)\}$ where each control is deformed by two independent systematic errors. A general two-parameter error model for the control $v_\mu(t) = f_\mu[\boldsymbol{u}(t); \epsilon_i, \epsilon_j]$ may be formally expanded as

$$f_\mu[\boldsymbol{u}(t); \epsilon_i, \epsilon_j] = f_\mu[\boldsymbol{u}(t); 0, 0] + \epsilon_i \frac{\partial}{\partial \epsilon_i} f_\mu[\boldsymbol{u}(t); 0, 0] + \epsilon_j \frac{\partial}{\partial \epsilon_j} f_\mu[\boldsymbol{u}(t); 0, 0] + 2\epsilon_i \epsilon_j \frac{\partial^2}{\partial \epsilon_i \partial \epsilon_j} f_\mu[\boldsymbol{u}(t); 0, 0] + \mathcal{O}(\epsilon_i^2 + \epsilon_j^2) \qquad (60)$$

In practice, we need only concern ourselves with the first few expansion terms, because in most physically relevant error models the higher-order derivatives are identically zero. Following the same reasoning employed in Section II, we may decompose the imperfect propagator as $V(\boldsymbol{u}(t)) = U(\boldsymbol{u}(t))U^I(\boldsymbol{v}(t) - \boldsymbol{u}(t))$, where $U(\boldsymbol{u}(t))$ is the ideal propagation in the absence of errors and the interaction frame propagator $U^I(\boldsymbol{v}(t) - \boldsymbol{u}(t))$ represents the evolution induced by the systematic errors. Formally $U^I(\boldsymbol{v}(t) - \boldsymbol{u}(t))$ may be studied using a Magnus expansion, however, this method is usually impeded by the complexity of the Magnus series. The determination of sequences that compensate simultaneous errors is currently an unresolved problem, although some progress has been made by considering concatenated pulse sequences [31]. As an example relevant to an NMR quantum computer, we now study error models where two systematic errors occur simultaneously and show that these errors can be compensated by concatenation of pulse sequences.

a. Simultaneous Amplitude and Detuning Errors. When studying the control of qubits based on coherent spectroscopy methods, it is natural to consider situations where systematic errors in the field amplitude and tuning are simultaneously present. From the NMR control Hamiltonian Eq. (9) it is straightforward to derive the error model, $v_{x/y} = u_{x/y}(t)(1 + \epsilon_A)$ and $v_z = u_z(t) + \epsilon_D$. In this case, the independent amplitude and detuning errors are decoupled and the parameters ϵ_A and ϵ_D do not affect the same control.

b. Simultaneous Pulse-Length and Detuning Errors. Likewise, simultaneous errors in the pulse length and field tuning may occur. Again, it is simple to derive the joint error model, $v_{x/y} = u_{x/y}(t)(1 + \epsilon_T)$ and $v_z = (u_z(t) + \epsilon_D)(1 + \epsilon_T)$. Unlike the previous case, this error model couples the parameters ϵ_T and ϵ_D, because they both effect the H_z control function.

It is sometimes possible to produce a sequence that compensates two simultaneous errors by pulse sequence concatenation. Recall that the Wimperis sequence B2 is a second-order compensation sequence for both amplitude and pulse-length errors. However, if only detuning errors are present, then the response of a B2 sequence to pure detuning errors is similar to that of a native pulse Similarly, a CORPSE sequence would eliminate the first-order term produced by the detuning offsets but not correct amplitude errors. The sequence B2CORPSE corrects both errors *independently* by concatenating the B2 and CORPSE sequences.

B2CORPSE is composed of three B2 subsequences that form a larger CORPSE sequence

$$M_{\text{B2C1}}(\theta, 0) = M_{\text{B2}}(\theta_3, 0)M_{\text{B2}}(\theta_2, \pi)M_{\text{B2}}(\theta_1, 0) \tag{61}$$

where the angles $\{\theta_1, \theta_2, \theta_3\}$ are given in Eq. (55). At the lower level of concatenation, B2 sequences are used to synthesize rotations robust to pure amplitude or pulse-length errors, whereas on the higher level the CORPSE construction compensates pure detuning errors. Figure 11a–d are plots of the infidelity of plain pulses, and the CORPSE, B2, and B2CORPSE sequences in the presence of simultaneous amplitude and detuning errors. Similarly, Fig. 11e–h are infidelity plots for simultaneous pulse-length and detuning errors. As expected, the CORPSE sequences improve the accuracy of gates with respect to detuning errors, but offers little improvement against either amplitude or pulse length errors. Unsurprisingly, this behavior is inverted for the B2 sequences. The B2CORPSE sequence performs well for either amplitude/pulse-length or detuning errors; in the presence of simultaneous errors the performance diminishes, yet the fidelity of the applied gate is still vastly improved over uncompensated pulses.

D. Shaped Pulse Sequences

Thus far, we have considered sequences composed of control functions $u_\mu(t)$, which are piecewise constant over the pulse interval. This construction is particularly convenient from a sequence design perspective, although in practice instrumental shortcomings will frequently distort the pulse profile. Also, in some applications, the rectangular profile is nonideal. For example, in the frequency domain, the square pulse corresponds to a sinc function whose local maxima may complicate the control of certain systems. Finally, we note that in some systems, such as in superconducting Josephson junction qubits, bandwidth requirements forbid the sudden switching of control fields (i.e., place a limit on the derivative $|\dot{u}_\mu| \leq \dot{u}_\mu^{\text{MAX}}$). In these cases, it is desirable to consider shaped pulse sequences composed by a set of continuous differentiable control functions [66]. In the present chapter, we shall only consider shaped pulse sequences, which are also fully compensating (class A), and therefore appropriate for use in a quantum processor. Specifically, we study shaped pulse sequences that also compensate systematic errors. We shall see that the additional flexibility admitted by shaped pulses frequently produces superior sequences.

Once given a pulse waveform, we may verify that it is compensating for a particular error model by computing the Magnus expansion in the appropriate interaction frame; however, we emphasize that these methods require the evaluation of successive nested integrals (e.g., Eq. (15)) and are difficult to analytically implement beyond the first few orders [67]. Furthermore, the inverse problem (solving for the control functions) is especially difficult except in the most simple

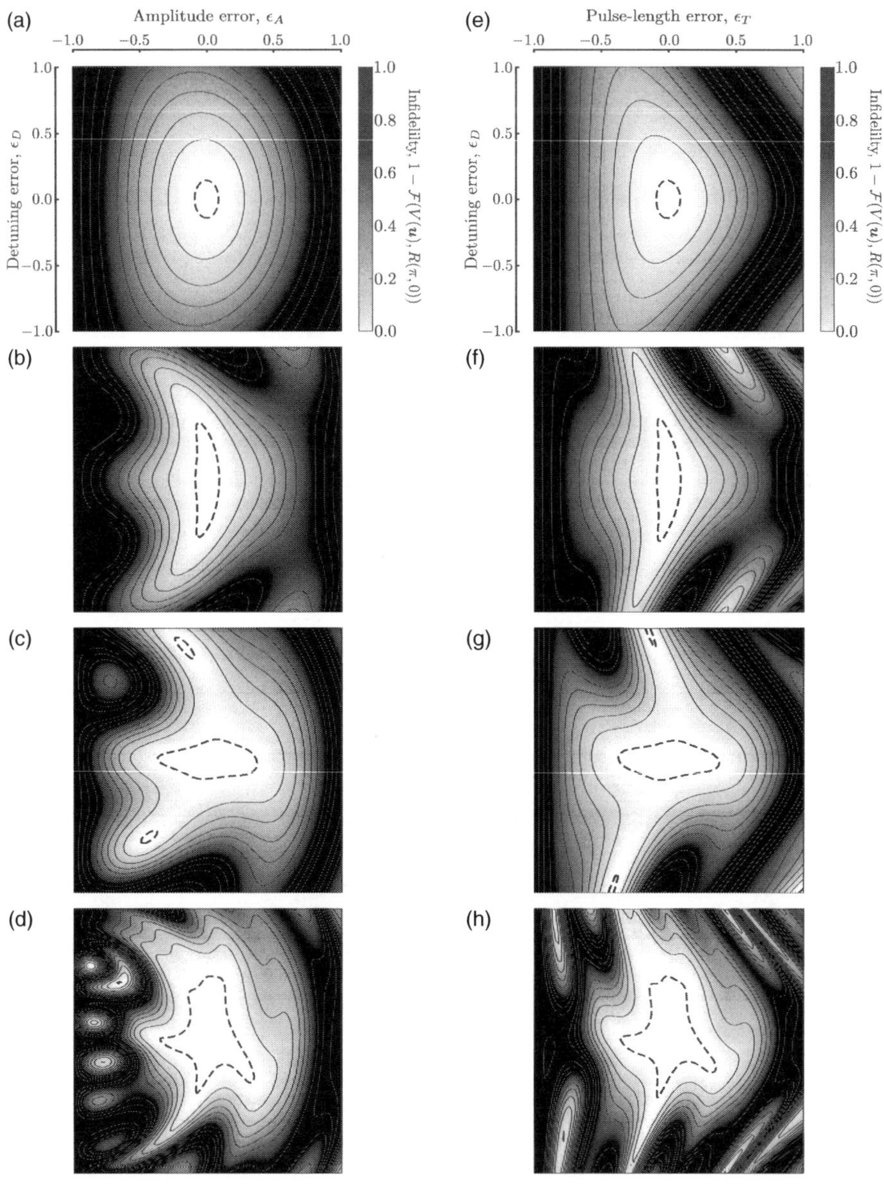

Figure 11. Infidelity of (**a, e**) plain pulses; (**b, f**) CORPSE; (**c, g**) B2; and (**d, h**) a concatenated B2CORPSE sequence in the presence of simultaneous amplitude and detuning (left column) and pulse-length and detuning (right column) errors. The target rotation is $U_T = R(\pi, 0)$. The dashed contour corresponds to an infidelity of 0.01, while the remaining contours are plotted at 10% intervals.

cases. For this reason, various numerical optimization methods have become popular, including gradient-accent techniques, optimal control methods [68–70], and simulated annealing [71,72]. Methods that use elements of optimal-control theory merit special attention; in recent years the GRAPE [73,74] and Krotov algorithms have been especially successful in pulse design, and has been applied to NMR [68], trapped ions [75], and ESR [76]. The main advantage of the GRAPE algorithm is an efficient estimation of the gradient of the fidelity as a function of the controls, which then enables optimization via a gradient-accent method. A recent review of these methods may be found in Refs [77,78].

To demonstrate the relative performance of shaped sequences, we consider the continuous analogs of the CORPSE sequence that modulate $u_x(t)$ to compensate detuning errors. From Eq. (53) observe that the first-order compensation condition $\epsilon_D \Omega_1(\tau, 0) = 0$ is met when $\int_0^\tau dt \cos(\vartheta(t)) = \int_0^\tau dt \sin(\vartheta(t)) = 0$. What remains is to solve for a set of control functions $u_x(t)$ that generate the target operation U_T while remaining robust to the detuning error. One popular scheme is to decompose the control functions as a Fourier series [72,79] over the pulse interval

$$u_x(t) = \omega \sum_{n=0}^{\infty} a_n \cos\left(n\omega\left(t - \frac{\tau}{2}\right)\right) + b_n \sin\left(n\omega\left(t - \frac{\tau}{2}\right)\right), \quad \omega = \frac{2\pi}{\tau} \tag{62}$$

This has the advantage of specifying the controls using only a few expansion parameters $\{a_n\}$ and $\{b_n\}$. Also, the series may be truncated to avoid high-frequency control modulations that are incompatible with some control systems. Moreover, by choosing each of the expansion terms $b_n = 0$, the resultant sequence may be made symmetric with respect to time reversal. We note that not all compensating sequences completely eliminate the first-order error term; in some sequences the leading order errors are highly suppressed rather than completely eliminated. This behavior is more common in sequences obtained from numerical methods.

Historically, the first class A shaped pulses developed belonged to the U-BURP (band-selective, uniform response, pure-phase) family designed by Geen and Freeman [25,72], using simulated-annealing, and gradient decent methods. Soon after, Abramovich and Vega attacked the same problem using approximation methods based on Floquet theory [80]. These sequences are designed to correct detuning errors over an extremely large range; however, they fail to completely eliminate the leading-order error terms. In the design of compensating sequences for quantum computing, the emphasis has been on the synthesis of composite rotations of extraordinary accuracy over a narrow window of errors. More recently, Steffen and Koch considered shaped Gaussian modulated controls specially designed for superconducting qubit manipulations [81]. Pryadko and Sengupta also derived a family of sequences ($S_1(\phi_0)$,$S_2(\phi_0)$,$Q_1(\phi_0)$,$Q_2(\phi_0)$) using a semianalytic method based on the Magnus expansion and average Hamiltonian theory [67,82]. By design, these sequences eliminate the leading-order error terms in a manner similar to the square-pulse sequences previously discussed.

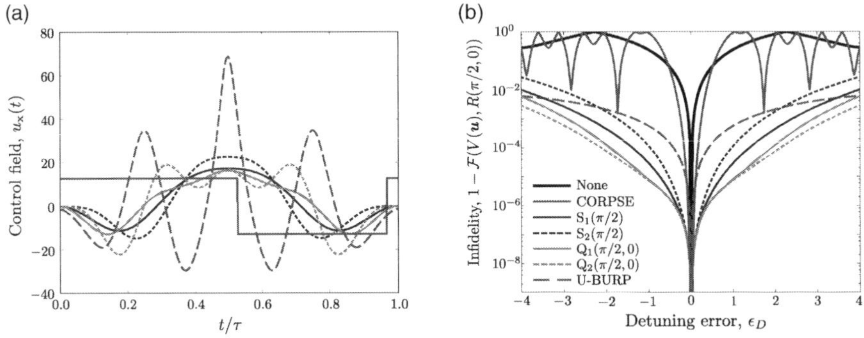

Figure 12. (**a**) Control function $u_x(t)$ for various pulse sequences designed to compensate systematic detuning errors. (**b**) Performance of several shaped pulses over a wide detuning range. (*See the color version of this figure in the Color Plates section.*)

It is interesting to compare the performance of shaped sequences to a sequence of square pulses, such as CORPSE. Many of these sequences compensate detuning errors by modulating the amplitude of the control function $u_x(t)$. In Fig. 12a, pulse shapes are given for CORPSE, U-BURP, $S_1(\pi/2)$, $S_2(\pi/2)$, $Q_1(\pi/2)$, and $Q_2(\pi/2)$ sequences that implement the target operation $U_T = R(\pi/2, 0)$. The relative performance of these sequence over a wide range of systematic detuning error is given in Fig. 12b. Clearly, shaped pulses outperform CORPSE over a wide range of field tunings; however, for very small detunings, CORPSE has more favorable scaling behavior. Figure 13a and b show the behavior of the shaped pulse $S_1(\pi/2)$ as trajectories on the Lie algebra and on the Bloch sphere, respectively.

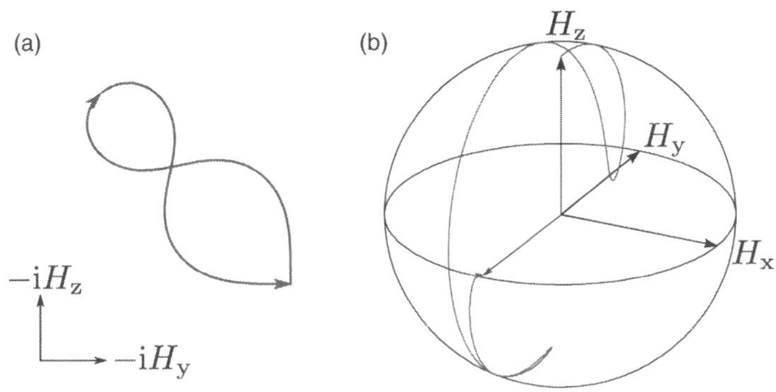

Figure 13. (**a**) Vector path on the Lie algebra traced out by the interaction frame Hamiltonian $H^I(t) = \epsilon_D H_z^I(t)$ for the shaped sequence $S_1(\pi/2)$. (**b**) Magnetization trajectory for a qubit under an $S_1(\pi/2)$ sequence for the target operation $U_T = R(\pi/2, 0)$, for the exceptionally large detuning error $\epsilon_D = 1$.

V. COMPOSITE PULSE SEQUENCES ON OTHER GROUPS

We have studied sequences that compensate imperfect single-qubit rotations (i.e., operations that form a representation of the group $\mathsf{SU}(2)$). Another class of problems of practical and fundamental interest is the design of sequences for other Lie groups, such as the group of n-qubit operations $\mathsf{SU}(2^n)$ [50]. Several compensating sequences exist for multi-qubit gates [57,58,83]. In general, these sequences rely on operations that form an $\mathsf{SU}(2)$ subgroup to perform compensation in a way analogous to the one-qubit case and is the topic of this section.

We note the design of compensating pulse sequences that do not rely on an $\mathsf{SU}(2)$ or $\mathsf{SO}(3)$ subgroup is a largely unexplored topic, and is an interesting subject for future study. For many cases, we can determine which errors can be compensated by examining the algebra [11], but it is unclear what the natural compensating sequences are for groups that are not equivalent to rotations in three dimensions.

A. Compensated Two-Qubit Operations

Thus far, we have shown how compensating sequences may be used to correct systematic errors in arbitrary one-qubit operations. Universal quantum computation also requires accurate two-qubit gates. The study of two-qubit compensating sequences is therefore of great fundamental and practical interest. In this section, we study two-qubit operations using Lie theoretic methods. The Cartan decomposition of the dynamical Lie algebra will be central to this approach. We begin by studying the properties and decompositions of the Lie group and its associated algebra. Then we consider systematic errors in two-qubit gates derived from the Ising interaction, and show how the Cartan decomposition may be used to construct compensating sequences.

Any two-qubit Hamiltonian (up to a global phase) may be written in the form

$$H(t) = \sum_{\mu} \sum_{\nu} u_{\mu\nu}(t) H_{\mu\nu} \tag{63}$$

where the controls are represented in the product-operator basis [84], corresponding to the control Hamiltonians $H_{\mu\nu} = 2H_\mu \otimes H_\nu$, where $H_1 = \frac{1}{2}\mathbb{1}$ and the indices μ, ν run over $H_\mu \in \{H_1, H_x, H_y, H_z\}$. As a matter of convention, we exclude the term proportional to the identity $H_{11} = \frac{1}{2}\mathbb{1} \otimes \mathbb{1}$ since it generates an unimportant global phase and otherwise does not contribute to the dynamics; it is implied in the following equations that this term never appears. The product-operator representation is particularly convenient because each control Hamiltonian is orthogonal under the Hilbert–Schmidt inner product, $\langle H_{\mu\nu}, H_{\rho\sigma}\rangle = \delta_{\mu,\rho}\delta_{\nu,\sigma}$. The family of all possible solutions to a control equation, for example, Eq. (1) with the Hamiltonian Eq. (63), forms a representation of the special unitary group $\mathsf{SU}(4)$. This group contains all possible single-qubit (local) operations that may be applied to each qubit among the pair as well as all two-qubit (nonlocal) operations.

1. Cartan Decomposition of Two-Qubit Gates

Associated with the group $\mathsf{SU}(4)$ is the corresponding Lie algebra $\mathfrak{su}(4) = \bigoplus_{\mu\nu} \mathrm{span}\{-\mathrm{i}H_{\mu\nu}\}$ excluding H_{11} (including H_{11} would make the group $\mathsf{U}(4)$). Consequently, $\mathfrak{su}(4)$ is the algebra of four-dimensional traceless skew-Hermitian matrices. Observe that $\mathfrak{su}(4)$ may be decomposed as $\mathfrak{su}(4) = \mathfrak{k} \oplus \mathfrak{m}$, where

$$\mathfrak{k} = \mathfrak{su}(2) \oplus \mathfrak{su}(2) = \mathrm{span}\{-\mathrm{i}H_{\mathrm{x1}}, -\mathrm{i}H_{\mathrm{y1}}, -\mathrm{i}H_{\mathrm{z1}}, -\mathrm{i}H_{\mathrm{1x}}, -\mathrm{i}H_{\mathrm{1y}}, -\mathrm{i}H_{\mathrm{1z}}\}$$
$$\mathfrak{m} = \mathrm{span}\{-\mathrm{i}H_{\mathrm{xx}}, -\mathrm{i}H_{\mathrm{xy}}, -\mathrm{i}H_{\mathrm{xz}}, -\mathrm{i}H_{\mathrm{yx}}, -\mathrm{i}H_{\mathrm{yy}}, -\mathrm{i}H_{\mathrm{yz}}, -\mathrm{i}H_{\mathrm{zx}}, -\mathrm{i}H_{\mathrm{zy}}, -\mathrm{i}H_{\mathrm{zz}}\}$$

It may be verified that $[\mathfrak{k}, \mathfrak{k}] \subseteq \mathfrak{k}$, $[\mathfrak{m}, \mathfrak{k}] = \mathfrak{m}$, and $[\mathfrak{m}, \mathfrak{m}] \subseteq \mathfrak{k}$. Therefore, the decomposition $\mathfrak{k} \oplus \mathfrak{m}$ is a Cartan decomposition of the algebra $\mathfrak{su}(4)$. Also, the subalgebra $\mathfrak{a} = \mathrm{span}\{-\mathrm{i}H_{\mathrm{xx}}, -\mathrm{i}H_{\mathrm{yy}}, -\mathrm{i}H_{\mathrm{zz}}\}$ is a maximal abelian subalgebra and $\mathfrak{a} \subset \mathfrak{m}$. Therefore, $\mathfrak{a}$ is a Cartan subalgebra. This admits the decomposition $U = K_2 A K_1$ for any element $U \in \mathsf{SU}(4)$. The operators

$$K_j = \exp\left(-\mathrm{i} \sum_{\mu \in \{\mathrm{x,y,z}\}} \alpha^{(j)}_{\mu 1} H_{\mu 1} + \alpha^{(j)}_{1\mu} H_{1\mu}\right) \tag{64}$$

are in the subgroup $\mathrm{e}^{\mathfrak{k}} = \mathsf{SU}(2) \otimes \mathsf{SU}(2)$ comprising all single-qubit operations for both qubits, whereas the operator

$$A = \exp\left(-\mathrm{i} \sum_{\mu \in \{\mathrm{x,y,z}\}} \alpha_{\mu\mu} H_{\mu\mu}\right) \tag{65}$$

is in the abelian group $\mathrm{e}^{\mathfrak{a}} = \mathsf{T}^3$, isomorphic to the 3-torus, generated by the two-qubit interaction terms H_{xx}, H_{yy}, and H_{zz}. In analogy with the Euler decomposition, the parameters $\alpha_{\mu\nu}$ may be regarded as rotation angles. This construction is a KAK Cartan decomposition for the propagator U, and may be used as a framework for a pulse sequence to generate any arbitrary two-qubit operation. Furthermore, if the operators K_2, A, and K_1 may be implemented using a compensating pulse sequence, robust for a given error model, then their product will also be a compensating sequence.

2. Operations Based on the Ising Interaction

We seek a method for implementing an arbitrary gate $A \in \mathrm{e}^{\mathfrak{a}}$ using a two-qubit coupling interaction. To be specific, we consider an NMR quantum computer of two heteronuclear spin qubits coupled by an Ising interaction. It is well known that propagators generated by the Ising interaction and single-qubit rotations are universal for SU(4) [85]; for completeness we explicitly show how arbitrary gates may be synthesized using the KAK form. Under the assumption of the qubits can be spectrally distinguished due to large differences in Larmor frequencies, the system Hamiltonian takes the form

$$H(t) = \left(\sum_{\mu \in \{x,y,z\}} u_{\mu 1}(t) H_{\mu 1} + u_{1\mu}(t) H_{1\mu} \right) + u_{zz}(t) H_{zz} \tag{66}$$

where the controls $u_{\mu 1}(t)$ and $u_{1\mu}(t)$ are applied by the appropriate rf fields and $u_{zz}(t) = 2\pi J(t)$ is the strength of the spin–spin coupling interaction. In some cases, it is useful to manipulate the strength of the scalar coupling $J(t)$ (e.g., by using spin decoupling techniques) [25]. We consider the simpler case of constant Ising couplings, and also assume that it is possible to apply hard pulses, where rf coupling amplitude greatly exceeds J, and the spin–spin coupling may be considered negligible. In terms of resource requirements for the quantum computer, this implies that single-qubit operations are fairly quick, whereas two-qubit operations driven by the Ising coupling are much slower.

Consider the propagators $U_{\mu\mu}(\alpha) = \exp(-\mathrm{i}\alpha H_{\mu\mu}) \in e^{\mathfrak{a}}$. Let us define the one-qubit rotation operators $R_1(\theta, \phi) = R(\theta, \phi) \otimes \mathbb{1}$ and $R_2(\theta, \phi) = \mathbb{1} \otimes R(\theta, \phi)$. In the absence of applied rf fields, the system evolves according to $U_{zz}(\theta_{zz})$, where $\theta_{zz} = u_{zz}\Delta t$ and Δt is the duration of the free-precession interval. Observe that for

$$\begin{aligned} K_x &= R_1(\pi/2, 0) R_2(\pi/2, 0) = \exp(-\mathrm{i}\pi/2(H_{x1} + H_{1x})) \\ K_y &= R_1(\pi/2, \pi/2) R_2(\pi/2, \pi/2) = \exp(-\mathrm{i}\pi/2(H_{y1} + H_{1y})) \end{aligned} \tag{67}$$

$U_{xx}(\alpha) = K_y U_{zz}(\alpha) K_y^{\dagger}$ and $U_{yy}(\alpha) = K_x^{\dagger} U_{zz}(\alpha) K_x$. Because the group $e^{\mathfrak{a}}$ is abelian, then any arbitrary group element A specified by the decomposition angles $\{\alpha_{xx}, \alpha_{yy}, \alpha_{zz}\}$, may be produced by the product $A = U_{xx}(\alpha_{xx}) U_{yy}(\alpha_{yy}) U_{zz}(\alpha_{zz})$. Then any $U = K_2 A K_1 \in$ SU(4) may be produced using single-qubit rotations and the Ising interaction.

a. Systematic Errors in Ising Control. In practice, systematic errors introduced by experimental imperfections prohibit the application of perfect two-qubit gates. In the case of gates produced by the Ising interaction, the errors may arise from several sources, such as experimental uncertainty in the strength of the coupling J. It is, therefore, desirable to design sequences that implement accurate Ising gates over a range of coupling strengths. The error model in this case is similar to the case of amplitude errors; the control u_{zz} is replaced by the imperfect analogue $v_{zz} = u_{zz}(1 + \epsilon_J)$ where the parameter ϵ_J is proportional to the difference between the nominal (measured) and actual coupling strengths. For now we assume the remaining controls are error free. In correspondence, the perfect propagators $U_{zz}(\theta_{zz})$ are replaced by $V_{zz}(\theta_{zz}) = U_{zz}(\theta_{zz}(1+\epsilon_J))$.

Jones was the first to study compensating pulse sequences for the Ising interaction [55,57], and proposed sequences closely related to the Wimperis sequences studied in Section IV.B. Here, we demonstrate that these sequences are easily derived using Lie algebraic techniques. Observe that several subalgebras are contained in $\mathfrak{su}(4)$, for instance $\mathfrak{j} = \text{span}\{-\mathrm{i}H_{x1}, -\mathrm{i}H_{yz}, -\mathrm{i}H_{zz}\}$, which are representations of $\mathfrak{su}(2)$. Furthermore, if it is feasible to produce any imperfect propagator in the group $\mathsf{J} = \mathrm{e}^{\mathfrak{j}}$, then the compensating sequences discussed in Section IV may be reused for this system.

This strategy can be implemented using accurate one-qubit rotations. For example, let

$$\mathcal{R}(\theta,\phi) = R_1^\dagger(\phi,0)U_{zz}(\theta)R_1(\phi,0) = \exp(-\mathrm{i}\theta(\cos\phi H_{zz} + \sin\phi H_{yz})) \quad (68)$$

Also, let us define the imperfect rotation $\mathcal{M}(\theta,\phi) = R_1^\dagger(\phi,0)V_{zz}(\theta)R_1(\phi,0) = \mathcal{R}(\theta(1+\epsilon_J),\phi)$. The two-qubit unitaries $\mathcal{R}(\theta,\phi)$ are isomorphic to single-qubit rotations $\mathcal{R}(\theta,\phi)$ and Jones used this similarity to construct an alternative sequence we refer to as B2-J [57,58]:

$$\begin{aligned}\mathcal{M}_{\text{B2-J}}(\theta,0) &= \mathcal{M}(\pi,\phi)\mathcal{M}(2\pi,3\phi)\mathcal{M}(\pi,\phi)\mathcal{M}(\theta,0)\\ &= R_1^\dagger(\phi,0)U_{zz}(\pi(1+\epsilon_J))R_1(\phi,0)\; R_1^\dagger(3\phi,0)U_{zz}(2\pi(1+\epsilon_J))\\ &\quad\times R_1(3\phi,0)R_1^\dagger(\phi,0)U_{zz}(\pi(1+\epsilon_J))R_1(\phi,0)\; U_{zz}(\theta(1+\epsilon_J))\\ &= U_{zz}(\theta) + \mathcal{O}(\epsilon_J^3)\end{aligned} \quad (69)$$

where again $\phi = \arccos(-\theta/4\pi)$. If the sequence B2-J is used in place of the simple Ising gate $U_{zz}(\theta)$ then the first- and second-order effects of the systematic error are eliminated. In this manner, any number of sequences designed for operations in $\mathsf{SU}(2)$ may be mapped into sequences that compensate Ising gates, including the higher-order Trotter–Suzuki sequences, which produce gates at an arbitrary level of accuracy. However, we note that because two-qubit gates occur so slowly, the practical utility of very long sequences is not clear, especially in systems where the two-qubit gate time is comparable to the qubit coherence lifetime. Substantial improvements in the minimum time requirements may be possible with shaped pulse sequences or using time-optimal control methods [86].

In this method, accurate one-qubit gates are used to transform inaccurate Ising gates into a representation of $\mathsf{SU}(2)$. Naturally, it is unimportant which qubit among the pair is rotated to perform this transformation (i.e., the subalgebra $\mathfrak{j}' = \text{span}\{-\mathrm{i}H_{1x}, -\mathrm{i}H_{zy}, -\mathrm{i}H_{zz}\}$ would serve just as well). Given a control with a systematic error and a perfect rotation operator that transforms the control

Hamiltonian H_μ to an independent Hamiltonian H_ν, it is possible to perform compensation if H_μ, and H_ν generate a representation of $\mathfrak{su}(2)$ [58].

b. Correcting Simultaneous Errors. Accurate single-qubit gates are required to compensate errors in the Ising coupling by transforming Ising gates into the larger dynamical Lie group. If accurate single-qubit operations are not available, then propagators of the form Eq. (68) can no longer be reliably prepared. However, if a compensating pulse sequence may be implemented in place of each imperfect single-qubit rotation, then the effect of this error may be reduced. This procedure was used in Section IV to produce a concatenated CORPSE and B2 sequence robust to simultaneous amplitude and detuning errors. A similar strategy can be employed to produce accurate Ising gates in the presence of simultaneous spin-coupling (two-qubit) and amplitude (one-qubit) systematic errors [58]. Consider the control system Eq. (66) under the influence of these two independent errors. The imperfect propagator takes the form

$$V(\boldsymbol{u}(t)) = U(\boldsymbol{u}(t) + \epsilon_A \boldsymbol{\delta u}_1(t) + \epsilon_J \boldsymbol{\delta u}_2(t)) \tag{70}$$

where $\epsilon_A \boldsymbol{\delta u}_1(t) = \epsilon_A u_x(t) + \epsilon_A u_y(t)$ is the amplitude error of the single-qubit controls, and $\epsilon_J \boldsymbol{\delta u}_2(t) = \epsilon_J u_{zz}(t)$ is the error in the Ising coupling. The systematic error on the one-qubit controls $\epsilon_A \boldsymbol{\delta u}_1(t)$ complicates the synthesis of compensated Ising gates, as unitary propagators of the form Eq. (68) can no longer be reliably prepared, that is, $\mathcal{M}(\theta,\phi) = R_1^\dagger(\phi(1+\epsilon_A),0)U_{zz}(\theta(1+\epsilon_J))R_1(\phi(1+\epsilon_A),0) = \mathcal{R}(\theta(1+\epsilon_J),\phi) + \mathcal{O}(\epsilon_A^2)$. This difficulty may be avoided if we use a B2 sequence to correct ϵ_A, before correcting ϵ_J using B2-J. Let $U_{\text{B2}}(\theta,\phi) = U_T \otimes \mathbb{1} + \mathcal{O}(\epsilon_A^3)$ represent the propagator produced by a B2 sequence for the target rotation $U_T = R(\theta,\phi)$ on the first qubit. The sequence B2-WJ is

$$\begin{aligned}\mathcal{M}_{\text{B2-WJ}}(\theta,0) = {} & M_{\text{B2}}^\dagger(\phi,0)U_{zz}(\pi(1+\epsilon_J))M_{\text{B2}}(\phi,0)\, M_{\text{B2}}^\dagger(3\phi,0)U_{zz}(2\pi(1+\epsilon_J)) \\ & \times M_{\text{B2}}(3\phi,0)M_{\text{B2}}^\dagger(\phi,0)U_{zz}(\pi(1+\epsilon_J))M_{\text{B2}}(\phi,0)\, U_{zz}(\theta(1+\epsilon_J))\end{aligned} \tag{71}$$

This sequence replaces imperfect $R_1(\phi(1+\epsilon_A),0)$ pulses with the compensated rotation produced by the B2 sequence. When $\epsilon_A = 0$, B2-WJ scales as $\mathcal{O}(\epsilon_J^3)$, and when $\epsilon_J = 0$ the sequence scales as $O(\epsilon_A^3)$. Figure 14a–c are plots of the infidelity of plain Ising pulses, and the B2-J and B2-WJ sequences in the presence of simultaneous amplitude and Ising coupling errors.

3. *Extension to* SU(2^n)

This sequence demonstrated for two-qubits can be naturally extended to compensate operations on a network of n qubits with single-qubit operations and Ising couplings. In this case, the dynamical Lie algebra for this system is SU(2^n). Khaneja

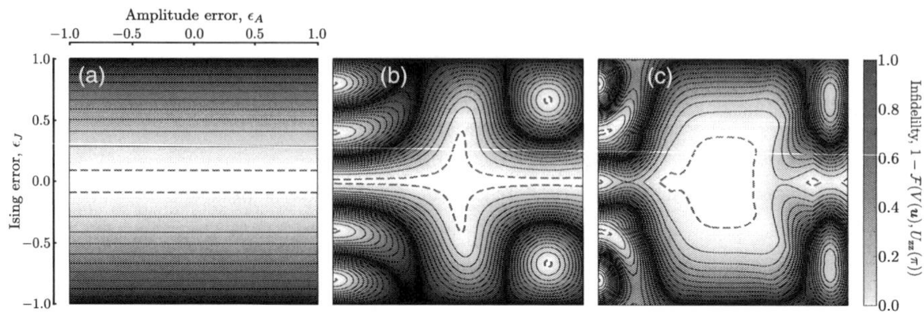

Figure 14. Infidelity of (**a**) plain Ising pulse, (**b**) B2-J, and (**c**) B2-WJ in the presence of simultaneous Ising and amplitude errors. The target rotation is $U_T = U_{ZZ}(\pi, 0)$. The dashed contour corresponds to an infidelity of 0.01, while the remaining contours are plotted at 10% intervals.

has identified a recursive Cartan decomposition for this group that allows any $U \in \mathsf{SU}(2^n)$ to be written as a product of single-qubit (local) and two-qubit (nonlocal) operations [50]. Given a pair of sequences for both single- and two-qubit operations that compensate for a particular error, it is always possible to produce a sequence for any $U \in \mathsf{SU}(2^n)$ using this decomposition. In particular, imperfect Ising gates may be replaced with a B2-J sequence, while imperfect single-qubit gates may be corrected using the methods described in Section IV. The surprising result is that one needs only a single accurate control to compensate an unlimited number of uncorrelated but systematic errors [58]. In practice, this concatenating scheme is expensive for the whole system, but it points to a method for minimizing the amount of calibration required when a few good controls can compensate nearby errors.

a. Computation on Subspaces. Several interesting proposals involve the encoding of logical qubits on subspaces of a larger Hilbert space, which may offer certain advantages over other encoding schemes. In decoherence-free subspace schemes, qubits are encoded on a subspace that is decoupled from environmental noise sources [87–89]. Also, several theoretical proposals involve the use of only two-qubit interactions to perform quantum computation [90]. In these schemes, the control algebra is chosen to be sufficiently large to allow universal computation on a subspace.

An interesting question is whether gates applied to an encoded qubit may be corrected by using a compensating sequence on the code space [58]. Recall that any gate in $\mathsf{SU}(2^n)$ may be decomposed as a product of single-qubit (encoded) and two-qubit gates; it is sufficient to consider these cases individually. We may reuse the sequences described in Section IV if the controls for the encoded gates are distorted by a similar error model. For example, given a set of two controls with correlated systematic errors $v_\mu(t) = (1+\epsilon)u_\mu(t)$ and $v_\nu(t) = (1+\epsilon)u_\nu(t)$, it is possible to perform compensation if the control Hamiltonians H_μ and H_ν

generate a representation of $\mathfrak{su}(2)$. In Ref. [58], it is shown that universal subspace computation can be compensated if the two-qubit Hamiltonians are of the XY model, $H = H_{\mathrm{xx}} + H_{\mathrm{yy}}$, but only single-qubit operations can be compensated if the two-qubit couplings are of the exchange type, $H = H_{\mathrm{xx}} + H_{\mathrm{yy}} + H_{\mathrm{zz}}$.

VI. CONCLUSIONS AND PERSPECTIVES

Recent advances in quantum information and quantum control have revitalized interest in compensating composite pulse sequences. Specifically for the case of systematic control errors these techniques offer a particularly resource-efficient method for quantum error reduction. As quantum information processing experiments continue to grow in both size and complexity, these methods are expected to play an increasingly important role.

In this chapter, we have presented a unified picture of compensating sequences based on control theoretic methods and a dynamic interaction picture. Our framework allows us to view each order of the error as a path in the dynamical Lie algebra, highlighting the geometric features. Correction of the first two orders has a natural geometric interpretation: the path of errors must be closed and the signed area enclosed by the path must be zero.

The geometric method helps illuminate the construction of arbitrarily accurate composite pulses. Currently arbitrarily accurate pulse sequences are of limited use because as the length of the sequence increases, other noise sources become important. For most experiments, decoherence and random errors limit the fidelity of second-order compensation sequences. Our review of the Solovay–Kitaev pulses resulted in a modest improvement of the sequence time by implementing a new geometric construction. Shorter sequences may be achieved through numeric operation and continuous controls.

Finally, we note that CORPSE and related pulse sequences are similar to dynamically corrected gates [4], in that both remove coupling to an external field while performing an operation. Combining the methods here with the developments in dynamically corrected gates and dynamic decoupling could lead to pulse sequences robust against environmental and control errors. These operations will lower the initial error and in the end limit the resources required for quantum error correction, which will ultimately determine the feasibility of performing large quantum chemistry calculations on a quantum computer [91,92].

Acknowledgments

This work was supported by the NSF-CCI on Quantum Information for Quantum Chemistry (CHE-1037992) and by IARPA through ARO contract W911NF-10-1-0231. JTM acknowledges the support of a Georgia Tech Presidential Fellowship.

REFERENCES

1. D. Gottesman, in *Quantum Information Science and Its Contributions to Mathematics*, American Mathematical Society, 2010, pp. 15–58.
2. L. Viola, E. Knill, and S. Lloyd, *Phys. Rev. Lett.* **82**, 2417 (1999).
3. K. Khodjasteh and D. A. Lidar, *Phys. Rev. Lett.* **95**, 180501 (2005).
4. K. Khodjasteh and L. Viola, *Phys. Rev. Lett.* **102**, 080501 (2009).
5. M. H. Levitt, *Prog. Nucl. Mag. Res. Spec.* **18**, 61 (1986).
6. G. M. Huang, T. Tarn, and J. W. Clark, *J. Math. Phys.* **24**, 2608 (1983).
7. S. G. Schirmer, A. I. Solomon, and J. V. Leahy, *J. Phys. A* **35**, 4125 (2002).
8. F. Albertini and D. D'Alessandro, *IEEE T. Automat. Contr.* **48**, 1399 (2003).
9. H. Mabuchi and N. Khaneja, *Int. J. Robust Nonlin. Contr.* **15**, 647 (2005).
10. D. D'Alessandro, *Introduction to Quantum Control and Dynamics*, Chapman & Hall/CRC, Boca Raton, 2008.
11. J.-S. Li and N. Khaneja, *Phys. Rev. A* **73**, 030302 (2006).
12. J. M. Taylor et al., *Phys. Rev. B* **76**, 035315 (2007).
13. I. Chiorescu, Y. Nakamura, C. J. P. M. Harmans, and J. E. Mooij, *Science* **299**, 1869 (2003).
14. J. M. Martinis, S. Nam, J. Aumentado, K. M. Lang, and C. Urbina, *Phys. Rev. B* **67**, 094510 (2003).
15. J. Clarke and F. K. Wilhelm, *Nature* **453**, 1031 (2008).
16. D. Leibfried, R. Blatt, C. Monroe, and D. Wineland, *Rev. Mod. Phys.* **75**, 281 (2003).
17. H. Häffner, C. Roos, and R. Blatt, *Phys. Rep.* **469**, 155 (2008).
18. F. J. Dyson, *Phys. Rev.* **75**, 1736 (1949).
19. W. Magnus, *Commun. Pure Appl. Math.* **7**, 649 (1954).
20. S. Blanes, F. Casas, J. A. Oteo, and J. Ros, *Phys. Rep.* **470**, 151 (2009).
21. P. Madhu and N. Kurur, *Chem. Phys. Lett.* **418**, 235 (2006).
22. R. M. Wilcox, *J. Math. Phys.* **8**, 962 (1967).
23. F. Albertini and D. D'Alessandro, *IEEE T. Automat. Contr.* **48**, 1399 (2003).
24. M. H. Levitt and R. Freeman, *J. Magn. Reson.* **43**, 65 (1981).
25. R. Freeman, *Spin Choreography: Basic Steps in High Resolution NMR*, Oxford University Press, USA, 1998.
26. M. H. Levitt and R. Freeman, *J. Magn. Reson.* **33**, 473 (1979).
27. R. A. de Graaf, *Magn. Reson. Med.* **53**, 1297 (2005).
28. R. S. Said and J. Twamley, *Phys. Rev. A* **80**, 032303 (2009).
29. F. Schmidt-Kaler et al., *Nature* **422**, 408 (2003).
30. B. Luy, K. Kobzar, T. E. Skinner, N. Khaneja, and S. J. Glaser, *J. Magn. Reson.* **176**, 179 (2005).
31. T. Ichikawa, M. Bando, Y. Kondo, and M. Nakahara, *Phys. Rev. A* **84**, 062311 (2011).
32. L. M. K. Vandersypen and I. L. Chuang, *Rev. Mod. Phys.* **76**, 1037 (2005).
33. J. A. Jones, *Prog. Nucl. Mag. Reson. Spec.* **59**, 91 (2011).
34. S. S. Ivanov and N. V. Vitanov, *Opt. Lett.* **36**, 1275 (2011).
35. R. Gilmore, *Lie Groups, Lie Algebras, and Some of Their Applications*, Krieger Publishing Company, FL, 1994.
36. P. Brumer and M. Shapiro, *Chem. Phys. Lett.* **126**, 541 (1986).

37. M. Demiralp and H. Rabitz, *Phys. Rev. A* **47**, 809 (1993).
38. V. Ramakrishna, M. V. Salapaka, M. Dahleh, H. Rabitz, and A. Peirce, *Phys. Rev. A* **51**, 960 (1995).
39. N. Khaneja, R. Brockett, and S. J. Glaser, *Phys. Rev. A* **63**, 032308 (2001).
40. N. Khaneja, S. J. Glaser, and R. Brockett, *Phys. Rev. A* **65**, 032301 (2002).
41. D. Grensing and G. Grensing, *Z. Phys. C* **33**, 307 (1986).
42. E. B. Dykin, *Dokl. Akad. Nauk SSSR* **57**, 323 (1947).
43. D. P. Burum, *Phys. Rev. B* **24**, 3684 (1981).
44. U. Haeberlen and J. S. Waugh, *Phys. Rev.* **175**, 453 (1968).
45. J. S. Waugh, in *Encyclopedia of Magnetic Resonance*, John Wiley & Sons, Chichester, 2007.
46. H. F. Trotter, *P. Am. Math. Soc.* **10**, 545 (1959).
47. M. Suzuki, *Phys. Lett. A* **165**, 387 (1992).
48. N. Wiebe, D. Berry, P. Høyer, and B. C. Sanders, *J. Phys. A* **43**, 065203 (2010).
49. C. M. Dawson and M. A. Nielsen, *Quant. Inf. Comput.* **6**, 81 (2006).
50. N. Khaneja and S. J. Glaser, *Chem. Phys.* **267**, 11 (2001).
51. K. R. Brown, A. W. Harrow, and I. L. Chuang, Phys. Rev. A **70**, 052318 (2004); K. R. Brown, A. W. Harrow, and I. L. Chuang, Phys. Rev. A **72**, 039005 (2005).
52. S. Wimperis, *J. Magn. Reson.* **109**, 221 (1994).
53. W. G. Alway and J. A. Jones, *J. Mag. Reson.* **189**, 114 (2007).
54. J. J. L. Morton et al., *Phys. Rev. Lett.* **95**, 200501 (2005).
55. L. Xiao and J. A. Jones, *Phys. Rev. A* **73**, 032334 (2006).
56. S. E. Beavan, E. Fraval, M. J. Sellars, and J. J. Longdell, *Phys. Rev. A* **80**, 032308 (2009).
57. J. A. Jones, *Phys. Rev. A* **67**, 012317 (2003).
58. Y. Tomita, J. T. Merrill, and K. R. Brown, *New J. Phys.* **12**, 015002 (2010).
59. M. H. Levitt, *J. Chem. Phys.* **128**, 052205 (2008).
60. M. Suzuki, *Phys. Lett. A* **180**, 232 (1993).
61. R. Tycko, *Phys. Rev. Lett.* **51**, 775 (1983).
62. H. K. Cummins and J. A. Jones, *J. Magn. Reson.* **148**, 338 (2001).
63. H. K. Cummins and J. A. Jones, *New J. Phys.* **2**, 6 (2000).
64. H. K. Cummins, G. Llewellyn, and J. A. Jones, *Phys. Rev. A* **67**, 042308 (2003).
65. E. Collin et al., *Phys. Rev. Lett.* **93**, 157005 (2004).
66. M. Steffen, J. M. Martinis, and I. L. Chuang, *Phys. Rev. B* **68**, 224518 (2003).
67. L. Pryadko and P. Sengupta, *Phys. Rev. A* **78**, 032336 (2008).
68. T. E. Skinner, *J. Magn. Reson.* **163**, 8 (2003).
69. T. E. Skinner, T. O. Reiss, B. Luy, N. Khaneja, and S. J. Glaser, *J. Magn. Reson.* **167**, 68 (2004).
70. J.-S. Li, J. Ruths, T.-Y. Yu, H. Arthanari, and G. Wagner, *Proc. Natl. Acad. Sci. USA* **108**, 1879 (2011).
71. H. Geen, S. Wimperis, and R. Freeman, *J. Magn. Reson.* **85**, 620 (1989).
72. H. Geen and R. Freeman, *J. Magn. Reson.* **93**, 93 (1991).
73. N. Khaneja, T. Reiss, C. Kehlet, T. Schulte-Herbrüggen, and S. J. Glaser, *J. Magn. Reson.* **172**, 296 (2005).
74. P. de Fouquieres, S. G. Schirmer, S. J. Glaser, and I. Kuprov, *J. Magn. Reson.* **212**, 412 (2011).

75. N. Timoney et al., *Phys. Rev. A* **77**, 052334 (2008).
76. J. S. Hodges, J. C. Yang, C. Ramanathan, and D. G. Cory, *Phys. Rev. A* **78**, 010303 (2008).
77. K. Singer et al., *Rev. Mod. Phys.* **82**, 2609 (2010).
78. S. Machnes et al., *Phys. Rev. A* **84**, 022305 (2011).
79. B. Pryor and N. Khaneja, in *46th IEEE Conference on Decision and Control*, IEEE, 2007, pp. 6340–6345.
80. D. Abramovich, *J. Magn. Reson.* **105**, 30 (1993).
81. M. Steffen and R. H. Koch, *Phys. Rev. A* **75**, 062326 (2007).
82. P. Sengupta and L. Pryadko, *Phys. Rev. Lett.* **95**, 037202 (2005).
83. M. J. Testolin, C. D. Hill, C. J. Wellard, and L. C. L. Hollenberg, *Phys. Rev. A* **76**, 012302 (2007).
84. O. W. Sørensen, G. W. Eich, and M. H. Levitt, *Prog. Nucl. Mag. Res. Spec.* **16**, 163 (1983).
85. M. A. Nielsen and I. L. Chuang, *Quantum Computation and Quantum Information*, Cambridge University Press, 2000.
86. M. Lapert, Y. Zhang, S. J. Glaser, and D. Sugny, *J. Phys. B* **44**, 154014 (2011).
87. D. A. Lidar and K. B. Whaley, in *Irreversible Quantum Dynamics*, Springer, 2003.
88. Y. S. Weinstein and C. S. Hellberg, *Phys. Rev. A* **72**, 022319 (2005).
89. M. J. Storcz et al., *Phys. Rev. B* **72**, 064511 (2005).
90. A. M. Childs, D. Leung, L. Mancinska, and M. Ozols, *Quant. Inf. Comput.* **11**, 19 (2011).
91. C. R. Clark, T. S. Metodi, S. D. Gasster, and K. R. Brown, *Phys. Rev. A* **79**, 062314 (2009).
92. I. Kassal, J. D. Whitfield, A. Perdomo-Ortiz, M.-H. Yung, and A. Aspuru-Guzik, *Annu. Rev. Phys. Chem.* **62**, 185 (2011).

REVIEW OF DECOHERENCE-FREE SUBSPACES, NOISELESS SUBSYSTEMS, AND DYNAMICAL DECOUPLING

DANIEL A. LIDAR

Departments of Electrical Engineering, Chemistry, and Physics, and Center for Quantum Information Science & Technology, University of Southern California, Los Angeles, CA 90089, USA

Advances in Chemical Physics, Volume 154: Quantum Information and Computation for Chemistry, First Edition. Edited by Sabre Kais.

I. INTRODUCTION

The protection of quantum information is a central task in quantum information processing [1]. Decoherence and noise are obstacles that must be overcome and managed before large-scale quantum computers can be built. This chapter provides an introduction to the theory of decoherence-free subspaces (DFSs), noiseless subsystems (NSs), and dynamical decoupling (DD), among the key tools in the arsenal of decoherence mitigation strategies. It is based on lectures given by the author at the University of Southern California as part of a graduate course on quantum error correction, and as such is not meant to be a comprehensive review, nor to supply an exhaustive list of references. Rather, the goal is to get the reader quickly up to speed on a subset of key topics in the field of quantum noise avoidance and suppression. For previous reviews overlapping with some of the theoretical topics covered here, see, for example, Refs [2–4].

The chapter is structured as follows. Section II introduces decoherence-free subspaces. Section III defines and analyzes the collective dephasing model, and explains how to combine the corresponding DFS encoding with universal quantum computation. Section IV considers the same problem in the context of the more general collective decoherence model, where general noise afflicts all qubits simultaneously. Section V introduces and analyzes noiseless subsystems, a key generalization of DFSs that underlies all known methods of quantum information protection. The NS structure is illustrated with the three-qubit code against collective decoherence, including computation over this code. We then proceed to dynamical decoupling. Section VI introduces the topic by analyzing the protection of a single qubit against pure dephasing and against general decoherence, using both ideal (zero-width) and real (finite-width) pulses. Section VII briefly discusses DD as a symmetrization procedure. Section VIII discusses combining DD with DFS in the case of two qubits. Section IX addresses concatenated dynamical decoupling (CDD), a method to achieve high-order decoupling. In the final technical Section X, we come full circle and connect dynamical decoupling to the representation theory ideas underlying noiseless subsystems theory, thus presenting a unified view of the approaches. Concluding remarks and additional literature entries are presented in Section XI.

II. DECOHERENCE-FREE SUBSPACES

Let us begin by assuming that we have two systems: S and B, defined by the Hilbert spaces $\mathcal{H}_S$ and $\mathcal{H}_B$, respectively. In general, the dynamics of these two systems are generated by

$$H = H_S + H_B + H_{SB} \tag{1}$$

where H_S and H_B are the Hamiltonians corresponding to the pure dynamics of systems S and B, respectively, and H_{SB} is the interaction between the two systems. Using the Kraus operator sum representation (OSR) [1], we can effectively study the reduced dynamics of system S for an initial state $\rho_S(0)$, where

$$\rho_S(0) \mapsto \rho_S(t) = \sum_\alpha K_\alpha(t)\rho_S(0)K_\alpha^\dagger(t) \tag{2}$$

after the partial trace over system B is completed. The Kraus operators $K_\alpha(t)$ satisfy the relation $\sum_\alpha K_\alpha^\dagger(t)K_\alpha(t) = I_S \; \forall t$, where I_S is the identity operator on the system S. The OSR generally results in nonunitary evolution in the system Hilbert space. Therefore, let us define decoherence as follows.

Definition 1 *An open system undergoes decoherence if its evolution is not unitary. Conversely, an open system that undergoes purely unitary evolution (possibly only in a subspace of its Hilbert space) is said to be decoherence-free.*

We would now like to study how we can avoid decoherence.

A. A Classical Example

Let us begin with a simple classical example. Assume we have three parties: Alice, Bob, and Eve. Alice wants to send a message to Bob, and evil Eve (the environment) wants to mess that message up. Let us also assume that the only way in which Eve can act to mess up the message is by, with some probability, flipping all of the bits of the message. If Alice were to send only one bit to Bob, there would be no way of knowing if that bit had been flipped. But let us say Alice is smarter than Eve and decides to send two bits. She also communicates with Bob beforehand and tells him that if he receives a 00 or 11 he should treat it as a “logical” 0 ($\bar{0}$), and if he receives a 01 or 10 to treat it as a “logical” 1 ($\bar{1}$). If this scheme is used, Eve’s ability to flip both bits has no effect on their ability to communicate:

$$\bar{0} = \begin{cases} 00 \\ 11 \end{cases} \mapsto \begin{cases} 11 \\ 00 \end{cases} = \bar{0} \tag{3a}$$

$$\bar{1} = \begin{cases} 01 \\ 10 \end{cases} \mapsto \begin{cases} 10 \\ 01 \end{cases} = \bar{1} \tag{3b}$$

In this example, we use parity conservation to protect information. The logical 0 is even parity and the logical 1 is odd parity. Encoding logical bits in parity in this way effectively hides the information from Eve’s bit flip error.

It is easy to see that the same strategy works for N bits when all Eve can do is to flip all bits simultaneously. Namely, Alice and Bob agree to encode their logical bits into the bit-string pairs $x_1x_2 \ldots x_N$ and $y_1y_2 \ldots y_N$, where $y_i = x_i \oplus 1$ (addition modulo 2), that is, $y_i = 0$ if $x_i = 1$ and $y_i = 1$ if $x_i = 0$. This encoding strategy yields $N - 1$ logical bits given N physical bits; that is, the code rate (defined as the number of logical bits divided by the number of physical bits) is $1 - 1/N$, which is asymptotically close to 1 in the large-N limit.

B. Collective Dephasing DFS

Let us now move to a genuine quantum example by analyzing in detail the operation of the simplest decoherence-free subspace. Suppose that a system of N qubits is coupled to a bath in a symmetric way, and undergoes a dephasing process. Namely, qubit j undergoes the transformation

$$|0\rangle_j \to |0\rangle_j, \qquad |1\rangle_j \to e^{i\phi}|1\rangle_j \tag{4}$$

which puts a random phase ϕ between the basis states $|0\rangle$ and $|1\rangle$ (eigenstates of σ_z with respective eigenvalues $+1$ and -1). Notice how the phase ϕ—by assumption—has no space (j) dependence; that is, the dephasing process is invariant under qubit permutations. Suppose the phases have a distribution p_ϕ and define the matrix $R_z(\phi) = \text{diag}\left(1, e^{i\phi}\right)$ acting on the $\{|0\rangle, |1\rangle\}$ basis. If each qubit is initially in an arbitrary pure state $|\psi\rangle_j = a|0\rangle_j + b|1\rangle_j$, then the random process outputs a state $R_z(\phi)|\psi\rangle_j$ with probability p_ϕ; that is, it yields a pure state ensemble $\{R_z(\phi)|\psi\rangle_j, p_\phi\}$, which is equivalent to the density matrix $\rho_j = \sum_\phi p_\phi R_z(\phi)|\psi\rangle_j \langle\psi| R_z^\dagger(\phi)$. Clearly, this is in the form of a

Kraus OSR, with Kraus operators $K_\phi = \sqrt{p_\phi} R_z(\phi)$; that is, we can also write $\rho_j = \sum_\phi K_\phi |\psi\rangle_j \langle\psi| K_\phi^\dagger$. Thus, each of the qubits will decohere. To see this explicitly, let us assume that ϕ is continuously distributed, so that

$$\rho_j = \int_{-\infty}^{\infty} p(\phi) R_z(\phi) |\psi\rangle_j \langle\psi| R_z^\dagger(\phi)\, \mathrm{d}\phi \tag{5}$$

where $p(\phi)$ is a probability density, and we assume the initial state of all qubits to be a product state. For a Gaussian distribution, $p(\phi) = (4\pi\alpha)^{-1/2} \exp(-\phi^2/4\alpha)$, it is simple to check that

$$\rho_j = \begin{pmatrix} |a|^2 & ab^* \mathrm{e}^{-\alpha} \\ a^* b \mathrm{e}^{-\alpha} & |b|^2 \end{pmatrix} \tag{6}$$

The decay of the off-diagonal elements in the computational basis is a signature of decoherence.

Let us now consider what happens in the two-qubit Hilbert space. The four basis states undergo the transformation

$$\begin{aligned} |0\rangle_1 \otimes |0\rangle_2 &\to |0\rangle_1 \otimes |0\rangle_2 \\ |0\rangle_1 \otimes |1\rangle_2 &\to \mathrm{e}^{i\phi} |0\rangle_1 \otimes |1\rangle_2 \\ |1\rangle_1 \otimes |0\rangle_2 &\to \mathrm{e}^{i\phi} |1\rangle_1 \otimes |0\rangle_2 \\ |1\rangle_1 \otimes |1\rangle_2 &\to \mathrm{e}^{2i\phi} |1\rangle_1 \otimes |1\rangle_2 \end{aligned} \tag{7}$$

Observe that the basis states $|0\rangle_1 \otimes |1\rangle_2$ and $|1\rangle_1 \otimes |0\rangle_2$ acquire the same phase, hence experience the same error. Let us define encoded states by $|0_L\rangle = |0\rangle_1 \otimes |1\rangle_2 \equiv |01\rangle$ and $|1_L\rangle = |10\rangle$. Then the state $|\psi_L\rangle = a|0_L\rangle + b|1_L\rangle$ evolves under the dephasing process as

$$|\psi_L\rangle \to a|0\rangle_1 \otimes \mathrm{e}^{i\phi}|1\rangle_2 + b\mathrm{e}^{i\phi}|1\rangle_1 \otimes |0\rangle_2 = \mathrm{e}^{i\phi}|\psi_L\rangle \tag{8}$$

and the overall phase thus acquired is clearly unimportant. This means that the two-dimensional subspace $\mathrm{DFS}_2(1) = \mathrm{Span}\{|01\rangle, |10\rangle\}$ of the four-dimensional Hilbert space of two qubits is *decoherence-free*. The subspaces $\mathrm{DFS}_2(2) = \mathrm{Span}\{|00\rangle\}$ and $\mathrm{DFS}_2(0) = \mathrm{Span}\{|11\rangle\}$ are also (trivially) DF, since they each acquire a global phase as well, 1 and $\mathrm{e}^{2i\phi}$, respectively. Since the phases acquired by the different subspaces differ, there is no coherence *between* the subspaces. You might want to pause at this point to think about the similarities and differences between the quantum case and the case of two classical coins.

For $N = 3$ qubits, a similar calculation reveals that the subspaces

$$\mathrm{DFS}_3(2) = \mathrm{Span}\{|001\rangle, |010\rangle, |100\rangle\} \tag{9a}$$

$$\mathrm{DFS}_3(1) = \mathrm{Span}\{|011\rangle, |101\rangle, |110\rangle\} \tag{9b}$$

are DF, as well the (trivial) subspaces $\mathrm{DFS}_3(3) = \mathrm{Span}\{|000\rangle\}$ and $\mathrm{DFS}_3(0) = \mathrm{Span}\{|111\rangle\}$.

By now it should be clear how this generalizes. Let λ_N denote the number of 0's in a computational basis state (i.e., a bit string) over N qubits. Then it is easy to check that any subspace spanned by states with constant λ_N is DF against collective dephasing, and can be denoted $\mathrm{DFS}_N(\lambda_N)$ in accordance with the above notation. The dimensions of these subspaces are given by the binomial coefficients: $d_N \equiv \dim[\mathrm{DFS}_N(\lambda_N)] = \binom{N}{\lambda_N}$ and they each encode $\log_2 d_N$ qubits. It might seem that we lost a lot of bits in the encoding process. However, consider the encoding rate $r \equiv (\log_2 d_N)/N$, defined as the number of output qubits divided by the number of input qubits. Using Stirling's formula $\log_2 x! \approx (x + \frac{1}{2}) \log_2 x - x$, we find, for the case of the highest-dimensional DFS ($\lambda_N = N/2$, for even N),

$$r \overset{N \gg 1}{\approx} 1 - \frac{1}{2} \frac{\log_2 N}{N} \tag{10}$$

where we neglected $1/N$ compared to $\log(N)/N$. Thus, the rate approaches 1 with only a logarithmically small correction.

C. Decoherence-Free Subspaces in the Kraus OSR

Let us now partition the Hilbert space into two subspaces $\mathcal{H}_S = \mathcal{H}_G \oplus \mathcal{H}_N$. The subspace of the Hilbert space not affected is denoted by $\mathcal{H}_G$, the "good" portion, and $\mathcal{H}_N$ denotes the decoherence-affected, "noisy" subspace. The point of this decomposition is to establish a general condition under which $\mathcal{H}_G$ remains unaffected by the open system evolution, and evolves unitarily. If we can do this, then we are justified in calling $\mathcal{H}_G$ a DFS. We now make two assumptions:

- Assume that it is possible to partition the Kraus operators as

$$K_\alpha(t) = g_\alpha U \oplus N_\alpha = \begin{pmatrix} g_\alpha U & 0 \\ 0 & N_\alpha \end{pmatrix} \tag{11}$$

  such that U defines a unitary operator acting solely on $\mathcal{H}_G$, $g_\alpha \in \mathbb{C}$, and N_α is an arbitrary (possibly nonunitary) operator acting solely on $\mathcal{H}_N$.
- Assume that the initial state is partitioned in the same manner, as

$$\rho_S(0) = \rho_G(0) \oplus \rho_N(0) = \begin{pmatrix} \rho_G(0) & 0 \\ 0 & \rho_N(0) \end{pmatrix} \tag{12}$$

  where $\rho_G(0) : \mathcal{H}_G \mapsto \mathcal{H}_G$ and $\rho_N(0) : \mathcal{H}_N \mapsto \mathcal{H}_N$.

Note that by the normalization of the Kraus operators,

$$\sum_{\alpha} |g_\alpha|^2 = 1, \quad \sum_{\alpha} N_\alpha^\dagger N_\alpha = I_N \tag{13}$$

Under these two assumptions, the Kraus OSR [Eq. (2)] then becomes

$$\begin{aligned}\rho_S \mapsto \rho_S' &= \sum_{\alpha} (g_\alpha U \oplus N_\alpha)\, \rho_S(0) \left(g_\alpha^* U^\dagger \oplus B_\alpha^\dagger\right) \\ &= \begin{pmatrix} U\rho_G(0)U^\dagger & 0 \\ 0 & \sum_\alpha N_\alpha \rho_N(0) N_\alpha^\dagger \end{pmatrix}\end{aligned} \tag{14}$$

The remarkable thing to notice about this last result is that ρ_G evolves purely unitarily; that is, it satisfies the definition of decoherence-freeness. If we disregard the evolution in the "noisy" subspace, we have the following result.

Theorem 1 *If the above two assumptions hold, then the evolution of an open system that is initialized in the "good" subspace $\mathcal{H}_G$ is decoherence-free; that is, $\mathcal{H}_G$ is a DFS.*

D. Hamiltonian DFS

Discussing DFS in terms of Kraus operators works well, but we would like to develop a bottom-up understanding of the DFS concept, using Hamiltonian evolution. Assume we are given a system in which our computation is occurring, and a bath that is connected to the system. The Hamiltonian governing the whole system can be written as usual as

$$H = H_S \otimes I_B + I_S \otimes H_B + H_{SB} \tag{15}$$

where H_S acts only on the system we are interested in, H_B acts only on the bath, and H_{SB} governs the interaction between the two. Assume also, without loss of generality, that the interaction Hamiltonian can be written as

$$H_{SB} = \sum_{\alpha} S_\alpha \otimes B_\alpha \tag{16}$$

where each S_α is a pure-system operator and each B_α is a pure-bath operator. The Hilbert space can be written as $\mathcal{H} = \mathcal{H}_S \otimes \mathcal{H}_B$, and $\mathcal{H}_S = \mathcal{H}_G \otimes \mathcal{H}_N$, where

$$\mathcal{H}_G = \text{Span}\{|\gamma_i\rangle\} \tag{17a}$$

$$\mathcal{H}_N = \text{Span}\{|\nu_k\rangle\} \tag{17b}$$

$$\mathcal{H}_B = \text{Span}\{|\beta_j\rangle\} \tag{17c}$$

To formulate a theorem, we also need the following assumptions:

1. The system state is initialized in the *good* subspace:

$$\rho_S = \rho_G \oplus 0 = \sum_{i,j} r_{ij} |\gamma_i\rangle\langle\gamma_j| \oplus 0 \tag{18}$$

2. The basis states of the good subspace are eigenvectors of the interaction Hamiltonian:

$$S_\alpha |\gamma_i\rangle = c_\alpha |\gamma_i\rangle, \quad c_\alpha \in \mathbb{C} \tag{19}$$

3. The basis states of the good subspace, when acted on by the system Hamiltonian, remain in the good subspace:

$$H_S |\gamma_i\rangle \in \mathcal{H}_G \tag{20}$$

With these assumptions in hand and with $U(t) = \mathrm{e}^{-iHt}$, we can posit the following theorem.

Theorem 2 *Assuming 1–3, the evolution of the open system can be written as*

$$\rho_S(t) = \mathrm{Tr}_B[U(t)\,(\rho_S(0) \otimes \rho_B(0))\,U^\dagger(t)] = U_S(t)\rho_G(0)\,U_S^\dagger(t) \tag{21a}$$

$$U_S(t)\,|\gamma_i\rangle \in \mathcal{H}_G \tag{21b}$$

where $U_S(t) = \mathrm{e}^{-iH_S t}$.

Proof. Using Eqs. (16) and (19), we can write

$$\begin{aligned}(I_S \otimes H_B + H_{SB})\,|\gamma_i\rangle_S \otimes |\beta_j\rangle_B &= |\gamma_i\rangle \otimes (H_B |\beta_j\rangle) + \sum_\alpha c_\alpha |\gamma_i\rangle \otimes B_\alpha |\beta_j\rangle \\ &= |\gamma_i\rangle \otimes \left(\sum_\alpha c_\alpha B_\alpha + H_B\right) |\beta_j\rangle \\ &= |\gamma_i\rangle \otimes H_{B'} |\beta_j\rangle \end{aligned}\tag{22}$$

where $H_{B'}$ acts only on the bath. Applying this to Eq. (15), we find that the complete Hamiltonian can be decomposed into a portion that acts only on the system and a portion that acts only on the bath.

$$H\,|\gamma_i\rangle \otimes |\beta_j\rangle = (H_S \otimes I_B + I_S \otimes H_{B'})\,|\gamma_i\rangle \otimes |\beta_j\rangle \tag{23}$$

If we plug this form of the Hamiltonian into the unitary evolution matrix, we get

$$U(t) = \mathrm{e}^{-i(H_S \otimes I_B + I_S \otimes H_{B'})t} = U_S(t) \otimes U_C(t) \tag{24}$$

where $U_x(t) = \exp(-itH_x)$, $x = S, C$.

To find $\rho(t)$, we apply this unitary to $\rho(0)$ with $\rho_S(0) = \rho_G(0) \oplus 0$.

$$\begin{aligned}\rho(t) &= U(\rho_S(0) \otimes \rho_B(0))U^\dagger \\ &= U_S(\rho_G(0) \oplus 0)U_S^\dagger \otimes U_C \rho_B(0) U_C^\dagger\end{aligned} \tag{25}$$

We find the state of our system of interest $\rho_S(t)$ by taking the partial trace.

$$\begin{aligned}\rho_S(t) &= \mathrm{Tr}_B[U_S(\rho_G(0) \oplus 0)U_S^\dagger \otimes U_C \rho_B(0) U_C^\dagger] \\ &= U_S(\rho_G(0) \oplus 0)U_S^\dagger \\ &= U_S^G \rho_G(0)(U_S^G)^\dagger\end{aligned} \tag{26}$$

where in the last equality we projected U_S to the good subspace, that is, $U_S^G \equiv U_S|_{\mathcal{H}_G}$. □

Thus, Theorem 2 guarantees that if its conditions are satisfied, a state initialized in the DFS will evolve unitarily.

E. Deutsch's Algorithm

As a first example of error avoidance utilizing the DFS construction, we can consider the first known algorithm that offers a quantum speedup. Deutsch's algorithm [1] presents a simple decision problem in which the goal is to decide whether a function is constant or balanced. Let $f(x) : \{0, 1\}^n \mapsto \{0, 1\}$ (in decimal notation $x \in \{0, 1, \ldots, 2^n - 1\}$) denote the function where if

$$f(x) = \begin{cases} 0, & \forall x \\ 1, & \forall x \end{cases} \tag{27}$$

then the function is called constant and if

$$f(x) = \begin{cases} 0, & \text{half the inputs} \\ 1, & \text{half the inputs} \end{cases} \tag{28}$$

then the function is called balanced. Classically, we find that in the worst case, making a decision on whether $f(x)$ is constant or balanced requires a minimum of $2^n/2 + 1$ total queries to f. Deutsch and Jozsa showed that the exponential cost in f-queries is drastically reduced—to just a single query!—by considering a quantum version of the algorithm.

The decision problem can be recast in terms of the following quantum circuit:

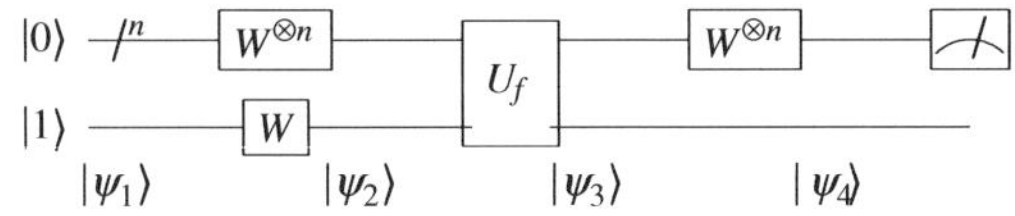

where each classical bit n corresponds to a qubit. The unitary operator U performs the query on $f(x)$ by

$$U_f : |x\rangle\,|y\rangle \mapsto |x\rangle\,|y \oplus f(x)\rangle \text{ (addition mod 2)} \tag{29}$$

where the first register (x) contains the first n qubits and the second register (y) contains the last qubit. Here W represents the Hadamard gate: $W\,|0\rangle = |+\rangle$ and $W\,|1\rangle = |-\rangle$, where $|\pm\rangle = (|0\rangle \pm |1\rangle)/\sqrt{2}$.

In order to illustrate Deutsch's algorithm for the above quantum circuit, consider the single-qubit version ($n = 1$). In this case, there are four functions, two of which are constant and two of which are balanced: $\{f_0(x) = 0,\ f_1(x) = 1\}$ (constant), $\{f_2(x) = x,\ f_3(x) = \bar{x}\}$ (balanced), where the bar denotes bit negation. Clearly, two classical queries to f are required to tell whether f is constant or balanced.

Initially, the total system state is given by $|\psi_1\rangle = |0\rangle\,|1\rangle$. Applying the Hadamard gate, the system state becomes $|\psi_2\rangle = |+\rangle\,|-\rangle$. Applying the unitary operator U_f, the resulting state is

$$\begin{aligned} |\psi_3\rangle = U_f\,|+\rangle\,|-\rangle &= U_f\left[\frac{1}{2}(|00\rangle - |01\rangle + |10\rangle - |11\rangle)\right] \\ &= \frac{1}{2}(|0, 0 \oplus f(0)\rangle - |0, 1 \oplus f(0)\rangle + |1, 0 \oplus f(1)\rangle \\ &\quad - |1, 1 \oplus f(1)\rangle) \\ &= \frac{1}{2}\left(|0, f(0)\rangle - \left|0, \bar{f}(0)\right\rangle + |1, f(1)\rangle - \left|1, \bar{f}(1)\right\rangle\right) \end{aligned} \tag{30}$$

Applying each constant and balanced function to $|\psi_3\rangle$, we find

$$|\psi_3\rangle = \begin{cases} f_0 : |00\rangle - |01\rangle + |10\rangle - |11\rangle = +\,|+-\rangle \\ f_1 : |01\rangle - |00\rangle + |11\rangle - |10\rangle = -\,|+-\rangle \\ f_2 : |00\rangle - |01\rangle + |11\rangle - |10\rangle = +\,|--\rangle \\ f_3 : |01\rangle - |00\rangle + |10\rangle - |11\rangle = -\,|--\rangle \end{cases} \tag{31}$$

for $f_j(x) \in \{0, 1, x, \bar{x}\}$ as defined by Eqs. (27) and (28). The remaining Hadamard gate yields the final state

$$|\psi_4\rangle = \begin{cases} f_0 : +|0-\rangle \\ f_1 : -|0-\rangle \\ f_2 : +|1-\rangle \\ f_3 : -|1-\rangle \end{cases} \tag{32}$$

and the characteristic of the function is determined by measuring the first qubit: a result of 0 indicates a constant function, and a result of 1 a balanced function. Thus, remarkably, we find that the quantum version only requires a *single* query to the function, while the classical case requires two queries (this scenario is the original Deutsch algorithm).

The circuit depicted above can be subjected to a similar analysis in the n-qubit case (the Deutsch–Jozsa algorithm) and the conclusion is that the quantum version of the algorithm still requires only a single f-query, thus resulting in an exponential speedup relative to its classical counterpart in the worst case.

F. Deutsch's Algorithm with Decoherence

To gain an understanding of how a DFS works, we can look at the Deutsch problem with added decoherence. We can consider the circuit diagram for the single-qubit Deutsch algorithm, but introduce a dephasing element as follows, where the dotted box denotes dephasing on the top qubit only:

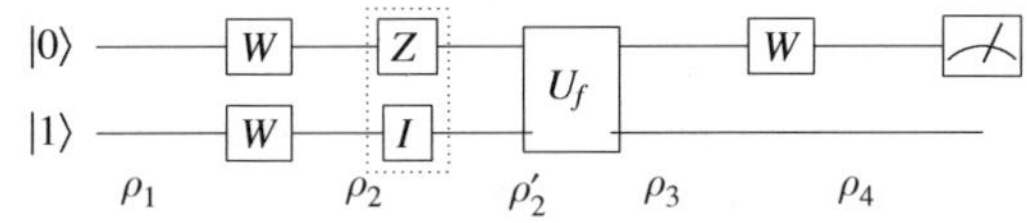

The Kraus operators governing the dephasing of ρ_2 are

$$K_0 = \sqrt{1-p}I_1 \otimes I_2 \tag{33a}$$
$$K_1 = \sqrt{p}Z_1 \otimes I_2 \tag{33b}$$

With probability $(1-p)$, nothing happens. However, with probability p, the first qubit experiences dephasing. If we follow the density matrix states through the algorithm, we can see the effect this dephasing has on our result. As before, we have $\rho_1 = |01\rangle\langle 01|$ and $\rho_2 = |+-\rangle\langle +-|$. By applying the Kraus operators, we find the state after dephasing to be

$$\begin{aligned} \rho_2' &= K_0\rho_2 K_0^\dagger + K_1\rho_2 K_1^\dagger \\ &= (1-p)\rho_2 + p(Z\otimes I)|+-\rangle\langle +-|(Z\otimes I)^\dagger \end{aligned} \tag{34}$$

It is easy to check that $Z\left|+\right\rangle = \left|-\right\rangle$; thus, we find

$$\rho_2' = (1-p)\left|+-\right\rangle\left\langle +-\right| + p\left|--\right\rangle\left\langle --\right| \tag{35}$$

Using Eq. (29), we can compute ρ_3:

$$\rho_3 = \begin{cases} (1-p)\left|+-\right\rangle\left\langle +-\right| + p\left|--\right\rangle\left\langle --\right| & f_0, f_1 \text{ (constant)} \\ (1-p)\left|--\right\rangle\left\langle --\right| + p\left|+-\right\rangle\left\langle +-\right| & f_2, f_3 \text{ (balanced)} \end{cases} \tag{36}$$

After applying the final Hadamard, we see that the state we will measure is

$$\rho_4 = \begin{cases} (1-p)\left|0-\right\rangle\left\langle 0-\right| + p\left|1-\right\rangle\left\langle 1-\right| & f_0, f_1 \text{ (constant)} \\ (1-p)\left|1-\right\rangle\left\langle 1-\right| + p\left|0-\right\rangle\left\langle 0-\right| & f_2, f_3 \text{ (balanced)} \end{cases} \tag{37}$$

If we now measure the first qubit to determine whether the function is constant or balanced, with probability p we will misidentify the outcome. For example, if we obtain the outcome 1, with probability p this could have come from the constant case. But, according to Eq. (32), the outcome 1 belongs to the balanced case.

It is possible to overcome this problem by use of a DFS. Let us again modify the original circuit design. We can add a third qubit and then define logical bits and gates.

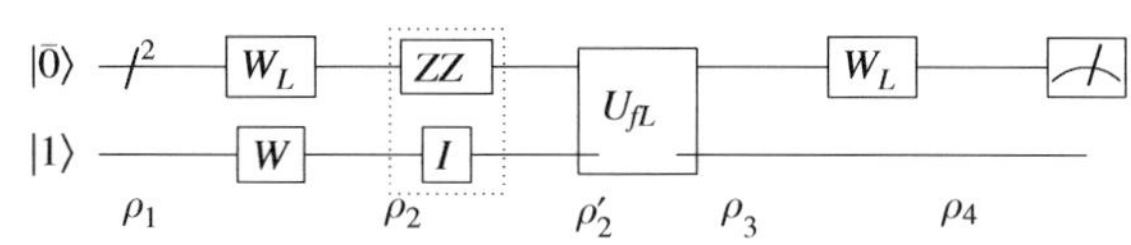

Now the Z dephasing acts simultaneously on both top qubits that comprise the logical qubit $\left|\bar{0}\right\rangle$. In this case, the Kraus operators are

$$K_0 = \sqrt{1-p}III, \qquad K_1 = \sqrt{p}ZZI \tag{38}$$

where $ZZI = \sigma^z \otimes \sigma^z \otimes I$. Recall the requirements for a DFS. The Kraus operators, as in Eq. (11), must be of the form

$$K_\alpha = \left(\begin{array}{c|c} g_\alpha U & 0 \\ \hline 0 & B_\alpha \end{array}\right) \tag{39}$$

and the state must be initialized in a *good* subspace; that is, $\rho_S = \rho_G \oplus \rho_N$, where the direct sum reflects the same block structure as in Eq. (39). If these conditions are met, then $\rho_S' = \sum K_\alpha \rho_S K_\alpha^\dagger = U\rho_G U^\dagger \oplus \rho_N'$. In other words, the evolution of ρ_G is entirely unitary. We start by checking the matrix form of the Kraus operators. K_0 is simply the identity and trivially satisfies this condition. We can check the

ZZ portion of K_1 since that is what will act on our logical qubit.

$$ZZ = \begin{pmatrix} 1 & 0 & 0 & 0 \\ 0 & -1 & 0 & 0 \\ 0 & 0 & -1 & 0 \\ 0 & 0 & 0 & 1 \end{pmatrix} \begin{matrix} 00 \\ 01 \\ 10 \\ 11 \end{matrix} \tag{40}$$

This obviously does not fit the required matrix format, in that there is no block of 1's like in K_0. However, with a simple reordering of the basis states, we obtain the following matrix:

$$ZZ = \begin{pmatrix} 1 & 0 & 0 & 0 \\ 0 & 1 & 0 & 0 \\ 0 & 0 & -1 & 0 \\ 0 & 0 & 0 & -1 \end{pmatrix} \begin{matrix} 00 \\ 11 \\ 01 \\ 10 \end{matrix} \tag{41}$$

Now we have a 2×2 matrix of 1's, so both ZZ and the identity matrix act as the same unitary on the subspace spanned by $|00\rangle$ and $|11\rangle$. The full matrix, K_1, then takes the form

$$K_1 = \sqrt{p} \begin{pmatrix} I_2 & 0 & 0 & 0 \\ 0 & I_2 & 0 & 0 \\ 0 & 0 & -I_2 & 0 \\ 0 & 0 & 0 & I_2 \end{pmatrix} \begin{matrix} 00i \\ 11i \\ 01i \\ 10i \end{matrix} \tag{42}$$

where I_2 denotes the 2×2 identity matrix, $i = \{0, 1\}$, and K_0 is the 8×8 identity matrix. Thus, we see that both K_0 and K_1 have the same upper block format, namely $U = I_{4\times4}$, where $g_0 = \sqrt{1-p}$ and $g_1 = \sqrt{p}$. Now we can define our logical bits $|\bar{0}\rangle = |00\rangle$ and $|\bar{1}\rangle = |11\rangle$. With these states, we can construct our logical Hadamard.

$$(W_L)_{4\times4} = \left(\begin{array}{c|c} W & 0 \\ \hline 0 & V \end{array}\right) \tag{43}$$

Here the logical Hadamard acts as a regular Hadamard on our logical qubits.

$$W_L\,|\bar{0}\rangle = |+_L\rangle \tag{44a}$$

$$W_L\,|\bar{1}\rangle = |-_L\rangle \tag{44b}$$

where $|\pm_L\rangle = \frac{1}{\sqrt{2}}\left(|\bar{0}\rangle \pm |\bar{1}\rangle\right)$. The other unitary action of H_L, namely V, we do not care about. Similarly, we can construct a logical U_f.

$$\left(U_{fL}\right)_{8\times 8} = \left(\begin{array}{c|c} U_f & 0 \\ \hline 0 & V' \end{array}\right) \tag{45}$$

Again, U_{fL} acts as U_f on our logical bits, and V' we do not care about. Neither V nor V' affects our logical qubits in any way. Now that we have set our system up, we can apply the Deutsch algorithm again to see if the DFS corrects the possibility of misidentifying the result. Our system begins in the state $\rho_1 = |\bar{0}\rangle |1\rangle \langle\bar{0}| \langle 1|$ and after applying the logical Hadamard we get $\rho_2 = |+_L\rangle |-\rangle \langle +_L| \langle -|$. Now we can apply the Kraus operators to see the effect of the decoherence. K_0 has no effect other than to multiply the state by $\sqrt{1-p}$ because it is proportional to the identity matrix. It is enough to examine the effect of K_1 on the state $|+_L\rangle$.

$$\begin{aligned} K_1 |+_L\rangle &= \frac{1}{\sqrt{2}}\left(K_1 |\bar{0}\rangle + K_1 |\bar{1}\rangle\right) \\ &= \frac{1}{\sqrt{2}}\sqrt{p}\left(I |\bar{0}\rangle + I |\bar{1}\rangle\right) \\ &= \sqrt{p}\,|+_L\rangle \end{aligned} \tag{46}$$

Therefore, $\rho_2' = \rho_2$. The decoherence has no effect on our system and the rest of the algorithm will proceed without any possibility of error in the end.

III. COLLECTIVE DEPHASING

A. The Model

Consider the example of a spin–boson Hamiltonian. In this example, the system of qubits could be the spins of N electrons trapped in the periodic potential well of a crystalline lattice. The bath is the phonons of the crystal (its vibrational modes). We also assume that the system–bath interaction has *permutation symmetry* in the sense that the interaction between the spins and phonons is the same for all spins because the phonon wavelength is long compared to the spacing between spins. This assumption is crucial for our purpose of demonstrating the appearance of a DFS. If the potential wells are deep enough, then the motional degrees of freedom of the electrons can be ignored. Let i denote the index for the set of N electrons in the system (the same as the index for the set of occupied potential wells in the solid), let k denote the vibrational mode index, $b_k^\dagger |n_1, \ldots, n_k, \ldots\rangle = \sqrt{n_k + 1}\,|n_1, \ldots, n_k + 1, \ldots\rangle$ is the action of the creation operator for mode k on a Fock state with occupation number n_k, $b_k |n\rangle = \sqrt{n_k}\,|n_1, \ldots, n_k - 1, \ldots\rangle$ is the action of the annihilation operator for mode k, and $b_k^\dagger b_k$ is the number operator, that is, $b_k^\dagger b_k \,|n_1, \ldots, n_k, \ldots\rangle = n_k \,|n_1, \ldots, n_k, \ldots\rangle$. With σ_i^z the Pauli-z spin operator acting on the ith spin, the system–bath Hamiltonian is

$$H_{SB} = \sum_{i,k} g_{i,k}^z \sigma_i^z \otimes (b_k + b_k^\dagger) + h_{i,k}^z \sigma_i^z \otimes b_k^\dagger b_k \tag{47}$$

The permutation symmetry assumption implies

$$g_{i,k}^z = g_k^z, \quad h_{i,k}^z = h_k^z \tag{48}$$

that is, the coupling constants do not depend on the qubit index. The system–bath Hamiltonian can then be written as

$$\begin{aligned} H_{SB} &= \sum_i \sigma_i^z \otimes \sum_k g_k^z (b_k + b_k^\dagger) + h_k^z b_k^\dagger b_k \\ &= S_z \otimes B_z \end{aligned} \tag{49}$$

where

$$S_z \equiv \sum_i \sigma_i^z, \quad B_z \equiv \sum_k g_k^z (b_k + b_k^\dagger) + h_k^z b_k^\dagger b_k \tag{50}$$

If these conditions are met, the bath acts identically on all qubits and system–bath Hamiltonian is invariant under permutations of the qubits' order. The operator S_z is a *collective* spin operator.

B. The DFS

We will now see that this model results in a DFS that is essentially identical to the one we saw in Section II.B.

First consider the case of $N = 2$. In light of the DFS condition [Eq. (19)],

$$N = 2 \Rightarrow S_z = Z \otimes I + I \otimes Z \tag{51}$$

Thus,

$$\begin{aligned} |00\rangle &\overset{S_z}{\rightarrow} 2 \cdot |00\rangle && \Rightarrow c_z = 2 \\ |01\rangle &\overset{S_z}{\rightarrow} |01\rangle - |01\rangle = 0 && \Rightarrow c_z = 0 \\ |00\rangle &\overset{S_z}{\rightarrow} |10\rangle - |10\rangle = 0 && \Rightarrow c_z = 0 \\ |00\rangle &\overset{S_z}{\rightarrow} -2 \cdot |11\rangle && \Rightarrow c_z = -2 \end{aligned} \tag{52}$$

It follows that the DFSs for the two spins are

$$\begin{aligned} \tilde{\mathcal{H}}_{N=2}(2) &= \{|00\rangle\} \\ \tilde{\mathcal{H}}_{N=2}(0) &= \mathrm{Span}\{|01\rangle\,, |10\rangle\} \\ \tilde{\mathcal{H}}_{N=2}(-2) &= \{|11\rangle\} \end{aligned} \tag{53}$$

where we used the notation $\tilde{\mathcal{H}}_N(c_z)$ to denote the "good" subspace $\mathcal{H}_G$ for N qubits, with eigenvalue c_z.

In the $\tilde{\mathcal{H}}_{N=2}(0)$ DFS, there are two states, so we have an encoded qubit:

$$|\bar{0}\rangle = |01\rangle \quad \text{logical } 0 \tag{54a}$$

$$|\bar{1}\rangle = |10\rangle \quad \text{logical } 1 \tag{54b}$$

For three spins, we have

$$N = 3 \Rightarrow S_z = Z \otimes I \otimes I \, + \, I \otimes Z \otimes I \, + \, I \otimes I \otimes Z \tag{55}$$

Similarly, it follows that the DFSs for the three spins are

$$\begin{aligned} \tilde{\mathcal{H}}_{N=3}(3) &= \{|000\rangle\} \\ \tilde{\mathcal{H}}_{N=3}(1) &= \mathrm{Span}\{|001\rangle\,, |010\rangle\,, |100\rangle\} \\ \tilde{\mathcal{H}}_{N=3}(-1) &= \mathrm{Span}\{|011\rangle\,, |101\rangle\,, |110\rangle\} \\ \tilde{\mathcal{H}}_{N=3}(-3) &= \{|111\rangle\} \end{aligned} \tag{56}$$

We find that there are two possible encoded qutrits for $N = 3$, one in $\tilde{\mathcal{H}}_{N=3}(1)$ and the other in $\tilde{\mathcal{H}}_{N=3}(-1)$.

In general, the DFS $\tilde{\mathcal{H}}_N(c_z)$ is the eigenspace of each eigenvalue of S_z. It is easy to see that the number of spin-ups (0's) and the number of spin-downs (1's) in each eigenstate are constant throughout a given eigenspace. This corresponds to the value of total spin projection along z. In fact, for arbitrary N,

$$c_z = \#0 - \#1 \tag{57}$$

Figure 1, known as the *Bratteli diagram*, shows the eigenvalues (y-axis) of S_z as N (x-axis) increases. It represents the constellation of DF subspaces in the parameter space (N, c_z). Each intersection point in the figure represents a DFS. Each upward stroke on the diagram indicates the addition of one new spin-up particle, $|0\rangle$, to the system, while each downward stroke indicates the addition of a new spin-down particle, $|1\rangle$, to the system. The number of paths from the origin to a given intersection point on the diagram is therefore exactly the dimension of the eigenspace with eigenvalue c_z, which is given by

$$\dim(\tilde{\mathcal{H}}_N(c_z)) = \binom{N}{\#0} \tag{58}$$

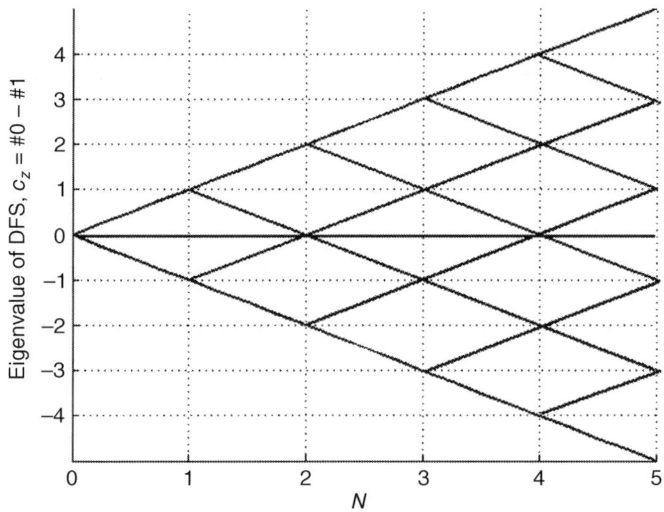

Figure 1. Bratteli diagram showing DFSs for each N. Each intersection point corresponds to a DFS. The number of paths leading to a given point is the dimension of the corresponding DFS. The diagram shows that there is a two-dimensional DFS for $N = 2$ at $c_z = 0$, yielding one encoded qubit.

and note that since $N = \#0 + \#1$, we have $\#0 = (N + c_z)/2$. The highest-dimensional DFS for each N is thus

$$\max_{c_z}\{\dim(\tilde{\mathcal{H}}_N(c_z))\} = \begin{cases} \binom{N}{\frac{N}{2}} & N \text{ even} \\ \binom{N}{\frac{N\pm 1}{2}} & N \text{ odd} \end{cases} \tag{59}$$

Given a D-dimensional DFS, $\tilde{\mathcal{H}}$:

$$\#\text{ of DFS qubits in } \mathcal{H} = \log_2 D \tag{60}$$

We can use this to calculate the rate of the DFS code, that is,

$$\begin{aligned} r &\equiv \frac{\#\text{ of DFS qubits in } \tilde{\mathcal{H}}}{\#\text{ of physical qubits}} = \frac{\log_2 D}{N} \\ &\overset{N\gg 1}{\approx} 1 - \frac{1}{2}\frac{\log_2 N}{N} \end{aligned} \tag{61}$$

where in the second line we used Eq. (10). For a DFS of given dimension, it may be preferable to think in terms of qudits rather than qubits. For example, for the DFS $\tilde{\mathcal{H}}_3(-1)$, the dimension $D = 3$, and so this DFS encodes one qutrit.

Any superposition of states in the same DFS will remain unaffected by the coupling to the bath, since they all share the same eigenvalue of S_z, and hence only acquire a joint overall phase under the action of S_z. But a superposition of states in different DFSs, that is, eigenspaces of S_z, will not evolve in a decoherence-free manner, since they will acquire relative phases due to the different eigenvalues of S_z.

We do not *have* to build a multiqubit DFS out of the largest good subspace. In some cases, it makes sense to sacrifice the code rate to gain simplicity or physical realizability. We shall see an example of this in the next subsection.

C. Universal Encoded Quantum Computation

From here on, encoded qubits will be called "logical qubits." To perform arbitrary single-qubit operations, we need to be able to apply any two of the Pauli operators on the logical qubits. In our example system, $\tilde{\mathcal{H}}_2(0)$, we have

$$\begin{aligned}
|\bar{0}\rangle &\equiv |01\rangle \\
|\bar{1}\rangle &\equiv |10\rangle \\
|\bar{\phi}\rangle &: \text{ DFS encoded logical qubit, } a|\bar{0}\rangle + b|\bar{1}\rangle \\
\bar{U} &: \text{ logical operator on the DFS qubits}
\end{aligned}$$

Thus, the logical Pauli-z operator is

$$\left.\begin{aligned} \bar{Z}|\bar{0}\rangle &= |\bar{0}\rangle \\ \bar{Z}|\bar{1}\rangle &= -|\bar{1}\rangle \end{aligned}\right\} \Rightarrow \bar{Z} = Z \otimes I \Rightarrow \left\{\begin{aligned} |01\rangle &\overset{Z\otimes I}{\rightarrow} |01\rangle = |\bar{0}\rangle \\ |10\rangle &\overset{Z\otimes I}{\rightarrow} -|10\rangle = -|\bar{1}\rangle \end{aligned}\right. \tag{62}$$

and the logical Pauli-x operator is

$$\left.\begin{aligned} \bar{X}|\bar{0}\rangle &= |\bar{1}\rangle \\ \bar{X}|\bar{1}\rangle &= |\bar{0}\rangle \end{aligned}\right\} \Rightarrow \bar{X} = X \otimes X \Rightarrow \left\{\begin{aligned} |01\rangle &\overset{X\otimes X}{\rightarrow} |10\rangle = |\bar{1}\rangle \\ |10\rangle &\overset{X\otimes X}{\rightarrow} |01\rangle = |\bar{0}\rangle \end{aligned}\right. \tag{63}$$

In general, suppose we have $2N$ physical qubits all experiencing collective dephasing. We can pair them into N logical qubits, each pair in $\tilde{\mathcal{H}}_2(0)$, and perform Z or X logical operations on the ith logical qubit using the following operators:

$$\bar{Z}_i \equiv Z_{2i-1} \otimes I_{2i} \tag{64a}$$

$$\bar{X}_i \equiv X_{2i-1} \otimes X_{2i} \tag{64b}$$

Note that by using this pairing we obtain a code whose rate is $N/(2N) = 1/2$, which is substantially less than the highest rate possible, when we use the codes specified by Eq. (59). However, the sacrifice is well worth it since we now have simple and physically implementable encoded logical operation involving at most two-body interactions. Moreover, it is easy to see that the tensor product of two-qubit DFSs is itself still a DFS (just check that all basis states in this tensor product space still satisfy the DFS condition, that is, have the same eigenvalue under the action of S_z).

We can define arbitrary rotations about the logical X or Z axis as $R_{\bar{X}}(\theta) = \exp[i\bar{X}\theta]$ and $R_{\bar{Z}}(\phi) = \exp[i\bar{Z}\phi]$. An arbitrary single logical qubit rotation (an arbitrary element of SU(2)) can then be obtained using the Euler angle formula, as a product of three rotations: $R_{\bar{X}}(\theta_2)R_{\bar{Z}}(\phi)R_{\bar{X}}(\theta_1)$.

To generate arbitrary operators on multiple qubits, we need to add another gate to the generating set: the controlled phase gate. The logical controlled phase gate can be generated from $\bar{Z}_i \otimes \bar{Z}_j \equiv Z_i \otimes Z_{2j-1}$. Thus, a Hamiltonian of the form

$$\bar{H}_S = \sum_i \omega_{Z_i}(t)\bar{Z}_i + \sum_i \omega_{X_i}(t)\bar{X}_i + \sum_{i \neq j} \Omega_{ij}(t)\bar{Z}_i \otimes \bar{Z}_j \tag{65}$$

not only does not take the encoded information outside the DFS $\tilde{\mathcal{H}}_2(0)$ of each of the N encoded qubits, that is, satisfies the DFS preservation condition [Eq. (20)], but is also sufficient to generate a universal set of logical gates over the logical DFS qubits. Moreover, this Hamiltonian is composed entirely of one- and two-body physical qubit operators, so it is physically implementable.

IV. COLLECTIVE DECOHERENCE AND DECOHERENCE-FREE SUBSPACES

The collective dephasing model can be readily modified to give the more general collective decoherence model. The interaction Hamiltonian has the following form:

$$H_{SB} = \sum_{i,k} \left[g^z_{i,k}\sigma^z_i \otimes (b_k + b^\dagger_k) + g^+_{i,k}\sigma^+_i \otimes b_k + g^-_{i,k}\sigma^-_i \otimes b^\dagger_k \right] \tag{66}$$

where

$$\sigma^{\pm} = \frac{1}{2}\left(\sigma_x \mp i\sigma_y\right) \tag{67}$$

corresponds to the raising (+) and lowering (−) operators, respectively, that is,

$$\sigma^+ \left|0\right\rangle = \left|1\right\rangle, \quad \sigma^+ \left|1\right\rangle = 0 \tag{68a}$$

$$\sigma^- \left|0\right\rangle = 0, \quad \sigma^- \left|1\right\rangle = \left|0\right\rangle \tag{68b}$$

where 0 here corresponds to the null vector and should not be confused with the $|0\rangle$ state. Thus, $\sigma^+ = |1\rangle\langle 0|/2$ and $\sigma^- = |0\rangle\langle 1|/2$ and the factor of $1/2$ is important for the rules of angular momentum addition we shall use below. Thus, all the Pauli matrices in this section also include factors of $1/2$.

The first term in the summation of Eq. (66) corresponds to an energy conserving (dephasing) term while the second and third terms correspond to energy exchange via, respectively, phonon absorption/spin excitation and spin relaxation/phonon emission.

By assuming that all qubits are coupled to the same bath, thereby introducing a permutation symmetry assumption, we have

$$g_{ik}^{\alpha} = g_k^{\alpha}, \quad \forall k, \quad \alpha \in \{+, -, z\}$$

$$\Rightarrow H_{SB} = \sum_i \sigma_i^+ \otimes \overbrace{\sum_k g_k^+ b_k}^{B_+} + \sum_i \sigma_i^- \otimes \overbrace{\sum_k g_k^- b_k^\dagger}^{B_-} + \sum_i \sigma_i^z \otimes \overbrace{\sum_k g_k^z (b_k + b_k^\dagger)}^{B_z}$$
$$= \sum_{\alpha \in \{+,-,z\}} S_\alpha \otimes B_\alpha \tag{69}$$

where

$$S_\alpha = \sum_{i=1}^{N} \sigma_i^\alpha \tag{70}$$

is the total spin operator acting on the entire system of N physical qubits. We can derive the following relations directly from the commutation relations of the Pauli matrices:

$$\left.\begin{aligned} [S_\pm, S_z] &= \pm 2 S_\pm \\ [S_-, S_+] &= S_z \end{aligned}\right\} \text{commutation relations for SL(2) triple} \tag{71}$$

where SL(2) is a Lie algebra [5].

We wish to define the total angular momentum operator $\vec{S}^2$ in terms of the angular momenta operators around each axis. It will be convenient to define the vector of angular momenta: $\vec{S} = (S_x, S_y, S_z)$, where $S_x \equiv S_+ + S_-$ and $S_y \equiv i(S_+ - S_-)$. We note that $\vec{S}^2 \equiv \vec{S} \cdot \vec{S} = \sum_{\alpha \in \{x,y,z\}} S_\alpha^2$ satisfies $[\vec{S}^2, S_z] = 0$. Since $\vec{S}^2$ and S_z commute and are both Hermitian, they are simultaneously diagonalizable; that is, they share a common orthonormal eigenbasis.

Recalling some basic results from the quantum theory of angular momentum, we note that for the basis $\{|S, m_S\rangle\}$, where S represents the total spin quantum number of N spin-$\frac{1}{2}$ particles and m_S represents the total spin projection quantum number onto the z-axis, we can show that

$$\vec{S}^2 \, |S, m_S\rangle \equiv S(S+1) \, |S, m_S\rangle \tag{72a}$$

$$S_z \, |S, m_S\rangle \equiv m_S \, |S, m_S\rangle \tag{72b}$$

where

$$S \in \left\{0, \frac{1}{2}, 1, \frac{3}{2}, 2, \ldots, \frac{N}{2}\right\} \tag{73}$$

and

$$m_S \in \{-S, -S+1, \ldots, S-1, S\} \tag{74}$$

Keeping in mind that the basis states of the good subspace are eigenvectors of the interaction Hamiltonian and also satisfy Eq. (19) for $\alpha \in \{+, -, z\}$, let us examine the cases $N = 1, 2, 3$, and 4 in turn.

A. One Physical Qubit

For a single physical qubit ($N = 1$), the basis $\{|0\rangle, |1\rangle\}$ corresponds to that of our familiar spin-$\frac{1}{2}$ particle, with $S = \frac{1}{2}$ and $m_S = \pm\frac{1}{2}$. We identify our logical zero and one states as follows:

$$|0\rangle = \left|S = \frac{1}{2}, m_S = \frac{1}{2}\right\rangle \tag{75a}$$

$$|1\rangle = \left|S = \frac{1}{2}, m_S = -\frac{1}{2}\right\rangle \tag{75b}$$

B. Two Physical Qubits

For two physical qubits ($N = 2$), which we label A and B, with individual spins $S_A = \frac{1}{2}$ and $S_B = \frac{1}{2}$, we first note that the prescription for adding angular momentum (or spin) given $\vec{S_A}$ and $\vec{S_B}$ is to form the new spin operator $\vec{S} = \vec{S_A} + \vec{S_B}$ with eigenvalues

$$S \in \{|S_A - S_B|, \ldots, S_A + S_B\} \tag{76}$$

with the corresponding spin projection eigenvalues

$$m_S \in \{-S, \ldots, S\} \tag{77}$$

Thus, for two physical qubits, we see that the total spin eigenvalues $S_{(N=2)}$ can only take the value 0 or 1. For $S_{(N=2)} = 0$, we see that m_S can only take the value 0 (singlet subspace) whereas when $S_{(N=2)} = 1$, m_S can take any one of the three values $-1, 0, 1$ (triplet subspace). For our singlet subspace,

$$\left|S_{(N=2)} = 0, m_S = 0\right\rangle = \frac{1}{\sqrt{2}}(|01\rangle - |10\rangle) \tag{78}$$

we see that $S_z\,|S_{(N=2)}=0, m_S=0\rangle = 0$ for our system operator $S_z = \sigma_1^z \otimes I + I \otimes \sigma_2^z$. In fact, $S_\alpha\,|S_{(N=2)}=0, m_S=0\rangle = 0$ for $\alpha \in \{+, -, z\}$, where $S_\alpha = \sum_i \sigma_i^\alpha$. Similarly, it can also be shown that $\vec{S}^2\,|S_{(N=2)}=0, m_S=0\rangle = 0$. Since the singlet state clearly satisfies condition (19), we conclude that $|S_{(N=2)}=0, m_S=0\rangle$ is by itself a one-dimensional DFS. However, we also note that the triplet states are not eigenstates of S_z, S_+, S_- and thus violate Eq. (19).

C. Three Physical Qubits

For three physical qubits ($N=3$), let us label the physical qubits A, B, and C, each with corresponding total spins $S_A=\frac{1}{2}$, $S_B=\frac{1}{2}$, and $S_C=\frac{1}{2}$. If we think of this system as a combination of a pair of spins (A and B) with another spin C, we can again apply our rule for adding angular momenta that gives us, from combining our pair of physical qubits into a $S_{(N=2)}=0$ system with a spin-$\frac{1}{2}$ particle, eigenvalues of the total spin operator of

$$S_{(N=3)} = \left|0-\frac{1}{2}\right|, \ldots, \left|0+\frac{1}{2}\right| = \frac{1}{2}$$

with corresponding spin projection eigenvalues $m_S=\pm\frac{1}{2}$. If, instead, we chose to combine our pair of physical qubits A and B into a $S_{(N=2)}=1$ system with a spin-$\frac{1}{2}$ particle, the eigenvalues of the total spin operator would be

$$S_{(N=3)} = \left|1-\frac{1}{2}\right|, \ldots, \left|1+\frac{1}{2}\right| = \frac{1}{2}, \frac{3}{2}$$

with corresponding spin projection eigenvalues $m_S=\pm\frac{1}{2}$ for $S_{(N=3)}=\frac{1}{2}$ or $m_S=\pm\frac{1}{2}, \pm\frac{3}{2}$ for $S_{(N=3)}=\frac{3}{2}$. These distinct cases arise because there are two distinct ways we can get a total spin of $S=\frac{1}{2}$ from a system with three physical qubits, either with two of the qubits combined as a spin-1 system and then combined with the spin-$\frac{1}{2}$ particle or alternatively with two qubits combined as a spin-0 system and subsequently combined with the remaining spin-$\frac{1}{2}$ particle.

D. Generalization to *N* Physical Qubits

The extension of this idea of combining spin angular momenta is straightforward. There is an inductive method of building up from the above procedure to higher N. Suppose we wish to build up the spin states of N physical qubits. We would first build up the states for a set of $N-1$ physical qubits and then couple the spin of the last qubit.

Suppose we consider the case with $N=4$ physical qubits. We can create a Bratteli diagram for this scenario (Fig. 2). The decoherence-free states lie on the

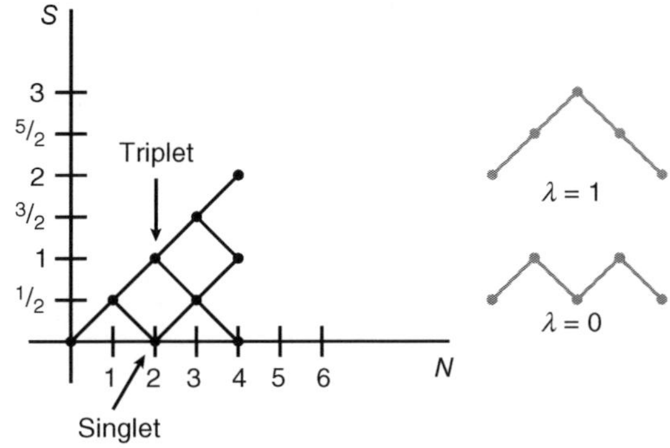

Figure 2. Example Bratteli diagram for $N = 4$ physical qubits. The decoherence-free states lie on the points of the axis where $S = 0$. There are two ways of getting to $S = 0$ when $N = 4$ because there are two possible paths starting from $N = 0$. So we can realize a qubit by setting the logical computational basis states $|\bar{0}\rangle$ and $|\bar{1}\rangle$ to these two decoherence-free states. The parameter λ indexes the two possible paths for $N = 4$ physical qubits.

axis where $S = 0$. There are two possible paths to build up the states from $N = 0$ to $N = 4$. So we can construct a qubit with each logical state $|\bar{0}\rangle$ and $|\bar{1}\rangle$ equal to a decoherence-free state indexed by the path label λ.

$|\bar{0}\rangle =$  $= |S = 0, m_S = 0, \lambda = 0\rangle$ (79)

$|\bar{1}\rangle =$  $= |S = 0, m_S = 0, \lambda = 1\rangle$ (80)

Before proceeding, let us define the singlet state $|s\rangle_{ij}$ over the ith and jth qubits as

$$|s\rangle_{ij} \equiv \frac{1}{\sqrt{2}}\left(|0_i1_j\rangle - |1_i0_j\rangle\right) \tag{81}$$

and the three triplet states as

$$|t^-\rangle_{ij} \equiv |1_i1_j\rangle = |S = 1, m_S = -1\rangle \tag{82a}$$

$$|t^0\rangle_{ij} \equiv \frac{1}{\sqrt{2}}\left(|0_i1_j\rangle + |1_i0_j\rangle\right) = |S = 1, m_S = 0\rangle \tag{82b}$$

$$|t^+\rangle_{ij} \equiv |0_i0_j\rangle = |S = 1, m_S = 1\rangle \tag{82c}$$

For $N = 4$ physical qubits, the logical zero is given by

$|\bar{0}\rangle =$ 

$$\begin{aligned} &= |\text{singlet}\rangle \otimes |\text{singlet}\rangle \\ &= |s\rangle_{12} \otimes |s\rangle_{34} \\ &= \frac{1}{2}\left(|0101\rangle - |0110\rangle - |1001\rangle + |1010\rangle\right) \end{aligned} \tag{83}$$

On the other hand, the logical one will be later seen to be given by

$$\left|\bar{1}\right\rangle =$$ 

$$= \frac{1}{\sqrt{3}}\left[\left|t^{+}\right\rangle_{12} \otimes \left|t^{-}\right\rangle_{34} + \left|t^{-}\right\rangle_{12} \otimes \left|t^{+}\right\rangle_{34} - \left|t^{0}\right\rangle_{12} \otimes \left|t^{0}\right\rangle_{34}\right]$$
$$= \frac{1}{\sqrt{3}}\left(|1100\rangle + |0011\rangle - \frac{1}{2}|0101\rangle - \frac{1}{2}|0110\rangle - \frac{1}{2}|1001\rangle - \frac{1}{2}|1010\rangle\right) \tag{84}$$

In a similar fashion, for $N = 6$ physical qubits, we have for the logical zero

$$\left|\bar{0}\right\rangle = |s\rangle_{12} \otimes |s\rangle_{34} \otimes |s\rangle_{56} \tag{85}$$

Note that permutations of qubit labels are permissible and can be used to define alternative basis states. Actually, we shall see in the next section that such permutations can be used to implement logical operations on the logical qubits.

E. Higher Dimensions and Encoding Rate

Clearly, more paths exist as N grows. We thus have more logical states available as we increase N because these states correspond to the distinct paths leading to each intersection point on the horizontal axis. There exists a combinatorial formula for the number of paths to each point in the Bratteli diagram with $S = 0$ for a given N and hence for the dimension d_N of the DFS $\tilde{\mathcal{H}}(N)$ of N spin-$\frac{1}{2}$ physical qubits:

$$d_N \equiv \dim\left(\tilde{\mathcal{H}}(N)\right) = \frac{N!}{(N/2)!\,(N/2+1)!} \tag{86}$$

As in the case of collective dephasing [Eq. (61)], we can determine the encoding rate from the above formula. The encoding rate r_N is the number of logical qubits N_L we obtain divided by the number of physical qubits N we put into the system. We can construct logical qubits from the logical states in the DFS $\tilde{\mathcal{H}}(N)$, and the number of logical qubits N_L is logarithmic in the number of logical states of $\tilde{\mathcal{H}}(N)$ with $N_L = \log_2(d_N)$. So the encoding rate r_N is

$$r_N \equiv \frac{\text{\# of DFS qubits in } \tilde{\mathcal{H}}(N)}{\text{\# of physical qubits}} = \frac{N_L}{N} = \frac{\log_2 d_N}{N} \tag{87}$$

It can be shown using Stirling's approximation

$$\log_2 N! \approx (N + 1/2)\log_2 N - N \tag{88}$$

that the rate

$$r_N \approx 1 - \frac{3}{2}\frac{\log_2 N}{N} \tag{89}$$

for $N \gg 1$ and hence that the rate r_N asymptotically approaches unity

$$\lim_{N \to \infty} r_N = 1 \tag{90}$$

This implies that when N is very large, remarkably we get about as many logical qubits out of our system as physical qubits we put into the system.

F. Logical Operations on the DFS of Four Qubits

How can we compute over a DFS? Suppose we group the qubits into blocks of length 4, and encode each block into the logical qubits given in Eqs. (83) and (84).

Now, $\forall x, y \in \{0, 1\}$, define the exchange operation E_{ij} on the state $|x\rangle_i \otimes |y\rangle_j$ to be

$$E_{ij}\left(|x\rangle_i \otimes |y\rangle_j\right) \equiv |y\rangle_i \otimes |x\rangle_j \tag{91}$$

Thus, E_{ij} has the following matrix representation in the standard basis of two qubits:

$$E_{ij} = \begin{pmatrix} 1 & 0 & 0 & 0 \\ 0 & 0 & 1 & 0 \\ 0 & 1 & 0 & 0 \\ 0 & 0 & 0 & 1 \end{pmatrix} \tag{92}$$

and it is easy to see that

$$\left[H_{SB}, E_{ij}\right] = 0 \qquad \forall i, j \tag{93}$$

for the collective decoherence case. The exchange operator has a natural representation using the so-called Heisenberg exchange Hamiltonian, namely

$$H_{\text{Heis}} = \sum_{ij} J_{ij} \vec{S}_i \cdot \vec{S}_j \tag{94}$$

where $\vec{S}_i \equiv (X_i, Y_i, Z_i)$, with X, Y, and Z the regular Pauli matrices (without the prefactor of $1/2$) and the J_{ij} are controllable coefficients that quantify the magnitude of the coupling between spin vectors $\vec{S}_i$ and $\vec{S}_j$. Indeed, we easily find by direct matrix multiplication that

$$E_{ij} = \frac{1}{2}(\vec{S}_i \cdot \vec{S}_j + I) \tag{95}$$

where I is the 4×4 identity matrix. Since the identity matrix will only give rise to an overall energy shift, it can be dropped from H_{Heis}. The nice thing about this realization is that the Heisenberg exchange interaction is actually very prevalent as it arises directly from the Coulomb interaction between electrons, and has been tapped for quantum computation, for example, in quantum dots [6,7].

We will now show that E_{ij}'s can be used to generate the encoded X, Y, and Z operators for the encoded qubits in our four-qubit DFS case.

Consider the operator $(-E_{12})$ and its action on the two encoded states given in Eqs. (83) and (84):

$$\begin{aligned}
(-E_{12})\,|\bar{0}\rangle &= -(E_{12})\frac{1}{2}\left(|0101\rangle - |0110\rangle - |1001\rangle + |1010\rangle\right) \\
&= -\frac{1}{2}\left(|1001\rangle - |1010\rangle - |0101\rangle + |0110\rangle\right) \\
&= |\bar{0}\rangle
\end{aligned} \tag{96a}$$

$$\begin{aligned}
(-E_{12})\,|\bar{1}\rangle &= -(E_{12})\frac{1}{\sqrt{3}}\Bigg(|1100\rangle + |0011\rangle - \frac{1}{2}|0101\rangle - \frac{1}{2}|0110\rangle \\
&\quad - \frac{1}{2}|1001\rangle - \frac{1}{2}|1010\rangle\Bigg) \\
&= -\frac{1}{\sqrt{3}}\Bigg(|1100\rangle + |0011\rangle - \frac{1}{2}|1001\rangle - \frac{1}{2}|1010\rangle \\
&\quad - \frac{1}{2}|0101\rangle - \frac{1}{2}|0110\rangle\Bigg) \\
&= -|\bar{1}\rangle
\end{aligned} \tag{96b}$$

Therefore, $(-E_{12})$ acts as a Z operator on the encoded qubits.

Similarly, we may check that $\frac{1}{\sqrt{3}}(E_{23} - E_{13})$ acts like an X operator on the encoded qubits.

Thus, we may define one set of X, Y, and Z operators for the DFS in our case to be

$$\bar{\sigma}^z \equiv (-E_{12}) \tag{97a}$$

$$\bar{\sigma}^x \equiv \frac{1}{\sqrt{3}}(E_{23} - E_{13}) \tag{97b}$$

$$\bar{\sigma}^y \equiv \frac{i}{2}\left[\bar{\sigma}^x, \bar{\sigma}^z\right] \tag{97c}$$

As we saw in Section III.C, with the $\bar{\sigma}^x$ and $\bar{\sigma}^z$ operations we can construct arbitrary qubit rotations via the Euler angle formula:

$$\exp(i\theta\hat{n} \cdot \overrightarrow{\bar{\sigma}}) = \exp(i\alpha\bar{\sigma}^x)\exp(i\beta\bar{\sigma}^z)\exp(i\gamma\bar{\sigma}^x) \tag{98}$$

To perform universal quantum computation, we also need to construct entangling logical operations between the logical qubits. This too can be done entirely using exchange operations. See Ref. [8] for the original construction of such a gate between the logical qubits of the four-qubit DFS code, and Ref. [9] for a more recent and efficient construction.

V. NOISELESS/DECOHERENCE-FREE SUBSYSTEMS

A. Representation Theory of Matrix Algebras

We begin this section by stating a theorem in representation theory of matrix algebras.

Recall the general form of the system–bath Hamiltonian, $H_{SB} = \sum_\alpha S_\alpha \otimes B_\alpha$. Let $\mathcal{A} = \{S_\alpha\}$ be the algebra generated by all the system operators S_α (all sums and products of such operators).

Theorem 3 ([10]) *Assume that $\mathcal{A}$ is $\dagger$-closed (i.e., $A \in \mathcal{A} \Rightarrow A^\dagger \in \mathcal{A}$) and that $I \in \mathcal{A}$. Then*

$$\mathcal{A} \cong \bigoplus_J I_{n_J} \otimes \mathcal{M}_{d_J}(\mathbb{C}) \tag{99}$$

The system Hilbert space can be decomposed as

$$\mathcal{H}_S = \bigoplus_J \mathbb{C}^{n_J} \otimes \mathbb{C}^{d_J} \tag{100}$$

Consequently, the subsystem factors $\mathbb{C}^{n_J}$'s are unaffected by decoherence.

Here $\mathcal{M}_d(\mathbb{C})$ denotes the algebra of complex-valued $d \times d$ irreducible matrices $\{M_d(\mathbb{C})\}$, while as usual I is the identity matrix. The number J is the label of an irreducible representation (irrep) of $\mathcal{A}$, n_J is the degeneracy of the Jth irrep, and d_J is the dimension of the Jth irrep. Irreducibility means that the matrices $\{M_d(\mathbb{C})\}$ cannot be further block diagonalized.

Each left factor $\mathbb{C}^{n_J}$ is called a "subsystem" and the corresponding right factor $\mathbb{C}^{d_J}$ is called a "gauge." Their tensor product forms a proper subspace of the system Hilbert space.

Note that the central conclusion of Theorem 3, that it is possible to safely store quantum information in each of the left factors, or "subsystems" $\mathbb{C}^{n_J}$, is a direct consequence of the fact that every term in $\mathcal{A}$ acts trivially (as the identity operator) on these subsystem factors. These components $\mathbb{C}^{n_J}$ are called *noiseless subsystems*.

The DFS case arises when $d_J = 1$: Then $\mathbb{C}^1$ is just a scalar and $\mathbb{C}^{n_J} \otimes \mathbb{C}^1 = \mathbb{C}^{n_J}$; that is, the summand $\mathbb{C}^{n_J} \otimes \mathbb{C}^1$ reduces to a proper subspace.

Also note that it follows immediately from Eq. (100) that the dimension of the full system Hilbert space $\mathcal{H}_S = (\mathbb{C}^2)^{\otimes N} = \mathbb{C}^{2^N}$ can be decomposed as

$$2^N = \sum_J n_J d_J \tag{101}$$

The technical conditions of the theorem are easy to satisfy. To ensure that $I \in \mathcal{A}$, just modify the definition of H_{SB} so that it includes also the pure-bath term $I \otimes H_B$. And to ensure that $\mathcal{A}$ is $\dagger$-closed, we can always redefine the terms in H_{SB}, if needed, as follows in terms of new Hermitian operators:

$$S'_\alpha \otimes B'_\alpha \equiv \frac{1}{2}(S_\alpha \otimes B_\alpha + S_\alpha^\dagger \otimes B_\alpha^\dagger) \tag{102a}$$

$$S''_\alpha \otimes B''_\alpha \equiv \frac{i}{2}(S_\alpha \otimes B_\alpha - S_\alpha^\dagger \otimes B_\alpha^\dagger) \tag{102b}$$

Then $S_\alpha \otimes B_\alpha = S'_\alpha \otimes B'_\alpha - iS''_\alpha \otimes B''_\alpha$ and by writing H_{SB} in terms of $S'_\alpha \otimes B'_\alpha$ and $S''_\alpha \otimes B''_\alpha$ we have ensured that $\mathcal{A}$ is $\dagger$-closed.

As an application of Theorem 3, we now know that in the right basis (the basis that gives the block-diagonal form (99)), every system operator S_α has a matrix representation in the form $S_\alpha = \bigoplus_J I_{n_J} \otimes M_{d_J}^\alpha$:

$$S_\alpha = \begin{bmatrix} Q_1^\alpha & 0 & 0 & \cdots \\ 0 & Q_2^\alpha & 0 & \cdots \\ 0 & 0 & Q_3^\alpha & \cdots \\ \vdots & \vdots & \vdots & \ddots \end{bmatrix} \tag{103}$$

Each block $Q_J^\alpha = I_{n_J} \otimes M_{d_J}^\alpha$ is $n_J d_J \times n_J d_J$ dimensional, and is of the form

$$Q_J^\alpha = \begin{bmatrix} M_{d_J}^\alpha & 0 & 0 & \cdots \\ 0 & M_{d_J}^\alpha & 0 & \cdots \\ 0 & 0 & M_{d_J}^\alpha & \cdots \\ \vdots & \vdots & \vdots & \ddots \end{bmatrix} \tag{104}$$

Since the system–bath interaction H_{SB} is just a weighted sum of the S_α (the weights being the B_α), it is also an element of $\mathcal{A}$, and hence has the same block-diagonal form. The same applies to any function of H_{SB} that can be written in terms of sums and products, so in particular $e^{-itH_{SB}}$. Thus, the system–bath unitary evolution operator also has the same block-diagonal form, and its action on the NS factors $\mathbb{C}^{n_J}$ is also trivial, that is, proportional to the identity operator.

B. Computation Over a NS

As we saw in our study of DFSs, operators that do not commute with H_{SB} will induce transitions outside of the DFS. Thus, we had to restrict our attention to system Hamiltonians H_S that preserve the DFS [Eq. (20)]. For the same reason, we now consider the *commutant* $\mathcal{A}'$ of $\mathcal{A}$, defined to be the set

$$\mathcal{A}' = \{X : [X, A] = 0, \forall A \in \mathcal{A}\} \tag{105}$$

This set also forms a †-closed algebra and is reducible to, over the same basis as $\mathcal{A}$,

$$\mathcal{A}' \cong \bigoplus_J \mathcal{M}_{n_J}(\mathbb{C}) \otimes I_{d_J} \tag{106}$$

These are the logical operations for performing quantum computation: they act nontrivially on the noiseless subsystems $\mathbb{C}^{n_J}$.

C. Example: Collective Decoherence Revisited

1. General Structure

Let us return to the collective decoherence model. Recall that collective decoherence on N qubits is characterized by the system operators $S_\alpha = \sum_{i=1}^N \sigma_i^\alpha$, for $\alpha \in \{x, y, z\}$. In this case, the system space is

$$\mathcal{H}_S = \bigoplus_{J=0(1/2)}^{N/2} \mathbb{C}^{n_J} \otimes \mathbb{C}^{d_J} \tag{107}$$

where J labels the total spin, and the sum is from $J = 0$ or $J = 1/2$ if N is even or odd, respectively. For a fixed J, there are $2J + 1$ different eigenvalues of m_J, and hence

$$d_J = 2J + 1 \tag{108}$$

By using angular momentum addition rules, one can prove that

$$n_J = \frac{(2J+1)N!}{(N/2+1+J)!(N/2-J)!} \tag{109}$$

which is equal to the number of paths from the origin to the vertex (N, J) on the Bratteli diagram (Fig. 2), and generalizes the DFS dimensionality formula [Eq. (86)].

We have

$$\mathcal{H}_S^{(N)} = \mathbb{C}^{n_0} \otimes \mathbb{C}^1 \oplus \mathbb{C}^{n_1} \otimes \mathbb{C}^3 \oplus \cdots \tag{110}$$

for N even, and

$$\mathcal{H}_S^{(N)} = \mathbb{C}^{n_{1/2}} \otimes \mathbb{C}^2 \oplus \mathbb{C}^{n_{3/2}} \otimes \mathbb{C}^4 \oplus \cdots \tag{111}$$

for N odd. For example, when $N = 3$: $n_{1/2} = \frac{2\cdot 3!}{3!1!} = 2$.

The DFS case arises when $J = 0$ (so that $d_J = 1$): Then $\mathbb{C}^1$ is just a scalar and $\mathbb{C}^{n_0} \otimes \mathbb{C}^1 = \mathbb{C}^{n_0}$; that is, the left (subsystem) factor has a dimension equal to the number of paths and the right (gauge) factor is just a scalar. In this case, the summand $\mathbb{C}^{n_0} \otimes \mathbb{C}^1$ reduces to a proper subspace.

The noiseless subsystems corresponding to different values of J for a given N can be computed by using the addition of angular momentum, as illustrated below.

2. *The Three-Qubit Code for Collective Decoherence*

The smallest N that encodes one qubit in a noiseless subsystem is $N = 3$. In this case,

$$\mathcal{H}_S^{(N=3)} = \mathbb{C}^2 \otimes \mathbb{C}^2 \oplus \mathbb{C}^1 \otimes \mathbb{C}^4 \tag{112}$$

Thus, we can encode one qubit in the first factor $\mathbb{C}^2$ of $J = 1/2$. The two paths of $|\bar{0}\rangle$ and $|\bar{1}\rangle$ are, respectively,  $(\lambda = 0)$ and 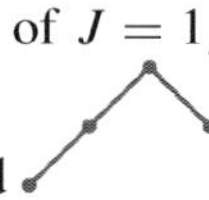 $(\lambda = 1)$. The end points of these two paths each have two spin projections $m_J = \pm 1/2$ (since they correspond to a total spin $J = 1/2$). Using the state notation $|J, \lambda, m_J\rangle$, we thus have

$$\begin{aligned} |\bar{0}\rangle &= \alpha\,|1/2, 0, -1/2\rangle + \beta\,|1/2, 0, 1/2\rangle \\ &= |1/2, 0\rangle \otimes (\alpha\,|-1/2\rangle + \beta\,|1/2\rangle) \end{aligned} \tag{113a}$$

$$\begin{aligned} |\bar{1}\rangle &= \alpha\,|1/2, 1, -1/2\rangle + \beta\,|1/2, 1, 1/2\rangle \\ &= |1/2, 1\rangle \otimes (\alpha\,|-1/2\rangle + \beta\,|1/2\rangle) \end{aligned} \tag{113b}$$

where α and β are completely arbitrary. Or using the vector form, we have

$$|\bar{0}\rangle = \begin{pmatrix} 1 \\ 0 \end{pmatrix} \otimes \begin{pmatrix} \alpha \\ \beta \end{pmatrix}, \quad |\bar{1}\rangle = \begin{pmatrix} 0 \\ 1 \end{pmatrix} \otimes \begin{pmatrix} \alpha \\ \beta \end{pmatrix} \tag{114}$$

Suppose we want to encode a state $|\psi\rangle = a|0\rangle + b|1\rangle$. The encoded state is

$$|\bar{\psi}\rangle = a|\bar{0}\rangle + \beta|\bar{1}\rangle = \begin{pmatrix} a \\ b \end{pmatrix} \otimes \begin{pmatrix} \alpha \\ \beta \end{pmatrix} \tag{115}$$

where we only care about the encoded information a and b. Notice how this last result precisely corresponds to the $\mathbb{C}^2 \otimes \mathbb{C}^2$ term in Eq. (112). Thus, α and β are "gauge amplitudes"; their precise values do not matter.

The interaction Hamiltonian restricted to the system S is of the form

$$\begin{aligned} H_{SB}|_S &= \bigoplus_{J=1/2}^{3/2} I_{n_J} \otimes \mathcal{M}_{d_J} \\ &= I_2 \otimes \mathcal{M}_2 \oplus I_1 \otimes \mathcal{M}_4 \\ &= \left[\begin{array}{c|c} I_2 \otimes \mathcal{M}_2 & \\ \hline & \mathcal{M}_4 \end{array} \right] \end{aligned} \tag{116}$$

It means that the term $I_2 \otimes \mathcal{M}_2$ acts on $|\bar{\psi}\rangle$ and leaves its first factor alone (this is good since that is where we store the qubit), but applies some arbitrary matrix M_2 to the second factor (we do not care). $\mathcal{M}_4$ acts on the $\mathbb{C}^1 \otimes \mathbb{C}^4$ subspace, where we do not store any quantum information.

We can check that the dimensions satisfy Eq. (101):

$$\sum_{J=1/2}^{3/2} n_J d_J = n_{1/2} d_{1/2} + n_{3/2} d_{3/2} = 2 \cdot 2 + 1 \cdot 4 = 8 = 2^3 \tag{117}$$

Let us now find explicit expressions for the basis state of the three-qubit noiseless subsystem. Recall that $|0\rangle = |J = 1/2, m_J = 1/2\rangle$, $|1\rangle = |1/2, -1/2\rangle$, the singlet state $|s\rangle = |0, 0\rangle = \frac{1}{\sqrt{2}}(|01\rangle - |10\rangle)$, and the triplet states are $|t_+\rangle = |1, 1\rangle = |00\rangle$, $|t_-\rangle = |1, 1\rangle = |11\rangle$, and $|t_0\rangle = |1, 0\rangle = \frac{1}{2}(|01\rangle + |10\rangle)$. We now derive the four $J = 1/2$ states by using the addition of angular momentum and Clebsch–Gordan coefficients.

$$\begin{aligned} |1/2, 0, -1/2\rangle &= |s\rangle \otimes |m_3 = -1/2\rangle \\ &= \frac{1}{\sqrt{2}}(|011\rangle - |101\rangle) \end{aligned} \tag{118a}$$

$$\begin{aligned}
|1/2, 0, 1/2\rangle &= |s\rangle \otimes |0\rangle \\
&= \frac{1}{\sqrt{2}}(|010\rangle - |100\rangle) &\text{(118b)}\\
|1/2, 1, -1/2\rangle &= \frac{1}{\sqrt{3}}(\sqrt{2}\,\big|J_{12} = 1, m_{J_{12}} = -1\big\rangle \otimes |m_3 = 1/2\rangle \\
&\quad - \big|J_{12} = 1, m_{J_{12}} = 0\big\rangle \otimes |m_3 = -1/2\rangle) \\
&= \frac{1}{\sqrt{6}}(2\,|110\rangle - |011\rangle - |101\rangle) &\text{(118c)}\\
|1/2, 1, 1/2\rangle &= \frac{1}{\sqrt{3}}(\big|J_{12} = 1, m_{J_{12}} = 0\big\rangle \otimes |m_3 = 1/2\rangle \\
&\quad - \sqrt{2}\,\big|J_{12} = 1, m_{J_{12}} = 1\big\rangle \otimes |m_3 = -1/2\rangle) \\
&= \frac{1}{\sqrt{6}}(|010\rangle + |100\rangle - 2\,|001\rangle) &\text{(118d)}
\end{aligned}$$

These are the basis states that appear in Eq. (113), so they complete the specification of the three-qubit code.

3. *Computation Over the Three-Qubit Code*

Consider the permutation operator $E_{ij} = \frac{1}{2}(I + \vec{\sigma}_i \cdot \vec{\sigma}_j)$ such that $E_{ij}\,|x\rangle_i\,|y\rangle_j = |y\rangle_i\,|x\rangle_j$ for $x, y \in \{0, 1\}$. We have

$$E_{12}\,|1/2, 0, -1/2\rangle = \frac{1}{\sqrt{2}}(-\,|011\rangle + |101\rangle) = -\,|1/2, 0, -1/2\rangle \qquad \text{(119a)}$$

$$E_{12}\,|1/2, 0, 1/2\rangle = \frac{1}{\sqrt{2}}(|100\rangle - |010\rangle) = -\,|1/2, 0, 1/2\rangle \qquad \text{(119b)}$$

$$E_{12}\,|1/2, 1, -1/2\rangle = \frac{1}{\sqrt{6}}(2\,|110\rangle - |011\rangle - |101\rangle) = |1/2, 1, -1/2\rangle \qquad \text{(119c)}$$

$$E_{12}\,|1/2, 1, 1/2\rangle = \frac{1}{\sqrt{6}}(|010\rangle + |100\rangle - 2\,|001\rangle) = |1/2, 1, 1/2\rangle \qquad \text{(119d)}$$

Thus, E_{12} works as a logical $-\sigma^z$, in the sense that

$$E_{12} = \begin{bmatrix} -1 & & & \\ & -1 & & \\ & & 1 & \\ & & & 1 \end{bmatrix} = -\sigma^z \otimes I = -\bar{\sigma}^z \qquad \text{(120)}$$

in the ordered basis of the four $J = 1/2$ states given in Eq. (119). Again, this agrees with the $\mathbb{C}^2 \otimes \mathbb{C}^2$ structure of the Hilbert subspace where we store our qubit.

Similarly, one can easily verify that

$$\frac{1}{\sqrt{3}}(E_{13} - E_{23}) = \sigma^x \otimes I = \bar{\sigma}^x \tag{121}$$

Then $\bar{\sigma}^y$ can be obtained from

$$2i\bar{\sigma}^y = [\bar{\sigma}^z, \bar{\sigma}^x] \tag{122}$$

Finding the explicit form of the encoded CNOT is a complicated problem. See Ref. [11] for a constructive approach using infinitesimal exchange generators, and Ref. [12] for a numerical approach that yields a finite and small set of exchange-based gates.

VI. DYNAMICAL DECOUPLING

As we saw in the discussion of noiseless subsystems, the error algebra $\mathcal{A} = \{S_\alpha\}$ is isomorphic to a direct sum of n_J copies of $d_J \times d_J$ complex matrix algebras: $\mathcal{A} \cong \bigoplus_J I_{n_J} \otimes \mathcal{M}_{d_J}(\mathbb{C})$, where n_J is the degeneracy of the Jth irrep and d_J is the dimension of the Jth irrep. We can store quantum information in a factor $\mathbb{C}^{n_J}$ when $n_J > 1$. However, from general principles (Noether's theorem) we know that degeneracy requires a symmetry, and in our case we would only have $n_J > 1$ when the system–bath coupling has some symmetry. When there is no symmetry at all, $n_J = 1$ for all J's, and a DFS or NS may not exist. Starting in this section, we discuss how to "engineer" the system–bath coupling to have some symmetry.

To sum up, the idea of a DFS/NS is powerful: we can use naturally available symmetries to encode and hide quantum information, and we can compute over the encoded, hidden information. But often such symmetries are imperfect, and we need additional tools to protect quantum information. Such an approach, which adds active intervention to the passive DFS/NS approach, is dynamical decoupling.

A. Decoupling Single-Qubit Pure Dephasing

1. *The Ideal Pulse Case*

Consider a single-qubit system with the pure dephasing system–bath coupling Hamiltonian

$$H_{SB} = \sigma^z \otimes B^z \tag{123}$$

and system Hamiltonian

$$H_S = \lambda(t)\sigma^x \tag{124}$$

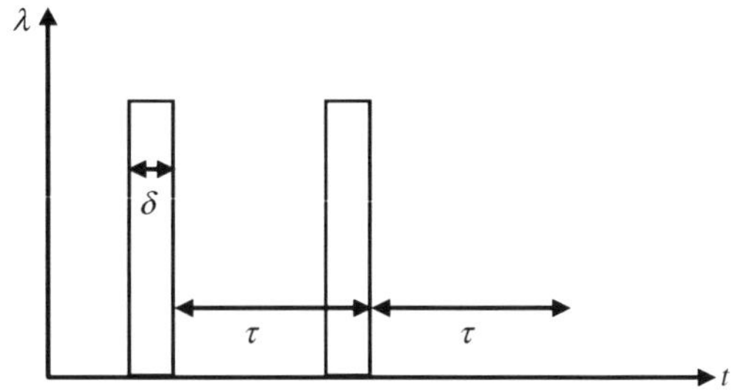

Figure 3. Schematic of a dynamical decoupling pulse sequence. Pulses have width δ and intervals of duration τ. The modulation function $\lambda(t)$ is responsible for switching the pulses on and off.

We assume that $\lambda(t)$ is a fully controllable field, for example, several pulses of a magnetic or electric field applied to the system. Assume these pulses last for a period of time δ, and with strength λ, and

$$\delta\lambda = \frac{\pi}{2} \tag{125}$$

Assume that at $t = 0$, we turn on the pulse for a period of time δ, then let the system and bath interact for a period of time τ, and repeat this procedure, as shown in Fig. 3. In the *ideal* case, $\delta \to 0$ and $\lambda \to \infty$ while still satisfying $\delta\lambda = \frac{\pi}{2}$, which means the pulses are a series of delta functions. For simplicity, we temporarily assume that $H_B = 0$.

To formalize this "ideal pulse" scenario, let us define the system–bath "pulse-free" evolution operator f_τ and the unitary transformation caused by the pulse, X, as follows:

$$f_\tau \equiv \mathrm{e}^{-i\tau H_{SB}} \tag{126a}$$

$$X \equiv \mathrm{e}^{-i\delta\lambda\sigma^x} \otimes I_B = \mathrm{e}^{-i\frac{\pi}{2}\sigma^x} \otimes I_B = -i\sigma^x \otimes I_B \tag{126b}$$

In the case of an ideal pulse ($\delta \to 0$, $\lambda \to \infty$), there is no system–bath interaction during the time the pulse is turned on, since the duration of the pulse is 0. Then the joint system–bath evolution operator at time $t = 2\tau$ is (dropping overall factors of i and minus signs)

$$Xf_\tau Xf_\tau = \sigma^x \mathrm{e}^{-i\tau H_{SB}} \sigma^x \mathrm{e}^{-i\tau H_{SB}} = \mathrm{e}^{-i\tau\sigma^x H_{SB}\sigma^x} \mathrm{e}^{-i\tau H_{SB}} \tag{127}$$

where in the second equality we used the identity

$$U\mathrm{e}^{A}U^\dagger = \mathrm{e}^{UAU^\dagger} \tag{128}$$

valid for any operator A and unitary U.

On the other hand, since the Pauli matrices are Hermitian and every pair of distinct Pauli matrices anticommutes,

$$\{\sigma_\alpha, \sigma_\beta\} = 0, \quad \alpha \neq \beta \tag{129}$$

where the anticommutator is defined as

$$\{A, B\} \equiv AB + BA \tag{130}$$

for any pair of operators A and B, it follows that the sign of H_{SB} is flipped:

$$\begin{aligned}\sigma^x H_{SB} \sigma^x &= \sigma^x \sigma^z \sigma^x \otimes B^z \\ &= -\sigma^z \otimes B^z \\ &= -H_{SB}\end{aligned} \tag{131}$$

This means that the evolution under H_{SB} has been effectively time-reversed!

Indeed, if we now substitute Eq. (131) into Eq. (127), we obtain

$$Xf_\tau Xf_\tau = \mathrm{e}^{+i\tau H_{SB}} \mathrm{e}^{-i\tau H_{SB}} = I \tag{132}$$

Thus, the bath has no effect on the system at the instant $t = 2\tau$. In other words, for a fleeting instant, at $t = 2\tau$, the system is completely decoupled from the bath. Clearly, if we were to repeat Eq. (127) over and over, the system would "stroboscopically" decouple from the bath every 2τ.

2. *The Real Pulse Case*

Unfortunately, in the real world, pulses cannot be described by δ functions, because that would require infinite energy. Generally, the pulse must be described by some continuous function $\lambda(t)$ in the time domain, which may or may not be a pulse. Then, during the period when the pulse is applied to the system, the system–bath Hamiltonian cannot be neglected, so we must take it into account. Keeping the assumption $H_B = 0$ for the time being, we have to modify the pulse to

$$X = \mathrm{e}^{-i\delta(\lambda\sigma^x + H_{SB})} \tag{133}$$

If $\lambda \gg \|H_{SB}\|$ and $\delta\lambda = \pi/2$ (we will define the norm momentarily), then it is true that $X \approx \sigma^x \otimes I_B$; that is, we can approximate the ideal pulse case of Eq. (126b). Let us now see how good of an approximation this is.

To deal with the real pulse case, we first recall the Baker–Campbell–Hausdorff (BCH) formula (see any advanced book on matrices, for example, Ref. [13])

$$\mathrm{e}^{\epsilon(A+B)} = \mathrm{e}^{\epsilon A} \mathrm{e}^{\epsilon B} \mathrm{e}^{(\epsilon^2/2)[A,B] + \mathcal{O}(\epsilon^3)} \tag{134}$$

for any pair of operators A and B. Now, set $\epsilon = -i\delta$, $A = \lambda\sigma^x$, and $B = H_{SB} = \sigma^z \otimes B^z$. Then the real pulse is

$$X = \underbrace{\mathrm{e}^{-i\delta\lambda\sigma^x}}_{\text{ideal pulse}} \underbrace{\mathrm{e}^{-i\delta H_{SB}}}_{\text{OK}} \underbrace{\mathrm{e}^{-\delta^2\lambda[\sigma^x, H_{SB}]/2 + \mathcal{O}(\delta^3)}}_{\text{does damage}} \tag{135}$$

The first exponential is just the ideal pulse, and the second is OK as well (we will see that shortly), but the third term will cause the pulse sequence to operate imperfectly. Let us analyze the pulse sequence subject to this structure of the real pulse.

First, let us define the operator norm [13]:

$$\|A\| \equiv \sup_{|\psi\rangle} \frac{\|A\,|\psi\rangle\|}{\||\psi\rangle\|} = \sup_{|\psi\rangle} \frac{\sqrt{\langle\psi|\,A^{\dagger}A\,|\psi\rangle}}{\||\psi\rangle\|} \tag{136}$$

that is, the largest singular value of A (the largest eigenvalue of $|A| = \sqrt{A^{\dagger}A}$), which reduces to the absolute value of the largest eigenvalue of A when A is Hermitian. The operator norm is an example of a unitarily invariant (ui) norm: If U and V are unitary, and A is some operator, a norm is said to be unitarily invariant if

$$\|UAV\|_{\text{ui}} = \|A\|_{\text{ui}} \tag{137}$$

Such norms are submultiplicative over products and distributive over tensor products [13]:

$$\|AB\|_{\text{ui}} \leq \|A\|_{\text{ui}}\|B\|_{\text{ui}}, \quad \|A \otimes B\|_{\text{ui}} = \|A\|_{\text{ui}}\|B\|_{\text{ui}} \tag{138}$$

Then, using $\|\sigma^{\alpha}\| = 1$ (the eigenvalues of σ^{α} are ± 1), we have

$$\|H_{SB}\| = \|\sigma^{x} \otimes B^{z}\| = \|B^{z}\| \tag{139}$$

Using this we find

$$\begin{aligned} \|\delta^2\lambda[\sigma^x, H_{SB}]/2\| &\leq \delta^2\lambda(\|\sigma^x H_{SB}\| + \|H_{SB}\sigma^x\|)/2 \\ &\leq (\pi/2)\delta\|\sigma^x\|\|H_{SB}\| \\ &= \mathcal{O}(\delta\|B^z\|) \end{aligned} \tag{140}$$

where we used the triangle inequality, submultiplicativity, and $\delta\lambda = \pi/2$. So, we arrive at the important conclusion that the pulse width should be small compared to the inverse of the system–bath coupling strength; that is,

$$\delta \ll 1/\|B^z\| \tag{141}$$

should be satisfied assuming $\|B^z\|$ is finite. This assumption will not always be satisfied (e.g., it does not hold for the spin–boson model), in which case different analysis techniques are required. In particular, operator norms will have to be replaced by correlation functions, which remain finite even when operator norms are formally infinite (see, for example, Ref. [14]). But, for now we shall simply assume that all operators norms we shall encounter are indeed finite.

Let us Taylor expand the "damage" term to lowest order:

$$\begin{aligned} e^{-\delta^2\lambda[\sigma^x, H_{SB}]/2} &= I - \delta^2\lambda[\sigma^x, H_{SB}]/2 + \mathcal{O}(\delta^3) \\ &= I + \mathcal{O}(\delta\|B^z\|) \end{aligned} \tag{142}$$

Putting everything together, including $e^{-i\delta\lambda\sigma^x} = -i\sigma^x$, the evolution subject to the real pulse is, from Eq. (135) (again dropping overall phase factors),

$$\begin{aligned} Xf_\tau Xf_\tau &= [\sigma^x e^{-i\delta H_{SB}}(I + \mathcal{O}(\delta\|B^z\|))]e^{-i\tau H_{SB}}[\sigma^x e^{-i\delta H_{SB}}(I + \mathcal{O}(\delta\|B^z\|))]e^{-i\tau H_{SB}} \\ &= e^{-i(\tau+\delta)\sigma^x H_{SB}\sigma^x} e^{-i(\tau+\delta)H_{SB}} + \mathcal{O}(\delta\|B^z\|) \\ &= e^{i(\tau+\delta)H_{SB}} e^{-i(\tau+\delta)H_{SB}} + \mathcal{O}(\delta\|B^z\|) \\ &= I + \mathcal{O}(\delta\|B^z\|) \end{aligned} \tag{143}$$

so we see that the real pulse sequence has a first-order pulse width correction.

Now let us recall that in fact $H_B \neq 0$. How does this impact the analysis? Both the free evolution and the pulse actually include H_B:

$$f_\tau = e^{-i\tau(H_{SB}+H_B)} \tag{144a}$$

$$X = e^{-i\delta(\lambda\sigma^x + H_{SB} + H_B)} \tag{144b}$$

so we need $\lambda \gg \|H_{SB} + H_B\|$. Set $H'_{SB} = H_B + H_{SB}$, and note that the ideal pulse commutes with H_B, so that

$$\sigma^x(H_{SB} + H_B)\sigma^x = -H_{SB} + H_B \tag{145}$$

Substituting Eqs. (144a) and (144b) into Eq. (143), we then have

$$\begin{aligned} Xf_\tau Xf_\tau &= e^{-i(\tau+\delta)\sigma^x H'_{SB}\sigma^x} e^{-i(\tau+\delta)H'_{SB}} + \mathcal{O}(\delta\|H'_{SB}\|) \\ &= e^{-i(\tau+\delta)(-H_{SB}+H_B)} e^{-i(\tau+\delta)(H_{SB}+H_B)} + \mathcal{O}(\delta\|H'_{SB}\|) \end{aligned} \tag{146}$$

Setting $A = H_{SB} + H_B$, $B = -H_{SB} + H_B$, and using the BCH formula (134) again in the form $e^{\epsilon A}e^{\epsilon B} = e^{\epsilon(A+B)}e^{-(\epsilon^2/2)[A,B]+\mathcal{O}(\epsilon^3)}$, we have $A + B = 2H_B$ and $\|[A, B]\|/2 \leq \|H_B - H_{SB}\|\|H_B + H_{SB}\| \leq (\|H_{SB}\| + \|H_B\|)^2$, so that Eq. (146) reduces to

$$\begin{aligned} Xf_\tau Xf_\tau = I_S \otimes e^{-2i(\tau+\delta)H_B} &+ \mathcal{O}[(\tau+\delta)^2(\|H_{SB}\| + \|H_B\|)^2] \\ &+ \mathcal{O}[\delta(\|H_{SB}\| + \|H_B\|)] \end{aligned} \tag{147}$$

Assuming that the pulses are very narrow, that is, $\delta \ll \tau$ (recall that we anyhow need this for ideal pulses), we can neglect δ relative to τ in the second term, and so the smallness conditions are

$$\delta \ll \tau \ll 1/(\|B^z\| + \|H_B\|) \tag{148}$$

which replaces the earlier $\delta \ll 1/\|B^z\|$ condition we derived when we ignored H_B.

B. Decoupling Single-Qubit General Decoherence

Let us now consider the most general one-qubit system–bath coupling Hamiltonian

$$H_{SB} = \sum_{\alpha=x,y,z} \sigma^\alpha \otimes B^\alpha \tag{149}$$

Using the anticommutation condition [Eq. (129)], we have

$$\sigma^x H_{SB} \sigma^x = \sigma^x \otimes B^x - \sigma^y \otimes B^y - \sigma^z \otimes B^z \tag{150}$$

so that the $Xf_\tau Xf_\tau$ pulse sequence should cancel both the y and z contributions. The remaining problem is how to deal with the σ^x term in H_{SB}.

Let us assume that the pulses are ideal ($\delta = 0$). We can remove the remaining σ^x term by inserting the sequence for pure dephasing into a second pulse sequence, designed to remove the σ_x term. This kind of recursive construction is very powerful, and we will see it again in Section IX.

Let the free evolution again be

$$f_\tau = \mathrm{e}^{-i\tau H_{SB}} \tag{151}$$

Then, after applying an X-type sequence,

$$Xf'_{2\tau} \equiv f_\tau X f_\tau = \mathrm{e}^{-i2\tau(\sigma^x \otimes B^x + H_B)} + \mathcal{O}(\tau^2) \tag{152}$$

To remove the remaining $\sigma^x \otimes B^x$, we can apply a Y-type sequence to $f'_{2\tau}$:

$$\begin{aligned} f''_{4\tau} &= Yf'_{2\tau}Yf'_{2\tau} \\ &= YXf_\tau Xf_\tau YXf_\tau Xf_\tau \\ &= Zf_\tau Xf_\tau Zf_\tau Xf_\tau \end{aligned} \tag{153}$$

where as usual we dropped overall phase factors. Clearly,

$$f''_{4\tau} = \mathrm{e}^{-i4\tau H_B} + \mathcal{O}(\tau^2) \tag{154}$$

so that at $t = 4\tau$ the system is completely decoupled from the bath. This pulse sequence is shown in Fig. 4, and is the *universal decoupling sequence* (for a single qubit), since it removes a general system–bath interaction.

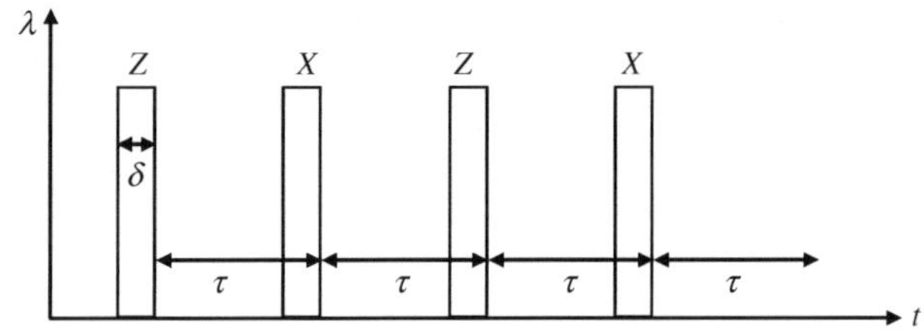

Figure 4. Schematic of the pulse sequence used to suppress general single-qubit decoherence. This pulse sequence is sometimes called XY-4, or the universal decoupling sequence.

VII. DYNAMICAL DECOUPLING AS SYMMETRIZATION

We saw in Eq. (153) that the universal decoupling sequence $Zf_\tau Xf_\tau Zf_\tau Xf_\tau$ decouples a single qubit from an arbitrary bath (to first order). We constructed this sequence using a recursive scheme. In this section, we would like to adopt a different perspective, which will help us generalize the theory beyond the single-qubit case. This perspective is based on symmetrization [15].

Up to a global phase, we have

$$Zf_\tau Xf_\tau Zf_\tau Xf_\tau = (Zf_\tau Z)(Yf_\tau Y)(Xf_\tau X)(If_\tau I) \tag{155}$$

On the right-hand side of (155), we see a clear structure: we are "cycling" over the group formed by the elements $\{I, X, Y, Z\}$. Note that because we are not concerned with global phases, this is not the Pauli group, which is the 16-element group $\{\pm I, \pm X, \pm Y, \pm Z, \pm iI, \pm iX, \pm iY, \pm iZ\}$. Rather, the four-element group is the abelian Klein group, whose multiplication table is given by

×	I	X	Y	Z
I	I	X	Y	Z
X	X	I	Z	Y
Y	Y	Z	I	X
Z	Z	Y	X	I

Returning to the decoupling discussion, to see why the sequence in Eq. (155) works, note that if we let $A^\alpha = \sigma^\alpha \otimes B^\alpha$ we have

$$If_\tau I = f_\tau = \mathrm{e}^{-i\tau(A^x + A^y + A^z + H_B)} \tag{156}$$

$$Xf_\tau X = \mathrm{e}^{-i\tau\sigma^x H\sigma^x} = \mathrm{e}^{-i\tau(A^x - A^y - A^z + H_B)} \tag{157}$$

$$Yf_\tau Y = \mathrm{e}^{-i\tau\sigma^y H\sigma^y} = \mathrm{e}^{-i\tau(-A^x + A^y - A^z + H_B)} \tag{158}$$

$$Zf_\tau Z = \mathrm{e}^{-i\tau\sigma^z H\sigma^z} = \mathrm{e}^{-i\tau(-A^x - A^y + A^z + H_B)} \tag{159}$$

Using the BCH expansion (134) again, we see that when we add all four exponents they cancel all A^α terms perfectly, so that the right-hand side of Eq. (155) is just

$$(Zf_\tau Z)(Yf_\tau Y)(Xf_\tau X)(If_\tau I) = \mathrm{e}^{-4i\tau H_B} + \mathcal{O}(\tau^2) \tag{160}$$

just like in Eq. (154). This is the first-order decoupling we were looking for.

From the right-hand side of Eq. (155), we also gain some intuition as to what our strategy should be beyond the single-qubit case. Again, define

$$f_\tau = \exp[-i\tau (H_{SB} + H_B)] \tag{161}$$

where now H_{SB} and H_B are completely general system–bath and pure-bath operators. Generalizing from Eq. (155), consider a group

$$\mathcal{G} = \{g_0, \ldots, g_K\} \tag{162}$$

(with $g_0 \equiv I$) of unitary transformations g_j acting purely on the system. Assuming that each such pulse g_j is effectively instantaneous, the pulse sequence shall consist of a full cycle over the group, lasting total time

$$T = (K + 1)\,\tau \tag{163}$$

More specifically, we apply the following *symmetrization sequence*:

$$\begin{aligned} U(T) &= \prod_{j=0}^{K} g_j^\dagger f_\tau g_j \\ &= \prod_{j=0}^{K} \mathrm{e}^{-i\tau\left(g_j^\dagger H_{SB} g_j + H_B\right)} \\ &= \mathrm{e}^{-i\tau\left(\sum_{j=0}^{K} g_j^\dagger H_{SB} g_j + (K+1) H_B\right)} + \mathcal{O}\left(T^2\right) \\ &= \mathrm{e}^{-iT\left(H'_{SB} + H_B\right)} + \mathcal{O}\left(T^2\right) \end{aligned} \tag{164}$$

where we used Eq. (128) in the second equality, the BCH formula in the third, and defined the effective or *average Hamiltonian*

$$H'_{SB} = \frac{1}{K+1} \sum_{j=0}^{K} g_j^\dagger H_{SB} g_j \tag{165}$$

Thus, the effect of the pulse sequence defined by $\mathcal{G}$ is to transform the original H_{SB} into the *group-averaged* H'_{SB}. If we can choose the *decoupling group* $\mathcal{G}$ so that H'_{SB} is harmless, we will have achieved our decoupling goal.

Thus, our strategy for general first-order decoupling could be one of the following:

1. Pick a group $\mathcal{G}$ such that $H'_{SB} = 0$.
2. Pick a group $\mathcal{G}$ such that $H'_{SB} = I_S \otimes B'$.

The first of these is precisely what we saw for decoupling a single qubit using the Pauli (or Klein) group, that is, Eq. (160). To see when we can achieve the second strategy (which obviously included the first as a special case with $B' = 0$), note that H'_{SB} belongs to the centralizer of the group $\mathcal{G}$, that is,

$$H_{SB} \stackrel{\mathcal{G}}{\longmapsto} H'_{SB} \in Z(\mathcal{G}) \equiv \{A \,|[A, g] = 0 \; \forall g \in \mathcal{G}\} \tag{166}$$

To prove this we only need to show that $g^\dagger H'_{SB} g = H'_{SB}$ for all $g \in \mathcal{G}$, since this immediately implies that $[H'_{SB}, g] = 0 \; \forall g \in \mathcal{G}$. Indeed,

$$\begin{aligned} g^\dagger H_{SB} g &= \frac{1}{K+1} \sum_{j=0}^{K} g^\dagger g_j^\dagger H_{SB} g_j g \\ &= \frac{1}{K+1} \sum_{j=0}^{K} \left(g_j g\right)^\dagger H_{SB} \left(g_j g\right) \\ &= H'_{SB} \end{aligned} \tag{167}$$

since by group closure $\{g_j g\}_{j=0}^{K}$ also covers all of $\mathcal{G}$.

The fact that H'_{SB} commutes with everything in $\mathcal{G}$ means that we can apply Schur's lemma [5].

Lemma 1 (Schur's lemma) *Let* $\mathcal{G} = \{g_i\}$ *be a group. Let* $\mathrm{T}(\mathcal{G})$ *be an irreducible* d*-dimensional representation of* $\mathcal{G}$ *(i.e., not all of the* $\mathrm{T}(g_i)$ *are similar to a block-diagonal matrix). If there is a* $d \times d$ *matrix* A *such that* $\left[A, g_i\right] = 0 \; \forall g_i \in \mathcal{G}$*, then* $A \propto I$.

Thus, it follows from this lemma that, provided we pick $\mathcal{G}$ so that its matrix representation over the relevant system Hilbert space is irreducible, indeed $H'_{SB} \propto I_S$, since it already commutes with every element of $\mathcal{G}$.

For example, $\mathcal{H}_S = \left(\mathbb{C}^2\right)^{\otimes n} = \mathbb{C}^{2^n}$ for n qubits; the dimension of the irrep should then be 2^n in this case. Which decoupling group has a 2^n-dimensional irrep over $\left(\mathbb{C}^2\right)^{\otimes n}$? An example is the n-fold tensor product of the Pauli group: $\mathcal{G} = \pm, \pm i\{I \otimes \cdots \otimes I, X \otimes I \otimes \cdots \otimes I, \ldots, Z \otimes \cdots \otimes Z\}$. And indeed, this decoupling group suffices to decouple the most general system–bath Hamiltonian in the case of n qubits:

$$H_{SB} = \sum_{\alpha} \sigma_1^{\alpha_1} \otimes \cdots \otimes \sigma_n^{\alpha_n} \otimes B^{\alpha} \tag{168}$$

where $\alpha = \{\alpha_1, \ldots, \alpha_n\}$, and $\alpha_i \in \{0, x, y, z\}$, with the convention that $\sigma^0 = I$. Fortunately, such a system–bath interaction is completely unrealistic, since it involves $(n+1)$-body interactions. "Fortunately," since the decoupling group we just wrote down has $K - 1 = 4^n$ elements, so that the time it would take to apply just once symmetrization sequence (164) grows exponentially with the number of qubits, and we would only achieve first-order decoupling (there is still a correction term proportional to T^2).

Actually, this approach using Schur's lemma is a bit too blunt. We have already seen that the Pauli group is too much even for a single qubit; the Pauli group has 16 elements, but the 4-element Klein group already suffices. Clearly, the approach suggested by Schur's lemma (looking for a group with a 2^n-dimensional irrep) is sufficient but not necessary. Moreover, as we shall see, it is possible to drastically reduce the required resources for decoupling, for example, by combining decoupling with DFS encoding, or by focusing on more reasonable models of system–bath interactions.

VIII. COMBINING DYNAMICAL DECOUPLING WITH DFS

We saw that to decouple the general system–bath interaction H_{SB} in Eq. (168) would require a group with an exponentially large number of elements. This not only is impractical, but might also destroy any benefit we hope to get from efficient quantum algorithms. Therefore, we now consider ways to shorten the decoupling sequence. As we will see, this is possible, at the expense of using more qubits. There will thus be a space–time trade-off. For an entry into the original literature on this topic, see Ref. [16].

A. Dephasing on Two Qubits: A Hybrid DFS–DD Approach

Consider a system consisting of two qubits that are coupled to a bath by the dephasing interaction

$$H_{SB} = \sigma_1^z \otimes B_1^z + \sigma_2^z \otimes B_2^z \tag{169}$$

This Hamiltonian is not invariant under swapping the two qubits since they couple to different bath operators. To make this more apparent, rewrite the interaction as

$$H_{SB} = \left(\frac{\sigma_1^z - \sigma_2^z}{2}\right) \otimes B_- + \left(\frac{\sigma_1^z + \sigma_2^z}{2}\right) \otimes B_+ \tag{170}$$

where the redefined bath operators are $B_\pm = B_1^z \pm B_2^z$. We find that $\left[\left(\sigma_1^z + \sigma_2^z\right)/2\right]$ is a "collective dephasing" operator that applies the same dephasing to both qubits, while the "differential dephasing" operator $\left[\left(\sigma_1^z - \sigma_2^z\right)/2\right]$ applies opposite dephasing to the two qubits. From our DFS studies, we already know that we can encode a single logical qubit as $|\bar{0}\rangle = |01\rangle$ and $|\bar{1}\rangle = |10\rangle$, just as in Eq. (54). Having chosen a basis that vanishes under the effect of one part of the interaction

Hamiltonian, this effectively reduces the interaction to

$$H_{SB}|_{\mathrm{DFS}} = \left(\frac{\sigma_1^z - \sigma_2^z}{2}\right) \otimes B_- = \bar{\sigma}^z \otimes B_- \tag{171}$$

If the initial interaction had been symmetric, choosing the DFS would have reduced it to zero. However, the interaction was not symmetric in this case, and we are left with the above differential dephasing term. We notice further that the residual term is the same as a $\bar{\sigma}^z$, or logical Z operating on the DFS basis [this is a symmetrized version of the logical Z operator in Eq. (64)]. We recall that dephasing acting on a single qubit was decoupled by pulses that implemented the X or Y operators, and hence expect that the $\bar{\sigma}^z$ interaction can be decoupled using a $\bar{X}$ or $\bar{Y}$ pulse. We will use the convention that logical/encoded terms in the Hamiltonian are denoted by $\bar{\sigma}^\alpha$, while the corresponding unitaries are denoted by $\bar{X}$, $\bar{Y}$, or $\bar{Z}$. Thus,

$$\bar{\sigma}^x = \frac{\sigma_1^x\sigma_2^x + \sigma_1^y\sigma_2^y}{2}, \quad \bar{\sigma}^y = \frac{\sigma_1^y\sigma_2^x - \sigma_1^x\sigma_2^y}{2} \tag{172}$$

Restricted to the DFS, the implementation of an $\bar{X}$ pulse using a $\bar{\sigma}^x$ is analogous to the implementation of an X pulse by applying σ^x for an appropriate period of time:

$$\begin{aligned}
\mathrm{e}^{-i\frac{\pi}{2}\bar{\sigma}^x} &= \mathrm{e}^{-i\frac{\pi}{4}(\sigma_1^x\sigma_2^x + \sigma_1^y\sigma_2^y)} \\
&= \mathrm{e}^{-i\frac{\pi}{4}\sigma_1^x\sigma_2^x}\mathrm{e}^{-i\frac{\pi}{4}\sigma_1^y\sigma_2^y} \quad (\text{using } [\sigma_1^x\sigma_2^x, \sigma_1^y\sigma_2^y] = 0) \\
&= \frac{1}{\sqrt{2}}[I - i\sigma_1^x\sigma_2^x]\frac{1}{\sqrt{2}}[I - i\sigma_1^y\sigma_2^y] \\
&= \frac{1}{2}[I - i\sigma_1^x\sigma_2^x - i\sigma_1^y\sigma_2^y + \sigma_1^z\sigma_2^z] \\
&= -\frac{i}{2}(\sigma_1^x\sigma_2^x + \sigma_1^y\sigma_2^y) = -i\bar{X}
\end{aligned} \tag{173}$$

where the term $I + \sigma_1^z\sigma_2^z$ in the fourth line was ignored since it vanishes on the DFS.

Hence, the dynamical decoupling process is effective in the sense that

$$\bar{X} f_\tau \bar{X} f_\tau|_{\mathrm{DFS}} = \bar{I} \otimes \exp(-2i\tau\tilde{B}) + \mathcal{O}[(2\tau)^2] \tag{174}$$

where $\tilde{B}$ is a bath operator whose exact form does not matter, since we have obtained a pure-bath operator up to a time $\mathcal{O}[(2\tau)^2]$. The notation $\bar{I}$ denotes the identity operator projected to the DFS. What have we learned from this example? That we do not need to remove every term in the system–bath Hamiltonian; instead, we can use a DFS encoding along with DD. Next we will see how this can save us some pulse resources.

B. General Decoherence on Two Qubits: A Hybrid DFS–DD Approach

We now consider the most general system–bath Hamiltonian on two qubits:

$$H_{SB} = \sum_{\alpha_1,\alpha_2} (\sigma_1^{\alpha_1} \otimes \sigma_2^{\alpha_2}) \otimes B^{\alpha_1\alpha_2} \tag{175}$$

where $\alpha_i \in \{0, x, y, z\}$.

Within the framework of the same DFS as earlier (DFS = Span$\{|01\rangle, |10\rangle\}$), we can classify all possible ($4^2 = 16$) system operators as either

- leaving system states unchanged (i.e., acting as proportional to $\bar{I}$),
- mapping system states to other states within the DFS; these correspond to logical operations (these are errors since they occur as a result of interaction with the bath), or
- transitions from the DFS to outside and vice versa ("leakage").

The operators causing these errors and their effects are given in Table I.

For example, $\sigma_1^z\sigma_2^z$ acts on $|\bar{0}\rangle = |01\rangle$ and $|\bar{1}\rangle = |10\rangle$ as $-\bar{I}$, while σ_1^x takes both $|\bar{0}\rangle$ and $|\bar{1}\rangle$ out of the DFS, to $|11\rangle$ and $|00\rangle$, respectively.

Along the same lines as single-qubit dynamical decoupling, we look for an operator that anticommutes with all the leakage operators and an operator that anticommutes with all the logical operators. Performing a calculation very similar to Eq. (173), we find that

- $\exp(-i\pi\bar{\sigma}^x) = Z_1Z_2$ and anticommutes with the *entire* leakage set, and
- $\exp(-i\frac{\pi}{2}\bar{\sigma}^z) = -i\bar{Z}$ and anticommutes with the logical error operators $\bar{\sigma}^x$ and $\bar{\sigma}^y$.

A combination of these operators, along with $\bar{X}$ that we used above, is sufficient to reduce the effect of the system–bath interaction Hamiltonian to that of a pure-bath

TABLE I
Classification of All Two-Qubit Error Operators on the DFS for Collective Dephasing

Effect on DFS States	Operators
Unchanged	$I, \sigma_1^z + \sigma_2^z, \sigma_1^z\sigma_2^z, \sigma_1^x\sigma_2^x - \sigma_1^y\sigma_2^y, \sigma_1^x\sigma_2^y + \sigma_1^y\sigma_2^x$
Logical operation	$\bar{\sigma}^z, \bar{\sigma}^y, \bar{\sigma}^x$
Leakage	$\sigma_1^x, \sigma_2^x, \sigma_1^y, \sigma_2^y, \sigma_1^x\sigma_2^z, \sigma_1^z\sigma_2^x, \sigma_1^y\sigma_2^z, \sigma_1^z\sigma_2^y$

operator that acts trivially on the system. First we apply $\bar{X}$ to decouple the logical error operators $\bar{\sigma}^z$ and $\bar{\sigma}^y$, giving us a net unitary evolution

$$\begin{aligned} U_1(2\tau) &= \bar{X} f_\tau \bar{X} f_\tau \\ &= \exp[-2i\tau(\bar{\sigma}^x \otimes \bar{B}^x + \sum_{j=1}^{8} \text{leak}_j) \otimes B^{(j)}] + \mathcal{O}[(2\tau)^2] \end{aligned} \tag{176}$$

where the sum is over the eight leakage operators shown in Table I and $\bar{B}^x$ and $B^{(j)}$ are bath operators. Thus, we still have to compensate for the logical σ^z error and the leakage errors. The order in which we do this does not matter to first order in τ, so let us remove the leakage errors next. This is accomplished by using a ZZ pulse:

$$\begin{aligned} U_2(4\tau) &= ZZ \cdot U_1(2\tau) \cdot ZZ \cdot U_1(2\tau) \\ &= \exp[-4i\tau\bar{\sigma}^x \otimes \bar{B}^x] + \mathcal{O}[(4\tau)^2] \end{aligned} \tag{177}$$

All that remains now is to remove the logical error operator $\bar{\sigma}^x$, since it commutes with both the $\bar{X}$ and ZZ pulses we have used so far. This can be performed using $\bar{Z}$, which anticommutes with $\bar{\sigma}^x$. Hence, the overall time evolution that compensates for all possible (logical and leakage) errors is of period 8τ and is of the form

$$\begin{aligned} U_3(8\tau) &= \bar{Z}U_2(4\tau)\bar{Z}U_2(4\tau) \\ &= \bar{I} \otimes e^{-8i\tau B'} + \mathcal{O}[(8\tau)^2] \end{aligned} \tag{178}$$

A qubit encoded into the $|\bar{0}\rangle = |01\rangle$ and $|\bar{1}\rangle = |10\rangle$ DFS is acted on (at time $T = 8\tau$) only by the innocuous operators in the first row of Table I. As a result, it is completely free of decoherence, up to errors appearing to $\mathcal{O}(T^2)$, while we used a pulse sequence that has length 8τ, shorter by a factor of 2 compared to the sequence we would have had to use without the DFS encoding (the full two-qubit Pauli or Klein group). This, then, illustrates the space–time trade-off between using full DD without DFS encoding and using a hybrid approach, where we use up twice the number of qubits, but gain a factor of 2 in time.

However, we could have, of course, also simply discarded one of the two qubits and used the length-4 universal decoupling sequence for a single qubit. In this sense, the current example is not yet evidence of a true advantage. Such an advantage emerges when one considers constraints on which interactions can be controlled. The method we have discussed here essentially requires only an "XY"-type interaction, governed by a Hamiltonian with terms of the form $\sigma^x \otimes \sigma^x + \sigma^y \otimes \sigma^y$ [17].

For a discussion of how to generalize the construction we have given here to an arbitrary number of encoded qubits, see Ref. [18].

IX. CONCATENATED DYNAMICAL DECOUPLING: REMOVING ERRORS OF HIGHER ORDER IN TIME

The dynamical decoupling techniques considered so far have all involved elimination of decoherence up to first order in time. We now consider the question of whether it is possible to improve upon these techniques and remove the effect of noise up to higher orders in time. We saw in the earlier sections that applying pulses corresponding to the chosen decoupling group effectively causes a net unitary evolution

$$U^{(1)}(T_1) = \prod_{i=0}^{K} g_i^\dagger U^{(0)}(\tau) g_i \tag{179}$$

where $U^{(0)}(\tau) \equiv U_f = \mathrm{e}^{-iH\tau}$ is the free unitary evolution operator, $T_1 = k\tau$, where we let $k \equiv K + 1$ and $\tau \equiv T_0$ is the free evolution duration.

A CDD sequence is defined recursively for $m \geq 1$ as

$$U^{(m)}(T_m) = \prod_{i=0}^{K} g_i^\dagger U^{(m-1)}(T_{m-1}) g_i \tag{180}$$

lasting total time

$$T_m = kT_{m-1} = k^m \tau \tag{181}$$

For example, we could concatenate the universal decoupling sequence $U^{(1)}(T_1) = Zf_\tau Xf_\tau Zf_\tau Xf_\tau$ [Eq. (155)], where $T_1 = 4\tau$, in this manner. The second-order sequence we would obtain is then

$$\begin{aligned} U^{(2)}(T_2) &= ZU^{(1)}(T_1)XU^{(1)}(T_1)ZU^{(1)}(T_1)XU^{(1)}(T_1) \\ &= Z[Zf_\tau Xf_\tau Zf_\tau Xf_\tau]X[Zf_\tau Xf_\tau Zf_\tau Xf_\tau]Z[Zf_\tau Xf_\tau Zf_\tau Xf_\tau]X[Zf_\tau Xf_\tau Zf_\tau Xf_\tau] \\ &= f_\tau Xf_\tau Zf_\tau Xf_\tau Yf_\tau Xf_\tau Zf_\tau Xf_\tau f_\tau Xf_\tau Zf_\tau Xf_\tau Yf_\tau Xf_\tau Zf_\tau Xf_\tau \end{aligned} \tag{182}$$

Note that while some of the pulses have been compressed using equalities such as $Z^2 = I$ and (up to a phase) $XZ = Y$, the total duration of the sequence is dictated by the number of free evolution intervals, which is 16 in this case. Note that while this second-order sequence is time-reversal symmetric, this is not a general feature; indeed, the first-order sequence is not, nor is the third-order sequence, as is easily revealed by writing it down.

To analyze the performance of CDD, we start by rewriting, without loss of generality, $H = H_{SB} + I_S \otimes H_B$ as

$$H \equiv H^{(0)} = H_C^{(0)} + H_{NC}^{(0)} \tag{183}$$

where $H_C^{(0)}$ commutes with the group $\mathcal{G}$ and $H_{NC}^{(0)}$ does not ($H_C^{(0)}$ includes $I_S \otimes H_B$ for sure, and maybe also a part of H_{SB}). This split is done in anticipation of our considerations below. We proceed by using the BCH formula, which yields

$$U^{(1)}(T_1) = \exp\left\{-i\tau \sum_i g_i^\dagger H^{(0)} g_i + \frac{\tau^2}{2} \sum_{i<j} [g_i^\dagger H^{(0)} g_i, g_j^\dagger H^{(0)} g_j] + \mathcal{O}(\tau^3)\right\}$$
$$\equiv e^{-i\tau H^{(1)}} \tag{184}$$

where the effective Hamiltonian $H^{(1)}$ can be decomposed as

$$H^{(1)} = H_C^{(1)} + \tau H_{NC}^{(1)} + \mathcal{O}(\tau^2) \tag{185a}$$
$$H_C^{(1)} \equiv \sum_i g_i^\dagger H^{(0)} g_i \tag{185b}$$
$$H_{NC}^{(1)} \equiv \frac{\tau}{2} \sum_{i<j} [g_i^\dagger H^{(0)} g_i, g_j^\dagger H^{(0)} g_j] \tag{185c}$$

and where, as we have already seen in Eq. (167), $H_C^{(1)}$ lies in the centralizer of $\mathcal{G}$, that is, $[H_C^{(1)}, g] = 0\ \forall g \in \mathcal{G}$. Thus, the first-order term in the BCH series is harmless. Note the important fact that the bad, noncommuting (with $\mathcal{G}$) term $H_{NC}^{(1)}$ is of $\mathcal{O}(\tau)$, while the good, commuting term $H_C^{(1)}$ is of $\mathcal{O}(1)$.

Next let us see how this plays out in $U^{(2)}$:

$$U^{(2)}(T_2) = \prod_{i=0}^{K} g_i^\dagger U^{(1)}(T_1) g_i$$
$$= \exp\left\{-i\tau \sum_i g_i^\dagger H^{(1)} g_i + \frac{\tau^2}{2} \sum_{i<j} [g_i^\dagger H^{(1)} g_i, g_j^\dagger H^{(1)} g_j] + \cdots\right\}$$
$$\equiv e^{-i\tau H^{(2)}} \tag{186}$$

where the effective Hamiltonian $H^{(2)}$ can be decomposed as

$$H^{(2)} = H_C^{(2)} + \tau^2 H_{NC}^{(2)} + \mathcal{O}(\tau^3) \tag{187}$$

It is tempting but slightly counterproductive to try to work out the exact form of the $H_C^{(2)}$ and $H_{NC}^{(2)}$ terms. Clearly, $H_C^{(2)}$ contains both $\sum_i g_i^\dagger H^{(1)} g_i$ and the terms involving commutators of only $H_C^{(1)}$ (without any $H_{NC}^{(1)}$). The key point is that the lowest-order term in the second line of Eq. (186) that does not commute with $\mathcal{G}$ (and hence is responsible for decoherence) is of the type $\tau^2 \sum_{i<j}[g_i^\dagger H_C^{(1)} g_i, g_j^\dagger H_{NC}^{(1)} g_j]$ and hence is of $\mathcal{O}(\tau^3)$.

Comparing Eq. (185a) and (187), we see that the order of the noncommuting term increased by 1, from τ to τ^2. This will clearly continue as we proceed to higher concatenation levels, so that

$$U^{(m)}(T_m) \equiv \mathrm{e}^{-i\tau H^{(m)}} \tag{188}$$

where

$$H^{(m)} = H_C^{(m)} + \tau^m H_{NC}^{(m)} + \mathcal{O}(\tau^{m+1}) \tag{189}$$

This is a remarkable result: it tells us that using concatenation we can push the order of the error term $H_{NC}^{(m)}$ to become arbitrarily high. Of course, the price is an exponentially growing pulse sequence length, but this is a price that may be worth paying if we can make the error shrink fast enough.

Note, however, that in principle the norm of $H_{NC}^{(m)}$ may grow with m and thus is it not yet obvious at this point that increasing orders of CDD concatenation implies a better performance. Let us show that in fact there is an optimal level of concatenation (see Ref. [14] for a rigorous analysis; here we adapt the more intuitive presentation in Ref. [19]).

Given an mth level CDD sequence, one can define a dimensionless "error phase"

$$\phi_{\mathrm{CDD}}(m) = \tau^{m+1} \| H_{NC}^{(m)} \| \tag{190}$$

Let

$$J \equiv \|H_{SB}\|, \quad \beta \equiv \|H_B\| \tag{191}$$

Assuming $J < \beta$ and specializing to the universal DD sequence [Eq. (155)] as the base sequence for the concatenation, this error phase can be bounded as [19, Eq. (46)]

$$\phi_{\mathrm{CDD}}(m) \leq T_m (2^{m^2} (\beta\tau)^m J) \tag{192}$$

where m is the degree of concatenation and now $T_m = 4^m \tau$. It follows then that for an m-level CDD sequence we have the following two possibilities:

- Assume that $T_m = T$ is fixed; that is, $\tau = T/4^m$ can be made arbitrarily small. Then $\phi_{\mathrm{CDD}}(m) \leq T(2^{m^2}(\beta T/4^m)^m J) = JT(\beta T/2^m)^m$, and thus the noise strength decreases monotonically as the concatenation level m increases, as soon as $2^m > \beta T$. In practice, this implies that for a fixed T, provided βT is small enough, the more concatenations of DD the better.

However, it is, of course, not possible in practice to continuously reduce the pulse interval.

- On the other hand, if τ has a minimum physically achievable value, such that a higher concatenation level corresponds to a longer pulse sequence, then there is an optimal level of concatenation. To see this, consider how the upper bound $4^m\tau(2^{m^2}(\beta\tau)^m J)$ behaves as m grows. We see that

$$\log \phi_{\text{CDD}}(m) \leq \log(2)m^2 + \log(4\beta\tau)m + \log(J\tau) \tag{193}$$

This is just a quadratic expression in m, a parabola with a minimum. Differentiating with respect to m, we easily find that the minimum is at $m_{\text{opt}} = -\log(4\beta\tau)/2\log(2)$. Thus, $m_{\text{opt}} > 0$ whenever $T_1\beta = 4\tau\beta < 1$. Thus, provided we choose the base pulse sequence length $T_1 < 1/\beta$, it helps to concatenate, up to the level $\lfloor m_{\text{opt}} \rfloor$. But for m larger than this, the CDD process loses its effectiveness.

Concatenation has been tested experimentally and the optimal concatenation level has been observed (see, for example, Ref. [20]).

As a final note, it is possible to obtain the arbitrary-order error suppression using DD without the exponential cost of CDD. For some entries into this literature, see, for example, Refs [21–23].

X. DYNAMICAL DECOUPLING AND REPRESENTATION THEORY

In this section, our goal is to illustrate the connections between first-order DD and the result from representation theory, Theorem 3, which is a theorem of fundamental importance in the theory of quantum error correction. See Refs [24–26] for entries into the original literature on this topic.

A. Information Storage and Computation Under DD

We define the *group algebra* $\mathbb{C}\mathcal{G}$ of the group $\mathcal{G}$ over the complex field as $\mathbb{C}\mathcal{G} \equiv$ {all linear combinations of the elements in $\mathcal{G}$, over $\mathbb{C}$}. For example, if $\mathcal{G} = \{I, \sigma^x, \sigma^y, \sigma^z\}$, the Pauli group, then

$$\mathbb{C}\mathcal{G} = \left\{ \sum_{\alpha=0,x,y,z} a_\alpha \sigma^\alpha \right\}, \quad a_\alpha \in \mathbb{C} \tag{194}$$

Now consider the group algebra of our decoupling group, $\mathbb{C}\mathcal{G}$. Clearly, it is a matrix algebra of dimension $d \times d$, where $d = \dim(\mathcal{H}_S)$. We can always choose the group in such a way that $\mathbb{C}\mathcal{G}$ is $\dagger$-closed. Since every group includes the identity, we can thus invoke Theorem 3, using which we have

$$\mathbb{C}\mathcal{G} \cong \bigoplus_J I_{n_J} \otimes \mathcal{M}_{d_J} \tag{195}$$

where J is the irrep (irreducible representation) label, n_J is the multiplicity of irrep J, and d_J is the dimension of the irrep labeled by J. The system Hilbert space $\mathcal{H}_S$ is correspondingly partitioned into a direct sum of product spaces, which we can write as

$$\mathcal{H}_S \cong \bigoplus_J \mathbb{C}^{n_J} \otimes \mathbb{C}^{d_J} \tag{196}$$

Hence, every DD pulse that we apply to the system acts like identity on $\mathbb{C}^{n_J}$ and as some nontrivial operation on $\mathbb{C}^{d_J}$. It is clear that if we store our quantum information in $\mathbb{C}^{n_J}$, then the pulses do not affect it (of course, the system–bath interaction can still affect information stored in $\mathbb{C}^{n_J}$).

Now consider the *commutant* of the group algebra $\mathbb{C}\mathcal{G}$ [see Eqs. (105) and (106)],

$$\begin{aligned}\mathbb{C}\mathcal{G}' &\equiv \{A | [A, \mathbb{C}\mathcal{G}] = 0\} \\ &= \bigoplus_J \mathcal{M}_{n_J} \otimes I_{d_J}\end{aligned} \tag{197}$$

We can immediately see from this definition and from Eq. (165) that, by linearity, the effective system–bath Hamiltonian, H'_{SB}, lies in the commutant of the group algebra, that is, $H'_{SB} \in \mathbb{C}\mathcal{G}'$. It follows that H'_{SB} can be represented as

$$H'_{SB} = \bigoplus_J (H'_{SB})_{n_J} \otimes I_{d_J} \tag{198}$$

that is, the effective system–bath Hamiltonian has this block-diagonal representation, with blocks labeled by the irrep index J, each of dimension $n_J d_J$, and where each nontrivial factor $(H'_{SB})_{n_J}$ is an $n_J \times n_J$ matrix. It is clear that in this case we can encode our quantum information in $\mathbb{C}^{d_J}$, since H'_{SB} will act as identity on it. Note that it does not necessarily matter that the pulses themselves have a nontrivial effect on $\mathbb{C}_{d_J}$: because we know everything about the pulses, they can be compensated for by applying appropriate transformations.

However, it is often desirable not to have to compensate for the action of the DD pulses. In that case, we might want to be able to store information in $\mathbb{C}^{n_J}$ rather than $\mathbb{C}^{d_J}$. If so, clearly we need to somehow make the effect of H'_{SB} trivial on $\mathbb{C}^{n_J}$ as well. Since H'_{SB} is determined by our group of pulses, $\mathcal{G}$, it boils down to choosing the appropriate set of pulses, that is, picking the group $\mathcal{G}$ such that

$$\begin{aligned} H'_{SB} &\in \mathbb{C}\mathcal{G}' \cap \mathbb{C}\mathcal{G} \\ &= \bigoplus_J \lambda_J I_{n_J} \otimes I_{d_J}, \ \lambda_J \in \mathbb{C} \end{aligned} \tag{199}$$

where the second line can be easily deduced by examining Eq. (195) and the second line of Eq. (197). This means that if we pick an appropriate $\mathcal{G}$, H'_{SB} will have the form

$$H'_{SB} = \begin{pmatrix} \lambda_1 I_{n_1 d_1} & & & \\ & \lambda_2 I_{n_2 d_2} & & \\ & & \lambda_3 I_{n_3 d_3} & \\ & & & \ddots \end{pmatrix} \tag{200}$$

We have already seen examples where H'_{SB} is of this form:

- When the conditions of Schur's lemma are satisfied, then $H'_{SB} \propto I_d$. Thus, that result was a special case resulting from this more general structure.
- If $\lambda_J = 0$ for all J, then $H'_{SB} = 0$ and the system–bath interaction is annihilated (to first order in T). This describes the kind of situation we obtained with the Klein group symmetrization [see Eq. (154)].

So, now we have established that it is possible to protect information stored in both $\mathbb{C}^{n_J}$ and $\mathbb{C}^{d_J}$. Information stored in the latter is protected since the effective system–bath interaction acts like identity on that space, though we will have to compensate for the influence of the pulses we would apply. Storing in the former requires us to more cleverly choose the pulse group $\mathcal{G}$ such that the effective system–bath interaction becomes a block-diagonal matrix, with each block proportional to identity.

Our choice of information storage location might depend on which of $\max_J n_J$ and $\max_J d_J$ is greater, since that would provide us with a larger-dimensional space and hence more qubits. Or it might depend on whether we wish to apply computation as well, in which case using $\mathbb{C}^{n_J}$ would be preferred. Indeed, when the second line of Eq. (199) is satisfied, we can make use of the commutant $\mathbb{C}\mathcal{G}'$ to perform computation!

We now discuss some examples that illustrate these points.

B. Examples

In all of the following examples, our system is a set of N qubits, that is, $\mathcal{H}_S = (\mathbb{C}^2)^{\otimes N} = \mathbb{C}^{2^N}$.

1. Example 1: $\mathcal{G} = (SU(2))^{\otimes N}$

In this case, the pulses are products of all N arbitrary single-qubit unitaries, and the only operator that commutes with $\mathcal{G}$ is the identity operator, that is, $H'_{SB} \in \mathbb{C}\mathcal{G}' = \mathbb{C}$. This is the Schur's lemma situation again. Thus, in this case we have the choice of storing information in either the left-hand ($\mathbb{C}^{n_J}$) or right-hand factor ($\mathbb{C}^{d_J}$), or both if we do not mind compensating for the action of the pulses on the right-hand factor. It is interesting that in spite of the huge cost of directly implementing this group, it can be well approximated by picking the SU(2) rotation on each qubit at random [27].

2. Example 2: $\mathcal{G}$ = Collective $SU(2)$

Let us add to Example 1 the constraint that the same unitary matrix acts on every qubit, but every element of SU(2) is implemented (again, this can be approximated using random elements of SU(2) [27]). Then $H'_{SB} \in \mathbb{C}\mathcal{G}' = \mathbb{C}S_N$, where S_N is the permutation group on N elements. Thus, the effective system–bath Hamiltonian is *not* proportional to identity, and instead we have here the case where H'_{SB} acts trivially only on the right-hand factors. Its action on the left-hand factors is to apply permutations.

The collective SU(2) group is generated by the sum of the Pauli matrices, that is, $\{\sum_i \sigma_i^\alpha\}_{\alpha=x,y,z}$. We already encountered these sums in the study of collective decoherence [see Eqs. (69) and (70)]. In this case, our pulse group has the same generating Hamiltonian as the system part of the system–bath interaction in collective decoherence: $H_{SB}^{\text{coll.dec.}} = \sum_{\alpha=x,y,z} \left(\sum_i \sigma_i^\alpha\right) \otimes B^\alpha$. Since the DD group acts as identity on the left-hand factors, and the DD group behaves like the collective decoherence operators, while the effective system–bath interaction H'_{SB} acts like the exchange operators [Eq. (91)] we encountered in our study of computation over the DFS for collective decoherence, we see that the current situation is the reverse (or dual) of the situation back in the DFS case. In other words, we can invoke the machinery we developed then [see Eqs. (107)–(109)] but we should flip the role of d_J and n_J: $\mathcal{H}_S = \bigoplus_{J=0(1/2)}^{N/2} \mathbb{C}^{n_J} \otimes \mathbb{C}^{d_J}$, where now

$$n_J = 2J + 1 \tag{201a}$$

$$d_J = \frac{(2J+1)N!}{(N/2+1+J)!(N/2-J)!} \tag{201b}$$

We can always pick an irrep J so that $d_J > n_J$, and so we have here the same code rates as in the case of a DFS.

3. *Example 3: $\mathcal{G} = S_n$*

In this case, $\mathcal{G}$ is the permutation group. As we saw in the last example, the permutation group is dual to collective SU(2). So in this case, $H'_{SB} \in \mathbb{C}\mathcal{G}' =$ collective decoherence. In fact, since the permutation group can be obtained by swaps (or transpositions), we could as well take $\mathcal{G} = \{\mathrm{SWAP}_{i,j}\}$. And we know that

$$\mathrm{SWAP}_{i,j} = \mathrm{e}^{-i\pi\vec{\sigma}_i\cdot\vec{\sigma}_j/4} \tag{202}$$

which is generated by the Heisenberg interaction, so the decoupling group is implementable in physical systems (such as quantum dots) where the exchange interaction is controllable (see Ref. [28] for a discussion of how to efficiently implement $\mathcal{G}$ in this case). The Hilbert space again splits as $\mathcal{H}_S = \bigoplus_{J=0(1/2)}^{N/2} \mathbb{C}^{n_J} \otimes \mathbb{C}^{d_J}$, where now we have the irrep dimension and multiplicity formulas we encountered during the DFS study:

$$n_J = \frac{(2J+1)N!}{(N/2+1+J)!(N/2-J)!} \tag{203a}$$

$$d_J = 2J + 1 \tag{203b}$$

We have the option of encoding into the right-hand factor, where H'_{SB} acts as identity, but then the space dimension is only $2J + 1$. Alternatively, we can encode into the left-hand factors, where the effective system–bath interaction has nontrivial action, but it acts as collective decoherence, so that our DFS encoding will completely hide the quantum information from the action of H'_{SB}. This has the significant advantage (over right-hand factor encoding) of providing us with a code space of dimension n_J. This example leads us to the interesting conclusion that in this case, in fact, the second line of Eq. (199) applies; that is, the effective system–bath interaction acts trivially everywhere.

4. *Example 4: Linear System–Bath Coupling*

We consider a system–bath interaction of the form

$$H_{SB} = \sum_{\alpha=x,y,z} \sum_i \sigma_i^\alpha \otimes B_i^\alpha \tag{204}$$

Each qubit in this case has its own bath. In this noise model, we do not consider bilinear terms in the system such as $\sigma_i^\alpha \sigma_j^\beta \otimes B_{ij}^{\alpha\beta}$, because this is a three-body interaction that is typically much weaker in nature and also very hard to engineer.

The decoupling group we select for this is

$$\mathcal{G} = \left\{ I^{\otimes N}, X^{\otimes N}, Y^{\otimes N}, Z^{\otimes N} \right\} \tag{205}$$

We choose N to be even. Therefore, $\mathcal{G}$ becomes abelian. And from representation theory, we know that all the irreps of an abelian group are one dimensional (scalars, so $d_J = 1\ \forall J$) and the number of irreps is the order of the group. Here $|\mathcal{G}| = 4$. The irreps are

J	$I^{\otimes N}$	$X^{\otimes N}$	$Y^{\otimes N}$	$Z^{\otimes N}$
1	1	1	1	1
2	1	1	-1	-1
3	1	-1	1	-1
4	1	-1	-1	1

(206)

The group algebra is

$$\begin{aligned} \mathbb{C}\mathcal{G} &\cong \bigoplus_J I_{n_J} \otimes \mathcal{M}_{d_J} \\ &= \bigoplus_{J=1}^{4} c_J I_{2^{N-2}} \end{aligned} \tag{207}$$

where in the second line $n_J = 2^{N-2}$ because $\sum_J n_J d_J = 2^N$ and all the n_J's are equal (by use of a standard multiplicity formula from group theory [5], or by noting the symmetry between the X, Y, and Z operators), we have $4n_J = 2^N$, making $n_J = 2^{N-2}$.

Thus, the group algebra can be represented as a block-diagonal matrix, each block being proportional to an identity of dimension 2^{N-2} with proportionality constant c_J; that is, we have $(N-2)$ qubits in each block that will be unaffected by the system–bath interaction.

Let us pick the first (trivial) irrep to encode our qubits into, that is, $\{1, 1, 1, 1\}$. In this irrep each pulse acts as 1, so we are looking for code states that are "stabilized" by the group (each group element acts as identity). After a bit of thought, it is clear that such states are of the form

$$|\psi_r\rangle \equiv \frac{1}{\sqrt{2}}\left(|r\rangle + |\bar{r}\rangle\right) \tag{208}$$

where $\bar{r} = \text{NOT}(r)$ and $r \in \{0, 1\}^N$ is an *even* weight binary string of N bits; that is, r contains an even number of 1's. Then it is easy to see that the action of any member of the decoupling group $\mathcal{G}$ leaves $|\psi_r\rangle$ unchanged, so indeed $|\psi_r\rangle$ belongs to the trivial irrep.

Why did we pick this decoupling group? Because it has a couple of very interesting and useful features that we list below.

Feature 1: We can show that $H'_{SB} = 0$.

Proof.

$$\begin{aligned} H'_{SB} &= \frac{1}{|\mathcal{G}|}\sum_{j=0}^{|\mathcal{G}|} g_j^\dagger H'_{SB} g_j \\ &= \frac{1}{4}\left[IH_{SB}I + X^{\otimes N}H_{SB}X^{\otimes N} + Y^{\otimes N}H_{SB}Y^{\otimes N} + Z^{\otimes N}H_{SB}Z^{\otimes N}\right] \\ &= \frac{1}{4}\Big[H_{SB} + \sum_i \left(\sigma_i^X \otimes B_i^X - \sigma_i^Y \otimes B_i^Y - \sigma_i^Z \otimes B_i^Z\right. \\ &\quad - \sigma_i^X \otimes B_i^X + \sigma_i^Y \otimes B_i^Y - \sigma_i^Z \otimes B_i^Z \\ &\quad \left. - \sigma_i^X \otimes B_i^X - \sigma_i^Y \otimes B_i^Y + \sigma_i^Z \otimes B_i^Z\right)\Big] \\ &= 0 \end{aligned} \tag{209}$$

where to arrive at the third line we used the properties of the Pauli group to do the multiplication. For example,

$$\begin{aligned} X^{\otimes N}\sigma_i^Y X^{\otimes N} &= (\sigma_1^X \otimes \cdots \otimes \sigma_i^X \otimes \cdots \otimes \sigma_N^X)(\sigma_i^Y)(\sigma_1^X \otimes \cdots \otimes \sigma_i^X \otimes \cdots \otimes \sigma_N^X) \\ &= -\sigma_i^Y \end{aligned} \tag{210}$$

and so on. □

Thus, this decoupling group eliminates the system–bath interaction completely, to first order. Recall that this means that we can use the left-hand factor to encode and store information. This allows us to perform computation using the commutant, which has nontrivial action on the left-hand factor.

Feature 2: We can do computation on the decoherence-protected qubits.

For $N = 2$, that is, for two physical qubits, we have no logical qubits, as the only one state possible according to Eq. (208) is $|\psi_r\rangle = \frac{1}{\sqrt{2}}(|00\rangle + |11\rangle)$. This agrees with the fact that, since each irrep provides us with $N - 2$ logical qubits, we have zero logical qubits for two physical qubits.

Let us list the possible states in the case of $N = 4$, using Eq. (208):

$$|\psi_{0000}\rangle = \frac{1}{\sqrt{2}}(|0000\rangle + |1111\rangle) \equiv |\bar{0}\bar{0}\rangle \tag{211a}$$

$$|\psi_{0011}\rangle = \frac{1}{\sqrt{2}}(|0011\rangle + |1100\rangle) \equiv |\bar{1}\bar{0}\rangle \tag{211b}$$

$$|\psi_{0101}\rangle = \frac{1}{\sqrt{2}}(|0101\rangle + |1010\rangle) \equiv |\bar{0}\bar{1}\rangle \tag{211c}$$

$$|\psi_{0110}\rangle = \frac{1}{\sqrt{2}}(|0110\rangle + |1001\rangle) \equiv |\bar{1}\bar{1}\rangle \tag{211d}$$

and since we have four orthonormal states, we can use them as two qubits. Again, this agrees with the fact that $N - 2 = 2$ in this case.

How do we perform computations on these states? For that we use the commutant of the group,

$$\begin{aligned} \mathbb{C}\mathcal{G}' &\cong \bigoplus \mathcal{M}_{n_J} \otimes I_{d_J} \\ &= \bigoplus_{J=1}^{4} c_J \mathcal{M}_{2^{N-2}} \end{aligned} \tag{212}$$

where c_J are scalars; they are the columns of Eq. (206) (e.g., for $X^{\otimes N}$ we have $c_1 = c_2 = 1$ and $c_3 = c_4 = -1$). We can check that the commutant can be generated by $\{X_1 X_{j+1}\}_{j=1}^{N-2} \cup \{Z_{j+1} Z_N\}_{j=1}^{N-2}$. For example, we do not require $Y_i Y_j$ to be in the generating set, since $Y_i Y_j = X_i X_j Z_i Z_j$. Therefore, for $N = 4$, the generating set for the commutant becomes $\{X_1 X_2, X_1 X_3, Z_2 Z_4, Z_3 Z_4\}$.

Let us check the action of $X_1 X_2$ on our first logical state $|\psi_{0000}\rangle \equiv |\bar{0}\bar{0}\rangle$,

$$\begin{aligned} X_1 X_2 |\bar{0}\bar{0}\rangle &= (X_1 \otimes X_2 \otimes I \otimes I)\left(\frac{1}{\sqrt{2}}(|0000\rangle + |1111\rangle)\right) \\ &= \frac{1}{\sqrt{2}}(|0011\rangle + |1100\rangle) \equiv |\bar{1}\bar{0}\rangle \end{aligned} \tag{213}$$

We can similarly check that $X_1 X_2 |\bar{0}\bar{1}\rangle = |\bar{1}\bar{1}\rangle$. Therefore, $X_1 X_2$ is logical Pauli-X on the first logical qubit. We write this as $X_1 X_2 = \bar{X}_1$, where the bar denotes a logical operator. Similarly, we can verify that

$$X_1 X_{j+1} = \bar{X}_j \tag{214}$$

$$Z_{j+1} Z_N = \bar{Z}_j \tag{215}$$

And having obtained logical-X and logical-Z (note that they anticommute, as they should), we can implement any one qubit gate we like using only two-body interactions. And moreover, these logical gates lie in the commutant of the decoupling group. Therefore, we can compute while at the same time applying DD.

But, in order to perform universal quantum computation, we also need to be able to perform entangling operations on two qubits, for example, the CNOT gate. Let us examine $\bar{X}_i\bar{X}_j$, which is a logical entangling operation:

$$\begin{aligned} \bar{X}_i\bar{X}_j &= X_1X_{i+1}X_1X_{j+1} \\ &= X_{i+1}X_{j+1} \end{aligned} \tag{216}$$

which is a physical two-body interaction. So, we have managed to implement a logical entangling operation using only a two-body interaction.

It is well known that if we can implement any Hamiltonian of the form

$$H_S = \sum_i \omega_i^X(t)\sigma_i^X + \sum_i \omega_i^Z(t)\sigma_i^Z + \sum_{\alpha\in\{X,Z\}}\sum_{i,j} J_{ij}(t)\sigma_i^\alpha\sigma_j^\alpha \tag{217}$$

then we can perform universal quantum computation [1]. Therefore, in our case, replacing the Pauli operators in Eq. (217) with their logical counterparts and expanding them in terms of their decompositions (214) in the physical qubit space, we obtain

$$\begin{aligned} \bar{H}_S &= \sum_{i=1}^{N-2} \bar{\omega}_i^X(t)\bar{\sigma}_i^X + \sum_{i=1}^{N-2} \bar{\omega}_i^Z(t)\bar{\sigma}_N^i + \sum_{\alpha\in\{X,Z\}}\sum_{i,j} \bar{J}_{ij}(t)\bar{\sigma}_i^\alpha\bar{\sigma}_j^\alpha \\ &= \sum_{i=1}^{N-2} \omega_{i+1}^X(t)\sigma_1^x\sigma_{i+1}^x + \sum_{i=1}^{N-2} \omega_{i+1}^Z(t)\sigma_{i+1}^z\sigma_N^z + \sum_{\alpha\in\{X,Z\}}\sum_{i,j} J_{i+1,j+1}(t)\bar{\sigma}_{i+1}^\alpha\bar{\sigma}_{j+1}^\alpha \end{aligned} \tag{218}$$

Remarkably, we have obtained *decoupling and universal quantum computation* on the logical qubits using only two-body interactions on the physical qubits. A similar approach has been studied numerically in the context of the four-qubit DFS code, with gates protected by CDD, showing evidence for a highly robust set of universal gates [29]. Related ideas apply in the context of adiabatic quantum computation [30].

XI. CONCLUSIONS

This chapter has covered a selection of topics in the theory of decoherence-free subspaces, noiseless subsystems, and dynamical decoupling. We have seen how these tools allow one to hide information from the environment, and when this hiding is imperfect, how dynamical decoupling allows us to suppress the remaining residual decoherence. Moreover, we have shown explicitly how universal quantum computation is compatible with decoherence avoidance and suppression.

Many important topics were left out in this brief chapter. For example, apart from CDD we did not address high-order decoupling methods, in particular, schemes based on optimized pulse intervals [31–34]. Nor did we address the filter function approach to DD [35], optimized continuous modulation [36,37], or randomized decoupling, which is well suited to strongly time-dependent baths [38]. It is important to stress that beyond decoherence avoidance and suppression, the theory of noiseless subsystems gave rise also to important advances in the theory of quantum error correcting codes, such as operator quantum error correction [39]. However, perhaps our greatest omission has been the abundance of experimental results that have both confirmed and driven the theoretical developments described here. For an entry into that literature, as well as many additional theoretical topics, see Ref. [4]. Nevertheless, hopefully we have given the reader the tools and inspiration to delve deeper into the large and fascinating literature on decoherence avoidance and suppression.

Acknowledgments

I am grateful to all the students in my Spring 2007 and Fall 2011 Quantum Error Correction courses, in particular, Jose Raul Gonzalez Alonso, Chris Cantwell, Kung-Chuan Hsu, Siddharth Muthu Krishnan, Ching-Yi Lai, Hokiat Lim, Osonde Osoba, Kristen Pudenz, Greg Quiroz, Bilal Shaw, Mark Wilde, Sunil Yeshwanth, and Yicong Zheng, for their meticulous notes, which formed the basis for this chapter. Thanks also to Gerardo Paz who helped with the section on CDD. I am indebted to Steve Huntsman who found and helped correct many typos. This work was supported by the NSF Center for Quantum Information and Computation for Chemistry, Grant No. CHE-1037992, NSF Grant No. CHE-924318, and by the ARO MURI Grant No. W911NF-11-1-0268.

REFERENCES

1. M. Nielsen and I. Chuang, *Quantum Computation and Quantum Information, Cambridge Series on Information and the Natural Sciences*, Cambridge University Press, Cambridge, 2000.
2. D. A. Lidar and K. B. Whaley, in *Irreversible Quantum Dynamics, Lecture Notes in Physics*, Vol. 622, F. Benatti and R. Floreanini, eds., Springer, Berlin, 2003, pp. 83–120.
3. W. Yang, Z.-Y. Wang, and R.-B. Liu, *Front. Phys.* **6**, 2 (2011).
4. D. Lidar and T. Brun, eds., *Quantum Error Correction*, Cambridge University Press, Cambridge, 2013.
5. M. Hammermesh, *Group Theory and Its Application to Physical Problems*, Dover Publications, 1989.
6. D. Loss and D. P. DiVincenzo, *Phys. Rev. A* **57**, 120 (1998).

7. A. Mizel and D. A. Lidar, *Phys. Rev. B* **70**, 115310 (2004).
8. D. Bacon, J. Kempe, D. A. Lidar, and K. B. Whaley, *Phys. Rev. Lett.* **85**, 1758 (2000).
9. R. Woodworth, A. Mizel, and D. A. Lidar, *J. Phys.: Condens. Matter* **18**, S721 (2006).
10. E. Knill, R. Laflamme, and L. Viola, *Phys. Rev. Lett.* **84**, 2525 (2000).
11. J. Kempe, D. Bacon, D. A. Lidar, and K. B. Whaley, *Phys. Rev. A* **63**, 042307 (2001).
12. D. P. DiVincenzo, D. Bacon, J. Kempe, G. Burkard, and K. B. Whaley, *Nature* **408**, 339 (2000).
13. R. Bhatia, *Matrix Analysis, Graduate Texts in Mathematics*, Vol. 169, Springer, New York, 1997.
14. H.-K. Ng, D. A. Lidar, and J. P. Preskill, *Phys. Rev. A* **84**, 012305 (2011).
15. P. Zanardi, *Phys. Lett. A* **258**, 77 (1999).
16. M. S. Byrd and D. A. Lidar, *Phys. Rev. Lett.* **89**, 047901 (2002).
17. D. A. Lidar and L. A. Wu, *Phys. Rev. Lett.* **88**, 017905 (2001).
18. L. A. Wu, M. S. Byrd, and D. A. Lidar, *Phys. Rev. Lett.* **89**, 127901 (2002).
19. K. Khodjasteh and D. A. Lidar, *Phys. Rev. A* **75**, 062310 (2007).
20. G. A. Álvarez, A. Ajoy, X. Peng, and D. Suter, *Phys. Rev. A* **82**, 042306 (2010).
21. G. S. Uhrig, *Phys. Rev. Lett.* **98**, 100504 (2007).
22. J. R. West, B. H. Fong, and D. A. Lidar, *Phys. Rev. Lett.* **104**, 130501 (2010).
23. Z.-Y. Wang and R.-B. Liu, *Phys. Rev. A* **83**, 022306 (2011).
24. P. Zanardi, *Phys. Lett. A* **258**, 77 (1999).
25. L. Viola, S. Lloyd, and E. Knill, *Phys. Rev. Lett.* **83**, 4888 (1999).
26. L. Viola, E. Knill, and S. Lloyd, *Phys. Rev. Lett.* **85**, 3520 (2000).
27. L. Viola and E. Knill, *Phys. Rev. Lett.* **94**, 060502 (2005).
28. L. A. Wu and D. A. Lidar, *Phys. Rev. Lett.* **88**, 207902 (2002).
29. J. R. West, D. A. Lidar, B. H. Fong, and M. F. Gyure, *Phys. Rev. Lett.* **105**, 230503 (2010).
30. D. A. Lidar, *Phys. Rev. Lett.* **100**, 160506 (2008).
31. G. S. Uhrig, *Phys. Rev. Lett.* **98**, 100504 (2007).
32. J. R. West, B. H. Fong, and D. A. Lidar, *Phys. Rev. Lett.* **104**, 130501 (2010).
33. Z.-Y. Wang and R.-B. Liu, *Phys. Rev. A* **83**, 022306 (2011).
34. Y. Xia, G. S. Uhrig, and D. A. Lidar, *Phys. Rev. A* **84**, 062332 (2011).
35. H. Uys, M. J. Biercuk, and J. J. Bollinger, *Phys. Rev. Lett.* **103**, 040501 (2009).
36. G. Gordon, G. Kurizki, and D. A. Lidar, *Phys. Rev. Lett.* **101**, 010403 (2008).
37. J. Clausen, G. Bensky, and G. Kurizki, *Phys. Rev. Lett.* **104**, 040401 (2010).
38. L. F. Santos and L. Viola, *Phys. Rev. Lett.* **97**, 150501 (2006).
39. D. Kribs, R. Laflamme, and D. Poulin, *Phys. Rev. Lett.* **94**, 180501 (2005).

FUNCTIONAL SUBSYSTEMS AND STRONG CORRELATION IN PHOTOSYNTHETIC LIGHT HARVESTING

DAVID A. MAZZIOTTI and NOLAN SKOCHDOPOLE

Department of Chemistry and the James Franck Institute, The University of Chicago, Chicago, IL 60637, USA

I. INTRODUCTION

Nature harvests solar energy with a remarkably high *quantum efficiency*, the percentage of charge carriers created by photons. Recent spectroscopic experiments [1–3] and theoretical models [4–13], provide strong evidence that efficient light harvesting in nature occurs by a quantum mechanism involving sustained electronic coherence [14] and entanglement [15,16] between chromophores. Quantum coherence in the Fenna–Matthews–Olson (FMO) antennae complex of green

Advances in Chemical Physics, Volume 154: Quantum Information and Computation for Chemistry, First Edition. Edited by Sabre Kais.

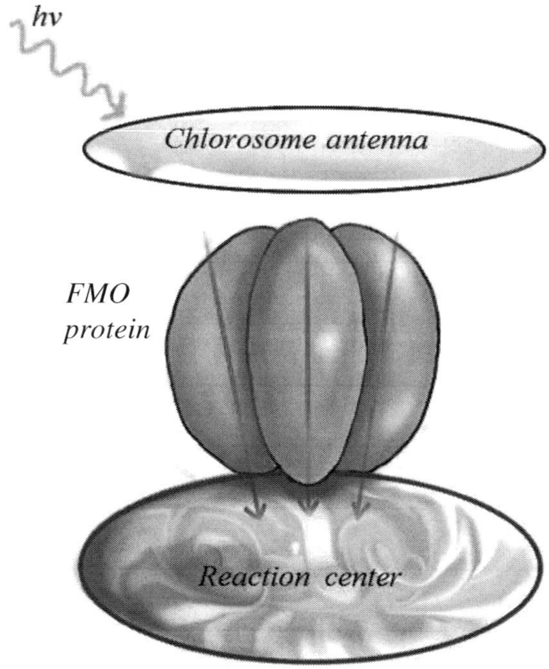

Figure 1. After harvesting a photon of energy, the Fenna–Matthews–Olson (FMO) antennae complex transfers it to the reaction center. Illustration by K. Naftchi-Ardebili, The University of Chicago. Reprinted with permission from Ref. [4]. Copyright 2011 American Chemical Society.

sulfur bacteria (Fig. 1) has been shown by two-dimensional Fourier-transform electronic spectroscopy to last over 500 fs at ambient temperatures and over 300 fs at physiological temperatures [1–3].

In this chapter, we study two key features of the photosynthetic light harvesting: (1) the effect of *strong correlation* of the electrons within the chromophores on the transfer efficiency of energy to the reaction center and (2) the role of *functional subsystems*—subsets of chromophores with the ability to transfer energy efficiently to the reaction center without the other chromophores—in providing a built-in *quantum redundancy* to keep photosynthesis "humming" along even with more than half of the chromophores temporarily or permanently shut down. The concepts of strong correlation and functional subsystems are important to not only improving our understanding photosynthetic light harvesting but also providing new design principles for the development of enhanced photovoltaic devices.

II. EFFECT OF STRONG ELECTRON CORRELATION

A. Overview

Although the chromophores are chlorophyll molecules containing large networks of conjugated carbon bonds that surround a charged magnesium ion, they have largely been represented in theoretical studies [4–13] by one-electron models that neglect the effects of electron correlation and entanglement within chromophores. Two advanced methods in electronic structure, density-matrix renormalization group [17] and two-electron reduced-density-matrix theory [18,19], have recently shown that networks of conjugated bonds as in acene chains [17,20,21], acene sheets [20,21], and chlorophyll are associated with polyradical character that cannot be adequately described without a strongly correlated many-electron quantum model.

In the first part of this chapter, we examine the efficiency of light harvesting where we represent each chromophore by a correlated N-electron model to treat strong electron correlation. Figure 2 illustrates the replacement of one-electron models for each of the seven chromophores in the FMO complex of green-sulfur bacteria by N-electron models of increasing complexity. Here, we employ the N-electron Lipkin–Meshkov–Glick (LMG) model [22] used extensively in electronic structure [23–31] and quantum information [16]. We find that strong electron correlation dramatically enhances the efficiency of the energy transfer to the reaction center by more than 100%.

Strong electron correlation, the results show, is likely employed by nature to enhance its energy-transfer efficiency as much as other factors such as (i) environmental noise [5–10] and (ii) entanglement between chromophores [11–13], which have been extensively studied in the recent literature. The correlation of electrons within a molecular subunit like a chromophore cannot be separated from their entanglement between molecular subunits. Furthermore, the results suggest a

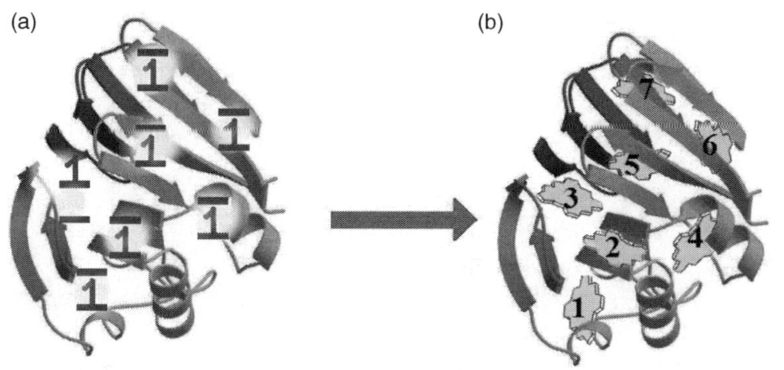

Figure 2. Single electrons or correlated chromophores. Each of the seven chromophores in the FMO complex is generally treated as a single electron in a two-state model (**a**), and yet the chromophores are constructed from chlorophyll molecules with many strongly correlated electrons (**b**). Here, we treat each of the chromophores by a correlated N-electron model by Lipkin, Meshkov, and Glick. Reprinted with permission from Ref. [32]. Copyright 2012 American Institute for Physics.

general design principle for human-made materials in which electron correlations and entanglements both within and between subunits are simultaneously tuned for achieving enhanced quantum efficiencies. In combination with other recent advances, including the study of functional subsystems of the FMO complex [4], this interplay of electron entanglements on different length scales may enable us to develop materials with quantum efficiencies approaching those found in natural processes from photosynthesis to bioluminescence.

B. Multi-Electron Chromophore Model

The FMO complex of green-sulfur bacteria contains three identical subunits, each with a network of seven chromophores. Theoretical models of an FMO complex's subunit typically represent each of M chromophore by a *one-electron model* in which the electron has access to two energy levels separated by the excitation energy of the chromophore. Interactions $\hat{U}$ between pairs of chromophores are modeled by the exchange of single-electron excitations (or *excitons*) between them:

$$\hat{H} = \hat{H}_0 + \hat{U} \tag{1}$$

where

$$\hat{H}_0 = \frac{1}{2} \sum_{s,m} m \epsilon_s a^{\dagger}_{s,m} a_{s,m} \tag{2}$$

$$\hat{U} = \sum_{s \neq t} U_{s,t} a^{\dagger}_{s,+1} a_{s,-1} a^{\dagger}_{t,-1} a_{t,+1} \tag{3}$$

The quantum numbers s and t denote the seven sites of the chromophores while the quantum number m, equal to $+1$ or -1, indicates one of the two energy levels within each chromophore. The second-quantized operator $a^{\dagger}_{s,m}$ ($a_{s,m}$) creates (annihilates) an electron on chromophore s in energy level m. The M parameters ϵ_s are the excitation energies of the chromophore, and the $M(M-1)/2$ parameters $U_{s,t}$ are the coupling energies between all pairs of chromophores. Typical values for the excitation and coupling energies of the seven-site FMO complex are given in the 7×7 Hamiltonian of Ref. [33].

Here, we generalize this representation by using the N-electron LMG model [22]. Each chromophore is modeled as N electrons in two energy levels that are each N-fold degenerate; a pairwise interaction scatters two electrons from the lower level to the upper level or from the upper level to the lower level [25]. The Hamiltonian of the M interacting LMG chromophores can be expressed as follows:

$$\hat{H} = \hat{H}_0 + \lambda \hat{U} + \hat{V} \tag{4}$$

where

$$\hat{H}_0 = \frac{1}{2} \sum_{s,m,p} m\epsilon_s a^\dagger_{s,m,p} a_{s,m,p} \tag{5}$$

$$\hat{U} = \frac{1}{N} \sum_{s \neq t,p} U_{s,t} a^\dagger_{s,+1,p} a_{s,-1,p} a^\dagger_{t,-1,p} a_{t,+1,p} \tag{6}$$

$$\hat{V} = \frac{1}{2} \sum_{s,m,p,q} V a^\dagger_{s,m,p} a^\dagger_{s,m,q} a_{s,-m,q} a_{s,-m,p} \tag{7}$$

where $\hat{V}$ accounts for the interactions *within* each chromophore and $\lambda \in [0, 1]$ is a screening parameter to be defined later. The new quantum number p (pr q) denotes the N degenerate states within each energy level, which are necessary to accommodate the N electrons. When the interaction strength V equals zero, the N electrons on each site are noninteracting, and the model behaves the same as the one-electron-per-site models; when V is unequal to zero, we have a generalized model for light harvesting with a tunable electron correlation on the chromophores.

Environmental effects of dephasing and dissipation as well as the transfer of energy to the reaction center (sink) can be incorporated by introducing a Lindblad operator $\hat{L}$ into the *quantum Liouville equation*

$$\frac{d}{dt} D = -\frac{i}{\hbar}[\hat{H}, D] + \hat{L}(D) \tag{8}$$

where D is the many-electron density matrix. The Lindblad operator can be divided into three operators that account for dephasing, dissipation, and the sink

$$\hat{L}(D) = \hat{L}_{\text{deph}}(D) + \hat{L}_{\text{diss}}(D) + \hat{L}_{\text{sink}}(D) \tag{9}$$

where

$$\hat{L}_{\text{deph}}(D) = \alpha \sum_k 2\langle k|D|k\rangle |k\rangle\langle k| - \{|k\rangle\langle k|, D\} \tag{10}$$

$$\hat{L}_{\text{diss}}(D) = \beta \sum_k 2\langle k|D|k\rangle |g\rangle\langle g| - \{|k\rangle\langle k|, D\} \tag{11}$$

$$\hat{L}_{\text{sink}}(D) = 2\gamma\langle 3|D|3\rangle |s\rangle\langle s| - \gamma\{|3\rangle\langle 3|, D\} \tag{12}$$

The state $|g\rangle$ denotes the ground eigenstate of the Hamiltonian in Eq. (4), the states $|k\rangle$ represent the excited eigenstates computed from this Hamiltonian where the interaction $\hat{U}$ between chromophores is set to zero, the state $|s\rangle$ denotes the reaction center (sink), and $|3\rangle$ indicates the first excited state of the third chromophore multiplied by the ground states of the other chromophores. Non-Markovian effects from the environment can also be added to the quantum Liouville equation (e.g., refer to Ref. [10]); however, they will not qualitatively change the effect of strong electron correlation within chromophores on energy-transfer efficiency.

C. Application of Multielectron Chromophore Model

1. *Model Parameters*

The values for the site excitation energies of the seven-site FMO complex are defined by the 7×7 Hamiltonian of Ref. [33]; when $V > 0$, the parameters $\{\epsilon_s\}$ are adjusted to ensure that the excitations of the correlated LMG models agree with the excitation energies of the seven chromophores. Selection of the coupling energies $U_{s,t}$ in Eq. (4) also requires adjustment. Because the coupling energies given in Ref. [33] are "dressed" dipole–dipole interactions that account implicitly for both the electron correlation within the chromophores and the protein environment surrounding the chromophores, they require adjustment for the LMG chromophore model that contains an explicit electron–electron interaction. Previous 2-RDM-based calculations of acene chains and sheets reveal the presence of strong electron correlation in conjugated systems of sizes similar to the bacteriochlorophyll molecules in the FMO complex. Based on acene-chain data [20,21], we estimate the interaction strength V with $N = 4$ to be 0.8, which gives an 80% probability of finding an electron in the highest occupied orbital (occupied in a mean-field treatment) and a 20% probability of finding an electron in the lowest unoccupied orbital. This estimate is conservative because (i) the acene chains of a similar length typically reveal a nearly biradical filling ($\approx$50% in the highest occupied orbital) and (ii) the presence of the Mg ion with its d orbitals is expected to enhance the degree of correlation. To prevent overcounting of the electron correction's effect on the coupling, we screen the coupling energies $U_{s,t}$ of Ref. [33] by selecting λ in Eq. (4) to be less than unity. Specifically, we set $\lambda = 0.629$ to match experimental data with the LMG model when $N = 4$ and $V = 0.8$. The rate parameters α, β, and γ in the Lindblad operators in Eq. (9) are chosen in atomic units to be 1.52×10^{-4}, 7.26×10^{-5}, and 1.21×10^{-8}, respectively. These definitions are similar to those employed in previous work when $V = 0$ [33].

2. *Enhanced Efficiency*

The exciton populations in chromophores 1, 2, and 3 as well as the sink population are shown as functions of time in femtoseconds (fs) in Fig. 3 for $N = 4$ and $\lambda = 0.629$ with (a) $V = 0.0$ and (b) $V = 0.8$. Population dynamics of the excitation are generated by evolving the Liouville equation in Eq. (8) from an initial density matrix with chromophore 1 in its first excited state and the other chromophores in their ground states. Correlating the four electrons on each chromophore significantly accelerates the increase in the sink population with time. By 1 ps the sink population for $V = 0.0$ is 0.114 while the population for $V = 0.8$ is 0.287. Correlating the excitons on each chromophore also has the effect of

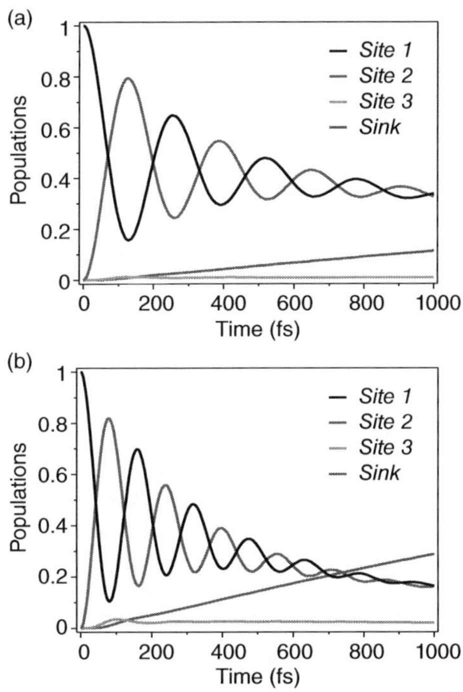

Figure 3. Populations of chromophores 1–3 and sink with (**b**) and without (**a**) electron correlation per site. The exciton populations in chromophores 1, 2, and 3 as well as the sink population are shown as functions of time in femtoseconds (fs) for $N = 4$ with (**a**) $V = 0.0$ and (**b**) $V = 0.8$. Correlating the four electrons on each chromophore significantly accelerates the increase in the sink population with time. Reprinted with permission from Ref. [32]. Copyright 2012 American Institute for Physics. (*See the color version of this figure in the Color Plates section.*)

shortening the periods of oscillation in chromophores 1 and 2 and accelerating the population decay in these chromophores, which is consistent with the change in the sink population.

The sink population as a function of time (fs) is shown in Fig. 4 for a range of V with $N = 4$ and $\lambda = 0.629$. Importantly, as V increases, we observe a dramatic acceleration of the increase of the sink population. For V increasing by the

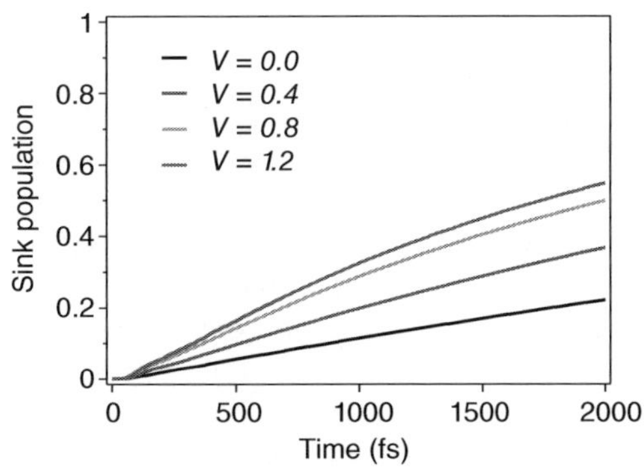

Figure 4. Correlation-enhanced transfer to the reaction center. The reaction center (sink) population as a function of time (fs) is shown in for a range of V with $N = 4$. Correlating the electrons within the chromophores significantly increases the efficiency of energy transfer to the reaction center (sink). Reprinted with permission from Ref. [32]. Copyright 2012 American Institute for Physics. (*See the color version of this figure in the Color Plates section.*)

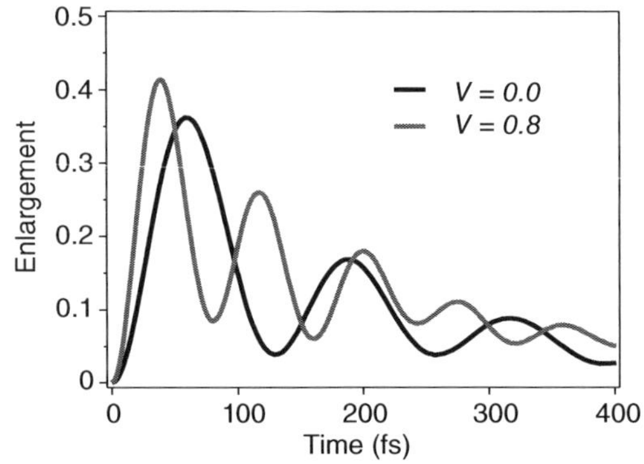

Figure 5. Entanglement of excitons with and without electron correlation. A measure of global entanglement is shown as a function of time (fs) for $N = 4$ with $V = 0.0$ and $V = 0.8$. The correlation of the excitons increases the degree of the entanglement between chromophores at early times and the frequency of its oscillation. Reprinted with permission from Ref. [32]. Copyright 2012 American Institute for Physics.

sequence 0.0, 0.4, 0.8, and 1.2, the sink population at 2 ps increases by the sequence 0.221, 0.367, 0.498, and 0.547. For $N = 4$ correlating the electrons within the chromophores significantly increases the efficiency of energy transfer to the reaction center (sink) by as much as 148%. While we choose $N = 4$, the number N of electrons per chromophore can model electron correlation for any $N > 2$. The precise value of $N > 2$ is unimportant because the effect of changing N can be related to a rescaling of the interaction V.

Figure 5 examines the entanglement of excitons between the LGM-model chromophores for $N = 4$ and $\lambda = 0.629$ with $V = 0.0$ and $V = 0.8$. *Entanglement* is a correlation that cannot occur in a classical system; Bell [34] defined entanglement as "a correlation that is stronger than any classical correlation." Parts of a quantum system become *entangled* when the total density matrix for the system cannot be expressed as a product of the density matrices for the parts [15,16,35]. We employ a measure of global entanglement in which the squared Euclidean distance between the quantum density matrix and its nearest classical density matrix is computed [28,36–38]:

$$\sigma(D) = ||D - C||^2 = \sum_{i,j} |D^i_j - C^i_j|^2 \tag{13}$$

In some cases like the entanglement of the chromophores, the squared Euclidean distance can be viewed as the sum of the squares of the *concurrences* [15], a measure of local entanglement. Within the mathematical framework of Bergmann distances, the squared Euclidean distance can also be related to *quantum relative entropy* [13,16], which is often applied as a global entanglement measure. The squared Euclidean distance $\sigma(D)$ is nonzero if and only if the excitons on different chromophores are entangled. The correlation of electrons increases the degree of the entanglement between chromophores at early times and the frequency of its oscillation. The greater entanglement at early times reflects the opening of additional channels between chromophores for quantum energy transfer.

D. Implications and Significance

The chromophores interact through intermolecular forces, both dipole–dipole and London dispersion forces. The present results imply that nature enhances these intermolecular forces through strong electron correlation in the π-bonded networks of the chromophores to achieve the observed energy-transfer efficiency. The two parameters of the LMG chromophore model provide the simplest approach to studying the effect of strong electron correlation V on the effective coupling between chromophores. The one-electron or dipole models with their coupling energies $U_{s,t}$ can mimic the efficiency from such correlation within chromophores through an empirical inflation of the one-electron U coupling, but they do not provide a mechanism for either isolating or estimating the magnitude of the enhanced coupling due to strong electron correlation. Orbital occupations from recent 2-RDM calculations of correlation in polyaromatic hydrocarbons [20,21] suggest $V = 0.8$ to be a conservative estimate of the correlation within the LMG models of the seven chromophores. Using this estimate with coupling energies screened to match experimental and computational data, we observe a greater than 100% enhancement from the strong electron correlation.

Correlation-enhanced energy transfer can be compared with noise-assisted transfer. Theoretical models [5–9,12,13] have shown that noise from the environment (dephasing) can assist energy transfer in the FMO complex by interfering with the coherence (resonance) between chromophores with similar energies, which facilitates the downhill flow of energy to the lowest-energy, third chromophore, connected to the reaction center. Electron correlation on each chromophore, we have shown, enhances transfer by opening additional coherent channels between chromophores, which also accelerates energy transfer to the third chromophore. Photosynthesis can draw from both of these sources, strong electron correlation within chromophores and environmental noise, to increase the rates of energy transfer. Some of the experimental energy-transfer efficiency attributed to noise in one-electron chromophores models may in fact be due to strong electron correlation.

Briggs and Eisfeld [12] recently examined whether the energy-transfer efficiency from quantum entanglement might be matched by a purely classical process. They conclude that if chromophores are approximated as dipoles, then quantum and classical treatments can achieve similar efficiency. While their result might also be extendable to other dipole or one-electron approximations of the chromophores, many-electron models of the chromophores cannot be represented within classical physics. Neither the electron correlation, present in the LMG model of the chromophores when $V > 0$, nor the associated enhancement of energy-transfer efficiency can be mapped onto an analogous classical process.

III. ROLE OF FUNCTIONAL SUBSYSTEMS

In the second part of this chapter, we show that photosynthetic light harvesting exhibits *quantum redundancy*. By quantum redundancy we mean that photosynthetic antennae complexes have multiple efficient quantum pathways for transferring energy to the reaction center. These pathways involve different subsets of the antennae's chromophores, which form *functional subsystems*. The antennae complex known as the FMO complex is a trimer where each monomer consists of seven bacteriochlorophyll-a chromophores [39]. Previous research has shown that in each monomer initial excitation can occur on either chromophore 1 or chromophore 6 and that this energy must move to chromophore 3, which is most closely coupled with the reaction center [40]. Here, we show that within each monomer of the FMO complex *there exist many subsets of the seven chromophores with efficiencies close to or even better than the efficiency of the entire set of seven chromophores*. Quantum redundancy contributes a robustness to nature's quantum device with potential survival benefits.

The computations presented here reveal that the functional subsystems achieve their efficiencies by a quantum mechanism similar to that of the full system including the roles of entanglement and environmental noise. We assess each subsystem's entanglement by a global entanglement measure [15,35] based on the squared Euclidean distance [28,36–38,41,42]. The study of these chromophore subsystems gives more information about the role of each chromophore in the energy transfer in the whole FMO complex [10,11,13,33,43–50]. This more in-depth understanding can provide insight to other antennae complexes and ultimately be applied to create synthetic solar cells that rival the efficiency of nature.

A. Application of Variable-M Chromophore Model

1. Model Parameters

In Figs. 6–8 and Table I, we study subsystems of the FMO complex, in which certain of the seven chromophores are "turned off" and the energy transfer is tracked within a smaller set of chromophores. For the full system of seven chromophores, we employ the 7×7 Hamiltonian given in Ref. [33]; chromophores are removed from the full system by deleting rows and columns of the Hamiltonian matrix. We denote the subsystem by the numbers of the chromophores retained; for example, the subsystem with chromophores 1, 2, and 3 is denoted as 123. We solve the Liouville equation in Eq. (8) with each of the M chromophores being represented by a one-electron model as in Eq. (1) and the effects of dissipation, dephasing, and transfer to the sink being represented by the Lindblad operator in Eq. (9). At $t = 0$, we initialize the density matrix with a single excitation (exciton) on either site 1 or 6. The rate parameters α, β, and γ in the Lindblad operator in Eq. (9) are chosen

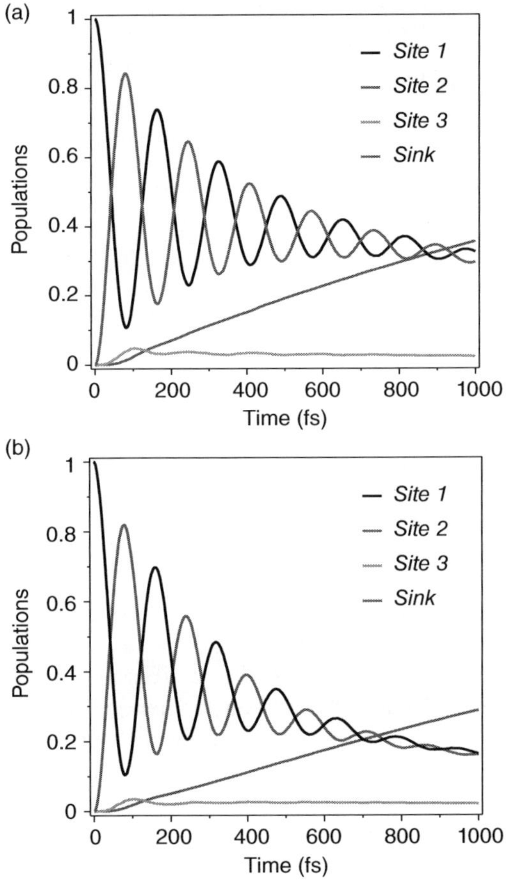

Figure 6. Populations of chromophores 1–3 and the reaction center for (**a**) the subsystem 123 and (**b**) the full FMO system 1234567. Subsystem 123 of the FMO complex transfers energy more efficiently from chromophore 1 to the reaction center (sink) than the full FMO system. Reprinted with permission from Ref. [4]. Copyright 2011 American Chemical Society. (*See the color version of this figure in the Color Plates section.*)

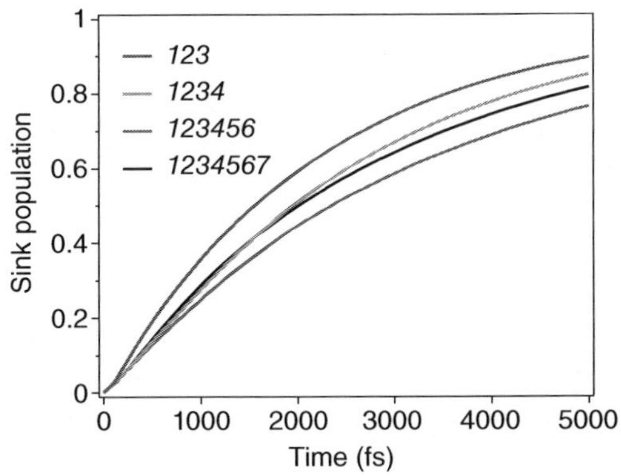

Figure 7. Population in the reaction center for several chromophore subsystems. The subsystems 123, 1234, and 12345 (not shown) are more efficient than the full FMO system at transferring energy from chromophore 1 to the reaction center (sink). Reprinted with permission from Ref. [4]. Copyright 2011 American Chemical Society. (*See the color version of this figure in the Color Plates section.*)

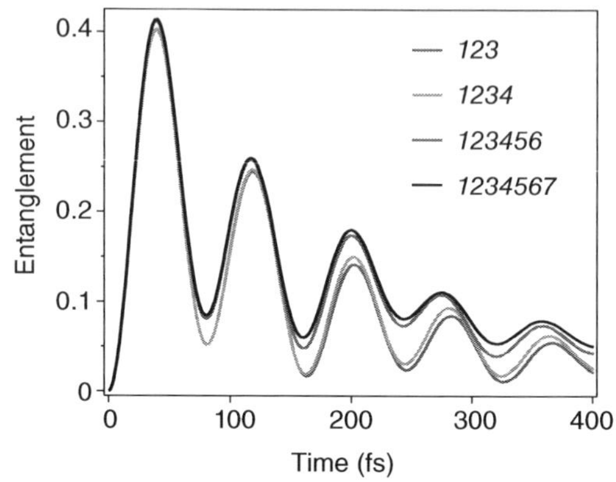

Figure 8. Entanglement of excitons for several chromophore subsystems as a function of time. A measure of global entanglement is shown as a function of time (fs) for several chromophore subsystems, 123, 1234, and 123456, as well as the full system 1234567. Like the full system, the subsystems exhibit entanglement of excitons between the chromophores. Reprinted with permission from Ref. [4]. Copyright 2011 American Chemical Society. (*See the color version of this figure in the Color Plates section.*)

in atomic units to be 1.52×10^{-4} , 7.26×10^{-5}, and 1.21×10^{-8}, respectively. These values are consistent with those employed in Ref. [33].

2. *Subsystem Efficiency*

Figure 6 shows the exciton populations of chromophores 1, 2, 3, and the reaction center for reduced system 123 and the full system. We observe that the two systems are quite similar, and yet they exhibit some important differences. The reduced system features a slightly slower decay of the exciton populations in chromophores 1 and 2, and most significantly, it transfers energy from chromophore 1 to the reaction center more quickly than the full system. Because chromophores 4–7 draw some of the excitation energy in the full system, the populations of chromophores 1–3 and the reaction center are lower in the full system than in the reduced system.

The efficiency of energy transfer from site 1 to the sink is compared in Fig. 7 for several subsystems, 123, 1234, 12345 (not shown because its efficiency is similar to 1234), and 123456, as well as the full system. We observe that subsystems 123, 1234, and 12345 are more efficient than the full system even though efficiency decreases as we add chromophores from 123 to 123456. The addition of chromophores 4, 5, and 6 increases the number of sites for the energy to enter and thereby decreases the population of every site, including the reaction center. The addition of chromophore 7 to 123456, however, improves efficiency; because chromophore 7 has a lower energy than chromophores 5 and 6, it offers a new downhill quantum path through chromophore 4, even lower in energy, to chromophore 3 and the reaction center. The importance of chromophore 7 can also be seen from Table I, which shows that the subsystem containing chromophores 12347 is the most efficient system, again because of the downhill path created from chromophore 7 to the reaction center.

TABLE I
Populations of the Reaction Center of Different Chromophore Subsystems at 2 ps

		Reaction Center Population	
Initial Excitation	System	No Dephasing	Dephasing
Chromophore 1	123	0.387	0.592
	1234	0.161	0.509
	12345	0.144	0.506
	123456	0.120	0.445
	1234567	0.459	0.498
	12347	0.347	0.564
	123467	0.314	0.514
Chromophore 6	346	0.010	0.090
	367	0.009	0.026
	3467	0.048	0.486
	3456	0.045	0.305
	34567	0.103	0.445
	234567	0.086	0.434
	1234567	0.139	0.433
	123467	0.074	0.456

Source: Reprinted with permission from Ref. [4]. Copyright 2011 American Chemical Society.

Table I reports the efficiencies of subsystems with the initial excitation on either chromophore 1 or chromophore 6. Subsystems with both chromophores 1 and 6 are included in both categories. Table I shows that most of these subsystems are more efficient than the full system at transferring energy from either chromophore 1 *or* chromophore 6 to the reaction center. However, only one subsystem 123467 is more efficient than the full system when efficiency is assessed with equal weight given to the excitation pathways from chromophores 1 *and* 6 to the reaction center. Although this model suggests that chromophore 5 is superfluous, it may have other necessary duties in nature that are not considered in the present model. The superfluity of chromophore 5 is predictable from the Hamiltonian of the system, as the site energy for chromophore 5 is much higher than the other chromophores [40]. However, it is worth noting that this is only the case in the FMO complex for *Prosthecochloris aestuarii*, which we are studying in this chapter, while in the FMO complex for *Chlorobium tepidum*, chromophore 5 is not the highest in energy and its exclusion would most likely make the system less efficient. Very recent crystallographic and quantum chemistry studies [51,52] indicate that an eighth chromophore is likely in the FMO of *P. aestuarii*, which due to sample preparation is not present in the ultrafast spectroscopic studies [1–3]. An eighth chromophore would further support the existence and importance of functional subsystems with the seven chromophores in recent ultrafast experiments serving as a functional subsystem of the full system in nature.

3. Subsystem Dephasing and Entanglement

The subsystems, we observe, function by a quantum mechanism similar to that observed in the full system. Two features of this mechanism, which have received significant attention for the full system in the literature [10,11,13,33,43–50], are (i) dephasing from environmental noise and (ii) entanglement of excitons. As in the full system, the dephasing assists transport of the excitons from chromophore 1 or 6 to the reaction center by facilitating the distribution of the exciton population from its initial chromophore. The efficiencies of different subsystems with and without dephasing are reported in Table I. The perceived enhancement is a characteristic of systems of heterogeneous chromophores, that is chromophores with different site energies. We also observe in Fig. 7, displaying the entanglement in several subsystems as well as the full system, that the entanglement of excitons exhibits similar behavior in the reduced systems as in the full system. We compute the entanglement from Eq. (13) to demonstrate that the transfer of energy for both reduced and full systems *within the model* occurs by the same quantum mechanism. The addition of each chromophore slightly increases the total entanglement of the system, as measured by the squared Euclidean distance, by adding extra off-diagonal elements to the density matrix. Each system shows that the entanglement lasts for a few hundred femtoseconds, which matches previous experimental data as well as predictions from other theoretical models [3,53].

IV. CONCLUDING REMARKS

Many theoretical models [5–9,12,13] have been designed to explore the energy transfer in recent light-harvesting experiments, but most of them treat each chromophore by a single electron with two possible energy states. In reality, however, the chromophores are assembled from chlorophyll molecules that contain an extensive network of conjugated carbon–carbon bonds surrounding magnesium ions, from which significant strong electron correlation, including polyradical character, has been shown to emerge [17,20]. In the first part of this chapter, we have examined the effect of strong electron correlation and entanglement *within* chromophores through an extension of single-electron models of the chromophores to N-electron models, based on the LMG model [16,22,25,28]. We find that increasing the degree of electron correlation of each LMG-model chromophore significantly enhances the efficiency with which energy is transferred to the reaction center (sink). This result, showing that nature likely uses strong electron correlation to achieve its energy-transfer efficiency, also has implications for the design of more energy- and information-efficient materials.

In the second part of this chapter, we have identified multiple functional subsystems of the FMO light-harvesting complex whose efficiencies are comparable to the efficiency of the full FMO complex. Many of these subsystems are, in fact, more efficient in transferring energy from either chromophore 1 or 6 to the reaction center, and one of these subsystems is even more efficient in transferring energy from both chromophores 1 and 6 to the reaction center. These functional subsystems provide evidence for quantum redundancy in photosynthetic light harvesting. There exists multiple quantum pathways or circuits that are capable of transferring energy efficiently to the reaction center. For example, subsystem 123 transfers energy from chromophore 1 to the reaction center with 59.2% efficiency (after 2 ps), and subsystem 3467 transfers energy from chromophore 6 to the reaction center with 48.6% efficiency. This built-in redundancy likely provides benefits to the natural system because damage to any chromophore save 3 will not disrupt light harvesting. The characterization of functional subsystems within the FMO complex offers a detailed map of the energy flow within the FMO complex with potential applications to the design of more efficient photovoltaic devices.

Acknowledgments

DAM expresses his appreciation to Dudley Herschbach, Herschel Rabitz, and Alexander Mazziotti for their support and encouragement. DAM also thanks Sabre Kais for the invitation to contribute this chapter. The authors gratefully recognizes support from the Army Research Office Grant No. W91 INF-1 1-1-0085, the National Science Foundation, the Henry-Camille Dreyfus Foundation, and the David-Lucile Packard Foundation.

REFERENCES

1. G. S. Engel, T. R. Calhoun, E. L. Read, T. K. Ahn, T. Mancal, Y. C. Cheng, R. E. Blankenship, and G. R. Fleming, *Nature* **446**, 782–786 (2007).
2. E. Collini, C. Y. Wong, K. E. Wilk, P. M. G. Curmi, P. Brumer, and G. D. Scholes, *Nature* **463**, 644 (2010).
3. G. Panitchayangkoon, D. Hayes, K. A. Fransted, J. R. Caram, E. Harel, J. Z. Wen, R. E. Blankenship, and G. S. Engel, *Proc. Natl. Acad. Sci. USA* **107**, 12766–12770 (2010).
4. N. Skochdopole and D. A. Mazziotti, *J. Phys. Chem. Lett.* **2**, 2989 (2011).
5. J. S. Cao and R. J. Silbey, *J. Phys. Chem. A* **113**, 13825–13838 (2009).
6. F. Caruso, A. W. Chin, A. Datta, S. F. Huelga, and M. B. Plenio, *Phys. Rev. A* **81**, 062346 (2010).
7. P. Huo and D. F. Coker, *J. Chem. Phys.* **133**, 184108 (2010).
8. D. P. S. McCutcheon and A. Nazir, *Phys. Rev. B* **83**, 165101 (2011).
9. B. Palmieri, D. Abramavicius, and S. Mukamel, *J. Chem. Phys.* **130**, 204512 (2009).
10. J. Zhu, S. Kais, P. Rebentrost, and A. Aspuru-Guzik, *J. Phys. Chem. B* **115**, 1531–1537 (2011).
11. K. Bradler, M. M. Wilde, S. Vinjanampathy, and D. B. Uskov, *Phys. Rev. A* **82**, 062310 (2010).
12. J. S. Briggs and A. Eisfeld, *Phys. Rev. E* **83**, 051911 (2011).
13. M. Sarovar, A. Ishizaki, G. R. Fleming, and K. B. Whaley, *Nat. Phys.* **6**, 462–467 (2010).
14. C. Brif, R. Chakrabarti, and H. Rabitz, *New J. Phys.* **12**, 075008 (2010).
15. S. Kais, *Adv. Chem. Phys.* **134**, 493 (2007).

16. R. Horodecki, P. Horodecki, M. Horodecki, and K. Horodecki, *Rev. Mod. Phys.* **81**(2), 865–942 (2009).
17. J. Hachmann, J. J. Dorando, M. Avilés, and G. K. Chan, *J. Chem. Phys.* **127** (2007).
18. D. A. Mazziotti, *Phys. Rev. Lett.* **106**, 083001 (2011).
19. R. M. Erdahl, *Adv. Chem. Phys.* **134**, 61–91 (2007).
20. G. Gidofalvi and D. A. Mazziotti, *J. Chem. Phys.* **129**, 134108 (2008).
21. K. Pelzer, L. Greenman, G. Gidofalvi, and D. A. Mazziotti, *J. Phys. Chem. A* **115**, 5632–5640 (2011).
22. H. J. Lipkin, N. Meshkov, and A. J. Glick, *Nucl. Phys.* **62**, 188 (1965).
23. J. Arponen and J. Rantakivi, *Nucl. Phys. A* **407**, 141 (1983).
24. R. Perez, M. C. Cambiaggio, and J. P. Vary, *Phys. Rev. C* **37**, 2194 (1988).
25. D. A. Mazziotti, *Phys. Rev. A* **57**, 4219–4234 (1998).
26. D. A. Mazziotti, *Chem. Phys. Lett.* **289**, 419 (1998).
27. J. Stein, *J. Phys. G* **26**, 377 (2000).
28. D. A. Mazziotti and D. R. Herschbach, *Phys. Rev. A* **62**, 043603 (2000).
29. K. Yasuda, *Phys. Rev. A* **65**, 052121 (2002).
30. D. A. Mazziotti, *Phys. Rev. A* **69**, 012507 (2004).
31. G. Gidofalvi and D. A. Mazziotti, *Phys. Rev. A* **74**, 012501 (2006).
32. D. A. Mazziotti, *J. Chem. Phys.* **137**, 074117 (2012).
33. A. W. Chin, A. Datta, F. Caruso, S. F. Huelga, and M. B. Plenio, *New J. Phys.* **12**, 065002 (2010).
34. J. S. Bell, *Physics* **1**, 195–200 (1964).
35. Z. Huang and S. Kais, *Chem. Phys. Lett.* **413**, 1–5 (2005).
36. A. O. Pittenger and M. H. Rubin, *Lin. Alg. Appl.* **346**, 47–71 (2002).
37. R. A. Bertlmann, H. Narnhofer, and W. Thirring, *Phys. Rev. A* **66**, 032319 (2002).
38. J. E. Harriman, *Phys. Rev. A* **17**, 1249–1256 (1978).
39. R. E. Fenna and B. W. Matthews, *Nature* **258**, 573–577 (1975).
40. J. Adolphs and T. Renger, *Biophys. J.* **91**, 2778–2797 (2006).
41. L. Greenman and D. A. Mazziotti, *J. Chem. Phys.* **133**, 164110 (2010).
42. T. Juhasz and D. A. Mazziotti, *J. Chem. Phys.* **125**, 174105 (2006).
43. P. F. Huo and D. F. Coker, *J. Phys. Chem. Lett.* **2**, 825–833 (2011).
44. A. Kelly and Y. M. Rhee, *J. Phys. Chem. Lett.* **2**, 808–812 (2011).
45. T. Mancal, V. Balevicius, and L. Valkunas, *J. Phys. Chem. A* **115**, 3845–3858 (2011).
46. M. Sener, J. Strumpfer, J. Hsin, D. Chandler, S. Scheuring, C. N. Hunter, and K. Schulten, *Chem. Phys. Chem.* **12**, 518–531 (2011).
47. X. T. Liang, *Phys. Rev. E* **82**, 051918 (2010).
48. P. Nalbach and M. Thorwart, in *Semiconductors and Semimetals*, Vol. 83, Academic Press, San Diego, 2010, pp. 39–75.
49. V. I. Novoderezhkin and R. van Grondelle, *Phys. Chem. Chem. Phys.* **12**, 7352–7365 (2010).
50. J. L. Wu, F. Liu, Y. Shen, J. S. Cao, and R. J. Silbey, *New J. Phys.* **12**, 105012 (2010).
51. M. Schmidt am Busch, F. Müh, M. E. Madjet, and T. Renger, *J. Phys. Chem. Lett.* **2**, 93 (2011).
52. D. E. Tronrud, J. Z. Wen., L. Gay, and R. E. Blankenship, *Photosynth. Res.* **100**, 79–87 (2009).
53. A. Ishizaki and G. R. Fleming, *Proc. Natl. Acad. Sci.* **106**, 17255–17260 (2009).

VIBRATIONAL ENERGY TRANSFER THROUGH MOLECULAR CHAINS: AN APPROACH TOWARD SCALABLE INFORMATION PROCESSING

C. GOLLUB, P. VON DEN HOFF, M. KOWALEWSKI, U. TROPPMANN, and R. DE VIVIE-RIEDLE

Department Chemie, Ludwig-Maximilians-Universität, Butenandt-Str. 11, 81377 München, Germany

I. INTRODUCTION

Quantum information processing is a rapidly developing field and has entered different areas in physics and chemistry. The first principal ideas came from the quantum optics community and considerable success was reported with cavity

Advances in Chemical Physics, Volume 154: Quantum Information and Computation for Chemistry, First Edition. Edited by Sabre Kais.

quantum electrodynamics [1], trapped ions [2,3], and nuclear magnetic resonance (NMR) [4,5]. Here, the quantum systems representing the qubits are photons, atoms, and the nuclear spin in molecules. In these approaches, the quantum information is encoded in well-isolated, identical qubits with equal transition frequencies. The great achievements in laser control during the past two decades have opened the connection between femtosecond lasers and quantum information processing. Modulated light fields in the femtosecond regime are able to precisely control the internal molecular degrees of motions such as vibration, rotations, or electronic transitions [6,7]. The step from molecular reaction control to the implementation of molecular quantum gates was in this sense quite natural and opened a new direction in the field of quantum information processing using internal molecular degrees of freedom as qubits [8–10]. This new approach has been followed by numerous studies working with internal motional states of molecules such as rovibrational states [11] and vibrational states in diatomic [12–14] and polyatomic systems [8–10,15,16].

In all proposals for quantum computing, basic gate operations could be demonstrated. Quantum algorithms and quantum circuits [17] were discussed and partly realized in the experiment [4,11,18]. Some experiments were even able to prove that quantum computing has become reality with an impressive number of qubits [2–5,18–21]. Nevertheless, large-scale quantum computation is still a vision that requires ongoing research. In the end, only a large-scale quantum computer will be able to efficiently solve some of the most difficult problems in computational science, such as integer factorization, quantum simulation, and modeling, intractable on any present or conceivable future classical computer. Two pathways for scaling the number of qubits for the construction of quantum registers can be followed: enlargement of a single quantum register, or a network with nodes consisting of smaller units.

For the concept of molecular quantum computing with vibrational qubits, the first possibility is to use more and more normal modes of a molecule to define the qubit modes and to set up a multiqubit system for a single polyatomic molecule. Quantum gates for up to six qubit modes have been demonstrated theoretically [22]. Phase encoding could be demonstrated experimentally for an even larger number of qubits [11,23] and can be regarded as a first step to precompiled quantum computing. Molecular quantum computing with vibrational modes prefers low-lying modes of polyatomic molecules [8,10] to avoid decoherence as far as possible. Because not all normal modes are strongly IR (or Raman) active and resolvable at the same time, the size of the qubit system is generally limited. A different idea we present here follows the network-type ansatz and is based on the connection of individual qubit systems through molecular chains. The study focuses on the key step, the laser-driven vibrational energy transfer across the bridging molecules from one qubit site to the other. In principle, such a building block can be augmented by linkage to a second molecuar chain a third qubit and so on. The process can be regarded as information transfer in the context of molecular quantum information processing with vibrational qubits. The idea is related to one-dimensional spin chains acting as quantum channels [24,25]. The generation of entanglement and the transport of quantum information has already been investigated in such systems [26,27].

In general, population transfer between vibrational eigenstates is an important process for several phenomena in chemistry and physics. Intramolecular vibrational redistribution (IVR) has been studied in many different systems, such as protein molecules [28–30], charge transport in molecular wires [31–33], transport of excitons in macromolecules [34,35], transport of heat and energy in chain molecules [36–39], or other molecular systems [40–43]. A comprehensive review on energy transfer dynamics can be found in Ref. [44].

After a local vibrational excitation of molecules, IVR processes can be studied with time-resolved pump-probe techniques. Particularly, multidimensional IR spectroscopy [30,45–50] facilitates an insight into the vibrational dynamics. Besides the time resolution, also technical advances have been reached in the spatial resolution of spectroscopic techniques (e.g., optical nearfield control in nanostructures) [51,52]. On the basis of all these new developments, the precise detection and control of vibrational energy transfer processes in molecules will become possible in future.

In the present study, the aim is to implement a laser-driven vibrational state transfer from one qubit site to another across the molecular chain states. A model system is set up based on the linear octatetrayne molecule, which is described by ab initio methods. Two qubit systems are coupled to the chain, and laser fields are calculated with optimal control theory (OCT), driving an efficient state transfer. In addition, vibrational relaxation to mimic possible IVR and dephasing processes is incorporated in the study.

II. FUNDAMENTALS OF QUANTUM DYNAMICS AND COHERENT CONTROL

The control prospects of vibrational quantum processes, such as state-to-state transitions, unitary transformations in individual moleculesare combined with the concept of long-range vibrational energy transfer through molecular chains (Section III.B). This first strategic idea for a quantum register relies exclusively on internal degrees of motions. The driving forces are external, electric laser fields controlling the vibrational quantum dynamics.

These quantum processes can be studied by solving the time-dependent Schrödinger equation, where usually numerical propagation schemes are applied. The calculations can be performed, based either on a grid [53] or on an eigenstate representation. For the latter case, the eigenfunctions and eigenvalues have to be known. Both approaches differ in the choice of the basis functions and can easily be transferred into each other. Generally, a quantum system cannot be regarded as completely isolated and different intra- and intermolecular effects may play a role in the dynamics. For the investigation of induced dissipative effects, the wave function description is transferred to the density matrix representation, where effects such as energy relaxation can be incorporated in the quantum dynamical calculation.

In the following, the basic quantum dynamical tools to solve the time-dependent Schrödinger equation in the eigenstate basis will be summarized. The description of dissipative effects in the density matrix formalism and the fundamentals of OCT are briefly reviewed.

A. Wavepacket Dynamics

The quantum dynamics of a molecular system with the Hamiltonian $\hat{H}_{\text{mol}}$ is governed by the time-dependent Schrödinger equation

$$i\frac{\partial}{\partial t}\Psi_{\text{mol}}(t) = \hat{H}_{\text{mol}}\Psi_{\text{mol}}(t) \tag{1}$$

(all equations are given in atomic units [au], i.e., $\hbar$=1 and m_e=1). The stationary, molecular wave function Ψ_{mol} depends on the nuclear coordinates R and on the electronic coordinates r. It can be separated according to

$$\Psi_{\text{mol}}(R, r) = \Psi_{\text{nuc}}(R)\Psi_{\text{el}}(r; R) \tag{2}$$

into a nuclear wave function $\Psi_{\text{nuc}}(R)$, depending on the nuclei coordinates only, and an electronic part $\Psi_{\text{el}}(r; R)$, with a parametric dependence on the nuclear arrangement. As the quantum dynamical calculations are performed within the Born–Oppenheimer approximation, the electronic Schrödinger equation is solved with standard quantum chemical program packages to obtain the potential energy curves. A comprehensive review on quantum chemistry can be found in Ref. [54]. A different approach is pursued in case of density functional theory (DFT) [55], where the many-body electronic wave function is replaced by the electronic density as the basic quantity.

For the present studies, DFT is used to calculate the potential energy surfaces $E(R)$ and the corresponding molecular properties.

The intramolecular motion of the nuclei is described by the quantum dynamics of the nuclear wave function $\Psi_{\text{nuc}}(R)$ on the calculated potential energy surfaces $E(R) \equiv \hat{V}_{\text{nuc}}$. From now, the subscript $_{\text{nuc}}$ is omitted for the nuclear wave function Ψ_{nuc}, and it will be denoted by Ψ. In case of the Hamiltonian, the label $_0$ indicates that it is time-independent. The time evolution of Ψ is governed by the time-dependent, nuclear Schrödinger equation:

$$i\frac{\partial}{\partial t}\Psi(t) = \hat{H}_0\Psi(t) \tag{3}$$

The time-independent Hamiltonian $\hat{H}_0$ includes the kinetic energy $\hat{T}_{\rm nuc}$ of the nuclei and the potential energy $\hat{V}_{\rm nuc}$. Integrating the time-dependent, nuclear Schrödinger equation [Eq. (3)] determines the equations of motion as the action of a propagator on the nuclear wave function

$$\Psi(t) = \hat{U}(t_j, t_i)\,\Psi(t_i) = e^{-i\hat{H}_0\,(t_j - t_i)}\Psi(t_i) \tag{4}$$

where the propagator of the time-independent Hamiltonian $\hat{H}_0$ is $\hat{U}(t_j, t_i)$. The nuclear wave function evolves in the time interval $\Delta t = t_j - t_i$ according to

$$\Psi(t_j) = e^{-i\hat{H}_0\Delta t}\Psi(t_i) \tag{5}$$

An additional, time-dependent perturbation of the Hamiltonian like an external, electric field $\varepsilon(t)$, interacting with the molecular system can be included as

$$\hat{H}(t) = \hat{H}_0 - \hat{\mu}\varepsilon(t) \tag{6}$$

The interaction is mediated by the molecular dipole moment μ. Now the Hamiltonian $\hat{H}(t)$ is time-dependent, and the propagation has to be performed in sufficiently small time steps, so the perturbation can be regarded as constant during the time interval Δt. The corresponding propagation equation is given by

$$\Psi(t_j) = e^{-i(\hat{H}_0 - \hat{\mu}\varepsilon(t_i))\Delta t}\,\Psi(t_i) \tag{7}$$

1. Propagation in the Eigenstate Basis

To describe the molecular chain dynamics the wave function is formulated in the eigenstate basis. The basis vectors correspond to the vibrational eigenfunctions Ψ_n. The wave function is represented by a n-dimensional vector $\mathbf{c}(t)$ with the complex, time-dependent elements $c_n(t)$, derived from the projection

$$c_n(t) = \langle \Psi_n | \Psi(t) \rangle \tag{8}$$

The corresponding matrix representation $\mathbf{H}_0$ of the time-independent Hamiltonian $\hat{H}_0$ is diagonal, where the matrix elements $H_{0\,(nn)}$ are given by the eigenenergies E_n.

$$H_{0\,(nn)} = \langle \Psi_n(R)|\hat{H}_0(R)|\Psi_n(R)\rangle = E_n, \quad H_{0\,(nm)} = 0 \tag{9}$$

For the interaction with the external field, the dipole matrix elements μ_{nm} of the matrix $\boldsymbol{\mu}$ are evaluated according to

$$\mu_{nm} = \langle \Psi_n(R)|\hat{\mu}(R)|\Psi_m(R)\rangle \tag{10}$$

The temporal evolution of the wave function in the eigenstate representation is formally determined by

$$\mathbf{c}(t_j) = e^{-i(\mathbf{H}_0 - \boldsymbol{\mu}\varepsilon(t_i))\Delta t}\mathbf{c}(t_i) = e^{-i(\mathbf{H}\Delta t)}\mathbf{c}(t_i) = \mathbf{U}(t_j, t_i)\mathbf{c}(t_i) \tag{11}$$

Different numerical approaches can be used to evaluate the propagation steps, given by the exponential equation [Eq. (11)]. Detailed reviews on quantum dynamical methods are presented in [56,57].

The numerical evaluation of the term $e^{-i\mathbf{H}\Delta t}\,\Psi$ can be performed efficiently with different techniques. Here, the Chebychev polynomial expansion is applied and summarized briefly. The Chebychev propagation scheme [58] is based on polynomial expansions of the time evolution operator. Here, the propagator $\mathbf{U}(t_j, t_i)$ is approached by a Chebychev series, taking the form

$$e^{-i\mathbf{H}t} \equiv \sum_{n=0}^{N} a_n(t)\Phi_n(-i\mathbf{H}) \tag{12}$$

where

$$a_n(t) = 2J_n(t) \text{ and } a_0(t) = J_0(t) \tag{13}$$

Φ_n are complex Chebychev polynomials, depending on the Hamiltonian and obeying the recursion relation

$$\Phi_{n+1} = -2i\mathbf{H}\Phi_n + \Phi_{n-1} \tag{14}$$

The time-dependent expansion coefficients $a_n(t)$ are determined by Bessel functions. For the implementation, the argument of Φ_n has to be mapped onto the interval $[-i, i]$. The eigenvalues of $\mathbf{H}$ are consequently shifted and scaled to the range $[-1, 1]$. The propagation is then performed with the normalized Hamiltonian and a shift parameter is introduced, compensating for the normalization. The order of expansion N has to be chosen large enough to ensure the convergence of the series [58].

2. *Calculation of Eigenstates*

The vibrational eigenfunctions Ψ_n have to be evaluated explicitly to set up the quantum dynamical calculations in the corresponding eigenstate representation. They can be determined by solving the stationary, vibrational Schrödinger equation:

$$\hat{H}_0(R)\Psi_n(R) = E_n\Psi_n(R) \tag{15}$$

A relaxation method [59,60] is applied, where an incident, vibrational wave packet $\Psi(t)$ is propagated in negative imaginary time.

$$\tilde{\Psi}_n(R) \equiv e^{-i\hat{H}_0(R)(-i\Delta t)}\Psi(R,t) = e^{-\hat{H}_0(R)\Delta t}\Psi(R,t) \tag{16}$$

The propagation methods just presented can be used, but here the wave functions and operators are constructed in the grid basis. The components of the wave packet with the highest energies are attenuated faster during the propagation period, and the wave packet is basically relaxed to the vibrational ground state in each time step. This procedure is applied several times, for each eigenfunction and for each time step the resulting approximated vibrational eigenfunction $\tilde{\Psi}_n$ is projected out. Afterwards, the Hamiltonian is set up in the basis of the approximated eigenfunctions and diagonalized to obtain very precise solutions of the stationary, nuclear Schrödinger equation [Eq. (15)].

B. Dissipative Dynamics

Often, a molecular system cannot be regarded as completely isolated such that environmental effects may play a role for the processes under investigation. Molecular collision or intramolecular vibrational redistribution can occur, changing the population (relaxation) or stochastically perturbing the phase (dephasing) of individual quantum states. The simulation of these effects can be incorporated in the quantum dynamical studies, using density matrix theory [61].

1. Density Matrix

The molecular system investigated here corresponds to a set of selected vibrational normal modes. They can be regarded as open quantum systems in the density matrix formalism and can interchange energy with the environment, where the environmental effects are described as distortions. The density matrix is a statistical operator, defined as

$$\hat{\rho} = |\Psi\rangle\langle\Psi| \tag{17}$$

with the vibrational state vector $|\Psi\rangle$. Expressing the wave function in the basis of vibrational eigenfunctions: $|\Psi\rangle = \sum_n a_n|\Psi_n\rangle$ and $\langle\Psi| = \sum_m a_m^*\langle\Psi_m|$ (in general, the coefficients are time-dependent) leads to the matrix representation $\boldsymbol{\rho}$ of the operator $\hat{\rho}$:

$$\boldsymbol{\rho} = \sum_{nm} a_n a_m^* |\Psi_n\rangle\langle\Psi_m| \tag{18}$$

The respective matrix elements of the density operator are given by

$$\rho_{nm} = \langle \Psi_n | \boldsymbol{\rho} | \Psi_m \rangle = a_n a_m^* \tag{19}$$

The diagonal elements $\rho_{nn} = |a_n|^2$ are equal to the probability that the system is in the state $|\Psi_n\rangle$ and the off-diagonal elements ρ_{nm} $(n \neq m)$ represent the coherences of the system. Due to the orthonormality of the basis functions, the trace of the density matrix is $\mathrm{Tr}(\boldsymbol{\rho}) = 1$.

2. *Liouville–von Neumann Equation*

The temporal evolution of a quantum system, represented by a density matrix, is governed by the quantum Liouville equation:

$$i\frac{\partial \boldsymbol{\rho}(t)}{\partial t} = [\mathbf{H}, \boldsymbol{\rho}(t)] \tag{20}$$

Equation (20) is known as the nondissipative Liouville–von Neumann equation, which can also be rewritten as

$$\dot{\boldsymbol{\rho}}(t) = \mathcal{L}_{\text{sys}} \boldsymbol{\rho}(t) = -i[\boldsymbol{H}, \boldsymbol{\rho}(t)] \tag{21}$$

where the Hamiltonian is expressed by the Liouvillian superoperator $\mathcal{L}_{\text{sys}}$. When the quantum system is interacting with the environment, the dynamics can be described by the dissipative Liouville–von Neumann equation under the Markov approximation (neglecting memory effects). The bath modes (environment) are not treated explicitly, but their influence on the quantum system is described according to

$$\dot{\boldsymbol{\rho}}(t) = \mathcal{L}\boldsymbol{\rho}(t) = (\mathcal{L}_{\text{sys}} + \mathcal{L}_D)\boldsymbol{\rho}(t) = -i[\mathbf{H}, \boldsymbol{\rho}(t)] + \mathcal{L}_D(\boldsymbol{\rho}(t)) \tag{22}$$

The Liouvillian superoperator $\mathcal{L}$ consists of a system part $\mathcal{L}_{\text{sys}}$ and a dissipative part $\mathcal{L}_D$. The dissipative correction $\mathcal{L}_D$ is a function of the density matrix. Here, the Lindblad approach [62–64] to the Markovian description of open quantum systems is applied with the mathematical form

$$\mathcal{L}_D(\boldsymbol{\rho}(t)) = \sum \left\{ \mathbf{C}_i \boldsymbol{\rho} \mathbf{C}_i^\dagger - \frac{1}{2}\left[\mathbf{C}_i^\dagger \mathbf{C}_i, \boldsymbol{\rho}\right]_+ \right\} \tag{23}$$

and the Lindblad operators $\mathbf{C}_i$. They correspond to raising and lowering operators of the ith two-level system $|a\rangle$, $|b\rangle$, which is set up for every relaxation channel

$$\hat{C}_i = \sqrt{\Gamma_{ab}}|a\rangle\langle b| = \sqrt{\Gamma_{ab}}\begin{pmatrix} 0 & 1 \\ 0 & 0 \end{pmatrix}, \quad \hat{C}_i^\dagger = \sqrt{\Gamma_{ba}}|b\rangle\langle a| = \sqrt{\Gamma_{ba}}\begin{pmatrix} 0 & 0 \\ 1 & 0 \end{pmatrix} \tag{24}$$

The energy relaxation rates are defined as $\Gamma_{ab} = \frac{1}{T_1}$. Additionally, pure dephasing effects can be taken into account with Lindblad operators of the type

$$\mathbf{C}_i = \sqrt{\gamma^*_{ab}}(|b\rangle\langle b| - |a\rangle\langle a|) = \sqrt{\gamma^*_{ab}}\begin{pmatrix} -1 & 0 \\ 0 & 1 \end{pmatrix} \tag{25}$$

In this case, γ^*_{ab} is associated with the pure dephasing time scale T_2^*: $\gamma^*_{ab} = \frac{1}{T_2^*}$. The total dephasing rate is given by $\frac{1}{T_2} = \frac{1}{T_2^*} + \frac{1}{2T_1}$. Inserting the Lindblad approach for the dissipative part of Eq. (23) into the Liouville–von Neumann equation [Eq. (22)] leads to the following equations of motion for the diagonal and off-diagonal elements of the density matrix:

$$\frac{\partial \rho_{nn}}{\partial t} = \sum_p -i[V_{np}(t)\rho_{pn} - \rho_{np}V_{pn}(t)] + \sum_p (\Gamma_{pn}\rho_{pp} - \Gamma_{np}\rho_{nn}) \tag{26}$$

$$\frac{\partial \rho_{mn}}{\partial t} = -i[(E_m - E_n)\rho_{mn} + \sum_p (V_{mp}(t)\rho_{pn} - \rho_{mp}V_{pn}(t))] - \gamma^*_{mn}\rho_{mn} \tag{27}$$

The matrix $\mathbf{V}$ is the laser-molecule interaction potential, defined as $\mathbf{V} = -\boldsymbol{\mu}\varepsilon(t)$ and E_i are the vibrational eigenenergies (i.e. the diagonal matrix elements of the molecular Hamiltonian matrix $\mathbf{H}_0$).

3. *Density Matrix Propagation*

The Liouville–von Neumann equation [Eq. (22)] is numerically solved using the Faber propagator [65–67]. This propagator scheme uses a polynomial expansion and is related to the Chebychev propagation scheme. The Faber polynomials can be used to approximate functions of variables, defined in the complex plane. The formal solution of Equation (22) is given by

$$\boldsymbol{\rho}(t) = e^{\mathcal{L}(t-t_0)}\boldsymbol{\rho}(t_0) \tag{28}$$

with the initial density matrix $\boldsymbol{\rho}(t_0)$ at the time $t = t_0$, where again the matrix representation is used for the quantum dynamical calculations. The Faber polynomial method is applied to approximate the exponential of the matrix $\mathcal{L}$:

$$\boldsymbol{\rho}(t) \equiv \sum_{k=0}^{n} b_k(t)F_k(\mathcal{L})\boldsymbol{\rho}(t_0) \tag{29}$$

The structure is equivalent to the corresponding Chebychev approximation [Eq. (12)], the time-dependent coefficients $b_k(t)$ depend on Bessel functions and the domain of the complex eigenvalues of the Liouvillian is determined by the strength of the dissipative versus the system part. The Faber polynomials can be constructed according to a recursion relation, similar to the Chebychev polynomials. In contrast to the Chebychev series, the Faber polynomials provide a more efficient scaling scheme for $\mathcal{L}$ in the complex plane. Detailed discussions on the Faber propagator can be found in Refs [65,66].

C. Optimal Control Theory

Quantum systems and quantum processes under the influence of light pulses have been studied for a long time. The first goal was to understand the observed system dynamics in real time [68]. Beyond mere observation, the central question is the controllability and efficient manipulation of quantum systems. Here, the aim is to steer quantum processes in a desired way, which is commonly referred to as quantum control [69–71]. The experiments are based on ultrashort laser pulses modulated in amplitude and phase.

The corresponding theoretical technique to quantum control experiments is OCT. In the context of this the objective is long-range vibrational energy transfer. In OCT an optimality criterion has to be achieved and the method finds an appropriate control law for it, the optimal laser field. Different OCT concepts for quantum control investigations were developed, predominantly in the groups of Rabitz [72,73], Tannor and Rice [74,75] based on the calculus of variations. In general, the following OCT functional in Eq. (30) has to be maximized.

$$J\left(\psi_i(t), \lambda(t), \varepsilon(t)\right) = F(\tau) - \int_0^T \alpha(t) \left|\varepsilon(t)\right|^2 dt - \int_0^T \lambda(t)\, G\left(\psi_i(t), \varepsilon(t)\right) dt \tag{30}$$

It includes three terms, the optimization aim $F(\tau)$, an integral over the laser field, penalizing the pulse fluence, and an ancillary constraint.

The optimization aim $F(\tau)$ is to transfer the initial wave function ψ_i into a final state ϕ_f after the laser excitation time T. The objective can be formulated as the square of the scalar product of the initial state, propagated in time with the target state:

$$F(\tau) = \left|\langle \psi_i(T) | \phi_f \rangle\right|^2 \tag{31}$$

Initial and target states can be chosen as eigenstates or arbitrary superpositions of eigenstates.

For the implementation of global quantum gates, it is necessary to perform several qubit basis transitions with the same laser pulse. In this sense, global means that independently of the initial qubit state the correct transition has to be performed forward and backward with the same pulse. For these calculations, the definition of the control aim is extended and for a N-dimensional qubit basis it takes the form

$$F(\tau) = \sum_{k=1}^{N} |\langle \psi_{ik}(T)|\phi_{fk}\rangle|^2 \tag{32}$$

A different formulation of the optimization aim [76–78] additionally facilitates the correct phase relation between the single transitions.

The second term of Eq. (30) is an integral over the laser field $\varepsilon(t)$ with a time-dependent factor $\alpha(t)$. In principle, high values of α assure low field intensities and complexities. Depending on the implementation, it is known as the penalty factor or Krotov change parameter. With the choice of $\alpha(t) = \alpha_0/s(t)$ and a sinusoidal shape function $s(t)$, for example, an envelope function can be impressed on the laser field [60,79]. This guarantees smooth switching on and off behavior of the pulse, instead of abrupt field intensity changes for the times $t = 0, T$.

The last term of the functional [Eq. (30)] comprises the time-dependent Schrödinger equation as an ancillary constraint, denoted by $G(\psi_i(t), \varepsilon(t))$, with the Lagrange multiplier $\lambda(t)$:

$$\int_0^T \lambda(t) G(\psi_i(t), \varepsilon(t))\, dt = 2\Re \left[C \int_0^T \left\langle \lambda(t) \middle| i \left[\hat{H}_0 - \hat{\mu}\varepsilon(t) \right] + \frac{\partial}{\partial t} \middle| \psi_i(t) \right\rangle dt \right] \tag{33}$$

Separable differential equations can be derived from this form due to the formulation $2\Re$ in Eq. (33) and a suitable choice of the factor C in dependence on the definition of the optimization aim.

In case, a single transition is chosen as control aim in Eq. (31), the factor C becomes $C = \langle \psi_i(t)|\psi_f(t)\rangle$. Multitarget optimal control theory (MTOCT) is essential for unitary transformations in the context of quantum computing. Here, the control objective equals Eq. (32) and the factor C in the ancillary constraint includes a sum, running over all k targets. The complete multitarget functional reads

$$\begin{aligned} J(\psi_{ik}(t), \lambda_k(t), \varepsilon(t)) = \sum_{k=1}^{N} \Bigg\{ & |\langle \psi_{ik}(t)|\phi_{fk}\rangle|^2 \\ & - 2\Re \left[\langle \psi_{ik}(T)|\phi_{fk}\rangle \int_0^T \langle \lambda_k(t)| i \left[\hat{H}_0 - \hat{\mu}\varepsilon(t) \right] + \frac{\partial}{\partial t} |\psi_{ik}(t)\rangle dt \right] \Bigg\} \\ & - \alpha_0 \int_0^T \frac{|\varepsilon(t)|^2}{s(t)} dt \end{aligned} \tag{34}$$

The calculation of optimal laser fields now relies on finding the extremum of the functional [Eq. (34)] with respect to the functions $\psi_{ik}(t)$, $\lambda_k(t)$, and $\varepsilon(t)$. The derivative of the functional with respect to $\lambda_k(t)$ and $\psi_{ik}(t)$ leads to the following coupled equations of motion and boundary conditions

$$i\frac{\partial}{\partial t}\psi_{ik}(t) = \left[\hat{H}_0 - \hat{\mu}\varepsilon(t)\right]\psi_{ik}(t), \quad \psi_{ik}(0) = \phi_{ik} \tag{35}$$

$$i\frac{\partial}{\partial t}\lambda_k(t) = \left[\hat{H}_0 - \hat{\mu}\varepsilon(t)\right]\lambda_k(t), \quad \lambda_k(T) = \phi_{fk} \tag{36}$$

The propagated wave functions $\psi_{ik}(t)$ have to correspond to the initial states ϕ_{ik} at the time $t = 0$, and the Lagrange multipliers are equal to the target states at the end of the propagation $\lambda(T) = \phi_{fk}$. According to Ref. [72], the functional [Eq. (34)] is also differentiated with respect to the laser field $\varepsilon(t)$, where only the linear terms are kept and terms containing $(\delta\varepsilon(t))^2$ are neglected.

$$\begin{aligned}\delta_{\varepsilon(t)}J &= J\left(\psi_{ik}(t), \lambda_k(t), \varepsilon(t) + \delta\varepsilon(t)\right) - J\left(\psi_{ik}(t), \lambda_k(t), \varepsilon(t)\right)\\ &\approx -\sum_{k=1}^{N}\int_0^T\left[2\alpha_0\frac{\varepsilon(t)}{s(t)} + 2\Im\langle\psi_{ik}(t)|\phi_{fk}\rangle\langle\psi_{fk}(t)|\hat{\mu}|\psi_{ik}(t)\rangle\right]\delta\varepsilon(t)dt\\ &= 0\end{aligned} \tag{37}$$

Because no incident condition is imposed on $\delta\varepsilon(t)$, Eq. (37) is fulfilled when the integrand turns zero and an equation constructing the electric field can be derived

$$\varepsilon(t) = -\frac{s(t)}{\alpha_0 N}\Im\left[\sum_{k=1}^{N}\langle\Psi_{ik}(t)|\Psi_{fk}(t)\rangle\left\langle\psi_{fk}(t)\middle|\hat{\mu}\middle|\psi_{ik}(t)\right\rangle\right] \tag{38}$$

The coupled Eqs. (35), (36), and (38) can be interpreted in different ways, and different methods to obtain the optimal field were proposed. The schemes can be based on gradient-type optimization of the laser fields [80,81]. Alternatively, the Krotov method, which is a global iterative procedure, was developed [75,77,82]. In this case, the $2N + 1$ coupled differential Eqs. (35), (36), and (38) are solved iteratively to self-consistency and proceeds in the following way. The target states $\Psi_{fk}(t)$ are propagated backward in time with the electric field $\varepsilon(t)$ from Eq. (36). Afterwards, simultaneous propagation forward in time of the wave functions and the target states takes place in Eqs. (35) and (36), where the new field is determined in each step as intermediate feedback according to Eq. (38). This new field is then used in the next iteration for back-propagation. Also, schemes using an immediate feedback from the control field in an entangled fashion were proposed, where quadratic convergence is reached [73].

According to Ref. [83], the constraint on the pulse fluence can also been chosen to take the following form:

$$\int_0^T \frac{\alpha_0}{s(t)}[\varepsilon(t) - \varepsilon'(t)]^2 \tag{39}$$

in the OCT functional [Eq. (34)], where $\varepsilon'(t)$ corresponds to the electric field from the previous iteration. The constraint restricts the change in pulse energy in each iteration with the Krotov change parameter α_0. In the next iteration step of MTOCT, the improved laser field $\varepsilon(t)^{n+1}$ is constructed as follows:

$$\varepsilon^{n+1}(t) = \varepsilon^n(t) + \frac{s(t)}{\alpha_0 N} \Im \left[\sum_{k=1}^{N} \langle \Psi_{ik}(t) | \Psi_{fk}(t) \rangle \langle \psi_{fk}(t) | \hat{\mu} | \psi_{ik}(t) \rangle \right] \tag{40}$$

This method is known as the modified Krotov OCT scheme.

D. OCT-Frequency Shaping

For the optimized transfer through the molecular chain molecule it turned out to be advantageous to gain control also over the laser pulse spectra within the OCT formalism. To this extent, an OCT technique [84] is applied that allows the combined optimization both in the time and in the frequency domain, thereby gaining complete control over the spectra of the laser pulses. Frequency filtering is realized during the optimization cycles by a modified implementation of OCT based on the Krotov method. The key point is to extend the control functional by an additional side condition

$$S = \gamma(t) \int_0^T |F(\epsilon(t))| = 0 \tag{41}$$

where the transformation $F(\epsilon(t))$ acts as a frequency filter on the electric field. The corresponding Lagrange multiplier is $\gamma(t)$. If F is chosen to remove all components from the spectrum representing a valid solution, the side condition $S = 0$ is fulfilled. The new extended functional now reads

$$K = J(\psi_k(t), \Phi_k(t), \varepsilon(t)) - \gamma(t) \int_0^T |F(\epsilon(t))| \tag{42}$$

The filter operation is formulated in the time domain and thus can be treated with a FIR filter [85], which can be regarded as a convolution with a frequency mask

$$F(\epsilon(t)) = \sum_{j=0}^{M} c_j \epsilon(t - j\Delta t) \tag{43}$$

where c_j are the FIR filter coefficients and Δt is the step size in the discrete time representation. From Eqs. (41) and (43) it is evident that the side condition is only linearly depenendent on $\epsilon(t)$. The functional derivative with respect to the electric field yields the Lagrange multiplier $\gamma(t)$

$$\frac{\delta S[\epsilon(t)]}{\delta \epsilon(t)} = \gamma(t) \tag{44}$$

If the complete functional derivative equation

$$\frac{\delta K[\epsilon(t)]}{\delta \epsilon(t)} = 0 \tag{45}$$

is now solved, only the expression for the electric field from Eq. (40) is extended by $\gamma(t)$

$$\epsilon^{n+1}(t) = \epsilon^{n}(t) + \Delta\varepsilon \tag{46}$$

$$\Delta\varepsilon = \frac{s(t)}{2N\alpha_0}\left(\sum_{k}^{N} C_k^n - \gamma(t)\right) \tag{47}$$

where C_k^n corresponds to

$$\begin{aligned} C_k^n = \Im[&\langle \Phi_k(t, \varepsilon^n)|\psi_k(t, \varepsilon^{n+1})\rangle \\ &\times \langle \Phi_k(t, \varepsilon^n)|\hat{\mu}|\psi_k(t, \varepsilon^{n+1})\rangle] \end{aligned} \tag{48}$$

The Lagrange multiplier $\gamma(t)$ can be interpreted as a correction field, which subtracts the undesired frequency components in each iteration. Because no additional equation has been introduced, $\gamma(t)$ cannot be determined in a direct fashion. Instead an educated guess from the correlation between the initial and the target state is generated

$$\begin{aligned} \gamma'(t) = \sum_k \Im[&\langle \Phi_k(t, \epsilon^n)|\psi_k(t, \epsilon^n)\rangle \\ &\times \langle \Phi_k(t, \epsilon^n)|\hat{\mu}|\psi_k(t, \epsilon^n)\rangle] \approx \sum_{k=1}^{N} C_k^n \end{aligned} \tag{49}$$

By removing the allowed frequencies from $\gamma'(t)$ with a simple Fourier filter, a sufficiently good guess is generated. Morever the side condition is additionally maintained by filtering the generated field after every iteration. The iterative procedure is monotonous convergent in practice. Another approach using a different strategy for a frequency-shaping algorithm has been recently reported in Ref. [86].

III. IMPLEMENTATION OF QUANTUM INFORMATION PROCESSING

Feynman first proposed the idea of processing data via a computer with the help of quantum phenomena [87]. The concept of a universal quantum computer was presented by Deutsch, suggesting that a theoretical quantum computing machine should combine principles from quantum mechanics together with the concept of a Turing machine [88].

The main difference from a classical computer establishes in the units of information, which are quantum bits (qubits), instead of classical bits. Whereas, a classical bit is always represented by either of the two states 0 and 1, a qubit can take any state of a linear, coherent superposition of both basis states $|0\rangle$ and $|1\rangle$ with the probability amplitudes α and β:

$$\psi = \alpha|0\rangle + \beta|a\rangle, \quad |\alpha|^2 + |\beta|^2 = 1; \quad \alpha, \beta \in \mathbb{C} \tag{50}$$

Measuring the qubit in the standard basis, the probability of the outcome $|0\rangle$ is $|\alpha|^2$ and for $|1\rangle$ it is $|\beta|^2$. The measurement leads to a collapse to a classical state. The state space for N classical bits is of the dimension $2N$, where a register of N qubits spans a 2^N-dimensional Hilbert space. Two major differences originate from the quantum nature of the qubits. One of them is quantum parallelism, which is the fundamental principle of the power of most modern quantum algorithms. It arises from the fact that the quantum register can exist in a superposition of basis states. The other is quantum correlation or quantum entanglement, which is a pure quantum mechanical phenomenon and gives rise to the speed up of quantum algorithms operating on pure states.

Deutsch formulated requirements that have to be fulfilled by quantum computers [88]. They include the preparation of a defined initial state and the implementation of a universal set of quantum gates. The quantum gates are the elementary, logic operations, which can be performed by a quantum computer on the qubits. Quantum gates are reversible and can be represented mathematically by unitary matrices. The set of universal quantum gates consists of a certain reduced amount of operations, which are the two-qubit controlled NOT (CNOT) gate together with all one-qubit gates NOT, Π, and Hadamard [89].

The corresponding Pauli matrices take the forms given in Eqs. (51) and (52) for the two-qubit basis $\{|00\rangle, |01\rangle, |10\rangle, |11\rangle\}$.

$$\mathbf{NOT} = \begin{pmatrix} 0 & 1 & 0 & 0 \\ 1 & 0 & 0 & 0 \\ 0 & 0 & 0 & 1 \\ 0 & 0 & 1 & 0 \end{pmatrix}, \quad \mathbf{CNOT} = \begin{pmatrix} 1 & 0 & 0 & 0 \\ 0 & 1 & 0 & 0 \\ 0 & 0 & 0 & 1 \\ 0 & 0 & 1 & 0 \end{pmatrix} \tag{51}$$

$$\Pi = \begin{pmatrix} 1 & 0 & 0 & 0 \\ 0 & 1 & 0 & 0 \\ 0 & 0 & -1 & 0 \\ 0 & 0 & 0 & -1 \end{pmatrix}, \quad \mathbf{H} = \frac{1}{\sqrt{2}} \begin{pmatrix} 1 & 0 & 1 & 0 \\ 0 & 1 & 0 & 1 \\ 1 & 0 & -1 & 0 \\ 0 & 1 & 0 & -1 \end{pmatrix} \tag{52}$$

The NOT and CNOT gates, given in Eq. (51), are qubit flip gates. In case of the NOT operation, the flip of the state of the second (active) qubit is performed independently of the state of the first qubit. The corresponding CNOT gate involves a control qubit (first qubit). Only, if it is in state $|1\rangle$, the active (second) qubit is flipped. A phase rotation of π of a qubit in state $|1\rangle$ is implemented by a Π operation in Eq. (52). The Hadamard gate in Eq. (52) involves phase rotations in combination with qubit flips and is essential for the preparation of superposition states. A corresponding set of gates exists, where the role of the first and second (passive and active) qubits are interchanged. In accordance with the definition that a universal set can be used to implement any quantum operation, the sequence $\mathrm{CNOT}_2\ \mathrm{CNOT}_1\ \mathrm{CNOT}_2$ composes a swap operation.

A further requirement of Deutsch is the preferably correct readout of the qubits after the quantum gate operations. Additionally, DiVincenzo formulated further essential requirements [90] for the realization of quantum computers. The defined qubit system needs to be scalable, the storage of quantum information data must be possible and stable; furthermore, a favorable relation of switching to decoherence times needs to exist. Related to molecular quantum computing with vibrational qubits (see Section III.A), we find a favorable relation of switching to decoherence time and the possibility to store information in vibrational or electronic states [10]. The issue of scalability will be addressed in Section III.B.

Only 10 years after the first description of the universal quantum computer, the first CNOT gate was realized by Monroe and Wineland [91] based on trapped ions, an implementation previously proposed by Cirac and Zoller [92]. In principle, any two-level system could be used as a qubit system, but multilevel systems are also suitable if it is possible to decouple the nonqubit basis states efficiently. The electronic states of the ions in an electromagnetic trap are used to code the qubit states. The quantum information can be processed through the collective motion of the ions in the trap. Lasers apply couplings between the qubit states or a coupling between the internal qubit states and the external motional states to perform quantum gates or generate entanglement. A strategy for scaling the ion trap approach to larger numbers of qubits was developed, based on arrays of ion traps [93]. In 2005, the first quantum byte was achieved [20] and larger and larger systems could be realized since then [21,94]. Another promising quantum computing technology was proposed in 1997 by Cory and is based on nuclear magnetic resonance (NMR) [4,95]. In the year 2000, a five-qubit NMR computer was presented using spin states of molecules as qubits [5]. The difference to other implementations is that the scheme is based on ensembles of molecules. The quantum gates are realized through radio frequency pulses.

Other concepts rely on electronic degrees of freedom in solid state, such as spins of electrons in quantum dots and superconducting flux (or charge) qubits in Josephson junctions [96–102]. A scalable setup for quantum computing with superconducting circuits has been recently presented [103]. Implementations based on photons, where, for example, the polarization of light codes of the qubit states are also investigated [104–106].

A. Molecular Quantum Computing

The first concept of molecular quantum computing with vibrational qubits was proposed in Refs [8,9]. In this proposal the qubit states are encoded in vibrational states of normal modes of polyatomic molecules. Mediated by the molecular dipole moment, the quantum gates can be realized by ultrashort, specially shaped femto- to picosecond pulses. The theoretical proof of principle was demonstrated for the molecule acetylene [8–10], where in case of a two- and three-qubit system, the universal set of quantum gates could be successfully implemented. Additionally, quantum algorithms, such as the Deutsch–Josza algorithm [15] or a QFT [78] were calculated for two-qubit systems. Further studies suggest selecting transition metal carbonyls as promising candidates, due to strong IR absorbance [16,107,108]. Studies on the effects of molecular parameters determining the vibrational modes, the intramode anharmonicity, and the anharmonic coupling reflected basic requirements on these parameters and the effects on the efficiencies and properties of the quantum gates [109]. Investigations of further molecular candidates with the same or similar qubit coding schemes were performed by several other groups [110–125]. A comprehensive review on this topic is given in Ref. [10]. First experiments, which may allow for the realization of the proposed quantum computing scheme, were presented recently [126–130]. Here, the shaping of mid-IR pulses and tracing the population mechanism are essential. In theory, the optimization of the quantum gates is based on MTOCT (see Section II.C).

B. Approach for Quantum Information Processing with Molecular Vibrational Qubits

Recently a register of ultracold NaCs polar molecules in a static electric field has been discussed as an approach to scalability [131] in the field of molecular quantum computing. Here, we follow a different strategy to communicate between spatially separated qubits, taking advantage of the fast vibrational motions. We investigate the vibrational energy transfer through a molecular chain as a possible approach to scalable quantum information. Our model consists of two individual molecular qubit systems attached at either end of a carbon chain. As a chain model

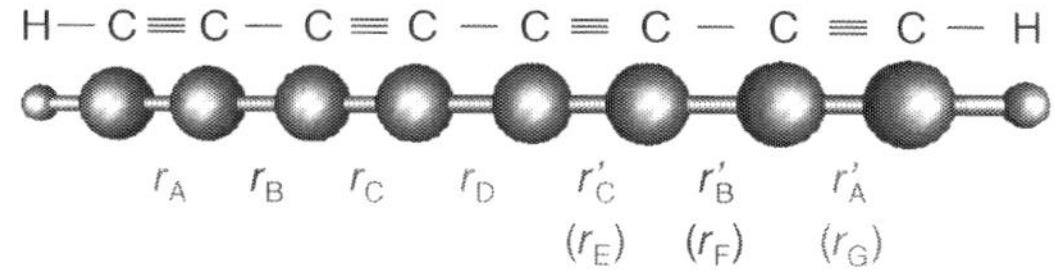

Figure 1. Octatetrayne chain with the local mode coordinates q_i referring to the displacement of the C—C bondings, with the equilibrium bond lengths r_i.

we selected octatetrayne (Fig. 1). The complete model is described in the eigenstate representation; for the octatetrayne chain, the stretching local modes are selected as basis. The labels assigned to the C—C distance such as r_X and r'_X indicate the symmetry of the molecule that can be used to reduce the number of interatomic interactions to be calculated.

The molecular geometry is optimized quantum chemically using DFT (bp86/6-31G(d,p)) [132]. Along each local mode (explicitly for the coordinates q_A, q_B, q_C, q_D, referring to the bond lengths r_A, r_B, r_C, r_D in Fig. 1), the 1D C—C potentials ($\hat{V}_\mathrm{C}^{1\mathrm{D}}$) are calculated and the corresponding vibrational eigenfunctions and eigenvalues are evaluated (see Section II.A.2) with the kinetic Hamiltonian operator $\hat{T}_\mathrm{C}^{1\mathrm{D}}$ and the mass m_C

$$\hat{T}_\mathrm{C}^{1\mathrm{D}} = -\frac{1}{m_\mathrm{C}}\frac{\partial}{\partial r_i^2} \tag{53}$$

Additionally, for pairs of local modes ($q_\mathrm{A}q_\mathrm{B}$, $q_\mathrm{B}q_\mathrm{C}$, $q_\mathrm{C}q_\mathrm{D}$, $q_\mathrm{A}q_\mathrm{C}$, $q_\mathrm{B}q_\mathrm{D}$, $q_\mathrm{C}q_\mathrm{E}$), the 2D potentials $\hat{V}_{\mathrm{C}_i\mathrm{C}_j}^{2\mathrm{D}}$ are set up. For neighboring oscillators ($q_\mathrm{A}q_\mathrm{B}$, $q_\mathrm{B}q_\mathrm{C}$, $q_\mathrm{C}q_\mathrm{D}$), the kinetic coupling has to be accounted for in the kinetic Hamiltonian $\hat{T}_{\mathrm{C}_i\mathrm{C}_j}^{2\mathrm{D}}$

$$\hat{T}_{\mathrm{C}_i\mathrm{C}_j}^{2\mathrm{D}} = -\frac{1}{2}\left(\frac{2}{m_\mathrm{C}}\frac{\partial}{\partial r_i^2} + \frac{2}{m_\mathrm{C}}\frac{\partial}{\partial r_j^2} - \frac{2}{m_\mathrm{C}}\frac{\partial}{\partial r_i}\frac{\partial}{\partial r_j}\right) \tag{54}$$

In contrast, for two nonneighboring oscillators (indicated by the label $\mathrm{C}_iX\mathrm{C}_j$), the Hamiltonian consists of the 2D potential part $\hat{V}_{\mathrm{C}_iX\mathrm{C}_j}^{2\mathrm{D}}$ and the kinetic Hamiltonian $\hat{T}_{\mathrm{C}_iX\mathrm{C}_j}^{2\mathrm{D}}$, which is assumed to be of cartesian type (i.e., uncoupled):

$$\hat{T}_{\mathrm{C}_iX\mathrm{C}_j}^{2\mathrm{D}} = -\frac{1}{2}\left(\frac{2}{m_\mathrm{C}}\frac{\partial}{\partial r_i^2} + \frac{2}{m_\mathrm{C}}\frac{\partial}{\partial r_j^2}\right) \tag{55}$$

In this case, the only coupling of the oscillators is due to an intermode anharmonicity of the potential energy surfaces ($\hat{V}_{\mathrm{C}_i\mathrm{C}_j}^{2\mathrm{D}}$ and $\hat{V}_{\mathrm{C}_iX\mathrm{C}_j}^{2\mathrm{D}}$).

The total wave function Ψ, representing the carbon stretching mode of the octatetrayne molecule, is expanded on the basis of the 1D local mode functions ϕ_n. The number of local wave functions with the vibrational coordinates

q_n is $n = 7$. The product function Ψ_m takes the form

$$\Psi_m = \Pi_n \phi_n(q_n) = \phi_1\phi_2\phi_3\phi_4\phi_5\phi_6\phi_7 \tag{56}$$

The Hamiltonian matrix **H** and the corresponding matrix elements are calculated as follows. The diagonal elements are equal to the sum of the eigenvalues of the corresponding local mode states, which will be shown next. For the mth product state (e.g., the total ground state $\Psi_0 \equiv 0000000$ is assumed as the state $m = 0$ and the state with one quantum of excitation in the q_A local mode $\Psi_1 \equiv 1000000$ corresponds to $m = 1$), the matrix element is calculated according to

$$H_{mm} = \langle \Psi_m | \sum_n \hat{H}_n^{1D}(q_n) | \Psi_m \rangle \tag{57}$$

$$= \langle \Pi_n^{(m)} \phi_n^{(m)}(q_n) | \sum_n \hat{H}_n^{1D}(q_n) | \Pi_n^{(m)} \phi_n^{(m)}(q_n) \rangle \tag{58}$$

$$= \sum_n \langle \phi_n^{(m)}(q_n) | \hat{H}_n^{1D}(q_n) | \phi_n^{(m)}(q_n) \rangle \tag{59}$$

$$H_{00} = 2 \cdot \epsilon_A^0 + 2 \cdot \epsilon_B^0 + 2 \cdot \epsilon_C^0 + \epsilon_D^0 \tag{60}$$

$$H_{11} = \epsilon_A^1 + \epsilon_A^0 + 2 \cdot \epsilon_B^0 + 2 \cdot \epsilon_C^0 + \epsilon_D^0 \tag{61}$$

Inserting the product function Eq. (56) into Eq. (57) leads to Eq. (58). This equation can be further simplified, yielding Eq. (59), as the 1D Hamiltonians $\hat{H}_n^{1D}(q_n)$ depend only on one local coordinate q_n. Two examples for diagonal matrix elements are given in Eqs. (60) and (61). Here, ϵ_n^x are the eigenenergies of the local modes q_n, and x denotes the degree of excitation of these modes. The kinetic coupling of next-neighbor oscillators as well as the intermode anharmonicity of the 2D potential energy surfaces contribute to the off-diagonal elements. The highest dimensionality of couplings taken into account here is 2D. Equivalently, one could also calculate the corresponding 3D or multidimensional potential energy surfaces and take the higher dimensional potential couplings into account. However, from test calculations on the smaller butadiyne system, it was found that the importance of these terms are lower and they can be neglected for the octatetrayne model system. The off-diagonal elements H_{lm} ($l \neq m$) of the Hamiltonian matrix **H** are calculated as

$$H_{lm} = \langle \Psi_l | \sum_k \sum_j \hat{H}_{kj}^{2D}(q_k q_j) | \Psi_m \rangle \tag{62}$$

$$= \langle \Pi_n^{(l)} \phi_n^{(l)}(q_n) | \sum_k \sum_j \hat{H}_{kj}^{2D}(q_k q_j) | \Pi_n^{(m)} \phi_n^{(m)}(q_n) \rangle \tag{63}$$

$$= \langle \Pi_n^{(l)} \phi_n^{(l)}(q_n) | \sum_k \sum_j (\hat{T}_{kj}^{2D}(q_k q_j) + \hat{V}_{kj}^{2D}(q_k q_j)) | \Pi_n^{(m)} \phi_n^{(m)}(q_n) \rangle \tag{64}$$

The indexes k and j refer to all 2D operators, taken into account. In this study, they denote the following pairs of local modes: $q_Aq_B, q_Bq_C, q_Cq_D, q_Aq_C, q_Bq_D, q_Cq_E$, and the corresponding mirrored parts (e.g., $q'_Aq'_B$ referring to the bond lengths r'_A and r'_B in Fig. 1). Equation (64) can be further simplified, because the 2D Hamiltonians $\hat{T}^{2D}_{kj}(q_kq_j)$ and $\hat{V}^{2D}_{kj}(q_kq_j)$ depend on two coordinates and they only act on the local mode wave functions, referring to these local coordinates (q_kq_j), for example,

$$H_{01} = \langle 0000000| \sum_k \sum_j \hat{H}^{2D}_{kj}(q_kq_j)|1000000\rangle \tag{65}$$

$$= \langle 0000000|\hat{H}^{2D}_{AB}(q_Aq_B) + \hat{H}^{2D}_{BC}(q_Bq_C) + \hat{H}^{2D}_{AC}(q_Aq_C) + \cdots |1000000\rangle \tag{66}$$

$$= \langle 00|\hat{H}^{2D}_{AB}(q_Aq_B)|10\rangle + \langle 00|\hat{H}^{2D}_{BC}(q_Bq_C)|00\rangle + \langle 00|\hat{H}^{2D}_{AC}(q_Aq_C)|10\rangle + \cdots \tag{67}$$

Although exclusively 2D couplings are accounted for in this setup of the Hamiltonian matrix **H**, the vibrational normal modes can be well approached, because delocalized states such as 1011101 are taken into account. The maximum number of states used in these calculations is on the order of 300.

The octatetrayne model, as previously described, is extended by vibrational qubit systems, which are assumed to be directly connected to the octatetrayne chain. Similar molecular systems, including sp-hybridized carbon chains, have been synthesized in the group of Gladysz [133,134]. In these structures, platinum complexes are directly linked to both ends of the molecular carbon wire. These platinum complexes are chemically related to the transition metal carbonyl compounds, investigated as two-qubit systems in previous work [16,107,108]. Here, the model is simplified as we assume only one-qubit systems attached at either end of the octatetrayne chain. Thus, one-qubit operations, like NOT or Hadamard, can be performed on the individual qubits, the individual nodes, and their outcome can be communicated between the nodes. For the calculations, the product wave function [Eq. (56)] is expanded by qubit normal mode states, and the system Hamiltonian is extended by the respective matrix elements. The qubit systems are supposed to be connected to the chain ends, instead of the hydrogen atoms in octatetrayne (Fig. 1). A model setup of this system, composed of a linear carbon chain, described in the local mode basis and two coupled qubit normal modes, is sketched in Fig. 2.

For the optimization of a state transfer from one qubit site to the other one across the chain molecule, two different qubit systems linked to the molecular bridge were constructed according to Fig. 3a and b. The qubit mode Q(A) (left side of Figs. 2 and 3a) is assumed to have a fundamental transition frequency of $\omega_{Q(A)} = 1400\,\text{cm}^{-1}$ and an anharmonicity of $\Delta_{Q(A)} = 43\,\text{cm}^{-1}$, whereas the

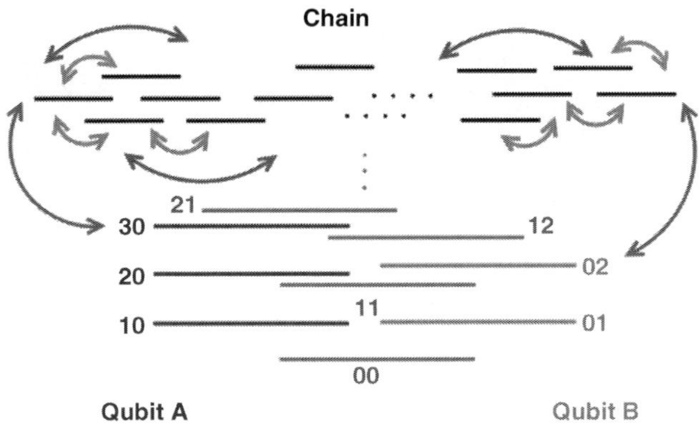

Figure 2. Model system for two qubits connected by local mode states of a molecular chain (black). As indicated by the green and blue qubit levels, the fundamental frequencies of the qubit modes differ in this setup. The red and purple arrows indicate the kinetic and potential couplings between the local mode chain states, and additionally a coupling between a qubit overtone state and one chain state at each site is assumed (brown arrows). (*See the color version of this figure in the Color Plates section.*)

parameters for the qubit mode Q(B) are $\omega_{Q(B)} = 2200\,\text{cm}^{-1}$ and $\Delta_{Q(B)} = 30\,\text{cm}^{-1}$. The dipole moment of the Q(A) mode is set to $0.18\,\text{au} = 0.45\,\text{Debye}$ and to $0.13\,\text{au} = 0.33\,\text{Debye}$ for the Q(B) mode. They are assumed to scale harmonically (i.e., with the square root of the vibrational quantum number).

Additionally, a coupling is introduced between the overtone state $v = 3$ of the qubit mode Q(A) and a chain state, where the vibrational excitation is located at the Q(A) site of the chain in the local coordinate q_A ($\Psi \equiv 2000000$, Fig. 3b). The size of the coupling element is selected to be 0.0008 au. Analogously, the second overtone state ($v = 2$) of the Q(B) qubit mode is coupled to the chain state $\Psi \equiv 0000002$.

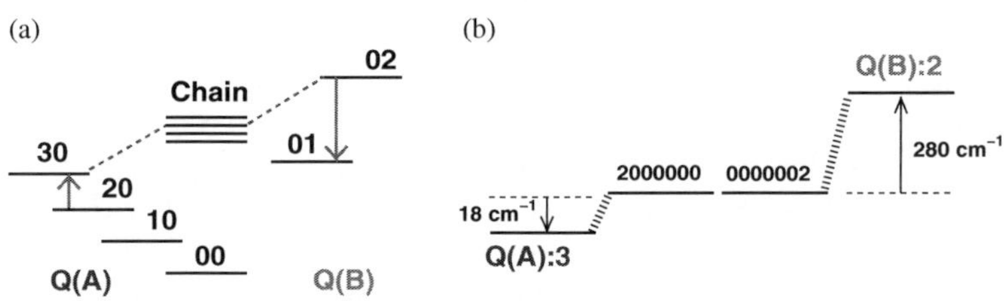

Figure 3. (**a**) Vibrational state-to-state transfer, which has to be driven from the second excited state of the qubit mode Q(A) to the first excited state of the qubit mode Q(B) by an optimized ultrashort laser field. The overtone and combination states are not shown, for reasons of simplicity, but they are taken into account in the calculations. (**b**) Couplings between the qubit mode and local chain states and energetics of the corresponding levels.

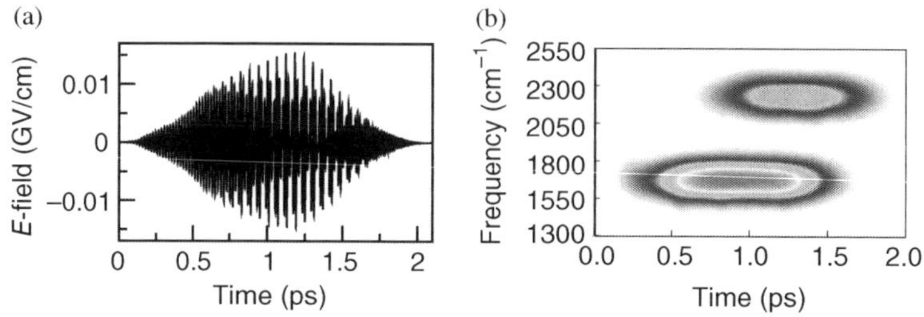

Figure 4. (**a**) Optimized laser field, driving the state-to-state transfer from $v = 2$ on the Q(A) qubit mode to $v = 1$ on the Q(B) qubit mode across the chain states. (**b**) Corresponding FROG representation of the laser pulse.

C. State Transfer and Quantum Channels

The optimization aim that simulates information exchange between different qubit sites is a vibrational state-to-state transfer from the state $v_{Q(A)} = 2$ to $v_{Q(B)} = 1$ as indicated by the arrows in Fig. 3a. For example, a successful NOT operation [Eq. (51), left] on qubit mode Q(A) can be communicated to qubit mode Q(B). The initial and target states of interest will be denoted as 20 and 01, only referring to the vibrational degrees of excitation in the qubit modes $q_{Q(A)}q_{Q(B)}$, neglecting the chain states. A laser pulse driving the vibrational population transfer $20 \rightarrow 01$ is optimized with OCT in Eq. (34) and an efficiency of 99.2 % is reached for a short pulse duration of ~2.1 ps. The corresponding laser field is depicted in Fig. 4a, and the FROG representation is shown in Fig. 4b. The FROG diagram reveals a simple pulse structure with two subpulses delayed in frequency and time. It can be traced that the center frequency of the subpulses does not match the transition frequencies $v_2 \rightarrow v_3$ of 1314 cm^{-1} for the qubit mode Q(A) and $v_2 \rightarrow v_1$ of 2170 cm^{-1} for the qubit mode Q(B), as indicated by the arrows in Fig. 3a.

The population transfer from qubit mode to local chain modes is investigated to understand the vibrational energy transfer process. The evolution of the population in the qubit mode states is shown in Fig. 5a. Initially, the complete population is in the state 20 (blue) and the chain local modes are in the ground state. After 0.5 ps the amount of population in the state 20 starts to decrease and is transferred into the target state 01 (dark green), but also intermediately to a small extent to the state 02 (light green), which couples directly to the chain at the qubit site Q(B). The mechanism of the local mode chain states is shown in Fig. 5b. At least 15 local mode states significantly participate in the transfer process and are intermediately populated.

The transition pathway of the population transfer from the initial qubit state 20 into the chain states and out of them into the target state 01 is not easy to interprete in the local modes picture. To clarify the mechanism the Hamiltonian, set up in the local mode representation, is diagonalized to obtain the normal modes of the coupled qubit-chain system. Certain local mode chain states mix with the

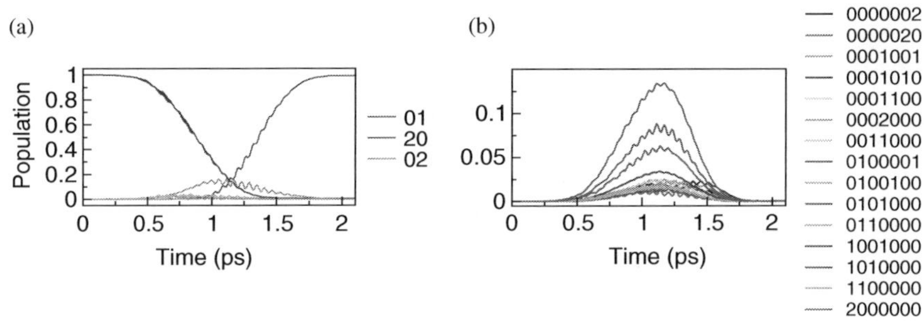

Figure 5. Mechanism induced by the laser field depicted in Fig. 4a. (**a**) The population in the qubit mode states are depicted, where the vibrational quantum numbers of the local chain states are 0. The blue line refers to the initial state 20 and the dark-green line to the target state 01 of the transfer process. (**b**) The intermediate population of the local chain states is shown. Here, only the vibrational quantum number of the local chain states are given. (*See the color version of this figure in the Color Plates section.*)

qubits mode states 30 and 02, which are directly coupled to the chain, and they are included in the normal modes of the system. The basis of the dipole moment matrix $\boldsymbol{\mu}$, referring to qubit mode transitions, is changed to the normal mode basis. This is performed with a transformation matrix **Y** [Eq. (68)] corresponding to the matrix of the normal mode eigenvectors, which were obtained from the diagonalization of the local mode Hamiltonian.

$$\boldsymbol{\mu}^{\text{normal}} = \mathbf{Y}\boldsymbol{\mu}^{\text{local}}\mathbf{Y}^{\dagger} \tag{68}$$

As a result, different transition pathways connecting the initial and target state of the qubit modes with the normal modes can be detected. These pathways can be associated with different transition dipole strengths and are illustrated in Fig. 6. The blue lines refer to the transitions from the initial state 20 into the normal mode states and the green lines back into the target state 01. Every blue transition line is related to a green transition path. The differing dipole strengths are indicated by the width of the transition lines. In general, the dipole moment matrix elements

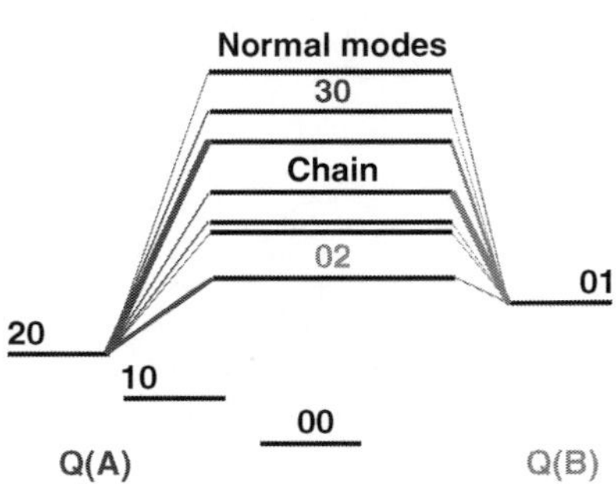

Figure 6. From the change of the local mode to the normal mode basis, different transition pathways from the initial and target states into the normal mode states can be found. They are indicated by the blue lines for the initial state and by the green lines for the target state. The varying width of the lines indicates that the transition dipole strengths differ. (*See the color version of this figure in the Color Plates section.*)

are larger when the contribution of the coupled local mode states 2000000 and 0000002 to the respective normal mode vectors is higher.

The transition pathways, as shown in Fig. 6, allow for the identification of the optically accessible normal mode states n (i.e., the transition matrix elements into these states is $\mu \neq 0$). The possible transitions from the initial (20, blue) and target (01, green) states into these normal mode states n are depicted in Fig. 7a. They

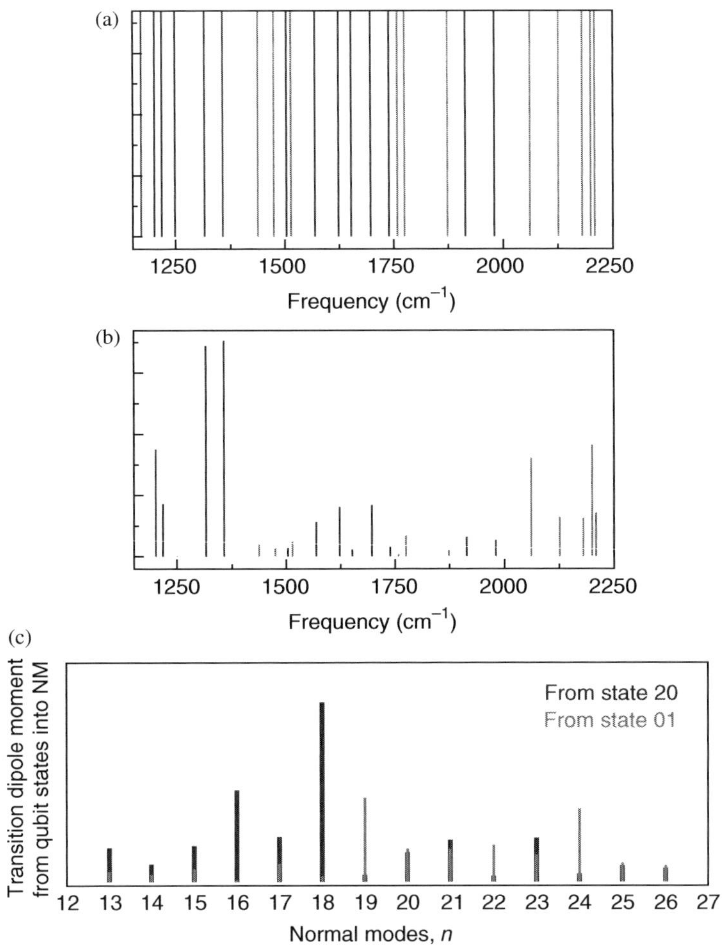

Figure 7. (**a**) Transition pathways from the initial (20, blue) and target (01, green) states into the normal mode states, plotted as a function of the transition frequency. The spectrum of the blue transitions is plotted for $n = 13 - 26$, for the green transitions for $n = 14 - 23$. (**b**) Transition pathways scaled by the size of the respective dipole matrix elements. (**c**) Pathways from both qubit sites, plotted against the normal modes n. (*See the color version of this figure in the Color Plates section.*)

are plotted versus the transition frequencies $20 \rightarrow n$ (blue) and $n \rightarrow 01$ (green). Figure 7b shows an equivalent graph, but here the size of the corresponding dipole matrix elements for the transitions are indicated by the height of the vertical lines. Each blue line ($20 \rightarrow n$) refers to a corresponding green line ($n \rightarrow 01$), where in both cases the same normal mode state n is used for the transfer. This can be visualized by plotting the transition dipole strengths against the number of the respective normal modes states n, as shown in Fig. 7c.

By means of these plots, the state transfer process induced by the optimized laser pulse (Fig. 4a) across the molecular chain can be understood. The pulse couples the qubit mode Q(A) to one or more optically addressable normal mode states n with the first subpulse and transfers the population from the state 20 into the local chain states, which form the respective normal modes. The second subpulse draws the population out into the target state 01. The carrier frequencies of the two subpulses (1677 cm^{-1} and 2233 cm^{-1} traced from the FROG representation of the optimized laser field in Fig. 4b) with respect to the transition pathways (Fig. 7b and c) reveal that predominantly two normal modes ($n = 21$ and $n = 23$) are used as transfer channels. As can be deduced from Fig. 7c several different quantum channels are available for the vibrational population transfer process.

Now a question arises on whether and how the other channels can be used, and if the vibrational energy transfer can be optimized through a single normal mode. For the OCT calculations of the state transfer processes through different normal modes, the initial guess laser fields have to be chosen properly; that is, a good starting laser field should provide the two frequency components matching the transition frequencies. A desired transfer channel can be selected (Fig. 7c) and the required frequencies can be extracted from Fig. 7b. A helpful OCT technique, which assists these kinds of optimizations, is the frequency filtering method [84] outlined in Section II.D.

In the population transfer optimizations through distinct quantum channels, only minor parts of the spectrum have to be suppressed, as long as the guess laser field is set into resonance with the normal mode transitions and the penalty factor α for the optimization is selected high enough. Corresponding optimizations were performed for the normal modes $n = 17, 20, 21$, and 23 (Fig. 7c) and the results are depicted in Fig. 8. The respective laser fields in the time domain are similarly simple as the one depicted in Fig. 4a. The spectra of the calculated laser pulses are mapped onto the available transition pathways, and one normal mode is used predominantly as a transfer channel in each case (Fig. 8a, $n = 17$; Fig. 8b, $n = 20$; Fig. 8c, $n = 21$; and Fig. 8d, $n = 23$). The transition frequencies for the excitation processes $20 \rightarrow n$ and $n \rightarrow 01$, with n referring to the transfer modes (17, 20, 21, and 23), exactly correspond to the carrier frequencies of the single subpulses. The relative size of the dipole matrix elements into and out of the quantum channels account for the differing intensities of the two subpulses. Because the transition dipole moment for the process $20 \rightarrow n$ is larger in case of $n = 17, 21, 23$ (blue

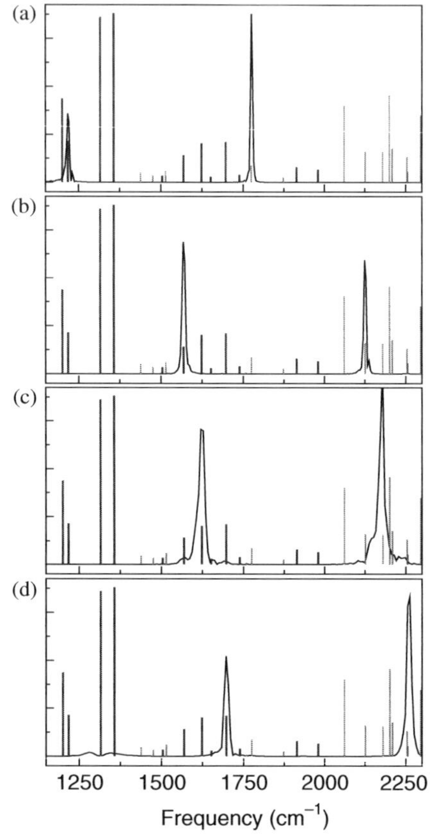

Figure 8. Transition pathways with spectra of the laser pulses driving the state transfer across the chain molecule through different quantum channels. (**a**) $n = 17$, (**b**) $n = 20$, (**c**) $n = 21$, and (**d**) $n = 23$ (Fig. 7c). (*See the color version of this figure in the Color Plates section.*)

lines in Fig. 7c), the spectral intensities of the first subpulses are lower (spectral parts associated with the blue lines in Fig. 8a, c, and d). The situation is reversed for $n = 20$ (Fig. 8b). The presented quantum channels ($n = 17$, 20, 21, and 23) turned out to be the most suited ones for the vibrational transfer process and when inspecting the corresponding transition dipole strengths (Fig. 7c) the reason becomes clear. All of them show similarly sized dipole moment elements for the excitations $20 \to n$ and $n \to 01$ (i.e., the green and blue lines are equally high). This is particularly the case for the normal mode 20, and accordingly, the spectral intensities of both subpulses in Fig. 8b are similar.

From these calculations, it is expectable that not only a state-to-state transfer but also a transfer of superposition states as they result from Hadamard transformations [Eq. (52), right] is possible. In this case, different quantum channels can be used for the process. This result may facilitate quantum information processing with vibrational qubits in future, where after quantum gate operations, the resulting eigenstates or superposition states can be communicated to other qubit units.

D. Dissipative Influence on Vibrational Energy Transfer Related to IVR

A crucial factor for the maximum possible efficiency in large molecular complexes or in condensed phase is the impact of dissipative effects. These effects are not always disadvantageous. Cooperative disspative effects have recently been reported in Ref. [135] in the context of quantum computing. Regarding our molecular chain model possible sources for decoherence are vibrational relaxation within the modes forming the quantum channel and couplings to other vibrational degrees of freedom in the molecular chain.

Based on our ab initio data a simple model for dissipation is set up by introducing IVR-processes to the octatetrayne chain. The perturbations due to IVR are also set up in the normal mode representation. Within this model the dissipative effects are studied for the vibrational state transfer process across the chain molecule as outlined in Section II.B.3. The complete population is assumed to be in the first vibrational level of the highest energy C—C stretching mode of octatetrayne ($\sim$2300 cm^{-1}). As observed in Section III.C, such a state can serve as a transfer channel for the vibrational population. For the dissipation, the vibrational energy relaxation within this normal mode is taken into account as well as a coupling to overtone and combination states of deformation modes, which are in resonance to the initially excited state. To find the resonant states, for each deformation mode an anharmonicity of 5 cm^{-1} is assumed and the energy of the overtone and combination states are extrapolated. Overall, three resonances were found, with energy differences below $\Delta E = 5\,\text{cm}^{-1}$ between the deformation states and the initially populated state. The time evolution of the respective deformational modes, abbreviated as def$_1$, def$_2$, and def$_3$, are shown in Fig. 9. Different relaxation rates for the IVR processes into these states are extrapolated from individual 2D ab initio

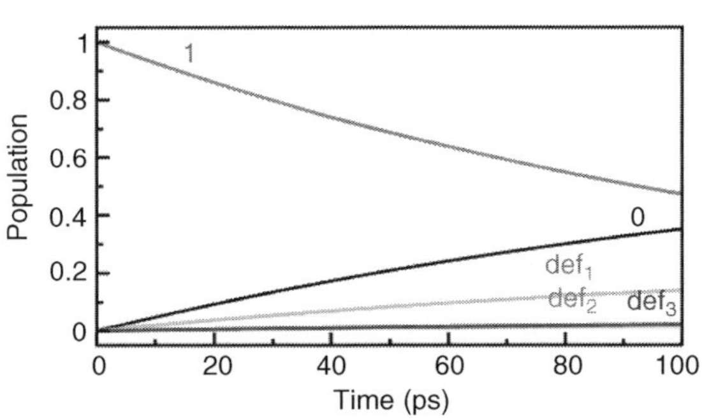

Figure 9. Mechanism of a dissipative propagation. The vibrational population in the first excited state of the highest energy normal mode of octatetrayne (1) relaxes to the ground state (0) and to several resonant deformation mode states (here: def$_1$, def$_2$, def$_3$). For short propagation times of $\sim$2 ps, as necessary for the vibrational state transfer, the decay of population is negligible.

models for the selected coupling modes. Together with the estimated relaxation into the ground state, taken in analogy to measured molecular T_1 times [128], a range of relaxation rates is scanned and a realistic example with a relaxation time of $\sim$200 ps is shown in Fig. 9. The calculations show that IVR and relaxation time scales for the selected octatetrayne chain are tolerable when they stay in the order of 200 ps. In this case high efficiencies of the vibrational state-to-state transfer processes (with durations of approximately 2 ps, as calculated in Section III.C) are guaranteed.

IV. CONCLUSIONS

With the idea of a linear, molecular chain connecting individual molecular qubit systems, a first effort toward the scalability of the molecular quantum computing approach with vibrational qubits has been made. In this chapter, two different qubit systems are assumed to share quantum information with the help of a laser-driven vibrational state-to-state information transfer. A model system, composed of an octatetrayne chain and two qubit modes at each site, was set up, where the bridging molecule was described in the local mode basis. As a first approach, a coupling is assumed between local chain states and an overtone state of each qubit mode. This description may be extended in the future, by incorporating different coupling elements, with the size depending on the energy gap between the coupled states. With OCT as an optimization tool, highly efficient laser fields providing very short pulse durations could be optimized, driving a fast information transfer. From the inspection of the transfer mechanisms, different available transition pathways corresponding to normal mode states could be detected. They can be regarded as quantum channels for the information transfer, and it was shown that several of these pathways can be used for the transfer. The suitability of the quantum channels depends on the relative size of the transition dipole matrix elements with regard to both qubit sites and the OCT strategy with spectral constraints [84] allows for the control of the transfer through the single channels. The existence of several equivalent channels establish the possibility for superposition state transfer, showing a path to extend the communication to all fundamental quantum gate operations. The advantage of the scheme lies in its ultrafast transfer rates that in themselves function as a protection against decoherence due to the otherwise unavoidable IVR processes.

Acknowledgment

The authors would like to thank the Deutsche Forschungsgemeinschaft for financial support through the Normalverfahren and through the excellence cluster "Munich Center for Advanced Photonics" (MAP).

REFERENCES

1. M. Brune, F. Schmidt-Kaler, A. Maali, J. Dreyer, E. Hagley, J. M. Raimond, and S. Haroche, *Phys. Rev. Lett.* **76**, 1800 (1996).
2. F. Schmidt-Kaler, H. Haffner, M. Riebe, S. Gulde, G. P. T. Lancaster, T. Deuschle, C. Becher, C. F. Roos, J. Eschner, and R. Blatt, *Nature* **422**, 408 (2003).
3. D. Leibfried et al., *Nature* **438**, 639 (2005).
4. J. A. Jones and M. Mosca, *J. Chem. Phys.* **109**, 1648 (1998).
5. R. Marx, A. F. Fahmy, J. M. Myers, W. Bermel, and S. J. Glaser, *Phys. Rev. A* **62**, 012310 (2000).
6. T. Brixner and G. Gerber, *ChemPhysChem* **4**, 418 (2003).
7. P. Nürnberger, G. Vogt, T. Brixner, and G. Gerber, *Phys. Chem. Chem. Phys.* **9**, 2470 (2007).
8. C. M. Tesch and R. de Vivie-Riedle, *Phys. Rev. Lett.* **89**, 157901 (2002).
9. C. M. Tesch, L. Kurtz, and R. de Vivie-Riedle, *Chem. Phys. Lett.* **343**, 633 (2001).
10. R. de Vivie-Riedle and U. Troppmann, *Chem. Rev.* **107**, 5082 (2007).
11. J. Vala, Z. Amitay, B. Zhang, S. R. Leone, and R. Kosloff, *Phys. Rev. A* **66**, 062316 (2002).
12. R. Zadoyan, D. Kohen, D. A. Lidar, and V. A. Apkarian, *Chem. Phys.* **266**, 323 (2001).
13. Z. Bihary, D. R. Glenn, D. A. Lidar, and V. A. Apkarian, *Chem. Phys. Lett.* **360**, 459 (2002).
14. Y. Ohtsuki, *Chem. Phys. Lett.* **404**, 126 (2005).
15. C. M. Tesch and R. de Vivie-Riedle, *J. Chem. Phys.* **121**, 12158 (2004).
16. B. Korff, U. Troppmann, K. Kompa, and R. de Vivie-Riedle, *J. Chem. Phys.* **123**, 244509 (2005).
17. A. Daskin and S. Kais, *J. Chem. Phys.* **134**, 144112 (2011).
18. L. M. K. Vandersypen, M. Steffen, G. Breyta, C. S. Yannoni, M. H. Sherwood, and I. L. Chuang, *Nature* **414**, 883 (2001).
19. I. L. Chuang, N. Gershenfeld, and M. Kubinec, *Phys. Rev. Lett.* **80**, 3408 (1998).
20. H. Häffner et al., *Nature* **438**, 643 (2005).
21. B. P. Lanyon et al., *Science* **334**, 57 (2011).
22. M. Schröder and A. Brown, *J. Chem. Phys.* **131**, 034101 (2009).
23. K. Ohmori, *Ann. Rev. Phys. Chem.* **60**, 487 (2009).
24. S. Bose, *Phys. Rev. Lett.* **91**, 207901 (2003).
25. M. Christandl, N. Datta, A. Ekert, and A. J. Landahl, *Phys. Rev. Lett.* **92**, 187902 (2004).
26. K. Audenaert, J. Eisert, M. B. Plenio, and R. F. Werner, *Phys. Rev. A* **66**, 042327 (2002).
27. B. Alkurtaas, G. Sadiek, and S. Kais, *Phys. Rev. A* **84**, 022314 (2011).
28. K. Moritsugu, O. Miyashita, and A. Kidera, *Phys. Rev. Lett.* **85**, 3970 (2000).
29. J. Antony, B. Schmidt, and C. Schutte, *J. Chem. Phys.* **122**, 014309 (2005).
30. M. Schade, A. Moretto, M. Crisma, C. Toniolo, and P. Hamm, *J. Phys. Chem. B* **113**, 13393 (2009).
31. A. Pecchia, M. Gheorghe, A. D. Carlo, and P. Lugli, *Phys. Rev. B.* **68**, 235321 (2003).
32. M. Gheorghe, R. Gutiérrez, N. Ranjan, A. Pecchia, A. D. Carlo, and G. Cuniberti, *Europhys. Lett.* **71**, 438 (2005).
33. A. Nitzan and M. A. Ratner, *Science* **300**, 1384 (2003).
34. M. Tommasini, G. Zerbi, V. Chernyak, and S. Mukamel, *J. Phys. Chem. A* **105**, 7057 (2001).
35. E. Atas, Z. Peng, and V. D. Kleiman, *J. Phys. Chem. B* **109**, 13553 (2005).
36. D. Schwarzer, P. Kutne, C. Schröder, and J. Troe, *J. Chem. Phys.* **121**, 1754 (2004).

37. Z. Wang, A. Pakoulev, and D. D. Dlott, *Science* **296**, 2201 (2002).
38. D. Schwarzer, C. Hanisch, P. Kutne, and J. Troe, *J. Phys. Chem. A* **106**, 8019 (2002).
39. D. Antoniou and S. D. Schwartz, *J. Chem. Phys.* **103**, 7277 (1995).
40. A. Zwielly, A. Portnov, C. Levi, S. Rosenwaks, and I. Bar, *J. Chem. Phys.* **128**, 114305 (2008).
41. T. Kim and P. M. Felker, *J. Phys. Chem. A* **111**, 12466 (2007).
42. Y. Yamada, Y. Katsumoto, and T. Ebata, *Phys. Chem. Chem. Phys.* **9**, 1170 (2007).
43. Y. Pang, J. C. Deak, W. T. Huang, A. Lagutchev, A. Pakoulev, J. E. Patterson, T. D. Sechler, Z. H. Wang, and D. D. Dlott, *Int. Rev. Phys. Chem.* **26**, 223 (2007).
44. V. May and O. Kühn, *Charge and Energy Transfer Dynamics in Molecular Systems*, Wiley VCH, Berlin, 2000.
45. P. Hamm, M. Lim, W. F. DeGrado, and R. M. Hochstrasser, *Proc. Natl. Acad. Sci. USA* **96**, 2036 (1999).
46. R. M. Hochstrasser, *Chem. Phys.* **266**, 273 (2001).
47. M. T. Zanni and R. M. Hochstrasser, *Curr. Opin. Struct. Biol.* **11**, 516 (2001).
48. S. Gnanakaran and R. M. Hochstrasser, *J. Am. Chem. Soc.* **123**, 12886 (2001).
49. Y. S. Kim and R. M. Hochstrasser, *J. Phys. Chem. B* **111**, 9697 (2007).
50. P. Hamm, L. H. Lim, and R. M. Hochstrasser, *J. Phys. Chem. B* **102**, 6123 (1998).
51. T. Brixner, F. J. García de Abajo, J. Schneider, and W. Pfeiffer, *Phys. Rev. Lett.* **95**, 093901 (2005).
52. M. Aeschlimann, M. Bauer, D. Bayer, T. Brixner, F. J. G. de Abajo, W. Pfeiffer, M. Rohmer, C. Spindler, and F. Steeb, *Nature* **446**, 301 (2007).
53. D. J. Tannor, *Introduction to Quantum Mechanics: A Time-Dependent Perspective*, University Science, Sausalito, 2007.
54. A. Szabo and N. S. Ostlund, *Modern Quantum Chemistry: Introduction to Advanced Electronic Structure Theory*, Dover Publications, Inc., Mineola, NY, 1996.
55. W. Koch and M. C. Holthausen, *A Chemist's Guide to Density Functional Theory*, Wiley-VCH, Weinheim, 2001.
56. R. Kosloff, *J. Phys. Chem.* **92**, 2087 (1988).
57. C. Leforestier et al., *J. Comp. Phys.* **94**, 59 (1991).
58. H. T. Ezer and R. Kosloff, *J. Chem. Phys.* **81**, 3967 (1984).
59. K. Sundermann, *Parallele Algorithmen zur Quantendynamik und optimalen Laserpulskontrolle chemischer Reaktionen*, PhD Thesis, Freie Universität Berlin, 1998.
60. J. Manz, K. Sundermann, and R. de Vivie-Riedle, *Chem. Phys. Lett.* **290**, 415 (1998).
61. K. Blum, *Density Matrix Theory and Applications*, Plenum, New York, 1981.
62. G. Lindblad, *Comm. Math. Phys.* **33**, 305 (1973).
63. G. Lindblad, *Comm. Math. Phys.* **39**, 111 (1974).
64. G. Lindblad, *Comm. Math. Phys.* **48**, 119 (1976).
65. L. Pesce, *Dissipative Quantum Dynamics of Elementary Chemical Processes at Metal Surfaces*, PhD Thesis, Freie Universität Berlin, 1998.
66. W. Huisinga, L. Pesce, R. Kosloff, and P. Saalfrank, *J. Chem. Phys.* **110**, 5538 (1999).
67. Y. Huang, D. Kouri, and D. Hoffman, *J. Chem. Phys.* **101**, 10493 (1994).
68. A. H. Zewail, *J. Phys. Chem. A* **104**, 5660 (2000).
69. H. Rabitz, R. de Vivie-Riedle, M. Motzkus, and K. Kompa, *Science* **288**, 824 (2000).

70. S. A. Rice and M. Zhao, *Optical Control of Molecular Dynamics*, Wiley-Interscience, Hoboken, 2000.
71. M. Shapiro and P. Brumer, *Quantum Control of Molecular Processes*, Wiley-VCH, Weinheim, 2012.
72. W. Zhu and H. Rabitz, *J. Chem. Phys.* **109**, 385 (1998).
73. W. Zhu, J. Botina, and H. Rabitz, *J. Chem. Phys.* **108**, 1953 (1998).
74. D. Tannor and S. A. Rice, *J. Chem. Phys.* **83**, 5013 (1985).
75. D. J. Tannor, V. Kazakov, and V. Orlov, *Time Dependent Quantum Molecular Dynamics*, Plenum, New York, 1992.
76. J. P. Palao and R. Kosloff, *Phys. Rev. Lett.* **89**, 188301 (2002).
77. J. P. Palao and R. Kosloff, *Phys. Rev. A* **68**, 062308 (2003).
78. U. Troppmann, C. Gollub, and R. de Vivie-Riedle, *New J. Phys.* **8**, 100 (2006).
79. K. Sundermann and R. de Vivie-Riedle, *J. Chem. Phys.* **110**, 1896 (1999).
80. P. Gross, D. Neuhauser, and H. Rabitz, *J. Chem. Phys.* **96**, 2834 (1992).
81. S. Shi, A. Woody, and H. Rabitz, *J. Chem. Phys.* **88**, 6870 (1988).
82. J. Somloi, V. A. Kazakov, and D. J. Tannor, *Chem. Phys.* **172**, 85 (1993).
83. C. P. Koch, J. P. Palao, R. Kosloff, and F. Masnou-Seeuws, *Phys. Rev. A* **70**, 013402 (2004).
84. C. Gollub, M. Kowalewski, and R. de Vivie-Riedle, *Phys. Rev. Lett.* **101**, 073002 (2008).
85. A. V. Oppenheimer, R. W. Schaffer, and J. Buck, *Discrete Time Signal Processing*, Prentice Hall, Upper Saddle River, NJ, 1999.
86. M. Lapert, R. Tehini, G. Turinici, and D. Sugny, *Phys. Rev. A* **79**, 063411 (2009).
87. R. Feynman, *Found. Phys.* **16**, 507 (1986).
88. D. Deutsch, *Proc. R. Soc. Lond. A* **400**, 97 (1985).
89. A. Barenco, C. H. Bennett, R. Cleve, D. P. DiVincenzo, N. Margolus, P. Shor, T. Sleator, J. A. Smolin, and H. Weinfurter, *Phys. Rev. A* **52**, 3457 (1995).
90. D. P. diVincenzo, *Science* **270**, 255 (1995).
91. C. Monroe, D. M. Meekhof, B. E. King, W. M. Itano, and D. J. Wineland, *Phys. Rev. Lett.* **75**, 4714 (1995).
92. J. I. Cirac and P. Zoller, *Phys. Rev. Lett.* **74**, 4091 (1995).
93. D. Kielpinski, C. Monroe, and J. Wineland, *Nature* **417**, 709 (2002).
94. J. P. Home, D. Hanneke, J. D. Jost, J. M. Amini, D. Leibfried, and D. J. Wineland, *Science* **325**, 1227 (2009).
95. D. G. Cory, A. F. Fahmy, and T. F. Havel, *Proc. Natl. Acad. Sci. USA* **94**, 1634 (1997).
96. P. Zanardi and F. Rossi, *Phys. Rev. Lett.* **81**, 4752 (1998).
97. E. Biolatti, R. C. Iotti, P. Zanardi, and F. Rossi, *Phys. Rev. Lett.* **85**, 5647 (2000).
98. I. D'Amico, E. Biolatti, E. Pazy, P. Zanardi, and F. Rossi, *Physica E* **13**, 620 (2002).
99. G. Burkard, H.-A. Engel, and D. Loss, *Fortschr. Phys.* **48**, 965 (2000).
100. V. Cerletti, W. A. Coish, O. Gywat, and D. Loss, *Nanotechnology* **16**, R27 (2005).
101. Y. Makhlin, G. Schön, and A. Shnirman, *Nature* **398**, 305 (1999).
102. C. H. van der Wal, F. K. Wilhelm, C. J. P. M. Harmans, and J. E. Mooij, *Eur. Phys. J. B* **31**, 111 (2003).
103. F. Helmer, M. Mariantoni, A. G. Fowler, J. von Delft, E. Solano, and F. Marquardt, *arXiv:0706.3625v1*, 2008.

104. A. Rauschenbeutel, G. Nagues, S. Osnaghi, P. Bertet, M. Brune, J. M. Raimond, and S. Haroche, *Phys. Rev. Lett.* **83**, 5166 (1999).

105. J. M. Raimond, M. Brune, and S. Haroche, *Rev. Mod. Phys.* **73**, 565 (2001).

106. A. Auffeves, P. Maioli, T. Meunier, S. Gleyzes, G. Nogues, M. Brune, J. M. Raimond, and S. Haroche, *Phys. Rev. Lett.* **91**, 230405 (2003).

107. C. Gollub, B. Korff, K. Kompa, and R. de Vivie-Riedle, *Phys. Chem. Chem. Phys.* **9**, 369 (2007).

108. B. Schneider, C. Gollub, K. Kompa, and R. de Vivie-Riedle, *Chem. Phys.* **338**, 291 (2007).

109. C. Gollub, U. Troppmann, and R. de Vivie-Riedle, *New J. Phys.* **8**, 48 (2006).

110. D. Babikov, *J. Chem. Phys.* **121**, 7577 (2004).

111. S. Suzuki, K. Mishima, and K. Yamashita, *Chem. Phys. Lett.* **410**, 358 (2005).

112. L. Bomble, D. Lauvergnat, F. Remacle, and M. Desouter-Lecomte, *J. Chem. Phys.* **128**, 064110 (2008).

113. K. Mishima, K. Takumo, and K. Yamashita, *Chem. Phys.* **343**, 61 (2008).

114. M. Ndong, L. Bomble, D. Sugny, Y. Justum, and M. Desouter-Lecomte, *Phys. Rev. A* **76**, 043424 (2007).

115. D. Weidinger and M. Gruebele, *Mol. Phys.* **105**, 1999 (2007).

116. K. Shioya, K. Mishima, and K. Yamashita, *Mol. Phys.* **105**, 1283 (2007).

117. M. Y. Zhao and D. Babikov, *J. Chem. Phys.* **126**, 204102 (2007).

118. M. Tsubouchi and T. Momose, *Phys. Rev. A* **77**, 052326 (2008).

119. M. Ndong, D. Lauvergnat, X. Chapuisat, and M. Desouter-Lecomte, *J. Chem. Phys.* **126**, 244505 (2007).

120. D. Sugny, C. Kontz, M. Ndong, Y. Justum, G. Dive, and M. Desouter-Lecomte, *Phys. Rev. A* **74**, 043419 (2006).

121. D. Sugny, M. Ndong, D. Lauvergnat, Y. Justum, and M. Desouter-Lecomte, *J. Photochem. Photobiol. A* **190**, 359 (2007).

122. T. W. Cheng and A. Brown, *J. Chem. Phys.* **124**, 034111 (2006).

123. P. Pellegrini and M. Desouter-Lecomte, *Eur. Phys. J. D.* **64**, 163 (2011).

124. R. R. Zaari and A. Brown, *J. Chem. Phys.* **135**, 044317 (2011).

125. K. Mishima and K. Yamashita, *Chem. Phys.* **379**, 13 (2011).

126. S.-H. Shim, D. B. Strasfeld, E. C. Fulmer, and M. T. Zanni, *Opt. Lett.* **31**, 838 (2006).

127. S.-H. Shim, D. B. Strasfeld, and M. T. Zanni, *Opt. Exp.* **14**, 13120 (2006).

128. D. B. Strasfeld, S.-H. Shim, and M. T. Zanni, *Phys. Rev. Lett.* **99**, 038102 (2007).

129. H. S. Tan and W. S. Warren, *Opt. Exp.* **11**, 1021 (2003).

130. M. Tsubouchi and T. Momose, *J. Opt. Soc. Am. B* **24**, 1886 (2007).

131. L. Bomble, P. Pellegrini, P. Ghesquiere, and M. Desouter-Lecomte, *Phys. Rev. A* **82**, 062323 (2010).

132. U. Troppmann, PhD thesis, Studien zur Realisierbarkeit von Molekularem Quantencomputing. Ludwig-Maximilians-Universität München, 2006.

133. J. Stahl, J. C. Bohling, E. B. Bauer, T. B. Peters, W. Mohr, J. M. Martín-Alvarez, F. Hampel, and J. A. Gladysz, *Angew. Chem. Int. Ed.* **41**, 1871 (2002).

134. L. de Quadras, E. B. Bauer, J. Stahl, F. Zhuravlev, F. Hampel, and J. A. Gladysz, *New J. Chem.* **31**, 1594 (2007).

135. R. Schmidt, A. Negretti, J. Ankerhold, T. Calarco, and J. Stockburger, *Phys. Rev. Lett.* **107**, 130404 (2011).

ULTRACOLD MOLECULES: THEIR FORMATION AND APPLICATION TO QUANTUM COMPUTING

ROBIN CÔTÉ

Department of Physics, U-3046, University of Connecticut, 2152 Hillside Road, Storrs, CT 06269-3046, USA

Advances in Chemical Physics, Volume 154: Quantum Information and Computation for Chemistry, First Edition. Edited by Sabre Kais.

I. INTRODUCTION

Historic developments in information technology have had an unprecedented impact on society. The past few decades have witnessed an explosion of computing capacity consistent with *Moore's law* asserting that processor power doubles every 18 months. This trend has been accompanied by a commensurate increase in processor density, one that cannot continue unchecked without dramatic changes to the basic circuit elements of modern processors. Indeed, existing circuit elements are reaching the *quantum limit*, the size scale where quantum phenomena begin to dominate the physics governing their behavior. While this prediction itself motivates the study of "quantum" information and computation, discoveries of the past decade have revealed that *quantum phenomena can be exploited to yield a qualitatively new computational paradigm* [1], *one that offers provable speedup over its classical counterpart for a wide variety of computational problems*. Indeed, efficient quantum algorithms have been designed for problems believed to be intractable for classical computation, such as integer factorization [2] or large database search [3].

Although this promise of a new, provably more efficient, frontier of computation is exciting, construction of general purpose quantum computing equipment remains a significant engineering challenge. In fact, it is essential to address coherently quantum states in a system and to perform reversible quantum logic operations. This can be achieved only if the qubits can interact via controlled coherent physical process; preserving coherence is crucial because quantum interference and entanglement are extremely fragile. Many systems are being studied to manipulate quantum information based mainly on atomic or molecular platforms, condensed matter systems, and nonlinear optical setups.

Polar molecules present a promising new platform for quantum computation [4,5], because they incorporate the prime advantage of both neutral atoms [6,7] and trapped ions [8–10], that is long coherence times and strong interactions, respectively [11,12]. For example, schemes using entanglement of vibrational eigenstates [13,14] and optimal control [15] have been proposed. Here, we describe schemes based on polar molecules, where the dipolar interaction is crucial to implement quantum gates. The advances in cooling [16–20] and storing [21–24] techniques are beginning to make the precise manipulation of single molecules possible. In addition, polar molecules could be integrated into condensed matter physics architectures, using, for example, molecule-chips [25–28] or microtraps connected to superconducting wires [11,12,29,30]. In a recent article [11], the experimental implementation of quantum information processing using superconducting stripline resonators has been studied in detail.

We first discuss the approaches used to obtain ultracold molecules. We give an overview of direct methods to cool molecules, such as buffer gas cooling, stark deceleration, and direct laser cooling. This overview is followed by a description of indirect methods to form ultracold molecules based on the photoassociation of ultracold atoms using one- and two-photon transitions. We then proceed to a detailed discussion of the use of Feshbach resonances to enhance the formation rate of ultracold molecules, and their role in efficient coherent transfer of atom pairs into molecules (and vice versa).

Second, after a brief review of some basic concepts in quantum information processing, such as entangled states and quantum logic gates, we list various systems considered to implement these ideas. We focus our attention on a particular one, namely polar molecules. After a brief overview of their properties, we show how the strong interaction between them can be used to implement universal two-qubit logic gates (so called *phase gates*), from which any quantum algorithm can be constructed. We describe two main schemes: one based on large permanent dipole moments, and one based on "dipolar switching," where dipole moments are switched on and off at will. Finally, we discuss some sources of decoherence and errors and how they can be controlled.

II. ULTRACOLD MOLECULE FORMATION

We begin with a short overview of the key ingredient for quantum information processing with ultracold molecules, namely how to obtain ultracold molecules. We briefly review existing techniques, and then focus on a particular approach based on photoassociation near a Feshbach resonance, and its extension to coherent atom-molecule transfer.

A. Overview

In recent years, we have witnessed rapid developments in the production of cold and ultracold molecules [31–34]. The approaches used can be divided in two broad categories: direct slowing and cooling techniques, and indirect building/assembling with already ultracold components.

1. Direct Methods

The three main approaches actively pursued experimentally are Stark deceleration, buffer gas cooling, and direct laser cooling.

- *Stark deceleration*: This technique relies on the Stark effect to decelerate molecules with a permanent dipole moment. It was first developed by Meijer's group [35–37], using time-varying inhomogeneous electric fields arranged in a linear array (see Fig. 1a). Typically, an incident molecular beam is prepared via supersonic expansion to cool both the internal and external degrees of freedom, and spatially inhomogeneous electric fields (acting via the

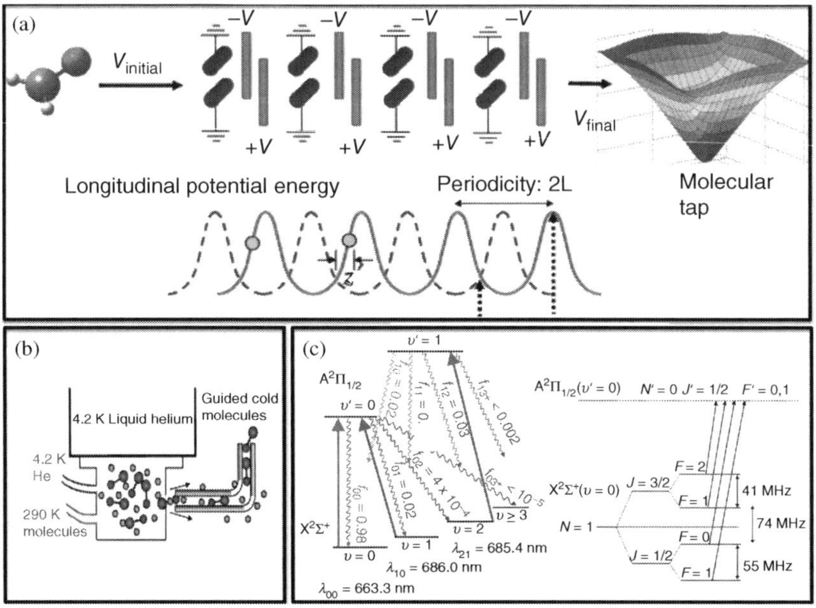

Figure 1. Direct cooling methods. (**a**) Stark decelerator. Top: a polar molecule passes through sets of electrodes with alternating voltage, and finally reaches a trap. Bottom: potential energy due to the stark energy shift of the molecules in a weak-field-seeking state moving in the spatially inhomogeneous electric fields. (**b**) Buffer gas cooling: molecules thermalize by elastic collisions with the cold He atoms, and are guided out of the chamber. (**c**) Laser cooling. Left: relevant electronic and vibrational structure in SrF (solid upward lines indicate laser-driven transitions at wavelengths λ). Right: Relevant rotational energy levels, splittings and transitions (vertical arrows) in the SrF cycling scheme. Parts (a) and (b) are reproduced from Ref. [32] by permission from the Institute of Physics (IOP), and (c) from Ref. [48] by permission from the Nature Publishing Group.

Stark energy shifts of the traveling molecules) are switched synchronously as the molecules travel. As a molecule prepared in its weak-field-seeking state moves toward an increasing field strength, its longitudinal kinetic energy is converted to potential energy, its kinetic energy decreases, and the molecule slows down. Before the molecule exits the high-field region and starts going down and accelerates, the electric field is switched so that the net effect is to remove energy from the molecule. This process is repeated with successive stages of electrodes until the molecules have been slowed to the desired velocity. At the output of the decelerator, the slowed molecules can be loaded into a trap. Current experiments typically decelerate molecules to a mean speed adjustable between 600 m/s to rest, with a translational temperature tunable from 10 mK to 1 K, corresponding to a longitudinal velocity spread from a few ~1–100 m/s. These slowed packets contain 10^4–10^6 molecules at a density of ~10^5–10^7 cm^{-3} in the beam and ~10^6 cm^{-3} in a trap. A variety of polar molecules (CO [35], OH [36], ND_3 and NH [37], LiH [38], H_2CO [39], YbF [40], and SO_2 [41]) have been decelerated. For molecules without electric dipole moment, but with a magnetic dipole moment, one can use decelerators based on the Zeeman effect, which employs inhomogeneous magnetic fields instead [42,43].

- *Buffer gas cooling*: In the approach pioneered by Doyle's group at Harvard [44], a cold gas of helium atoms is used to cool atoms or molecules *via* elastic collisions. Figure 1b illustrates the principle of this powerful and versatile method; warm molecules are "injected" (e.g., by laser ablation or other sources) into a chamber where they collide elastically with cryogenically cooled He atoms, resulting in momentum transfer and thus cooling of the molecule. The cold molecules can be trapped inside the chamber (e.g., by a magnetic field if they possess a magnetic moment) or guided out of the chamber (e.g., through an orifice with curved electric or magnetic field guides so that only sufficiently slow molecules make it to the target final location). This technique has been used to cool a large variety of molecular species [34,44,45], such as CaH, CaF, and NH, at temperatures in the range of 250 mK to 1 K, and molecular densities reaching values larger than 10^9 cm^{-3}. It also is used to form van der Waals molecules [46] and cool atomic species [47].
- *Laser cooling*: This method has been extremely successful to cool atoms and reach the sub-mK regime, using optical cycling transitions. However, because of their complex internal structure, laser cooling of molecules has been only recently demonstrated for strontium monofluoride (SrF) [48]. Although this approach requires particular molecular level structure (see Fig. 1c for SrF), for example, with large Franck–Condon factors between only a few levels, it provides an alternative route to obtain ultracold molecules, bridging the gap between ultracold (sub-mK) temperatures reached by indirect methods (see below), and the 1 K temperatures attainable with the other direct methods. In addition, this technique should allow the production of large samples of molecules at ultracold temperatures for species that are chemically distinct from bi-alkalis.

2. *Indirect Method: Photoassocation*

The indirect method consists in photoassociating atoms that are already ultracold to produce ultracold molecules. Although in principle it could be used to form trimers from three atoms (or from a dimer and an atom), and so on, it has been successful in producing mostly diatomic molecules of alkali and alkaline earth

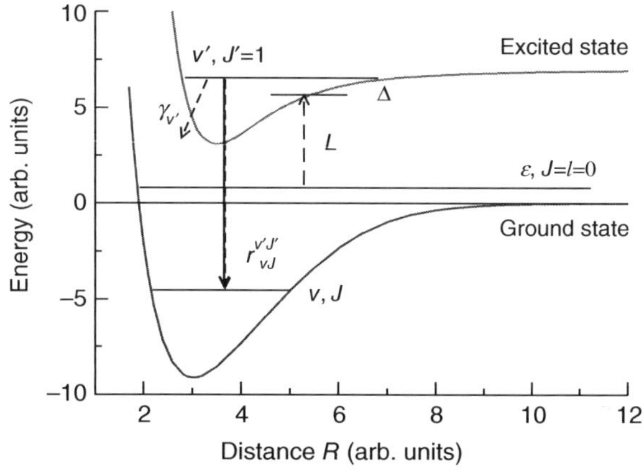

Figure 2. Schematics of PA: two colliding atoms of energy ε make a transition to a bound level v' (of width $\gamma_{v'}$) induced by a laser L of intensity I and detuning Δ from v'. This excited ultracold molecule can decay to a lower level v, J with a branching ratio $r_{vJ}^{v'J'}$.

atoms. This approach leads to a range of variants to produce ultracold molecules, some relying on spontaneous decay, others on Raman-like excitations or stimulated Raman adiabatic passage (STIRAP). We first describe the one- and two-photon photoassociation (PA) process for the formation of diatomic molecules.

In general, a pair of atoms approaching each other along the molecular ground state in the presence of an optical field can make a transition into a molecular bound level v' of an excited electronic state (see Fig. 2). The corresponding one-photon photoassociation rate coefficient $\mathsf{K}_{v'}^{(1)}$ from a laser $L = \{I, \Delta\}$ (with intensity I and detuning Δ from a bound level (v', J') is given by Refs [49–52]

$$\mathsf{K}_{v'}^{(1)}(T, L) = \left\langle \frac{\pi v_{\text{rel}}}{k^2} \sum_{\ell=0}^{\infty} (2\ell + 1) |S_{\ell,v'}(\varepsilon, L)|^2 \right\rangle \tag{1}$$

where $\varepsilon = \hbar^2 k^2/2\mu = \mu v_{\text{rel}}^2/2$, μ is the reduced mass, v_{rel} is the relative velocity of the colliding pair, and $S_{\ell,v'}$ represents the scattering matrix element for producing the state v' from the continuum state. Averaging over v_{rel} is implied by $\langle \ldots \rangle$.

At ultracold temperatures, only the s-wave ($\ell = 0$) scattering is relevant (for distinguishable atoms or identical bosons). Assuming a nondegenerate gas, keeping only the $\ell = 0$ contribution, and averaging over a Maxwellian velocity distribution characterized by the temperature T, we obtain [50–52]

$$\mathsf{K}_{v'}^{(1)}(T, L) = \frac{1}{hQ_T} \int_0^{\infty} d\varepsilon \; e^{-\varepsilon/k_B T} |S_{\ell=0,v'}(\varepsilon, L)|^2 \tag{2}$$

where $Q_T = \left(\frac{2\pi\mu k_B T}{h^2}\right)^{3/2}$. We note that only $(v', J' = 1)$ rovibrational levels can be populated from the $\ell = 0$ partial wave. The scattering matrix is well approximated by [50,52]

$$|S_{\ell,v'}(\varepsilon, L)|^2 = \frac{\gamma_{v'}\gamma_s(\varepsilon, \ell)}{(\varepsilon - \Delta)^2 + (\gamma/2)^2} \tag{3}$$

where $\gamma = \gamma_{v'} + \gamma_s$. Here, $\gamma_{v'}$ is the natural width of the bound level v' and γ_s is the stimulated width from the continuum initial state $|\varepsilon, \ell = 0\rangle$ to the target state $|v', J' = 1\rangle$. The stimulated width γ_s can be expressed using the Fermi golden rule as [50–52]

$$\gamma_s \simeq \frac{\pi I}{\epsilon_0 c}|D_{v'}(\varepsilon)|^2 \equiv 2\pi\hbar^2\Omega^2_{\varepsilon v'} \tag{4}$$

where ϵ_0 and c are the vacuum permittivity and speed of light, respectively, and $D_{v'}(\varepsilon) \equiv \langle\phi_{v',J'=1}|D|\Psi_{\varepsilon,\ell=0}\rangle$ is the dipole transition matrix element, which depends on the molecular dipole transition moment $D(R)$ connecting the ground and excited electronic states, and $\phi_{v',J'=1}(R)$ and $\Psi_{\varepsilon,\ell=0}(R)$ stand for the wave functions of the excited level v' and the colliding pair of atoms, respectively (note that those three quantities vary with R). The expression above also defines a Rabi frequency $\Omega_{\varepsilon v'}$ between the continuum and bound states (following the convention of Ref. [50]). We note that the dipole matrix element appearing in Eq. (4) follows Wigner's threshold law [53], that is, $|D_{v'}(\varepsilon)|^2 = C_{v'}\sqrt{\varepsilon}$, where the coefficient $C_{v'}$ depends on the details of the wave functions [54–56]. It is proportional to the Franck–Condon overlap between the initial continuum scattering and the target bound level wave functions; the better the overlap, the larger the $C_{v'}$.

If $\gamma_s/\gamma_{v'} \ll 1$, we can approximate $|S_{\ell,v'}|^2$ by $2\pi\gamma_s(\varepsilon, \ell)\delta(\varepsilon - \Delta)$ [50–52,55,56]. Substituting this expression in Eq. (2), and using $\gamma_s \propto C_{v'}\sqrt{\varepsilon}$, we find $\mathsf{K} \propto e^{-\Delta/k_BT}\sqrt{\Delta}$, with a maximum value for $\Delta = k_BT/2$. The rate coefficient then takes the simple form [55,56]

$$\mathsf{K}^{(1)}_{v'}(T, I) = \frac{2\pi^2}{h}\frac{I}{\epsilon_0 c}\frac{e^{-1/2}}{Q_T}C_{v'}\sqrt{k_BT/2} \tag{5}$$

Note that we have not accounted for the light polarization in the preceding expression. The numerical prefactors are simple to evaluate. For example, if $C_{v'}$ is in a.u., I in W/cm^2, and T in Kelvin, $\mathsf{K}^{(1)}_{v'}$ for Li_2 and LiH is simply given by

$$\mathsf{K}^{(1)}_{v'}(T, I) = C_{v'}\frac{I}{T} \times \begin{cases} 1.1 \times 10^{-24}\ \text{cm}^3/\text{s}, & \text{for Li}_2 \\ 9.0 \times 10^{-24}\ \text{cm}^3/\text{s}, & \text{for LiH} \end{cases} \tag{6}$$

and can be found for any other diatomic system because it scales as $\mu^{-3/2}$ according to the expression for Q_T.

The level $(v', J' = 1)$ of the excited electronic state will undergo spontaneous radiative decay into allowed states of the lower electronic state, in a distribution of the continuum and rovibrational levels, characterized by the lifetime

$$\tau_{v'} = \frac{1}{A_{v'}^{\text{tot}}} \tag{7}$$

The branching ratio for radiative decay from an initial $(v', J' = 1)$ into a specific bound level (v, J) of the electronic ground state is simply given by

$$r_{vJ}^{v'J'}(\alpha) \equiv \frac{A_{v'v}^{\alpha}}{A_{v'}^{\text{tot}}} \tag{8}$$

where α stands for the specific allowed branch (P, Q, or R-branch).

We can describe a two-photon stimulated Raman photoassociation process by the rate coefficient $\mathsf{K}^{(2)}$ (Fig. 3). If the value of Δ is large when compared to the natural width $\gamma_{v'}$ of the intermediate level v', we can use an effective Rabi frequency formulation [57] and write the two-photon Raman rate coefficient $\mathsf{K}^{(2)}$ in terms of the one-photon photoassociation rate $\mathsf{K}^{(1)}$ to v' and the ratio of the bound–bound Rabi frequency $\Omega_{vv'}$ and the detuning Δ

$$\mathsf{K}_{vv'}^{(2)}(T, \{L\}) = \mathsf{K}_{v'}^{(1)}(T, I_1, \delta)\left(\frac{\Omega_{vv'}}{\Delta}\right)^2 \tag{9}$$

Here, $\{L\} \equiv \{L_1, L_2\}$ stands for the various laser parameters. Note that $\mathsf{K}^{(1)}$ is computed using the two-photon detuning δ. This approximation clearly fails when

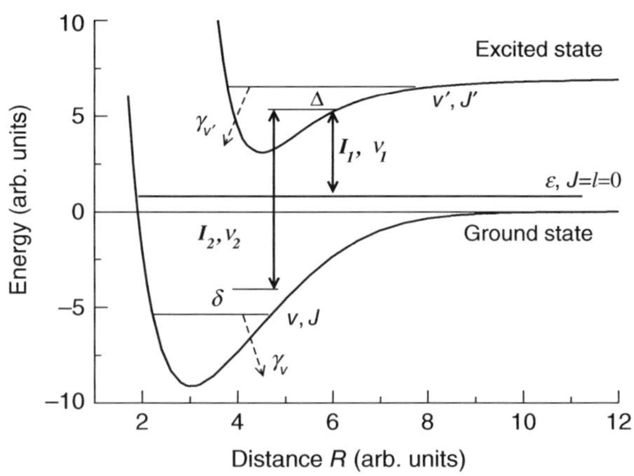

Figure 3. Two-photon PA between the continuum state and a bound level v, J of the ground electronic state using an intermediate level v', J' of an excited electronic state.

$\Delta \to 0$; if $\Delta/\gamma_{v'}$, the ratio of the detuning to the spontaneous decay width of level v', is not large enough, Autler–Townes splittings and large spontaneous decays would need to be taken into account. Note also that it is straightforward to rederive the previous result using an effective Rabi frequency for the two-photon process (when Δ is far detuned from v'): $\Omega_{\rm eff} = \Omega_{\varepsilon v'}\Omega_{v'v}/\Delta$. Writing $\hbar^2\Omega^2_{v'v} = (I_2/2\epsilon_0 c)|D_{v'v}|^2$, where $D_{v'v} \equiv \langle \phi_{v',J'=1}(R)|D(R)|\Psi_{v,J}(R)\rangle$ is the dipole transition matrix element between the intermediate level (v', J') and the target state (v, J), and using the result of Eq. (5) for $\mathsf{K}^{(1)}$, Eq. (9) becomes

$$\mathsf{K}^{(2)}_{vv'}(T, \{L\}) = \frac{4\pi^4}{h^3}\frac{I_1 I_2}{\epsilon_0^2 c^2}\frac{e^{-1/2}}{Q_T}\sqrt{\frac{k_B T}{2}}\,C_{v'}\frac{|D_{v'v}|^2}{\Delta^2} \tag{10}$$

Finally, a rate R of molecules formed per second can be obtained if we multiply the appropriate photoassociation rate coefficient K by the densities of each of the atomic species, n_1 and n_2, and by the volume V illuminated by the laser beam(s). For both one- and two-photon processes, we write

$$\mathsf{K}_v = \begin{cases} r^{v'J'}_{vJ}(\alpha)\mathsf{K}^{(1)}_{v'}(T, I_1) & \text{one-photon} \\ \mathsf{K}^{(2)}_{vv'}(T, \{L\}) & \text{two-photon} \end{cases} \tag{11}$$

and the formation rate of molecules simply becomes

$$\mathsf{R}_v = \chi n_1 n_2 \mathsf{K}_v V \tag{12}$$

Here, χ is the fraction of collision with the appropriate symmetry (e.g., for nonspin polarized alkali atoms colliding, 1/4 of the collisions are along the singlet manifold and 3/4 along the triplet manifold; if the hyperfine structure is accounted for, χ will depend on the exact nuclear spin as well); to form bi-alkali molecules in their ground state (singlet symmetry), 1/4 of the collisions will have the right symmetry.

We note that the number of molecules formed is often not very large, because it depends on the overlap of the initial continuum wave function with the intermediate state v', and of the overlap of v' with the lower states v. Usually, the Franck–Condon overlap between the continuum and target bound level wave functions is good only for high-lying excited rovibrational levels, that is, for very extended bound levels; for more tightly bound states, $C_{v'}$ are usually many orders of magnitude smaller, leading to poor photoassociation rates. Similarly, the branching ratio $r^{v'J'}_{vJ}$ is rarely above 0.01, and if large laser intensities are used with the two-photon approach, back-stimulation into the continuum becomes important [56]. In the following sections, we describe approaches to circumvent these problems.

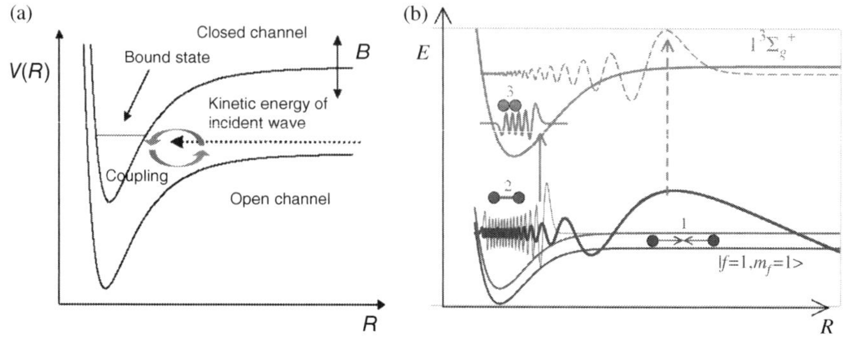

Figure 4. (**a**) A Feshbach resonance (tuned by the magnetic B-field) between two coupled channels occurs when the energy of a pair of scattering atoms (in an open channel) is degenerate with the energy of a bound state (in a closed channel). (**b**) FOPA: Colliding atoms (1) interact via open and closed channels due to hyperfine interactions. A Feshbach resonance occurs when a bound level (2) wave function coincides with the continuum state (1) wave function. A photon (wavelength λ) can associate the atoms into a more deeply bound level v (3) of the excited state potential, as opposed to the extended high-lying level (dashed line). Part (b) is reproduced from Ref. [63] by permission of the Institute of Physics (IOP).

B. Feshbach Optimized Photoassociation (FOPA)

As mentioned in the previous section, it is difficult to find levels with the appropriate overlaps in order to form large amount of ultracold molecules. To enhance the probability density at short range, hence increasing $C_{v'}$, $\mathsf{K}_{v'}^{(1)}$, and $\mathsf{K}_{vv'}^{(2)}$, one can use a magnetically induced Feshbach resonance [58]. A Feshbach resonance occurs when the energy of the colliding atoms coincides with the energy of a bound state in a closed channel [59] (see Fig. 4). FOPA allows transitions to deeply bound levels .

We determine $|\Psi_{\varepsilon,\ell=0}\rangle$ in Eq. (4) by solving the Hamiltonian for two colliding atoms in a magnetic field [58,59]

$$H = \frac{p^2}{2\mu} + V_C + \sum_{j=1}^{2} H_j^{\text{int}} \tag{13}$$

where $V_C = V_0(R)P^0 + V_1(R)P^1$ is the Coulomb interaction, decomposed into singlet (V_0) and triplet (V_1) molecular potentials, with the associated projection operator P^0 and P^1. The internal energy of atom j, $H_j^{\text{int}} = \frac{a_{\text{hf}}^{(j)}}{\hbar^2}\vec{s}_j \cdot \vec{i}_j + (\gamma_e \vec{s}_j - \gamma_n \vec{i}_j) \cdot \vec{B}$, consists of the hyperfine and Zeeman contributions, respectively. Here, $\vec{s}_j$ and $\vec{i}_j$ are the electronic and nuclear spins of atom j with hyperfine constant

$a_{\rm hf}^{(j)}$, and $\vec{B}$ is the magnetic field. We solve for $|\Psi_{\varepsilon,\ell=0}\rangle$, which can be expanded onto the basis constructed from the hyperfine states of both atoms,

$$|\Psi_{\varepsilon,\ell=0}\rangle = \sum_{\sigma=1}^{N} \psi_\sigma(R)\{|f_1, m_1\rangle \otimes |f_2, m_2\rangle\}_\sigma \tag{14}$$

where $\vec{f}_j = \vec{i}_j + \vec{s}_j$ is the total spin of atom j, and m_j its projection on the magnetic axis. Here, $\psi_\sigma(R)$ stands for the radial wave function associated with channel σ labeled by the quantum numbers f_i, m_i.

As an example, we consider PA to the $1^3\Sigma_g^+$ excited molecular state of Li_2 from a gas of ultracold ^{7}Li atoms initially prepared in the $|f = 1, m = 1\rangle$ hyperfine state: this system has been explored in recent experiments of Hulet and coworkers [60]. We note that the scattering properties for these entrance channels are well understood and have been experimentally investigated [60–62]. We used the ^{7}Li potential curves and transition dipole moment described in Côté et al. [54], adjusted to reproduce the Feshbach resonance at 736 G for the $|f = 1, m = 1\rangle$ entrance channel [60] (see Fig. 5). With these curves, we calculated $\mathsf{K}_{v'}$ to $v' = 83$ of the $1^3\Sigma_g^+$ state as a function of the B-field at $T = 10\ \mu$K [63]. We choose these values of v and T to compare with a recent measurement [60] at higher I (see next). Figure 5 shows $\mathsf{K}_{v'=83}$ for $I = 1$ mW/cm^2, which displays a maximum at resonance ($B \sim 736$ G) and a minimum near 710 G, spanning over eight orders of magnitude. These results are similar to those of our original work [58], where we considered forming LiNa in the ground $X^1\Sigma^+$ state directly from the continuum, starting with

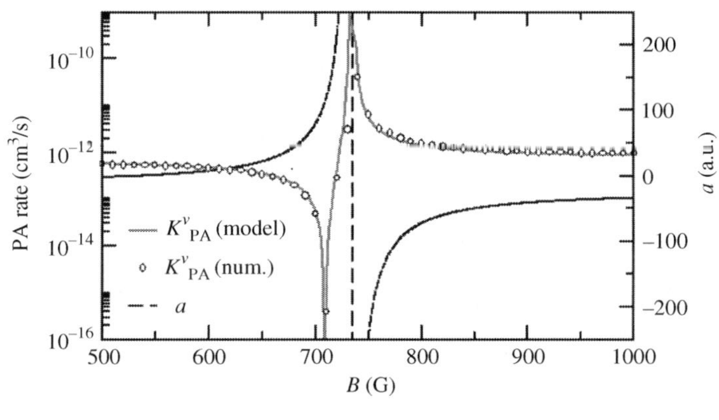

Figure 5. $\mathsf{K}_{v'}$ into $v' = 83$ versus B for $I = 1$ mW/cm^2 and $T = 10\ \mu$K (circle: full coupled-channel, solid curve: two-channel model). $\mathsf{K}_{v'=83}$ increases by more than four orders of magnitude near the resonance. The right axis shows the scattering length of two ^{7}Li atoms in $|f = 1, m = 1\rangle$, which is well fitted by $a = a_{\rm bg}\left(1 - \frac{\Delta}{B - B_{\rm res}}\right)$ with $a_{\rm bg} = -18$ a.u., $\Delta = -200$ G, and $B_{\rm res} = 736$ G. Reproduced from Ref. [63] by permission of the Institute of Physics (IOP).

^{6}Li($f=\frac{1}{2}, m=-\frac{1}{2}$) and ^{23}Na($f=1, m=-1$), with eight channels with total projection $M=m_1+m_2=-\frac{3}{2}$ being coupled by the Hamiltonian (13).

The numerical results shown in Fig. 5 can also be understood in terms of a simple two-coupled channel model. Following [58], the stimulated width γ_s can be expressed in terms of the resonant s-wave phase shift δ as

$$\gamma_s(I, k, B) = \gamma_s^{\rm off}\, |1 + C_1 \tan\delta + C_2 \sin\delta|^2 \tag{15}$$

where $\gamma_s^{\rm off}(I, k)$ is the stimulated rate off-resonance ($\delta = 0$) given by Eq. (4), and C_1 and C_2 are ratios of dipole transition matrix elements [58]. Ignoring saturation effects, the rate coefficient is then given by

$$\mathsf{K}_{v'} = \mathsf{K}_{v'}^{\rm off}\, |1 + C_1 \tan\delta + C_2 \sin\delta|^2 \tag{16}$$

where $\mathsf{K}_{v'}^{\rm off}$ is the off-resonance rate coefficient ($\delta = 0$) given by Eq. (5). To the first order in k, the resonant and background s-wave phase shifts are related to the scattering length a by $\tan(\delta + \delta_{\rm bg}) = -ka$, with $\delta_{\rm bg} = -ka_{bg}$ and [59]

$$a = a_{bg}\left(1 - \frac{\Delta_B}{B - B_{\rm res}}\right) \tag{17}$$

where $a_{\rm bg}$ is the background scattering length of the pair of atoms (which can vary slowly with B), $B_{\rm res}$ is the position of the resonance, and Δ_B is its width [59]. In Fig. 5, we compare the coupled channel numerical results with the two-channel model using $\gamma_s^{\rm off} = 0.8$ kHz, and the fitted values $C_1 = 6$, and $C_2 = 1$ (see Fig. 5 for the values of $a_{\rm bg}$, Δ, and $B_{\rm res}$); the agreement is excellent.

Figure 6 illustrates $\mathsf{K}_{v'}$ for many excited levels v' of the $1^3\Sigma_g^+$ state with $I = 1$ mW/cm^2: they basically behave the same way. The figure also shows another characteristic of FOPA: the sensitivity of $\mathsf{K}_{v'}$ to the exact level v' being probed. In fact, whereas the maximum of $\mathsf{K}_{v'}$ near the resonance at 736 G does not vary much, the exact position of the minimum is very sensitive to v: it follows a "croissant trajectory" as v' decreases, first moving to lower B at large v' and then to higher B at smaller v', even passing to the right of the resonance at lower v'. The reason for this sensitivity resides in the nature of the minimum: it comes from poor overlap of the target level v' and scattering wave functions, reflected in C_1 and C_2 of Eq. (16). Any small difference in the target wave functions will be amplified and will result in a shift of the minimum. It is worth noting that such a shift was observed in recent measurements for $v = 82 - 84$ [60], consistent with our results. This sensitivity to minute changes in the target level wave function could actually be used as a spectroscopic tool for high-precision measurements of both the scattering and bound states.

The preceding discussion does not account for saturation, which can become a dominant effect near a Feshbach resonance. If we consider $\mathsf{K}_{v'}$ of Eq. (2) with

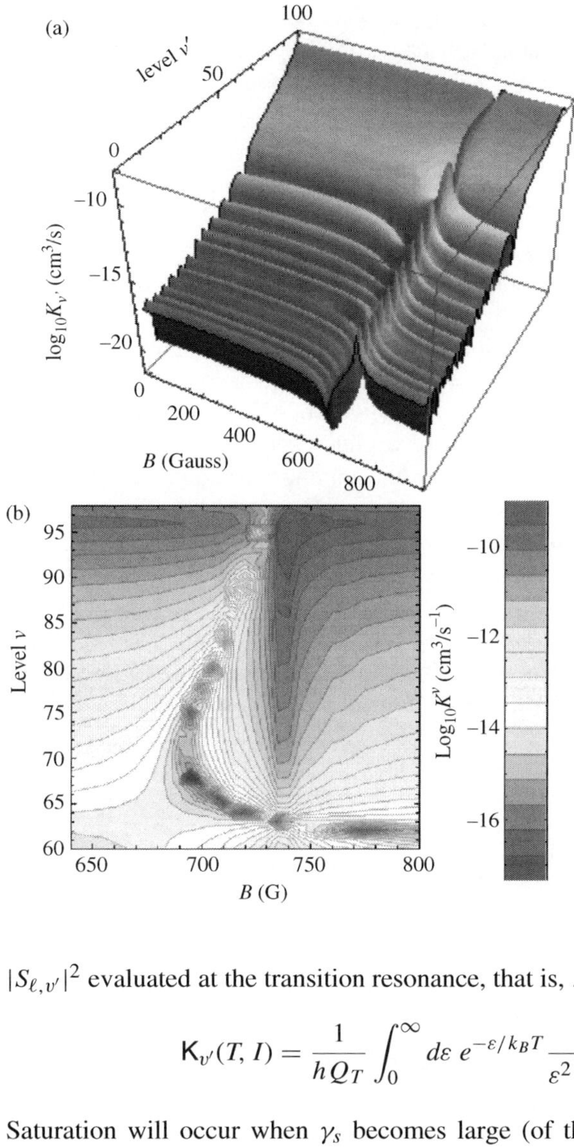

Figure 6. (**a**) $\mathsf{K}_{v'}$ into $1^3\Sigma_g^+$ of $^7Li_2$ versus B for all v'. (**b**) Contour plot showing the maximum always at $B_{\rm res}$ and the minimum following a"croissant shape" trajectory. Part (b) reproduced from Ref. [63] by permission of the Institute of Physics (IOP).

$|S_{\ell,v'}|^2$ evaluated at the transition resonance, that is, $\Delta = 0$ in Eq. (3), we have

$$\mathsf{K}_{v'}(T, I) = \frac{1}{hQ_T}\int_0^\infty d\varepsilon\, e^{-\varepsilon/k_BT}\frac{\gamma_{v'}\gamma_s}{\varepsilon^2 + (\frac{\gamma_{v'}+\gamma_s}{2})^2} \tag{18}$$

Saturation will occur when γ_s becomes large (of the order of $\gamma_{v'}$), which can take place because one increases the laser intensity I, or if one is near a Feshbach resonance. For a pair of atoms experiencing s-wave scattering, we have $\gamma_s^{\rm off}(I, k) \propto Ik$ as $k \to 0$, in agreement with Wigner's threshold law [53], while $\tan\delta(k) = -ka_{\rm bg}\frac{\Delta}{B-B_{\rm res}}$; although $\gamma_s^{\rm off}$ becomes small as $k \to 0$, the $\tan\delta(k)$ term in γ_s can become very large if $B \sim B_{\rm res}$. The maximum PA rate coefficient reachable for a given temperature is obtained when the scattering matrix element in Eq. (18)

is equal to its maximum value of one: $\gamma_{v'}\gamma_s/[\varepsilon^2 + (\frac{\gamma_{v'}+\gamma_s}{2})^2] = 1$. The integration over the Maxwell–Boltzmann velocity distribution can then be performed so that Eq. (18) gives

$$\mathsf{K}_{v'}^{\text{limit}}(T) = \frac{k_B T}{h Q_T} = \frac{h^2}{(2\pi\mu)^{3/2}} \frac{1}{\sqrt{k_B T}} \tag{19}$$

$\mathsf{K}_{v'}^{\text{limit}}(T)$ for a thermal gas increases as the temperature drops.

Saturation near a Feshbach resonance, as well as the other features discussed can be seen in a recent experiment by Hulet and coworkers [60]. We compare our calculation with measurements obtained in Ref. [60]; Figure 7 shows our multichannel results at $T = 10$ μK and $I = 1.6$ W/cm^2 together with experimental measurements, where the atom loss rate K_P is twice our $\mathsf{K}_{v'}$, since two atoms are lost during the PA process. Our results are within or close to the uncertainty of Hulet and coworkers [60]; when taking into account the variation in temperatures between data points (~9–18 μK [60]), the discrepancy vanishes. Figure 7 also shows enlargements around the minimum and the maximum. We also note that the experimental points at saturation near the resonance are the least accurate, and that the point at 745 G was particularly challenging to obtain [60], consistent with the large and negative scattering length a (i.e., attractive interaction, see Fig. 5). The extremely good agreement between our results (obtained without any adjustable parameters) and the experimental data shows that this treatment captures all the relevant physics in these binary processes.

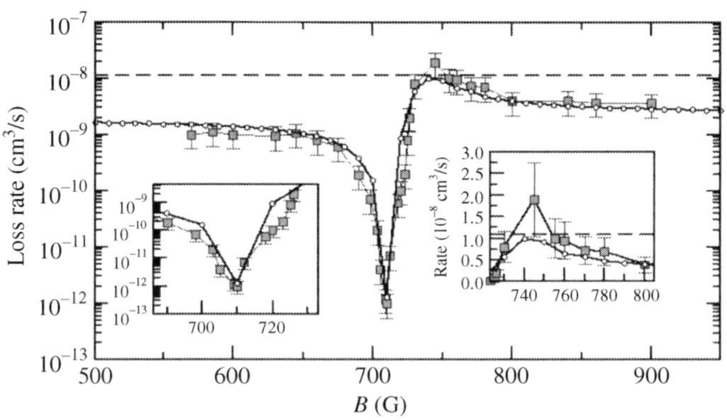

Figure 7. Comparison of $2\mathsf{K}_{v'=83}$ with experimental data [60] at 10 μK and 1.6 W/cm^2. The insets show the minimum and maximum (linear scale); only the point at 745 G is outside the uncertainty of 45% [60]. The dashed horizontal line corresponds to twice the saturation limit of 5.5×10^{-9} cm^3/s. Reproduced from Ref. [63] by permission of the Institute of Physics (IOP).

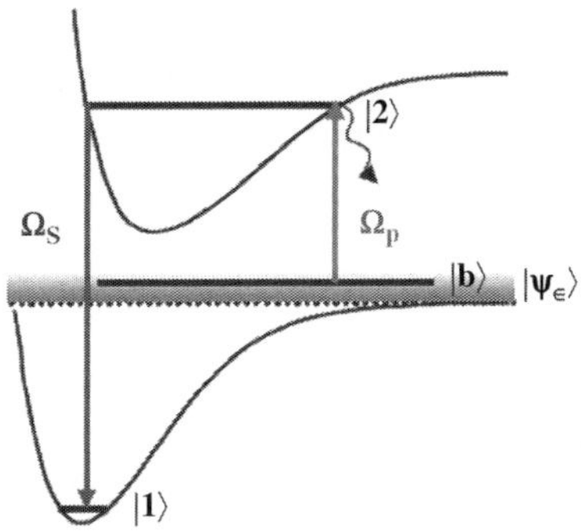

Figure 8. Schematics: population from the initial state $|\Psi_\epsilon\rangle$ is transferred to a final target state $|1\rangle$ via an intermediate state $|2\rangle$. Both $|\Psi_\epsilon\rangle$ and $|1\rangle$ are coupled to $|2\rangle$ by a pump and a Stokes pulse, respectively labeled Ω_P and Ω_S. A bound level $|b\rangle$ corresponding to a closed channel can be imbedded in the continuum. Reproduced from Ref. [64] by permission of the Institute of Physics (IOP).

C. STIRAP with FOPA

As mentioned previously, the large increase in γ_s near a Feshbach resonance could be used to obtain large molecule formation rates via one- or two-photon processes. Another advantage of FOPA relates to the possibility of performing STIRAP directly from the continuum. We review here a simple model exhibiting all relevant features.

We consider a three-level system plus a continuum as shown in Fig. 8, representing scattering states of two colliding atoms and bound states of a molecule [64]. The ground level labeled $|1\rangle$ is the final product state to which a maximun of population must be transfered (typically the lowest vibrational level ($v = 0$, $J = 0$) of a ground molecular potential). This ground level is coupled to an excited bound level $|2\rangle$ of an excited molecular potential via a "Stokes" pulse depicted by the down-arrow in Fig. 8. This level $|2\rangle$ is itself coupled via a pump pulse (up-arrow) to an initial continuum of unbound scattering states $|\Psi_\epsilon\rangle$ of energies ϵ (shaded area in Fig. 8). If we denote $C_1(t)$, $C_2(t)$, and $C(\epsilon, t)$ the time dependent amplitudes associated to the final, intermediate, and initial states $|1\rangle$, $|2\rangle$, and $|\Psi_\epsilon\rangle$, respectively, then the total wave function $|\Phi\rangle$ of the system is given by

$$|\Phi\rangle = C_1(t)\,|1\rangle + C_2(t)\,|2\rangle + \int d\epsilon\; C(\epsilon, t)\,|\Psi_\epsilon\rangle \qquad (20)$$

We assume that the levels associated with states $|1\rangle$ and $|2\rangle$ are well-isolated, and no off-resonant laser couplings to other levels: this ensures the sufficient accuracy of the three-state model (see, e.g., Refs [65–67]).

We consider the continuum state $|\Psi_\epsilon\rangle$ to be a multichannel scattering state in which a bound level $|b\rangle$ associated to a closed channel is embedded in the continuum of scattering states $|\epsilon'\rangle$ of an open channel. As described in the previous section, when the energy of $|\epsilon'\rangle$ coincides with that of $|b\rangle$, a Feshbach resonance [59] occurs. Following the Fano theory presented in Ref. [68], the scattering state $|\Psi_\epsilon\rangle$ can be expressed as [64]:

$$|\Psi_\epsilon\rangle = a(\epsilon)\,|b\rangle + \int d\epsilon'\; b(\epsilon,\epsilon')\,|\epsilon'\rangle \tag{21}$$

with

$$a(\epsilon) = \sqrt{\frac{2}{\pi\Gamma(\epsilon)}}\,\sin\delta \tag{22}$$

$$b(\epsilon,\epsilon') = \frac{1}{\pi}\sqrt{\frac{\Gamma(\epsilon')}{\Gamma(\epsilon)}}\frac{\sin\delta}{\epsilon-\epsilon'} - \cos\delta\;\cdot\;\delta(\epsilon-\epsilon') \tag{23}$$

Here, $\delta = -\arctan(\frac{\Gamma}{2(\epsilon-\epsilon_R)}) \in [-\pi/2,\pi/2]$ is the phase shift due to the interaction between $|b\rangle$ and the scattering state $|\epsilon\rangle$ of the open channel. The width of the Feshbach resonance, $\Gamma = 2\pi|V(\epsilon)|^2$, is weakly dependent on the energy, while $V(\epsilon)$ is the interaction strength between the open and closed channels. The position of the resonance, $\epsilon_R = E_b + P\int\frac{|V(\epsilon')|^2 d\epsilon'}{\epsilon-\epsilon'}$, includes an interaction-induced shift from the energy of the bound state E_b.

If we label E_i the energy of the state $|i\rangle$, the total Hamiltonian H is given by

$$H = \sum_{i=1,2} E_i|i\rangle\langle i| + \int d\epsilon\;\epsilon|\Psi_\epsilon\rangle\langle\Psi_\epsilon| + V_{\text{light}} \tag{24}$$

The light–matter interaction Hamiltonian V_{light} takes the form:

$$\begin{aligned} V_{\text{light}} = &- \int d\epsilon\;\left[\vec{\mu}_{2\Psi_\epsilon}\,|2\rangle\,\langle\Psi_\epsilon| + \text{H.c.}\right]\cdot\left(\vec{\mathcal{E}}_p + \vec{\mathcal{E}}_S + \text{c.c.}\right) \\ &- \left[\vec{\mu}_{21}|2\rangle\langle 1| + \text{H.c.}\right]\cdot\left(\vec{\mathcal{E}}_p + \vec{\mathcal{E}}_S + \text{c.c.}\right) \end{aligned} \tag{25}$$

where $\vec{\mathcal{E}}_{p,S} = \hat{e}_{p,S}\mathcal{E}_{p,S}\exp(-i\omega_{p,S}t)$ are the pump and Stokes laser fields of polarization $\hat{e}_{p,S}$, respectively, while $\vec{\mu}_{21}$ and $\vec{\mu}_{2\Psi_\epsilon}$ are the dipole transition moments between the states $|2\rangle$ and $|1\rangle$, and $|2\rangle$ and $|\Psi_\epsilon\rangle$, respectively. The Schrödinger equation describing the STIRAP conversion of two atoms into a molecule gives

$$i\hbar\frac{\partial C_1}{\partial t} = E_1\,C_1 - \vec{\mu}^*_{21}\cdot\vec{\mathcal{E}}^*_S\,C_2 \tag{26}$$

$$\begin{aligned} i\hbar\frac{\partial C_2}{\partial t} &= E_2\,C_2 - \vec{\mu}_{21}\cdot\vec{\mathcal{E}}_S\,C_1 \\ &\quad - \int_{\epsilon_{th}}^{\infty} d\epsilon\;\vec{\mu}_{2\Psi_\epsilon}\cdot\vec{\mathcal{E}}_p\,C(\epsilon,t) \end{aligned} \tag{27}$$

$$i\hbar\frac{\partial C(\epsilon,t)}{\partial t} = \epsilon\,C(\epsilon,t) - \vec{\mu}^*_{2\Psi_\epsilon}\cdot\vec{\mathcal{E}}^*_p\,C_2 \tag{28}$$

Setting the origin of the energy to be the position of the ground state $|1\rangle$, and using the rotating wave approximation with $C_1 = c_1$, $C_2 = c_2 e^{-i\omega_S t}$, and $C(\epsilon, t) = c(\epsilon, t)e^{-i(\omega_S - \omega_P)t}$, Eqs. (26)–(28) become

$$i\frac{\partial c_1}{\partial t} = -\Omega_S c_2 \tag{29}$$

$$i\frac{\partial c_2}{\partial t} = \delta_2 c_2 - \Omega_S c_1 - \int_{\epsilon_{\text{th}}}^{\infty} d\epsilon\ \Omega_\epsilon c(\epsilon, t) \tag{30}$$

$$i\frac{\partial c(\epsilon, t)}{\partial t} = \Delta_\epsilon c(\epsilon, t) - \Omega_\epsilon^* c_2 \tag{31}$$

where $\delta_2 = E_2/\hbar - \omega_S$, $\Delta_\epsilon = \epsilon/\hbar - (\omega_S - \omega_p)$, and ϵ_{th} is the dissociation energy of the ground electronic potential with respect to the state $|1\rangle$. The Rabi frequencies of the fields are $\Omega_S = \vec{\mu}_{21} \cdot \hat{e}_S \mathcal{E}_S/\hbar$ (assumed real), $\Omega_\epsilon = \vec{\mu}_{2\Psi_\epsilon} \cdot \hat{e}_p \mathcal{E}_p/\hbar$.

This system of three equations can be reduced into a two-equation system by eliminating the continuum amplitude $c(\epsilon, t)$; introducing $c(\epsilon, t) = s(\epsilon, t)\exp(-i\Delta_\epsilon t)$ into Eq. (31), we get

$$s = i\int_0^t dt'\ \Omega_\epsilon^*(t')c_2(t')e^{i\Delta_\epsilon t'} + s_0(\epsilon) \tag{32}$$

where $s_0(\epsilon) \equiv s_0(\epsilon, t = 0)$ with $t = 0$ some moment before the collision of the two atoms. This initial amplitude of the continuum wave function $s_0(\epsilon)$ has been discussed in various contributions [69–71]. The resulting continuum amplitude is

$$c(\epsilon, t) = i\int_0^t dt'\ \Omega_\epsilon^*(t')c_2(t')e^{i\Delta_\epsilon(t'-t)} + s_0(\epsilon)e^{-i\Delta_\epsilon t}$$

Inserting this result into Eq. (30), we obtain a final system of equations for the amplitudes of the bound states:

$$i\frac{\partial c_1}{\partial t} = -\Omega_S c_2 \tag{33}$$

$$i\frac{\partial c_2}{\partial t} = (\delta_2 - i\gamma)c_2 - \Omega_S c_1 - S + iT \tag{34}$$

where we introduced a spontaneous decay term γc_2 in Eq. (34), and the two functions

$$S \equiv \int_{\epsilon_{\text{th}}}^{\infty} d\epsilon\ \Omega_\epsilon(t)s_0(\epsilon)e^{-i\Delta_\epsilon t}$$

$$T \equiv \int_{\epsilon_{\text{th}}}^{\infty} d\epsilon\ \Omega_\epsilon(t)\int_0^t dt'\ \Omega_\epsilon^*(t')c_2(t')e^{i\Delta_\epsilon(t'-t)} \tag{35}$$

S corresponds to the source function, whereas T corresponds to the "back-stimulation" term (or back-conversion), which accounts for the transfer of the bound molecules back into the continnum.

To describe the two-atom collision, a Gaussian wave packet characterized by an energy bandwidth δ_ϵ can be used

$$s(\epsilon, t = 0) = \frac{1}{(\pi\delta_\epsilon^2)^{1/4}} e^{-\frac{(\epsilon-\epsilon_0)^2}{2\delta_\epsilon^2} + \frac{i}{\hbar}(\epsilon-\epsilon_0)t_0} \tag{36}$$

where t_0 is the moment of the collision and ϵ_0 is the central energy of the wave packet. The results for two different cases are shown in Fig. 9; a wide Feshbach

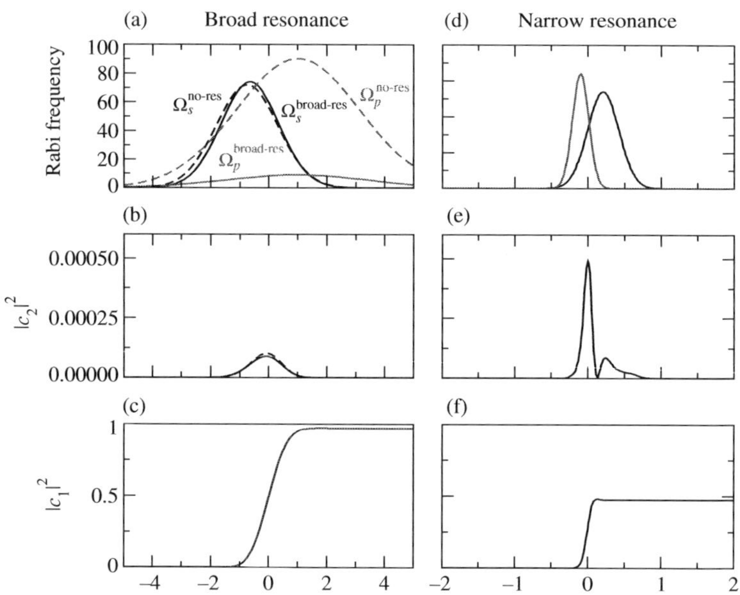

Figure 9. Time-dependence of the Stokes and pump pulses (**a, d**) and population in state $|2\rangle$ (**b, e**) and target state $|1\rangle$ (**c, f**) for the STIRAP transfer of a pair of atoms within the center of the thermal distribution. The left column is for a broad Feshbach resonance, while the right column is for a narrow resonance (using parameters in Table I). The dashed lines in the left column are the results obtained without resonance, when the parameters are adjusted to obtain the same overall transfer efficiency as for the broad resonance. Ω_S is in units of $10^6\ \text{s}^{-1}$, while Ω_p is in dimensionless units $(16\pi/\delta_\epsilon)^{1/4}\, \vec{\mu}_{2\epsilon} \cdot \vec{e}_p \mathcal{E}_p$ in the broad resonance limit and $(2\pi/\Gamma)^{1/2}\, \vec{\mu}_{2\epsilon} \cdot \vec{e}_p \mathcal{E}_p$ in the narrow resonance limit. The scale for the Rabi frequencies in the narrow resonance case is 40 times the scale for the broad resonance, and the magnitude of Ω_p is enlarged 10 times for better visibility. Reproduced from Ref. [64] by permission of the Institute of Physics (IOP).

TABLE I
Parameters of the Pulses Used in Fig. 9; $q = 10$, $\gamma = 10^8$ s^{-1}, and $\mu_{2b} = \mu_{21} = 0.1$ D (1 D = 0.3934 ea_0)

	δ_ϵ	Γ	Ω_S^0	I_S	I_p	T_S	T_p	τ_S	τ_p
Resonance	μK		10^8s^{-1}	W/cm^2		μs			
None	10	–	0.72	62	4×10^5	1.5	3	0.75	1.0
Broad	10	1000	0.74	65	4×10^3	1.4	3.4	0.65	1.0
Narrow	100	1	2.24	600	4×10^2	0.157	0.3	0.1	0.207

Rabi frequencies are modeled by Gaussians $\Omega_{S,p} = \Omega_{S,p}^0 \exp(-(t - t_0 \pm \tau_{S,p}))/T_{S,p}^2$, where $\pm$ refers to the Stokes and pump pulse, respectively.

resonance with $\Gamma \gg \delta_\epsilon$, where the width Γ of the Feshbach resonance is much larger than the thermal energy spread δ_ϵ of the colliding atoms, and a narrow resonance with $\Gamma \ll \delta_\epsilon$. The exact parameters used to obtain the results are listed in Table I; the corresponding Fano parameter q has the value 10. The q factor is essentially the ratio of the dipole matrix elements from the state $|2\rangle$ to the bound state $|b\rangle$ (modified by the continuum) and to an unperturbed continuum state $|\epsilon\rangle$.

For the broad case, we considered a Feshbach resonance with a width $\Gamma = 1$ mK (typical for broad resonances), and a thermal atomic ensemble with an energy bandwidth $\delta_\epsilon = 10$ μK. We see that the transfer efficiency can reach ~97% of the continuum state into the target state $|1\rangle$ (see Fig. 9c). When comparing these results to that of the unperturbed continuum (i.e., far from the resonance), we find that all populated continuum states experience the same transition dipole matrix element enhancement factor to the state $|2\rangle$, so that the system essentially reduces to the case of a flat continuum with an uniformly enhanced transition dipole matrix element. In this limit, one expects the adiabatic passage to be efficient. This is clearly demonstrated in Fig. 9 (left column, dashed lines): to reach the same ~ 97% transfer efficiency achieved with the broad resonance, a very large pump laser intensity (~100 times larger) is required if there is no resonance in the continuum (Fig. 9a), while the Stoke laser intensity is basically the same. Considering the intensity used in this particular example, this would lead to intensities in the range of 5×10^5 W/cm^2, making STIRAP from the continuum technically impossible to achieve without a resonance, consistent with the analysis of Ref. [71].

The results for a narrow resonance (typical width $\Gamma = 1$ μK and ensemble energy bandwidth $\delta_\epsilon = 100$ μK) are also shown in Fig. 9. In this limit, the transfer efficiency is lower: in the specific case analyzed here, it does not exceed 47%. The reason for this lower efficiency is the destructive quantum interference, which leads to electromagnetically induced transparency [72] in the transition from the continuum to the excited state. This mechanism is similar to the Fano interference effect, the difference is that the continuum is initially populated. One can therefore view it as an inverse Fano effect [64].

We note that during the transfer an initial incoherent mixture of atomic scattering states is converted into a pure internal state, which seems to decrease the entropy of the system. However, the entropy is transferred to the center-of-mass motion of the created molecules, which can lead to a slight translational heating of the sample.

D. STIRAP for Additional Intermediate States

In this section, we briefly present cases in which additional intermediate states are used to increase the efficiency of the STIRAP transfer to deeply bound molecular levels [73]. Figure 10 illustrates the simplest five-level model molecular system with states chainwise coupled by optical fields. The states $|g_1\rangle$, $|g_2\rangle$, and $|g_3\rangle$ are vibrational levels of a ground electronic molecular state, while $|e_1\rangle$ and $|e_2\rangle$ are vibrational states of an excited electronic molecular state.

Molecules are initially in the state $|g_1\rangle$, which could be a high-lying bound level, a Feshbach molecular bound state, or a continuum couple of a closed channel as in the previous section. The state $|g_3\rangle$ is the target final state, assumed to be the deepest bound vibrational state $v = 0$, and $|g_2\rangle$ is an intermediate vibrational state. To efficiently transfer population from the state $|g_1\rangle$ to state $|g_3\rangle$, at least two vibrational levels $|e_1\rangle$ and $|e_2\rangle$ in an excited electronic state are required; one with a good Franck–Condon overlap with $|g_3\rangle$, and the other with the initial state $|g_1\rangle$. In the states $|e_1\rangle$ and $|e_2\rangle$, molecules decay due to spontaneous emission and collisions, and in the state $|g_2\rangle$ they experience fast inelastic collisions with background atoms leading to loss of molecules from a trap. Thus, one must avoid populating intermediate states to have an efficient and coherent transfer process.

We give here the results if we neglect all decays (details including decays are presented in Ref. [73]). The wave function of the system is

$$|\Psi\rangle = \sum_i C_i \exp\left(-i\phi_i(t)\right)|i\rangle \tag{37}$$

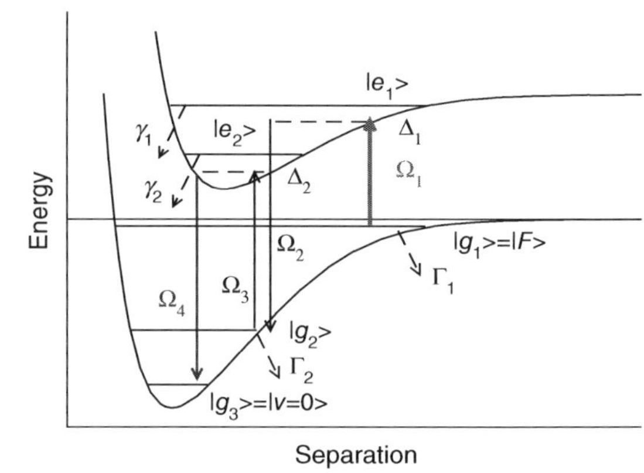

Figure 10. Schematic showing the multistate chainwise STIRAP transfer of population from the Feshbach $|g_1\rangle$ to the ground $|g_3\rangle$ vibrational state. Reproduced from Ref. [73] by permission of the American Physical Society (APS).

where $i = g_1, e_1, g_2, e_2, g_3$; $\phi_{g_1} = 0$, $\phi_{e_1} = \nu_1 t$, $\phi_{g_2} = (\nu_2 - \nu_1)t$, $\phi_{e_2} = (\nu_3 + \nu_2 - \nu_1)t$, $\phi_{g_3} = (\nu_4 - \nu_3 + \nu_2 - \nu_1)t$; ν_i is the frequency of the ith optical field. The evolution is then governed by the Schrödinger equation

$$i\hbar \frac{\partial |\Psi\rangle}{\partial t} = H(t) |\Psi\rangle \tag{38}$$

where the time-dependent Hamiltonian of the system in the rotating wave approximation is

$$H(t) = - \begin{pmatrix} 0 & \Omega_4(t) & 0 & 0 & 0 \\ \Omega_4(t) & -\Delta_2 & \Omega_3(t) & 0 & 0 \\ 0 & \Omega_3(t) & 0 & \Omega_2(t) & 0 \\ 0 & 0 & \Omega_2(t) & -\Delta_1 & \Omega_1(t) \\ 0 & 0 & 0 & \Omega_1(t) & 0 \end{pmatrix} \tag{39}$$

where $\Omega_1(t) = \mu_1 \mathcal{E}_1(t)/2\hbar$, $\Omega_2(t) = \mu_2 \mathcal{E}_2(t)/2\hbar$, $\Omega_3(t) = \mu_3 \mathcal{E}_3(t)/2\hbar$, and $\Omega_4(t) = \mu_4 \mathcal{E}_4(t)/2\hbar$ are the Rabi frequencies of optical fields; $\mathcal{E}_i$ is the amplitude of ith optical field, μ_i is the dipole matrix element along the respective transition, $\Delta_1 = \omega_1 - \nu_1$ and $\Delta_2 = \omega_4 - \nu_4$ are one-photon detunings of the fields, and the ω_i are the molecular frequencies along transition i.

If we assume that pairs of fields coupling two neighboring ground state vibrational levels are in a two-photon (Raman) resonance, the system has a dark state given by

$$\left|\Phi^0\right\rangle = \frac{\Omega_2\Omega_4 |g_1\rangle - \Omega_4\Omega_1 |g_2\rangle + \Omega_1\Omega_3 |g_3\rangle}{\sqrt{\Omega_4^2\Omega_1^2 + \Omega_1^2\Omega_3^2 + \Omega_2^2\Omega_4^2}} \tag{40}$$

In "classical" STIRAP, where the optical fields are applied in a counterintuitive way, that is, at $t = -\infty$ only a combination of the $\Omega_4, \Omega_3, \Omega_2$ fields, and at $t = +\infty$ only of Ω_3, Ω_2, and Ω_1 is present, the dark state is initially associated with the $|g_1\rangle$ and finally with the $|g_3\rangle$ states. Adiabatically changing the Rabi frequencies of the optical fields so that the system stays in the dark state during evolution, one can transfer the system from the initial $|g_1\rangle$ to the ground vibrational $|g_3\rangle$ state with unit efficiency, defined as the population of the $|g_3\rangle$ state at $t = +\infty$. The dark state does not have contributions from the $|e_1\rangle$ and $|e_2\rangle$ excited states, and thus the decay from these states does not affect the transfer efficiency. Decay from the $|g_1\rangle$, $|g_2\rangle$, $|g_3\rangle$ states will, however, degrade the coherent superposition (40) and result in population loss from the dark state and reduction of the transfer efficiency. We note that Ω_2 and Ω_3 could be left on (i.e., cw-lasers) or varying, while Ω_1 and Ω_4 need to follow the standard STIRAP sequence. Naturally, this approach can be generalized to many more intermediate states [74].

III. QUANTUM INFORMATION WITH MOLECULES

In this section, we briefly describe the foundations of quantum information and computing. We compare classical and quantum units of information, and list various quantum algorithms. We define entangled states and quantum logic phase gates, and enumerate various properties needed to realize quantum computing. We list some physical systems that have been explored to create them. We finally show why ultracold molecules represent an attractive candidate for the realization of quantum logic gates, especially using hybrid platforms.

A. Overview of QI

The individual unit of classical information is the bit: an object that can take either one of two values, say 0 or 1. The corresponding unit of quantum information is the quantum bit or *qubit*. It describes a state in the simplest possible quantum system [1]. The smallest nontrivial Hilbert space is two-dimensional, and we may write an orthonormal basis for the vector space as $|0\rangle$ and $|1\rangle$. A single bit or qubit can represent at most two numbers, but qubits can be put into infinitely many other states by a superposition:

$$|Q^{(1)}\rangle = c_0|0\rangle + c_1|1\rangle \qquad \text{with} \qquad |c_0|^2 + |c_1|^2 = 1 \tag{41}$$

where c_0 and c_1 are complex numbers. A system of N classical bits can take 2^N possible values (e.g., from 0 to $2^N - 1$), but a particular classical register is described only by a binary string of length N, and therefore contains N bits of information. A quantum register of size N is described by a vector in the 2^N-dimensional Hilbert space: $|Q^{(N)}\rangle = \sum_{x=0}^{2^N-1} c_x|x\rangle$, where c_x are complex numbers such that $\sum_{x=0}^{2^N-1} |c_x|^2 = 1$. The decimal notation relates to the binary notation for encoding the states. For example, with $N = 3$, there are eight three-qubit states, $|0\rangle \equiv |0\rangle|0\rangle|0\rangle$, $|1\rangle \equiv |0\rangle|0\rangle|1\rangle$, $|2\rangle \equiv |0\rangle|1\rangle|0\rangle, \ldots, |7\rangle \equiv |1\rangle|1\rangle|1\rangle$ [75].

If prepared in a general superposition state, quantum registers (or memories) can store 2^N bits of information simultaneously, as compared to classical registers where only N bits of information are stored. So, in principle, quantum memories are extremely large: a system of $N = 100$ qubits will be described by $2^{100}\sim 10^{30}$ complex numbers, an impressive number (by comparison, the classical register would be described by a binary string of length 100). However, not all the information contained in quantum memories can be accessed by physical measurements [1,75].

The reason why quantum computers can be very fast is the so-called *quantum parallelism*: they can process quantum superpositions of many numbers in one computational step. Each computational step is a unitary transformation of quantum registers. So, instead of operating on a single string of N bits (classical computer), 2^N bits of information can be acted on with a single operation, and the speedup can be exponential. To achieve this speedup, a universal quantum computer should be able to perform an arbitrary unitary transformation on any superposition of states, which requires, in particular, the ability to generate the states that cannot be written as a direct product of individual qubit states. The existence of such *entangled states* constitute one of the most intriguing features of quantum mechanics and leads to many unusual phenomena such as the EPR paradox [76].

Like classical computers, quantum computers can be built out of quantum circuits composed of a set of elementary logic gates [77]. Such a set is universal if it can be used to design any possible computation. It is by now known that all universal gates require a two-qubit nonlocal (entangled) state and a local operation (on a single qubit). One example of a nontrivial two-qubit operation is a so-called *phase gate* [1,75]. Consider, for example, two atoms A and B, each in a superposition state (or qubit): $|A\rangle = c_0^A|0\rangle_A + c_1^A|1\rangle_A$ and $|B\rangle = c_0^B|0\rangle_B + c_1^B|1\rangle_B$. The combined $A + B$ system is described by the two-qubit state

$$|Q^{(2)}\rangle = |A\rangle|B\rangle = c_0^A c_0^B|0\rangle_A|0\rangle_B + c_0^A c_1^B|0\rangle_A|1\rangle_B + c_1^A c_0^B|1\rangle_A|0\rangle_B + c_1^A c_1^B|1\rangle_A|1\rangle_B \tag{42}$$

If a unitary transformation $\hat{U}_\phi$ changes the phase of the last pair, the new two-qubit state is

$$\hat{U}_\phi|Q^{(2)}\rangle = c_0^A c_0^B|0\rangle_A|0\rangle_B + c_0^A c_1^B|0\rangle_A|1\rangle_B + c_1^A c_0^B|1\rangle_A|0\rangle_B + e^{i\phi}c_1^A c_1^B|1\rangle_A|1\rangle_B \tag{43}$$

For example, if $\phi \neq 0$, such a state cannot be written as the product of any single atom states. Two qubits in such a state are strongly (and nonlocally) correlated. By using a sequence of such phase gates (i.e., operations $\hat{U}_\phi$) and simple unitary operations on a single qubit, any quantum computation can be completed [1,75].

A similar example is the *controlled not* (CNOT) gate. In this operation, the state of the *target* qubit $|B\rangle$ is flipped (i.e., $|0\rangle_B \leftrightarrow |1\rangle_B$) if the *control* qubit $|A\rangle$ is in the state $|1\rangle_A$; otherwise qubit $|B\rangle$ is not affected. Just as for the phase gate, the combination of CNOT gates and single-qubit gates are sufficient to perform any arbitrary computation. The conditional nature of the CNOT gate is transparently similar to the minimal requirements for classical computation, where a single conditional two-bit logic gate (e.g., NAND) is sufficient.

To take advantage of parallelism and entanglement, quantum algorithms are being devised. Three of the best-known quantum algorithms illustrate the various speedups expected when compared to classical computing times. Grover's algorithm [3,78] for the search of unsorted data base is demonstrably faster than the best classical algorithm, but not exponentially faster. The Deutsch–Josza algorithm [79] for the oracle problem, where the function of a black box must be determined from the inputs and outputs alone, is an example of *relativized* exponential speedup. Finally, Shor's algorithm [2,80] for factoring large numbers is the perfect example of a truly exponential speedup: the problem can be solved in polynomial time instead of the exponential time required for classical computers. Note that the adaptation of known classical algorithms to their quantum mechanical counterparts is also a promising research direction; for example, algorithms featuring quantum random walks [81] have recently been shown to have exponential speedups [82]. Although not all algorithms rely on entanglement (e.g., Grover's does not), most do: for that reason, entanglement is a critical property that quantum computers should possess.

B. Requirements to Implement Quantum Computers

The information contained in qubits and entangled states include the phase: any error in the phase has important implications (e.g., changing a nonlocal to a local state). To perform complex quantum computations, we need to prepare reliably a delicate superposition of states of a relatively large quantum system that cannot be perfectly isolated from the environment; hence the superpositions always decay. The decoherence of an entangled state is even faster: the nonlocal correlations are extremely fragile and decay rapidly. Also, applications of unitary transformations to qubits will not be flawless, and errors can accumulate.

Recent theoretical developments in quantum error correction [83–85] have addressed these points, and it was shown that quantum computing can be fault tolerant [86–88]. A rapidly expanding experimental effort to process coherent quantum information is taking place. To build hardware for a quantum computer, technology that enables the manipulation of qubits is needed. The requirements for that technology are given, for example, by DiVincenzo [89] and of which the most elusive ones are:

1. A set of individually addressable qubits in which coherence can be stored long enough to complete interesting computations; hence, qubits should be isolated from the environment to minimize decoherence
2. Quantum gates to perform nontrivial two-qubit interactions; this can be achieved only via controlled, coherent interactions among qubits
3. Reliable and efficient readout methods

Many systems are being studied to manipulate quantum information. Some are using individual atoms: cold-trapped ions [8–10,90,91], neutral atoms in optical lattices [6,7,92,93], atoms excited to high Rydberg states [94], or atoms in crystals [95]. Others involve spin of the particle [96–98], such as nitrogen-valence (NV) centers in diamond [99–101], or photons in cavity QED or nonlinear optical setups [102,103], as well as more exotic ones where geometric combinations of elementary excitations are defined as qubits, such as in *topological quantum computing*

[104]. However, none of these systems has yet emerged as a definitive way to build a quantum information processor. Quantum computers require a coherent, controlled evolution for a period of time necessary to complete the computation. The essential dichotomy is that we need weak coupling to the environment to avoid decoherence, but also strong coupling to at least some external modes to manipulate qubits and have controlled interactions. As we will see in the next section, ultracold molecules, especially polar molecules, represent a good candidate to obtain quantum logic gates, in which the two required features can peacefully coexist.

To implement quantum information processing with polar molecules, they could be stored in a 1D or 2D array so that their dipole moments could be aligned by an electric field perpendicular to the array. We assume the full development of the storage and addressing capabilities of two recently proposed architectures (see Fig. 11). The first is an optical lattice with a lattice spacing of about 1 μm, as suggested in Ref. [4]. Using a DC field for dipole alignment during trapping naturally allows the repulsive dipole–dipole interaction to aid with homogeneous distribution in the lattice. In this case, addressing single qubits can be accomplished by either using the inhomogeneous DC electric fields proposed in Ref. [4] to create individualized transition frequencies, or by individually addressing with light in the visible part of the frequency spectrum. The second architecture is based on a "stripline wire" architecture, as suggested at Yale and Harvard [11]. Here,

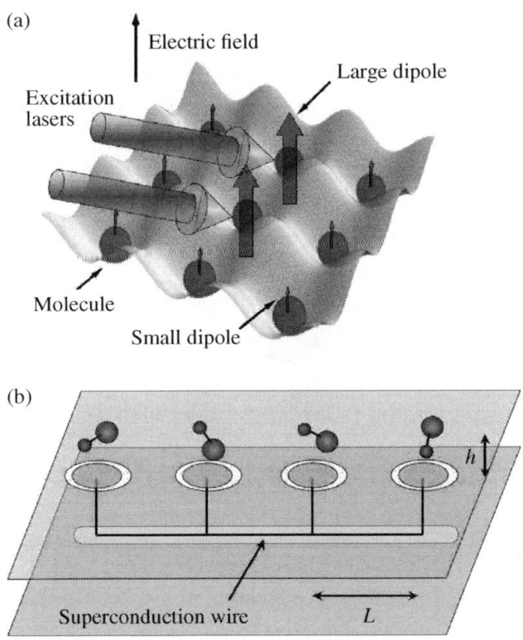

Figure 11. Setups: (**a**) molecules individually addressable by lasers are stored in an optical lattice. (**b**) Superconducting wires are used to "deliver" the interaction. In both, molecules are selectively excited, and interact only if both are in $|e\rangle$. Reproduced from Ref. [73] by permission of the American Physical Society (APS).

molecules sit on their own small traps, which also serve for addressing, and are connected via a superconducting transmission line using microwave fields to allow for long-range dipole–dipole interaction.

C. Wishlist: Properties of Polar Molecules

Polar molecules have a permanent dipole moment with respect to their molecular axis, but which averages to zero in the laboratory frame. An electric field with a given orientation in the laboratory frame will orient the molecules and align the two frames. This means that rotational states are mixed, leading to observable Stark shifts of the energy levels of the molecules.

Since the dipole moment in the molecular frame $\vec{\mu}(R)$ is a function of the separation R of the atoms forming the molecules, the dipole moment depends on the exact internal rovibrational state of the molecule. Usually, since $\vec{\mu}(R)$ is small at large separation, the upper rovibrational states have small dipole moments, while the deeper ones have larger dipole moments. To store the information, a variety of internal molecular states can be used, such as specific rotational and/or vibrational states, or hyperfine states (for molecules with nuclear and electronic spins).

Overall, we can list five requirements on properties required of polar molecules that would ensure a realistic implementation of quantum gates (as described in the next section):

1. Choice of qubits: We need long-lived states to store the information. These should interact as little as possible with the environment to minimize decoherence. Good candidates are hyperfine and rotational states of a ground electronic molecular state.
2. Coupling strength: Fast one- and two-qubit gates require large interaction strengths, while storage needs long lifetimes. So, transitions between the storage states $|0\rangle$ and $|1\rangle$ should have long lifetime. Either direct transitions, or Raman transitions via intermediate states, could be used to perform one-qubit operations.
3. Robust dipole–dipole interaction: For two-qubit gates, strong interactions are required. The interaction strength can be maximized by choosing the smallest possible (effective) distance between qubits, and also by using molecules with large dipole moments.
4. Cooling and trapping: To fulfill the previous requirements, molecules must be cooled down to sub-Kelvin temperatures. This prevents the population of undesired rotational states. Also, tightly trapped molecules help to minimize errors in their position and so on gates and readout operations.
5. Decoherence: In order to store the information for long durations, we need to minimize the interaction with the trapping potential, the environment, and all other fields used to manipulate the qubits. It helps to use states with negligible dipole moments to store the information, since these states interact

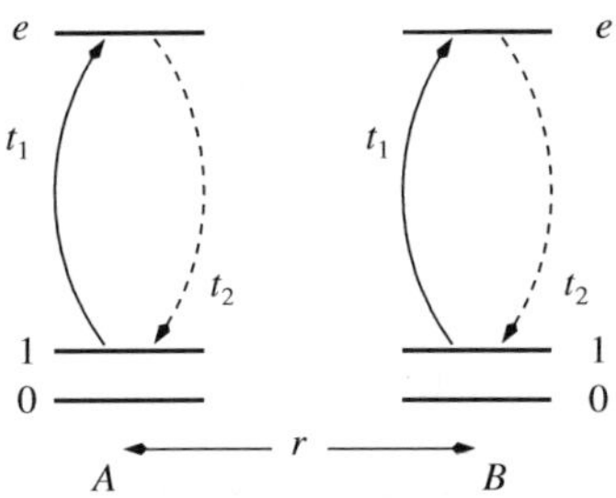

Figure 12. Phase gate: two molecules A and B separated by r are prepared in a superposition of states $|0\rangle$ and $|1\rangle$. At $t_1 = 0$, we excite $|1\rangle$ of both into $|e\rangle$: both interact via dipole–dipole interactions, and acquire a phase ϕ. At time $t_2 = \tau$ such that $\phi = \pi$, we stimulate coherently both $|e\rangle$ back to $|1\rangle$. Reproduced from Ref. [73] by permission of the American Physical Society (APS).

only weakly with the each other, the environment, or other fields present in the setup. Switching on and off strong dipoles at the appropriate time is conceptually attractive for controlling decoherence, but the switching itself must be performed carefully lest it become the largest source of decoherence.

D. Schemes with Switchable Dipoles

As mentioned before, the original proposal to use polar molecules is based on their permanent dipole [4], which has been extensively studied also in Ref. [105]. Other approaches using molecules are based on internal molecular states in single molecules [13,14,106,107]. Here, we instead focus our attention on schemes where the dipole–dipole interaction is switched on and off [108,109].

We first describe the generic setup to obtain a phase gate, or universal two-qubit operation, in Fig. 12. We assume that the molecules are individually addressable by optical or microwave fields. We choose $|0\rangle$ and $|1\rangle$ as the hyperfine states within a zero-dipole-moment manifold, in a level with a long coherence time; and $|e\rangle$ as a metastable state in a large-dipole-moment manifold.

Single-qubit rotations can be accomplished with optical or microwave fields. The initial states of two individual sites A and B can be prepared in a superposition state, for example, using $\pi/2$ Raman pulses. To perform a two-qubit gate, a one- or two-photon transition couples $|1\rangle$ and $|e\rangle$ coherently, but not $|0\rangle$ and $|e\rangle$. This can always be accomplished by either polarization or frequency selection. The molecules interact via a dipole–dipole interaction only if both are in the $|e\rangle$ state; in this case the system acquires a phase $\phi(t)$. After a time $t = \tau$ such that $\phi = \pi$, we coherently stimulate the state $|e\rangle$ back to $|1\rangle$. This process can be summarized by [108,109]

$$\begin{array}{ccccccc}
|00\rangle & \stackrel{\pi-\text{pulse}}{\longrightarrow} & |00\rangle & \stackrel{\text{dip}-\text{dip}}{\longrightarrow} & |00\rangle & \stackrel{\pi-\text{pulse}}{\longrightarrow} & |00\rangle \\
|01\rangle & \longrightarrow & |0e\rangle & \longrightarrow & |0e\rangle & \longrightarrow & |01\rangle \\
|10\rangle & \longrightarrow & |e0\rangle & \longrightarrow & |e0\rangle & \longrightarrow & |10\rangle \\
|11\rangle & \longrightarrow & |ee\rangle & \longrightarrow & -|ee\rangle & \longrightarrow & -|11\rangle
\end{array}$$

The resulting transformation corresponds to a phase gate.

The π-phase shift produced in the time τ between the exciting and deexciting π-pulses is given by

$$\phi = \pi = \frac{1}{\hbar}\int_0^{\tau} d\tau' \frac{\mu}{r^3}\left(3\cos^2\theta - 1\right)\rho_e^2(\tau') \tag{44}$$

where μ and ρ_e are the dipole moment and fractional population in the excited state, r is the distance between molecules A and B, and θ is the angle between the dipole moments. This formulation allows for finite excitation and deexcitation times and imperfect π-pulses.

We now describe three possible setups utilizing variations of our switchable phase-gate scheme [108,109]. The first system is based on carbon monoxide (CO). As far as dipolar molecules are concerned, CO is an anomaly; while its electronic ground state $X^1\Sigma^+$ has a very small dipole moment ($\mu \approx 0.1$ D in the vibrational ground state, which is expected to be the easiest to trap), there exists a very long-lived ($\tau_{\text{life}} \approx 10-1000$ ms) excited electronic state $a^3\Pi$ with a large dipole moment, $\mu \approx 1.5$ D. As $|0\rangle$ and $|1\rangle$, one can choose, for example, the two nuclear spin projection states of $X^1\Sigma^+$, $v = 0$, $N = 0$, $I = 1/2$, $F = 1/2$ of ^{13}CO. With a magnetic-field-induced Zeeman splitting, selective excitation from $|1\rangle$ to $|e\rangle$ is possible. The transition frequency between $X^1\Sigma^+$ and $a^3\Pi$ is in the UV (about 48,000 cm^{-1}), and the optical lattice architecture would be the ideal choice. With a coherence time in an optical lattice of a few seconds [4] and a necessary dipole–dipole interaction time of several milliseconds, about 10^3 operations are possible. This scheme is straightforward, the techniques are in place or nearly so, and CO is a well-studied molecule.

A more common situation can be found in molecules such as mixed bi-alkali molecules, (e.g., LiCs or RbCs). These molecules have large permanent dipole moments μ (as large as 7 D) in their ground electronic state $X^1\Sigma^+$ (for $|0\rangle$ and $|1\rangle$), and a metastable electronic state $a^3\Sigma^+$ that can be accessed from $X^1\Sigma^+$ due to spin-orbit coupling (for $|e\rangle$); the $a^3\Sigma^+$ potential well is located at large nuclear separation and supports at bound states; in most cases, these triplet states have permanent dipole moments close to zero. These properties can be used to implement a scheme in all important points similar to the CO scheme, except for three details. First, the phase gate would be "inverted" (i.e., $|00\rangle \rightarrow -|00\rangle$, $|01\rangle \rightarrow |01\rangle$, $|10\rangle \rightarrow |10\rangle$, and $|11\rangle \rightarrow |11\rangle$). Second, it requires the molecules to be stored, with the help of an aligning DC electric field, in the large-dipole state, which would most likely lead to seriously shortened coherence times. In addition, the interaction would happen for *all* molecules, not just the two we wish to be coupled by a phase gate. However, this issue can be mitigated, for example by switching on an aligning DC field only during interaction times, and for exactly a 2π phase shift, as measured for the lower states. By adding together this 2π phase shift using a DC field and the "negative" π phase shift for the molecules in the $|e\rangle$ state, the phase gate is given by

$$\begin{array}{ccccccccc}
|00\rangle & \stackrel{\text{exc+DC}}{\longrightarrow} & |00\rangle & \stackrel{\pi}{\longrightarrow} & -|00\rangle & \stackrel{\text{de-exc}}{\longrightarrow} & -|00\rangle & \stackrel{DC}{\longrightarrow} & |00\rangle \\
|01\rangle & \longrightarrow & |0e\rangle & \longrightarrow & |0e\rangle & \longrightarrow & |01\rangle & \longrightarrow & -|01\rangle \\
|10\rangle & \longrightarrow & |e0\rangle & \longrightarrow & |e0\rangle & \longrightarrow & |10\rangle & \longrightarrow & -|10\rangle \\
|11\rangle & \longrightarrow & |ee\rangle & \longrightarrow & |ee\rangle & \longrightarrow & |11\rangle & \longrightarrow & -|11\rangle
\end{array}$$

Note that the scheme described for CO could be adapted for these molecules by using two vibrational, rotational, or hyperfine states of $a^3\Sigma^+$ for the doublet $|0\rangle$ and $|1\rangle$, and a low-lying vibrational state of $X^1\Sigma^+$ as $|e\rangle$.

The last setup we describe here is the "rotational scheme," which is based on the zero dipole moment μ of a molecule in the rotational ground state $N = 0$ (or any pure rotational state). We consider rotational states of the ground electronic and vibrational state; $|0\rangle$ *and* $|1\rangle$ are hyperfine states in the rotational ground state $N = 0$, while $|e\rangle$ is the *superposition* of neighboring rotational states $|e\rangle = |e_1\rangle + |e_2\rangle$ (see Fig. 13). This superposition state can be obtained by coupling two neighboring rotation states via microwaves [108,109]. Because both $|0\rangle$ and $|1\rangle$ are in the absolute ground state with exactly zero dipole moment, this system has several advantages: maximum coherence time, ease of storage, and no residual dipole–dipole interaction. Moreover, any polar molecule can be used with this scheme, as long as it has at least two hyperfine states. Interesting choices would be CaF or NaCl with a dipole moment of up to 5–10 D.

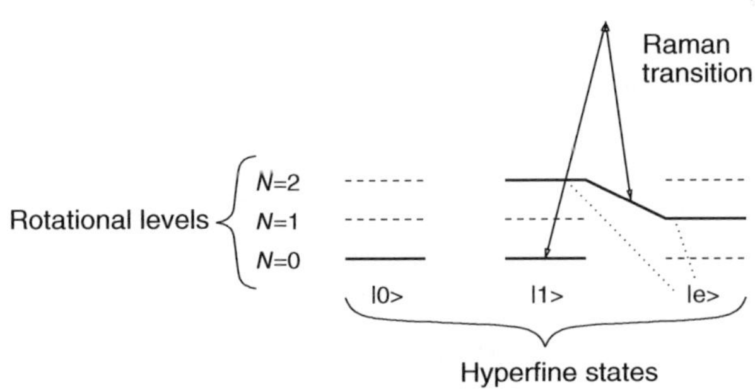

Figure 13. Level system for "rotational scheme": all states are part of the electronic and vibrational ground state. For example, a laser Raman π-pulse can transfer $|1\rangle$ to a storage state $|s\rangle$ that is a sublevel of, for example, the $N = 2$ state. Then a microwave $\pi/2$ pulse transfers $|s\rangle$ to the superposition state $|e\rangle \propto |e_1\rangle + |e_2\rangle$, where $|e_2\rangle = |s\rangle$ and $|e_1\rangle$ is a sublevel of the $N = 1$ manifold. Alternatively, if lasers are not needed for spatial selectivity, $|1\rangle$ could simply be transferred to the superposition state $|e\rangle \propto |1\rangle + |e_1\rangle$, again with $|e_1\rangle$ in the $N = 1$ manifold. Reproduced from Ref. [73] by permission of the American Physical Society (APS).

If the sites can be addressed individually and the dipole–dipole interactions are strong, the previous schemes could take advantage of the so-called dipole *blockade* mechanism. This mechanism has been introduced for quantum information processing with Rydberg atoms, and generalized to mesoscopic ensembles as well as van der Waals interactions. The original dipole blockade proposal [110–112] relies on a rapid "hopping" of the excitation between the energy levels of two Rydberg atoms, leading to an effective splitting of the doubly excited state. When this splitting is sufficiently large, the energy levels are shifted far away from the unperturbed atomic resonance, effectively eliminating the transition to this doubly excited state; one atom can be excited into a Rydberg state, but additional Rydberg excitations are prevented by the large energy shifts.

In a similar fashion, the blockade mechanism can be generalized to polar molecules. If the dipole–dipole interaction is strong enough (i.e., larger than the bandwidth of the excitation field), the doubly excited state corresponding to $|ee\rangle$ will be shifted out of resonance and never excited. If both sites A and B are addressable individually, the ability to drive a 2π transition in site B depends on whether site A is excited (see Fig. 14). At t_1, we apply a π pulse to molecule A and populate the state $|e\rangle$. At t_2, we apply a second pulse (2π) to molecule B: if A is already in $|e\rangle_A$, the dipole–dipole interaction shifts the state $|e\rangle_B$, and the photon is off-resonance, hence no transition. If A is not in $|e\rangle_A$, B acquires a phase of π after the process. At t_3, we de-excite A with another π pulse; in summary,

$$
\begin{array}{ccccccc}
 & t_1 & & t_2 & & t_3 & \\
|00\rangle & & |00\rangle & & |00\rangle & & |00\rangle \\
|01\rangle & & |01\rangle & \longrightarrow & -|01\rangle & & -|01\rangle \\
|10\rangle & \longrightarrow & i\,|e0\rangle & & i\,|e0\rangle & \longrightarrow & -|10\rangle \\
|11\rangle & \longrightarrow & i\,|e1\rangle & \not\longrightarrow & i\,|e1\rangle & \longrightarrow & -|11\rangle
\end{array}
$$

This scheme is robust with respect to the separation between the molecules; as long as the excitation is blockaded, the exact separation is not important.

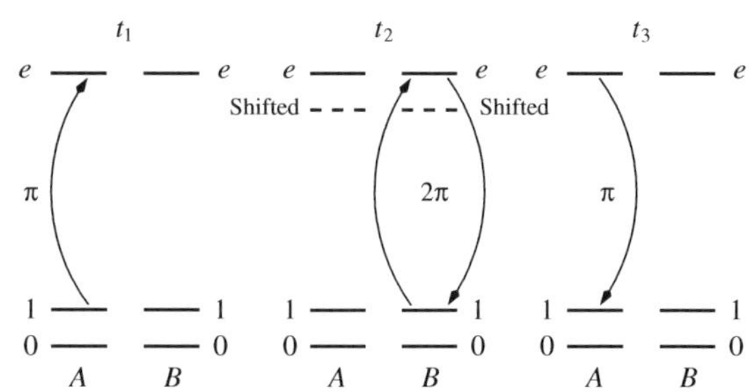

Figure 14. Principle of the dipole blockade (see text). Reproduced from Ref. [73] by permission of the American Physical Society (APS).

A key operation at the end of several-qubit operations is the readout of the quantum registers. Several approaches could be employed with polar molecules. As mentioned earlier, selective ionization of one of the states (0 or 1) and the detection of molecular ions can be readily accomplished. However, this method is destructive, because the molecule is lost after the readout, and the site would need to be refilled. A different method uses a "cycling" fluorescent transition in which the molecules decay after irradiation directly back into the state from which they came. Although this process is substantially more difficult for molecules than for atoms because of the large number of molecular levels, it offers the advantage of being "nondestructive." Another approach based on evanescent-wave mirrors for polar molecules might yield promising results [113,114]; both states "0" and "1" would feel a different trapping potential far away from the wall, and could in principle be imaged separately. Because reflection takes place far away from the surface of the mirror, it might help to minimize decoherence due to shorter range interactions with the surface. Finally, as discussed before [11,115], for molecules trapped near a microwave stripline cavity, the qubit state can be read out by monitoring the dispersive phase shift of photons transmitted through the cavity.

IV. AN ATOM-MOLECULE HYBRID PLATFORM

In recent years, hybrid systems taking advantage of specific characteristics of given platforms for quantum information processing [30,115,116] have generated increasing interest. Neutral-atom based quantum computing [117,118] offers long-lived atomic states (e.g., hyperfine sublevels to encode qubits and well-mastered techniques to manipulate atomic transitions), but usually suffers from weak coupling between atoms. As described in previous sections, while polar molecules offer a variety of long-lived states for qubit encoding (rotational, spin, hyperfine, etc.) and have long-range dipole–dipole interaction due to their permanent electric dipole moment, control and manipulation over molecular states are not yet as developed as for neutral atoms.

We describe next an atom–molecule platform that combines the principal advantages of the neutral atom and polar molecule-based approaches; it employs atoms with long-lived states to encode qubits, initialize and read out their states, and polar molecules to perform two-qubit gates [119].

A. Overview of the Atom–Molecule Platform

Starting with an optical lattice containing two atomic species per site in a Mott insulating phase, a qubit can be encoded in hyperfine states of one atom, while the second atom serves as an "enabler" to realize a two-qubit gate. This allows qubit initialization, storage, readout, and one-qubit operations, using well-developed techniques of atomic state manipulation. To execute a two-qubit gate, a pair of atoms at one site is selectively converted into a stable molecule with a large dipole moment, which can interact with a molecule formed at another site via a long-range dipole–dipole interaction. This hybrid atom–molecule platform could also provide an alternative scheme for readout of molecular qubits; qubits could be stored in long-lived molecular states as in Ref. [4], and conditionally converted into atoms for readout. The atom–molecule conversion is a nondestructive alternative to state-selective ionization; after readout atoms can be converted back into molecules that can be used again.

The characteristics needed for the hybrid platform can be summarized in the following list:

1. Long-lived atomic states
2. Elastic collisions between two atoms in a lattice site
3. Possibility to control and reversibly convert atoms into stable molecules
4. Molecules with large dipole moments

Many systems exhibit these properties. In particular, alkali atoms have hyperfine states with long coherence times, and most molecules formed from mixed alkali atoms have a large permanent electric dipole moment in their ground state. The atom–molecule conversion could be achieved using the STIRAP+FOPA approach described in the previous section.

B. A Specific Example

We illustrate the scheme using an atomic pair of ^{87}Rb and ^{7}Li (or ^{6}Li), since the LiRb molecule has one of the largest permanent dipole moments among bi-alkali diatomic molecules (4.2 D in its ground electronic state), and Rb is a popular atom in neutral atom quantum computing. In addition, a Feshbach resonance has been observed at 649 G for the mixed ^{87}Rb+^{7}Li system between the $|f.m\rangle$ atomic states $|1,1\rangle^{\mathrm{Rb}}$ and $|1,1\rangle^{\mathrm{Li}}$ [120].

- *Qubit choice:* A qubit can be encoded into hyperfine sublevels $|f, m\rangle$ of one atom, while the other atom is in a state stable to collisions. In Rb–Li, ^{87}Rb can store the information, while ^{7}Li can serve as the "enabler": $|2,2\rangle$ and $|1,1\rangle$ states of ^{87}Rb can be used for qubit encoding, and ^{7}Li can be kept in $|2,2\rangle$ during storage and one-qubit operations (see Fig. 15). At ultralow temperatures, scattering of Rb and Li in these states is elastic, preventing atom loss and qubit decoherence [64,119]. For the two-qubit gate (see below), Li is transferred into the "enabling" state $|1,1\rangle$, where a Feshbach resonance occurs at $B \sim 649$ G for the $|1,1\rangle^{\mathrm{Rb}} + |1,1\rangle^{\mathrm{Li}}$ channel, strongly enhancing the atom–molecule conversion efficiency [64]. Qubits during storage and

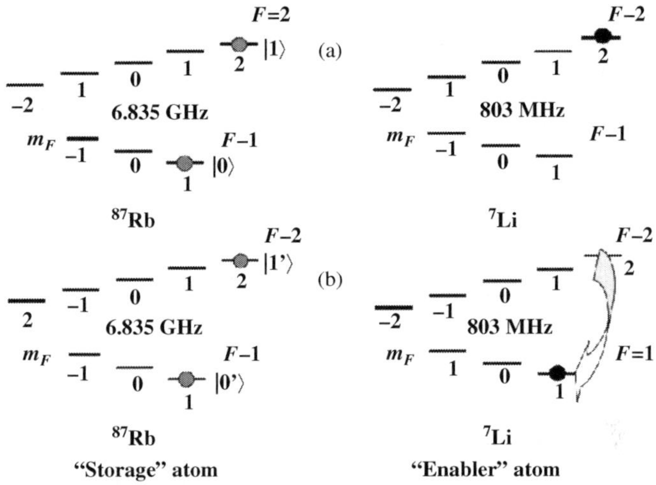

Figure 15. (a) Storage and one-qubit operations: information is encoded into hyperfine sublevels $|f, m\rangle = |2, 2\rangle$ and $|1, 1\rangle$ of ^{87}Rb, while ^{7}Li is in an "inert" state $|2, 2\rangle$, leading to the storage qubits $|0\rangle$ and $|1\rangle$. (b) For two-qubit operations the "enabler" ^{7}Li atom is transferred into $|1, 1\rangle$, leading to the "enabled" qubits $\left|0'\right\rangle$ and $\left|1'\right\rangle$. Note that the spacing is not to scale. Reproduced from Ref. [119] by permission of the American Physical Society (APS).

one-qubit operations, and "enabled" (primed) qubits during a two-qubit gate are

$$\left.\begin{aligned} |0\rangle &\equiv |1, 1\rangle^{\mathrm{Rb}} \otimes |2, 2\rangle^{\mathrm{Li}} \\ |1\rangle &\equiv |2, 2\rangle^{\mathrm{Rb}} \otimes |2, 2\rangle^{\mathrm{Li}} \\ \left|0'\right\rangle &\equiv |1, 1\rangle^{\mathrm{Rb}} \otimes |1, 1\rangle^{\mathrm{Li}} \\ \left|1'\right\rangle &\equiv |2, 2\rangle^{\mathrm{Rb}} \otimes |1, 1\rangle^{\mathrm{Li}} \end{aligned}\right\} \tag{45}$$

One-qubit rotations can be implemented using optical Raman pulses resonant with the qubit transition of Rb while not affecting the "enabler," which is far-detuned.

- *Two-qubit phase gate:* For a two-qubit phase gate, we conditionally transfer a pair of atoms into a stable molecular state $|g\rangle$ with a large permanent electric dipole moment. An external electric field orients a dipole moment in a laboratory-fixed frame and molecules in two different sites A and B can interact via their dipole–dipole interaction. After a π phase shift is accumulated, the molecules are converted back into atoms, realizing a phase gate. Combining the phase gate with two $\pi/2$ rotations applied to a target qubit implements a CNOT gate. Two unbound atoms can be coherently transferred into a molecular state by two-color photoassociation (e.g., using the FOPA +

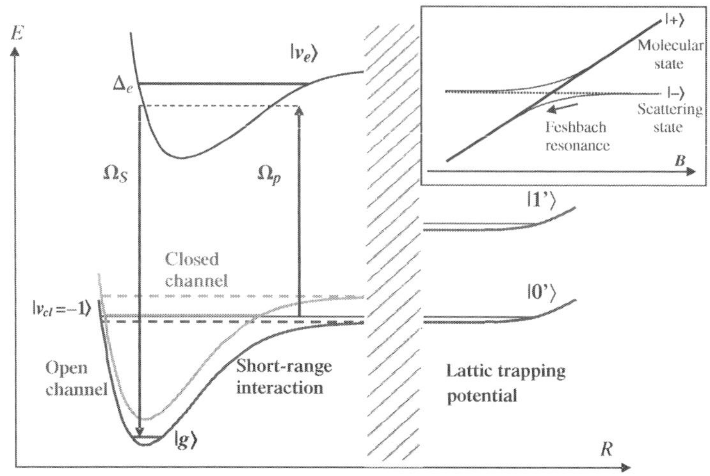

Figure 16. Conversion of atom pairs into ground-state molecules. Atoms in a qubit state $|i\rangle$ ($|0'\rangle$ or $|1'\rangle$), are selectively and coherently transferred from the lowest motional state of the trap into a deeply bound molecule $|g\rangle$, by setting pump (Ω_p) and Stokes (Ω_S) laser fields in a two-photon resonance with the $|i\rangle - |g\rangle$ transition. Ω_p is enhanced by a Feshbach resonance due to a closed channel ($|v_{cl} = -1\rangle$: see text for details). The dashed lines show the molecular thresholds. The inset illustrates the formation of the $|\pm\rangle$ mixed states close to a Feshbach resonance. Reproduced from Ref. [128] by permission Springer.

STIRAP approach discussed previously, see Fig.16). Conditional transfer can be realized by setting the laser fields in a two-photon resonance with the $|0'\rangle - |g\rangle$ transition, in this case the $|1'\rangle - |g\rangle$ transition will be far-detuned and no transfer will take place. Once the storage states $|0\rangle$ and $|1\rangle$ have been transferred into the enabling states $|0'\rangle$ and $|1'\rangle$, the sequence of operation to implement a phase gate can be summarized as

$$\begin{array}{lllllll}
|00\rangle' & \stackrel{\text{FOPA}}{\longrightarrow} & |gg\rangle' & \stackrel{\pi}{\longrightarrow} & -|gg\rangle' & \stackrel{\text{FOPA}^{-1}}{\longrightarrow} & -|00\rangle' \\
|01\rangle' & \longrightarrow & |g1\rangle' & \longrightarrow & |g1\rangle' & \longrightarrow & |01\rangle' \\
|10\rangle' & \longrightarrow & |1g\rangle' & \longrightarrow & |1g\rangle' & \longrightarrow & |10\rangle' \\
|11\rangle' & \longrightarrow & |11\rangle' & \longrightarrow & |11\rangle' & \longrightarrow & |11\rangle'
\end{array}$$

Here, FOPA and FOPA^{-1} refer to the STIRAP+FOPA conversion of atoms into molecules and the reverse process, respectively, and $|ij\rangle' \equiv |i'\rangle_A \otimes |j'\rangle_B$ and $|gg\rangle' \equiv |g\rangle_A \otimes |g\rangle_B$; after the sequence, the $|0'\rangle$ and $|1'\rangle$ states need to be returned to their storage states $|0\rangle$ and $|1\rangle$. The resulting state entangles the Rb atoms in site A and B.

- *Molecular qubit readout:* A different application of the conditional atom–molecule conversion relates to the readout of molecular qubits. Molecules possess long-lived rotational and spin sublevels in the ground electronic state that can encode qubits. Specifically, molecules with a singlet $^1\Sigma^+$ ground state, which is typical for alkali dimers, have zero electronic spin. Nuclear spins of constituent atoms are therefore decoupled from the electronic spin [121,122], making them insensitive to magnetic field fluctuations. It makes hyperfine or Zeeman sublevels of the molecular ground state an attractive choice for qubit encoding. However, initialization and readout are expected to be more difficult for molecules than for atoms. Due to the complex molecular level structure and the lack of closed ("cycling") transitions, the optical pumping and fluorescence used for initialization and readout in atoms are not in general available for molecules. Conditional conversion of a molecule into a pair of atoms followed by atomic detection represents a possible solution to this problem. In fact, coherent conversion followed by atomic imaging has been used recently to detect ground-state KRb molecules [65]. For rotational qubits, conditional conversion can be realized by selectively exciting a specific rotational state with a laser field, since rotational splittings are typically in the GHz range and the states are well-resolved. For spin qubits, one can first map the qubit state onto rotational sublevels and then apply the readout sequence.

C. Phase Gate Implementation

We discuss details of the scheme focusing on the implementation of the conditional atom–molecule conversion, assuming that atoms are transferred to the stable ground rovibrational state with a large permanent dipole moment. The transfer can be realized by an optical Raman π pulse or STIRAP pulses, as illustrated in Fig.16, where a pair of atoms in $\left|0'\right\rangle$ qubit state and lowest motional state of the lattice is transferred into the molecular ground state $|g\rangle$. Choosing the Rabi frequencies of the pulses much smaller than the qubit transition frequency avoids excitation of the $\left|1'\right\rangle$ state. It is in general difficult to transfer atoms from a delocalized scattering state, even when confined in a lattice, to a tightly localized, deeply bound molecular state. Namely, the wave function overlap of the intermediate molecular state $|v_e\rangle$ with either the initial atomic or the final molecular state will be small, requiring large intensity of the laser pulse to provide Rabi frequency sufficiently large for fast transfer. As discussed previously, unfavorable wave function overlap can be alleviated by using FOPA [64], which mixes the scattering state of an open collision channel and a bound vibrational state of a closed channel and strongly enhances the dipole moment for the transition to the molecular state $|v_e\rangle$. For ^{87}Rb+^{7}Li, the wide resonance at 649 G in the $|1, 1\rangle^{\mathrm{Rb}} + |1, 1\rangle^{\mathrm{Li}}$ collision channel, recently observed in [120], can be used. For this, Li has to be transferred to the "enabled" $|1, 1\rangle$ state, the conditional conversion will then take place from the $|0'\rangle$ qubit state.

First, we denote the eigenstates of a system of two different atoms in a deep lattice site interacting via a short range van der Waals interaction as $|\omega_{\rm op}^n\rangle$ if they correspond to the open channel of two unbound atoms, and as $|v_{\rm cl}\rangle$ if they correspond to a closed bound molecular state; the eigenenergies and eigenstates depend on the interaction strength via a background s-wave scattering length $a_{\rm bg}$. These have been calculated in Refs [123,124]; ω^n stands for the oscillation frequency of the $n^{\rm th}$ motional state in the lattice, and v for a set of rovibrational quantum numbers of the molecular bound state. At a Feshbach resonance, a bound molecular state $|v_{\rm cl}\rangle$ of a closed channel overlaps in energy a lowest motional state of two unbound atoms of an open channel $|\omega_{\rm op}^0\rangle$, and both become mixed by the hyperfine interaction. The corresponding Hamiltonian can be written as

$$H = \epsilon_{\rm cl} |v_{\rm cl}\rangle \langle v_{\rm cl}| + \sum_n \left(n + \frac{1}{2}\right) \hbar\omega \left|\omega_{\rm op}^n\right\rangle + \sum_n \left(V^n |v_{\rm cl}\rangle \left\langle \omega_{\rm op}^n\right| + H.c.\right) \tag{46}$$

where $\epsilon_{\rm cl} = \epsilon_{\rm cl}(B)$ is the energy of $|v_{\rm cl}\rangle$ (tunable by an external magnetic field B), and V^n is the interaction strength between the nth motional state of the open channel and the bound state of the closed channel. If the motional frequency ω is much larger than the interaction strength, we can neglect the coupling of $|v_{\rm cl}\rangle$ to all motional states but the lowest $|\omega_{\rm op}^0\rangle$, in which case $|\omega_{\rm op}^0\rangle$ and $|v_{\rm cl}\rangle$ form "mixed" or "dressed" states $|\pm\rangle$ given by

$$|\pm\rangle = c_\pm^{\rm cl} |v_{\rm cl}\rangle + c_\pm^{\rm op} \left|\omega_{\rm op}^0\right\rangle \tag{47}$$

with amplitudes

$$c_\pm^{\rm cl} = \frac{|V^0|}{\sqrt{2|\epsilon_\pm|}\left\{|V^0|^2 + \frac{1}{4}\epsilon_{\rm cl}^2\right\}^{1/4}} \tag{48}$$

$$c_\pm^{\rm op} = \pm \frac{|\epsilon_\pm|}{\sqrt{2}\left\{|V^0|^2 + \frac{1}{4}\epsilon_{\rm cl}^2\right\}^{1/4}} \tag{49}$$

with corresponding eigenenergies (see the inset to Fig. 16)

$$\epsilon_\pm = \frac{1}{2}\epsilon_{\rm cl} \pm \frac{1}{2}\sqrt{\epsilon_{\rm cl}^2 + 4|V^0|^2} \tag{50}$$

For large detunings ($|\epsilon_{\rm cl}| \geq |V^0|$) the "dressed" states become

$$|\pm\rangle \approx |v_{\rm cl}\rangle \pm \frac{|V^0|}{\epsilon_{\rm cl}} \left|\omega_{\rm op}^0\right\rangle \tag{51}$$

$$|\mp\rangle \approx \frac{|V^0|}{\epsilon_{\rm cl}} |v_{\rm cl}\rangle \mp \left|\omega_{\rm op}^0\right\rangle \tag{52}$$

where the top/bottom sign refers to positive/negative detunings, $\epsilon_{\rm cl} > 0$ and $\epsilon_{\rm cl} < 0$, respectively. By choosing the $|+\rangle$ or $|-\rangle$ states for positive or negative detunings, respectively, with significant fraction of the bound state $|v_{\rm cl}\rangle$ near the resonance, the dipole moment from the "dressed" state to an excited molecular state $|v_e\rangle$ is strongly enhanced by a factor $\sim |\langle v_e|\vec{\mu}\,|v_{\rm cl}\rangle|/|\langle v_e|\vec{\mu}\left|\omega_{\rm op}^0\right\rangle|$ as compared to the case without resonance.

Using an optical Raman π pulse for illustration purposes, the population of the molecular state $|g\rangle$ during the atom–molecule conversion is

$$|c_g|^2 = \frac{|\Omega_R|^2}{|\Omega_R|^2 + \delta^2} \sin^2\left(\sqrt{|\Omega_R|^2 + \delta^2}\,\frac{t}{2}\right) \tag{53}$$

where $\Omega_R = \Omega_p\Omega_S/2\Delta_e$ is the Rabi frequency of the pulse, $\Omega_{p,S}$ are the Rabi frequencies of the pump and Stokes fields, Δ_e is the common one-photon detuning of the fields from the excited molecular state, and δ is the two-photon detuning. Molecules at each site will interact via dipole–dipole interaction, accumulating the phase

$$\phi(t) = \frac{1}{\hbar}\int_0^t V_{\rm dd}(t')|c_g(t')|^4 dt' \tag{54}$$

for the $|g, g\rangle$ state (the other combinations do not acquire a phase, because the electric dipole moment is null in at least one of the states). Here, $V_{\rm dd} = \mu_{\rm ind}^2/r^3$ is the dipole–dipole interaction strength, $\mu_{\rm ind} \approx \mu(\mu F/3\hbar B_{\rm rot})$ is the molecular dipole moment induced by an electric field of magnitude F, μ is the permanent dipole moment, $\hbar B_{\rm rot}$ is the molecular rotational energy, and r is the distance between the molecules. After a specific time we can reverse the process and convert the molecules back into atoms $|g\rangle \rightarrow \left|0'\right\rangle$ such that we get the total accumulated phase $\phi = \pi$ and finally rotate the "enabler" atoms ^{7}Li back to the "inert" state $|2, 2\rangle$, resulting in a phase gate.

D. Realistic Estimates, Decoherence, and Errors

We discuss here the switchable dipole schemes, including the atom–molecule conversion scheme; a thorough analysis of schemes based on the permanent dipole moment of polar molecules [4] is given in Ref. [105]. To estimate the duration of one- and two-qubit gates, we need to account for the interaction of two atoms in a lattice site, whose strength is proportional to the background s-wave scattering length $a_{\rm bg}$. Because it is in general different for the $|0\rangle$ and $|1\rangle$ qubit states, the motional states will differ as well. To avoid undesirable entanglement between internal and motional states, one-qubit rotations have to be performed adiabatically compared to the oscillation period of the lattice, limiting the one-qubit rotation time in the tens microseconds for traps with $\sim$100 kHz oscillation frequency. The duration of the optical Raman π pulse transferring the Li atom to the "enabled" $|1, 1\rangle$ state is limited by the same requirement.

If we assume that the dipole moment induced by an external electric field is of the order of the permanent dipole moment ($\mu = 4.2$ D for LiRb [125]), the dipole–dipole interaction between two in neighboring lattice sites separated by $r = \lambda/2 = 400$ nm (λ is the wavelength of the optical field forming the lattice) is $V_{\rm dd} = \mu^2/r^3 \sim 2.7 \times 10^5\ {\rm s}^{-1}$. To allow $\sim 10^3$–10^4 operations during the coherence time, the atom–molecule conversion pulses have to be short enough; a careful choice of hyperfine states for qubit encoding can achieve coherence times $\sim$170 ms. The phase gate duration, therefore, has to be in the tens to several hundreds microseconds range. These gate durations correspond to the conversion pulses Rabi frequencies $\Omega_R \geq 10^4 - 10^5\ {\rm s}^{-1}$ (i.e., of the same order or larger than $V_{\rm dd}$), making it difficult to use the dipole blockade. Thus, direct dipole–dipole interaction between molecules must be used instead to realize the phase gate, which requires $\Omega_R \gg V_{\rm dd}$.

Assuming that the two-photon detuning is $\delta = V_{\rm dd}/\hbar$, and $|\Omega_R| \gg \delta$, the phase accumulated during two Raman π pulses and interaction time $\tau_{\rm int}$ is

$$\phi \approx \frac{V_{\rm dd}}{\hbar}\tau_{\rm int} \tag{55}$$

so that the interaction time to accumulate the π phase shift is $\tau_{\rm int} \approx \pi\hbar/V_{\rm dd} \sim 5\ \mu{\rm s}$.

Molecule-atom conversion for molecular qubit readout is limited by the requirement of state selectivity, that is, the bandwidth and the Rabi frequency of the laser field must be much smaller than the qubit transition frequency. Hyperfine splittings in the ground rovibrational state of $^1\Sigma^+$ electronic ground state are expected to be in the kiloHertz range [121,122], resulting in readout pulses of microseconds duration. These can be shortened to nanoseconds by mapping the spin qubit states onto rotational sublevels, having splittings in gigaHertz range.

The following sections contain a few potential sources of error that can take place in either the switchable schemes or the atom–molecule hybrid platform, briefly analyze them, and show ways to mitigate them.

1. Dipole–Dipole Interaction Strength

The fidelity of two-qubit gates depends on the separation and orientation of the molecular dipoles, which can lead to errors. The error per gate can, however, be reduced below a threshold value at which fault-tolerant quantum computing can be realized [126,127]. A typically cited desirable error threshold is 0.01%, or 1% with error correction, while errors per gate demonstrated experimentally to date are 3%. We analyze the error in a phase gate, assuming that two molecules are in a state having a large dipole–dipole interaction matrix element. While the interaction is "switched on," ideally a π phase shift accumulates.

To analyze the phase dependence on the relative distance and orientation of the dipole moments, we assume that each molecule is in a ground translational state of a 1D or 2D optical lattice potential. We use a translational ground state wave function of the 3D isotropic harmonic potential for a simple estimate of the mean distance a molecule travels in the ground state of an optical lattice potential. We make a harmonic approximation of the potential $V_0 k^2 \sum_i x_i^2$, giving the frequency $\omega = k\sqrt{2V_0/m}$ with the corresponding translational ground state wave function width $w = \sqrt{\hbar/m\omega}$ for a molecule of mass m. A typical potential depth a molecule experiences in a lattice is $V_0 = \eta E_R$, where $E_R = h^2/(2m\lambda^2)$ is the molecular recoil energy; the dimensionless parameter η defines the depth of the optical lattice (typically $\eta = 10 - 50$). The width w can then be related to the lattice field wavelength λ and the lattice depth η as $w = (2\eta)^{-1/4}\lambda/2\pi$. For molecules in neighboring lattice sites $R = \lambda/2$ this gives the ratio $w/R \approx 0.1$. While with adiabatic excitation this will not result in any uncertainty of the phase (just a renormalization), nonadiabaticity can result in errors of the order of w/R. For the typical case of $w/R \approx 0.1$, the error would too big to tolerate, and dipole-blockade would have to be used. If static dipoles are aligned by the same (static) electric field, errors in orientation will be negligible. However, for the "rotational" scheme described earlier, the dipole rotates with a frequency given by the rotational energy splitting. Hence, errors in relative timing of the excitation pulses are equivalent to orientation errors, and so the timing must be tightly controlled.

2. *Molecular State Decoherence*

If qubits are stored in hyperfine sublevels of the ground rovibrational electronic state the phase is relatively insensitive to local fluctuations of DC and AC electric fields, but more sensitive to magnetic field fluctuations, which should be minimized. The time scale of this minimization should be of the order of the coherence time of the trap. The states used to "switch on" the dipole–dipole interaction have to be long-lived to minimize decoherence from spontaneous emission. With gate operation times of ≤ 100 μs and metastable excited state lifetimes of several hundred microseconds, decoherence due to spontaneous emission will be small.

Interaction of molecules in the large dipole states can give rise to mechanical forces between molecules, leading to motional decoherence if the gates are not performed adiabatically. In the dipole-blockade mechanism, however, molecules are never actually transferred to the large dipole moment states simultaneously, thus this source of motional decoherence is minimized. A similar problem could be excitation into higher translational states due to momentum transfer from laser Raman pulses, for example, or forces from an electric field gradient. Again, however, typical pulse lengths (i.e., gate operation times) are expected to be larger than the translational oscillation period, allowing adiabatic transfers that minimize this decoherence source sufficiently.

3. Trap Induced Decoherence

Lifetimes of ultracold molecules in a far-detuned optical lattice of ~1 s have been obtained by minimizing the scattering of lattice photons. Given that the lifetime of a nuclear spin state of a single molecule, isolated from the environment, can be as long as hours, the lifetime of a molecule in a lattice (i.e., Raman scattering into other translational states), will be the major decoherence mechanism in this context. Similar numbers are expected in other trap schemes. Where our scheme relies on the dipole-blockade effect, it is not critical for the gate operation to cool molecules to the translational ground state of a lattice. The spatial dependence of driving fields, however, could cause decoherence due to excitation of higher translational states. The finite width of the translational ground state in a lattice potential does result in a spatially varying Rabi frequency of optical pulses experienced by the molecule. In particular, for a Gaussian Raman drive beam of width σ, the variation of the Rabi frequency over the molecular wave function can be of the order of w^2/σ^2, where w is the width of the molecular state. Beams, however, usually will not be focused down below the individual trap size (e.g., the size of one lattice site); hence, this error will be at most on the order of 1%.

V. CONCLUSIONS AND OUTLOOK

We discussed the formation of ultracold molecules, paying special attention to polar diatomic molecules, and how they could be used for quantum information processing.

The brief overview of the methods now being used for cooling molecules is far from complete. Instead, it illustrates the rapid developments of those and new techniques, and the string of achievements accomplished by a growing number of experimental and theoretical teams. We presented a more detailed description of the indirect approach to obtain ultracold molecules, using one- and two-photon photoassociation of pairs of atoms, and how Feshbach resonances could be employed to increase the formation rate, in an approach labeled FOPA. In addition, we discussed how FOPA could be used in tandem with STIRAP to coherently transfer pairs of atoms directly from the continuum into deeply bound levels of the electronic molecular ground state, sometimes using more than one intermediate state. All these advances are only the beginning of a successful effort to obtain ultracold molecules. Rapid progress is taking place in the generalization of photoassociation to produce larger ultracold molecules, for example, by using an atom and a diatomic molecule, or two diatomic molecules. A promising route is to manipulate the long-range interaction between the partners to create better Franck–Condon overlaps between ground and excited molecular electronic states at shorter range, thus allowing efficient production of more complex molecules. This could be achieved by orienting molecules with external fields.

The second aspect of this chapter relates to quantum information processing using polar molecules. In addition to a short review of concepts relevant to quantum information, such as qubits, gates, and algorithms, we listed the properties that polar molecules should possess to be a viable platform. We also described schemes in which the dipole–dipole interaction between two polar molecules could be switched on and off by pumping the molecules into appropriate states.

We presented a combined atomic–molecular system for quantum computation, which uses the principal advantages of neutral atom and polar molecule-based platforms. Encoding a qubit in atomic states allows easy initialization, readout, and one-qubit operations, as well as mapping the qubit state onto a photon for quantum communication. On the other hand, conditional conversion of a pair of atoms into a polar molecule gives rise to strong dipole–dipole interaction, resulting in fast two-qubit gates. The conditional conversion can also be used for reading out of molecular qubit states, which presents an efficient nondestructive alternative to state-selective ionization. We also give a short discussion of sources of decoherence and error, their effects, and estimates using realistic atomic–molecular systems.

Finally, we note that these fields of research are evolving rapidly, and that they may impact several other subfields of physics and chemistry. For example, creating ultracold molecules and controlling their long-range interaction may give insight in state-to-state chemical reactions or open new avenues in which the statistics of the interacting particles (composite bosons/fermions) may affect the results. The combined atomic–molecular system shows interesting many-body physics due to the presence of three species with independently tunable interactions, including short-range interatomic and long-range dipole–dipole intermolecular ones, allowing studies of exotic quantum phases such as a checkerboard solid and a pair supersolid. The same is true with the coherent control of molecules by atoms, and vice versa, which can become a new control tool in addition to traditional laser field-based control. And naturally, quantum information science is another field in which molecules, and hybrid platforms using molecules, provide a rich field of research. The next few years promise to be very exciting.

Acknowledgments

The author thanks the many co-authors of the papers referenced in this chapter. Funding from the National Science Foundation, the U.S. Department of Energy Office of Basic Sciences, and the Air Force Office of Scientific Research is gratefully acknowledged.

REFERENCES

1. M. A. Nielsen and I. L. Chuang, *Quantum Computation and Quantum Information*, Cambridge University Press, Cambridge, 2000.
2. P. W. Shor, *SIAM J. Comput.* **26**, 1484 (1997).
3. L. K. Grover, *Phys. Rev. Lett.* **79**, 4709 (1997).
4. D. DeMille, *Phys. Rev. Lett.* **88**, 067901 (2002).
5. C. Lee and E. A. Ostrovskaya, *Phys. Rev. A* **72**, 062321 (2005).
6. D. Jaksch, H.-J. Briegel, J. I. Cirac, C. W. Gardiner, and P. Zoller, *Phys. Rev. Lett.* **82**, 1975 (1999).
7. O. Mandel, M. Greiner, A. Widera, T. Rom, T. W. Hänsch, and I. Bloch, *Nature* **425**, 937 (2003).
8. J. I. Cirac and P. Zoller, *Phys. Rev. Lett.* **74**, 4091 (1995).
9. C. Monroe, D. M. Meekhof, B. E. King, W. M. Itano, and D. J. Wineland, *Phys. Rev. Lett.* **75**, 4714 (1995).
10. B. E. King, C. S. Wood, C. J. Myatt, Q. A. Turchette, D. Leibfried, W. M. Itano, C. Monroe, and D. J. Wineland, *Phys. Rev. Lett.* **81**, 1525 (1998).
11. A. André, D. DeMille, J. M. Doyle, M. D. Lukin, S. E. Maxwell, P. Rabl, R. J. Schoelkopf, and P. Zoller, *Nat. Phys.* **2**, 636 (2006).
12. R. Côté, *Nat. Phys.* **2**, 583 (2006).
13. U. Troppmann, C. M. Tesch, R. de Vivie-Riedle, *Chem. Phys. Lett.* **378**, 273 (2003).
14. D. Babikov, *J. Chem. Phys.* **121**, 7577 (2004).
15. J. P. Palao and R. Kosloff, *Phys. Rev. Lett.* **89**, 188301 (2002).
16. K. Maussang, D. Egorov, J. S. Helton, S. V. Nguyen, and J. M. Doyle, *Phys. Rev. Lett.* **94**, 123002, (2005).
17. A. J. Kerman, J. M. Sage, S. Sainis, T. Bergeman, and D. DeMille, *Phys. Rev. Lett.* **92**, 153001 (2004).
18. E. Hodby, S. T. Thompson, C. A. Regal, M. Greiner, A. C. Wilson, D. S. Jin, E. A. Cornell, and C. E. Wieman, *Phys. Rev. Lett.* **94**, 120402 (2005).
19. D. Wang, J. Qi, M. F. Stone, O. Nikolayeva, H. Wang, B. Hattaway, S. D. Gensemer, P. L. Gould, E. E. Eyler, and W. C. Stwalley, *Phys. Rev. Lett.* **93**, 243005 (2004).
20. C. Haimberger, J. Kleinert, M. Bhattacharya, and N. P. Bigelow, *Phys. Rev. A* **70**, 021402(R) (2004).
21. J. G. E. Harris, R. A. Michniak, S. V. Nguyen, W. C. Campbell, D. Egorov, S. E. Maxwell, L. D. van Buuren, and J. M. Doyle, *Rev. Sci. Instrum.* **75**, 17 (2004).
22. J. van Veldhoven, H. L. Bethlem, and G. Meijer, *Phys. Rev. Lett.* **94**, 083001 (2005).
23. S. Y. T. van de Meerakker, P. H. M. Smeets, N. Vanhaecke, R. T. Jongma, and G. Meijer, *Phys. Rev. Lett.* **94**, 023004 (2005).
24. T. Junglen, T. Rieger, S. A. Rangwala, P. W. H. Pinkse, and G. Rempe, *Phys. Rev. Lett.* **92**, 223001 (2004).
25. M. Vengalattore, R. S. Conroy, and M. G. Prentiss, *Phys. Rev. Lett.* **92**, 183001 (2004).
26. S. Groth, P. Krüger, S. Wildermuth, R. Folman, T. Fernholz, J. Schmiedmayer, D. Mahalu, and I. Bar-Joseph, *Appl. Phys. Lett.* **85**, 2980 (2004).
27. C. D. J. Sinclair, E. A. Curtis, I. Llorente Garcia, J. A. Retter, B. V. Hall, S. Eriksson, B. E. Sauer, and E. A. Hinds, *Phys. Rev. A* **72**, 031603(R) (2005).

28. J. Janis, M. Banks, and N. P. Bigelow, *Phys. Rev. A* **71**, 013422 (2005).
29. A. Wallraff, D. I. Schuster, A. Blais, L. Frunzio, J. Majer, M. H. Devoret, S. M. Girvin, and R. J. Schoelkopf, *Phys. Rev. Lett.* **95**, 060501 (2005).
30. A. S. Sørensen, C. H. van der Wal, L. I. Childress, and M. D. Lukin, *Phys. Rev. Lett.* **92**, 063601 (2004).
31. R. V. Krems, W. C. Stwalley, and B. Friedrich, *Cold Molecules: Theory, Experiment, Applications*, CRC Press, Boca Raton, FL, 2009.
32. L. D. Carr and J. Ye, *New J. Phys.* **11**, 055009 (2009), and references therein.
33. L. D. Carr, D. DeMille, R. V. Krems, and J. Ye, *New J. Phys.* **11**, 055049 (2009).
34. J. M. Doyle, B. Friedrich, R. V. Krems, and F. Masnou-Seeuws, *Eur. Phys. J. D* **31**, 149 (2004).
35. H. L. Bethlem, G. Berden, and G. Meijer, *Phys. Rev. Lett.* **83**, 1558 (1999).
36. J. R. Bochinski, E. R. Hudson, H. J. Lewandowski, G. Meijer, and J. Ye, *Phys. Rev. Lett.* 91 243001 (2003).
37. S. Y. T. Van De Meerakker, H. L. Bethlem, and G. Meijer, *Nat. Phys.* **4**, 595 (2008).
38. S. K. Tokunaga, J. M. Dyne, E. A. Hinds, and M. R. Tarbutt, *New J. Phys.* **11**, 055038 (2009).
39. E. R. Hudson, C. Ticknor, B. C. Sawyer, C. A. Taatjes, H. J. Lewandowski, J. R. Bochinski, J. L. Bohn, and J. Ye, *Phys. Rev. A* **73**, 063404 (2006).
40. M. R. Tarbutt, H. L. Bethlem, J. J. Hudson, V. L. Ryabov, V. A. Ryzhov, B. E. Sauer, G. Meijer, and E. A. Hinds, *Phys. Rev. Lett.* **92**, 173002 (2004).
41. S. Jung, E. Tiemann, and C. Lisdat, *Phys. Rev. A* **74**, 040701 (2006).
42. S. D. Hogan, D. Sprecher, M. Andrist, N. Vanhaeck, and F. Merkt, *Phys. Rev. A* **76**, 023412 (2007).
43. E. Narevicius, A. Libson, C. G. Parthey, I. Chavez, J. Narevicius, U. Even, and M. G. Raizen, *Phys. Rev. Lett.* **100**, 109902 (2008).
44. J. D. Weinstein, R. deCarvalho, T. Guillet, B. Friedrich, and J. M. Doyle, *Nature* **395**, 148 (1998).
45. D. Patterson D and J. M. Doyle, *J. Chem. Phys.* **126**, 154307 (2007).
46. N. Brahms, T. V. Tscherbul, P. Zhang, J. Klos, H. R. Sadeghpour, A. Dalgarno, J. M. Doyle, and T. G. Walker, *Phys. Rev. Lett.* **105**, 033001 (2010).
47. J. G. E. Harris, R. A. Michniak, S. V. Nguyen, N. Brahms, W. Ketterle, and J. M. Doyle, *Europhys. Lett.* **67**, 198 (2004).
48. E. S. Shuman, J. F. Barry, and D. DeMille, *Nature* **467**, 820 (2010).
49. H. R. Thorsheim, J. Weiner, and P. S. Julienne, *Phys. Rev. Lett.* **58**, 2420 (1987).
50. J. L. Bohn and P. S. Julienne, *Phys. Rev. A* **54**, R4637 (1996).
51. R. Côté and A. Dalgarno, *Phys. Rev. A* **58**, 498 (1998).
52. K. M. Jones, E. Tiesinga, P. D. Lett, and P. S. Julienne, *Rev. Mod. Phys.* **78**, 483 (2006).
53. E. P. Wigner, *Phys. Rev.* **73**, 1002 (1948).
54. R. Côté, A. Dalgarno, Y. Sun, and R. G. Hulet, *Phys. Rev. Lett.* **74**, 3581 (1995).
55. E. Juarros, P. Pellegrini, K. Kirby, and R. Côté, *Phys. Rev. A* **73**, 041403(R) (2006).
56. E. Juarros, K. Kirby, and R. Côté, *J. Phys. B* **39**, S965 (2006).
57. E. Taylor-Juarros, R. Côté, and K. Kirby, *Eur. Phys. J. D* **31**, 213 (2004).
58. P. Pellegrini, M. Gacesa, and R. Côté, *Phys. Rev. Lett.* **101**, 053201 (2008).
59. T. Köhler, K. Góral, and P. S. Julienne, *Rev. Mod. Phys.* **78**, 1311 (2006).

60. M. Junker, D. Dries, C. Welford, J. Hitchcock, Y. P. Chen, and R. G. Hulet, *Phys. Rev. Lett.* **101**, 060406 (2008).
61. K. E. Strecker, G. B. Partridge, A. G. Truscott, and R. G. Hulet, *Nature* **417**, 150 (2002).
62. I. D. Prodan, M. Pichler, M. Junker, R. G. Hulet, and J. L. Bohn, *Phys. Rev. Lett.* **91**, 080402 (2003).
63. P. Pellegrini and R. Côté, *New J. Phys.* **11**, 055047 (2009).
64. E. Kuznetsova, M. Gacesa, P. Pellegrini, S. F. Yelin, R. Côté, *New J. Phys.* **11**, 055028 (2009).
65. K.-K. Ni, S. Ospelkaus, M. H. G. de Miranda, A. Pe'er, B. Neyenhuis, J. J. Zirbel, S. Kotochigova, P. S. Julienne, D. S. Jin, and J. Ye, *Science* **322**, 231 (2008).
66. J. G. Danzl, M. J. Mark, E. Haller, M. Gustavsson, R. Hart, J. Aldegunde, J. M. Hutson, H. C. Nagerl, *Nat. Phys.* **6**, 265 (2010).
67. F. Lang, K. Winkler, C. Strauss, R. Grimm, and J. Hecker Denschlag, *Phys. Rev. Lett.* **101**, 133005 (2008);
68. U. Fano, *Phys. Rev.* **124**, 1866 (1961).
69. A. Vardi, D. Abrashkevich, E. Frishman, and M. Shapiro, *J. Chem. Phys.* **107**, 6166 (1997).
70. A. Vardi, M. Shapiro, and K. Bergmann, *Optics Express* **4**, 91 (1999).
71. E. A. Shapiro, M. Shapiro, A. Pe'er, and J. Ye, *Phys. Rev. A* **75**, 013405 (2007).
72. M. Fleischhauer, A. Imamoglu, and J. P. Marangos, *Rev. Mod. Phys.* **77**, 633 (2005).
73. E. Kuznetsova, P. Pellegrini, R. Côté, M. D. Lukin, and S. F. Yelin, *Phys. Rev. A* **78**, 021402(R) (2008).
74. B. W. Shore, K. Bergemann, J. Oreg, and S. Rosenwaks, *Phys. Rev. A* **44**, 7442 (1991).
75. A. Ekert, Quantum Computation in *Atomic Physics*, Vol. 14, Fourteenth International Conference on Atomic Physics, Boulder, CO (1994), D. J. Wineland, C. E. Wieman, and S. J. Smith, eds., American Institute of Physics, New York, 1995, p. 450.
76. A. Einstein, B. Podolsky, and N. Rosen, *Phys. Rev.* **47**, 777 (1935).
77. D. Deutsch, *Proc. R. Soc. Lond. A* 425, 73 (1989).
78. L. K. Grover, *Phys. Rev. Lett.* **79**, 325 (1997).
79. D. Deutsch and R. Jozsa, *Proc. R. Soc. Lond. A* **439**, 553 (1992).
80. A. Ekert and R.Jozsa, *Rev. Mod. Phys.* **68**, 733 (1996).
81. D. Aharonov, A. Ambainis, J. Kempe, and U. Vazirani, *Proc. 33rd ACM STOC*, 50–59 (2001).
82. A. M. Childs, R. Cleve, E. Deotto, E. Farhi, S. Gutmann, and D. A. Spielman, *Proc. 35th ACM STOC*, 59–68 (2003)
83. A. M. Steane, *Phys. Rev. Lett.* **77**, 793 (1996).
84. A. Steane, *Proc. R. Soc. Lond. A* **452**, 2551 (1996).
85. A. R. Calderbank and P. W. Shor, *Phys. Rev. A* **54**, 1098 (1996).
86. E. Knill, R. Laflamme, and W. H. Zurek, *Science* **279**, 342 (1998).
87. J. Preskill, *Proc. R. Soc. Lond. A* **454**, 385 (1998).
88. J. Preskill, *Phys. Today* **52**(6), 24 (1999).
89. D. P. DiVincenzo, *Fortschr. Phys.* **48**, 771 (2000).
90. J. F. Poyatos, J. I. Cirac, and P. Zoller, *Phys. Rev. Lett.* **81**, 1322 (1998).
91. D. F. V. James, *Phys. Rev. Lett.* **81**, 317 (1998).
92. A. Sørensen and K. Mølmer, *Phys. Rev. Lett.* **83**, 2274 (1999).
93. G. K. Brennen, C. M. Caves, P. S. Jessen, and I. H. Deutsch, *Phys. Rev. Lett.* **82**, 1060 (1999).

94. M. Saffman, T. G. Walker, and K. Mølmer, *Rev. Mod. Phys.* **82**, 2313 (2010).
95. M. D. Lukin and P. R. Hemmer, *Phys. Rev. Lett.* **84**, 2818 (2000).
96. I. L. Chuang, N. Gershenfeld, and M. Kubinec, *Phys. Rev. Lett.* **80**, 3408 (1998).
97. D. Loss and D. P. DiVincenzo, *Phys. Rev. A* **57**, 120 (1998).
98. B. E. Kane, *Nature* **393**, 133 (1998).
99. J. Wrachtrup and F. Jelezko, *J. Phys. C* **18**, S807 (2006).
100. L. Childress, M. V. Gurudev Dutt, J. M. Taylor, A. S. Zibrov, F. Jelezko, J. Wrachtrup, P. R. Hemmer, and M. D. Lukin, *Science* **314**, 281 (2006).
101. D. Marcos, M. Wubs, J. M. Taylor, R. Aguado, M. D. Lukin, and A. S. Sørensen, *Phys. Rev. Lett.* **105**, 210501 (2010).
102. H. J. Kimble, *Phys. Scripta* **76**, 127 (1998).
103. Q. A. Turchette, C. J. Hood, W. Lange, H. Mabuchi, and H. J. Kimble, *Phys. Rev. Lett.* **75**, 4710 (1995).
104. C. Nayak, S. H. Simon, A. Stern, M. Freedman, and S. Das Sarma, *Rev. Mod. Phys.* **80**, 1083 (2008).
105. Q. Wei, S. Kais, B. Friedrich, and D. Herschbach, *J. Chem. Phys.* 134, 124107 (2011).
106. B. M. R. Korff, U. Troppmann, K. L. Kompa, and R. de Vivie-Riedle, *J. Chem. Phys.* **123**, 244509 (2005).
107. L. Bomble, D. Lauvergnat, F. Remacle, and M. Desouter-Lecomte, *J. Chem. Phys.* **128**, 064110 (2008).
108. S. F. Yelin, K. Kirby, and R. Côté, *Phys. Rev. A* **74**, 050301(R) (2006).
109. E. Kuznetsova, R. Côté, K. Kirby, and S. F. Yelin, *Phys. Rev. A* **78**, 012313 (2008).
110. D. Jaksch, J. I. Cirac, P. Zoller, S. L. Rolston, R. Côté, and M. D. Lukin, *Phys. Rev. Lett.* **85**, 2208 (2000).
111. M. D. Lukin, M. Fleischhauer, R. Côté, L. M. Duan, D. Jaksch, J. I. Cirac, and P. Zoller, *Phys. Rev. Lett.* **87**, 037901 (2001).
112. D. Tong, S. M. Farooqi, J. Stanojevic, S. Krishnan, Y. P. Zhang, R. Côté, E. E. Eyler, and P. L. Gould, *Phys. Rev. Lett.* **93**, 063001 (2004).
113. S. Kallush, B. Segev, and R. Côté, *Phys. Rev. Lett.* **95**, 163005 (2005).
114. S. Kallush, B. Segev, and R. Côté, *Eur. Phys. J. D* **35**, 3 (2005).
115. K. Tordrup and K. Mølmer, *Phys. Rev. A* **77**, 020301 (2008).
116. P. Rabl, D. DeMille, J. M. Doyle, M. D. Lukin, R. J. Schoelkopf, and P. Zoller, *Phys. Rev. Lett.* **97**, 033003 (2006).
117. D. Jaksch, *Contemp. Phys.* **45**, 367 (2004).
118. P. Treutlein, T. Steinmetz, Y. Colombe, B. Lev, P. Hommelhoff, J. Reichel, M. Greiner, O. Mandel, A. Widera, T. Rom, I. Bloch, and T. W. Hänsch, *Fortschr. Phys.* **54**, 702 (2006).
119. E. Kuznetsova, M. Gacesa, S. F. Yelin, and R. Côté, *Phys. Rev. A* **81**, 030301 (2010).
120. C. Marzok, B. Deh, C. Zimmermann, Ph. W. Courteille, E. Tiemann, Y. V. Vanne, and A. Saenz, *Phys. Rev. A* **79**, 012717 (2009).
121. J. Aldegunde, B. A. Rivington, P. S. Zuchowski, and J. M. Hutson, *Phys. Rev. A* **78**, 033434 (2008).
122. J. Aldegunde, and J. M. Hutson, *Phys. Rev. A* **79** 013401 (2009).
123. J. F. Bertelsen and K. Mølmer, *Phys. Rev. A* **76**, 043615 (2007).

124. F. Deuretzbacher, K. Plassmeier, D. Pfannkuche, F. Werner, C. Ospelkaus, S. Ospelkaus, K. Sengstock, and K. Bongs, *Phys. Rev. A* **77**, 032726 (2008).

125. M. Aymar and O. Dulieu, *J. Chem. Phys.* **122**, 204302 (2005).

126. E. Knill, *Nature* **434**, 39 (2005).

127. P. Aliferis and J. Preskill, *Phys. Rev. A* **79**, 012332 (2009).

128. E. Kuznetsova, S. F. Yelin, and R. Côté, *Quant. Inf. Process* **10**, 821 (2011).

DYNAMICS OF ENTANGLEMENT IN ONE- AND TWO-DIMENSIONAL SPIN SYSTEMS

GEHAD SADIEK,[1,2] QING XU,[3] and SABRE KAIS[4]

[1] *Department of Physics, King Saud University, Riyadh, Saudi Arabia*
[2] *Department of Physics, Ain Shams University, Cairo 11566, Egypt*
[3] *Department of Chemistry, Purdue University, 560 Oval Drive, West Lafayette, IN 47907, USA*
[4] *Department of Chemistry and Physics, Purdue University, 560 Oval Drive, West Lafayette, IN 47907, USA; Qatar Environment & Energy Research Institute (QEERI), Doha, Qatar; Santa Fe Institute, Santa Fe, NM 87501, USA*

Advances in Chemical Physics, Volume 154: Quantum Information and Computation for Chemistry, First Edition. Edited by Sabre Kais.

I. INTRODUCTION

The state of a classical composite system is described in the phase space as a product of its individual constituents, separate states. On the other hand, the state of a composite quantum system is expressed in the Hilbert space as a superposition of tensor products of its individual subsystems, states. In other words, the state of the quantum composite system is not necessarily expressible as a product of the individual quantum subsystem states. This peculiar property of quantum systems is called *entanglement*, which has no classical analog [1]. The phenomenon of entanglement was first introduced by Schrödinger [2] who called it "Verschrankung" and stated "For an entangled state the best possible knowledge of the whole does not include the best possible knowlege of its parts." Quantum entanglement is a nonlocal correlation between two (or more) quantum systems such that the description of their states has to be done with reference to each other even if they are spatially well separated. In the early days of the quantum theory, the notion of entanglement was first noted and introduced by Einstein et al. [3] as a paradox in the formalism of the quantum theory. Einstein et al. in their famous EPR paper proposed a thought experiment to demonstrate that the quantum theory is not a complete physical theory because it lacks the elements of reality needed for such a theory. It needed about three decades before performing an experiment that invalidated the EPR argument and guaranteed victory to the quantum theory. The experiment was based on a set of inequalities derived by John Bell [4], which relate correlated measurements of two physical quantities that should be obeyed by any local theory. He demonstrated that the outcomes in the case of quantum entangled states violate the Bell inequality. This result emphasizes that entanglement is a quantum mechanical property that cannot be simulated using a classical formalism [5].

Recently, the interest in studying quantum entanglement was sparked by developments in the field of quantum computing, initiated in the 1980s by the pioneering work of Benioff, Bennett, Deutsch, Feynman, and Landauer [6–12]. This interest gained a huge boost in 1994 after the distinguished work of Peter Shor and the development of a quantum computer algorithm for efficiently prime factorizing composite integers [13]. Other fields in which entanglement plays a major role are quantum teleportation [14,15], dense coding [16,17], quantum communication [18], and quantum cryptography [19].

Different physical systems have been proposed as reliable candidates for the underlying technology of quantum computing and quantum information processing [20–28]. The basic idea in each one of these systems is to define certain quantum degrees of freedom to serve as a qubit, such as the charge, orbital, or spin angular momentum. This is usually followed by finding a controllable mechanism to form an entanglement between a two-qubit system in such a way to produce a fundamental quantum computing gate such as an exclusive Boolean XOR. In addition, we have to be able to coherently manipulate such an entangled state to provide an efficient computational process. Coherent manipulation of entangled states has been observed in different systems such as isolated trapped ions [29] and superconducting junctions [30]. The coherent control of a two-electron spin state in a coupled quantum dot was achieved experimentally in which the coupling mechanism was the Heisenberg exchange interaction between the electron spins [31–33].

Particularly, the solid-state systems have been the focus of interest as they facilitate the fabrication of large integrated networks that would be able to implement realistic quantum computing algorithms on a large scale. On the other hand, the strong coupling between a solid-state system and its complex environment makes it challenging to achieve the high coherence control required to manipulate the system. Decoherence is considered one of the main obstacles toward realizing an effective quantum computing system [34–37]. The main effect of decoherence is to randomize the relative phases of the possible states of the isolated system as a result of coupling to the environment. By randomizing the relative phases, the system loses all quantum interference effects and its entanglement character and may end up behaving classically.

The interacting Heisenberg spin systems, in one, two, and three dimensions represent reliable models for constructing quantum computing schemes in different solid-state systems and a rich model for studying the novel physics of localized spin systems [38–41]. These spin systems can be experimentally realized, for instance, as a one-dimensional chain and lattices of coupled nano quantum dots.

Multiparticle systems are of central interest in the field of quantum information, because a quantum computer is considered as many-body system by itself. Understanding, quantifying, and exploring entanglement dynamics may provide answers many questions regarding the behavior of complex quantum systems [42]. Quantum phase transitions and critical behavior as entanglement are considered to be the physical property responsible for the long-range quantum correlations accompanying these phenomena [43–47].

A. Entanglement Measures

One of the central challenges in the theory of quantum computing and quantum information and their applications is the preparation of entangled states and quantifying them. An enormous number of approaches have been taken to tackle this problem, both experimentally and theoretically. The term *entanglement measure* is used to describe any function that can be used to quantify entanglement. Unfortunately we are still far from a complete theory that can quantify entanglement of a general multipartite system in pure or mixed state [48–51]. A limited number of cases have achieved successful entanglement measures. Two of these cases are (1) a bipartite system in a pure state, and (2) a bipartite system of two spin $1/2$ in a mixed state. Because these two particular cases are of special interest for our studies in spin systems we discuss them in greater detail in the two following sections.

To quantify entanglement, that is, to find out how much entanglement is contained in a quantum state, Vedral et al. introduced the axiomatic approach to quantify entanglement [52]. They introduced the basic axioms that are necessary for an entanglement measure to satisfy. Before introducing these axioms one should first discuss the most common operations that can be performed on quantum systems and their effects on entanglement [53]. First, local operation is an operation that when applied to a quantum system consisting of two subsystems, each subsystem evolves independently. Therefore, possible preexisting correlations, whether classical or quantum, will not be affected. Hence, their entanglement also will not be affected. Second a global operation, when applied to a quantum system consisting of two subsystems, has subsystems that will evolve while interacting with each other. Therefore, correlations, both classical and quantum, may change under the effect of such operations. Hence, their entanglement will be affected by this operation. Third, local operations with classical communications (LOCC) are a special kind of global operations in which the subsystems evolve independently with classical communications allowed between them. Information about local operations can be shared using the classical communications and then further local operations may be done according to the shared information. Therefore, classical, but not quantum, correlations may be changed by LOCC. It is reasonable then to require that an entanglement measure should not increase under LOCC.

Now let us briefly introduce a list of the most commonly accepted axioms that a function E must obey to be considered an entanglement measure:

1. E is a mapping from the density matrix of a system to a positive real number $\rho \rightarrow E(\rho) \in \mathbf{R}$.
2. E does not increase under LOCC only [53].
3. E is invariant under local unitary transformations.
4. for a pure state $\rho = |\psi\rangle \langle\psi|$, E reduces to the entropy of entanglement [54] (discussed in greater detail later).

5. $E = 0$ iff the states are separable [55].
6. E takes its maximum for maximally entangled states (normalization).
7. E is continous [56].
8. E should be a convex function (i.e., it cannot be obtained by mixing states $[\rho_i$ [56], $E(w_i\rho_i) \leq w_i E(\rho_i)]$).
9. E is additive (i.e., given two pairs of entangled particles in the total state $\sigma = \sigma_1 \otimes \sigma_2$, then we have [57] $E(\sigma) = E(\sigma_1) + E(\sigma_2)$).

1. *Pure Bipartite State*

An effective approach to handle pure bipartite state is the Schmidt decomposition. For a pure state $|\psi\rangle$ of a composite system consists of two subsystems A and B, with two orthonormal basis $\{|\phi_{a,i}\rangle\}$ and $\{|\phi_{b,i}\rangle\}$, respectively, the Schmidt decomposition of the state $|\psi\rangle$ is defined by

$$|\psi\rangle = \sum_i \lambda_i \, |\phi_{a,i}\rangle \, |\phi_{b,i}\rangle \tag{1}$$

where λ_i are positive coefficients satisfying $\sum \lambda_i^2 = 1$ and are called Schmidt coefficients. Evaluating the reduced density operators

$$\rho_{A/B} = tr_{B/A}(|\psi\rangle) = \sum_i \lambda_i^2 |\phi_{(a/b),i}\rangle\langle\phi_{(a/b),i}| \tag{2}$$

shows that the two operators have the same spectrum λ_i, which means that the two subsystems will have many properties in common. If the state $|\psi\rangle$ is a cross-product of pure states, say $|\phi_{a,k}\rangle$ and $|\phi_{b,k}\rangle$, of the subsystems A and B, respectively, then $|\psi\rangle$ is a disentangled state and all Schmidt coefficients vanish except $\lambda_k = 1$ in that case. In general, the coefficients λ_i can be used to quantify the entanglement in the composite system.

Entropy plays a major role in the classical and quantum information theory. Entropy is a measure of our uncertainty (lack of information) of the state of the system. The Shannon entropy quantifies the uncertainty associated with a classical distribution $\{P_i\}$ and is defined as $H(x) = -\sum_x P_x \log P_x$. The quantum analog of the Shannon entropy is the Von Neumann entropy where the classical probability distribution is replaced with the density operators. Considering the density operator ρ to represent the state of a quantum system, the Von Neumann entropy is defined as $S(\rho) \equiv -tr(\rho \log \rho) = -\sum_i \alpha_i \log \alpha_i$ where $\{\alpha_i\}$ are the eigenvalues of the matrix ρ . For a bipartite system, the Von Neumann entropy of the reduced density matrix $\rho_{A/B}$, namely $S(\rho_{A/B}) = -\sum_i \lambda_i^2 \log \lambda_i^2$ is a measure of entanglement that satisfies all the axioms. Although the entanglement content in a multipartite system is difficult to quantify, the bipartite entanglement of the different constituents of the system can provide good insight about the entanglement of the whole system.

2. *Mixed Bipartite State*

When the composite physical system is in a mixed state, which is more common, different entanglement measures are needed. Contrary to the pure state, which has only quantum correlations, the mixed state contains both classical and quantum correlations; the entanglement measure should discriminate between these correlations. Developing an entanglement measure for mixed multipartite systems is a challenging task because it is difficult to discriminate between the quantum and classical correlations in that case [54,55]. Nevertheless for bipartite systems, different entanglement measures were introduced to overcome the mathematical difficulty, particularly for subsystems with only two degrees of freedom. Among the most common entanglement measures in this case are the relative entropy E_R [57,58]; entanglement of distillation E_D [58]; and negativity and logarithmic negativity [59].

One of the most widely used measures is entanglement of formation, which was the first measure to appear in 1996 in work by C. Bennett et al. [60]. For a mixed state, the entanglement of formation is defined as the minimum amount of entanglement needed to create the state. Any mixed state $|\psi\rangle$ can be decomposed into a mixture of pure states $|\psi_i\rangle$ with different probabilities p_i, which is called "ensemble." The entanglement of formation of a mixed state can be obtained by summing the entanglements of each pure state after multiplying each one by its probability $p_i E(|\psi_i\rangle)$. The entanglement of each pure state is expressed as the entropy of entanglement of that state. Hence, for an ensemble Γ of pure states $\{p_i, |\psi_i\rangle\}$ we have [61]

$$E_F(\epsilon) = \sum_i p_i E_{vN}(|\psi_i\rangle) \tag{3}$$

Because a mixed state can be decomposed into many different ensembles of pure states with different entanglements, the entanglement of formation is evaluated using what is called the most "economical" ensemble [60], that is,

$$E_F(\rho) = \inf_{\Gamma} \sum_i p_i E_{vN}(|\psi_i\rangle) \tag{4}$$

where the infimum is taken over all possible ensembles. E_F is called a convex roof, and the decomposition leading to this *convex roof* is called the *optimal decomposition* [62].

The minimum is selected because in the case of a decomposition where the average entanglement is zero, then this state can be created locally without the need for entangled pure states and therefore, $E_F = 0$ [62].

Performing minimization over all decompositions is a difficult task because of the large number of terms involved [51]. Nevertheless, it was shown that only limited number of terms is sufficient to perform the minimization. However, finding an explicit formula that does not need performing the minimization would simplify the evaluation of E_F significantly. Bennett et al. [60] evaluated E_F for a mixture of Bell's states, which are completely entangled qubits. Hill and Wootters [63] provided a closed form of E_F as a function of the density matrix for two-level bipartite systems having only two non-zero eigenvalues in terms of the *concurrence*, which was extended later to the case of all two-level bipartite systems (i.e., two qubits) [64]. The entanglement of formation satisfies the previously discussed axioms [54,56].

B. Entanglement and Quantum Phase Transitions

Quantum phase transition (QPT) in many-body systems, in contrary to classical phase transition, takes place at zero temperature. QPT are driven by quantum fluctuations as a consequence of Heisenberg uncertainty principle [65,66]. Examples of quantum phase transition are quantum Hall transitions, magnetic transitions of cuprates, superconductor–insulator transitions in two dimension, and metal–insulator transitions [66,67]. Quantum phase transition is characterized by a singularity in the ground-state energy of the system as a function of an external parameter or coupling constant λ [68]. In addition, QPT is characterized by a diverging correlation length ξ in the vicinity of the quantum critical point defined by the parameter value λ_c. The correlation length diverges as $\xi^{-1} \sim J|\lambda - \lambda_c|^{\nu}$, where J is an inverse length scale and ν is a critical exponent. Quantum phase transitions in many-body systems are accompanied by a significant change in the quantum correlations within the system. This led to a great interest in investigating the behavior of quantum entanglement close the critical points of transitions, which may shed some light on the different properties of the ground state wave function as it goes through the transition critical point. On the other hand, these investigations may clarify the role entanglement plays in quantum phase transitions and how it is related to the different properties of that transition, such as its order and controlling parameters. The increasing interest in studying the different properties of entangled states of complex systems motivated a huge amount of research in that area. One of the main consequences of the research is the consideration of entanglement as a physical resource that can be utilized to execute specific physical tasks in many-body systems [60,69]. Osborne and Nielsen have argued that the physical property responsible for long-range quantum correlations accompanying quantum phase transitions in complex systems is entanglement, and that it becomes maximum at the critical point [70]. The renormalization group calculations demonstrated that quantum phase transitions have a universal character independent of the dynamical properties of the system and is only affected by specific global properties such as the symmetry of the system [71]. In order to test whether the entanglement would show the same universal properties in the similar systems, the pairwise entanglement was studied in the XY spin model and its special case of the Ising model [44] where it was shown that the entanglement reaches a maximum value at the critical point of the phase transition in the Ising system. Also, the entanglement was proved to obey a scaling behavior in the vicinity of transition critical point for a class of one-dimensional magnetic system, XY model in a transverse magnetic field [72]. Furthermore, the scaling properties of entanglement in XXZ and XY spin-$1/2$ chains near and at the transition critical point were investigated, and the resemblance between the critical entanglement in spin system and entropy in conformal field theories was emphasized [73]. Several works have discussed the relation between entanglement and correlation functions and, as a consequence, the notion of localizable entanglement was introduced, which enabled the definition of entanglement correlation length bounded from above by entanglement assistance and from below by classical functions and diverges at a quantum phase transition [74–77]. Quantum discord, which measures the total amount of correlations in a quantum state and discerns it from the classical ones, first introduced by Olliver and Zurek [78], was used to study quantum phase transition in XY and XXZ spin systems [79]. It was demonstrated that while the quantum correlations increase close to the critical points, the classical correlations decreases, in XXZ model, and is monotonous in Ising model, in the vicinity of the critical points.

C. Dynamics of Entanglement

In addition to the interest in the static behavior of entanglement in many-body systems, its dynamical behavior has attracted great attention as well, where different aspects of these dynamics have been investigated recently. One of the most important aspects is the propagation of entanglement through a many-body system starting from a specific part within the system. The speed of propagation of entanglement through the system depends on different conditions and parameters such as the initial setup of the system, impurities within the system, the coupling strength among the system constituents, and the external magnetic field [80–82].

In most treatments, the system is prepared in an initial state described by an initial Hamiltonian H_i, then its time evolution is studied under the effect of different parameters, internal and external, which causes creation, decay, vanishing, or just transfer of entanglement through the system. In many cases, the system is abruptly changed from its initial state to another one, causing sudden change in entanglement as well.

The creation of entanglement between different parts of a many-body system rather than the transfer of entanglement through it was also investigated. The creation of entanglement between the end spins of a spin-$1/2$ XY chain was studied

[83]. A global time-dependent interaction between the nearest neighbor spins on the chain was applied to an initial separable state and the creation of entanglement between the end spins on the chain was tested. It was demonstrated that the amount of entanglement created dynamically was significantly larger than that created statically.

As heat can be extracted from a many-body solid-state system and be used to create heat, it was shown that entanglement can be extracted from a many-body system by means of external probes and be used in quantum information processing [84]. The idea is to scatter a pair of independent noninteracting particles simultaneously by an entangled manybody solid-state system (e.g., solid-state spin chain) or optical lattice with cold atoms where each incident particle interacts with a different entangled system particle. It was demonstrated that the entanglement was extracted from the many-body system and transferred to the incident probes and the amount of entanglement between the probes pair is proportional to the entanglement within the many-body system and vanishes for a disentangled system. Recently the time evolution of entanglement between an incident mobile particle and a static particle was investigated [85]. It was shown that the entanglement increases monotonically during the transient but then saturates to a steady-state value. The results were general for any model of two particles, where it was demonstrated that the transient time depends only on the group velocity and the wave packet width for the incident quasi-monochromatic particle and independent of the type and strength of the interaction. On the other hand, entanglement information extraction from spin-boson environment using noninteracting multi-qubit systems as a probe was considered [86]. The environment consists of a small number of quantum-coherent two-level fluctuators (TLFs) with a damping caused by independent bosonic baths. Special attention was devoted to the quantum correlations (entanglement) that build up in the probe as a result of the TLF-mediated interaction.

The macroscopic dynamical evolution of spin systems was demonstrated in what is known as quantum domino dynamics. In this phenomenon, a one-dimensional spin-1/2 system with nearest neighbor interaction in an external magnetic field is irradiated by a weak resonant transverse field [87,88]. It was shown that a wave of spin flip can be created through the chain by an initial single spin flip. This can be utilized as a signal amplification of spin-flipping magnetization.

II. DYNAMICS OF ENTANGLEMENT IN ONE-DIMENSIONAL SPIN SYSTEMS

A. Effect of a Time-Dependent Magnetic Field on Entanglement

We consider a set of N localized spin-1/2 particles coupled through exchange interaction J and subject to an external magnetic field of strength h. We investigate the dynamics of entanglement in the systems in the presence of a time-dependent magnetic fields. The Hamiltonian for such a system is given by Huang and Kais [89]

$$H = -\frac{J}{2}(1+\gamma)\sum_{i=1}^{N}\sigma_i^x\sigma_{i+1}^x - \frac{J}{2}(1-\gamma)\sum_{i=1}^{N}\sigma_i^y\sigma_{i+1}^y - \sum_{i=1}^{N}h(t)\sigma_i^z \quad (5)$$

where J is the coupling constant, $h(t)$ is the time-dependent external magnetic field, σ^a are the Pauli matrices ($a = x, y, z$), γ is the degree of anisotropy, and N is the number of sites. We can set $J = 1$ for convenience and use periodic boundary conditions. Next, we transform the spin operators into fermionic operators, so that, the Hamiltonian assumes the following form:

$$H = \sum_{p=1}^{N/2}\alpha_p(t)[c_p^+c_p + c_{-p}^+c_{-p}] + i\delta_p[c_p^+c_{-p}^+ + c_pc_{-p}] + 2h(t) = \sum_{p=1}^{N/2}\tilde{H}_p \quad (6)$$

where, $\alpha_p(t) = -2\cos\phi_p - 2h(t)$, $\delta_p = 2\gamma\sin\phi_p$, and $\phi_p = 2\pi p/N$. It is easy to show $[\tilde{H}_p, \tilde{H}_q] = 0$, which means the space of $\tilde{H}$ decomposes into noninteracting subspace, each of four dimensions. No matter what $h(t)$ is, no transitions will occur among those subspaces. Using the following basis for the pth subspace $(|0>; c_p^+c_{-p}^+|0>; c_p^+|0>; c_{-p}^+|0>)$, we can explicitly get

$$\tilde{H}_p(t) = \begin{pmatrix} 2h(t) & -i\delta_p & 0 & 0 \\ i\delta_p & -4\cos\phi_p - 2h(t) & 0 & 0 \\ 0 & 0 & -2\cos\phi_p & 0 \\ 0 & 0 & 0 & -2\cos\phi_p \end{pmatrix} \quad (7)$$

We only consider the systems that at time $t = 0$ are in the thermal equilibrium at temperature T. Let $\rho_p(t)$ be the density matrix of the pth subspace, we have $\rho_p(0) = e^{-\beta\tilde{H}_p(0)}$, where $\beta = 1/kT$ and k is the Boltzmann constant. Therefore, using Eq. (7), we can have $\rho_p(0)$. Let $U_p(t)$ be the time-evolution matrix in the pth subspace, namely($\hbar = 1$): $i\frac{dU_p(t)}{dt} = U_p(t)\tilde{H}_p(t)$, with the boundary condition $U_p(0) = I$. Now, the Liouville equation of this system is

$$i\frac{d\rho(t)}{dt} = [H(t), \rho(t)] \quad (8)$$

and can be decomposed into uncorrelated subspaces and solved exactly. Thus, in the pth subspace, the solution of Liouville equation is $\rho_p(t) = U_p(t)\rho_p(0)U_p(t)^\dagger$.

As a first step to investigate the dynamics of the entanglement, we can take the magnetic field to be a step function, and then generalize it to other relevant functional forms such as an oscillating one [89]. Figure 1 shows the results for

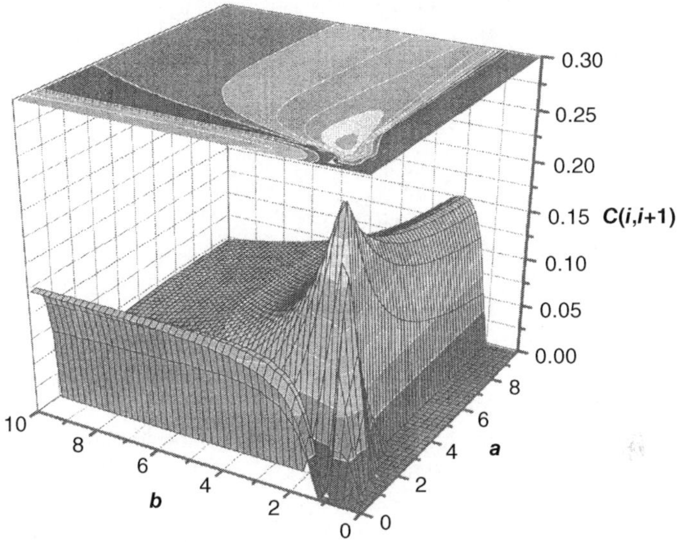

Figure 1. Nearest-neighbor concurrence C at zero temperature as a function of the initial magnetic field a for the step function case with final field b.

nearest-neighbor concurrence $C(i, i+1)$ at temperature $T = 0$ and $\gamma = 1$ as a function of the initial magnetic field a for the step function case with final field b. For $a < 1$ region, the concurrence increases very fast near $b = 1$ and reaches a limit $C(i, i+1) \sim 0.125$ when $b \to \infty$. It is surprising that the concurrence will not disappear when b increases with $a < 1$. This indicates that the concurrence will not disappear as the final external magnetic field increases at infinite time. It shows that this model is not in agreement with the obvious physical intuition, because we expect that increasing the external magnetic field will destroy the spin–spin correlation functions and make the concurrence vanish. The concurrence approaches maximum $C(i, i+1) \sim 0.258$ at ($a = 1.37, b = 1.37$), and decreases rapidly as $a \neq b$. This indicates that the fluctuation of the external magnetic field near the equilibrium state will rapidly destroy the entanglement. However, in the region where $a > 2.0$, the concurrence is close to zero when $b < 1.0$ and maximum close to 1. Moreover, it disappear in the limit of $b \to \infty$.

Now, let us examine the system size effect on the entanglement with three different external magnetic fields changing with time t [90]:

$$h_I(t) = \begin{Bmatrix} a & t \leq 0 \\ b + (a - b)e^{-Kt} & t > 0 \end{Bmatrix} \tag{9}$$

$$h_{II}(t) = \begin{Bmatrix} a & t \leq 0 \\ a - a\sin(Kt) & t > 0 \end{Bmatrix} \tag{10}$$

$$h_{III}(t) = \begin{Bmatrix} 0 & t \leq 0 \\ a - a\cos(Kt) & t > 0 \end{Bmatrix} \tag{11}$$

where a, b, and K are varying parameters.

We have found that the entanglement fluctuates shortly after a disturbance by an external magnetic field when the system size is small. For larger system size, the entanglement reaches a stable state for a long time before it fluctuates. However, this fluctuation of entanglement disappears when the system size goes to infinity. We also show that in a periodic external magnetic field, the nearest-neighbor entanglement displays a periodic structure with a period related to that of the magnetic field. For the exponential external magnetic field, by varying the constant K we have found that as time evolves, $C(i, i+1)$ oscillates but it does not reach its equilibrium value at $t \to \infty$. This confirms the fact that the nonergodic behavior of the concurrence is a general behavior for slowly changing magnetic field. For the periodic magnetic field $h_{II} = a(1 - \sin[-Kt])$ the nearest-neighbor concurrence is at maximum at $t = 0$ for values of a close to 1, because the system exhibits a quantum phase transition at $\lambda_c = J/h = 1$, where in our calculations we fixed $J = 1$. Moreover, for the two periodic $\sin[-Kt]$ and $\cos[-Kt]$ fields the nearest-neighbor concurrence displays a periodic structure according to the periods of their respective magnetic fields [90]. For the periodic external magnetic field $h_{III}(t)$, we show in Figure 2 that the nearest-neighbor concurrence $C(i, i+1)$

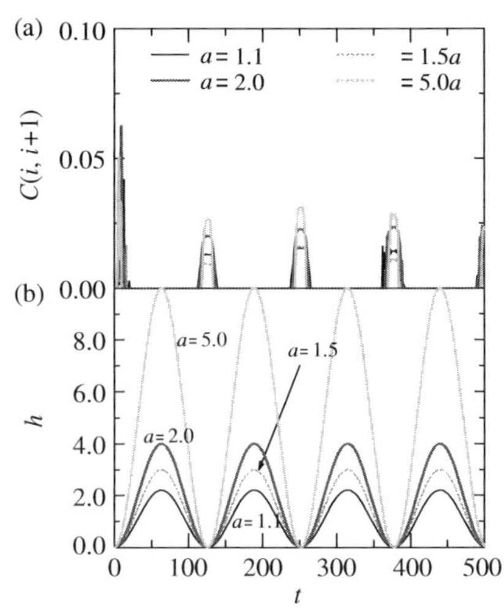

Figure 2. The nearest-neighbor concurrence $C(i, i+1)$ (**a**) and the periodic external magnetic field $h_{III}(t) = a(1 - \cos[Kt])$, see Eq. (14) in the text (**b**) for $K = 0.05$ with different values of a as a function of time t.

is zero at $t = 0$, because the external magnetic field $h_{III}(t = 0) = 0$ and the spins align along the x-direction: the total wave function is factorizable. By increasing the external magnetic field we see the appearance of nearest-neighbor concurrence but very small. This indicates that the concurrence cannot be produced without background external magnetic field in the Ising system. However, as time evolves one can see the periodic structure of the nearest-neighbor concurrence according to the periodic structure of the external magnetic field $h_{III}(t)$ [90].

B. Decoherence in a One-Dimesional Spin System

Recently, there has been a special interest in solid-state systems as they facilitate the fabrication of large integrated networks that would be able to implement realistic quantum computing algorithms on a large scale. On the other hand, the strong coupling between a solid-state system and its complex environment makes it a significantly challenging mission to achieve the high coherence control required to manipulate the system. Decoherence is considered as one of the main obstacles toward realizing an effective quantum computing system [34–37]. The main effect of decoherence is to randomize the relative phases of the possible states of the isolated system as a result of coupling to the environment. By randomizing the relative phases, the system loses all quantum interference effects and may end up behaving classically.

As a system of special interest, great efforts have been made to study the mechanism of electron phase decoherence and determine the time scale for such process (the decoherence time), in solid-state quantum dots both theoretically [36,91–94] and experimentally [31,33,95–97]. The main source of electron spin decoherence in a quantum dot is the inhomogeneous hyperfine coupling between the electron spin and the nuclear spins.

In order to study the decoherence of a two-state quantum system as a result of coupling to a spin bath, we examined the time evolution of a single spin coupled by exchange interaction to an environment of interacting spin bath modeled by the XY-Hamiltonian. The Hamiltonian for such system is given by Huang et al. [98]

$$H = -\frac{1+\gamma}{2}\sum_{i=1}^{N} J_{i,i+1}\sigma_i^x\sigma_{i+1}^x - \frac{1-\gamma}{2}\sum_{i=1}^{N} J_{i,i+1}\sigma_i^y\sigma_{i+1}^y - \sum_{i=1}^{N} h_i\sigma_i^z \qquad (12)$$

where $J_{i,i+1}$ is the exchange interaction between sites i and $i+1$, h_i is the strength of the external magnetic field on site i, σ^a are the Pauli matrices ($a = x, y, z$), γ is the degree of anisotropy, and N is the number of sites. We consider the centered spin on the lth site as the single spin quantum system and the rest of the chain as its environment, where in this case $l = (N+1)/2$. The single spin directly interacts with its nearest-neighbor spins through exchange interaction $J_{l-1,l} = J_{l,l+1} = J'$. We assume exchange interactions between spins in the environment are uniform,

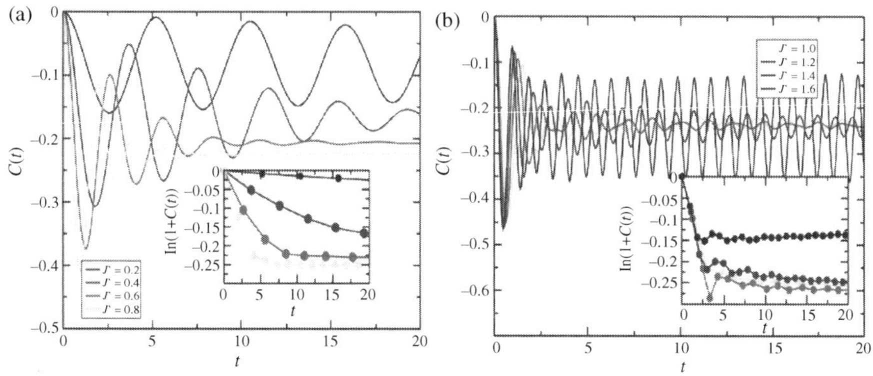

Figure 3. The spin correlation function $C(t)$ of centered spin for $N = 501$, $h = 0.5$, and $\gamma = 1.0$ versus time t for different values of the coupling (**a**) $J' \leq J$; (**b**) $J' \geq J$ at zero temperature. The decay profile for each case is shown in the inner panel. (*See the color version of this figure in the Color Plates section.*)

and simply set it as $J = 1$. The centered spin is considered as inhomogeneously coupled to all the spins in the environment by being directly coupled to its nearest neighbors and indirectly to all other spins in the chain through its nearest neighbors.

By evaluating the spin correlator $C(t)$ of the single spin, the jth site [98]

$$C_j(t) = \rho_j^z(t, \beta) - \rho_j^z(0, \beta) \tag{13}$$

we observed that the decay rate of the spin oscillations strongly depends on the relative magnitude of the exchange coupling between the single spin and its nearest neighbor J' and coupling among the spins in the environment J. The decoherence time varies significantly based on the relative couplings magnitudes of J and J'. The decay rate law has a Gaussian profile when the two exchange couplings are of the same order $J' \sim J$ but converts to exponential and then a power law as we move to the regimes of $J' > J$ and $J' < J$ as shown in Fig. 3. We also showed that the spin oscillations propagate from the single spin to the environmental spins with a certain speed as depicted in Fig. 4. Moreover, the amount of saturated decoherence induced into the spin state depends on this relative magnitude and approaches maximum value for a relative magnitude of unity. Our results suggest that setting the interaction within the environment in such a way that its magnitude is much higher or lower than the interaction with the single spin may reduce the decay rate of the spin state. The reason behind this phenomenon could be that the variation in the coupling strength along the chain at one point (where the single spin exits) blocks the propagation of decoherence along the chain by reducing the entanglement among the spins within the environment, which reduces its decoherence effect on

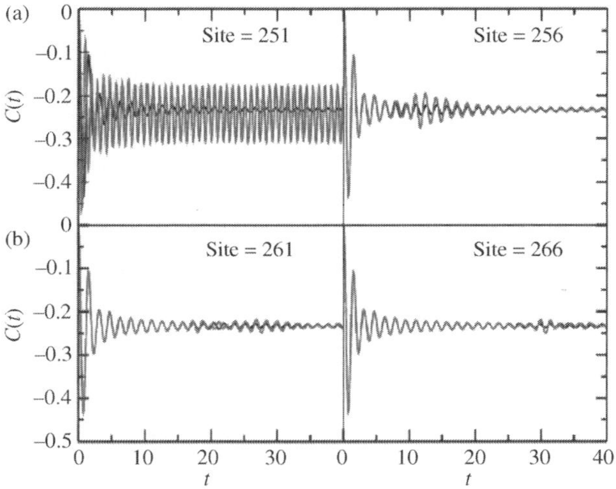

Figure 4. The spin correlation function $C(t)$ versus time t with $J' = 1.0$ (black) and $J' = 1.5$ (grey) for the centered spin (**a**, left) and the spin at site 256 (**a**, right), 261 (**b**, left), and 266 (**b**, right) in the environment. $N = 501, h = 0.5$, and $\gamma = 1.0$.

the single spin in return [98]. This result might be applicable in general to similar cases of a centered quantum system coupled inhomogeneously to an interacting environment with large degrees of freedom.

C. An Exact Treatment of the System with Step Time-Dependent Coupling and Magnetic Field

The obvious demand in quantum computation for a controllable mechanism to couple the qubits, led to one of the most interesting proposals in that regard, which is to introduce a time-dependent exchange interaction between the two valence spins on a doubled quantum dot system as the coupling mechanism [38,39]. The coupling can be pulsed over definite intervals resulting in a swap gate, which can be achieved by raising and lowering the potential barrier between the two dots through controllable gate voltage. The ground state of the two-coupled electrons is a spin singlet, which is a highly entangled spin state.

Many studies have focused on the entanglement at zero and finite temperature for isotropic and anisotropic Heisenberg spin chains in presence and absence of an external magnetic field [45,99–105]. Particularly, the dynamics of thermal entanglement have been studied in an XY spin chain considering a constant nearest-neighbor exchange interaction, in the presence of a time-varying magnetic field represented by a step, exponential, and sinusoidal functions of time we discussed already [89,90].

Recently, the dynamics of entanglement in a one-dimensional Ising spin chain at zero temperature was investigated numerically where the number of spins was seven at most [106]. The generation and transportation of the entanglement through the chain under the effect of an external magnetic field and irradiated by a weak resonant field were studied. It was shown that the remote entanglement between the spins is generated and transported though only nearest-neighbor coupling was considered. Later the anisotropic XY model for a small number of spins, with a time-dependent nearest-neighbor coupling at zero temperature was studied too [107]. The time-dependent spin–spin coupling was represented by a dc part and a sinusoidal ac part. It was found that an entanglement resonance occurs through the chain whenever the ac coupling frequency is matching the Zeeman splitting.

Here, we investigate the time evolution of quantum entanglement in an infinite one-dimensional XY spin chain system coupled through nearest-neighbor interaction under the effect of a time varying magnetic field $h(t)$ at zero and finite temperature. We consider a time-dependent nearest-neighbor Heisenberg coupling $J(t)$ between the spins on the chain. We discuss a general solution for the problem for any time dependence form of the coupling and magnetic field and present an exact solution for a particular case of practical interest, namely a step function form for both the coupling and the magnetic field. We focused on the dynamics of entanglement between any two spins in the chain and its asymptotic behavior under the interplay of the time-dependent coupling and magnetic field. Moreover, we investigated the persistence of quantum effects especially close to critical points of the system as it evolves in time and as its temperature increases.

The Hamiltonian for the XY model of a one-dimensional lattice with N sites in a time-dependent external magnetic field $h(t)$ with a time-dependent coupling $J(t)$ between the nearest-neighbor spins on the chain is given by

$$H = -\frac{J(t)}{2}(1+\gamma)\sum_{i=1}^{N}\sigma_i^x\sigma_{i+1}^x - \frac{J(t)}{2}(1-\gamma)\sum_{i=1}^{N}\sigma_i^y\sigma_{i+1}^y - \sum_{i=1}^{N}h(t)\sigma_i^z \tag{14}$$

where σ_is are the Pauli matrices and γ is the anisotropy parameter.

Following the standard procedure to treat the Hamiltonian (14), we transform the Hamiltonian into the form [108]

$$H = \sum_{p=1}^{N/2}\tilde{H}_p \tag{15}$$

with $\tilde{H}_p$ given by

$$\tilde{H}_p = \alpha_p(t)[c_p^\dagger c_p + c_{-p}^\dagger c_{-p}] + iJ(t)\delta_p[c_p^\dagger c_{-p}^\dagger + c_p c_{-p}] + 2h(t) \tag{16}$$

where $\alpha_p(t) = -2J(t)\cos\phi_p - 2h(t)$ and $\delta_p = 2\gamma\sin\phi_p$.

Writing the matrix representation of $\tilde{H}_p$ in the basis $\{|0\rangle, c_p^\dagger c_{-p}^\dagger |0\rangle, c_p^\dagger |0\rangle, c_{-p}^\dagger |0\rangle\}$, we obtain

$$\tilde{H}_p = \begin{pmatrix} 2h(t) & -iJ(t)\delta_p & 0 & 0 \\ iJ(t)\delta_p & -4J(t)\cos\phi_p - 2h(t) & 0 & 0 \\ 0 & 0 & -2J(t)\cos\phi_p & 0 \\ 0 & 0 & 0 & -2J(t)\cos\phi_p \end{pmatrix} \tag{17}$$

Initially the system is assumed to be in a thermal equilibrium state and therefore its initial density matrix is given by

$$\rho_p(0) = e^{-\beta \tilde{H}_p(0)} \tag{18}$$

where $\beta = 1/kT$, k is Boltzmann constant and T is the temperature.

Because the Hamiltonian is decomposable we can find the density matrix at any time t, $\rho_p(t)$, for the pth subspace by solving Liouville equation given by

$$i\dot{\rho}_p(t) = [H_p(t), \rho_p(t)] \tag{19}$$

which gives

$$\rho_p(t) = U_p(t)\rho_p(0)U_p^\dagger(t) \tag{20}$$

where $U_p(t)$ is time evolution matrix that can be obtained by solving the equation

$$i\dot{U}_p(t) = U_p(t)\tilde{H}_p(t) \tag{21}$$

Because $\tilde{H}_p$ is block diagonal, U_p should take the form

$$U_p(t) = \begin{pmatrix} U_{11}^p & U_{12}^p & 0 & 0 \\ U_{21}^p & U_{22}^p & 0 & 0 \\ 0 & 0 & U_{33}^p & 0 \\ 0 & 0 & 0 & U_{44}^p \end{pmatrix} \tag{22}$$

Fortunately, Eq. (21) may have an exact solution for a time-dependent step function form for both exchange coupling and the magnetic field we adopt in this work. Other time-dependent function forms will be considered in a future work where other techniques can be applied. The coupling and magnetic field are represented, respectively, by

$$J(t) = J_0 + (J_1 - J_0)\theta(t) \tag{23}$$

$$h(t) = h_0 + (h_1 - h_0)\theta(t) \tag{24}$$

where $\theta(t)$ is the usual mathematical step function. With this setup, the matrix elements of U_p can be evaluated. The reduced density matrix of any two spins is evaluated in terms of the magnetization defined by

$$M = \frac{1}{N}\sum_{j=1}^{N}\langle S_j^z\rangle = \frac{1}{N}\sum_{p=1}^{1/N} M_p \tag{25}$$

and the spin–spin correlation functions defined by

$$S_{l,m}^x = \langle S_l^x S_m^x \rangle, \quad S_{l,m}^y = \langle S_l^y S_m^y \rangle, \quad S_{l,m}^z = \langle S_l^z S_m^z \rangle \tag{26}$$

Using the obtained density matrix elements, one can evaluate the entanglement between any pair of spins using the Wootters method [109].

1. *Transverse Ising Model*

The completely anisotropic XY model, Ising model, is obtained by setting $\gamma = 1$ in the Hamiltonian (14). Defining a dimensionless coupling parameter $\lambda = J/h$, the ground state of the Ising model is characterized by a quantum phase transition that takes place at λ close to the critical value $\lambda_c = 1$ [44]. The order parameter is the magnetization $\langle \sigma^x \rangle$ that differs from zero for $\lambda \geq \lambda_c$ and zero otherwise. The ground state of the system is paramagnetic when $\lambda \to 0$ where the spins get aligned in the magnetic field direction, the z direction. For the other extreme case when $\lambda \to \infty$, the ground state is ferromagnetic and the spins are all aligned in the x direction.

We explored the dynamics of the nearest-neighbor concurrence $C(i, i+1)$ at zero temperature while the coupling parameter (and/or the magnetic field) is a step function in time. We found that the concurrence $C(i, i+1)$ shows a nonergodic behavior. This behavior follows from the nonergodic properties of the magnetization and the spin–spin correlation functions as reported by previous studies [89,110,111]. At higher temperatures, the nonergodic behavior of the system sustains but with reduced magnitude of the asymptotic concurrence (as $t \to \infty$).

We studied the behavior of the nearest-neighbor concurrence $C(i, i+1)$ as a function of λ for different values of J and h at different temperatures. As can be seen in Fig. 5a, the behavior of $C(i, i+1)$ at zero temperature depends only on the ratio J/h (i.e., λ) rather than their individual values. Studying entanglement at nonzero temperatures shows that the maximum value of $C(i, i+1)$ decreases as the temperature increases. Furthermore, $C(i, i+1)$ shows a dependence on the individual values of J and h, not only their ratio as illustrated in Fig. 5b, c, and d. We investigated the dependence of the asymptotic behavior (as $t \to \infty$) of the nearest-neighbor concurrence on the magnetic field and coupling parameters h_0, h_1, J_0, and J_1 at zero temperature. In Fig. 6a, we present a three-dimensional plot for the concurrence versus J_0 and J_1 where we set the magnetic field at $h_0 = h_1 = 1$,

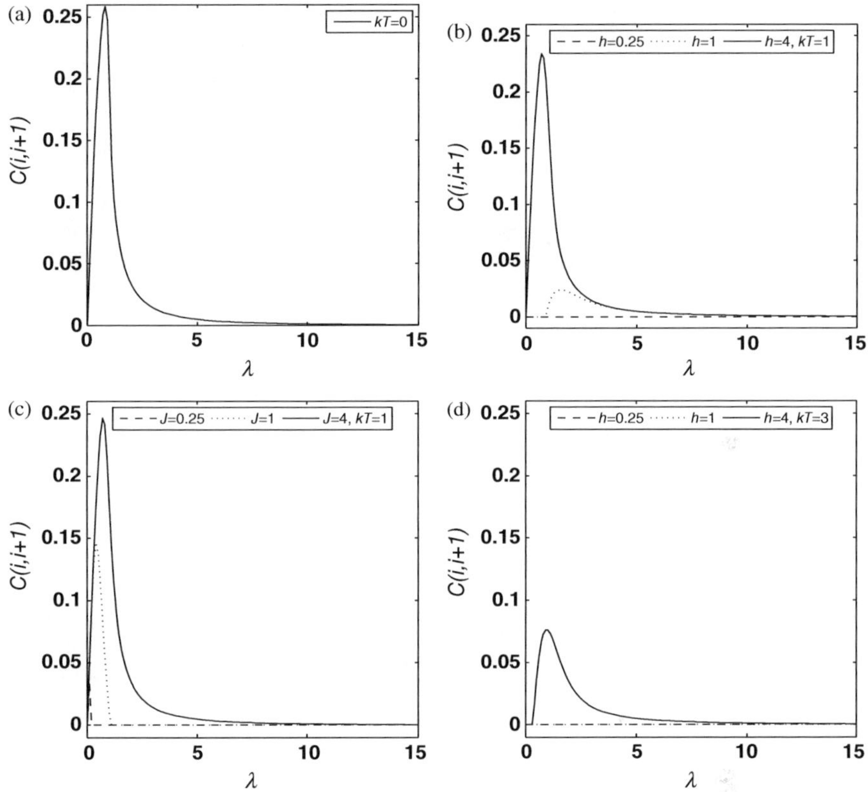

Figure 5. $C(i, i+1)$ as a function of λ for $h = h_0 = h_1$ and $J = J_0 = J_1$ at (**a**) $kT = 0$ with any combination of J and h; (**b**) $kT = 1$ with $h_0 = h_1 = 0.25, 1, 4$; (**c**) $kT = 1$ with $J_0 = J_1 = 0.25, 1, 4$; (**d**) $kT = 3$ with $h_0 = h_1 = 0.25, 1, 4$.

while Fig. 6b shows the asymptotic behavior of the nearest-neighbor concurrence as a function of h_0 and h_1, while fixing the coupling parameter at $J_0 = J_1 = 2$. In both cases, the entanglement reaches its maximum value close to the critical value $\lambda_c = 1$. Also, the behavior of the asymptotic concurrence is much more sensitive to the change in the magnetic field parameters compared to the coupling ones.

Interest continues to grow in investigating the effect of temperature on the quantum entanglement and the critical behavior of many-body systems and particularly spin systems [43,44,68,112,113]. Osborne and Nielsen have studied the persistence of quantum effects in the thermal state of the transverse Ising model as temperature increases [44]. Here, we investigate the persistence of quantum effects under both temperature and time evolution of the system in the presence of the time-dependent coupling and magnetic field. Interestingly, the time evolution

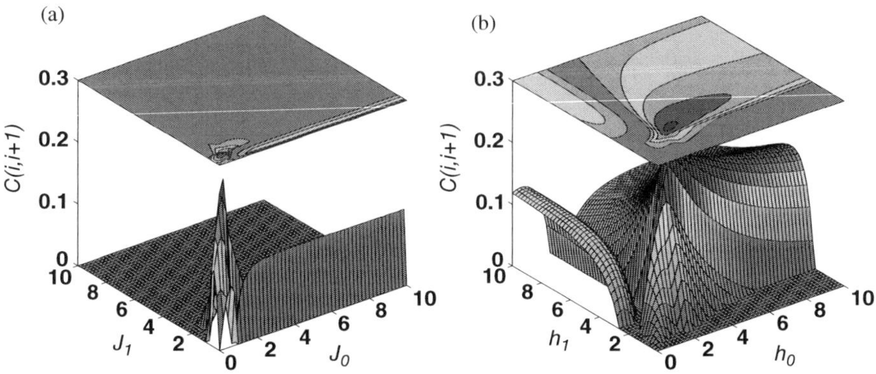

Figure 6. The asymptotic behavior of $C(i, i+1)$ as a function of (**a**) J_0 and J_1 with $h_0 = h_1 = 1$; (**b**) h_0 and h_1 with $J_0 = J_1 = 2$, where h_0, h_1, and J_0 are in units of J_1, $kT = 0$ and $\gamma = 1$. (*See the color version of this figure in the Color Plates section.*)

of entanglement shows a similar profile to that manifested in the static case (i.e., the system evolves in time preserving its quantum character in the vicinity of the critical point and $kT = 0$ under the time varying coupling) as shown in Fig. 7. Studying this behavior at different values of J_0 and h_0 shows that the threshold temperature, at which $C(i, i+1)$ vanishes, increases as λ_0 increases.

2. *Partially Anisotropic XY Model*

We now turn to the partially anisotropic system where $\gamma = 0.5$. First, we studied the time evolution of nearest-neighbor concurrence for this model and it showed

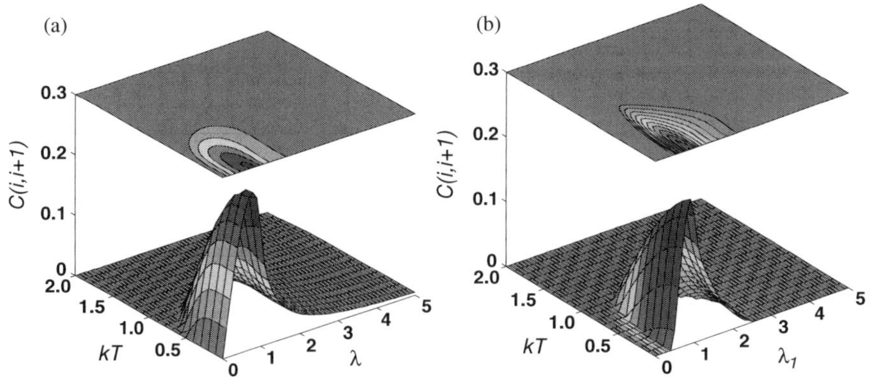

Figure 7. The asymptotic behavior of $C(i, i+1)$ as a function of (**a**) λ and kT, in units of J_1, with $\gamma = 1$, $h_0 = h_1 = 1$, and $J_0 = J_1$; (**b**) λ_1 and kT with $\gamma = 1$, $h_0 = h_1 = 1$, and $J_0 = 1$.

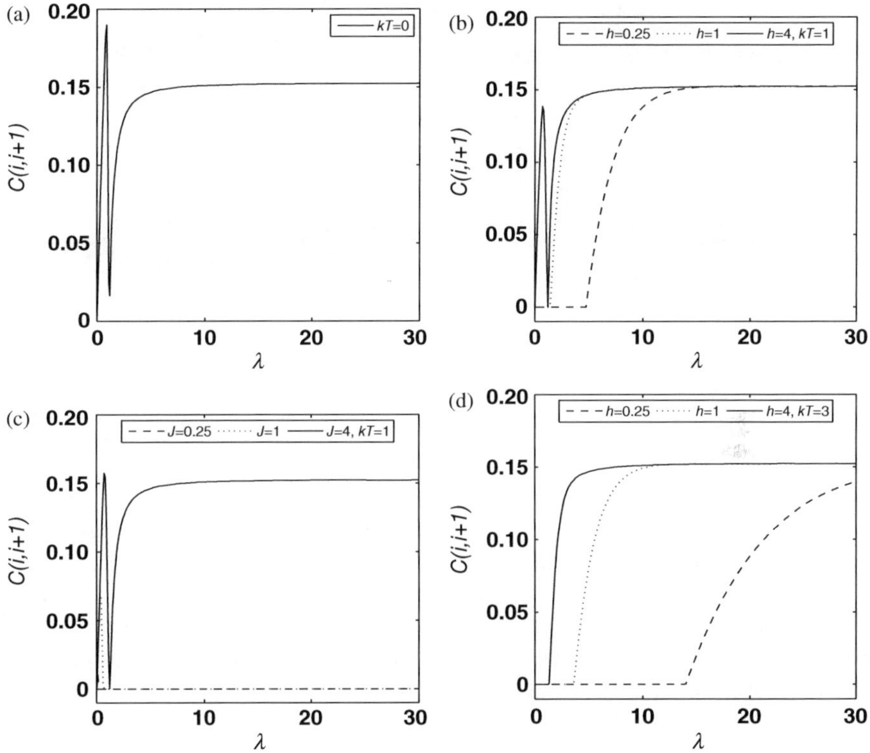

Figure 8. The asymptotic behavior of $C(i, i+1)$ with $\gamma = 0.5$ as a function of λ when $h_0 = h_1$ and $J_0 = J_1$ at (**a**) $kT = 0$ with any combination of constant J and h; (**b**) $kT = 1$ with $h_0 = h_1 = 0.25, 1, 4$; (**c**) $kT = 1$ with $J_0 = J_1 = 0.25, 1, 4$; (**d**) $kT = 3$ with $h_0 = h_1 = 0.25, 1, 4$.

a nonergodic behavior, similar to the isotropic case, which also follows from the nonergodic behavior of the spin correlation functions and magnetization of the system. Nevertheless, the equilibrium time in this case is much longer than the isotropic. We have investigated the behavior of the nearest-neighbor concurrence $C(i, i+1)$ as a function of λ for different values of J and h and at different temperatures as shown in Fig. 8. We first studied the zero temperature case at different constant values of J and h. For this particular case, $C(i, i+1)$ depends only on the ratio of J and h, similar to the isotropic case, rather than their individual values as depicted in Fig. 8a. Interestingly, the concurrence shows a complicated critical behavior in the vicinity of $\lambda = 1$, where it reaches a maximum value first and immediately drops to a minimum (very small) value before raising again to its equilibrium value. The raising of the concurrence from zero as J increases, for $\lambda < 1$, is expected as when the spins, which were originally aligned in the

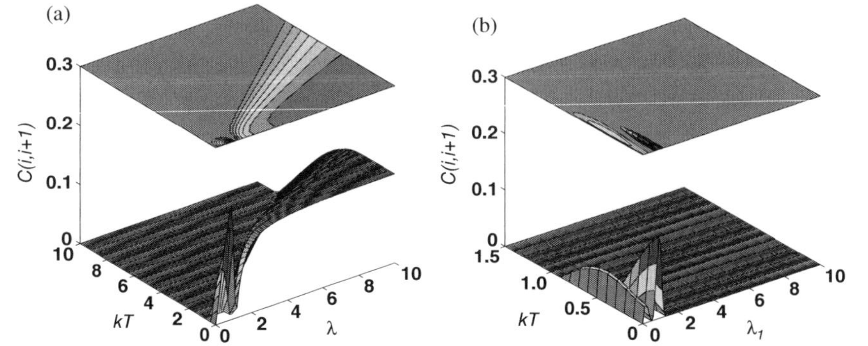

Figure 9. The asymptotic behavior of $C(i, i+1)$ as a function of (**a**) λ and kT, in units of J_1, with $\gamma = 0.5$, $h_0 = h_1 = 1$, and $J_0 = J_1$; (**b**) λ_1 and kT with $\gamma = 0.5$, $h_0 = h_1 = 1$, and $J_0 = 1$.

z-direction, change directions into the x- and y-directions. The sudden drop of the concurrence in the vicinity of $\lambda = 1$, where λ is slightly larger than 1, suggests that significant fluctuations is taking place and the effect of J_x is dominating over both J_y and h, which aligns most of the spins into the x-direction leading to a reduced entanglement value. Studying the thermal concurrence in Fig. 8b, c, and d, we note that the asymptotic value of $C(i, i+1)$ is not affected as the temperature increases. However, the critical behavior of the entanglement in the vicinity of $\lambda = 1$ changes considerably as the temperature is raised and the other parameters are varied. The effect of higher temperature is shown in Fig. 8d where the critical behavior of the entanglement disappears completely at all values of h and J, which confirms that the thermal excitations destroy the critical behavior due to suppression of quantum effects. The persistence of quantum effects as temperature increases and time elapses in the partially anisotropic case is examined and presented in Fig. 9. As demonstrated, the concurrence shows the expected behavior as a function of λ and decays as the temperature increases. As shown, the threshold temperature where the concurrence vanishes is determined by the value of λ; it increases as λ increases. The asymptotic behavior of the concurrence as a function λ_1 and kT is illustrated. Clearly the nonzero concurrence shows up at small values of kT and λ_1. The concurrence has two peaks versus λ_1, but as the temperature increases, the second peak disappears.

3. *Isotropic XY Model*

Now we consider the isotropic system where $\gamma = 0$ (i.e., $J_x = J_y$). We started with the dynamics of nearest-neighbor concurrence, we found that $C(i, i+1)$ takes a constant value that does not depend on the final value of the coupling J_1 and magnetic field h_1. This follows from the dependence of the spin correlation functions

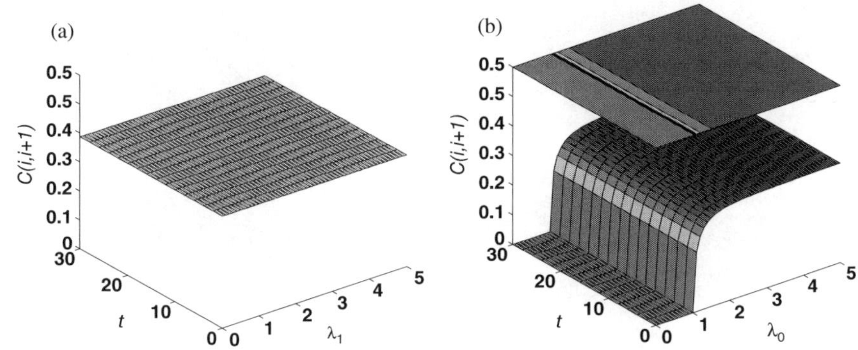

Figure 10. (**a**) $C(i, i+1)$ as a function of λ_1 and t, in units of J_1^{-1}, at $kT = 0$ with $\gamma = 0$, $h_0 = h_1 = 1$, and $J_0 = 5$; (**b**) $C(i, i+1)$ as a function of λ_0 and t at $kT = 0$ with $\gamma = 0$, $h_0 = h_1 = 1$, and $J_1 = 5$.

and the magnetization on the initial state only as well. The initial coupling parameters J_x and J_y, which are equal, force the spins to be equally aligned into the x- and y-directions, apart from those in the z-direction, causing a finite concurrence. Increasing the coupling parameters, strength would not change that distribution or the associated concurrence at constant magnetic field. The time evolution of nearest-neighbor concurrence as a function of the time-dependent coupling is explored in Fig. 10. Clearly, $C(i, i+1)$ is independent of λ_1. Studying $C(i, i+1)$ as a function of λ_0 and t with $h_0 = h_1 = 1$ at $kT = 0$ for various values of J_1, we noticed that the results are independent of J_1. Again, when $J_0 < h_0$, the magnetic field dominates and $C(i, i+1)$ vanishes. For $J_0 \geq h_0$, however, $C(i, i+1)$ has a finite value. Finally we explored the asymptotic behavior of the nearest-neighbor and next nearest-neighbor concurrence in the λ–γ phase space of the one dimensional XY spin system under the effect of a time-dependent coupling $J(t)$ as shown in Fig. 11. As one can notice, the nonvanishing concurrences appear in the vicinity of $\lambda = 1$ or lower and vanish for higher values. One interesting feature is that the maximum achievable nearest-neighbor concurrence takes place at $\gamma = 1$ (i.e., in a completely anisotropic system), while the maximum next nearest-neighbor concurrence is achievable in a partially anisotropic system, where $\gamma \approx 0.3$.

D. Time Evolution of the Driven Spin System in an External Time-Dependent Magnetic Field

We investigate the time evolution of quantum entanglement in a one-dimensional XY spin chain system coupled through nearest-neighbor interaction under the effect of an external magnetic field at zero and finite temperature. We consider

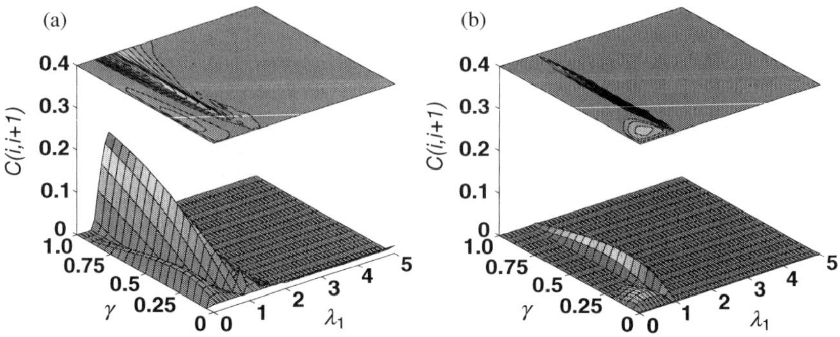

Figure 11. (**a**) The asymptotic behavior of $C(i, i+1)$ as a function of λ_1 and γ with $h_0 = h_1 = 1$ and $J_0 = 1$ at $kT = 0$. (**b**) The asymptotic behavior of $C(i, i+2)$ as a function of λ_1 and γ with $h_0 = h_1 = 1$ and $J_0 = 1$ at $kT = 0$.

both time-dependent nearest-neighbor Heisenberg coupling $J(t)$ between the spins on the chain and magnetic field $h(t)$, where the function forms are exponential, periodic, and hyperbolic in time. Particularly, we focus on the concurrence as a measure of entanglement between any two adjacent spins on the chain and its dynamical behavior under the effect of the time-dependent coupling and magnetic fields. We apply both analytical and numerical approaches to tackle the problem. The Hamiltonian of the system is given by

$$H = -\frac{J(t)}{2}(1+\gamma)\sum_{i=1}^{N}\sigma_i^x\sigma_{i+1}^x - \frac{J(t)}{2}(1-\gamma)\sum_{i=1}^{N}\sigma_i^y\sigma_{i+1}^y - \sum_{i=1}^{N}h(t)\sigma_i^z \quad (27)$$

where σ_is are the Pauli matrices and γ is the anisotropy parameter.

1. Numerical and Exact Solutions

Following the standard procedure to treat the Hamiltonian (27) [108], we again obtain the Hamiltonian in the form

$$H = \sum_{p=1}^{N/2}\tilde{H}_p \quad (28)$$

with $\tilde{H}_p$ given by

$$\tilde{H}_p = \alpha_p(t)[c_p^\dagger c_p + c_{-p}^\dagger c_{-p}] + iJ(t)\delta_p[c_p^\dagger c_{-p}^\dagger + c_p c_{-p}] + 2h(t) \quad (29)$$

where $\alpha_p(t) = -2J(t)\cos\phi_p - 2h(t)$ and $\delta_p = 2\gamma \sin\phi_p$. Initially the system is assumed to be in a thermal equilibrium state and therefore its initial density matrix is given by

$$\rho_p(0) = \frac{e^{-\beta \tilde{H}_p(0)}}{Z}, \qquad Z = Tr(e^{-\beta \tilde{H}_p(0)}) \tag{30}$$

where $\beta = 1/kT$, k is Boltzmann constant, and T is the temperature.

Because the Hamiltonian is decomposable we can find the density matrix at any time t, $\rho_p(t)$, for the pth subspace by solving Liouville equation, in the Heisenberg representation by following the same steps we applied in Eqs. (19)–(22).

To study the effect of a time-varying coupling parameter $J(t)$, we consider the following forms

$$J_{\exp}(t) = J_1 + (J_0 - J_1)\, e^{-Kt} \tag{31}$$
$$J_{\cos}(t) = J_0 - J_0 \cos(Kt) \tag{32}$$
$$J_{\sin}(t) = J_0 - J_0 \sin(Kt) \tag{33}$$
$$J_{\tanh}(t) = J_0 + \frac{J_1 - J_0}{2}\left[\tanh\left(K(t - \frac{5}{2})\right) + 1\right] \tag{34}$$

Note that Eq. (21), in the current case, gives two systems of coupled differential equations with variable coefficients. Such systems can only be solved numerically, which we carried out in this work. Nevertheless, an exact solution of the system can be obtained using a general time-dependent coupling $J(t)$ and a magnetic field in the following form:

$$J(t) = \lambda\, h(t) \tag{35}$$

where λ is a constant. Using Eqs. (21) and (35) we obtain the nonvanishing matrix elements

$$i\begin{pmatrix} \dot{u}_{11} & \dot{u}_{12} \\ \dot{u}_{21} & \dot{u}_{22} \end{pmatrix} = \begin{pmatrix} u_{11} & u_{12} \\ u_{21} & u_{22} \end{pmatrix}\begin{pmatrix} \frac{2}{\lambda} & -i\delta_p \\ i\delta_p & -4\cos\phi_p - \frac{2}{\lambda} \end{pmatrix} J(t) \tag{36}$$

and

$$i\,\dot{u}_{33} = -2\,\cos\phi_p\, J(t)\, u_{33}, \quad u_{44} = u_{33} \tag{37}$$

This system of coupled differential equations can be exactly solved to yield

$$u_{11} = \cos^2\theta e^{-i\lambda_1 \int_0^t J(t')dt'} + \sin^2\theta e^{-i\lambda_2 \int_0^t J(t')dt'} \tag{38}$$

$$u_{12} = -i \sin\theta \cos\theta \left\{ e^{-i\lambda_1 \int_0^t J(t')dt'} - e^{-i\lambda_2 \int_0^t J(t')dt'} \right\} \tag{39}$$

$$u_{21} = -u_{12} \tag{40}$$

$$u_{22} = \sin^2\theta e^{-i\lambda_1 \int_0^t J(t')dt'} + \cos^2\theta e^{-i\lambda_2 \int_0^t J(t')dt'} \tag{41}$$

$$u_{33} = u_{44} = e^{2i\cos\phi_p \int_0^t J(t')dt'} \tag{42}$$

where

$$\sin\theta = \sqrt{\frac{\sqrt{\delta_p^2 + (2\cos\phi_p + \frac{2}{\lambda})} - (2\cos\phi_p + \frac{2}{\lambda})}{2\sqrt{\delta_p^2 + (2\cos\phi_p + \frac{2}{\lambda})}}} \tag{43}$$

$$\cos\theta = \sqrt{\frac{\sqrt{\delta_p^2 + (2\cos\phi_p + \frac{2}{\lambda})} + (2\cos\phi_p + \frac{2}{\lambda})}{2\sqrt{\delta_p^2 + (2\cos\phi_p + \frac{2}{\lambda})}}} \tag{44}$$

where the angles ϕ and θ were found to be

$$\phi = (n+1)\pi, \quad \tan 2\theta = \frac{\delta_p}{2\cos\phi_p + \frac{2}{\lambda}} \tag{45}$$

where $n = 0, \pm 1, \pm 2, \ldots$, therefore

$$\sin 2\theta = \frac{\delta_p}{\sqrt{\delta_p^2 + (2\cos\phi_p + \frac{2}{\lambda})}}, \quad \cos 2\theta = \frac{2\cos\phi_p + \frac{2}{\lambda}}{\sqrt{\delta_p^2 + (2\cos\phi_p + \frac{2}{\lambda})}} \tag{46}$$

As usual, upon obtaining the matrix elements of the evolution operators, either numerically or analytically, one can evaluate the matrix elements of the two-spins density operator with the help of the magnetization and the spin–spin correlation funtions of the system and finally produce the concurrence using Wootters method [109].

2. *Constant Magnetic Field and Time-Varying Coupling*

First, we studied the dynamics of the nearest-neighbor concurrence $C(i, i+1)$ for the completely anisotropic system, $\gamma = 1$, when the coupling parameter is J_{exp} and the magnetic field is a constant using the numerical solution. As can be noticed in Fig. 12, the asymptotic value of the concurrence depends on K in addition to the coupling parameter and magnetic field. The larger the transition constant is, the lower is the asymptotic value of the entanglement and the more rapid the decay. This result demonstrates the nonergodic behavior of the system, where the

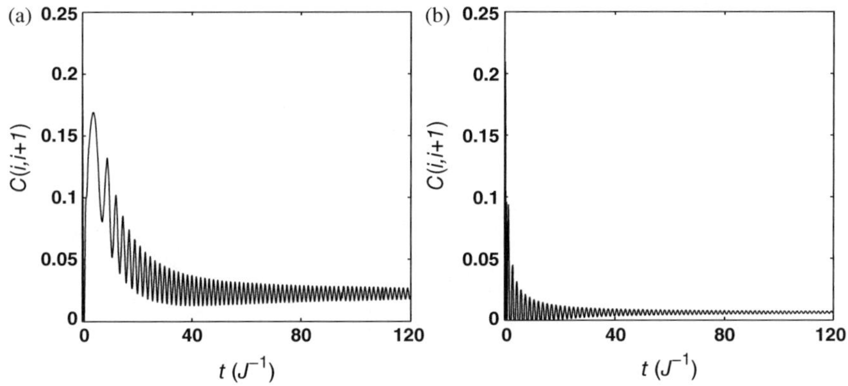

Figure 12. $C(i, i+1)$ as a function of t with $J_0 = 0.5$, $J_1 = 2$, $h = 1$, $N = 1000$ at $kT = 0$ and (**a**) $J = J_{\text{exp}}$, $K = 0.1$; (**b**) $J = J_{\text{exp}}$, $K = 10$ (**c**) $J = J_{\text{tanh}}$, $K = 0.1$ (**d**) $J = J_{\text{tanh}}$, $K = 10$.

asymptotic value of the entanglement is different from the one obtained under constant coupling J_1.

We have examined the effect of the system size N on the dynamics of the concurrence, as depicted in Fig. 13. We note that for all values of N the concurrence reaches an approximately constant value but then starts oscillating after some critical time t_c, that increases as N increases, which means that the oscillation will disappear as we approach an infinite one-dimensional system. Such

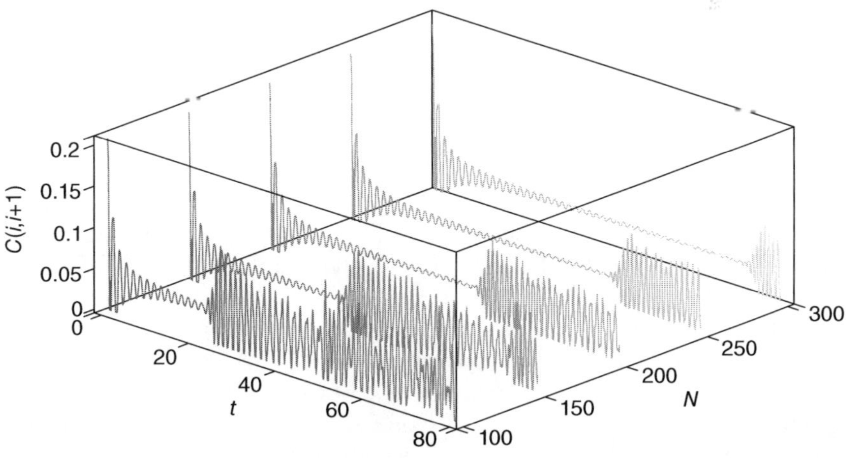

Figure 13. $C(i, i+1)$ as a function of t (units of J^{-1}) with $J = J_{\text{exp}}$, $J_0 = 0.5$, $J_1 = 2$, $h = 1$, $K = 1000$ at $kT = 0$, and N varies from 100 to 300.

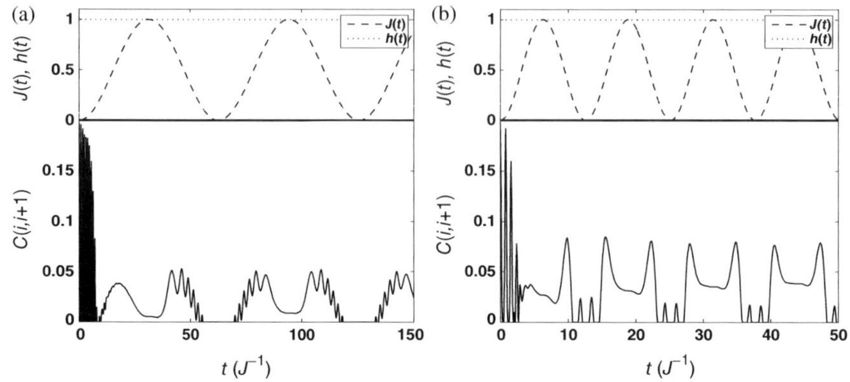

Figure 14. Dynamics of nearest-neighbor concurrence with $\gamma = 1$ for $J_{\cos}$ where $J_0 = 0.5$, $h = 1$ at $kT = 0$ and (**a**) $K = 0.1$; (**b**) $K = 0.5$.

oscillations are caused by the spin-wave packet propagation [90]. We next study the dynamics of the nearest-neighbor concurrence when the coupling parameter is $J_{\cos}$ with different values of K (i.e., different frequencies), shown in Fig. 14. We first note that $C(i, i+1)$ shows a periodic behavior with the same period of $J(t)$. We have shown in our previous work [114] that for the considered system at zero temperature, the concurrence depends only on the ratio J/h. When $J \approx h$, the concurrence has a maximum value. While when $J \gg h$ or $J \ll h$, the concurrence vanishes. In Fig. 14, one can see that when $J = J_{\max}$, $C(i, i+1)$ decreases because large values of J destroy the entanglement, while $C(i, i+1)$ reaches a maximum value when $J = J_0 = 0.5$. As $J(t)$ vanishes, $C(i, i+1)$ decreases because of the magnetic field domination.

3. *Time-Dependent Magnetic Field and Coupling*

Here, we use the exact solution to study the concurrence for four different forms of coupling parameter $J_{\exp}$, $J_{\tanh}$, $J_{\cos}$, and $J_{\sin}$ when $J(t) = \lambda h(t)$, where λ is a constant. We have compared the exact solution results with the numerical ones and they have shown coincidence. Figure 15 show, the dynamics of $C(i, i+1)$ when $h(t) = J(t) = J_{\cos}$ and $h(t) = J(t) = J_{\sin}$, respectively, where $J_0 = 0.5$ and $K = 1$. Interestingly, the concurrence in this case does not show a periodic behavior as it did when $h(t) = 1$ (i.e., for a constant magnetic field), in Fig. 14. In Fig. 16a we study the behavior of the asymptotic value of $C(i, i+1)$ as a function of λ at different values of the parameters J_0, J_1, and K, where $J(t) = \lambda h(t)$. Obviously, the asymptotic value of $C(i, i+1)$ depends only on the initial conditions, not on the form or behavior of $J(t)$ at $t > 0$. This result demonstrates the sensitivity of the concurrence evolution to its initial value. Testing the concurrence at nonzero

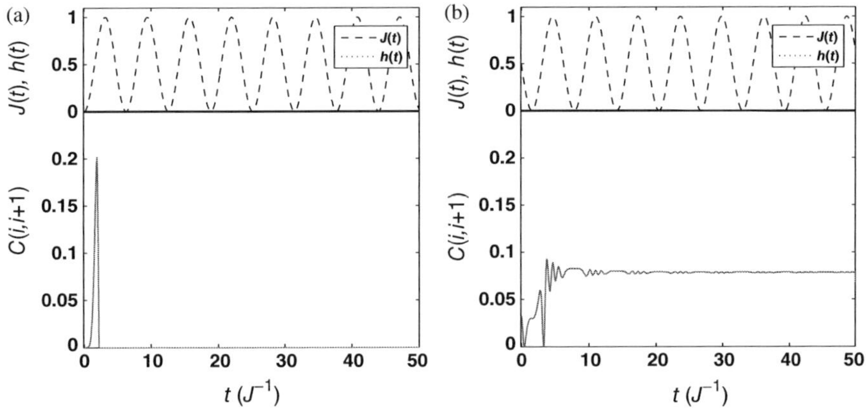

Figure 15. Dynamics of nearest-neighbor concurrence with $\gamma = 1$ at $kT = 0$ with $J_0 = h_0 = 0.5$, $K = 1$ for (**a**) $J_{\cos}$ and $h_{\cos}$; (**b**) $J_{\sin}$ and $h_{\sin}$.

temperatures demonstrates that it maintains the same profile but with reduced value with increasing temperature as can be concluded from Fig. 16b. Also the critical value of λ at which the concurrence vanishes decreases with increasing temperature as can be observed, which is expected as thermal fluctuations destroy the entanglement. Finally, in Fig. 17 we study the partially anisotropic system, $\gamma = 0.5$, and the isotropic system $\gamma = 0$ with $J(0) = 1$. We note that the behavior of $C(i, i+1)$ in this case is similar to the case of constant coupling parameters studied previously [114]. We also note that the behavior depends only on the initial coupling $J(0)$ and not on the form of $J(t)$ where different forms of $J(t)$

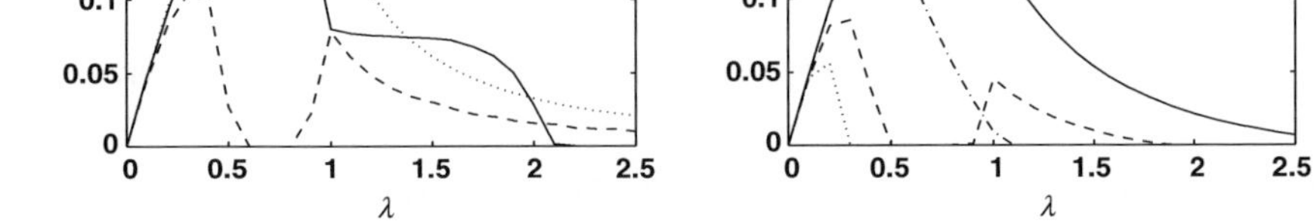

Figure 16. The behavior asymptotic value of $C(i, i+1)$ as a function of λ with $\gamma = 1$ at (**a**) $kT = 0$; (**b**) $kT = 0.5, 1$.

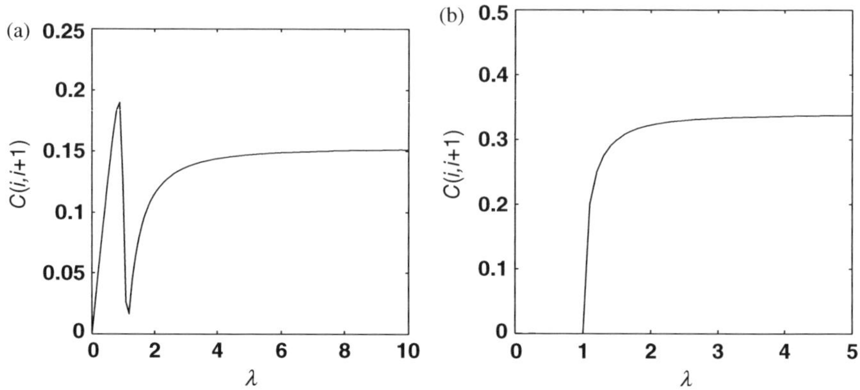

Figure 17. The behavior asymptotic value of $C(i, i+1)$ as a function of λ at $kT = 0$ with (**a**) $\gamma = 0.5$; (**b**) $\gamma = 0$.

have been tested. It is interesting to notice that the results of Figs. 15, 16, and 17 confirm one of the main results of the previous works [89,90,114], namely that the dynamic behavior of the spin system, including entanglement, depends only on the parameter $\lambda = J/h$, not on the individual values of h and J for any degree of anisotropy of the system.

In these previous works both the coupling and magnetic field were considered time-independent, while in this work we have assumed $J(t) = \lambda h(t)$, where $h(t)$ can take any time-dependent form. This explains why the asymptotic value of the concurrence depends only on the initial value of the parameters regardless of their function form. Furthermore in the previous works, it was demonstrated that for finite temperatures, the concurrence turnsout to be dependent not only on λ but also on the individual values of h and J, while according to Fig. 16b even at finite temperatures the concurrence still depends only on λ where $J(t) = \lambda J(t)$ for any form of $h(t)$.

III. DYNAMICS OF ENTANGLEMENT IN TWO-DIMENSIONAL SPIN SYSTEMS

A. An Exact Treatment of Two-Dimensional Transverse Ising Model in a Triangular Lattice

Entanglement close to quantum phase transitions was originally analyzed by Osborne and Nielsen [44] and Osterloh et al. [72] for the Ising model in one dimension. We studied before a set of localized spins coupled through exchange interaction and subject to an external magnetic filed [46,47,89,90]. We demonstrated for such a class of one-dimensional magnetic systems, that entanglement can be controlled and tuned by varying the anisotropy parameter in the Hamiltonian and by introducing impurities into the systems. In particular, under certain conditions, the entanglement is zero up to a critical point λ_c, where a quantum phase transition occurs, and is different from zero above λ_c [42].

In two and higher dimensions, nearly all calculations for spin systems were obtained by means of numerical simulations [115]. The concurrence and localizable entanglement in two-dimensional quantum XY and XXZ models were considered using quantum Monte Carlo [116,117]. The results of these calculations were qualitatively similar to the one-dimensional case, but entanglement is much smaller in magnitude. Moreover, the maximum in the concurrence occurs at a position closer to the critical point than in the one-dimensional case [62].

In this section, we introduce how to use the trace minimization algorithm [118,119] to carry out an exact calculation of entanglement in a 19-site two-dimensional transverse Ising model. We classify the ground-state properties according to entanglement for a certain class on two-dimensional magnetic systems and demonstrate that entanglement can be controlled and tuned by varying the parameter $\lambda = h/J$ in the Hamiltonian and by introducing impurities into the systems. We discuss the relationship of entanglement and quantum phase transition, and the effects of impurities on the entanglement.

1. Trace Minimization Algorithm

Diagonalizing a $2^{19} \times 2^{19}$ Hamiltonian matrix and partially tracing its density matrix is a numerically difficult task. We propose to compute the entanglement of formation, first by applying the trace minimization algorithm (Tracemin) [118,119] to obtain the eigenvalues and eigenvectors of the constructed Hamiltonian. Then, we use these eigenpairs and new techniques detailed in the appendix to build a partially traced density matrix.

The trace minimization algorithm was developed for computing a few of the smallest eigenvalues and the corresponding eigenvectors of the large sparse generalized eigenvalue problem

$$AU = BU\Sigma \tag{47}$$

where matrices $A, B \in \mathbb{C}^{n\times n}$ are Hermitian positive definite, $U = [u_1, ..., u_p] \in \mathbb{C}^{n\times p}$ and $\Sigma \in \mathbb{R}^{p\times p}$ is a diagonal matrix. The main idea of Tracemin is that minimizing $\mathrm{Tr}(X^* AX)$, subject to the constraints $X^* BX = I$, is equivalent to finding the eigenvectors U corresponding to the p smallest eigenvalues. This consequence of Courant–Fischer theorem can be restated as

$$\min_{X^* BX=I} \mathrm{Tr}(X^* AX) = \mathrm{Tr}(U^* AU) = \sum_{i=1}^{p} \lambda_i \tag{48}$$

where I is the identity matrix. The following steps constitute a single iteration of the Tracemin algorithm:

- $G = X_k^* B X_k$ (compute G)
- $G = V\Omega V^*$ (compute the spectral decomposition of G)
- $H = \tilde{Q}^* A \tilde{Q}$ (compute H, where $\tilde{Q} = X_k V \Omega^{-1/2}$)
- $H = W\Omega W^*$ (compute the spectral decomposition of H)
- $\bar{X}_k = \tilde{Q} W$ (now $\bar{X}_k^* A \bar{X}_k = \Lambda$ and $\bar{X}_k^* B \bar{X}_k = I$)
- $\bar{X}_{k+1} = \bar{X}_k - D$ (D is determined s.t. $\mathrm{Tr}(X_{k+1}^* A X_{k+1}) < \mathrm{Tr}(X_k^* A X_k)$)

In order to find the optimal update D in the last step, we enforce the natural constraint $\bar{X}_k^* BD = 0$, and obtain

$$\begin{pmatrix} A & B\bar{X}_k \\ X_k^* B & 0 \end{pmatrix} \begin{pmatrix} D \\ L \end{pmatrix} = \begin{pmatrix} A\bar{X}_k \\ 0 \end{pmatrix} \tag{49}$$

.

Considering the orthogonal projector $P = B\bar{X}_k (X_k^* B^2 X_k)^{-1} \bar{X}_k^* B$ and letting $D = (I - P)\bar{D}$, the linear system (49) can be rewritten in the following form:

$$(I - P)A(I - P)\bar{D} = (I - P)A\bar{X}_k \tag{50}$$

Notice that the conjugate gradient method can be used to solve (50), because it can be shown that the residual and search directions $r, p \in \mathrm{Range}(P)^{\perp}$. Also, notice that the linear system (50) needs to be solved only to a fixed relative precision at every iteration of Tracemin.

A reduced-density matrix, built from the ground state, which is obtained by Tracemin, is usually constructed as follows: diagonalize the system Hamiltonian $H(\lambda)$, retrieve the ground state $|\Psi>$ as a function of $\lambda = h/J$, build the density matrix $\rho = |\Psi><\Psi|$, and trace out contributions of all other spins in the density matrix to get reduced-density matrix by $\rho(i, j) = \sum_p < u_i(A)| < v_p(B)|\rho|u_j(A) > |v_p(B) >$, where $u_i(A)$ and $v_p(B)$ are bases of subspaces $\epsilon(A)$ and $\epsilon(B)$. It includes creating a $2^{19} \times 2^{19}$ density matrix ρ followed by permutations of rows, columns, and some basic arithmetic operations on the elements of ρ. Instead of operating on a huge matrix, we pick up only certain elements from $|\Psi>$, performing basic algebra to build a reduced-density matrix directly.

2. *General Forms of Matrix Representation of the Hamiltonian*

By studying the patterns of $\sum_{<i,j>} I \otimes \cdots \sigma_i^x \otimes \cdots \sigma_j^x \otimes \cdots I$ and $\sum_i I \otimes \cdots \sigma_i^z \otimes \cdots I$, one finds the following rules.

Scheme		Rule	Example $N = 3$
N		N	3
N–2		N–2	1
N	–2	N–2	1
N–2	–2	N–4	–1
N	–2	N–2	1
N–2	–2	N–4	–1
N–2	–2	N–4	–1
N–4	–2	N–6	–3
and so on		and so on	

Figure 18. Diagonal elements of $\sum_i \sigma_i^z$ for N spins.

a. $\sum_i \sigma_i^z$ *for N Spins.* The matrix is $2^N \times 2^N$; it has only 2^N diagonal elements. Elements follow the rules shown in Fig. 18.

If one stores these numbers in a vector and initializes $v = (N)$, then the new v is the concatenation of the original v and the original v with two subtracted from each of its elements. We repeat this N times, that is,

$$v = \begin{pmatrix} v \\ v - 2 \end{pmatrix} \tag{51}$$

$$v = (N) \tag{52}$$

$$\Rightarrow v = \begin{pmatrix} N \\ N - 2 \end{pmatrix} \tag{53}$$

$$\Rightarrow v = \begin{pmatrix} N \\ N - 2 \\ N - 2 \\ N - 4 \end{pmatrix} \tag{54}$$

b. $\sum_{<i,j>} I \otimes \cdots \sigma_i^x \otimes \cdots \sigma_j^x \otimes \cdots I$ *for N Spins.* Because $\begin{pmatrix} 1 & 0 \\ 0 & 1 \end{pmatrix}$ and $\begin{pmatrix} 0 & 1 \\ 1 & 0 \end{pmatrix}$ exclude each other, for matrix $I \otimes \cdots \sigma_i^x \otimes \cdots \sigma_j^x \otimes \cdots I$, every row/column contains only a single 1, then the matrix owns 2^N 1s and only 1

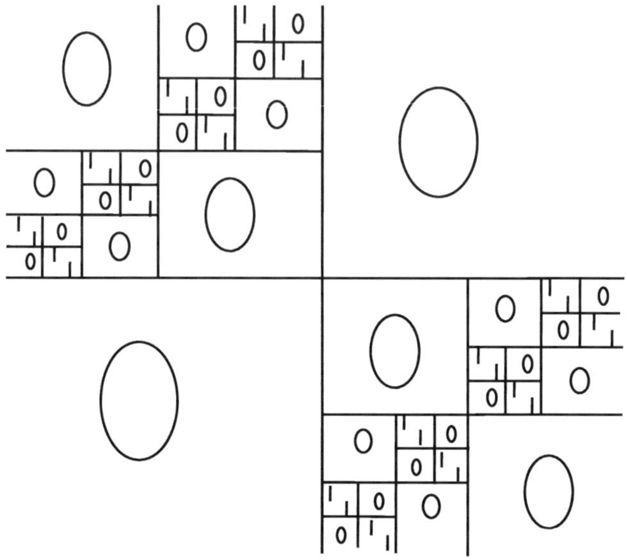

Figure 19. Scheme of matrix $I \otimes \sigma_3^x \otimes \sigma_2^x \otimes I \otimes I$.

in it. If we know the position of 1s, it turns out that we can set a $2^N \times 1$ array "col" to store the column position of 1s corresponding to the 1st $\to 2^N$th rows. In fact, the nonzero elements can be located by the properties stated next. For clarity, let us number N spins in the reverse order as: $N-1, N-2, \ldots, 0$, instead of $1, 2, \ldots, N$. The string of nonzero elements starts from the first row at: $1 + 2^i + 2^j$; with string length 2^j; and number of such strings 2^{N-j-1}. For example, Fig. 19 shows these rules for a scheme of $I \otimes \sigma_3^x \otimes \sigma_2^x \otimes I \otimes I$.

Again, because of the exclusion, the positions of nonzero element 1 of $I \otimes \cdots \sigma_i^x \otimes \cdots \sigma_j^x \otimes \cdots I$ are different from those of $I \otimes \cdots \sigma_l^x \otimes \cdots \sigma_m^x \otimes \cdots I$. So $\sum_{<i,j>} I \otimes \cdots \sigma_i^x \otimes \cdots \sigma_j^x \otimes \cdots I$ is a $2^N \times 2^N$ matrix with only 1 and 0.

After storing array "col," we repeat the algorithm for all the nearest pairs $\langle i, j \rangle$, and concatenate "col" to position matrix "c" of $\sum_{<i,j>} I \otimes \cdots \sigma_i^x \otimes \cdots \sigma_j^x \otimes \cdots I$. In the next section, we apply these rules to our problem.

3. *Specialized Matrix Multiplication*

Using the diagonal elements array v of $\sum_i \sigma_i^z$ and position matrix of nonzero elements c for $\sum_{<i,j>} I \otimes \cdots \sigma_i^x \otimes \cdots \sigma_j^x \otimes \cdots I$, we can generate matrix **H**, representing the Hamiltonian. However, we only need to compute the result of the matrix–vector multiplication $\mathbf{H}^*\boldsymbol{Y}$ in order to run Tracemin, which is the advantage of Tracemin, and consequently do not need to explicitly obtain **H**. Since matrix–vector multiplication is repeated many times throughout iterations, we propose an efficient implementation to speed up its computation specifically for Hamiltonian of Ising model (and XY by adding one term).

For simplicity, first let Y in $\mathbf{H}*Y$ be a vector and $J = h = 1$ (in general Y is a tall matrix and $J \neq h \neq 1$). Then

$$
\begin{aligned}
\mathbf{H} * Y &= \sum_{<i,j>} \sigma_i^x \sigma_j^x * Y + \sum_i \sigma_i^z * Y \\
&= \begin{pmatrix} 1 & & 1 & & 1 & \\ & 1 & & 1 & & 1 \\ & \cdots & & \cdots & & \\ & & & & & \\ & \cdots & & \cdots & & \\ 1 & & & & 1 & 1 \end{pmatrix} * \begin{pmatrix} Y(1) \\ Y(2) \\ \vdots \\ \vdots \\ Y(2^N) \end{pmatrix} \\
&+ \begin{pmatrix} v(1) & & & & \\ & v(2) & & & \\ & & \ddots & & \\ & & & \ddots & \\ & & & & v(2^N) \end{pmatrix} * \begin{pmatrix} Y(1) \\ Y(2) \\ \vdots \\ \vdots \\ Y(2^N) \end{pmatrix} \\
&= \begin{pmatrix} Y(c(1,1)) + Y(c(1,2)) + \ldots + Y(c(1, \text{ \# of pairs}) \\ \vdots \\ Y(c(k,1)) + Y(c(k,2)) + \ldots + Y(c(k, \text{ \# of pairs}) \\ \vdots \\ Y(c(2^N,1)) + Y(c(2^N,2)) + \ldots + Y(c(2^N, \text{\# of pairs}) \end{pmatrix} \\
&+ \begin{pmatrix} v(1) * Y(1) \\ v(2) * Y(2) \\ \vdots \\ \vdots \\ v(2^N) * Y(2^N) \end{pmatrix}
\end{aligned} \tag{55}
$$

where $p =$ # of pairs.

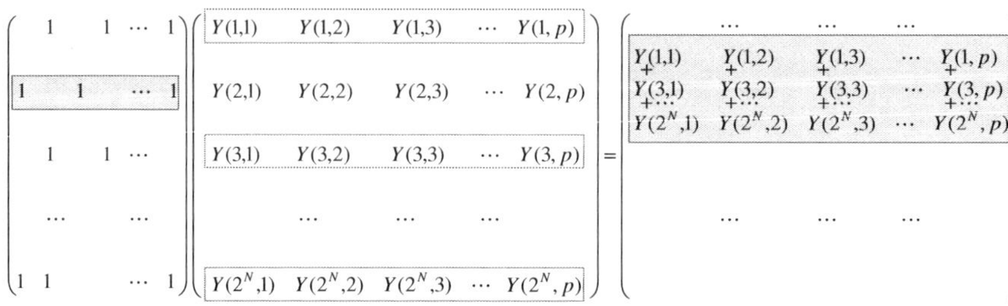

Figure 20. The illustration of $\mathbf{H}*Y$.

When Y is a matrix, we can treat Y (2^N by p) column by column for $\sum_{<i,j>} I \otimes \cdots \sigma_i^x \otimes \cdots \sigma_j^x \otimes \cdots I$. Also, we can accelerate the computation by treating every row of Y as a vector and adding these vectors at once. Figure 20 visualized the process.

Notice that the result of the multiplication of the xth row of $\sum_{<i,j>} \sigma_i^x \sigma_j^x$ (delineated by the shaded box) and Y, is equivalent to the sum of rows of Y, whose row numbers are the column indices of nonzero elements' of the xth row. Such that we transform a matrix operation to straight-forward summation and multiplication of numbers.

4. *Exact Entanglement*

We examine the change of concurrence between the center spin and its nearest-neighbor as a function of $\lambda = h/J$ for both the 7-site and 19-site systems. In Fig. 21, the concurrence of the 7-site system reaches its maximum 0.15275 when $\lambda = 2.61$. In the 19-site system, the concurrence reaches 0.0960 when $\lambda = 3.95$

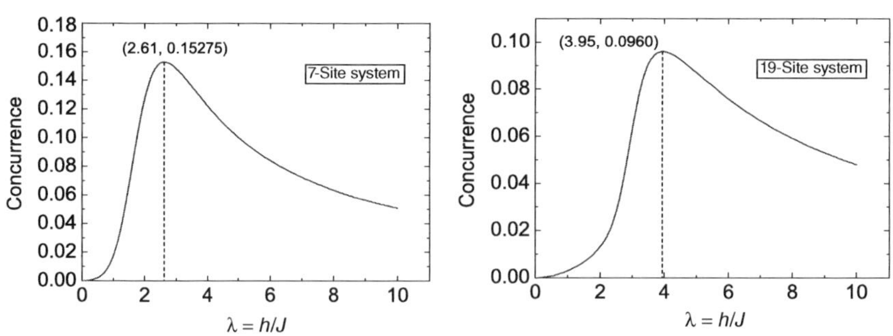

Figure 21. Concurrence of center spin and its nearest-neighbor as a function of λ for both 7-site and 19-site system. In the 7-site system, concurrence reaches maximum 0.15275 when $\lambda = 2.61$. In the 19-site system, concurrence reaches the maximum 0.0960 when $\lambda = 3.95$.

[120]. The maximum value of concurrence in the 19-site model, where each site interacts with six neighbors, is roughly 1/3 of the maximum concurrence in the one-dimensional transverse Ising model with size $N = 201$ [46], where it has only two neighbors for each site. It is the monogamy [121,122] that limits the entanglement shared among the number of neighboring sites. This property is also shown in the fact that the fewer the number of neighbors of a pair the larger the entanglement among other nearest-neighbors. Our numerical calculation shows that the maximum concurrence of next-nearest-neighbor is less than 10^{-8}. It shows that the entanglement is short ranged, though global.

B. Time Evolution of the Spin System

Decoherence is considered as one of the main obstacles toward realizing an effective quantum computing system [34]. The main effect of decoherence is to randomize the relative phases of the possible states of the considered system. Quantum error correction [123] and decoherence-free subspace [35,124] have been proposed to protect the quantum property during the computation process. Still, offering a potentially ideal protection against environmentally induced decoherence is difficult. In NMR quantum computers, a series of magnetic pulses were applied to a selected nucleus of a molecule to implement quantum gates [125]. Moreover, a spin-pair entanglement is a reasonable measure for decoherence between the considered two-spin system and the environmental spins. The coupling between the system and its environment leads to decoherence in the system and vanishing entanglement between the two spins. Evaluating the entanglement remaining in the considered system helps us to understand the behavior of the decoherence between the considered two spins and their environment [126].

The study of quantum entanglement in two-dimensional systems possesses a number of extra problems compared with systems of one dimension. The particular one is the lack of exact solutions. The existence of exact solutions has contributed enormously to the understanding of the entanglement for 1D systems [68,89,108,114]. Studies can be carried out on interesting but complicated properties, applied to infinitely large systems, and use finite scaling methods, to eliminate the size effects. Some approximation methods, such as density matrix renormalization group (DMRG), are also only workable in one dimension [125,127–129]. So when we carry out this two-dimensional study, no methods can be inherited from previous researches. They heavily rely on numerical calculations, resulting in severe limitations on the system size and properties. For example, dynamics of the system is a computational-costing property. We have to think of a way to improve the effectiveness of computation in order to increase the size of research objects. The physics in the observable systems may tell us the direction of less resource-costing large-scale calculations.

To tackle the problem, we introduce two calculation methods: step-by-step time-evolution matrix transformation and step-by-step projection. We compared them side by side, and besides giving the exactly same results, the step-by-step projection method turned out to be 20 times faster than the matrix transformation.

1. *The Evolution Operator*

According to quantum mechanics the transformation of $|\psi_i(t_0)\rangle$, the state vector at the initial instant t_0, into $|\psi_i(t)\rangle$, the state vector at an arbitrary instant, is linear [130]. Therefore, there exists a linear operator $U(t, t_0)$ such that

$$|\psi_i(t)\rangle = U(t, t_0)\,|\psi_i(t_0)\rangle \tag{56}$$

This is, by definition, the evolution operator of the system. Substituting Eq. (56) into the Schrödinger equation, we obtain

$$i\hbar\frac{\partial}{\partial t}U(t, t_0)|\psi(t_0)\rangle = H(t)U(t, t_0)|\psi(t_0)\rangle \tag{57}$$

which means

$$i\hbar\frac{\partial}{\partial t}U(t, t_0) = H(t)U(t, t_0) \tag{58}$$

Further, taking the initial condition

$$U(t_0, t_0) = \mathbb{I} \tag{59}$$

the evolution operator can be condensed into a single integral equation

$$U(t, t_0) = \mathbb{I} - \frac{i}{\hbar}\int_{t_0}^{t} H(t')U(t', t_0)dt' \tag{60}$$

When the operator H does not depend on time, Eq. (60) can easily be integrated and finally gives out

$$U(t, t_0) = e^{-iH(t-t_0)/\hbar} \tag{61}$$

2. *Step by Step Time-Evolution Matrix Transformation*

To unveil the behavior of concurrence at time t, we need to find the density matrix of the system at that moment, which can be obtained from

$$\rho(t) = U(t)\rho(0)U^{\dagger}(t) \tag{62}$$

Although Eq. (60) gives a beautiful expression for the evolution operator, in reality U is hard to obtain because of the integration involved. In order to overcome this

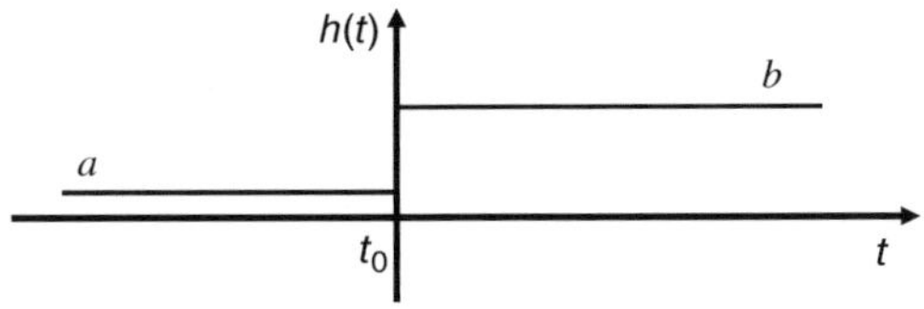

Figure 22. The external magnetic field in a step function form $h(t) = a$ @ $t \leq t_0$, $h(t) = b$ @ $t > t_0$.

obstacle, let us first consider the simplest time-dependent magnetic field: a step function of the form (Fig. 22)

$$h(t) = a + (b - a)\theta(t - t_0) \tag{63}$$

where $\theta(t - t_0)$ is the usual mathematical step function defined by

$$\theta(t - t_0) = \begin{cases} 0 & t \leq t_0 \\ 1 & t > t_0 \end{cases} \tag{64}$$

At t_0 and before, the system is time-independent since $H_a \equiv H(t \leq t_0) = -\sum_{<i,j>} \sigma_i^x \sigma_j^x - a\Sigma_i \sigma_i^z$. Therefore, we are capable of evaluating its ground state and density matrix at t_0 straightforwardly. For the interval t_0 to t, the Hamiltonian $H_b \equiv H(t > t_0) = -\sum_{<i,j>} \sigma_i^x \sigma_j^x - b\Sigma_i \sigma_i^z$ does not depend on time either, so Eq. (61) enables us to write out

$$U(t, t_0) = e^{-iH(t>t_0)(t-t_0)/\hbar} \tag{65}$$

and therefore

$$\rho(t) = U(t, t_0)\rho(t_0)U^\dagger(t, t_0) \tag{66}$$

Starting from here, it is not hard to think of breaking an arbitrary magnetic function into small time intervals, and treating every neighboring intervals as a step function. Comparing the two graphs in Fig. 23, the method has just turned a ski sliding into

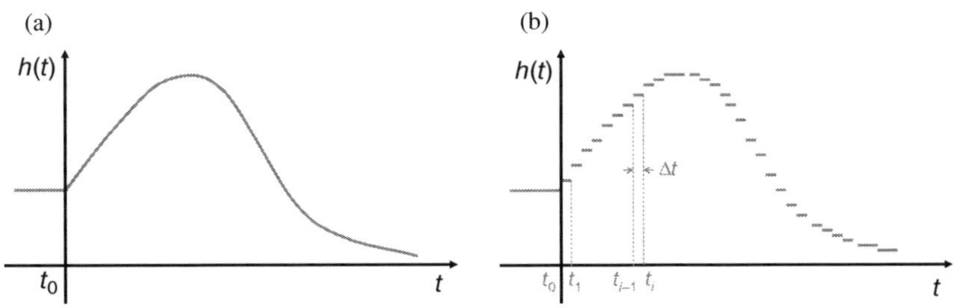

Figure 23. Divide an arbitrary magnetic field function into small time intervals. Every time step is Δt. Treat the field within the interval as a constant. In all, we turn (**a**) a smooth function into (**b**) a collection of step functions, which makes the calculation of dynamics possible.

a mountain climbing. Assuming each time interval is Δt, setting $\hbar = 1$ then

$$U(t_i, t_0)|\psi_0\rangle = U(t_i, t_{i-1})U(t_{i-1}, t_{i-2})...U(t_1, t_0)|\psi_0\rangle \tag{67}$$

$$U(t_i, t_0) = \prod_{k=1}^{i} \exp[-iH(t_k)\Delta t] \tag{68}$$

$$U(t_i, t_0) = \exp[-iH(t_i)\Delta t]U(t_{i-1} - t_0) \tag{69}$$

Here, we avoided integration, instead we have chain multiplications that can be easily realized as loops in computational calculations. This is a common numerical technique; desired precisions can be achieved via proper time step length adjustment.

3. *Step-by-Step Projection*

Step-by-step matrix transformation method successfully breaks down the integration, but still involves matrix exponential, which is numerically resource costing. We propose a projection method to accelerate the calculations. Let us look at the step magnetic field again in Fig. 22. For H_a, after enough long time, the system at zero temperature is in the ground state $|\phi\rangle$ with energy, say, ε. We want to ask how will this state evolve after the magnetic field is turned to the value b? Assuming the new Hamiltonian H_b has N eigenpairs E_i and $|\psi_i\rangle$. The original state $|\phi\rangle$ can be expanded in the basis $\{|\psi_i\rangle\}$

$$|\phi\rangle = c_1|\psi_1\rangle + c_2|\psi_2\rangle + ... + c_N|\psi_N\rangle \tag{70}$$

where

$$c_i = \langle\psi_i|\phi\rangle \tag{71}$$

When H is independent of time between t and t_0 then we can write

$$U(t,\ t_0)\,|\psi_{i,t_0}\rangle = e^{-iH(t>t_0)(t-t_0)/\hbar}|\psi_{i,t_0}\rangle = e^{-iE_i(t-t_0)/\hbar}|\psi_{i,t_0}\rangle \tag{72}$$

Now the exponent in the evolution operator is a number, no longer a matrix. The ground state will evolve with time as

$$\begin{aligned} |\phi(t)\rangle &= c_1|\psi_1\rangle e^{-iE_1(t-t_0)} + c_2|\psi_2\rangle e^{-iE_2(t-t_0)} + ... + c_N|\psi_N\rangle e^{-iE_N(t-t_0)} \\ &= \sum_{i=1}^{N} c_i|\psi_i\rangle e^{-iE_i(t-t_0)} \end{aligned} \tag{73}$$

and the pure state density matrix becomes

$$\rho(t) = |\phi(t)\rangle\langle\phi(t)| \tag{74}$$

Again any complicated function can be treated as a collection of step functions. When the state evolves to the next step, just repeat the procedure to get the following results. Our test shows, for the same magnetic field both methods give the same results, but projection is much more (about 20 times faster) than matrix transformation, which is a great advantage when the system size increases. But this is not the end of the problem. The summation is over all the eigenstates. Extending one layer out to 19 sites, fully diagonalizing thc $2^{19} \times 2^{19}$ Hamiltonian and summing over all of them in every time step is breathtaking.

4. *Dynamics of the Spin System in a Time-Dependent Magnetic Field*

We consider the dynamics of entanglement in a two-dimensional spin system, where spins are coupled through an exchange interaction and subject to an external time-dependent magnetic field. Four forms of time-dependent magnetic field are considered: step, exponential, hyperbolic, and periodic.

$$h(t) = a + (b - a)\theta(t - t_0) \tag{75}$$

$$h(t) = \begin{cases} a & t \leq t_0 \\ b + (a - b)e^{-\omega t} & t > t_0 \end{cases} \tag{76}$$

$$h(t) = \begin{cases} a & t \leq t_0 \\ \frac{(b-a)}{2}[\tanh(\omega t) + 1] + a & t > t_0 \end{cases} \tag{77}$$

$$h(t) = \begin{cases} a & t \leq t_0 \\ a - a\sin(\omega t + \phi) & t > t_0 \end{cases} \tag{78}$$

We show in Figs. 24–27 that the system entanglement behaves in an ergodic way in contrary to the one-dimensional Ising system. The system shows great controllability under all forms of external magnetic field except the step function, which creates rapidly oscillating entanglement. This controllability is shown to be breakable as the different magnetic field parameters increase. Also it will be shown that the mixing of even a few excited states by small thermal fluctuation is devastating to the entanglement of the ground state of the system. These can be explained by Fermi's golden rule and adiabatic approximation.

C. Tuning Entanglement and Ergodicity of Spin Systems Using Impurities and Anisotropy

Great interest is shown in studying the different sources of errors in quantum computing and their effect on quantum gate operations [131,132]. Different

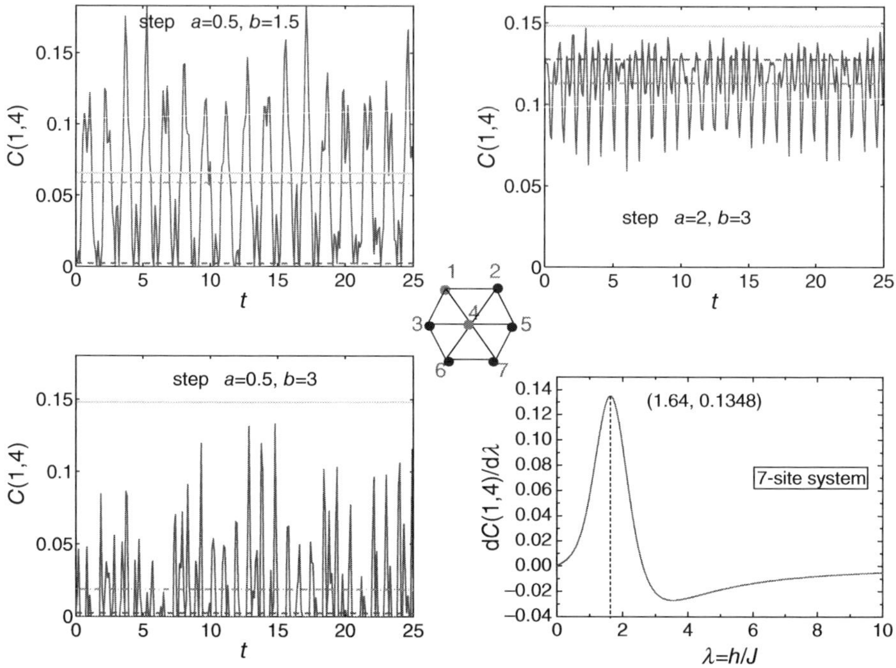

Figure 24. Dynamics of $C(1,4)$ (solid line) in the 7-site system when the step magnetic field is changed from $a = 0.5$ to $b = 1.5$ (before the "critical point" $h = 1.64$), from $a = 2$ to $b = 3$ (after) and from $a = 0.5$ to $b = 3$ (big step cross the "critical point"), where time t is in the unit of J^{-1}, the dashed line stands for the concurrence corresponding to a constant magnetic field $h = a$, the straight solid line for $h = b$ and the dot-dashed line for the average value of the oscillating concurrence.

approaches have been proposed for protecting quantum systems during the computational implementation of algorithms such as quantum error correction [123] and decoherence-free subspace [35]. Nevertheless, realizing a practical protection against the different types of induced decoherence is still a hard task. Therefore, studying the effect of naturally existing sources of errors such as impurities and lack of isotropy in coupling between the quantum systems implementing the quantum computing algorithms is a must. Furthermore, considerable efforts should be devoted to utilizing such sources to tune the entanglements rather than eliminating them. The effect of impurities and anisotropy of coupling between neighbor spins in a one-dimensional spin system has been investigated [46]. It was demonstrated that the entanglement can be tuned in a class of one-dimensional systems by varying the anisotropy of the coupling parameter as well as by introducing impurities into the spin system. For a physical quantity to be eligible for an equilibrium statistical mechanical description, it has to be ergodic, which means that its time average coincides with its ensemble average. To test ergodicity for a

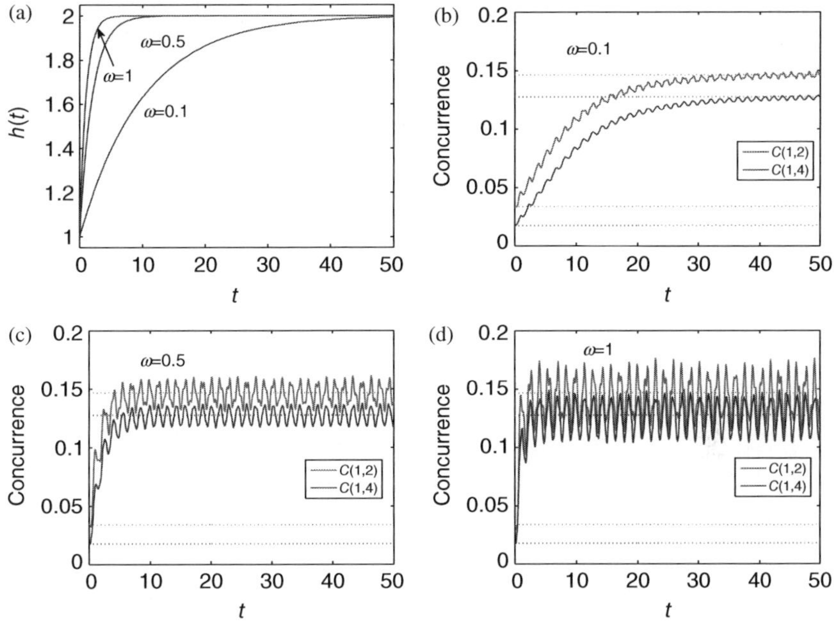

Figure 25. Dynamics of the concurrences $C(1,2)$ (solid upper line) and $C(1,4)$ (solid lower line) in applied exponential magnetic fields of various frequencies $\omega = 0.1, 0.5$, and 1, with strength $a = 1$, $b = 2$, where time t is in the unit of J^{-1}. The straight dotted upper lines are concurrences $C(1,2)$ under constant magnetic field $a = 1$ and $b = 2$; so are the lower lines for $C(1,4)$.

physical quantity one has to compare the time evolution of its physical state to the corresponding equilibrium state. Intensive efforts have been made to investigate ergodicity in one-dimensional spin chains where it was demonstrated that the entanglement, magnetization, spin–spin correlation functions are nonergodic in Ising and XY spin chains for finite number of spins as well as at the thermodynamic limit [90,111,114,133].

In this part, we consider the entanglement in a two-dimensional XY triangular spin system, where the nearest-neighbor spins are coupled through an exchange interaction J and subject to an external magnetic field h. We consider the system at different degrees of anisotropy to test its effect on the system entanglement and dynamics. The number of spins in the system is 7, where all of them are identical except one (or two), which are considered impurities. The Hamiltonian for such a system is given by

$$H = -\frac{(1+\gamma)}{2} \sum_{<i,j>} J_{i,j}\sigma_i^x\sigma_j^x - \frac{(1-\gamma)}{2} \sum_{<i,j>} J_{i,j}\sigma_i^y\sigma_j^y - h(t)\sum_i \sigma_i^z \qquad (79)$$

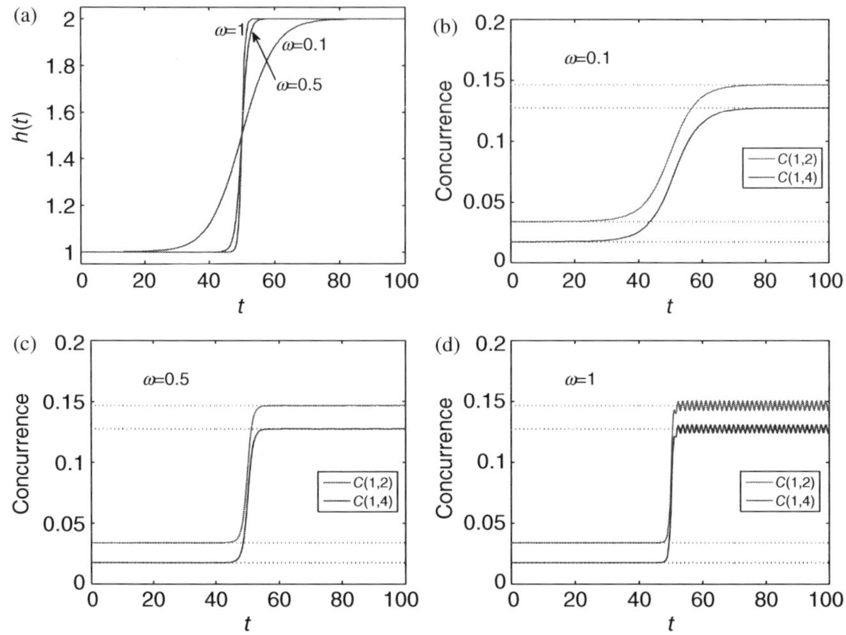

Figure 26. Dynamics of the concurrences $C(1,2)$ (solid upper line) and $C(1,4)$ (solid lower line) in applied hyperbolic magnetic fields of various frequencies $\omega = 0.1, 0.5$, and 1, with strength $a = 1$, $b = 2$, where time t is in the unit of J^{-1}. The straight dotted upper lines are concurrences $C(1,2)$ under constant magnetic field $a = 1$ and $b = 2$, so are the lower lines for $C(1,4)$.

where $\langle i, j\rangle$ is a pair of nearest-neighbors sites on the lattice, $J_{i,j} = J$ for all sites except the sites nearest to an impurity site. For a single impurity, the coupling between the impurity and its neighbors $J_{i,j} = J' = (\alpha + 1)J$, where α measures the strength of the impurity. For double impurities $J_{i,j} = J' = (\alpha_1 + 1)J$ is the coupling between the two impurities and $J_{i,j} = J'' = (\alpha_2 + 1)J$ is the coupling between any one of the two impurities and its neighbors while the coupling is just J between the rest of the spins. For this model it is convenient to set $J = 1$. Exactly solving Schrödinger equation of the Hamiltonian (79), yielding the system energy eigenvalues E_i and eigenfunctions ψ_i. The density matrix of the system is defined by

$$\rho = |\psi_0\rangle\langle\psi_0| \tag{80}$$

where $|\psi_0\rangle$ is the ground-state energy of the entire spin system. We confine our interest to the entanglement between two spins, at any sites i and j [72]. All

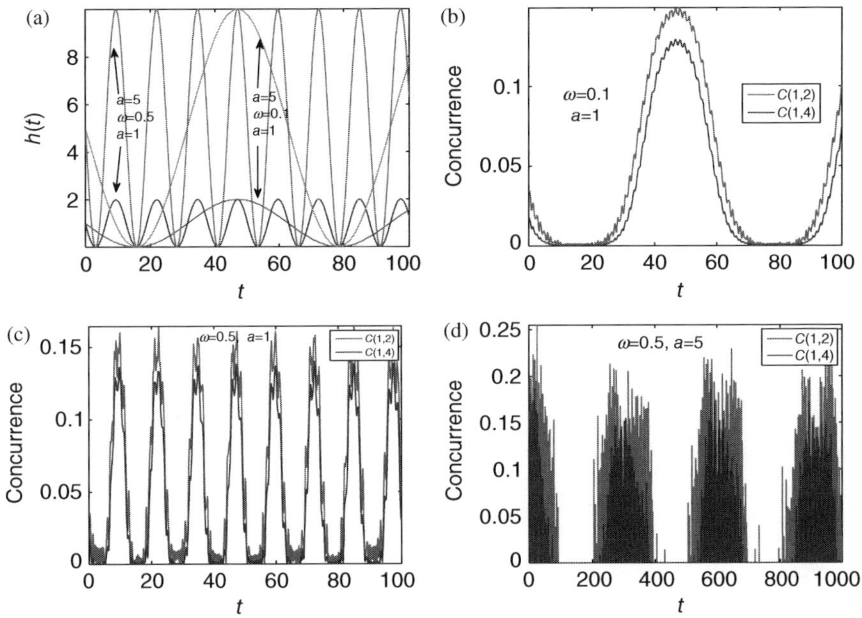

Figure 27. Dynamics of the concurrences $C(1,2)$ (upper line) and $C(1,4)$ (lower line) in applied sine magnetic fields of various frequencies and field strength $\omega = 0.1$ and $a = 1$, $\omega = 0.5$ and $a = 1$, and $\omega = 0.5$ and $a = 5$, where time t is in the unit of J^{-1}.

the information needed in this case, at any moment t, is contained in the reduced-density matrix $\rho_{i,j}(t)$, which can be obtained from the entire system density matrix by integrating out all the spins states except i and j. We adopt the entanglement of formation (or equivalently the concurrence C), as a well-known measure of entanglement [109]. The dynamics of entanglement is evaluated using the the step-by-step time-evolution projection technique introduced previously [134].

1. Single Impurity

We start by considering the effect of a single impurity located at the border site 1. The concurrence between the impurity site 1 and site 2, $C(1, 2)$, versus the parameter λ for the three different models, Ising ($\gamma = 1$), partially anisotropic ($\gamma = 0.5$), and isotropic XY ($\gamma = 0$) at different impurity strengths ($\alpha = -0.5, 0, 0.5, 1$) is in Fig. 28. First, the impurity parameter α is set to zero. For the corresponding Ising model, the concurrence $C(1, 2)$, in Fig. 28a, demonstrates the usual phase transition behavior where it starts at zero value and increases gradually as λ increases, reaching a maximum at $\lambda \approx 2$, then decays as λ increases further. As the

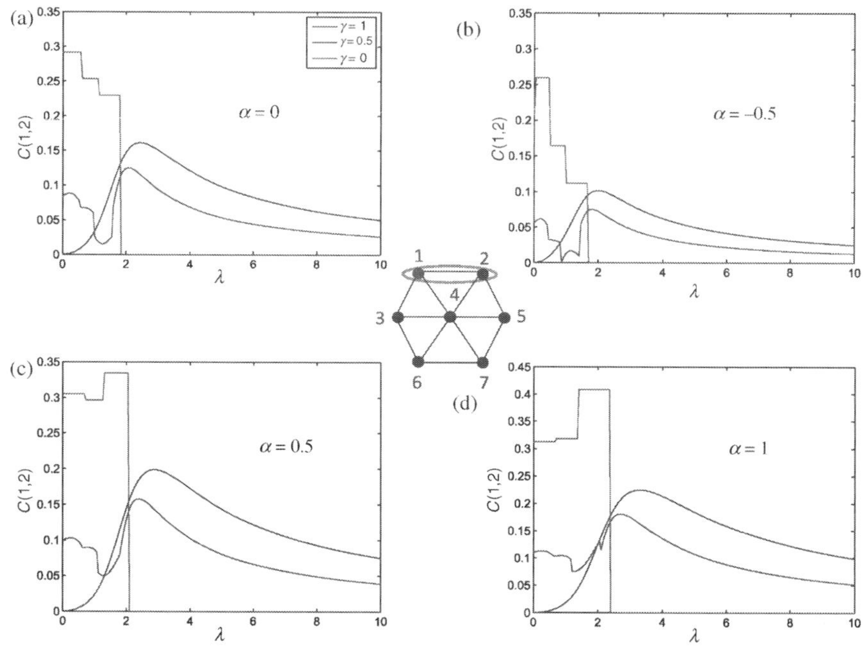

Figure 28. The concurrence $C(1, 2)$ versus the parameter λ with a single impurity at the border site 1 with different impurity coupling strengths $\alpha = -0.5, 0, 0.5, 1$ for different degrees of anisotropy $\gamma = 1, 0.5, 0$ as shown. The legend for all subfigures is as shown in part (a). (*See the color version of this figure in the Color Plates section.*)

degree of anisotropy decreases, the behavior of the entanglement changes, where it starts with a finite value at $\lambda = 0$ and then shows a step profile for the small values of λ. For the partially anisotropic case, the step profile is smooth and the entanglement mimics the Ising case as λ increases but with smaller magnitude. The entanglement of isotropic XY system shows a sharp step behavior, and then suddenly vanishes before reaching $\lambda = 2$. Interestingly, the entanglement behavior of the two-dimensional spin system at the different degrees of anisotropy mimics the behavior of the one-dimensional spin system at the same degrees of anisotropy at the extreme values of the parameter λ. Comparing the entanglement behavior in the two-dimensional Ising spin system with the one-dimensional system, one can see a great resemblance except that the critical value becomes $h/J \approx 2$ in the two-dimensional case as shown in Fig. 28. On the other hand, for the partially anisotropic and isotropic XY systems, the entanglements of the two-dimensional and one-dimensional systems agree at the extreme values of λ where it vanishes for $h >> J$ and reaches a finite value for $h << J$. The former case corresponds to an alignment of the spins in the z-direction, paramagnetic state, while the latter case corresponds to alignment in the x- and y-directions, which are a ferromagnetic state. The effect of a weak impurity ($J' < J$), $\alpha = -0.5$, is shown in Fig. 28b where the entanglement behavior is the same as before except that the entanglement magnitude is reduced compared with the pure case. On the other hand, considering the effect of a strong impurity ($J' > J$), where $\alpha = 0.5$ and 1, as shown in Fig. 28c and d, respectively, one can see that the entanglement profile for $\gamma = 1$ and 0.5 have the same overall behavior as in the pure and weak impurity cases except that the entanglement magnitude becomes higher as the impurity gets stronger and the peaks shift toward higher λ values. Nevertheless, the isotropic XY system behaves differently from the previous cases where it starts to increase first in a step profile before suddenly dropping to zero again, which will be explained later.

To explore the effect of the impurity location we investigate the case of a single impurity spin located at site 4, instead of site 2, where we plot the concurrences $C(1, 2)$ in Fig. 29. Interestingly, while changing the impurity location has almost

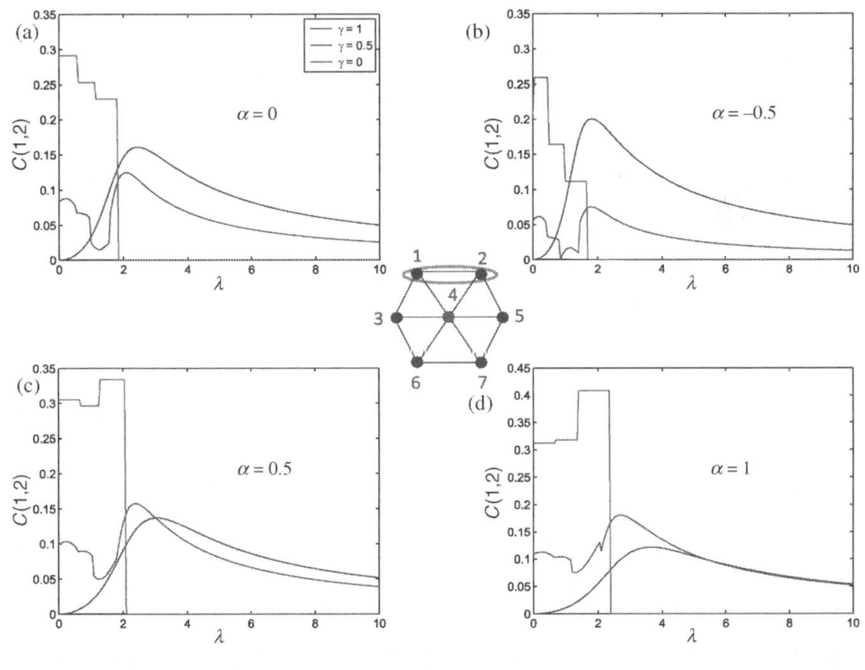

Figure 29. The concurrence $C(1, 2)$ versus the parameter λ with a single impurity at the central site 4 with different impurity coupling strengths $\alpha = -0.5, 0, 0.5, 1$ for different degrees of anisotropy $\gamma = 1, 0.5, 0$ as shown. The legend for all parts is as shown in part (a). (*See the color version of this figure in the Color Plates section.*)

no effect on the behavior of the entanglement $C(1, 2)$ of the partially anisotropic and isotropic XY systems, it has a great impact on that of the Ising system where the peak value of the entanglement increases significantly in the weak impurity case and decreases as the impurity gets stronger as shown in Fig. 29.

Now we turn to the dynamics of the two-dimensional spin system under the effect of a single impurity and different degrees of anisotropy. We investigate the dynamical reaction of the system to an applied time-dependent magnetic field with exponential form $h(t) = b + (a - b)e^{-wt}$ for $t > 0$ and $h(t) = a$ for $t \leq 0$. We start by considering the Ising system, $\gamma = 1$ with a single impurity at the border site 1, which is explored in Fig. 30. For the pure case, $\alpha = 0$ shown in Fig. 30a, the results confirm the ergodic behavior of the system that was demonstrated in our previous work [134], where the asymptotic value of the entanglement coincide with the equilibrium state value at $h(t) = b$. As can be noticed from Fig. 30b, c, and d, neither the weak nor strong impurities have effect on the ergodicity of

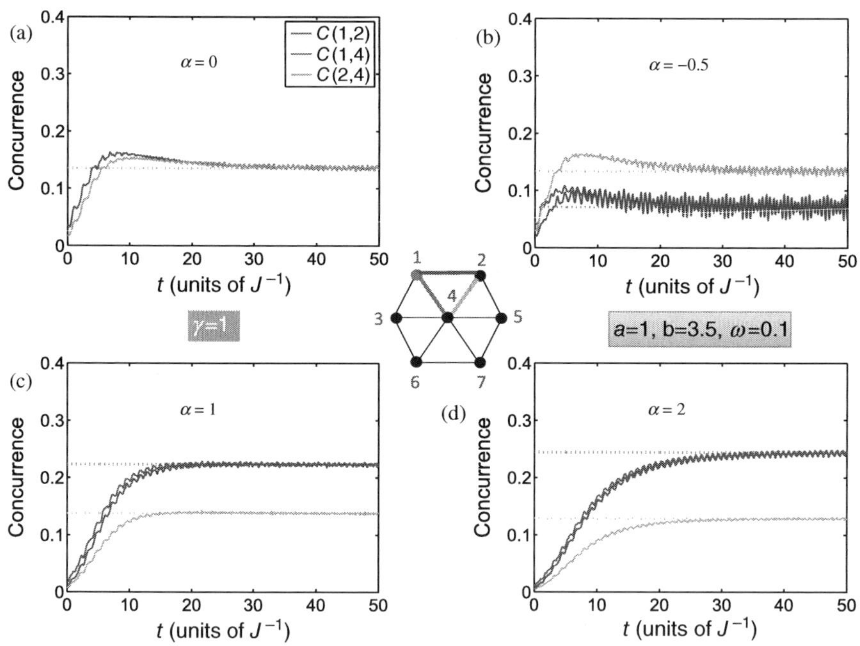

Figure 30. Dynamics of the concurrences $C(1, 2)$, $C(1, 4)$, $C(2, 4)$ with a single impurity at the border site 1 with different impurity coupling strengths $\alpha = -0.5, 0, 1, 2$ for the two-dimensional Ising lattice ($\gamma = 1$) under the effect of an exponential magnetic field with parameter values $a = 1$, $b = 3.5$, and $\omega = 0.1$. The straight dashed lines represent the equilibrium concurrences corresponding to constant magnetic field $h = 3.5$. The legend for all parts is as shown in part (a).

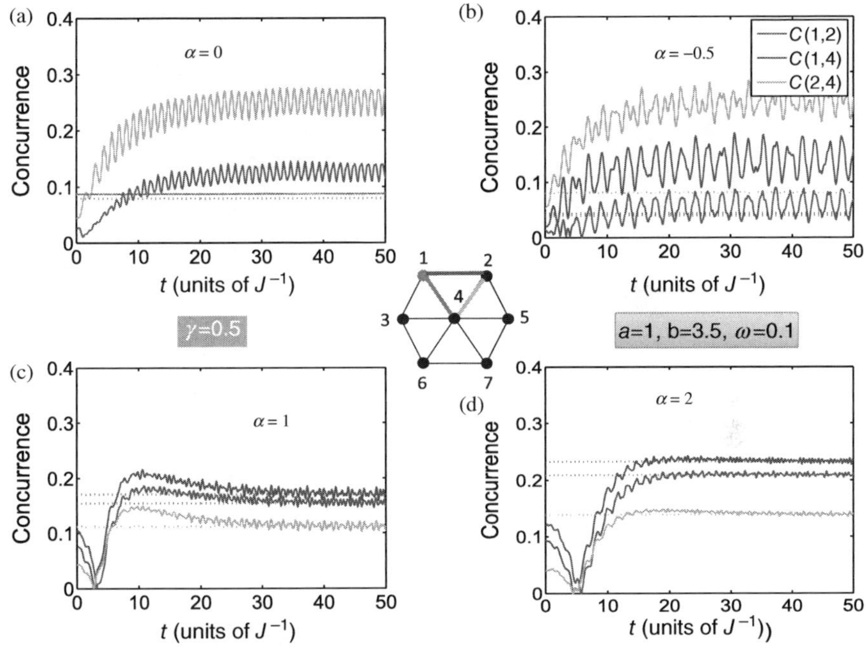

Figure 31. Dynamics of the concurrences $C(1, 2)$, $C(1, 4)$, $C(2, 4)$ with a single impurity at the border site 1 with different impurity coupling strengths $\alpha = -0.5, 0, 1, 2$ for the two-dimensional partially anisotropic lattice ($\gamma = 0.5$) under the effect of an exponential magnetic field with parameters values $a = 1, b = 3.5$, and $\omega = 0.1$. The straight dashed lines represent the equilibrium concurrences corresponding to constant magnetic field $h = 3.5$. The legend for all parts is as shown in part (b).

the Ising system. Nevertheless, there is a clear effect on the asymptotic value of entanglements $C(1, 2)$ and $C(1, 4)$ but not on $C(2, 4)$, which relates two regular sites. The weak impurity, $\alpha = -0.5$ reduces the asymptotic value of $C(1, 2)$ and $C(1, 4)$, while the strong impurities, $\alpha = 1, 2$ raise it compared to the pure case.

The dynamics of the partially anisotropic XY system under the effect of exponential magnetic field with parameters $a = 1$, $b = 3.5$, and $\omega = 0.1$, is explored in Fig. 31. It is remarkable to see that while for both the pure and weak impurity cases, $\alpha = 0$ and -0.5 , the system is nonergodic as shown in Fig. 31a and b, and it is ergodic in the strong impurity cases $\alpha = 1$ and 2 as illustrated in Fig. 31c and d.

2. *Double Impurities*

In this section, we study the effect of double impurity, where we start with two located at the border sites 1 and 2. We set the coupling strength between the two

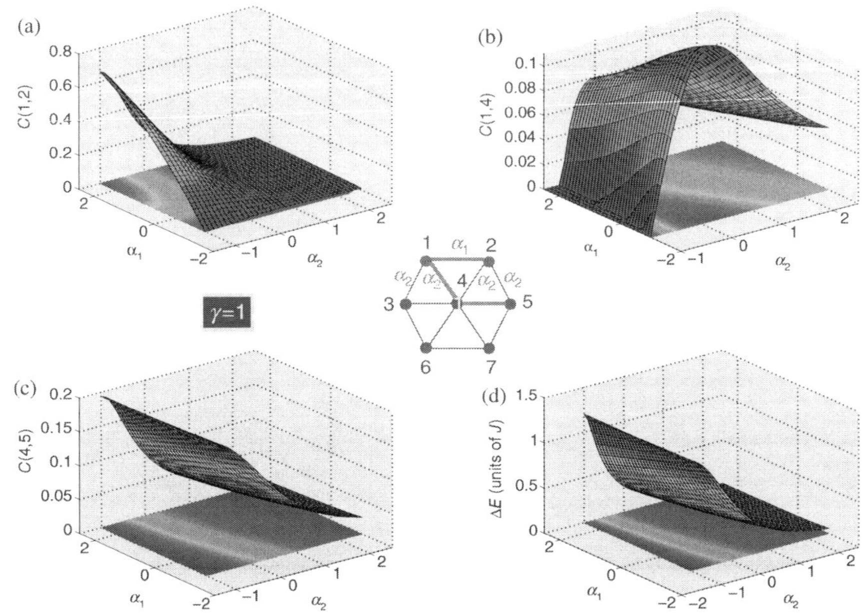

Figure 32. The concurrence $C(1,2)$, $C(1,4)$, $C(4,5)$ versus the impurity coupling strengths α_1 and α_2 with double impurities at sites 1 and 2 for the two-dimensional Ising lattice ($\gamma = 1$) in an external magnetic field $h = 2$.

impurities as $J' = (1 + \alpha_1)J$, between any one of the impurities and its regular nearest-neighbors as $J'' = (1 + \alpha_2)J$, and between the rest of the nearest-neighbor sites on the lattice as J. The effect of the impurities, strength on the concurrence between different pairs of sites for the Ising lattice is shown in Fig. 32. In Fig. 32a we consider the entanglement between the two impurity sites 1 and 2 under a constant external magnetic field $h = 2$. The concurrence $C(1,2)$ takes a large value when the impurity strengths α_1, controlling the coupling between the impurity sites, is large and when α_2, controlling coupling between impurities and their nearest-neighbors, is weak. As α_1 decreases and α_2 increases, $C(1,2)$ decreases monotonically until it vanishes. As one can conclude, α_1 is more effective than α_2 in controlling the entanglement in this case. On the other hand, the entanglement between the impurity site 1 and the regular central site 4 is illustrated in Fig. 32b, which behaves completely different from $C(1,2)$. The concurrence $C(1,4)$ is mainly controlled by the impurity strength α_2 where it starts with a very small value when the impurity is very weak and increases monotonically until it reaches a maximum value at $\alpha_2 = 0$ (i.e., with no impurity), and decays again as the impurity strength increases. The effect of α_1 in that case is less significant and makes the concurrence slowly decrease as α_1 increases, which is expected since as the coupling between the two border sites 1 and 2 increases, the entanglement between 1 and 4 decreases. It is important to note that in general $C(1, 2)$ is much larger than $C(1, 4)$ because the border entanglement is always higher than the central one as the entanglement is shared by many sites. The entanglement between two regular sites is shown in Fig. 32c where the concurrence $C(4,5)$ is depicted against α_1 and α_2, the entanglement decays gradually as α_2 increases while α_1 has a very small effect on the entanglement, which slightly decreases as α_1 increases as shown. Interestingly, the behavior of the energy gap between the ground state and the first excited state of the Ising system ΔE versus the impurity strengths α_1 and α_2, which is explored in Fig. 32d has a strong resemblance to that of the concurrence $C(4, 5)$ except that the decay of ΔE against α_2 is more rapid. The partially anisotropic system, $\gamma = 0.5$, with double impurity at sites 1 and 2 and under the effect of the external magnetic field $h = 2$ is explored in Fig. 33. As one can see, the overall behavior at the border values of the impurity strengths is the same as observed in the Ising case except that the concurrences suffer a local minimum within a small range of the impurity strength α_2 between 0 and 1 while corresponding to the whole α_1 range. The change of the entanglement around this local minimum takes a step-like profile, which is clear in the case of the concurrence $C(1, 4)$ shown in Fig. 33b. Remarkably, the local minima in the plotted concurrences coincide with the line of vanishing energy gap as shown in Fig. 33d.

3. *Entanglement and Quantum Phase Transition*

Critical quantum behavior in a many-body system happens either when an actual crossing takes place between the excited state and the ground state or a limiting avoided level-crossing between them exists (i.e., an energy gap between the two states that vanishes in the infinite system size limit at the critical point) [68]. When a many-body system crosses a critical point, significant changes in both its wave function and ground-state energy takes place, which are manifested in the behavior of the entanglement function. The entanglement in one-dimensional infinite spin systems, Ising and XY, was shown to demonstrate scaling behavior in the vicinity of critical points [72]. The change in the entanglement across the critical point was quantified by considering the derivative of the concurrence with respect to the parameter λ. This derivative was explored versus λ for different system sizes and although it did not show divergence for finite system sizes, it showed clear anomalies that developed to a singularity at the thermodynamic limit. The ground state of the Heisenberg spin model is known to have a double degeneracy for an odd number of spins, which is never achieved unless the thermodynamic limit is reached [68]. Particularly, the Ising 1D spin chain in an external transverse magnetic field has doubly degenerate ground state in a ferromagnetic phase that

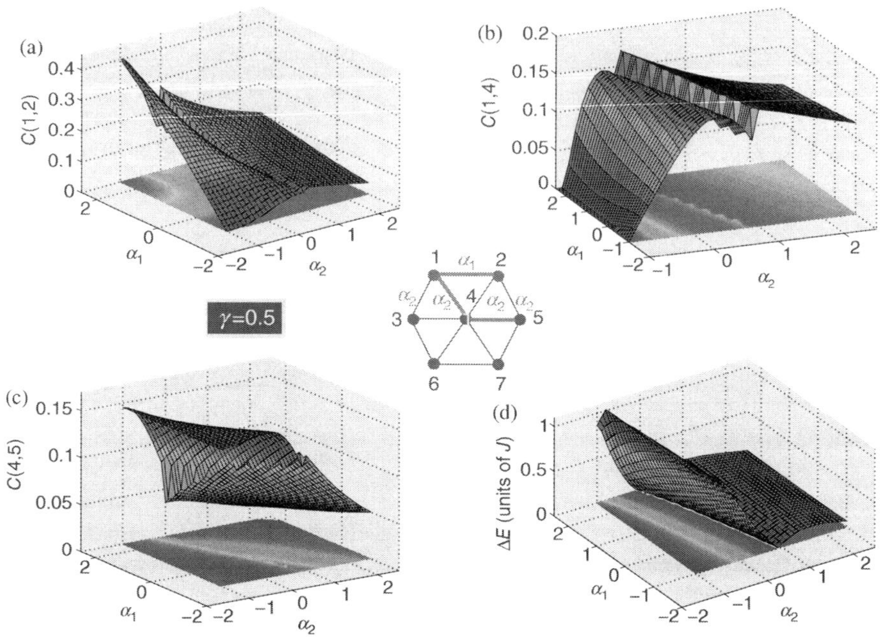

Figure 33. The concurrence $C(1,2)$, $C(1,4)$, $C(4,5)$ versus the impurity coupling strengths α_1 and α_2 with double impurities at sites 1 and 2 for the two-dimensional partially anisotropic lattice ($\gamma = 0.5$) in an external magnetic field $h = 2$.

is gapped from the excitation spectrum by $2J(1 - h/J)$, which is removed at the critical point and the system becomes in a paramagnetic phase. Now let us first consider our two-dimensional finite-size Ising spin system. The concurrence C_{14} and its first derivative are depicted versus λ in Fig. 34a and b, respectively. As one can see, the derivative of the concurrence shows strong tendency of being singular at $\lambda_c = 1.64$. The characteristics of the energy gap between the ground state and the first excited state as a function of λ are explored in Fig. 34c. The system shows strict double degeneracy, zero energy gap, only at $\lambda = 0$ (i.e., at zero magnetic field), but once the magnetic field on the degeneracy is lifted an extremely small energy gap develops, which increase very slowly for small magnetic field values but increases abruptly at certain λ value. It is important to emphasis here that at $\lambda = 0$, regardless of which one of the double ground states is selected for evaluating the entanglement, the same value is obtained. The critical point of a phase transition should be characterized by a singularity in the ground-state energy, and an abrupt change in the energy gap of the system as a function of the system parameter as it crosses the critical point. To better understand the behavior of the energy gap across

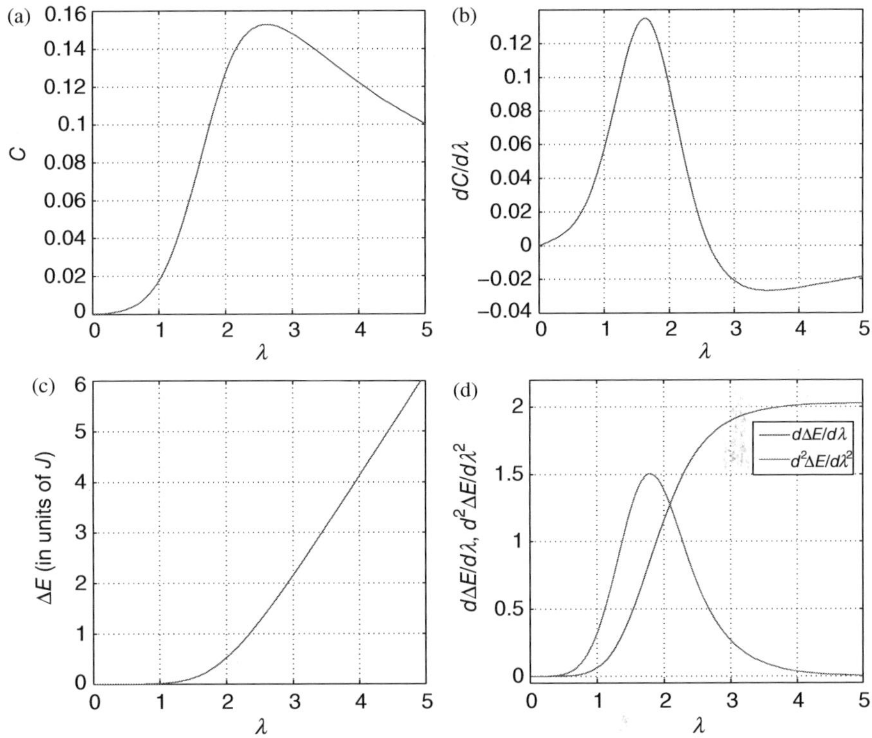

Figure 34. (**a**) The concurrence C_{14} versus λ; (**b**) the first derivative of the concurrence C_{14} with respect to λ versus λ; (**c**) the energy gap between the ground state and first excited state versus λ; (**d**) the first derivative (in units of J) and second derivative (in units of J^2) of the energy gap with respect to λ versus λ for the pure Ising system ($\gamma = 1$ and $\alpha = 0$).

the prospective critical point and identify it, we plot the first and second derivatives of the energy gap as a function of λ in Fig. 34d. Interestingly, the first derivative $d\Delta E/d\lambda$, which represents the rate of change of the energy gap as a function of λ, starts with a zero value at $\lambda = 0$ and then increases very slowly before it shows a great rate of change and finally reaches a saturation value. This behavior is best represented by the second derivative $d^2\Delta E/d\lambda^2$, which shows strong tendency of being singular at $\lambda_c = 1.8$, indicating the highest rate of change in the energy gap as a function of λ. The reason for the small discrepancy between the two values of the λ_c extracted from the $dC/d\lambda$ plot and the one of $d^2\Delta E/d\lambda^2$ is that the concurrence C_{14} is only between two sites and does not represent the whole system, contrary to the energy gap. One can conclude that the rate of change of

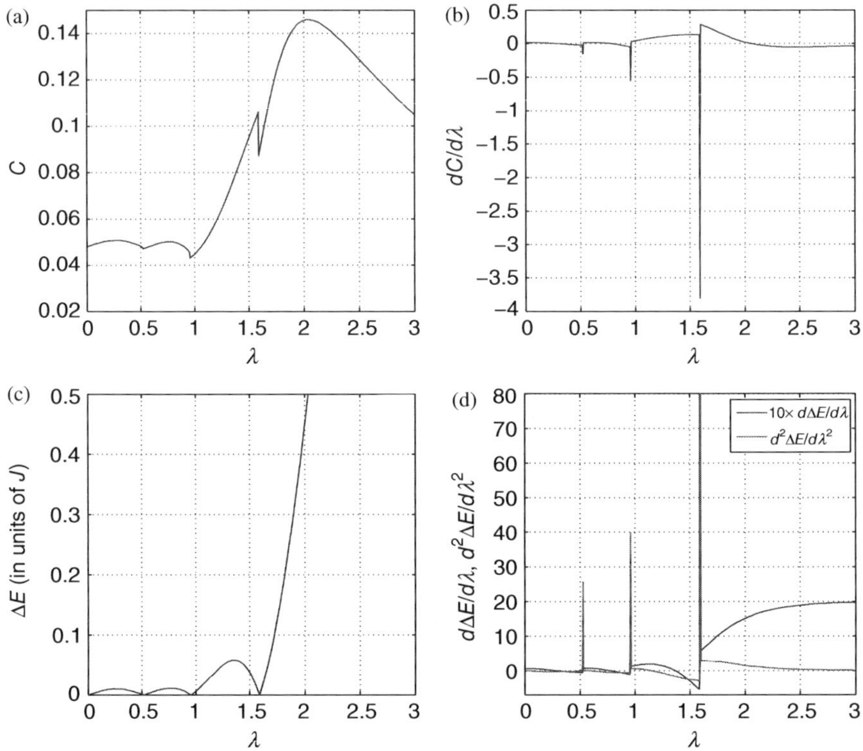

Figure 35. (**a**) The concurrence C_{14} versus λ; (**b**) the first derivative of the concurrence C_{14} with respect to λ versus λ; (**c**) the energy gap between the ground state and first excited state versus λ; (**d**) the first derivative (in units of J) and second derivative (in units of J^2) of the energy gap with respect to λ versus λ for the pure partially anisotropic system ($\gamma = 0.5$ and $\alpha = 0$). Notice that the first derivative of energy gap is enlarged 10 times its actual scale for clearness.

the energy gap as a function of the system parameter, λ in our case, should be maximum across the critical point. Turning to the case of the partially anisotropic spin system, $\gamma = 0.5$, presented in Fig. 35, one can notice that the concurrence (Fig. 35a) shows few sharp changes that are reflected in the energy gap plot as an equal number of minima Fig. 35b. Nevertheless, again there is only one strict double degeneracy at $\lambda = 0$ while the other three energy gap minima are nonzero and in the order of 10^{-5}. It is interesting to notice that the anomalies in both $dC/d\lambda$ and $d^2\Delta E/d\lambda^2$ are much stronger and sharper compared with the Ising case (Fig. 35c and d). Finally the isotropic system, which is depicted in Fig. 36, shows even sharper energy gap changes as a result of the sharp changes in the

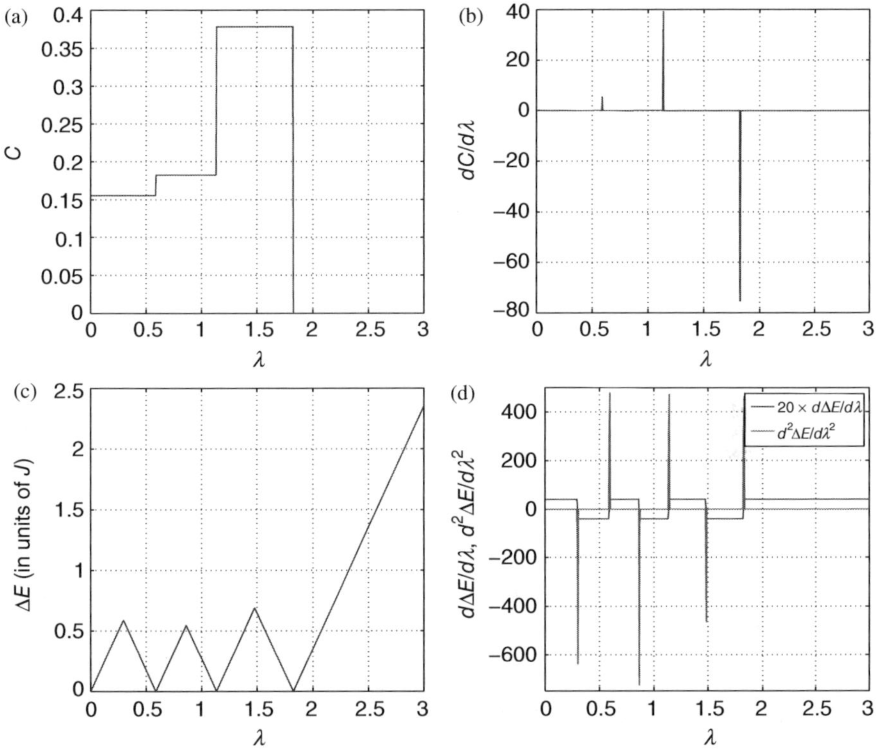

Figure 36. (**a**) The concurrence C_{14} versus λ; (**b**) the first derivative of the concurrence C_{14} with respect to λ versus λ; (**c**) the energy gap between the ground state and first excited state versus λ; (**d**) the first derivative (in units of J) and second derivative (in units of J^2) of the energy gap with respect to λ versus λ for the pure isotropic system ($\gamma = 0$ and $\alpha = 0$). Notice that the first derivative of energy gap is enlarged 20 times its actual scale for clearness.

concurrence, and the anomalies in the derivatives $dC/d\lambda$ and $d^2\Delta E/d\lambda^2$ are even much stronger than the previous two cases.

Acknowledgments

We would like to thank our collaborators Bedoor Alkurtass, Dr. Zhen Huang, and Dr. Omar Aldossary for their contributions to the studies of entanglement of formation and dynamics of one-dimensional magnetic systems.

We would also like to acknowledge the financial support of the Saudi NPST (project no. 11-MAT1492-02) and the deanship of scientific research, King Saud University . We are also grateful to the U.S. Army research office for partial support of this work at Purdue.

REFERENCES

1. A. Peres, Quantum Theory: Concepts and Methods, Kluwer, Dordrecht, The Netherlands, 1993.
2. E. Schrödinger, *Naturewissenschafen* **23**, 807 (1935).
3. A. Einstein, B. Podolsky, and N. Rosen, *Phys. Rev.* **47**, 777 (1935).
4. J. S. Bell, *Physics* **1**, 195 (1964).
5. N. Gisin, *Phys. Lett. A* **154**, 201 (1991).
6. P. Benioff, *J. Math. Phys.* **22**, 495 (1981).
7. P. Benioff, *Int. J. Theoret. Phys.* **21**, 177 (1982).
8. C. H. Bennett and R. Landauer, *Scientifc Am.* **253**, 48 (1985).
9. D. Deutsch, *Proc. R. Soc. A* **400**, 97 (1985).
10. D. Deutsch, *Proc. R. Soc. A* **425**, 73 (1989).
11. R. P. Feynman, *Int. J. Theoret. Phyis.* **21**, 467 (1982).
12. R. Landauer, *IBM J. Res. Develop.* **3**, 183 (1961).
13. P. W. Shor, in *Proceedings of the 35th Annual Symposium on Foundations of Computer Science*, IEEE Computer Society Press, Los Alamitos, CA, 1994.
14. D. Bouwmeester, J. W. Pan, K. Mattle, M. Eibl, H. Weinfurter, and A. A. Zeilinger, *Nature* **390**, 575 (1997).
15. D. Bouwmeester, K. Mattle, J.-W. Pan, H. Weinfurter, A. Zeilinger, and M. Zukowski, *Appl. Phys. B* **67**, 749 (1998).
16. C. H. Bennett and S. J. Wiesner, *Phys. Rev. Lett.* **69**, 2881 (1992).
17. K. Mattle, H. Weinfurter, P. G. Kwiat, and A. Zeilinger, *Phys. Rev. Lett.* **76**, 4546 (1996).
18. B. Schumacher, *Phys. Rev. A* **51**, 2738 (1995).
19. C. H. Bennett, G. Brassard, C. Crépeau, R. Jozsa, A. Peres, and W. K. Wootters, *Phys. Rev. Lett.* **70**, 1895 (1993).
20. A. Barenco, D. Deutsch, A. Ekert, and R. Jozsa, *Phys. Rev. Lett.* **74**, 4083 (1995).
21. L. Vandersypen, M. Steffen, G. Breyta, C. Yannoni, M. Sherwood, and I. Chuang, *Nature* **414**, 883 (2001).
22. I. L. Chuang, N. Gershenfeld, and M. Kubinec, *Phys. Rev. Lett.* **80**, 3408 (1998).
23. J. Jones, M. Mosca, and R. Hansen, *Nature* **393**, 344 (1998).
24. J. I. Cirac and P. Zoller, *Phys. Rev. Lett.* **74**, 4091 (1995).
25. C. Monroe, D. M. Meekhof, B. E. King, W. M. Itano, and D. J. Wineland, *Phys. Rev. Lett.* **75**, 4714 (1995).
26. Q. A. Turchette, C. J. Hood, W. Lange, H. Mabuchi, and H. J. Kimble, *Phys. Rev. Lett.* **75**, 4710 (1995).
27. D. V. Averin, *Solid State Commun.* **105**, 659 (1998).
28. A. Shnirman, G. Schön, and Z. Hermon, *Phys. Rev. Lett.* **79**, 2371 (1997).
29. J. e. a. Chiaverini, *Science* **308**, 997 (2005).
30. D. Vion, A. Aassime, A. Cottet, H. Joyez, P. Pothier, C. Urbina, D. Esteve, and M. H. Devoret, *Science* **296**, 886 (2002).
31. A. C. Johnson, J. R. Petta, J. M. Taylor, A. Yacoby, M. D. Lukin, C. M. Marcus, M. P. Hanson, and A. C. Gossard, *Nature* **435**, 925 (2005).

32. F. H. L. Koppens, J. A. Folk, J. M. Elzerman, R. Hanson, L. H. Willems van Beveren, I. T. Vink, H. P. Tranitz, W. Wegscheider, L. P. Kouwenhoven, and L. M. K. Vandersypen, *Science* **309**, 1346 (2005).
33. J. R. Petta, A. C. Johnson, J. M. Taylor, E. A. Laird, A. Yacoby, M. D. Lukin, C. M. Marcus, M. P. Hanson, and A. C. Gossard, *Science* **309**, 2180 (2005).
34. W. Zurek, *Phys. Today* **44**, 36 (1991).
35. D. Bacon, J. Kempe, L. D. A., and K. B. Whaley, *Phys. Rev. Lett.* **85**, 1758 (2000).
36. N. Shevni, R. de Sousa, and K. B. Whaley, *Phys. Rev. B* **71**, 224411 (2005).
37. R. de Sousa and S. Das Sarma, *Phys. Rev. B* **68**, 115322 (2003).
38. D. Loss and D. P. Divincenzo, *Phys. Rev. A* **57**, 120 (1998).
39. G. Burkard, D. Loss, and D. P. DiVincenzo, *Phys. Rev. B* **59**, 2070 (1999).
40. B. E. Kane, *Nature (London)* **393**, 133 (1998).
41. A. Sorensen, L. M. Duan, J. I. Cirac, and P. Zoller, *Nature (London)* **409**, 63 (2011).
42. S. Kais, in *Advances in Chemical Physics*, Vol. 134, John Wiley & Sons Inc, New York, 2007, pp. 493–535.
43. S. L. Sondhi, S. M. Girvin, J. P. Carini, and D. Shahar, *Rev. Mod. Phys.* **69**, 315 (1997).
44. T. J. Osborne and M. A. Nielsen, *Phys. Rev. A* **66**, 032110 (2002).
45. J. Zhang, F. M. Cucchietti, C. M. Chandrashekar, M. Laforest, C. A. Ryan, M. Ditty, A. Hubbard, J. K. Gamble, and R. Laflamme, *Phys. Rev. A* **79**, 012305 (2009).
46. O. Osenda, Z. Huang, and S. Kais, *Phys. Rev. A* **67**, 062321 (2003).
47. Z. Huang, O. Osenda, and S. Kais, *Phys. Lett. A* **322**, 137 (2004).
48. M. A. Nielsen and I. L. Chuang, *Quantum Computation and Quantum Information*, Cambridge University, Cambridge, 2000.
49. M. Horodecki, *Quant. Inf. Comput.* **1**, 3 (2001).
50. P. Horodecki and R. Horodecki, *Quant. Inf. Comput.* **1**, 45 (2001).
51. W. K. Wootters, *Quant. Inf. Comput.* **1**, 27 (2001).
52. V. Vedral, M. B. Plenio, M. A. Rippin, and P. L. Knight, *Phys. Rev. Lett.* **78**, 2275 (1997).
53. F. Mintert, A. R. R. Carvalho, K. M., and A. Buchleitner, *Phys. Rep.* **415**, 207 (2005).
54. P. MB and V. S, *Quant. Inf. Comput.* **7**, 1 (2007).
55. R. Horodecki, P. Horodecki, M. Horodecki, and K. Horodecki, *Quantum* **2**, 0702225 (2007).
56. M. Keyl, *Phys. Rep.* **369**, 431 (2002).
57. M. B. Plenio and V. Vedral, *Contemp. Phys.* **39**, 431 (1998), quant-ph/9804075.
58. B. Schumacher and M. D. Westmoreland, *Quantum* **1**, 0004045v1 (2000).
59. G. Vidal and R. F. Werner, *Phys. Rev. A.* **65**, 032314 (2002).
60. C. Bennett, D. P. Divincenzo, J. A. Smolin, and W. K. Wootters, *Phys. Rev. A.* **54**, 3824 (1996).
61. G. O. Myhr, *Quantum* **1**, 0408094 (2004).
62. L. Amico, R. Fazio, A. Osterloh, and V. Vedral, *Rev. Mod. Phys.* **80**, 517 (2008).
63. S. Hill and W. K. Wootters, *Phys. Rev. Lett.* **78**, 5022 (1997).
64. W. K. Wootters, *Phys. Rev. A.* **63**, 052302 (2001).
65. W. Kutzelnigg, G. Del Re, and G. Berthier, *Phys. Rev.* **49**, 172 (1968).
66. N. L. Guevara, S. R. P., and R. O. Esquivel, *Phys. Rev. A* **67**, 012507 (2003).
67. P. Ziesche, V. H. Smith, M. Hô, S. P. Rudin, P. Gersdorf, and M. Taut, *J. Chem. Phys.* **110**, 6135 (1999).

68. S. Sachdev, *Quantum Phase Transitions*, Cambridge University Press, Cambridge, 2001.
69. C. H. Bennett, H. J. Bernstein, S. Popescu, and B. Schumacher, *Phys. Rev. A* **53**, 2046 (1996).
70. T. J. Osborne and M. A. Nielsen, *Quant. Inf. Process.* **1**, 45 (2002).
71. J. Cardy, *Scaling and Renormalization in Statistical Physics*, Cambridge University Press, Cambridge, 1996.
72. A. Osterloh, L. Amico, G. Falci, and R. Fazio, *Nature* **416**, 608 (2002).
73. G. Vidal, J. I. Latorre, E. Rico, and A. Kitaev, *Phys. Rev. Lett.* **90**, 227902 (2003).
74. F. Verstraete, M. Popp, and J. I. Cirac, *Phys. Rev. Lett.* **92**, 027901 (2004).
75. M. Popp, F. Verstraete, M. A. Martin-Delgado, and J. I. Cirac, *Phys. Rev. A* **71**, 042306 (2005).
76. S. Popescu and D. Rohrlich, *Phys. Lett. A* **166**, 293 (1992).
77. A. Sen(De), U. Sen, M. Wiesniak, D. Kaszlikowski, and M. Zukowski, *Phys. Rev. A* **68**, 062306 (2003).
78. H. Olliver and W. H. Zurek, *Phys. Rev. Lett.* **88**, 017901 (2001).
79. R. Dillenschneider, *Phys. Rev. B* **78**, 224413 (2008).
80. L. Amico and A. Osterloh, *J. Phys. A* **37**, 291 (2004).
81. M. Christandl, N. Datta, A. Ekert, and A. Landahl, *Phys. Rev. Lett.* **92**, 187902 (2004).
82. M. E. Hartmann, M. J. Reuter and M. B. Plenio, *New J. Phys.* **8**, 94 (2006).
83. F. Glave, D. Zueco, S. Kohler, E. Lutz, and P. Hanggi, *Phys. Rev. A* **79**, 032332 (2009).
84. G. De Chiara, C. Brukner, F. R., G. M. Palma, and V. Vedral, *New J. Phys.* **8**, 95 (2006).
85. F. Ciccarello, M. Paternostro, G. M. Palma, and M. Zarcone, *Phys. Rev. B* **80**, 165313 (2009).
86. N. P. Oxtoby, A. Rivas, S. F. Huelga, and R. Fazio, *New J. Phys.* **11**, 063028 (2009).
87. J.-S. Lee and A. K. Khitrin, *Phys. Rev. A* **71**, 062338 (2005).
88. G. B. Furman, S. D. Goren, J.-S. Lee, A. K. Khitrin, V. M. Meerovich, and V. L. Sokolovsky, *Phys. Rev. B* **74**, 054404 (2006).
89. Z. Huang and S. Kais, *Int. J. Quant. Inf.* **3**, 483 (2005).
90. Z. Huang and S. Kais, *Phys. Rev. A* **73**, 022339 (2006).
91. A. Khaetskii, L. Daniel, and G. Leonid, *Phys. Rev. B* **67**, 195329 (2003).
92. J. M. Elzerman, H. R., L. H. W. Beveren, B. Witkamp, L. M. K. Vandersypen, and L. P. Kouwenhoven, *Nature (London)* **430**, 431 (2004).
93. M. Florescu and P. Hawrylak, *Phys. Rev. B* **73**, 045304 (2006).
94. W. A. Coish and L. Loss Daniel, *Phys. Rev. B* **70**, 195340 (2004).
95. A. K. Huttel, J. Weber, A. W. Holleitner, D. Weinmann, E. K., and R. H. Blick, *Phys. Rev. B* **69**, 073302 (2004).
96. A. M. Tyryshkin, S. A. Lyon, A. V. Astashkin, and A. M. Raitsimring, *Tyryshkin2003* **68**, 193207 (2003).
97. E. Abe, K. M. Itoh, I. J., and Y. S., *Phys. Rev. B* **70**, 033204 (2004).
98. Z. Huang, G. Sadiek, and S. Kais, *J. Chem. Phys.* **124**, 144513 (2006).
99. Z. M. Wang, K. Holmes, Y. I. Mazur, and G. J. Salamo, *Appl. Phys. Lett.* **84**, 1931 (2004).
100. M. Asoudeh and V. Karimipour, *Phys. Rev. A.* **73**, 062109 (2006).
101. R. Rossignoli and N. Canosa, *Phys. Rev. A* **72**, 012335 (2005).
102. M. S. Abdalla, E. Lashin, and G. Sadiek, *J. Phys. B* **41**, 015502 (2008).
103. G. Sadiek, E. Lashin, and M. S. Abdalla, *Phys. B* **404**, 1719 (2009).
104. H. Wichterich and S. Bose, *Phys. Rev. A* **79**, 060302(R) (2009).

105. P. Sodano, A. Bayat, and S. Bose, *Phys. Rev. B* **81**, 100412 (2010).
106. G. B. Furman, V. M. Meerovich, and V. L. Sokolovsky, *Phys. Rev. A* **77**, 062330 (2008).
107. B. Alkurtass, G. Sadiek, and S. Kais, *Phys. Rev. A* **84**, 022314 (2011).
108. E. Lieb, T. Schultz, and D. Mattis, *Ann. Phys.* **16**, 407 (1961).
109. W. K. Wootters, *Phys. Rev. Lett.* **80**, 2245 (1998).
110. P. Mazur, *Physica* **43**, 533 (1969).
111. E. Barouch, *Phys Rev. A* **2**, 1075 (1970).
112. M. C. Arnesen, S. Bose, and V. Vedral, *Phys. Rev. Lett.* **87**, 017901 (2001).
113. D. Gunlycke, V. M. Kendon, V. Vedral, and S. Bose, *Phys. Rev. A* **64**, 042302 (2001).
114. G. Sadiek, B. Alkurtass, and O. Aldossary, *Phys. Rev. A* **82**, 052337 (2010).
115. A. W. Sandvik and J. Kurkijärvi, *Phys. Rev. B* **43**, 5950 (1991).
116. O. F. Syljuåsen, *Phys. Rev. A* **68**, 060301 (2003).
117. O. F. Syljuåsen, *Phys. Lett. A* **322**, 25 (2004).
118. A. Sameh and J. Wisniewski, *SIAM J. Numer. Anal.* **19**, 1243 (1982).
119. A. Sameh and Z. Tong, *J. Comput. Appl. Math.* **123**, 155 (2000).
120. Q. Xu, S. Kais, M. Naumov, and A. Sameh, *Phys. Rev. A* **81**, 022324 (2010).
121. V. Coffman, J. Kundu, and W. K. Wootters, *Phys. Rev. A* **61**, 052306 (2000).
122. T. J. Osborne and F. Verstraete, *Phys. Rev. Lett.* **96**, 220503 (2006).
123. P. W. Shor, *Phys. Rev. A* **52**, R2493 (1995).
124. D. P. DiVincenzo, D. Bacon, J. Kempe, G. Burkard, and K. B. Whaley, *Nature (London)* **408**, 339 (2000).
125. S. Doronin, E. Fel'dman, and S. Lacelle, *Chem. Phys. Lett.* **353**, 226 (2002).
126. J. Lages, V. V. Dobrovitski, M. I. Katsnelson, H. A. De Raedt, and B. N. Harmon, *Phys. Rev. E* **72**, 026225 (2005).
127. J. C. Xavier, *Phys. Lett. B* **81**, 224404 (2010).
128. J. Silva-Valencia, J. Xavier, and E. Miranda, *Phys. Rev. B* **71**, 024405 (2005).
129. F. Capraro and C. Gros, *Eur. Phys. J. B* **29**, 35 (2002).
130. C. Cohen-Tannoudji, B. Diu, and F. Laloë, *Quantum Mechanics*, John Wiley & Sons Inc., 2005.
131. J. A. Jones, *Phys. Rev. A* **67**, 012317 (2003).
132. H. K. Cummins, G. Llewellyn, and J. A. Jones, *Phys. Rev. A* **67**, 042308 (2003).
133. A. Sen(De), U. Sen, and M. Lewenstein, *Phys. Rev. A* **70**, 060304 (2004).
134. Q. Xu, G. Sadiek, and S. Kais, *Phys. Rev. A* **83**, 062312 (2011).

FROM TOPOLOGICAL QUANTUM FIELD THEORY TO TOPOLOGICAL MATERIALS

PAUL WATTS,[1,2] GRAHAM KELLS,[3] and JIŘÍ VALA[1,2]

[1]*Department of Mathematical Physics, National University of Ireland Maynooth, Maynooth, Co. Kildare, Ireland*
[2]*School of Theoretical Physics, Dublin Institute for Advanced Studies, 10 Burlington Road, Dublin 4, Ireland*
[3]*Dahlem Center for Complex Quantum Systems, Fachbereich Physik, Freie Universität Berlin, Arminallee 14, D-14195 Berlin, Germany*

Advances in Chemical Physics, Volume 154: Quantum Information and Computation for Chemistry, First Edition. Edited by Sabre Kais.

I. INTRODUCTION

A *topological phase* is a phase of matter in exactly the same sense that "liquid" is a phase of matter, or "superconducting" is a phase of matter. They are, nevertheless, very unusual phases, because they are invariant under any local physical interactions or perturbations. Their quantum states are actually *completely decoupled* from these local physical processes and thus reflect purely global, topological properties of the space on which they are realized. They are effectively described by topological quantum field theories [1] from which their essential properties can be derived. They can be given concrete microscopic descriptions in the context of quantum lattice models (e.g., [2,3]); they are very likely to exist in condensed matter systems and solid-state materials; and they may eventually be engineered using cold atomic and molecular systems or superconducting electronics.

Interest in topological phases of matter has literally exploded in recent years for several reasons: they have attractive applications in quantum information science and technology [4] where they offer fault-tolerance built into quantum computing hardware, thus bypassing serious engineering challenges associated with fault-tolerant quantum computation. They are highly relevant to materials known as "topological insulators" and "topological superconductors" [5], which have been discovered recently.

Plus, let us be honest: they are really cool. By which we mean that they offer a rich variety of exciting problems for various areas of physics, a small list of which would include mathematical physics, string theory, both quantum and conformal field theory, statistical physics, and condensed matter. We believe that they are also quite relevant to chemical physics and material science.

Our objective is to present the theory of topological phases, from the basics of topological quantum field theory (TQFT) to recently developed quantum lattice models, and to outline their connection with realistic physical systems and materials. In order to do so, we first have to discuss what we mean by "topological."

How can we tell if two "spaces"—which can be anything from material objects to abstract sets of points—have the same topology or not? A nice (and very intuitive) rule-of-thumb answers this question: a space $\mathbb{X}$ is topologically equivalent to another space $\mathbb{Y}$ if we get $\mathbb{Y}$ from $\mathbb{X}$ purely by squashing, stretching, or distorting it *without* any cutting or puncturing. The classic example illustrating this is the doughnut-and-coffee mug: the former can be smoothly changed into the latter purely by remolding it, with the doughnut hole turning into the hole inside the mug's handle. Another way of saying this is that a doughnut can be *continuously deformed* into a coffee cup. Thus, the two are *topologically equivalent*. However, a solid rubber ball and a doughnut are *not* topologically equivalent; the ball has no holes, so it would have to be punctured in order to change it into a doughnut, which is not allowed in our rule-of-thumb. In essence, two spaces are topologically equivalent if they have the same general *shape*.

The preceding example is also useful in introducing the *topological invariants* of a space. A topological invariant of a space is a number that we assign to the space such that it *does not change* under any continuous deformation. The number of holes in the space, or "genus," is such an invariant: both a doughnut and coffee mug have one hole in them, and any continuous deformation of either will result in a space that has a single hole. A rubber ball, however, has no holes, so cannot possibly be topologically equivalent to any shape that has one or more holes in it. So characterizing topological spaces by their invariants is a convenient way of seeing quickly whether they are topologically equivalent to one another.

From the point of view of physical theories, these invariants may play another, more practical, role: they assign characteristic quantities to any state that determines their behavior. As an illustrative analogy, an isotropic, homogeneous gas looks the same at every point and in any direction, so the physics describing its behavior must be the same if we either translate or rotate the gas. This means that the quantities we use to describe the gas (state variables such as pressure, volume, temperature, etc.) must be translational and rotational invariants. Therefore, if we are to consider instead a physical system that behaves the same if we continuously deform it, then the quantities we want to use to describe it must be topological invariants.

What sort of physical systems would we expect to be topologically invariant? From what we have described in the preceding, the behavior of such systems would have to remain the same no matter how much we squash or stretch the system, providing we do not cut or puncture it. We can see from this requirement that a large number of familiar physical theories are immediately *excluded* from being topologically invariant.

For example, any theory in which interactions depend on the distances between objects cannot be topological, as stretching, squashing, or otherwise distorting the space will result in different behavior. Such theories are based on the existence of a *metric* g—a generalization of the usual vector dot-product—which gives the distance between the points x and y as $g(x - y, x - y)$ (e.g., $g(\vec{x}, \vec{y}) = \vec{x} \cdot \vec{y}$ in three-dimensional Euclidean space). So if the behavior of a topological theory cannot depend on distance, it must be *metric-independent*. Perhaps the most famous metric-dependent theory is general relativity, which describes how matter interacts gravitationally, but many other theories (such as classical electromagnetism and the Standard Model of particle physics) also depend on which metric we choose.

As a result, none of these theories are topologically invariant. (In fact, because the presence of a metric is the most common reason that a theory fails to be topologically invariant, the terms *topological* and *metric-independent* are usually interchangable.)

A topological theory can be purely classical: we can easily conceive of metric-independent systems that can be treated purely in terms of nonquantum physics, such as statistical mechanics and thermodynamics, for example. Despite the abundance of such theories (for an excellent review of many of them, see Ref. [6] and references therein), when we include quantum mechanics, this metric-independence has important consequences, not least of which is that a topologically invariant theory must have a *vanishing Hamiltonian*. We can see this by considering both quantum mechanics and relativity. The former tells us that time evolution of our theory is governed by the Hamiltonian, and the latter that our choice of metric tells us how we measure time. But if there is no metric, there is no way to measure time, and thus no time-dependence! The Hamiltonian therefore must be zero, and all of its eigenvalues—the energies available to the system—must also vanish.

So any theory that is explicitly topological has zero energy. Conversely, if a physical system is to have a topologically invariant phase, it must only appear in a state of zero total energy. So we already know one of the eigenvalues for states in a topological quantum theory; the challenge is in finding the values of the other state variables and the total degeneracy of the zero-energy state.

What kind of physical systems are relevant to these TQFTs? We can find them effectively realized in condensed matter systems. They can explain the fractional quantum Hall (FQH) effect [7,8], for example. This effect occurs when a two-dimensional electron gas is confined between two layers of semiconductors and is then exposed to a strong perpendicular magnetic field. These FQH systems exhibit a series of stable phases that are characterized by zero longitudinal resistivity (indicating dissipationless longitudinal current) and a stable nonzero value of the transverse Hall resistivity. This resistivity is precisely quantized and given as $R_{xy} = \frac{2\pi}{e^2}\frac{1}{\nu}$, where the *filling factor* ν is N_ϕ / N_e, the ratio of the number of electrons N_e in the gas to the number of magnetic flux lines N_ϕ penetrating the layer.

Phases of these FQH systems are treated as incompressible fluids in the bulk, which are gapped and thus stable to some extent under variations of the external parameters (e.g., the magnetic field). However, the edges are gapless, and the electric current is carried by quasiparticles localized there. These features—incompressibility of the bulk and gapless edge modes—reflect their topological character.

The systems with topological phases may also be discrete, like a system of spin-$1/2$ particles on a lattice that interact through some Hamiltonian H. We say that their ground states (i.e., their quantum phases) are topological if their *effective* description is given in terms of some TQFT. We say "effective" because such systems may have states of nonzero energy, and these phases cannot be topological as $H = 0$ is required for a TQFT to exist. However, the ground states, which *do* have vanishing Hamiltonian, may indeed be effectively described by a TQFT. (This is, in a way, similar to the effective description of a superconducting phase in terms of Ginzburg–Landau theory [9].)

The important basic properties of any such topological phases can be derived from their effective descriptions. For example, if we embody a TQFT in a lattice model, its ground state will have a finite degree of degeneracy determined by topological invariants, such as the genus of the surface on which the topological phase is realized. This has one immediate consequence: since the ground state of a lattice system has an infinite number of degrees of freedom at the thermodynamic limit (where we take the number of lattice sites to infinity), the finiteness of the ground state degeneracy implies that the ground state is separated from the rest of the energy spectrum of the system by an energy gap.

Our goal is to present a review of many of the ideas of topology as they arise out of, and are applied to, physical systems. The next sections give an introduction to some of the more theoretical and mathematical aspects of TQFT. This background material is on the more formal side, and so we have supplemented Appendices A and B: the first defines all the terms we use, and the second gives an outline of the axiomatic approach to the subject.

We will then concentrate our focus on spin lattice models because they not only serve as further, and more transparent, examples of TQFTs, but they provide important insights into the microscopic details of topological phases while abstracting themselves from irrelevant physical complications. They also provide natural connections with the realizations of these phases in atomic and molecular systems and materials. Because these models cover more recent results, we examine them in greater detail that we include in Section II.

We conclude by explaining the relevance of the theoretical results to experiment, by discussing current progress in realizing some of the abstract models presented herein in the laboratory, both to test the theory's predictions and to open the way for new research.

II. TOPOLOGICAL QUANTUM FIELD THEORY

Before we state any definitions or examples, we must first clarify what we mean when we say something is a *topological quantum field theory*, or TQFT for short.

- *Topological*—The theory is topologically invariant as already described: the behavior of the system in consideration does not change if we continuously deform it, and all variables that characterize any state of the system are topological invariants.
- *Quantum*—We describe the system using the ideas of quantum mechanics: states of the system are described as vectors in a Hilbert space, and

measureable quantities (observables) are the eigenvalues of Hermitian operators that act on this Hilbert space.

- *Field*—The quantities we measure can depend on where and when we measure them: they are, in principle, functions of both position and time (e.g., the electric and magnetic fields in electromagnetism).

Although this word-by-word breakdown of "topological quantum field theory" gives the general picture of what one might be, to fully formalize the definition, a more mathematically accurate approach is preferable. The standard five-axiom definition can be found in Ref. [1] and in our less formal version in Appendix B, but the two essential properties we need for a TQFT in this review are the following:

- The theory must be topologically invariant, namely, metric-independent. Physical measurements give results depending only on the global *shape* of the space, not its size or other local geometric features.
- The states the system can occupy must be determined entirely by its *boundary*: if a space $\mathbb{X}$ has boundary Σ, the states of a TQFT on $\mathbb{X}$ must come from some Hilbert space $\mathcal{H}(\Sigma)$ of finite dimension N_Σ. In the pathological case where there is no boundary ($\Sigma = \emptyset$), states are described by complex numbers.

A. The General Approach Toward Building TQFTs

So even though we now have a more definite idea of what we will call a TQFT, the axioms defining one seem more useful in telling us whether we already have one, not how we go about finding one. So let us look at TQFTs from a somewhat more constructive angle.

We start with the "field theory" part of TQFT: we have some underlying manifold $\mathbb{X}$ on which our system lives, and the system is described by the values of some function $A(x)$, where x is a generic point of $\mathbb{X}$. The way in which A evolves is determined by some equations of motion we assume are derived (via Hamilton's principle) from some action $S[A]$, which, in general, is expressible as an integral over $\mathbb{X}$.

Now for the "quantum" part: instead of treating $A(x)$ as a classical field, we assume that it is a second-quantized operator-valued object that acts on some Hilbert space. Alternatively, we may use the Feynman path integral to take into account the quantum nature of the system. The two approaches are entirely equivalent; both say that a classical A-dependent observable $\mathcal{O}[A]$ is replaced by its expectation value $\left\langle \mathcal{O}[A] \right\rangle$, but differ in how it is computed

$$\text{Path integral:} \quad \mathcal{O}[A] \mapsto \left\langle \mathcal{O}[A] \right\rangle = \int [\mathrm{d}A]\, \mathcal{O}[A] e^{iS[A]}$$

$$\text{Second quantization:} \quad \mathcal{O}[A] \mapsto \left\langle \mathcal{O}[A] \right\rangle = \left\langle 0 \middle| \mathcal{O}[A] \middle| 0 \right\rangle$$

where $|0\rangle$ is the vacuum state of the system's Hilbert space. (For a highly readable and entertaining introduction to QFTs, see Ref. [10].)

Finally, the "topological" part is brought in by saying that the behavior of the system is invariant under continuous deformations of $\mathbb{X}$. So a change of $\mathbb{X}$ to any topologically equivalent space cannot change the quantum field theory describing the system. At first glance, this looks like it might be quite unlikely: $S[A]$ is an integral over $\mathbb{X}$, so it seems possible that changing $\mathbb{X}$ will change S. Thus, the challenge in finding a TQFT lies primarily in finding an action whose change under a continuous deformation does not affect the expectation values of the system's observables. We would now like to illustrate this approach with a concrete example.

B. Chern–Simons Theory

One of the simplest theories that satisfies all the TQFT axioms was devised by Shiing-Shen Chern and James Simons in 1974 [11], and has since been called *Chern–Simons theory*. It not only serves as an illustrative example of the types of properties TQFTs have, it has also been shown to have deep connections with the theory of knots in three dimensions [12]. It also appears in some of the lattice models we discuss later in this chapter. Before we introduce it, however, we present a brief review of one of its major building blocks.

1. Gauge Theories

One of the major innovations of physics in the past half-century has been the invention of *gauge theories*. These theories introduce particles, or *gauge fields*, that describe symmetries of nature. The standard model of particle physics is perhaps the most famous example, but gauge theories are found in all areas of the physical sciences.

However, even though their study is relatively recent, their history goes back to the 1800s, when James Clerk Maxwell introduced his famous equations describing electromagnetism. When no electric charges or currents are present, these take the form

$$\vec{\nabla} \cdot \vec{B} = 0, \quad \vec{\nabla} \times \vec{E} = -\frac{\partial \vec{B}}{\partial t}$$
$$\vec{\nabla} \cdot \vec{E} = 0, \quad \vec{\nabla} \times \vec{B} = \frac{1}{c^2}\frac{\partial \vec{E}}{\partial t} \tag{1}$$

The first two equations can be automatically satisfied by introducing the scalar potential Φ and the vector potential $\vec{A}$ via

$$\vec{E} = -\vec{\nabla}\Phi - \frac{\partial \vec{A}}{\partial t}, \quad \vec{B} = \vec{\nabla} \times \vec{A}$$

However, these potentials are not unique; if we change them by the *gauge transformation*

$$\Phi \mapsto \Phi + \frac{\partial \chi}{\partial t}, \quad \vec{A} \mapsto \vec{A} - \vec{\nabla}\chi \tag{2}$$

where χ is any function, then the electric and magnetic fields are unchanged; we say that electromagnetism is a *gauge-invariant* theory.

Another property of Eq. (1) is that if we make the transformation

$$\vec{E} \mapsto c\vec{B}, \quad \vec{B} \mapsto -\vec{E}/c \tag{3}$$

we get back exactly the same equations. This transformation, which simply flips the quantities of interest without changing the underlying physics, is called a *duality* transformation, and we say the electric and magnetic fields are *dual* to each other. These types of flips can sometimes provide great insight into the behavior of a system, and so duality transformations are used whenever possible. (As we shall see later, they are particularly helpful in lattice models.)

Note that the scalar potential and the three components of the vector potential are ordinary functions: at a spacetime point $x = (x^0, x^1, x^2, x^3) = (ct, x, y, z)$, they just return real numbers. The leap to gauge theory comes when we think of them as *matrix-valued* objects. In this case, the gauge transformation is given not in terms of a scalar function $\chi(x)$, but a matrix function $U(x)$, and the three components of the vector potential would transform as

$$A_m \mapsto A'_m = U \cdot A_m \cdot U^{-1} - i\frac{\partial U}{\partial x^m} \cdot U^{-1}$$

Different gauge theories are thus determined by what types of matrices we use for U.

For example, if we think of A_m and U as 1×1 matrices (i.e., scalar functions), then $U = e^{-i\chi}$ gives the second of Eq. (2). This 1×1 matrix is *unitary*, meaning that its Hermitian adjoint $U^\dagger$ is its own inverse, so we say that electromagnetism is a $U(1)$ gauge theory.

But suppose that we take A_m to be a 2×2 matrix, and define our gauge transformation matrices U to be 2×2 unitary ones with $\det U = 1$. The set of such matrices is called $SU(2)$, and this defines a $SU(2)$ gauge theory. In this case, each component of the vector potential may be written in terms of the *Pauli matrices*

$$\sigma^x = \begin{pmatrix} 0 & 1 \\ 1 & 0 \end{pmatrix}, \quad \sigma^y = \begin{pmatrix} 0 & -i \\ i & 0 \end{pmatrix}, \quad \sigma^z = \begin{pmatrix} 1 & 0 \\ 0 & -1 \end{pmatrix}$$

as

$$A_m = \frac{1}{2}\left(A_m^x \sigma^x + A_m^y \sigma^y + A_m^z \sigma^z\right)$$

One of the fundamental differences between these two different gauge theories is that the $U(1)$ theory is *abelian* and the $SU(2)$ theory is *non-abelian*. That is to say, all $U(1)$ matrices commute, but two $SU(2)$ matrices do not necessarily commute with each other. Whether a gauge theory is a abelian can have a major influence on how the gauge fields behave.

So to summarize, a gauge theory is a physical system described by a gauge field $\vec{A}$ (which may be thought of as a matrix-valued vector potential) such that the behavior of the system does not change when $\vec{A}$ is transformed according to Eq. (2), with U being an invertible matrix taken from a set of matrices G. G is then called the *gauge group* or *symmetry group* of the theory. Electromagnetism is a $U(1)$ gauge theory; the standard model is a $SU(3) \times SU(2) \times U(1)$ gauge theory. And there are many more, one of which is the subject of the next section.

2. *The Chern–Simons Action*

We now wish to illustrate what we have discussed so far by introducing a theory of gauge fields that is both gauge- and topologically invariant: Chern–Simons theory. Let $\mathbb{X}$ be a three-dimensional manifold with boundary Σ on which we have an $SU(N)$ gauge theory, namely, matrix-valued fields A_1, A_2, and A_3, which transform as Eq. (2) under the group of $N \times N$ unitary matrices with determinant 1. The *Chern–Simons action* is defined as

$$S_{\mathrm{CS}}[A] = \frac{k}{4\pi} \sum_{\ell,m,n} \int_{\mathbb{X}} \mathrm{d}^3x\, \epsilon_{\ell mn} \mathrm{tr}\left(A_\ell \cdot \partial_m A_n + \frac{2i}{3} A_\ell \cdot A_m \cdot A_n\right)$$

where $\partial_m = \partial/\partial x^m$, tr is the usual matrix trace, $\epsilon_{\ell mn}$ the totally antisymmetric Levi-Civita symbol, and k is an as-yet-unspecified constant.

Why do we consider this particular action? Well, it has two important properties: first, it is invariant under coordinate transformations, as any action must be (the physics cannot change if we decide to work in spherical coordinates instead of Cartesian coordinates!). Second, it makes no reference at all to a metric; nowhere in this action is any mention made of lengths or distances. Therefore, it is *topologically invariant* and will have the same value on any manifold that is a continuous deformation of $\mathbb{X}$. So we are on the right track for a TQFT.

However, a gauge theory must be gauge-invariant, and at first glance, this one is not; under a gauge transformation, S_{CS} changes to $S_{\mathrm{CS}} + \Delta S_{\mathrm{CS}}$ with

$$\Delta S_{\mathrm{CS}}[A] = -\frac{ik}{12\pi}\sum_{\ell,m,n}\int_{\mathbb{X}} \mathrm{d}^3x\, \epsilon_{\ell mn}\mathrm{tr}\left(U^{-1}\cdot\partial_\ell U\cdot U^{-1}\cdot\partial_m U\cdot U^{-1}\cdot\partial_n U\right) - \frac{ik}{4\pi}\sum_{\ell,m,n}\int_{\Sigma} \mathrm{d}\sigma_\ell\, \epsilon_{\ell mn}\mathrm{tr}\left(A_m\cdot U^{-1}\cdot\partial_n U\right) \tag{4}$$

where $\mathrm{d}\sigma_\ell$ is the area element normal to the two-manifold Σ.

But note that when $\mathbb{X}$ has no boundary, the second integral in Eq. (4) vanishes; for this reason, we limit our CS theories to ones that live on *boundaryless* spaces only.[1] Furthermore, when $\mathbb{X}$ is closed and bounded (i.e., *compact*), it turns out that for *any* matrix U in $SU(N)$,

$$\sum_{\ell,m,n}\int_{\mathbb{X}} \mathrm{d}^3x\, \epsilon_{\ell mn}\mathrm{tr}\left(U^{-1}\cdot\partial_\ell U\cdot U^{-1}\cdot\partial_m U\cdot U^{-1}\cdot\partial_n U\right) = 24\pi^2 i \times \text{ integer}$$

The integer in the preceding equation depends on the manifold $\mathbb{X}$ and how the matrix U depends on the coordinates on $\mathbb{X}$, but it is *always* an integer. Therefore, the CS action will change by an integer multiple of $2\pi k$ under a gauge transformation.

This would still seem to preclude gauge-invariance, but it does not; when we quantize the theory, it is the path integral

$$Z_{\mathrm{CS}}(\mathbb{X}) = \int [\mathrm{d}A]\; e^{iS_{\mathrm{CS}}[A]} \tag{5}$$

that gives the physics, not the action by itself. But notice that the shift in S_{CS} will not change Z_{CS} if the constant k takes on only *integer* values, so gauge-invariance of this theory requires $k \in \mathbb{Z}$. Different values of k give different theories, so we define the *level* of the CS theory by specifying which integer we take, and say that we have a $SU(N)_k$ (read "$SU(N)$ level k") Chern–Simons theory on the space $\mathbb{X}$.

So any given CS theory is labeled by the level k, but what quantum numbers can we associate with the states that actually appear? We have argued that they must be topological invariants, but how can we find any of them? Well, we have one already: the CS action, which we already know is manifestly topologically invariant. But to get an actual number from it, we must evaluate it at a particular gauge field $A_m(x)$.

But the number we want from it should characterize a particular state of the system, and in order to find such a state, we must solve the Euler–Lagrange

[1] Chern–Simons theories also exist for manifolds with nonempty boundaries, but for illustrative purposes, we take $\Sigma = \emptyset$ here in this example.

equations of motion that come from S_{CS}. It can be shown that if the *field strength* is defined as

$$F_{mn} = \partial_m A_n - \partial_n A_m - iA_m \cdot A_n + iA_n \cdot A_m$$

then the equations of motion are nicely written as

$$F_{mn} = 0 \tag{6}$$

(Because $F_{mn} \mapsto U \cdot F_{mn} \cdot U^{-1}$ under a gauge transformation, we see these equations of motion are manifestly gauge-invariant, as expected.) A state of the system is thus obtained by solving these first-order partial differential equations, which may be difficult because $3\left(N^2 - 1\right)$ separate equations are found in Eq. (6), but we will assume that solutions can be found.

The simplest topological invariant that can be constructed from a solution is called the *first Chern number* ν_1: given a closed loop Γ in $\mathbb{X}$, we simply integrate the trace of A over it, that is,

$$\nu_1 = \frac{1}{4\pi i} \sum_{m=1}^{3} \oint_{\Gamma} \mathrm{d}x^m \operatorname{tr}(A_m) \tag{7}$$

However, if we are looking at an $SU(N)$ gauge theory, the gauge fields have zero trace (as in Eq. (4) when $N = 2$), and thus $\nu_1 = 0$ for all states. So in this case, it is somewhat trivial.

A more useful state variable for CS theory comes from evaluating the CS action at a solution: we define the *second Chern number* ν_2 by

$$\nu_2 = \frac{1}{8\pi^2} \sum_{\ell,m,n} \int_{\mathbb{X}} \mathrm{d}^3x\, \epsilon_{\ell mn} \operatorname{tr}\left(A_\ell \cdot \partial_m A_n + \frac{2i}{3} A_\ell \cdot A_m \cdot A_n\right)$$

Note that this differs from the CS action only by a factor of $2\pi k$, so it is also a topological invariant. But this additional factor also guarantees that ν_2 is an *integer*. So a nice quantumlike description arises, where each of the states of our theory is labeled by an integer.

Now to relate everything to TQFT: first of all, note that the path integral $Z_{CS}(\mathbb{X})$, which determines all of the behavior of the system, is topologically invariant. By construction, it is metric-independent and so does not change if $\mathbb{X}$ is deformed continuously. Second, the state of the system is given by evaluating $Z_{CS}(\mathbb{X})$ at a solution A_m of the equations of motion. But this is in $\mathbb{C}$, which is the Hilbert space for TQFTs on spaces without boundaries. Thus, this $SU(N)_k$ Chern–Simons theory satisfies both essential properties of a TQFT, and provides us with our first concrete example of one.

C. Boundaries and Conformal Field Theory

The Chern–Simons theory just discussed gave an explicit example of a TQFT on a space with no boundary; we had to assume this in order to ensure that the path integral (5) was both gauge- and topologically invariant. However, instances in which we have a space that *does* have a nonempty boundary are certainly possible, and so we now consider such cases.

1. TQFTs and Boundaries

One of the key features of a TQFT on a space is that the physical states come from a Hilbert space *entirely* determined by that space's boundary. The interior of the space (the "bulk") does not effect the states available to the system; it only determines which particular state the system is in. In more axiomatic language, if $\mathbb{X}$ is the space our TQFT lives on and Σ its boundary, then the system is in a state in the Hilbert space $\mathcal{H}(\Sigma)$. A TQFT on a different space $\mathbb{Y}$ with the same boundary might be in a different state, but this state would also be in $\mathcal{H}(\Sigma)$.

Theories on spaces with no boundary exist, of course; a physical theory in which all states are confined to a circle, sphere, or torus would be an example. However, "no boundary" is just another way of saying "with boundary $\emptyset$," and such cases are covered by our axioms, which require the Hilbert space describing a TQFT on any of these spaces to be $\mathbb{C}$, the complex numbers. As we discussed in the previous section, this is exactly what happened in Chern–Simons theory: the three-manifold $\mathbb{X}$ was assumed to have no boundary, and its state was described simply by the complex number $Z_{\text{CS}}(\mathbb{X})$. Thus, all states of the system are described by numerical topological invariants.

However, if $\mathbb{X}$ has a nonempty boundary Σ, then the situation may be quite different. The TQFT axioms no longer restrict the Hilbert space $\mathcal{H}(\Sigma)$, apart from it being of finite dimension N_Σ. This means that the possible states of the system are not given simply by complex numbers, but as linear combinations of some basis kets $\{|\phi_1^\Sigma\rangle, \ldots, |\phi_{N_\Sigma}^\Sigma\rangle\}$. Finding these "edge modes" is then, in a major sense, the main goal in the analysis of a TQFT on any space with a boundary, as all systems that have Σ as a boundary are described by them.

Without any more specific information about the TQFT in question, we cannot say much more about these edge modes; however, we can make two definite statements: first, because all TQFTs must have vanishing Hamiltonian, these kets (and therefore any linear combinations of them) must be zero-energy states. Second, N_Σ is the degeneracy of these states, and because it is the dimension of $\mathcal{H}(\Sigma)$—a topological invariant—it is determined entirely by the global topology of the system and not by any local properties it may have.

We will see examples later in this review of models where we can determine these modes and their degeneracies exactly, but we will see in the next section that, when our TQFT lives in a two- or three-dimensional spacetime, we can extract more structure without having to look at any specific systems.

2. *Conformal Field Theories and Two-Dimensional TQFTs*

Many TQFTs have connections to a somewhat different type of quantum field theory called *conformal field theory*, or CFT (cf. Ref. [13]). A CFT over a space $\mathbb{X}$ is a theory that is invariant under *conformal transformations* (i.e., deformations of $\mathbb{X}$ that *perserve angles*). For example, if $\vec{x}$ and $\vec{y}$ are the position vectors of two points in $\mathbb{R}^3$, then the angle $\theta(\vec{x}, \vec{y})$ between them can be found via

$$\cos\theta(\vec{x}, \vec{y}) = \frac{\vec{x} \cdot \vec{y}}{|\vec{x}\,||\,\vec{y}\,|}$$

A transformation of $\mathbb{R}^3$ that does not change this angle for any pair of vectors is a conformal transformation. More generally, if $\mathbb{X}$ is a space with a metric g, then conformal transformations are those changes of $\mathbb{X}$ that leave $\cos\theta(x, y) = g(x, y)/\sqrt{g(x, x)g(y, y)}$ unchanged for any pair of points x and y.

As an example of a conformal transformation, consider a uniform "expansion" of $\mathbb{R}^3$ in which we increase all distances between points by the same positive constant factor λ. In other words, if $\vec{x}$ and $\vec{y}$ are the position vectors of two points, then this transformation takes $|\,\vec{x} - \vec{y}\,|$ to $\lambda\,|\,\vec{x} - \vec{y}\,|$. It is easily seen that although this transformation increases the length of all position vectors by λ, it does not change the *angle* between any of them. Such "rescalings" are therefore conformal transformations. Thus, if we were to formulate a CFT on $\mathbb{R}^3$ and rescale the space, the behavior of the theory would have to be independent of λ, a property known as *scale-invariance*. (For a general metric g, this transformation would be written as $g(x, y) \mapsto \lambda^2 g(x, y)$.)

These are not the only types of transformation that leave angles unchanged; rotations and translations are others. In total, if our space is D-dimensional, there are $(D + 1)(D + 2)/2$ distinct types of conformal translations.

At least, that is the case when $D \neq 2$. In two dimensions, the situation is much different: there are an *infinite* number of distinct conformal transformations [14]. This is primarily due to a property of two-dimensional metric spaces, that they are always *conformally flat*. Put another way, any metric on a two-dimensional space can be made proportional to any other metric by making a suitable change of coordinates. A consequence of this is that not only are all two-dimensional CFTs scale-invariant, but all unitary scale-invariant theories in two dimensions are CFTs [15,16].

A rescaling of a space is simply a kind of "stretching"; it does not create any new holes in the space, nor does it rip it. This means that rescaling is a continuous deformation, and so a TQFT is a scale-invariant theory. This must be true for any dimension, but we now see the first link between the two types of theories: in two dimensions, a (unitary) TQFT is also a CFT, and so all the tools and techniques which have been developed for CFTs can be used for TQFTs as well.

As an example of the intimate connection between geometry and topology in two dimensions, perhaps the most famous is the remarkable *Gauss–Bonnet theorem* (cf. Ref. [17]): let Σ be a compact, orientable, boundaryless two-dimensional manifold with metric g and genus (number of "holes") G. The former is a *geometric* property of Σ, the latter a *topological* one. Then if $R[g]$ is the curvature of Σ, which is obtained from g,

$$2 - 2G = \frac{1}{4\pi} \int_{\Sigma} \mathrm{d}^2 x \sqrt{|\det g|} R[g]$$

The left-hand side is a topological quantity, because the genus of a space does not change under continuous deformations. The right-hand side is a geometric quantity that depends on the metric g. However, it is invariant under the rescaling $g \mapsto \lambda^2 g$, and is conformally invariant. This theorem gives the flavor of the $D = 2$ TQFT $\Rightarrow$ CFT relation.

3. *CFTs and Three-Dimensional TQFTs*

In the previous section, when we said "two-dimensional," we meant two *total* dimensions. We only ever needed two real numbers to specify the position of any point in our bulk space. However, many systems that appear two-dimensional but may be thought of as existing in *three* dimensions. For example, a particle moving in a plane needs not just two numbers to say where it is, but a third (a time coordinate) to say when it is there. Similarly, although you only need two angles to describe a point on the surface of a sphere, you need an additional number—the sphere's radius—to say where the point is if the sphere is sitting in three dimensions.

TQFTs in three-dimensional space exist (Chern–Simons theory was one example), but their connection to CFTs is less straightforward than it was for two-dimensional TQFTs. A scale-invariant theory is not automatically conformally invariant for $D \neq 2$. Luckily, in some cases we can relate a three-dimensional TQFT to a two-dimensional CFT in a straightforward and natural way.

a. Two-Dimensional Subspaces with Boundary. Consider a TQFT in two spatial dimensions and one time dimension (often called "2 + 1 dimensions"), and call the region of two-dimensional space that system occupies $\mathbb{M}$. Since time can take any value, the full spacetime is the three-manifold $\mathbb{X} = \mathbb{M} \times \mathbb{R}$. If $\mathbb{M}$ is boundaryless, then so is $\mathbb{X}$, and we can say little more about the TQFT without more information.

But suppose $\mathbb{M}$ has nonempty boundary $\partial\mathbb{M}$; then $\mathbb{X}$ has boundary $\Sigma = \partial\mathbb{M} \times \mathbb{R}$, a $(1 + 1)$-dimensional manifold. For example, if $\mathbb{M}$ is a circular disc, then $\mathbb{X}$ is a solid, infinitely long (in the time direction) cylinder. $\partial\mathbb{M}$ is the circle S^1, so Σ is the $(1+1)$-dimensional infinite hollow tube $S^1 \times \mathbb{R}$.

Our TQFT is still scale-invariant even when restricted to Σ, and thus is a CFT there. The states of this TQFT on a space must live in a Hilbert space associated with its boundary, and this CFT is a natural choice. So even though the full TQFT on the three-dimensional space $\mathbb{X}$ is not necessarily a CFT, there *is* a CFT on Σ that can play the role of $\mathcal{H}(\Sigma)$.

As a specific example of this TQFT/CFT relation, we consider the *fractional quantum Hall effect* [7]. This effect results from a current flowing through a magnetic field along a planar strip: the electrons are confined to move in a two-dimensional region with edges, a $(2+1)$-dimensional theory with a boundary. This system exhibits phases conjectured to be given by a TQFT in the $(2+1)$-dimensional bulk whose states are given by a CFT on the $(1+1)$-dimensional "edges" [18].

For example, one phase—the one with filling factor 5/2—is thought to be described by a CFT called the *critical Ising model* because of its relation to a system solved by Ernst Ising, one consisting of spins arranged in a circle and interacting only with its nearest neighbors [19]. The states of this CFT are given in terms of three particles: the lowest-energy vacuum state **1**, a fermion, ψ and a "spin-field" σ. Like most all CFTs, the critical Ising model is categorized in terms of how these states can combine with one another. These *fusion rules* describe what happens when any two particles interact with each other. In this particular case, they are written

$$\begin{aligned} \mathbf{1} \times \mathbf{1} &= \mathbf{1}, & \mathbf{1} \times \psi &= \psi, \\ \mathbf{1} \times \sigma &= \sigma, & \psi \times \psi &= \mathbf{1}, \\ \sigma \times \psi &= \sigma, & \sigma \times \sigma &= \mathbf{1} + \psi \end{aligned}$$

The first three of these tell us that any particle that interacts with a **1** is unchanged. Two ψs that meet each other combine to give a **1**, and a ψ and a σ give a σ. The last says that the *two* possible outcomes of a σ–σ interaction are a **1** or a ψ.

The CFT thought to describe the FQH effect is only one possible TQFT in $2+1$ dimensions, of course. But the point is still the same, that a TQFT on a $(2+1)$-dimensional space with an edge can be naturally described by a CFT living on its $(1+1)$-dimensional boundary.

b. Two-Dimensional Subspaces Without Boundary. Another situation in which we can relate two-dimensional CFTs to three-dimensional TQFTs is when we have a two-manifold Σ that is compact and boundaryless. Because of these properties, three-manifolds can have Σ as a boundary. $\Sigma = S^2$, the two-sphere, is such a manifold; the solid ball we obtain by "filling in" this sphere is a three-manifold with Σ as a boundary.

Suppose we are told that there is a CFT on such a Σ; is it possible that this CFT arises from a TQFT on $\mathbb{X}$, a three-manifold with Σ as boundary? In general, no; not all CFTs can be extended to TQFTs. However, certain types of CFTs can indeed be extended in this way. One that can is the *Wess–Zumino–Witten (WZW) model* [20,21]; this gauge theory is a CFT that is precisely the one derived from a Chern–Simons theory, a TQFT defined on $\mathbb{X}$.

For example, a WZW theory with gauge group $SU(2)$ on the surface of a torus gives a particular CFT. This torus is the boundary of a solid doughnut in which we have an $SU(2)_2$ CS theory (recall that the subscript refers to the level of the CS theory). We may therefore use either description as a CFT or as a TQFT to describe the system. For instance, using the TQFT view allows us to show that the system can be in one of three linearly independent states, and because the Hamiltonian must vanish for a TQFT, the ground state is threefold-degenerate [22].

III. TOPOLOGICAL PHASES IN LATTICE MODELS

In the previous section, we discussed TQFTs embodied in two-dimensional physical systems. The cases we considered were continuous (e.g., the FQH states of a two-dimensional electron gas in a strong perpendicular magnetic field).

They may also be discrete: spins stuck to the vertices of a polygonal lattice, for instance. Such models may not be TQFTs a priori, but many of them exhibit *topological phases*, solutions to the model whose ground states are robust under continuous deformations of the lattice. Because the ground states will have zero energy (or whose energy can be shifted to zero by an additive constant), they may be described by *effective* TQFTs. In this case, we could have a range of topological invariants at our disposal with which to classify the states of this effective TQFT.

One criterion that we need for this to have any chance of being true, however, is that the ground state must be *gapped*. In other words, the energy gap between the second-lowest energy and the ground-state energy must remain positive for all choices of the parameters that characterize the model. If not, the degeneracy of the ground state, which is invariant for a TQFT, would change as the energy gap goes to zero. We will assume a nonzero energy gap in all the spin lattice models we consider.

We will begin by describing the spin lattice model, which provides the simplest example of an abelian topological phase, that is, a phase such that the exchange of two particles results in the multiplication of their wavefunction by $e^{i\phi}$. This phenomenon is referred to as *abelian fractional statistics*. (If the multiplication were by a matrix instead, the statistics are *non-abelian*.) This model was originally proposed by Kitaev [2] as a stabilizer quantum error-correcting code [23], and has become known as the toric code.

We then discuss the class of trivalent spin lattice models also introduced by Kitaev [3]. Specifically, we will see that these models can always be mapped to lattice fermion models with an extra degrees of freedom, and that the toric code states play the role of the fermionic vacuum in this construction [24]. These models allow us simple access to some of the properties of topological phases and will serve as an ideal bridging point into many of the ideas in the previous TQFT discussion. We will briefly review similar mappings for other more exotic variations of the Kitaev model, namely the Yao–Kivelson model [25,26] and the square–octagon lattice; these models are particularly interesting in this context, as they exhibit a rich zoology of topological phases [28].

A. The Toric Code

1. *Original Definition*

The simplest model to exhibit topological order is the *toric code*. Its main advantage is that it is exactly solvable, and thus it allows us to analytically access all eigenvalues and all eigenvectors explicitly. In its original incarnation due to Kitaev [2], it consists of *qubits*, two-level quantum systems like spin-1/2 particles, located on the edges of a square lattice and interacting through the Hamiltonian

$$H = -\sum_{s \in \text{star}} A_s - \sum_{p \in \text{plaquette}} B_p$$

where A_s and B_p are operators that are defined as four-qubit interaction terms of two distinct kinds: the operator A_s represents the interaction between four qubits located on a *star* s (i.e., on four edges connected to a given vertex of the lattice). The operator B_p embodies the interaction between four spins on the edges of a square *plaquette* p. They are explicitly defined as shown in Fig. 1, as follows:

$$A_s = \prod_{i \in \text{star}} \sigma_i^x, \quad B_p = \prod_{i \in \text{plaquette}} \sigma_i^z$$

where σ_i^x and σ_i^z are the usual Pauli matrices acting on the qubit i. These definitions imply that all of the A-operators square to unity and mutually commute with one another, and all the B-operators do as well.

We can view B_p as the operator associated with a small loop around a single square, labeled by p, of the lattice. Their products (or more precisely, monomials) would then correspond to loops formed by products of the Pauli matrices σ^z of various lengths and configurations.

The star operators A_s are associated with the four "rays" coming out of a vertex labeled by s, and may be thought of as plaquette operators on the *dual lattice*, the lattice whose vertices are located in the centers of the plaquettes of the original.

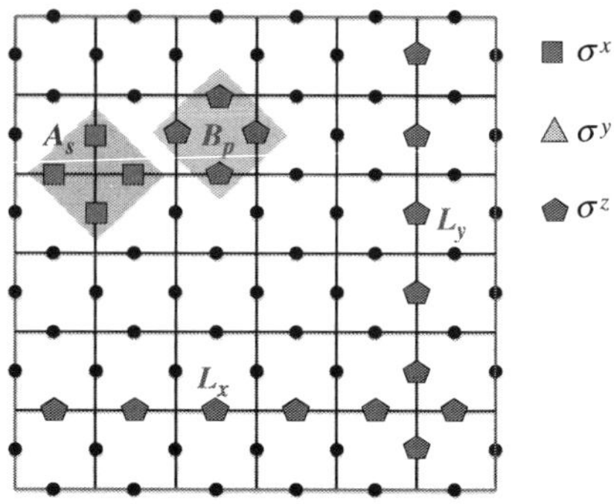

Figure 1. The toric code in the original formulation. On a torus, the opposite sides of the rectangle are to be identified, and two new symmetry operators are enabled.

Consequently, monomials of the operators A_s will appear as loops on the dual lattice.

2. *The Model on a Torus*

Any two operators A_s and B_p share either two (if s is a vertex of p) or zero (if not) qubits and therefore they also mutually commute. This means that they can be simultaneously diagonalized, and because they all square to unity, their only possible eigenvalues a_s and b_p (respectively) are either $+1$ or -1. On the torus, not all of these operators are independent because they must also satisfy the additional conditions

$$\prod_s A_s = 1, \quad \prod_p B_p = 1$$

It is worth pointing out that the number of spin-1/2 particles on a square lattice on a torus is the number of plaquette operators plus the number of star operators, so any basis state of the original model can now be rewritten as a configuration of the eigenvalues a_s and b_p. The ground state $|\text{g.s.}\rangle$ corresponds to the state that is characterized by all $a_s = +1$ and $b_p = +1$:

$$A_s \left| \text{g.s.} \right\rangle = B_p \left| \text{g.s.} \right\rangle = \left| \text{g.s.} \right\rangle$$

for all s and p. In this case, we say that the ground state is *stabilized* by all the operators A_s and B_p and we call the operators the *stabilizers*. (This connects us with the usual definition of "stabilizer" in quantum error correction.)

At this point we note that the ground state space is spanned by all spin configurations with even spin parity on the plaquettes (i.e., an even number of spins pointing up and an even number pointing down). For example, if we start from the top spin on a given plaquette and go clockwise, the set of spin configurations on the plaquette that contribute to the ground state is

$$\Big\{ \Big|\uparrow\uparrow\uparrow\uparrow\Big\rangle, \Big|\downarrow\downarrow\uparrow\uparrow\Big\rangle, \Big|\uparrow\downarrow\downarrow\uparrow\Big\rangle, \Big|\uparrow\uparrow\downarrow\downarrow\Big\rangle,$$
$$\Big|\downarrow\uparrow\uparrow\downarrow\Big\rangle, \Big|\downarrow\uparrow\downarrow\uparrow\Big\rangle, \Big|\uparrow\downarrow\uparrow\downarrow\Big\rangle, \Big|\downarrow\downarrow\downarrow\downarrow\Big\rangle \Big\}$$

These configurations will be mixed with similar configurations at the neighboring plaquettes by the star operators A_s, while forming the ground state as an equal superposition of all even parity plaquette configurations.

On a torus, two additional operators can be defined that commute with all the plaquette and star operators, and thus with the Hamiltonian: these are the *loop operators* L_x and L_y defined by

$$L_x = \prod_{i \in c_x} \sigma_i^z, \quad L_y = \prod_{i \in c_y} \sigma_i^z$$

where c_x and c_y are loops that wrap, respectively around the longitude (the x-direction) and meridian (the y-direction) of the torus; they are *topologically nontrivial*, meaning neither loop can be contracted to a single point without cutting and reconnecting. Because these operators commute with the Hamiltonian, they are symmetries that reflect the topology of the underlying manifold; in this case, a torus. Their eigenvalues, ℓ_x and ℓ_y, are either $+1$ or -1 and thus provide four distinct combinations of values to label the states of the toric code on a torus; therefore, any given eigenstate may be written as

$$\Big| \{a_s\}, \{b_p\}, \ell_x, \ell_y \Big\rangle$$

The ground state is then fourfold-degenerate on the torus and is given by the four kets

$$\Big| \text{g.s.} \Big\rangle = \Big| \{a_s = +1\}, \{b_p = +1\}, \ell_x = \pm 1, \ell_y = \pm 1 \Big\rangle$$

3. *Quasiparticle Excitations*

Note that, if we start from the ground state, applying the operator σ^x to the qubit on edge i will invert the qubit and thus also the parity of products of the spin operators around the two plaquettes p and p', which have i as an edge. This flips the eigenvalues of both plaquette operators B_p and $B_{p'}$ but commutes with all the A-operators, and so increases the energy of the system by 4, creating two "quasiparticle" excitations. The term *quasiparticle* is used to denote a collective mode—in this case, a spin configuration—rather than a single localized particle, in a sense similar to the way a collection of normal modes is called a "phonon" in condensed matter physics. Since these quasiparticles live on the plaquettes, we can call them *magnetic fluxes* in an analogy to classical electromagnetism.

Similarly, applying σ^z to a single qubit of the ground-state configuration will excite two quasiparticles associated with the star operators while commuting with all the B-operators. These excitations will also have energy 2, and we call them *electric charges*.

Both types of quasiparticles may be moved around the lattice by applying the appropriate operator (σ^x for fluxes, σ^z for charges) to a neighboring qubit. This means that a quasiparticle pair can be created and separated only by applying a "string" of operators acting on all the edges between them. It also gives us the ability to move particles in a closed loop by applying the product of the corresponding operators along the loop's edges. This allows us to now investigate the quasiparticles' statistical properties (i.e., the changes in the state of two particles under exchange).

For example, we can move an electric charge around a magnetic flux by applying a closed loop of σ^x operators, which encircles one magnetic flux. In this case, the loop of σ^x operators must cross—at some qubit k—the string of σ^z operators that connects a pair of magnetic fluxes, as in Fig. 2. Because σ_k^x and σ_k^z anticommute, the wave function changes by -1. This closed loop corresponds to two consecutive exchanges, so the statistical phase for one exchange of an electric charge and a magnetic flux is given by the imaginary unit i. (The result is the same if we encircle an electric charge by magnetic flux.) The charges and fluxes of our model thus behave as *abelian anyons*: particles whose statistics are neither bosonic nor fermonic, but whose exchange results in a multiplication by a numerical statistical phase i.

(We point out that this situation is quite analogous to the Bohm–Aharonov effect, where the phase of an electron wave function is rotated when we take it around an infinite magnetic flux line. In a further similarity to classical

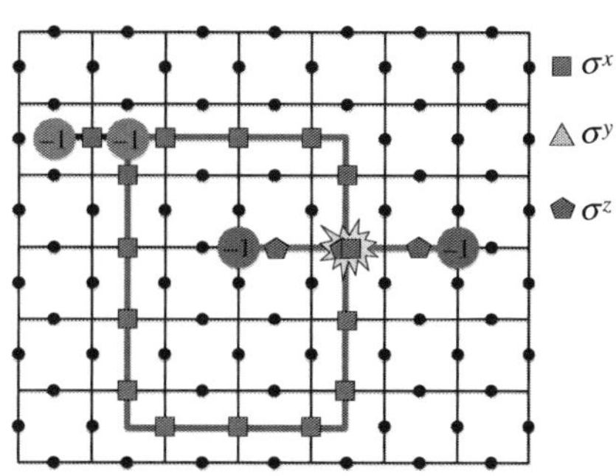

Figure 2. The toric code quasiparticles with electric and magnetic charges. The abelian statistical phase associated with an exchange of magnetic fluxes and electric charge emerges from the properties of the Pauli matrices.

electromagnetism, the electric and magnetic charges of the toric code model are dual to each other in the manner of Eq. (3).)

4. An Alternative Formulation

The toric code can be recast in a form that connects more naturally with the honeycomb lattice model [3], whose abelian phase is equivalent to that of the toric code on the fourth-order perturbation theory. In preparation for our discussion of this model, we now present this reformulation.

We can redraw the square lattice on which the qubits are located on the edges in a way that they become located on the vertices. We note that the stabilizer operators—defined on the stars and plaquettes of the original lattice—can be converted into each other by a simple unitary transformation; the square root of this operation will convert them into an equivalent form, which allows us to rewrite the Hamiltonian as a single sum of these new plaquette terms. Specifically, $H \mapsto U^\dagger H U$, with the transformation matrix given by

$$U = \left(\prod_{\substack{\text{horizontal}\\\text{links}}} X_i \right) \left(\prod_{\substack{\text{vertical}\\\text{links}}} Y_j \right)$$

where

$$X = \begin{pmatrix} 1 & 0 \\ 0 & -i \end{pmatrix}, \quad Y = \frac{1}{\sqrt{2}} \begin{pmatrix} i & 1 \\ i & -1 \end{pmatrix}$$

The transformed Hamiltonian is

$$H = -\sum_{\mathbf{p} \in \text{plaquettes}} Q_{\mathbf{p}}$$

with the new operators $Q_{\mathbf{p}}$ defined by

$$Q_{\mathbf{p}} = \sigma^z_{\mathbf{p}} \sigma^y_{\mathbf{p}+\mathbf{n}_x} \sigma^y_{\mathbf{p}+\mathbf{n}_y} \sigma^z_{\mathbf{p}+\mathbf{n}_x+\mathbf{n}_y}$$

where $\mathbf{p}$ is the location of the lowest corner of the plaquette and $\mathbf{n}_x$ and $\mathbf{n}_y$ are the unit vectors in the new lattice (see Fig. 3). Quasiparticles now appear as excitations of the same type of plaquette operators, but we keep in mind that, depending on whether they are of electric or magnetic type, they live on one or the other checkerboard patterns of the new square lattice. On a torus, each energy eigenstate is uniquely characterized by the eigenvalues $\zeta_{\mathbf{p}}$ $(= \pm 1)$ of the operators $Q_{\mathbf{p}}$ for all plaquettes $\mathbf{p}$, and by the two topologically nontrivial loop symmetries:

$$\left| \{\zeta_{\mathbf{p}}\}, \ell_x = \pm 1, \ell_y = \pm 1 \right\rangle$$

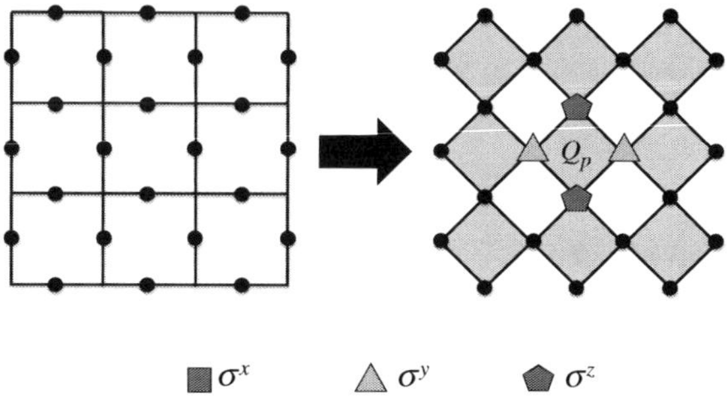

Figure 3. The alternative formulation of the toric code. The spins are located on vertices, and the underlying plaquette operators are unitarily equivalent to the star and plaquette operators of the model in its original formulation.

where the last two numbers are the eigenvalues of the operators defined on the new lattice by

$$L_x = \prod_{i \in c_x} \sigma_i^x, \tag{8}$$

$$L_y = \prod_{i \in c_y} \sigma_i^y \tag{9}$$

The ground state is again stabilized by all toric code stabilizers $Q_{\mathbf{p}}$:

$$\left| \{\zeta_{\mathbf{p}} = +1\}, \ell_x = \pm 1, \ell_y = \pm 1 \right\rangle \tag{10}$$

Its fourfold degeneracy arises purely from the topologically nontrivial symmetries of the torus; the ground state of the toric code on an infinite plane, which has no topologically nontrivial loop symmetries, is nondegenerate.

B. The Kitaev Honeycomb Model

1. Definition

The *Kitaev honeycomb model* [3] is a spin-1/2 lattice model in which the spins are located on the vertices of a hexagonal honeycomb lattice. Its basic Hamiltonian is given as follows:

$$H_0 = -J_x \sum_{x-\text{links}} K_{i,j}^x - J_y \sum_{y-\text{links}} K_{i,j}^y - J_z \sum_{z-\text{links}} K_{i,j}^z \tag{11}$$

where spins i and j interact through two-body nearest-neighbor interactions characterized by operators $K_{i,j}^\alpha = \sigma_i^\alpha \sigma_j^\alpha$ and coupling constants J_α, with $\alpha = x, y, z$ referring to the orientations of the links underlying the interactions as in Fig. 4.

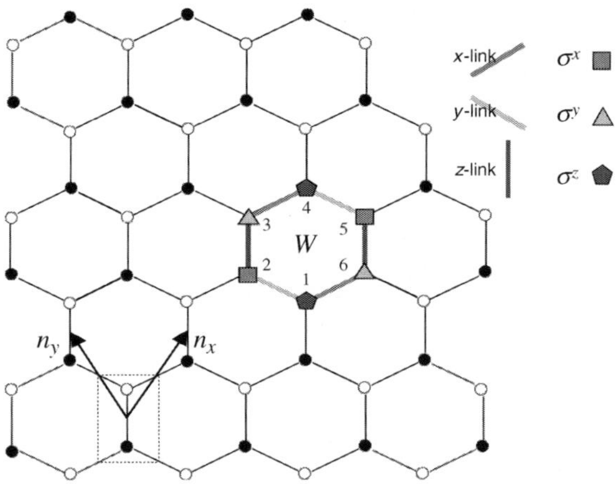

Figure 4. The Kitaev honeycomb lattice model. The unit cell is defined as two vertices connected by a z-link. The lattice is constructed by replicating the unit cell along the directions defined by the unit vectors $\mathbf{n}_x$ and $\mathbf{n}_y$. Note that it can be regarded as a square lattice of these unit cells. The bipartite character of the lattice is emphasized by distinct shading of the vertices of the two disjoint sublattices. The vortex operator $W_\mathbf{p}$ is defined as a product of six Pauli operators around plaquette $\mathbf{p}$ or, equivalently, as a product of the operators $K^\alpha_{i,j} = \sigma^\alpha_i \sigma^\alpha_j$, where i and j label the plaquette vertices and α the link between them.

We can also add to the Hamiltonian H_0 a time-reversal and parity-breaking potential, which is a sum of three-body spin terms. The contribution for each plaquette (labeled by $\mathbf{p}$, the position of their lowermost vertex) takes the form

$$\begin{aligned}V_\mathbf{p} = \sum_{l=1}^{6} P^l_\mathbf{p} &= P^1_\mathbf{p} + P^2_\mathbf{p} + P^3_\mathbf{p} + P^4_\mathbf{p} + P^5_\mathbf{p} + P^6_\mathbf{p} \\ &= \kappa^1_\mathbf{p}\, \sigma^x_1 \sigma^y_6 \sigma^z_5 + \kappa^2_\mathbf{p}\, \sigma^z_2 \sigma^y_3 \sigma^x_4 + \kappa^3_\mathbf{p}\, \sigma^y_1 \sigma^x_2 \sigma^z_3 + \kappa^4_\mathbf{p}\, \sigma^y_4 \sigma^x_5 \sigma^z_6 \\ &\quad + \kappa^5_\mathbf{p}\, \sigma^x_3 \sigma^z_4 \sigma^y_5 + \kappa^6_\mathbf{p}\, \sigma^y_2 \sigma^z_1 \sigma^x_6 \end{aligned} \tag{12}$$

so that $V = \sum_\mathbf{p} V_\mathbf{p}$ is the total potential. These additional terms can be linked at the third order of perturbation theory to the time-reversal breaking effect of a weak magnetic field:

$$V_m = -\sum_j \left(h_x \sigma^x_j + h_y \sigma^y_j + h_z \sigma^z_j \right) \tag{13}$$

If we take $J = J_x = J_y = J_z$, then the coupling constants $\kappa^l_\mathbf{p}$ in Eq. (12) are related to the magnetic field by $\kappa \sim h_x h_y h_z / J^2$.

In addition to the Hamiltonian, we define for each plaquette **p** an observable called the *vortex operator* $W_{\mathbf{p}}$:

$$\begin{aligned} W_{\mathbf{p}} &= K^y_{1,2}K^z_{2,3}K^x_{3,4}K^y_{4,5}K^z_{5,6}K^x_{6,1} \\ &= \sigma^z_1\sigma^x_2\sigma^y_3\sigma^z_4\sigma^x_5\sigma^y_6 \end{aligned}$$

These vortex operators commute with the total Hamiltonian $H = H_0 + V$ and thus can be simultaneously diagonalized yielding two eigenvalues, $+1$ or -1, for each plaquette **p**. We say that there is a vortex on the plaquette **p** if the eigenvalue of the corresponding vortex operator is -1. Because the vortex operators are symmetries of the model, the Hilbert space $\mathcal{H}$ splits into the subspaces with different numbers and configurations of vortices characterized by the eigenvalues $\{w_{\mathbf{p}}\}$ of the vortex operator for all plaquettes, namely,

$$\mathcal{H} = \bigoplus_{\{w_{\mathbf{p}}\}} \mathcal{H}_{\{w_{\mathbf{p}}\}}$$

As a consequence of a theorem due to Lieb [29], the ground state of the model will be characterized by the state with no vortices. We point out that in contrast to the toric code, where a state can be fully characterized by the eigenvalues of all the plaquette operators on the torus, this is *not* the case in the honeycomb model: all $\{w_{\mathbf{p}}\}$ configurations do give a complete basis.

2. *The Honeycomb Model Phase Diagram*

Kitaev solved the model using a mapping of the spin degrees of freedom into an enlarged Hilbert space of four fermionic operators [3]. These are represented by creation and annihilation operators satisfying the relations

$$\gamma_k\gamma_\ell + \gamma_\ell\gamma_k = 2\delta_{k\ell}1, \quad \gamma_j^\dagger = \gamma_j$$

where the first relation indicates that different fermionic operators anticommute, and the second tells us that each fermion is its own antiparticle, also called a *Majorana* fermion.

This mapping allows transformation of the model to a quadratic fermionic Hamiltonian that can be solved, and it is found that the model exhibits four distinct quantum phases (characterized by different ground states). Three phases, which are equivalent up to some global unitary transformation, are *gapped*: they correspond to ground states that are separated from the lowest excited state by a spectral gap $\Delta E > 0$ robust in the thermodynamic limit (an infinite number of spins). They occur when the coupling coefficients satisfy any one of the following conditions:

$$|J_x| > |J_y| + |J_z|$$
$$|J_y| > |J_x| + |J_z|$$
$$|J_z| > |J_x| + |J_y|$$

Using perturbation theory, Kitaev showed that these phases are equivalent to the abelian topological phase of the toric code. To briefly review this result, we start by taking only one of the couplings, say J_z, to be nonzero. In this case, the system is a collection of independent pairs of spins. Any two spins i and j are located on opposite vertices of z-links and are coupled through the interaction $-J_z\sigma_i^z\sigma_j^z$. This interaction is diagonal in the standard spin basis and has two degenerate excited states, $|\uparrow\downarrow\rangle$ and $|\downarrow\uparrow\rangle$, and two degenerate ground states, $|\uparrow\uparrow\rangle$ and $|\downarrow\downarrow\rangle$.

Here, we are interested in the low-energy behavior of the model, so we focus on the ground states, which we redefine in terms of the effective spins:

$$|\Uparrow\rangle = |\uparrow\uparrow\rangle, \quad |\Downarrow\rangle = |\downarrow\downarrow\rangle$$

As each pair of spins on the z-link collapses to one of these effective spins, we have a square lattice of the weakly interacting effective spins.

We can now treat the contribution from the other two coupling coefficients J_x and J_y as a perturbation to this low-energy theory; the first nonconstant term in the perturbative Hamiltonian emerges at fourth order and is equivalent to the toric code Hamiltonian in its alternative formulation:

$$H_{\text{eff}} = -\frac{J_x^2 J_y^2}{16|J_z|^3} \sum_{\mathbf{p}\in\text{plaquettes}} Q_{\mathbf{p}}$$

where the effective spins are now arranged on a square lattice. The four-spin plaquette operators in this effective-spin basis are

$$Q_p = \tau_{\mathbf{p}}^z \tau_{\mathbf{p}+\mathbf{n}_x}^y \tau_{\mathbf{p}+\mathbf{n}_y}^y \tau_{\mathbf{p}+\mathbf{n}_x+\mathbf{n}_y}^z \tag{14}$$

where $\tau_{\mathbf{p}}^{\alpha}$, $\alpha = x, y, z$, are the Pauli operators for the effective spin at vertex $\mathbf{p}$. This effective Hamiltonian therefore describes the A_z phase in Fig. 5.

The exact same argument would have applied if we had taken either J_x or J_y as strong and treated the other two couplings as weak; by doing so, we can establish the remaining gapped phases (the A_x and A_y) as abelian topological phases equivalent to the toric code at the fourth-order perturbation theory. Each will be characterized by a fourfold-degenerate ground state on the torus.

The remaining B phase is gapless; however, if we include the additional terms to the Hamiltonian that break time-reversal and sublattice symmetry of the model, a spectral gap opens in this phase. Kitaev identified this phase as the non-abelian

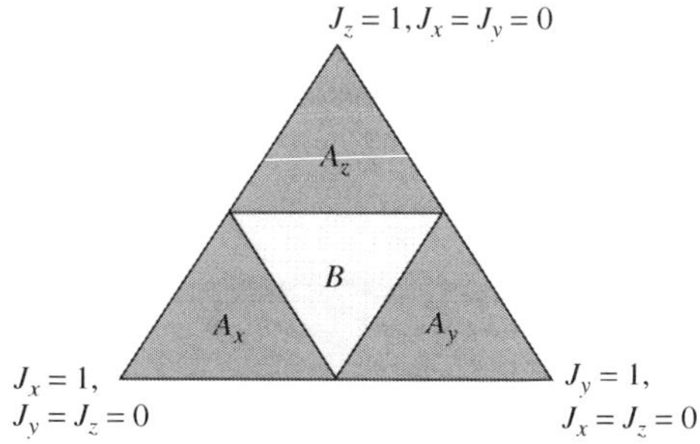

Figure 5. The phase diagram of the Kitaev model exhibits three equivalent abelian topological phases and one phase that, in the absence of the time-reversal and parity-breaking interaction V, is gapless. In the presence of V, this phase acquires a gap and becomes a non-abelian topological phase of the Ising universality class.

topological phase of the Ising universality class by calculating a topological invariant known in the FQH theory as the *spectral Chern number* or the *TKNN invariant* [30]. As we mentioned earlier, this phase is characterized by a threefold-degenerate ground state.

3. *The Effective Spin/Hardcore Boson Representation and Fermionization*

We now present an alternative solution of the honeycomb model, which relies on an intermediate mapping of the model onto a representation given by effective spins and hardcore bosons [31] and a subsequent Jordan–Wigner-type fermionization [24].

As in Fig. 4, we take the unit cell of the model to consist of two spins connected by the z-link. We start by fixing the position of one of these (the origin), and then construct the lattice by translating this cell by integer superpositions of the unit vectors $\mathbf{n}_x$ and $\mathbf{n}_y$, which connect to the nearest z-links in the next row of honeycomb plaquettes. Thus, all cells will be located at $q_x\mathbf{n}_x + q_y\mathbf{n}_y$ for some integers q_x and q_y; we therefore label them by the coordinate $\mathbf{q} = (q_x, q_y)$.

We will take advantage of the fact that the honeycomb lattice is bipartite, namely, its vertices can be divided into two disjoint sets; the sites at the top of a z-link will be denoted by $\circ$ and the ones at the bottom by $\bullet$. We then rewrite the state of any z-link as follows: the orientation of the spin on the black vertex defines the orientation of the effective spin, and we interpret the alignment of the white vertex as the absence (parallel/ferromagnetic) or presence (antiparallel/antiferromagnetic) of a *hardcore boson* on the link. Explicitly,

$$\left|\uparrow_\bullet\uparrow_\circ\right\rangle = \left|\Uparrow, 0\right\rangle, \quad \left|\downarrow_\bullet\downarrow_\circ\right\rangle = \left|\Downarrow, 0\right\rangle$$
$$\left|\uparrow_\bullet\downarrow_\circ\right\rangle = \left|\Uparrow, 1\right\rangle, \quad \left|\downarrow_\bullet\uparrow_\circ\right\rangle = \left|\Downarrow, 1\right\rangle \tag{15}$$

So if both spins are up, we say the cell has effective spin up and no hardcore boson present; if the black spin is down and the white spin up, this state has a

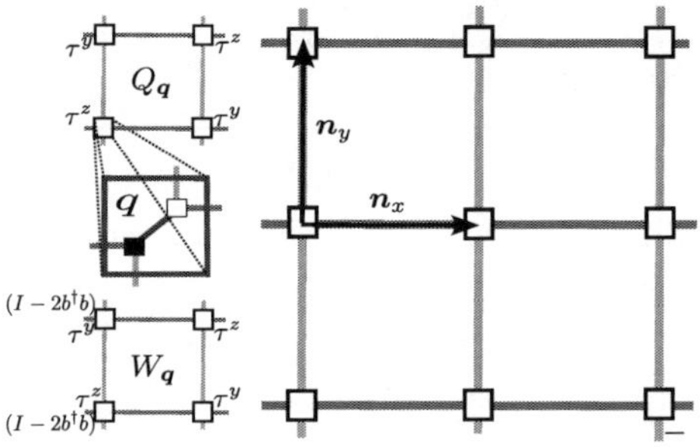

Figure 6. The effective spin/hardcore boson representation of the Kitaev model. The z-links of the original honeycomb lattice are contracted to vertices of the square lattice, which carry both effective spin and hardcore boson degrees of freedom. The vortex operators factorize into the product of the toric code plaquette operators and the hardcore boson operators. Reproduced from Fig. 3 of Ref. [24].

single hardcore boson with effective spin down. This transformation is illustrated in Fig. 6.

Although this may at first appear somewhat strange, it is simply a change of basis of our two-spin state. For instance, you can show that the operator $b_{\mathbf{q}}$, which changes a 1 to a 0 on the right-hand side of expression (15), is

$$b_{\mathbf{q}} = \frac{\sigma^x_{\mathbf{q},\circ} + i\sigma^z_{\mathbf{q},\bullet}\sigma^y_{\mathbf{q},\circ}}{2} \tag{16}$$

and the one that changes a 0 to a 1 is

$$b^\dagger_{\mathbf{q}} = \frac{\sigma^x_{\mathbf{q},\circ} - i\sigma^z_{\mathbf{q},\bullet}\sigma^y_{\mathbf{q},\circ}}{2} \tag{17}$$

Similarly, the Pauli matrix $\tau^x_{\mathbf{q}}$ must be the one that exchanges $\Uparrow$ and $\Downarrow$ in the effective-spin representation.

In summary, the transformations that give the two-spin representation from the effective-spin representation are

$$\begin{aligned} \sigma^x_{\mathbf{q},\bullet} &= \tau^x_{\mathbf{q}}(b^\dagger_{\mathbf{q}} + b_{\mathbf{q}}), \quad \sigma^x_{\mathbf{q},\circ} = b^\dagger_{\mathbf{q}} + b_{\mathbf{q}} \\ \sigma^y_{\mathbf{q},\bullet} &= \tau^y_{\mathbf{q}}(b^\dagger_{\mathbf{q}} + b_{\mathbf{q}}), \quad \sigma^y_{\mathbf{q},\circ} = i\tau^z_{\mathbf{q}}(b^\dagger_{\mathbf{q}} + b_{\mathbf{q}}) \\ \sigma^z_{\mathbf{q},\bullet} &= \tau^z_{\mathbf{q}}, \quad \sigma^z_{\mathbf{q},\circ} = \tau^z_{\mathbf{q}}(I - 2b^\dagger_{\mathbf{q}}b_{\mathbf{q}}) \end{aligned} \tag{18}$$

For example, we see that flipping the spin on the white vertex by application of $\sigma^x_{\mathbf{q},\circ}$ translates into creating or annihilating one hardcore boson, since it flips either from antiferromagnetic, to ferromagnetic, or vice versa.

The term *hardcore boson* may seem to be completely unmotivated, but using the properties of the Pauli matrices, it is easily shown from Eqs. (16) and (17) that for $\mathbf{q} \neq \mathbf{q}'$,

$$\left[b^\dagger_{\mathbf{q}}, b_{\mathbf{q}'}\right] = \left[b^\dagger_{\mathbf{q}}, b^\dagger_{\mathbf{q}'}\right] = \left[b_{\mathbf{q}}, b_{\mathbf{q}'}\right] = 0$$

but

$$\left\{b_{\mathbf{q}}^{\dagger}, b_{\mathbf{q}}\right\} = I$$

Because the operators *commute* when we compare different cells, they are bosonic. However, the *anticommutation* relation for a single cell has a fermionic form. This means that we cannot put more than one of these bosons at a single point; it has a *hard core*. (All of these facts will become relevant when we apply a Jordan–Wigner-type fermionization to the model in the effective spin/hardcore boson representation.)

One of the important consequences of this change of representation is a change of the lattice. Each unit cell of the original honeycomb system, which carried two spins, can be contracted to a single vertex that now carries an effective spin and is occupied by either zero or one hardcore boson. These four new degrees of freedom are now located on the vertices of a square lattice. This procedure is similar to the one we employed to relate the abelian phase of the model to the toric code, although in that case we restricted ourselves only to the low-energy part of the energy spectrum, which contained only the ferromagnetic states. This new mapping ensures that *all* of the original spin states are represented. We point out that the effective spins and hardcore bosons are independent degrees of freedom, so the operators that will represent them will commute.

Using (18), we can now write the original Hamiltonian H_0 from Eq. (11) in the effective spin/hardcore boson representation; it is

$$\begin{aligned} H = & -J_x \sum_{\mathbf{q}} \left(b_{\mathbf{q}}^{\dagger} + b_{\mathbf{q}}\right) \tau_{\mathbf{q}+\mathbf{n}_x}^{x} \left(b_{\mathbf{q}+\mathbf{n}_x}^{\dagger} + b_{\mathbf{q}+\mathbf{n}_x}\right) \\ & - j_y \sum_{\mathbf{q}} i\tau_{\mathbf{q}}^{z} \left(b_{\mathbf{q}}^{\dagger} - b_{\mathbf{q}}\right) \tau_{\mathbf{q}+\mathbf{n}_y}^{y} \left(b_{\mathbf{q}+\mathbf{n}_y}^{\dagger} + b_{\mathbf{q}+\mathbf{n}_y}\right) \\ & - J_z \sum_{\mathbf{q}} \left(I - 2b_{\mathbf{q}}^{\dagger} b_{\mathbf{q}}\right) \end{aligned}$$

The vortex operators in the new representation are

$$W_{\mathbf{q}} = \left(I - 2N_{\mathbf{q}}\right)\left(I - 2N_{\mathbf{q}+\mathbf{n}_y}\right) Q_{\mathbf{q}}$$

where $N_{\mathbf{q}} = b_{\mathbf{q}}^{\dagger} b_{\mathbf{q}}$ is the hardcore boson number operator and the operator $Q_{\mathbf{q}}$ is the toric code operator defined for the square plaquette with the vertices ($\mathbf{q}$, $\mathbf{q} + \mathbf{n}_x$, $\mathbf{q} + \mathbf{n}_y$, $\mathbf{q} + \mathbf{n}_x + \mathbf{n}_y$):

$$Q_{\mathbf{q}} = \tau_{\mathbf{q}}^{z} \tau_{q+\mathbf{n}_x}^{y} \tau_{\mathbf{q}+\mathbf{n}_y}^{y} \tau_{\mathbf{q}+\mathbf{n}_x+\mathbf{n}_y}^{z}$$

The boson number and the plaquette operators allow us to choose a convenient basis for the states of the Kitaev model within the effective spin/hardcore boson picture: because these operators represent independent degrees of freedom and thus can be simultaneously diagonalized, a basis can be written [24,32] in terms of their eigenstates. For instance, on an infinite plane, this basis is

$$\mathcal{B}_{\text{plane}} = \left\{ \left| \{\zeta_{\mathbf{q}}\}, \{\eta_{\mathbf{q}}\} \right\rangle \right\}$$

where $\zeta_{\mathbf{q}} \in \{-1, 1\}$ and $\eta_{\mathbf{q}} \in \{0, 1\}$ are the eigenvalues of $Q_{\mathbf{q}}$ and $N_{\mathbf{q}}$, respectively.

On a torus, the situation is quite similar, although in addition to the boson number and plaquette operators, we also have topologically nontrivial loop operators. These are formed by chains of spin operators that wrap around the meridian or longitude of a torus and commute with the Hamiltonian much like the vortex operators do. We denote these by $L_x^{(0)}$ and $L_y^{(0)}$, where the superscript reflects the fact that we define them along loops that pass through the vertex $\mathbf{q} = (0, 0)$, the origin. The basis states in this case are

$$\mathcal{B}_{\text{torus}} = \left\{ \left| \{\zeta_{\mathbf{q}}\}, \{\eta_{\mathbf{q}}\}, \ell_x^{(0)}, \ell_y^{(0)} \right\rangle \right\}$$

where $\ell_x^{(0)}$ and $\ell_y^{(0)}$ are the eigenvalues of the operators $L_x^{(0)}$ and $L_y^{(0)}$, respectively. Their possible values (± 1) reflect whether the state on the torus is characterized by symmetric or antisymmetric periodic boundary conditions in one or the other topologically nontrivial direction.

In this basis, the ground state of the model in the abelian sector of the phase diagram is inherited from the four kets

$$\left\{ \left| \{1\}, \{0\}, \pm 1, \pm 1 \right\rangle \right\} \tag{19}$$

The ground state is therefore be labeled by the four possible choices for the eigenvalues $(\ell_x^{(0)}, l_y^{(0)})$ of the topologically nontrivial symmetry operators, and so is fourfold-degenerate on the torus, a genus $G = 1$ surface. On a plane, whose genus is $G = 0$, the ground state is nondegenerate due to the absence of the topologically nontrivial symmetries. We see explicitly that the properties of the abelian topological phase of the Kitaev model depend on the topology of the surface on which the phase is realized.

4. *Fermionization with Jordan–Wigner Strings*

We now proceed to "fermionize" the model using the Jordan–Wigner-type procedure; the objective is to turn our hardcore bosons into fermions satisfying the fermionic commutation relations

$$\{c_{\mathbf{q}}^{\dagger}, c_{\mathbf{q}'}\} = \delta_{\mathbf{q}\mathbf{q}'}, \quad \{c_{\mathbf{q}}^{\dagger}, c_{\mathbf{q}'}^{\dagger}\} = \{c_{\mathbf{q}}, c_{\mathbf{q}'}\} = 0 \tag{20}$$

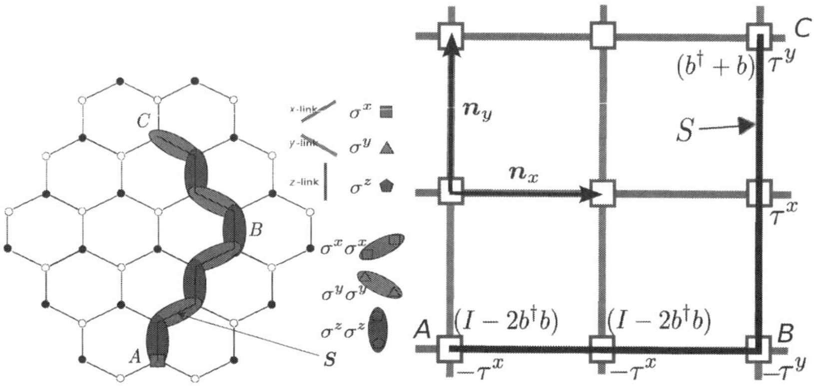

Figure 7. The string operators in the original and the effective spin/hardcore boson representations. Choosing a string convention is equivalent to choosing a gauge; this particular choice results in a Landau gauge. Reproduced from Figs. 1 and 3 of Ref. [24].

by attaching a suitable string of operators to the hardcore bosons. The final result will be a Hamiltonian, which is quadratic in the fermionic operators; we will then be able to solve the model using the Bogoliubov–de Gennes (BdG) technique developed within the theory of superconductivity, the subject of Appendix C.

In order to construct the string operators, we first start with the original honeycomb model: we choose one vertex as the origin $\mathbf{0} = (1, 1)$ and apply the Pauli matrix $\sigma_{\mathbf{0}}^x$ to the spin at this vertex. The rest of the string is made by applying the operators $K_{i,j}^{\alpha}$: we start first with alternating $K_{i,j}^z$ and $K_{j,k}^x$ and, after reaching the required length, we then apply alternating $K_{\ell,m}^z$ and $K_{m,n}^y$ until we reach the black site of the unit cell at $\mathbf{q} = (q_x, q_y)$ (see Fig. 7).

Each resulting string operator $S_{\mathbf{q}}$ squares to unity while different operators $S_{\mathbf{q}}$ and $S_{\mathbf{q}'}$ anticommute with each other. This leads us to define fermionic creation and annihilation operators $c_{\mathbf{q}}^\dagger$ and $c_{\mathbf{q}}$ by

$$S_{\mathbf{q}} = c_{\mathbf{q}}^\dagger + c_{\mathbf{q}} = \left(b_{\mathbf{q}}^\dagger + b_{\mathbf{q}}\right) S_{\mathbf{q}}'$$

where $S_{\mathbf{q}}'$ is simply the string $S_{\mathbf{q}}$ with the bosonic dependence of the end-point C removed (see Table I). Individually, our fermionic canonical creation and annihilation operators are

$$c_{\mathbf{q}}^\dagger = b_{\mathbf{q}}^\dagger S_{\mathbf{q}}', \quad c_{\mathbf{q}} = b_{\mathbf{q}} S_{\mathbf{q}}'$$

where the strings now insure that the operators $c^\dagger$ and c obey the canonical fermionic anticommutation relations (20).

TABLE I
The String S as Four Unique Segments

String Segment	Two-Spin Basis	Effective Spin/Hardcore Boson Basis
$[A, B)$	$\sigma^x_\circ \sigma^z_\circ \sigma^z_\bullet \sigma^x_\bullet$	$-\tau^x(I - 2b^\dagger b)$
B	$\sigma^y_\circ \sigma^z_\circ \sigma^z_\bullet \sigma^x_\bullet$	$-\tau^y$
(B, C)	$\sigma^y_\circ \sigma^z_\circ \sigma^z_\bullet \sigma^y_\bullet$	τ^x
C	$\sigma^y_\bullet$	$\tau^y(b^\dagger + b)$

Although bosons are only created/destroyed at the endpoint C of the string, the sites in the $[A, B)$ interval also have nontrivial bosonic dependence.

The Hamiltonian for the bare system can now be rewritten in a quadratic fermonic form by using the relations $b^\dagger_{\mathbf{q}} = c^\dagger_{\mathbf{q}} S'_{\mathbf{q}}$ and $b_{\mathbf{q}} = c_{\mathbf{q}} S'_{\mathbf{q}}$:

$$
\begin{aligned}
H_0 = {} & J_x \sum_{\mathbf{q}} X_{\mathbf{q}} \left(c^\dagger_{\mathbf{q}} - c_{\mathbf{q}}\right) \left(c^\dagger_{\mathbf{q}+\mathbf{n}_x} + c_{\mathbf{q}+\mathbf{n}_x}\right) \\
& + J_y \sum_{\mathbf{q}} Y_q \left(c^\dagger_{\mathbf{q}} - c_{\mathbf{q}}\right) \left(c^\dagger_{\mathbf{q}+\mathbf{n}_y} + c_{\mathbf{q}+\mathbf{n}_y}\right) \\
& + J_z \sum_{\mathbf{q}} \left(2c^\dagger_{\mathbf{q}} c_{\mathbf{q}} + I\right)
\end{aligned} \tag{21}
$$

(The time-reversal breaking potential V from Eq. (12) can also be rewritten in terms of these fermions; see Appendix D.)

The operators $X_{\mathbf{q}}$ and $Y_{\mathbf{q}}$ carry information about vortices and, on a torus, the topologically nontrivial symmetries. The Xs result from the product of two underlying strings $S'_{\mathbf{q}}$ leading to the sites that are connected by x-links on the original lattice. As the two strings are the same up to the turning point of the shorter string, their product results in a closed loop that can be expressed as a product of underlying vortex operators. On the plane, the $X_{\mathbf{q}}$ are thus given in terms of the vortex operators as

$$
X_{(q_x, q_y)} = \prod_{i_y=1}^{q_y - 1} W_{(q_x, i_y)} \tag{22}
$$

and all the Y-operators are equal to the identity.

To specify the Hamiltonian for a particular vortex configuration, one replaces the vortex operators in the preceding equation by their eigenvalues. Of course, the particular Jordan–Wigner convention used to define the fermions is directly responsible for how vorticity is encoded in the system; for the string convention chosen in Ref. [24], the vorticity is encoded in the fermionic Hamiltonian through Eq. (22).

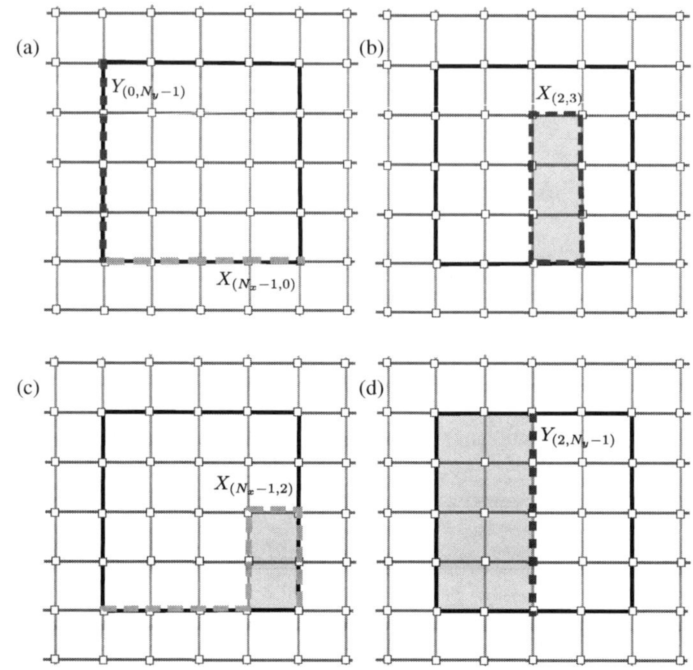

Figure 8. Vortex operators on a torus. The condition $\prod_{\mathbf{q}} W_{\mathbf{q}} = 1$ implies that we have $N/2 - 1$ independent vortex operators, where N is the number of spins of the original lattice. In addition, two topologically nontrivial loops are given by operators $L_x^{(0)}$ and $L_x^{(0)}$. (**a**) $X_{(N_x-1,0)} = -l_x^{(0)}$, $Y_{(0,N_y-1)} = -l_y^{(0)}$; (**b**) $X_{(2,3)} = \prod_{q_y=0}^{2} W_{(2,q_y)}$; (**c**) $X_{(N_x-1,2)} = -l_x^{(0)} \prod_{q_y=0}^{1} W_{(N_x-1,q_y)}$; (**d**) $Y_{(2,N_y-1)} = -l_y^{(0)} \prod_{q_x=0}^{1} \prod_{q_y=0}^{N_y-1} W_{(q_x,q_y)}$. Reproduced from Fig. 4 of Ref. [24].

On a torus, additional topologically nontrivial degrees of freedom also need to be incorporated in a manner consistent with Eq. (22). These degrees of freedom are encoded, via the eigenvalues of the loop operators $L_x^{(0)}$ and $L_x^{(0)}$, into the $X_{\mathbf{q}}$ and $Y_{\mathbf{q}}$ values only at the boundary of the system [24]; some examples appear in Fig. 8. These consistency relations have an interesting pictorial representation: for any vortex arrangement, we see that there are lines of $X_{\mathbf{q}} = -1$ and $Y_{\mathbf{q}} = -1$, or *branch cuts*, that together connect vortices in pairs. Figure 9 illustrates this phenomenon.

If the torus is an $N_x \times N_y$ lattice, the edges $q_x = 1$ and $q_x = N_x + 1$ are identified, as are $q_y = 1$ and $q_y = N_y + 1$. This periodicity suggests replacing the operators by their Fourier transforms $c_{\mathbf{q}} = (N_x N_y)^{-1/2} \sum_{\mathbf{k}} c_{\mathbf{k}} e^{i\mathbf{k}\cdot\mathbf{q}}$, where $\mathbf{k} = (k_x, k_y)$ is the momentum. If we make the further assumption that all the coupling

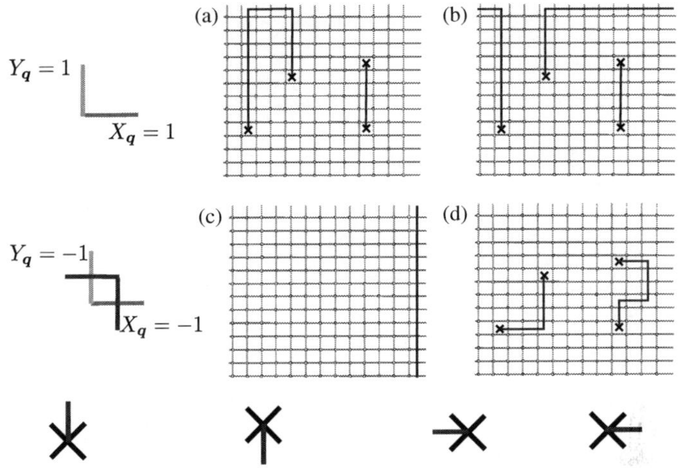

Figure 9. Vortices always appear at the end of a branch cut. Parts (a), (b), and (c) are real vortex configurations. Note that on a torus the topological sectors can also be encoded as branch cuts and their values dictate which vortices are connected to each other. Reproduced from Fig. 2 of Ref. [27].

coefficients in V are position-independent and equal to κ, then the Hamiltonian becomes (after antisymmetrization)

$$H = \frac{1}{2}\sum_{\mathbf{k}} \begin{pmatrix} c_{\mathbf{k}}^{\dagger} & c_{-\mathbf{k}} \end{pmatrix} H_{\mathbf{k}} \begin{pmatrix} c_{\mathbf{k}} \\ c_{-\mathbf{k}}^{\dagger} \end{pmatrix}$$

where $H_{\mathbf{k}}$ is the matrix

$$H_{\mathbf{k}} = \begin{pmatrix} \xi_{\mathbf{k}} & \Delta_{\mathbf{k}} \\ \Delta_{\mathbf{k}}^{*} & -\xi_{-\mathbf{k}} \end{pmatrix}$$

and

$$\begin{aligned} \Delta_{\mathbf{k}} &= 4\kappa\left[\sin(k_x) - \sin(k_y) - \sin(k_x - k_y)\right] + i\left[2J_x \sin(k_x) + 2J_y \sin(k_y)\right] \\ \xi_{\mathbf{k}} &= 2J_x \cos(k_x) + 2J_y \cos(k_y) + 2J_z \end{aligned} \tag{23}$$

At this point, we can use the BdG method outlined in Appendix C by defining

$$u_{\mathbf{k}} = \sqrt{\frac{1}{2}\left(1 + \frac{\xi_{\mathbf{k}}}{E_{\mathbf{k}}}\right)}, \qquad v_{\mathbf{k}} = -e^{i\arg(\Delta_{\mathbf{k}})}\sqrt{\frac{1}{2}\left(1 - \frac{\xi_{\mathbf{k}}}{E_{\mathbf{k}}}\right)}$$

where $E_{\mathbf{k}} = \sqrt{\xi_{\mathbf{k}}^2 + |\Delta_{\mathbf{k}}|^2}$, we can transform c and $c^\dagger$ to

$$\gamma_{\mathbf{k}} = -v_{\mathbf{k}} c_{-\mathbf{k}}^\dagger + u_{\mathbf{k}} c_{\mathbf{k}}, \quad \gamma_{\mathbf{k}}^\dagger = -v_{\mathbf{k}}^* c_{-\mathbf{k}} + u_{\mathbf{k}} c_{\mathbf{k}}^\dagger$$
$$\gamma_{-\mathbf{k}} = v_{\mathbf{k}} c_{\mathbf{k}}^\dagger + u_{\mathbf{k}} c_{-\mathbf{k}}, \quad \gamma_{-\mathbf{k}}^\dagger = v_{\mathbf{k}}^* c_{\mathbf{k}} + u_{\mathbf{k}}^* c_{-\mathbf{k}}^\dagger$$

to obtain

$$H = \sum_{\mathbf{k}} E_{\mathbf{k}} \left(\gamma_{\mathbf{k}}^\dagger \gamma_{\mathbf{k}} - \frac{1}{2} \right) \tag{24}$$

The usefulness of this Hamiltonian is not only in its particularly simple form, but also in the fact that it allows us to confirm that the ground states of the system are given in terms of (19) as Bardeen–Cooper–Schreiffer (BCS) superconductor states [33]:

$$\left| \text{g.s.} \right\rangle = \prod \left(u_{\mathbf{k}} + v_{\mathbf{k}} c_{\mathbf{k}}^\dagger c_{-\mathbf{k}}^\dagger \right) \left| \{1\}, \{0\}, \pm 1, \pm 1 \right\rangle \tag{25}$$

where the product is over distinct ($\mathbf{k}$, $-\mathbf{k}$) pairs. This state is annihilated by the operator $\gamma_{\mathbf{k}}$, and thus it has the lowest energy possible, $E_{\text{g.s.}} = -\sum_{\mathbf{k}} E_{\mathbf{k}}/2$.

This expression is a closed form expression of the ground state that does not require any additional modifications. It is noteworthy because it combines two powerful wave function descriptors: the BCS product and the stabilizer formalism. In Eq. (25), which is valid everywhere in the model's parameter space, the fermionic vacuum is fixed to be the toric code ground state.

Even though this implies that any mechanism for switching between the abelian and non-abelian topological phases must be contained exclusively within the BCS product, we should also recognize that the abelian phase is fourfold-degenerate due to the two loop symmetries. (This is referred to as a "$\mathbb{Z}_2 \times \mathbb{Z}_2$" phase to reflect the fact that both $\ell_x^{(0)}$ and $\ell_y^{(0)}$ can be either $+1$ or -1.) This degeneracy is *not* because of any mysterious property of the BCS product; to see this more clearly, note that in the A_z phase with $J_z = 1$ and $J_{x,y} \to 0$, we have $u_{\mathbf{k}} \to 1$ and $v_{\mathbf{k}} \to 0$. The ground state of $\left| \text{g.s.} \right\rangle$ therefore approaches (19), where all the Q-operators, and hence all the vortex operators $W_{\mathbf{q}}$, are unity. This is, of course, what one expects from the perturbation theory (see, e.g., Ref. [34]).

5. *Topological Invariants of the Model*

So far, we have analyzed the Kitaev honeycomb model using well-known techniques from quantum field theory, but even though we have referred to the different topologies of our system (e.g., an infinite plane versus a torus), it may not be obvious how this model connects to our earlier, more mathematical, discussion of

TQFTs. We now want to briefly discuss two numerical quantities (topological invariants) that will make this connection more concrete.

a. The Spectral Chern Invariant. Different topological phases of a model will be described by different topological invariants, and so finding such invariants can be a major step in identifying phase transitions and to getting important insights into the nature of these phases. One such invariant for the Kitaev honeycomb model is the *spectral Chern invariant* [3,30]. It is an integer-valued invariant defined as

$$\nu = \frac{1}{2\pi i}\int d^2\mathbf{k}\,\mathrm{tr}\left[P(\mathbf{k})\cdot\left(\frac{\partial P}{\partial k_x}\cdot\frac{\partial P}{\partial k_y}-\frac{\partial P}{\partial k_y}\cdot\frac{\partial P}{\partial k_x}\right)\right]$$

where $P(\mathbf{k})$ is a matrix that projects an arbitrary state onto the negative-energy eigenvectors of $H_{\mathbf{k}}$. Chern invariants calculated in this way are in complete agreement with those calculated using Kitaev's fermionization procedure [3].

The demonstrations that this number is both integer-valued and a topological invariant are similar to the analogous proofs we made for the action and path integral of Chern–Simons theory in Section II.B.2. In fact, this spectral Chern invariant is closely related to the first Chern number (7) we discussed in that section, so this is no coincidence.

As to how we can use this quantity, we note that if the form of $H_{\mathbf{k}}$ is such that there are several negative energy bands, then each band contributes its own Chern number to the total Chern invariant. (The nonzero eigenvalues of the Hamiltonian come in positive and negative pairs, so the Chern numbers of the completely positive energy bands are the same up to the sign and thus contain the same information.)

Thus, one of the ways we can distinguish topologically inequivalent honeycomb models is by looking at their energy bands; if two models have bands giving different spectral Chern invariants, they cannot be related by a simple continuous deformation of one into the other. They are truly topologically distinct.

b. The Ground State Degeneracy on a Torus. Basic quantum mechanics tells us that if our honeycomb model is on a torus, the periodic or antiperiodic boundary conditions that a state must satisfy restrict the values that the momentum $\mathbf{k}$ can have. The boundary conditions are encapsulated by the loop symmetries: $\ell_x^{(0)} = -1$, for example, requires that a state be periodic in the x-direction, while $\ell_y^{(0)} = +1$ says it is antiperiodic in the y-direction. For a $(N_x \times N_y)$-dimensional torus, the allowed values for k_x and k_y are

$$k_\alpha = \frac{\pi}{N_\alpha}\left(\frac{1+\ell_\alpha^{(0)}}{2}+2n_\alpha\right), \quad n_\alpha = 0, 1, \ldots N_\alpha - 1$$

In the abelian topological phases of the model on torus, we have four distinct copies of the BCS ground state, one for each vacuum state on which the product acts. The energies of these ground states, though not exactly the same due to the different momenta, converge as the system approaches the thermodynamic limit $N_{x,y} \to \infty$. Explicitly, the ground states are given by Eq. (25), where we emphasize that the vacuum states therein are the toric code states of the previous section defined on the *effective* spins.

In the non-abelian phase, the situation is more involved. For simplicity, we will focus on the regime where J_z is negative and both J_x and J_y are positive. In the doubly periodic sector where $\ell_x^{(0)} = \ell_y^{(0)} = -1$, $\mathbf{k} = \mathbf{0}$ is a possible value of the momentum and is therefore included in the product. But notice that the gap function $\Delta_{\mathbf{k}}$ from Eq. (23) vanishes at zero momentum. Moreover, the ratio $\xi_{\mathbf{0}}/E_{\mathbf{0}}$ flips abruptly from $+1$ to -1 as we cross this phase transition. This implies that $u_{\mathbf{0}} = 0$ and, because $c_{\mathbf{0}}^\dagger c_{\mathbf{0}}^\dagger = 0$ by the Pauli exclusion principle, the BCS ground state will be identically zero. This outcome is exactly what happens in the countinuum-model treatment of the spinless p-wave superconductor in Ref. [35].

The $\mathbf{k} = \mathbf{0}$ fermions are the ones behind this, so excluding them might allow us to remedy the situation [24]. Hence, we define a new state

$$\left|\psi\right\rangle = \prod_{\mathbf{k} \neq \mathbf{0}} (u_{\mathbf{k}} + v_{\mathbf{k}} c_{\mathbf{k}}^\dagger c_{-\mathbf{k}}^\dagger) \left| \{1\}, \{0\}, -1, -1 \right\rangle$$

What is the energy of this state? It is annihilated by all the $\gamma_{\mathbf{k}}$ operators for $\mathbf{k} \neq 0$ but is *not* annihilated by γ_0; thus, Eq. (24) gives its energy as

$$\begin{aligned} E_\psi &= \sum_{\mathbf{k} \neq 0} E_{\mathbf{k}} \left(0 - \frac{1}{2}\right) + E_0 \left(1 - \frac{1}{2}\right) \\ &= E_{\text{g.s.}} + E_0 \end{aligned}$$

Our assumption that the system is gapped means that the second-lowest energy differs from the lowest by a nonzero amount, so no matter what E_0 might be, $E_\psi > E_{\text{g.s.}}$.

Therefore, the B-phase of this model has only a *threefold* degeneracy in the ground state, because the $\ell_x^{(0)} = \ell_y^{(0)} = -1$ BCS state is "lifted" to a higher energy. It is a nice rediscovery of the TQFT/CFT result from Ref. [22], which we discussed in Section II.C.3.

This ground-state degeneracy is precisely the dimension N_Σ of the Hilbert space $\mathcal{H}(\Sigma)$ that is required in any TQFT and is a topological invariant. We have just shown that $N_\Sigma = 4$ for an A-phase honeycomb model and $N_\Sigma = 3$ for a B-phase one. Therefore, our claim that the phase transition between the two is a topological one is justified: different topological invariants imply different topologies.

C. Other Trivalent Kitaev Lattice Models

The key feature of the Kitaev honeycomb model is that every site interacts with exactly three other sites; it is *trivalent*. This follows directly from the hexagonal shape of the lattice.

There are, unsurprisingly, other lattices that have this property; is it possible that the same techniques that are used in solving the honeycomb model can be extended to such lattices? In this section, we look at two other trivalent lattices—the Yao–Kivelson and square–octagon models—and show that we can.

These models are characterized not just by their trivalence, but also by the fact that they are all two-dimensional and all sites have spin-$1/2$. We point out that by allowing different spins, we can extend these techniques to higher dimensions and higher-valence lattices. However, in this brief review, we will focus on the two-dimensional lattice models where the extension of the methodology presented has already been established.

1. *The Yao–Kivelson Model*

By an appropriate tessellation of the plane by triangles and dodecagons, we can construct a trivalent lattice, which is known as the *Yao–Kivelson (YK) lattice model* [25]. This model is remarkable because it exhibits spontaneous breaking of time-reversal symmetry: although the Hamiltonian itself is time-reversal invariant, Yao and Kivelson established the existence of abelian and non-abelian phases of particular vortex configurations, which are *not* symmetric under time-reversal.

Using pertubation theory, Ref. [36] showed that the abelian phase was yet again mappable to a toric code-like system and that the non-abelian phase was mappable to the non-abelian phase of the honeycomb lattice, with the time-reversal symmetry breaking terms appearing at higher orders in the theory.

The Hamiltonian consists of directional spin–spin interactions on the lattice. We use the representation of the model introduced in Ref. [36] as it provides a straightforward route, by contracting the Z-links, to the definition of the fermions as toric code states. In this representation, the Hamiltonian can be written as

$$\begin{aligned} H &= H_Z + H_J + H_K + H_L \\ &= -Z \sum_{Z\text{-links}} \sigma_i^z \sigma_j^z - J \sum_{J\text{-links}} \sigma_i^x \sigma_j^y \\ &\quad - K \sum_{K\text{-links}} \sigma_i^x \sigma_j^y - L \sum_{L\text{-links}} \sigma_i^x \sigma_j^y \end{aligned}$$

where the different links are illustrated in Fig. 10. It should be understood that the Z-links connect separate triangles and the J-, K-, and L-links within the triangles are the positive, zero, and negative slopes, respectively. We refer to triangles that

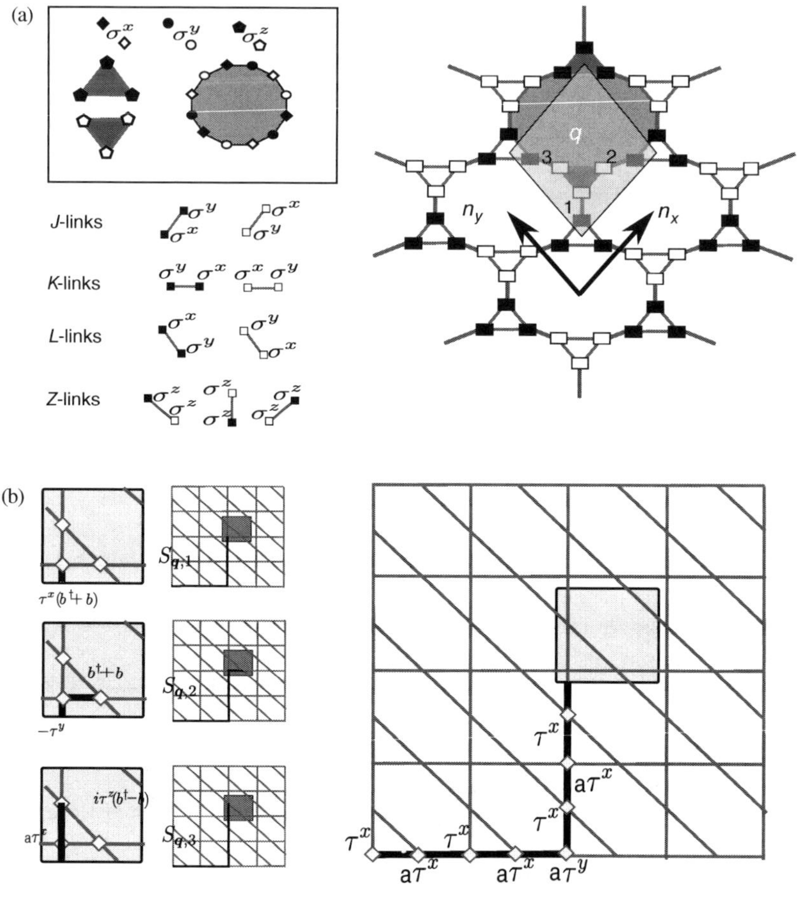

Figure 10. The Yao–Kivelson lattice. (**a**) The star lattice is a hexagonal lattice with each vertex replaced by a triangle, so the resulting lattice is triangle–dodecagon. The triangular symmetries are responsible for the spontaneous breaking of time-reversal symmetry [3,25]. (**b**) The lattice in the effective spin/hardcore boson representation with the definition of the fermionic string operators. Reproduced from Figs. 1 and 3 of Ref. [26].

point up as “black” and those that point down as “white” triangles; the sites on these triangles in the original lattice are colored black and white accordingly.

We define a basic unit cell of the lattice around a white triangle (so in a system of $6N$ spins, there are N unit cells). We label the Z-link at the bottom of the triangle with $n = 1$, the Z-link from the top right with $n = 2$ and the Z-link from the top left with $n = 3$. Each spin site can be specified using the position vector $\mathbf{q}$, the index n, and whether it is on a $\circ$ or a $\bullet$ site.

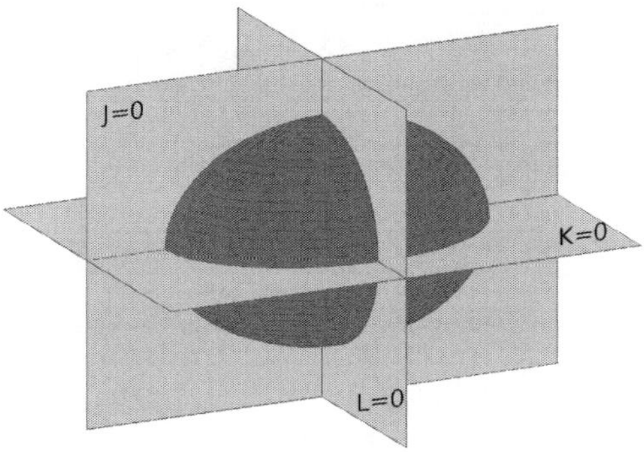

Figure 11. Schematic of the system phase diagram. The surface of the sphere of radius Z indicates the critical boundary between the abelian and non-abelian phases. Inside the sphere is a gapped abelian phase. Outside the sphere we are in a gapped non-abelian phase, provided we are not on the $J = 0$, $K = 0$, or $L = 0$ planes, where the system is gapless. Reproduced from Fig. 7 of Ref. [26].

The model can be solved in much the same way as the honeycomb model, via either the Majorana fermion method or by a Jordan–Wigner transformation; the second of these shows that the time-reversal symmetry is spontaneously broken at the level of the vacuum [26]. To summarize the results, the phase diagram and spectral gap of the model are shown in Figs. 11 and 12, respectively.

2. *The Square–Octagon Model*

The trivalent *square–octagon lattice model* was first discussed in Ref. [37], where it was found that the model exhibits two abelian phases with vanishing spectral Chern number and also $\nu = \pm 1$ phases that open up near the critical boundary between the abelian ones when one includes terms breaking the time-reversal symmetry. A much richer phase diagram with an abundance of abelian and non-abelian phases, with spectral Chern numbers $0, \pm 1, \pm 2, \pm 3, \pm 4$ (Fig. 13) was

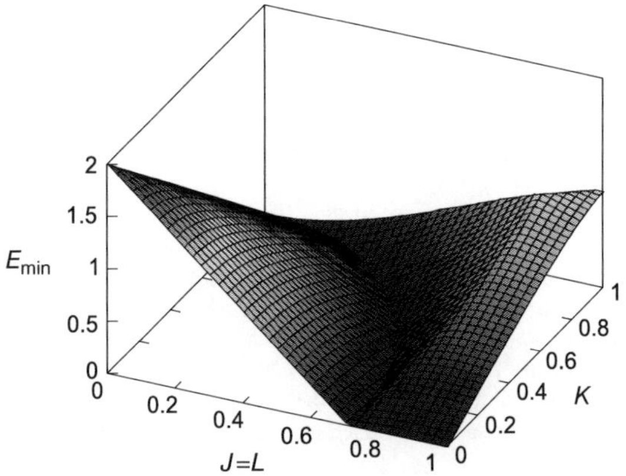

Figure 12. The minimum energy gap of the vortex-free sector for $Z = 1$. The critical point can be seen along the $J^2 + K^2 + L^2 = 1$ line. The system is gapless when $J = L > 1/\sqrt{2}$ and $K = 0$. More generally, if any of the parameters J, K, or L vanish, the system is gapless beyond the phase transition. Reproduced from Fig. 5 of Ref. [26].

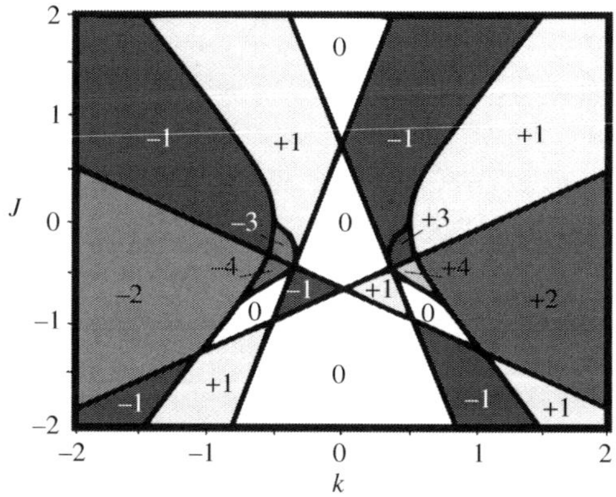

Figure 13. The extended phase diagram of the square–octagon model. Here, the parameters of the model are $J_z = 1$, $J = J_x = J_y$, and $\kappa_1 = \kappa_2$. The phases with odd spectral Chern number are non-abelian topological phases: $\nu = \pm 1$ characterizes the Ising topological phase and $\nu = \pm 3$ is the $SU(2)_2$ Chern–Simons phase. (*See the color version of this figure in the Color Plates section.*)

observed later [28] . Furthermore, the abelian $\nu = 0$ phases can both be mapped to toric code-like Hamiltonians.

The square–octagon lattice is a planar trivalent graph whose unit cells are formed by square plaquettes, with each unit cell connected corner-to-corner to four neighboring ones, as in Fig. 14a. The model has spins located at the vertices of the lattice, which interact through a Hamiltonian that is formally the same as

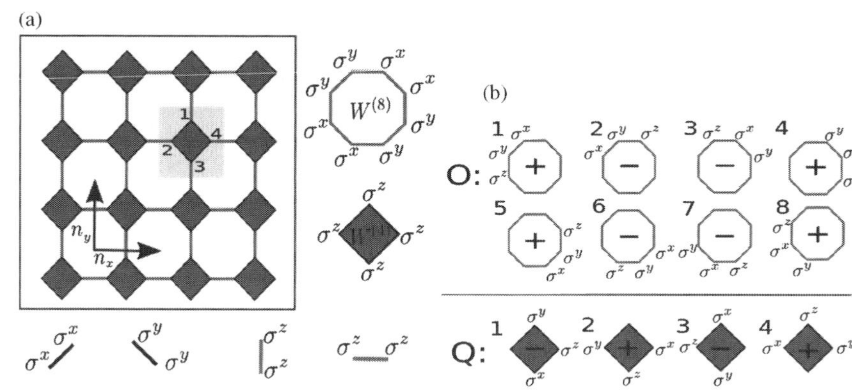

Figure 14. The square–octagon model. (**a**) The spins are located on the vertices of the model and are coupled via $\sigma^z\sigma^z$ interaction along the horizontal and vertical links (z-links) and via alternating $\sigma^x\sigma^x$ and $\sigma^y\sigma^y$ interactions along the diagonal links. The unit cell consists of four spins, that are located at the vertices of a square plaquette; their numbering, from the top counterclockwise, is relevant to the fermionization of the model. The vortex operators, defined on square and octagon plaquettes as shown, commute with the lattice Hamiltonian. (**b**) The time-reversal breaking terms of the square–octagon model are three-body spin terms constructed by overlapping the two interaction terms of the Hamiltonian on the underlying links with the sign of each term fixed as indicated. Reproduced from Figs. 1 and 2 of Ref. [28].

that of the Kitaev honeycomb model and the Yao–Kivelson model:

$$H_0 = -J_z \sum_{z-\text{links}} \sigma_i^z \sigma_j^z - J_x \sum_{x-\text{links}} \sigma_i^x \sigma_j^x - J_y \sum_{y-\text{links}} \sigma_i^y \sigma_j^y$$

where the z-links are those connecting the squares, while the x- and y-links make up the square plaquettes. In analogy with the honeycomb model, one defines the vortex operators $W^{(8)}$ and $W^{(4)}$ as the product of the appropriate Pauli operators around the octagonal and square plaquettes, respectively. These commute with the Hamiltonian, so energy eigenvectors will also have vortex operator eigenvalues ± 1.

As in the honeycomb system, we must augment the square–octagon Hamiltonian with time-reversal symmetry-breaking terms. Eight "octagon operators" O and four "square operators" Q, all of the general form $\sigma_i^\alpha \sigma_j^\beta \sigma_k^\gamma$ and defined in Fig. 14b, contribute to the total Hamiltonian via coupling constants κ_1 and κ_2:

$$H = H_0 + 2\kappa_1 \sum O + \kappa_2 \sum Q$$

This can be transformed into a quadratic fermionic form using the same techniques as the other models: by direct mapping onto Majorana fermions [37,38] or via a Jordan–Wigner fermionization [24,39,40].

The Jordan–Wigner fermionization method follows that of the Kitaev honeycomb model [24]: the model is first transformed into the effective spin/hardcore boson representation, which contracts the z-links and turns the original lattice into a square lattice whose vertices now carry the new degrees of freedom. One difference from the honeycomb model is that if we do not include the time-reversal breaking terms, the model permits *two* abelian topological phases of the toric code type at the fourth order of perturbation theory [28,37].

The hardcore bosons can again be converted to fermions, although we must do so with *two* separate types of string operators, which are illustrated in Fig. 15. The ground states can then be obtained in the same way as before as BCS-like condensates over toric code vacua, and the excited states are obtained as BdG quasiparticles over these condensates.

D. Effective Hamiltonians on Edges and Vortices

The most exciting—and arguably the *defining*—feature of the non-abelian phases found in the Kitaev and related superconducting models is the existence of Majorana fermions that can get "stuck" at the domain walls separating different topological phases.

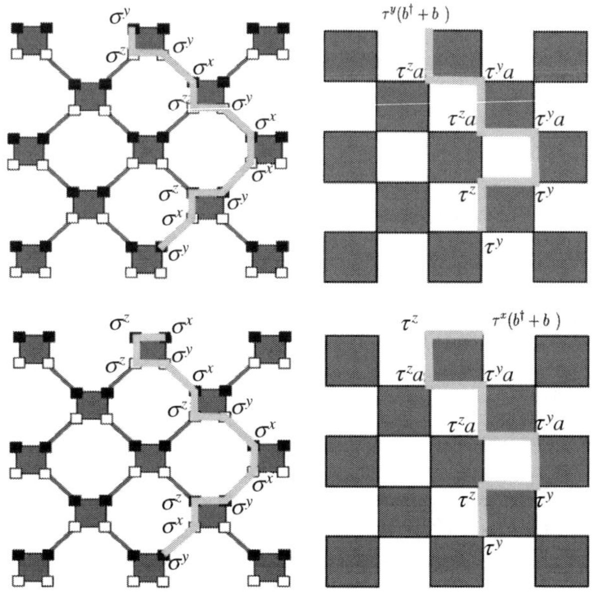

Figure 15. Fermionization of the square–octagon model. One possible string convention for the model in its original formulation and in the effective-hardcore boson representation where the z-links have been contracted to vertices. Two string operators are required for this process.

1. *Majorana Fermions*

The Majorana fermion itself is an electron/hole-like superposition that sits at the mid-gap ($E = 0$) point of certain particle–hole symmetric systems. The vanishing of the Hamiltonian for these states suggests that a TQFT may be describing them: this would fix the ground-state degeneracy (since it is a topological invariant) and imply that any effective Hamiltonian for the system would be metric-independent.

Majorana modes can occur around vortex excitations or between domain walls between abelian and non-abelian phases. In the case where the system contains $2N$ well-separated vortices, there are $2N$ zero-energy fermionic modes; N of these are created by the operators $\gamma_n^\dagger$ and N by their adjoints γ_n. One remarkable consequence of the relation (C2), which gives the relation between these operators and the c-fermions, is that one can *always* choose a superposition of these operators such that the resulting modes are fully localized around the vortex excitations:

$$\begin{aligned}\bar{\gamma}_j &= \sum_{n=1}^{N} \left(\alpha_{jn} \gamma_n^\dagger + \alpha_{j,n+N} \gamma_n \right) \\ &= \sum_{\mathbf{q}} \left(u_{\mathbf{q},j} c_{\mathbf{q}}^\dagger + v_{\mathbf{q},j} c_{\mathbf{q}} \right)\end{aligned}$$

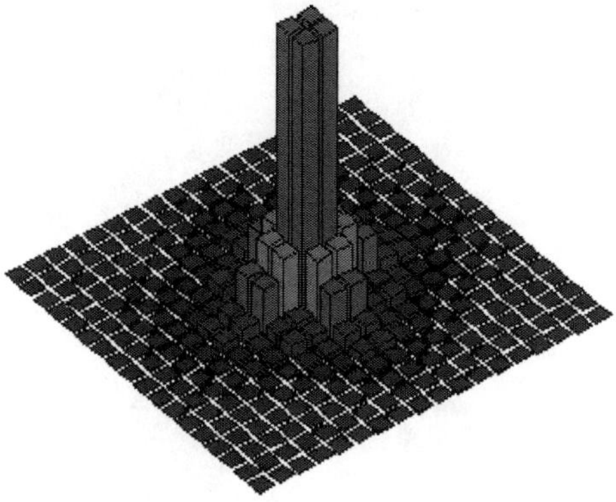

Figure 16. Position dependence of Majorana zero modes. $|u_{\mathbf{q}}|$ is plotted with $\mathbf{q}$ for $J_\alpha = 1$ and $\kappa = 0.5$. Reproduced from Fig. 3 of Ref. [27].

A full Dirac fermion can be made from a superposition of two of these localized modes, so each is, in a sense, "half" of a Dirac fermion. Similarly, the supporting vortices are sometimes called "half quantum vortices." Because the Dirac fermion mode is split between well-separated locations, it cannot interact with local processes that may alter the state; this is the feature that makes these systems potentially useful as quantum information-processing architecture.

It is interesting that this localization also enforces the condition $u_{\mathbf{q},j} = e^{i\Omega_n} v^*_{\mathbf{q},j}$. However, the fermion will only be a Majorana mode if $\bar{\gamma}_j = \bar{\gamma}_j^\dagger$, so it is necessary to multiply the states $(u, v)^T$ by the overall phase $e^{-i\Omega_n/2}$ such that $u_{\mathbf{q},j} = v^*_{\mathbf{q},j}$. This allows us to analyze the position-space structure of these Majorana modes: Fig. 16 plots $|u_{\mathbf{q}}|$ (and thus $|v_{\mathbf{q}}|$) as a function of $\mathbf{q}$ for $J_\alpha = 1$ and $\kappa = 0.5$.

2. *Majorana Edge States*

The Kitaev lattice models are what are known as *topological superconductors*. Roughly speaking, this means that even though an electric current cannot flow through the bulk of the lattice, the edges are conducting and current can flow along them.

This can be explained in terms of the *Fermi energy* E_F of the lattice, the total energy of the electrons bound in the material. An electron with an energy different from E_F cannot flow into the material because no open energy level is available, and electrons, as fermions, cannot enter an occupied state. Electrons with energies equal to E_F can, because those levels are unoccupied. Because $E_F = 0$ for these lattices, only zero modes will be able to contribute to a current.

The bulk spectra of these Kitaev models do not allow zero-energy states, so they are "bulk insulators." However, the energy spectrum on the edges includes $E = 0$, so the edges can conduct [3].

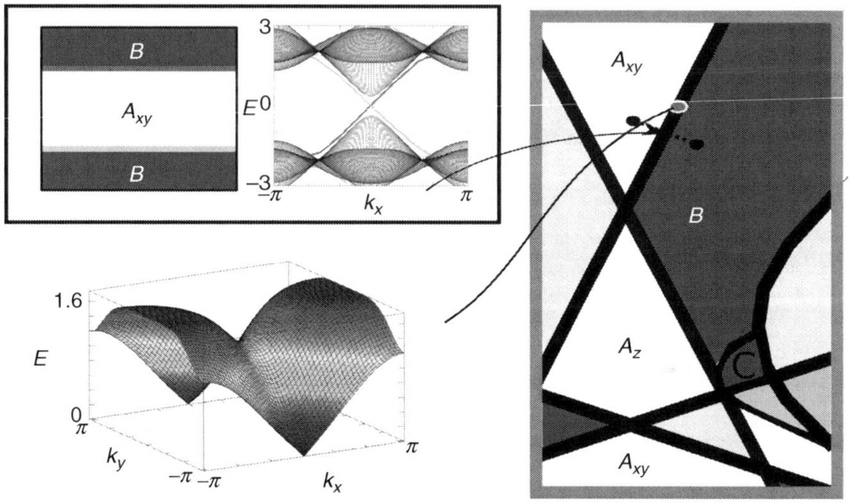

Figure 17. Edge modes of the square–octagon model on a cylinder. Plots of the energies associated with the edge modes at a phase boundary. (*See the color version of this figure in the Color Plates section.*)

For a careful choice of edge conditions, it is possible to analytically treat the conducting edge modes: for example, the type of spectrum one expects to see along a domain wall between an abelian and a non-abelian phase is shown in Fig. 17. The Majorana mode, in this case at $k_x = 0$, is accompanied by a continuum of other states that connect the insulating and bulk bands. If the system is on a torus (so that k_x is discretized), we can only have the exact zero energy Majorana mode if the boundary condition is periodic.

If we surround a non-abelian domain with an abelian domain, we have no zero-energy states if there are no vortices inside the non-abelian domain. If we place an odd number of vortices inside the non-abelian domain, then there is single-zero energy edge mode even though an odd number of branch cuts intersect the domain wall. This somewhat counterintuitive result comes about because phases are also picked up when the wall direction is changed, and these phases all add up to π, canceling the branch cut phase [26].

This picture can also immediately be applied to a domain of abelian phase inside a non-abelian one: if there is no vortex inside this abelian domain, then there is no branch cut. If there are an odd number of vortices in the abelian domain, then we have an odd number of branch cuts and a zero mode can exist. A zero mode due to a *single* vortex in the non-abelian domain is the limiting case of this scenario where the domain edge has been reduced to a single plaquette [35].

IV. TOPOLOGICAL PHASES AND PHYSICAL MATERIALS

The spin lattice systems presented in the previous sections are still rather abstract, and one might wonder whether these models and the topological phases they exhibit are manifested in physical systems, particularly in those relevant to chemical physics and material science. It seems most likely that any systems with topological phases will be found in condensed matter. There are also theoretical proposals to engineer these phases using atomic/molecular systems and superconducting electronics. Last, but not least, they are also relevant to solid-state materials.

Particular progress in the experimental investigation of topological phases has been made in the context of the FQH effect [7,8]. While most of the observed phases are likely to be abelian topological phases, there are also phases that are believed to be non-abelian. As mentioned in Section II.C.3 indications are that the state with filling factor $5/2$ is a non-abelian topological phase of the Ising universality class [18], and a number of experimental results are consistent with this expectation [43–46]. FQH states have also recently been observed in graphene under a magnetic field [47], thus offering a potential new platform for the study of topological phases.

Another direction toward the realization of topological phases is based on engineering topological phases using atomic and molecular systems. Condensed matter systems lend themselves naturally to simulation using trapped ultracold atoms [48], which combine a high level of controllability with the fact that the physics involved is understood in considerable detail. It is therefore hardly surprising that the first scheme for the realization of the Kitaev honeycomb lattice model was proposed using neutral atoms trapped in an optical honeycomb (super) lattice [49]. In this system, the trapped atoms carry spin-like degrees of freedom whose anisotropic interaction is mediated by suitably tuned and polarized laser fields.

The simulation of the Kitaev model is also possible using polar molecules trapped in an optical lattice via a recently proposed toolbox for the simulation of condensed matter systems [50]. This toolbox can also be augmented to include three-body interactions, which are required to break time-reversal symmetry [51]. Topologically protected quantum information systems, similar to lattices, can also be engineered in arrays of Josephson junctions, as predicted long ago [52] and observed recently [53].

Several types of solid-state materials may exhibit the topological phases we have discussed in the context of the lattice models: superconductors with p-wave pairing [35], like the experimentally studied strontium ruthenate (Sr_2RuO_4) [54,55], belong to the universality class characterized by the Ising topological field theory and are of the same nature as the superconducting phase revealed by the Jordan–Wigner fermionization of the Kitaev model described in Section III.B.

Another important class of materials are topological insulators [5]: these were predicted first in graphene [56] and then later in mercury telluride (CdTe/HgTe) heterostructure [57], followed by their experimental observation [58]. These solid-state materials are insulators in the bulk but have conducting edges or surfaces (in two and three dimensions, respectively) that are protected by time-reversal symmetry and are characterized by topological invariants [59,60]. The first three-dimensional topological insulators [61] were theoretically predicted [62,63], and experimentally observed [64] in a bismuth antimonide ($Bi_{1-x}Sb_x$) alloy using angle-resolved photoemission spectroscopy (ARPES), which can probe the surface states.

More recently, a pure stochiometric compound of Bi_2Se_3 [65] has been theoretically predicted and experimentally observed. Because it is a pure substance rather than an alloy, it should have an advantage over $Bi_{1-x}Sb_x$ in that it not only allows good access to surface state structure but also includes a large band gap equivalent to 3600 K [66], which suggests that topological properties of this insulator are robust at room temperature. This contrasts strikingly with FQH systems that require extremely high purity (and thus mobility) of samples and operational temperatures of milliKelvin order.

In addition, a family of related topological insulators that includes not only Bi_2Te_3 but also Sb_2Te_3 has been predicted using ab initio electronic structure calculations [67]. Similar calculations also predict topological insulators in other compounds, including ternary Heusler materials with band inversions similar to HgTe heterostructures [68]. Also predicted is a new class of three-dimensional topological insulators in thallium-based chalcogenides [69]. The advantage of three-dimensional topological insulators is that they offer gapless surface states. A spectral gap can then be induced in this two-dimensional electron gas through proximity with an ordinary superconducting material [70]. The result is a topological superconductor whose edge and vortex modes are predicted to carry Majorana fermions.

Topological insulators and superconductors have been classified into a periodic table [5] where their positions are based on 10 distinct symmetry classes [71], according to time-reversal symmetry, particle–hole symmetry, and chiral symmetry and according to the number of spatial dimensions. The lattice models we have presented in this chapter, and particularly the topological phases they exhibit, belong to the class of two-dimensional topological superconductors or superfluids [72–74], characterized by a integer topological invariant, the spectral Chern number. They are predicted within this classification to have gapless boundary states that reflect the bulk/boundary correspondence. The lattice models offer a microscopic realization that has allowed us to see these properties explicitly.

Acknowledgments

The authors thank Mark Howard for his comments and suggestions. We would also like to acknowledge funding from Science Foundation Ireland under the Principal Investigator Award 10/IN.1/I3013.

APPENDIX A: DEFINITIONS

Although many of these mathematical terms will be familiar to the reader, we include this glossary to define those that may not be.

Mathematicians say that a set of points $\mathbb{X}$—a "space"—has a *topology* if there is a consistent way to define open sets on it. For example, the standard topology on the real number line $\mathbb{R}$ is defined in the following way: if (a, b) is the interval of points between the numbers a and b (but *not* including a and b themselves), then an open set on $\mathbb{R}$ is any set that is either the union of an arbitrary number of these intervals or the intersection of a finite number of them. The full set $\mathbb{R}$ and the empty set $\emptyset = \{\}$ are also defined to be open. Although this example is reasonably familiar and straightforward, much more abstract topological spaces can be defined, and the rich mathematical field of topology is dedicated to studying such spaces and their properties.

However, for any given space, there may be several distinct ways to define open sets on it—a topology over a space is, in general, not unique. The two spaces $\mathbb{X}$ and $\mathbb{Y}$ may be able to be identified with each other point-by-point, but what we call open sets on one may not be in a one-to-one correspondence with what we call open sets on the other, in which case the two spaces are *topologically inequivalent*.

A *manifold* is defined to be a smooth topological space, a collection of points with no cusps or kinks in it and with a well-defined notion of open sets on it. The *dimension* of the manifold is the number of real-valued parameters you need to specify precisely where a point is. For example, the circle S^1 is a one-dimensional manifold (or "one-manifold"), because you only need a single angle between 0 and 2π to label every point on it. The sphere S^2, however, is two-dimensional; in spherical coordinates, the angles θ and ϕ label every point. $\mathbb{R}^3$ is a three-dimensional, with the Cartesian coordinates (x, y, z) being only one of the ways we can specify points in three spatial dimensions. (In mathematical language, $\mathbb{X}$ is a D manifold if it is locally diffeomorphic to $\mathbb{R}^D$.)

Manifolds are some of the most fundamental objects in any physical theory; they describe the space (or space–time) on which everything takes place. Furthermore, transformations of the system that depend on continuous parameters (e.g., translations and rotations) all have manifolds underlying them.

If there is a notion of "inside" and "outside" for a manifold—namely, that you can move through the manifold away from points that are in it toward ones that are not—then we say the manifold has a *boundary*. A disc, for example, is a two-dimensional manifold with a circle as a boundary, and moving out from the disc's center gets you closer to points not on the disc. But a circle is a one-dimensional manifold *without* a boundary: you never approach a point not on the circle as you go around it. In mathematical notation, if $\mathbb{X}$ is a manifold, then we denote its boundary by $\partial\mathbb{X}$, with $\partial\mathbb{X} = \emptyset$ indicating that the manifold has no boundary.

Note that the boundary of a manifold is itself a manifold of one dimension less (as with the two-dimensional disc having boundary S^1), and a manifold that is a boundary has no boundary itself: $\partial(\partial\mathbb{X}) = \emptyset$. Also, it is worth noting that the boundary of a manifold may not actually be in the manifold: for example, the open interval (0, 1) has boundary points 0 and 1, neither of which are in the interval itself.

The behavior of a system will almost always depend on its boundaries. Anyone who has ever solved a differential equation knows that the solution depends explicitly on what the boundaries are. And they are particularly important in TQFTs (as we will see in Appendix B); they determine what states can actually exist in a physical theory.

Somewhat related to the idea of a boundary is *compactness* of a topological space. Among a myriad of different, but equivalent, definitions for compactness, the best (and probably most intuitive) one for our purposes is that a compact set is *closed* and *bounded*. The former means that all the set's boundary points are in the set, and the latter means that the set—when thought of as a subset of $\mathbb{R}^D$—is finite in extent. As just mentioned, (0, 1) does not contain its boundary points, so although it is bounded, it is not closed, and so is not compact. The interval [0, 1], which *does* include 0 and 1, is compact.

Note that sets without boundaries are closed; they contain "all zero" of their boundary points, so technically fit the definition. Thus, sets without boundaries need only be bounded to be compact; the circle and sphere are such sets. However, not all closed sets are boundaryless, with [0, 1] being an example. We bring this up because some texts take boundarylessness to be a *requirement* for closedness, and thus compactness; however, we will not do so here, and will explicitly state when needed if compact sets have a boundary or not.

Compact manifolds have several properties that noncompact ones lack; for example, a compact space always has a finite volume. Physical states on compact manifolds tend to be described by *discrete* quantum numbers: the fact that ℓ and m, the eigenvalues of quantum-mechanical orbital angular momentum, take on only integer values is entirely due to the fact that the sphere S^2 is compact. Many of the topological invariants we will use to describe states in our TQFTs will also depend on whether if our space is compact.

A key concept we need in axiomatizing TQFT is *orientability*. An orientable manifold is one in which we can consistently define a "handedness" of our coordinates over the entire manifold. For example, the usual angular coordinate ϕ on a circle is such that we go counterclockwise as ϕ increases (e.g., $\phi = 0$ on the positive x-axis, $\phi = \pi/2$ on the positive y-axis, etc.). This is true everywhere on the circle, so the circle is an orientable one-manifold. In $\mathbb{R}^3$, we can define the Cartesian coordinates (x, y, z) so that we have right-handed axes (i.e., the unit direction vectors satisfy $\hat{e}_x \times \hat{e}_y = \hat{e}_z$ everywhere: it is an orientable three-manifold). Basically, any manifold on which there is a version (however abstract) of the "right-hand rule" is an orientable one.

But we could have just as well taken the opposite orientation for both of these examples: on the circle, $\phi' = -\phi$ is a perfectly good coordinate, but it increases in the *clockwise* direction everywhere. In $\mathbb{R}^3$, $(x', y', z') = (x, y, -z)$ are also good coordinates, but the primed axes give a *left*-handed Cartesian coordinate system. In general, any orientable manifold has *two* possible orientations, and which one we take to be "positive" or "right-handed" is a matter of convention. So a circle where we go around counterclockwise is the positively oriented S^1, whereas going around clockwise gives the negatively oriented circle, denoted $(S^1)^*$. In general, if $\mathbb{X}$ is an oriented space, then $\mathbb{X}^*$ is the same space with the opposite orientation.

But not all manifolds are orientable. Perhaps the most famous example of a nonorientable manifold is the Möbius strip: if you move a set of perpendicular axes—one pointing along the strip, the other pointing toward the edge—completely around the strip, the first axis stays the same, while the second flips direction. The same coordinate system is simultaneously right- *and* left-handed, which is not possible on an orientable manifold.

Why is the orientability of a manifold important? The main reason is that it allows us to integrate over the manifold; without a notion of handedness, we cannot consistently define an integral over the space. And in a physical theory, it is the integral of the Lagrangian over the underlying space—the action—that leads to the equations of motion giving its behavior. So it is not just desirable, but necessary, that we require orientability in any field theory.

APPENDIX B: THE AXIOMS OF TQFT

Using the definitions from Appendix A, we now state precisely what axioms must hold for anything we call a TQFT. We follow here the general approach of Ref. [1], in somewhat less mathematical terms and with some notational changes.

A D-dimensional TQFT consists of the following: for any compact, boundaryless, orientable D-manifold Σ, there is a Hilbert space $\mathcal{H}(\Sigma)$ of finite dimension N_Σ. Then for every $(D+1)$-dimensional compact manifold $\mathbb{X}$ that has Σ as a boundary, a map Z assigns a vector in $\mathcal{H}(\Sigma)$ to $\mathbb{X}$. These must satisfy the following axioms:

1. The Hilbert space assigned to the empty manifold is the complex numbers: $\mathcal{H}(\emptyset) = \mathbb{C}$.

   The empty set is compact, boundaryless, and orientable for any D, so it will always be a permissible choice for Σ. This axiom takes this into account by immediately assigning the one-dimensional Hilbert space $\mathbb{C}$ to it.
2. If Σ is positively oriented, then $\mathcal{H}(\Sigma)$ is spanned by N_Σ orthonormal kets $\{|\phi_i^\Sigma\rangle \, | i = 1, \ldots, N_\Sigma\}$, and the Hilbert space for the negatively oriented manifold Σ^* is spanned by the corresponding bras $\{\langle\phi_i^\Sigma|\}$.

This axiom simply relates the vector spaces of manifolds, which differ only in their orientation. Mathematically, we say that $\mathcal{H}(\Sigma^*)$ is *dual* to $\mathcal{H}(\Sigma)$, so we can take the inner product of a vector in one with a vector in the other.

3. If Σ_1 and Σ_2 are disjoint, then

$$\mathcal{H}(\Sigma_1 \cup \Sigma_2) = \mathcal{H}(\Sigma_1) \otimes \mathcal{H}(\Sigma_2)$$

The boundary of a space $\mathbb{X}$ will not always be a single connected set; for example, a hollow cylinder has two separate circles as its boundary. This axiom states that if this is the case, the overall Hilbert space is simply the tensor product of the individual spaces, namely, linear combinations of $\left|\phi_i^{\Sigma_1}\right\rangle \left|\phi_j^{\Sigma_2}\right\rangle$. So we only really need to define $\mathcal{H}(\Sigma)$ for *connected* manifolds, and use these to build up the Hilbert spaces for unconnected ones.

4. If Σ is part of the boundary of $\mathbb{X}_1$, we may "glue" it to another manifold $\mathbb{X}_2$ that has Σ^* as part of its boundary to obtain a new manifold denoted $\mathbb{X}_1 \cup_\Sigma \mathbb{X}_2$.

It makes perfect sense that we can glue two manifolds together only along boundaries that are topologically equivalent; we would not expect to be able to stitch a sphere to a torus, for instance.

The requirement that the orientations be opposite to each other is perhaps not as obvious, but a good way to think about it is that after the gluing, we should not be able to tell that there had ever been two boundaries there. Thus, their properties must "cancel out" in the gluing process (see Fig. 18).

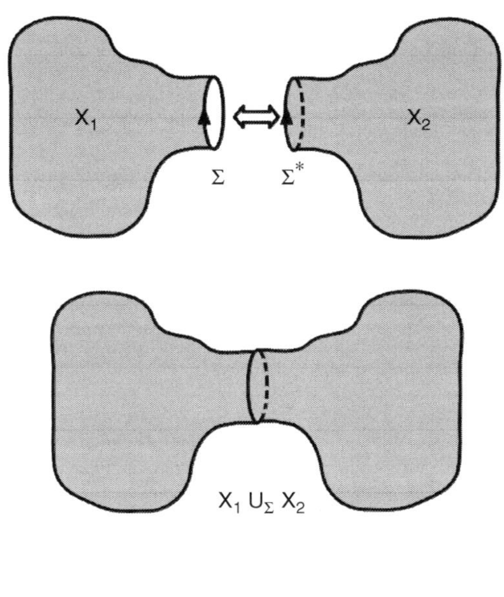

Figure 18. Gluing/cutting of manifolds along a shared boundary Σ. The arrows on Σ and Σ^* denote their orientations.

Put a bit more mathematically, no trace of the vector space $\mathcal{H}(\Sigma)$ can remain after the gluing, so we need to get rid of the kets $\left|\phi_i^\Sigma\right\rangle$ in $Z(\mathbb{X}_1)$ somehow. A natural way to do this is to take their inner product with the bras in $\mathcal{H}(\Sigma^*)$, which will just give numbers. Thus, $Z(\mathbb{X}_2)$ has to be expressible in terms of the bras $\left\langle\phi_i^\Sigma\right|$, indicating that Σ^* must be part of its boundary. The vectors assigned to the two spaces must therefore have the forms

$$Z(\mathbb{X}_1) = \sum_{i=1}^{N_\Sigma} a_i \left|\phi_i^\Sigma\right\rangle, \quad Z(\mathbb{X}_2) = \sum_{i=1}^{N_\Sigma} b_j \left\langle\phi_j^\Sigma\right|$$

for some coefficients a_i and b_i. So, if we use $\circ_\Sigma$ to denote the vector product when we glue along Σ, then

$$\begin{aligned} Z(\mathbb{X}_1) \circ_\Sigma Z(\mathbb{X}_2) &= Z(\mathbb{X}_1 \cup_\Sigma \mathbb{X}_2) \\ &= \sum_{i,j=1}^{N_\Sigma} a_i b_j \left\langle\phi_j^\Sigma \middle| \phi_i^\Sigma\right\rangle \\ &= \sum_{i=1}^{N_\Sigma} a_i b_i \end{aligned}$$

because we have an orthonormal basis. But this sum must be independent of Σ, because we should not be able to "see the joins" in the final manifold $\mathbb{X}_1 \cup_\Sigma \mathbb{X}_2$. This must work in reverse as well: we should be able to slice any space $\mathbb{X}$ along some subspace Σ and "unglue" it into two separate spaces $\mathbb{X}_1$ and $\mathbb{X}_2$ with boundaries Σ and Σ^*, respectively, but because we could pick any Σ we like, $Z(\mathbb{X})$ cannot depend on it.

5. If $\mathbb{I}$ is the closed interval [0, 1], then the vector assigned to the $(D+1)$-manifold $\Sigma \times \mathbb{I}$ is

$$Z(\Sigma \times \mathbb{I}) = \sum_{i=1}^{N_\Sigma} \left|\phi_i^\Sigma\right\rangle \left\langle\phi_i^\Sigma\right|$$

which, because the basis for $\mathcal{H}(\Sigma)$ is complete, is the identity operator.

One way to think of the meaning of this axiom is by realizing that $\Sigma \times \mathbb{I}$ is a sort of "cylinder" or "tube" whose cross-section is Σ. The boundaries of this tube are Σ on one side and Σ^* on the other, since they must have opposite orientations. Thus, $Z(\Sigma \times \mathbb{I})$ has to live in $\mathcal{H}(\Sigma) \otimes \mathcal{H}(\Sigma^*)$, which tells us that it must have the general form (ket $\otimes$ bra).

If Σ is part of the boundary of a space $\mathbb{X}$, then we can glue the Σ^* end of this tube to it; however, the effect is merely to continuously "pull" Σ a bit further away from $\mathbb{X}$. Because it is a continuous deformation, topological invariance tells us that the vector assigned by Z to $(\Sigma \times \mathbb{I}) \cup_{\Sigma} \mathbb{X}$ should be exactly the same as that assigned to $\mathbb{X}$. Therefore, $Z(\Sigma \times \mathbb{I})$ is the identity operator.

One useful consequence of this axiom is a quick way of computing the dimension of $\mathcal{H}(\Sigma)$; because the two ends of $\Sigma \times \mathbb{I}$ are the same manifold with opposite orientations, they can be glued together, resulting in a circular "doughnut" with cross-section Σ, or $\Sigma \times S^1$. This new manifold has no boundary, so $Z(\Sigma \times S^1)$ is just a complex number.

But recall that gluing requires us to take the inner product of the kets in $\mathcal{H}(\Sigma)$ with the bras in $\mathcal{H}(\Sigma^*)$, so that

$$Z\left(\Sigma \times S^1\right) = \sum_{i=1}^{N_\Sigma} \left\langle \phi_i^\Sigma \,\middle|\, \phi_i^\Sigma \right\rangle = \sum_{i=1}^{N_\Sigma} 1$$

since we have assumed our basis vectors are orthonormal. Therefore, $Z(\Sigma \times S^1) = N_\Sigma$, giving us a quick way of finding the dimension of $\mathcal{H}(\Sigma)$.

APPENDIX C: SUPERCONDUCTING HAMILTONIANS AND THE BOGOLIBOV–DEGENNES EQUATIONS

In Appendix C, we introduce the generic Bogolibov–deGennes (BdG) Hamiltonian for a lattice and some associated formalism. This Hamiltonian is quadratic in the fermionic operators and gives the energy cost for fermions to hop from one lattice point to another and the cost of creating and destroying fermionic pairs. The method for obtaining this Hamiltonian is a standard one, and so we only present a brief overview here; a fuller discussion can be found in Ref. [75].

In its most general form, a Hamiltonian that is quadratic in fermionic operators c and $c^\dagger$ may be written

$$H = \frac{1}{2} \sum_{jk} \left(\xi_{jk} c_j^\dagger c_k - \xi_{kj} c_j c_k^\dagger + \Delta_{jk} c_j^\dagger c_k^\dagger + \Delta_{jk}^* c_j c_k \right)$$

where the j and k indices label both the position on the lattice and any internal degrees of freedom. The ξs and Δs can be operators (the former being Hermitian), but if so, we assume that they commute with the fermionic operators.

By defining the matrices ξ and Δ to be those with elements ξ_{ij} and Δ_{ij}, and taking c_i and $c_i^\dagger$ as elements of vectors c and $c^\dagger$, the Hamiltonian can be rewritten in the following convenient matrix form:

$$H = \frac{1}{2}\begin{pmatrix} c^\dagger & c \end{pmatrix}\begin{pmatrix} \xi & \Delta \\ \Delta^\dagger & -\xi^T \end{pmatrix}\begin{pmatrix} c \\ c^\dagger \end{pmatrix} \tag{C1}$$

The system is diagonalized by solving the BdG eigenvalue problem

$$\begin{pmatrix} \xi & \Delta \\ \Delta^\dagger & -\xi^T \end{pmatrix} = \begin{pmatrix} U & V^* \\ V & U^* \end{pmatrix}\begin{pmatrix} E & \mathbf{0} \\ \mathbf{0} & -E \end{pmatrix}\begin{pmatrix} U & V^* \\ V & U^* \end{pmatrix}^\dagger$$

where E is a diagonal matrix whose entries are all nonnegative and the matrix conjugating diag$(E, -E)$ is unitary. (The form of the Hermitian matrix on the left-hand side guarantees there is always a solution to the BdG problem.)

We may now define new operators γ_n and $\gamma_n^\dagger$, which are the components of an column vector, by

$$\begin{pmatrix} \gamma \\ \gamma^\dagger \end{pmatrix} = \begin{pmatrix} U & V^* \\ V & U^* \end{pmatrix}^\dagger \begin{pmatrix} c \\ c^\dagger \end{pmatrix} \tag{C2}$$

or equivalently

$$\gamma_n = \sum_j \left(V_{jn}^* c_j^\dagger + U_{jn}^* c_j\right), \quad \gamma_n^\dagger = \sum_j \left(V_{jn} c_j + U_{jn} c_j^\dagger\right)$$

However, not all the eigenvalues of E are necessarily nonzero; if $E_1, \ldots, E_M$ are the M positive eigenvalues, they are the only ones that contribute to the Hamiltonian (C1). Therefore, it can be written as

$$H = \sum_{m=1}^{M} E_m \left(\gamma_m^\dagger \gamma_m - \frac{1}{2}\right)$$

The vacuum state $\left|\,\text{vac}\right\rangle$, which is annihilated by the c-operators, is *not* the ground state $\left|\,\text{g.s.}\right\rangle$ of this Hamiltonian. Rather, the ground state can usually be written down as the projection

$$\left|\,\text{g.s.}\right\rangle = \prod_{m=1}^{M} \gamma_m \left|\,\text{vac}\right\rangle$$

The energy of this ground state is $-\sum_{m=1}^{M} E_m/2$.

If the indices i refer only to the position $\mathbf{q}$ on a $N_x \times N_y$ lattice (e.g., they do not include other degrees of freedom such as spin), then we can perform the Fourier transform

$$c_{\mathbf{q}} = \frac{1}{\sqrt{N_x N_y}} \sum_{\mathbf{k}} c_{\mathbf{k}} e^{i\mathbf{k}\cdot\mathbf{q}}$$

to obtain the momentum-space Hamiltonian

$$H = \frac{1}{2} \sum_{\mathbf{k}} \begin{pmatrix} c_{\mathbf{k}}^{\dagger} & c_{-\mathbf{k}} \end{pmatrix} H_{\mathbf{k}} \begin{pmatrix} c_{\mathbf{k}} \\ c_{-\mathbf{k}}^{\dagger} \end{pmatrix}$$

where

$$H_{\mathbf{k}} = \begin{pmatrix} \xi_{\mathbf{k}} & \Delta_{\mathbf{k}} \\ \Delta_{\mathbf{k}}^{\dagger} & -\xi_{-\mathbf{k}}^{T} \end{pmatrix}$$

This is just a 2×2 matrix, so the BdG problem reduces to finding 2×2 matrices $\mathcal{W}_{\mathbf{k}}$ and E_D, satisfying

$$H_{\mathbf{k}} \mathcal{W}_{\mathbf{k}} = \mathcal{W}_{\mathbf{k}} E_D$$

The solution is

$$E_D = \begin{pmatrix} E_{\mathbf{k}} & 0 \\ 0 & -E_{\mathbf{k}} \end{pmatrix}, \quad \mathcal{W}_{\mathbf{k}} = \begin{pmatrix} u_{\mathbf{k}} & v_{\mathbf{k}} \\ -v_{\mathbf{k}}^{*} & u_{\mathbf{k}} \end{pmatrix}$$

where $E_{\mathbf{k}} = \sqrt{\xi_{\mathbf{k}}^2 + |\Delta_{\mathbf{k}}|^2}$ and

$$u_{\mathbf{k}} = \sqrt{\frac{1}{2}\left(1 + \frac{\xi_{\mathbf{k}}}{E_{\mathbf{k}}}\right)}, \quad v_{\mathbf{k}} = -e^{i\arg(\Delta_{\mathbf{k}})} \sqrt{\frac{1}{2}\left(1 - \frac{\xi_{\mathbf{k}}}{E_{\mathbf{k}}}\right)}$$

with $v_{-\mathbf{k}} = -v_{\mathbf{k}}$.

The quasiparticle excitations are then given by

$$\begin{pmatrix} \gamma_{\mathbf{k}}^{\dagger} & \gamma_{-\mathbf{k}} \end{pmatrix} = \begin{pmatrix} c_{\mathbf{k}}^{\dagger} & c_{-\mathbf{k}} \end{pmatrix} \mathcal{W}_{\mathbf{k}}, \quad \begin{pmatrix} \gamma_{\mathbf{k}} \\ \gamma_{-\mathbf{k}}^{\dagger} \end{pmatrix} = \mathcal{W}_{\mathbf{k}}^{\dagger} \begin{pmatrix} c_{\mathbf{k}} \\ c_{-\mathbf{k}}^{\dagger} \end{pmatrix}$$

or

$$\begin{aligned} \gamma_{\mathbf{k}} &= -v_{\mathbf{k}} c_{-\mathbf{k}}^{\dagger} + u_{\mathbf{k}} c_{\mathbf{k}} & \gamma_{\mathbf{k}}^{\dagger} &= -v_{\mathbf{k}}^{*} c_{-\mathbf{k}} + u_{\mathbf{k}} c_{\mathbf{k}}^{\dagger} \\ \gamma_{-\mathbf{k}} &= v_{\mathbf{k}} c_{\mathbf{k}}^{\dagger} + u_{\mathbf{k}} c_{-\mathbf{k}} & \gamma_{-\mathbf{k}}^{\dagger} &= v_{\mathbf{k}}^{*} c_{\mathbf{k}} + u_{\mathbf{k}}^{*} c_{-\mathbf{k}}^{\dagger} \end{aligned}$$

We then have $H = \sum E_{\mathbf{k}}(\gamma_{\mathbf{k}}^{\dagger}\gamma_{\mathbf{k}} - 1/2)$. The BCS-type ground state, annihilated by all $\gamma_{\mathbf{k}}$, is

$$\left|\text{g.s.}\right\rangle = \prod \left(u_{\mathbf{k}} + v_{\mathbf{k}} c_{\mathbf{k}}^{\dagger} c_{-\mathbf{k}}^{\dagger}\right) \left|\text{vac}\right\rangle$$

where the product is over distinct $(\mathbf{k}, -\mathbf{k})$ pairs. The energy of this state is

$$E_{\text{g.s.}} = -\frac{1}{2}\sum_{\mathbf{k}} E_{\mathbf{k}}$$

APPENDIX D: SYMMETRY-BREAKING TERMS IN THE FERMIONIZED MODEL

We present here for the interested reader the full form of the time-reversal and parity-invariance-breaking terms of the potential V in the Kitaev honeycomb model in the effective spin/hardcore boson representation. After using Eq. (18), the individual terms in Eq. (12) become

$$\begin{aligned}
P_{\mathbf{q}}^{1} &= \kappa_{\mathbf{q}}^{1}\left(b_{\mathbf{q}}^{\dagger} + b_{\mathbf{q}}\right) i\tau_{\mathbf{q}+\mathbf{n}_x}^{x}\left(b_{\mathbf{q}+\mathbf{n}_x}^{\dagger} - b_{\mathbf{q}+\mathbf{n}_x}\right)\\
P_{\mathbf{q}}^{2} &= \kappa_{\mathbf{q}}^{2} i\left(b_{\mathbf{q}+\mathbf{n}_y}^{\dagger} - b_{\mathbf{q}+\mathbf{n}_y}\right)\tau_{\mathbf{q}+\mathbf{n}_x+\mathbf{n}_y}^{x}\left(b_{\mathbf{q}+\mathbf{n}_x+\mathbf{n}_y}^{\dagger} + b_{\mathbf{q}+\mathbf{n}_x+\mathbf{n}_y}\right)\\
P_{\mathbf{q}}^{3} &= \kappa_{\mathbf{q}}^{3}\tau_{\mathbf{q}}^{z}\left(b_{\mathbf{q}}^{\dagger} - b_{\mathbf{q}}\right)\tau_{\mathbf{q}+\mathbf{n}_y}^{y}\left(b_{\mathbf{q}+\mathbf{n}_y}^{\dagger} - b_{\mathbf{q}+\mathbf{n}_y}\right)\\
P_{\mathbf{q}}^{4} &= \kappa_{\mathbf{q}}^{4}\,\tau_{\mathbf{q}+\mathbf{n}_x+\mathbf{n}_y}^{y}\left(b_{\mathbf{q}+\mathbf{n}_x+\mathbf{n}_y}^{\dagger} + b_{\mathbf{q}+\mathbf{n}_x+\mathbf{n}_y}\right)\tau_{\mathbf{q}+\mathbf{n}_x}^{z}\left(b_{\mathbf{q}+\mathbf{n}_x}^{\dagger} + b_{\mathbf{q}+\mathbf{n}_x}\right)\\
P_{\mathbf{q}}^{5} &= \kappa_{\mathbf{q}}^{5} i\left(b_{\mathbf{q}+\mathbf{n}_y}^{\dagger} + b_{\mathbf{q}+\mathbf{n}_y}\right)\tau_{\mathbf{q}+\mathbf{n}_x+\mathbf{n}_y}^{z}\tau_{\mathbf{q}+\mathbf{n}_x}^{z}\left(b_{\mathbf{q}+\mathbf{n}_x}^{\dagger} - b_{\mathbf{q}+\mathbf{n}_x}\right)\\
P_{\mathbf{q}}^{6} &= \kappa_{\mathbf{q}}^{6}\tau_{\mathbf{q}+\mathbf{n}_y}^{y}\left(b_{\mathbf{q}+\mathbf{n}_y}^{\dagger} + b_{\mathbf{q}+\mathbf{n}_y}\right)\tau_{\mathbf{q}}^{z}\left(I - 2b_{\mathbf{q}}^{\dagger}b_{\mathbf{q}}\right)\tau_{\mathbf{q}+\mathbf{n}_x}^{x}\left(b_{\mathbf{q}+\mathbf{n}_x}^{\dagger} + b_{\mathbf{q}+\mathbf{n}_x}\right)
\end{aligned}$$

After fermionization, we obtain

$$\begin{aligned}
P_{\mathbf{q}}^{1} &= -\kappa_{\mathbf{q}}^{1}\, iX_{\mathbf{q}}\left(c_{\mathbf{q}}^{\dagger} - c_{\mathbf{q}}\right)\left(c_{\mathbf{q}+\mathbf{n}_x}^{\dagger} - c_{\mathbf{q}+\mathbf{n}_x}\right)\\
P_{\mathbf{q}}^{2} &= -\kappa_{\mathbf{q}}^{2} iX_{\mathbf{q}+\mathbf{n}_y}\left(c_{\mathbf{q}+\mathbf{n}_y}^{\dagger} + c_{\mathbf{q}+\mathbf{n}_y}\right)\left(c_{\mathbf{q}+\mathbf{n}_x+\mathbf{n}_y}^{\dagger} + c_{\mathbf{q}+\mathbf{n}_x+\mathbf{n}_y}\right)\\
P_{\mathbf{q}}^{3} &= -\kappa_{\mathbf{q}}^{3}\, iY_{\mathbf{q}}\left(c_{\mathbf{q}}^{\dagger} - c_{\mathbf{q}}\right)\left(c_{\mathbf{q}+\mathbf{n}_y}^{\dagger} - c_{\mathbf{q}+\mathbf{n}_y}\right)\\
P_{\mathbf{q}}^{4} &= -\kappa_{\mathbf{q}}^{4} iY_{\mathbf{q}+\mathbf{n}_x}\left(c_{\mathbf{q}+\mathbf{n}_x}^{\dagger} + c_{\mathbf{q}+\mathbf{n}_x}\right)\left(c_{\mathbf{q}+\mathbf{n}_x+\mathbf{n}_y}^{\dagger} + c_{\mathbf{q}+\mathbf{n}_x+\mathbf{n}_y}\right)\\
P_{\mathbf{q}}^{5} &= \kappa_{\mathbf{q}}^{5} iX_{\mathbf{q}+\mathbf{n}_y}Y_{\mathbf{q}+\mathbf{n}_x}\left(c_{\mathbf{q}+\mathbf{n}_y}^{\dagger} - c_{\mathbf{q}+\mathbf{n}_y}\right)\left(c_{\mathbf{q}+\mathbf{n}_x}^{\dagger} - c_{\mathbf{q}+\mathbf{n}_x}\right)\\
P_{\mathbf{q}}^{6} &= \kappa_{\mathbf{q}}^{6} iX_{\mathbf{q}}Y_{\mathbf{q}}\left(c_{\mathbf{q}+\mathbf{n}_y}^{\dagger} + c_{\mathbf{q}+\mathbf{n}_y}\right)\left(c_{\mathbf{q}+\mathbf{n}_x}^{\dagger} + c_{\mathbf{q}+\mathbf{n}_x}\right)
\end{aligned}$$

These are also summarized in Fig. 19.

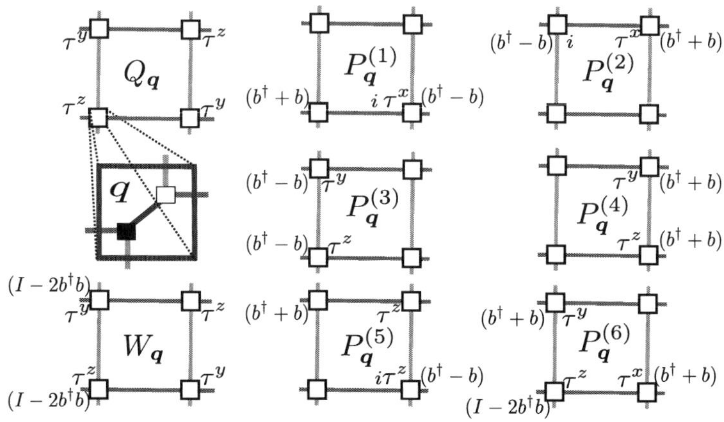

Figure 19. The time-reversal and parity-breaking terms of the potential V in the fermionic representation. Reproduced from Fig. 2 of Ref. [24].

REFERENCES

1. M. Atiyah, *The Geometry and Physics of Knots*, Cambridge University Press, Cambridge, 1990.
2. A. Kitaev, *Ann. Phys.* **303**, 2 (2003).
3. A. Kitaev, *Ann. Phys.* **321**, 2 (2006).
4. C. Nayak, S. H. Simon, A. Stern, M. H. Freedman, and S. Das Sarma, *Rev. Mod. Phys.* **80**, 1083 (2008).
5. M. Z. Hasan and C. L. Kane, *Rev. Mod. Phys.* **82**, 3045 (2010).
6. D. Birmingham, M. Blau, M. Rakowski, and G. Thompson, *Phys. Rep.* **209**, 129 (1991).
7. D. C. Tsui, H. L. Störmer, and A. C. Gossard, *Phys. Rev. Lett.* **48**, 1559 (1982).
8. J. P. Eisenstein and H. L. Störmer, *Science* **248**, 4962 (1990).
9. V. L. Ginzburg and L. D. Landau, *Zh. Eksp. Teor. Fiz.* **20**, 1064 (1950).
10. A. Zee, *Quantum Field Theory in a Nutshell*, Princeton University Press, Princeton, 2003.
11. S.-S. Chern and J. Simons, *Ann. Math.* **99**, 48 (1974).
12. E. Witten, *Commun. Math. Phys.* **121**, 351 (1989).
13. P. Di Francesco, P. Mathieu, and D. Sénéchal, *Conformal Field Theory*, Springer, New York, 1997.
14. A. A. Belavin, A. M. Polyakov, and A. B. Zamolodchikov, *Nucl. Phys. B* **241**, 333 (1984).
15. A. B. Zamolodchikov, *JETP Lett.* **43**, 730 (1986).
16. J. Polchinski, *Nucl. Phys. B* **303**, 226 (1988).
17. M. Spivak, *A Comprehensive Guide to Differential Geometry*, Vol. 5, Publish or Perish, Berkeley, 1979.
18. G. Moore and N. Read, *Nucl. Phys. B* **360**, 362 (1991).
19. E. Ising, *Z. Phys.* **25**, 253 (1925).
20. J. Wess and B. Zumino, *Phys. Lett. B* **37**, 95 (1971).

21. E. Witten, *Nucl. Phys. B* **223**, 422 (1983).
22. E. Verlinde, *Nucl. Phys. B* **300**, 360 (1988).
23. A. Kitaev, A. H. Shen, and M. N. Vyalyi, *Classical and Quantum Computation*, American Mathematical Society, Providence, 2002.
24. G. Kells, J. K. Slingerland, and J. Vala, *Phys. Rev. B* **80**, 125415 (2009).
25. H. Yao and S. A. Kivelson, *Phys. Rev. Lett.* **99**, 247203 (2007).
26. G. Kells, D. Mehta, J. K. Slingerland, and J. Vala, *Phys. Rev. B* **81**, 104429 (2010).
27. G. Kells and J. Vala, *Phys. Rev. B* **82**, 125122 (2010).
28. G. Kells, J. Kailasvuori, J. K. Slingerland, and J. Vala, *New J. Phys.* **13**, 095014 (2011).
29. E. H. Lieb, *Phys. Rev. Lett.* **73**, 2158 (1994).
30. D. J. Thouless, M. Kohmoto, M. P. Nightingale, and M. den Nijs, *Phys. Rev. Lett.* **49**, 405 (1982).
31. J. Vidal, K. P. Schmidt, and S. Dusuel, *Phys. Rev. B* **78**, 245121 (2008).
32. A. Bolukbasi, *Non-Abelian Anyons in the Kitaev Honeycomb Model*, Ph.D. thesis, National University of Ireland Maynooth, 2010.
33. J. Bardeen, L. N. Cooper, and J. R. Schrieffer, *Phys. Rev.* **106**, 162 (1957).
34. G. Kells et al., *Phys. Rev. Lett.* **101**, 240404 (2008).
35. N. Read and D. Green, *Phys. Rev. B* **61**, 10267 (2000).
36. S. Dusuel, K. P. Schmidt, and J. Vidal, *Phys. Rev. Lett.* **100**, 177204 (2008).
37. S. Yang, D. L. Zhou, and C. P. Sun, *Phys. Rev. B* **76**, 180404 (2007).
38. G. Baskaran, G. Santhosh, and R. Shankar, *arXiv:0908.1614*, 2009.
39. X.-Y. Feng, G.-M. Zhang, and T. Xiang, *Phys. Rev. Lett.* **98**, 087204 (2007).
40. H. D. Chen and Z. Nussinov, *J. Phys. A* **41**, 075001 (2008).
41. M. H. Freedman, A. Kitaev, M. J. Larsen, and Z. Wang, *Bull. Am. Math. Soc.* **40**, 31 (2003).
42. Y. Hatsugai, *Phys. Rev. Lett.* **71**, 3697 (1993).
43. M. Dolev et al., *Nature* **452**, 829 (2008).
44. R. L. Willett, L. N. Pfeiffer, and K. W. West, *Proc. Natl. Acad. Sci.* **106**, 8853 (2009).
45. R. L. Willett, L. N. Pfeiffer, and K. W. West, *Phys. Rev. B* **82**, 205301 (2010).
46. S. An et al., *arXiv:1112.3400*, 2011.
47. K. I. Bolotin et al., *Nature* **475**, 122 (2011).
48. M. Lewenstein et al., *Adv. Phys.* **56**, 243 (2007).
49. L.-M. Duan, E. Demler, and M. D. Lukin, *Phys. Rev. Lett.* **91**, 090402 (2003).
50. A. Micheli, G. K. Brennen, and P. Zoller, *Nat. Phys.* **2**, 341 (2006).
51. H. P. Büchler, A. Micheli and P. Zoller, *Nat. Phys.* **3**, 726 (2007).
52. L. B. Ioffe et al., *Nature* **415**, 503 (2002).
53. S. Gladchenko et al., *Nat. Phys.* **5**, 48 (2008).
54. Y. Maeno, T. M. Rice, and M. Sigrist, *Phys. Today* **54**, 42 (2001).
55. K. D. Nelson, Z. Q. Mao, Y. Maeno, and Y. Liu, *Science* **306**, 1151 (2004).
56. C. L. Kane and G. Mele, *Phys. Rev. Lett.* **95**, 226801 (2005).
57. B. A. Bernevig, T. L. Highes, and S.-C. Zhang, *Science* **314**, 1757 (2006).
58. M. König et al., *Science* **318**, 766 (2007).
59. C. L. Kane and G. Mele, *Phys. Rev. Lett.* **95**, 146802 (2005).
60. J. E. Moore and L. Balents, *Phys. Rev. B* **75**, 121306R (2007).

61. L. Fu and C. L. Kane, *Phys. Rev. Lett.* **98**, 106803 (2007).
62. L. Fu and C. L. Kane, *Phys. Rev. B* **76**, 045302 (2007).
63. J. C. Y. Teo, L. Fu, and C. L. Kane, *Phys. Rev. B* **78**, 045426 (2008).
64. D. Hsieh et al., *Nature* **452**, 970 (2008).
65. Y.-Q. Xia et al., *Nat. Phys.* **5**, 298 (2009).
66. J. Moore, *Nat. Phys.* **5**, 378 (2009).
67. H. Zhang et al., *Nat. Phys.* **5**, 438 (2009).
68. S. Chadov et al., *arXiv:1003.0193*, 2010.
69. B. Yan et al., *Europhys. Lett.* **90**, 37002 (2010).
70. L. Fu and C. L. Kane, *Phys. Rev. Lett.* **102**, 216403 (2009).
71. A. Altman and M. R. Zirnbauer, *Phys. Rev. B* **55**, 1142 (1997).
72. R. Roy, *arXiv:0803.2868*, 2008.
73. A. P. Schnyder et al., *Phys. Rev. B* **78**, 195125 (2008).
74. A. Kitaev, *arXiv:0901.2686*, 2009.
75. P. Ring and P. Schuck, *The Nuclear Many-Body Problem*, Springer-Verlag, Berlin, 2004.

TENSOR NETWORKS FOR ENTANGLEMENT EVOLUTION

SEBASTIAN MEZNARIC[1] and JACOB BIAMONTE[2,3]

[1] *Clarendon Laboratory, University of Oxford, Parks Road, Oxford OX1 3PU, UK*
[2] *ISI Foundation, Via Alassio 11/c, 10126, Torino, Italy*
[3] *Centre for Quantum Technologies, National University of Singapore, Block S15, 3 Science Drive 2, Singapore 117543, Singapore*

I. INTRODUCTION

Increasing recent interest in the topic of evolution of quantum entanglement was inspired by the work [1]. Konrad et al. considered the evolution equation for quantum entanglement for qubits [1] and its later extension to higher dimensions can be found in Refs [2,3]. The multipartite case was characterized by Gour [4]. Before [1], people generally dealt with entanglement evolution on a case-by-case basis. In this spirit, several authors considered non-Markovian entanglement evolution [5–8]. Similarly, upper and lower bounds can be obtained for special classes of states such as graph states [9].

Our approach aims to develop a tensor network diagrammatic method and to show which aspects of entanglement evolution can be described in those terms. We found that key results of Tiersch [2] can be reproduced in the language of tensor networks with several appealing features. Konrad et al. developed and relied on a

Advances in Chemical Physics, Volume 154: Quantum Information and Computation for Chemistry, First Edition. Edited by Sabre Kais.

sequence of diagrams that aided in their exposition of the topic at hand [1]. These diagrams were utilitarian in nature and used to convey a point. However, there exists a diagrammatic graphical language, of which Hilbert spaces are a complete model, known as Penrose tensor networks or string diagrams [10]. These diagrams can aid in intuition and also represent mathematical equations [11].

The diagrammatic language built on to describe the tensor networks dates back to Penrose [10]. Recently, we have also done work on the theory and expressiveness of tensor network states [12–15]. What we add to the literature on tensor network states is the development of some theory that connects these ideas to a specific version of channel-state duality. This offered an advancement and extension of the prior graphical aid of Ref. [1]. Through this rigorous graphical language a unified common topological structure of the networks representing the evolution of entanglement also emerges.

The introduction of a tensor network theory of entanglement evolution opens the door to apply numerical methods from the rapidly evolving area known as tensor network states [16–18]. In recent times, tensor networks states have been successfully used to simulate a wide class of quantum systems using classical computer algorithms. In addition, theory has been developed to use these methods to simulate classical stochastic systems [18]. At the heart of these methods are tensor network contraction algorithms, pairing physical insight with a new theory behind making approximations to truncate and hence minimize the classical data needed to represent a quantum state. The key properties and functions of these algorithms are known to be described in the graphical language we have adapted to entanglement evolution [10,12,13].

We will begin in Section II by reviewing those tensor network components that we will use. This introduction paves the way for a review of the so-called snake equation. This equation is then used to prove a graphical version of map-state duality, inspired by and tailored specifically to the problem at hand. Section III then applies these building blocks and associated tools from tensor network states to the theory of entanglement evolution. We begin by recalling the fundamental definitions and a quick outline of the known theory, we then cast these methods into our framework and pinpoint a wide class of measures that all share the same network topology in the tensor network language. The conclusion presents an overview of potential future directions by listing some open problems and new directions.

II. PENROSE GRAPHICAL NOTATION AND MAP-STATE DUALITY

We begin by showing how quantum states and operators may be represented graphically using the tensor network language. We then proceed to more complex topics, including the state-map duality and its graphical equivalents.

It is typical in quantum physics to fix the computational basis and consider a tensor as an indexed multiarray of numbers. For instance,

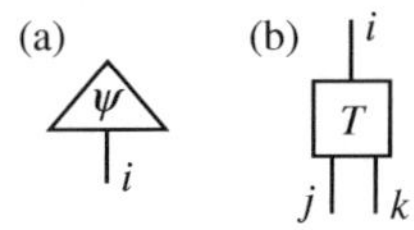

represent the tensor (a) ψ_i and (b) $T^i{}_{jk}$. The scenario readily extends.

Graphically a two-party quantum state $|\psi\rangle$ would have one wire for each index. We can write $|\psi\rangle$ in abstract index notation as ψ_{ij} or in terms of the Dirac notation convention (which we mainly adopt here) as $|\psi\rangle = \sum_{ij} c_{ij} |i\rangle |j\rangle$. Appropriately joining a flipped (transpose) and conjugated (star or overbar) copy of the quantum state $|\psi\rangle$ allows one to represent the density operator $\rho = \sum c_{ij}\overline{c}^{kl} |ij\rangle \langle kl|$ as follows [10] (Fig. 1).

Here, a box with two inputs (i, j) and two outputs (k, l) would be used to represent a general density operator $\rho = \sum \rho_{ij}{}^{kl} |ij\rangle \langle kl|$, with no known additional structure. Those familiar with quantum circuits will recognize this graphical language as their generalization [13].

Remark 1 (Tensor valence) *A tensor with n indices up and m down is called a valence-(n, m) tensor and sometimes a valence-k tensor for $k = n + m$.*

Remark 2 (Diagram convention—top to bottom, or right to left) *Open wires pointing toward the top of the page correspond to upper indices (bras); open wires pointing toward the bottom of a page correspond to lower indices (kets). For ease of presentation, we will often rotate this convention* 90 *degrees counterclockwise.*

We are interested in the following operations: (i) tensor index contraction by connecting legs of different tensors, (ii) raising and lowering indices by bending a leg upward or downwards, respectively, and thus taking the appropriate conjugate-transpose, and (iii) a duality between maps, states, and linear maps in general, called *Penrose wire bending duality*. This duality will play a major role in our application of the language to entanglement evolution. The duality is obtained by noticing that states are tensors with all open wires pointing in the same direction.

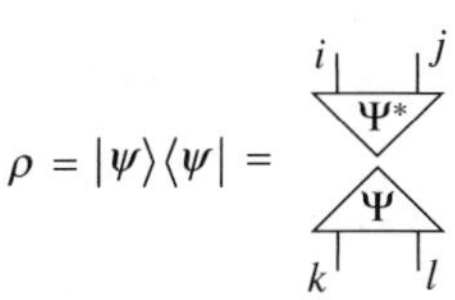

Figure 1. Penrose's graphical representation of a pure density matrix.

Bending some of the wires in the opposite direction makes various linear maps out of a given state. Given the computational basis, this amounts to turning all kets belonging to one of the Hilbert spaces into bras and vice versa. A bipartite state in the fixed standard basis

$$|\psi\rangle = \sum_{i,j} A_{i,j} |i\rangle |j\rangle \tag{1}$$

for example, is dual to the operator

$$[\psi] = \sum_{i,j} A_{i,j} |i\rangle \langle j| \tag{2}$$

This result can be shown by considering components of the equation $(A \otimes \mathbf{1}) \left|\Phi^+\right\rangle$ where $\left|\Phi^+\right\rangle$ is the Bell state (5).

The Choi–Jamio lkowski isomorphism can be cast as a case of Penrose duality. To do this we must recall Penrose's cups, caps, and identity wires. As in Ref. [10], these three tensors are given diagrammatically as

By thinking of these tensors now in terms of components (e.g., δ_{ij} is 1 whenever $i = j$ and 0 otherwise), we note that

$$\mathbf{1} = \sum_{ij} \delta^i_j |i\rangle \langle j| = \sum_k |k\rangle \langle k| \tag{3}$$

$$\langle 00| + \langle 11| + \cdots + \langle nn| = \sum_{ij} \delta^{ij} \langle ij| = \sum_k \langle kk| \tag{4}$$

$$|00\rangle + |11\rangle + \cdots + |nn\rangle = \sum_{ij} \delta_{ij} |ij\rangle = \sum_k |kk\rangle \tag{5}$$

where the identity map (a) corresponds to Eq. (3), the cup (b) to Eq. (4), and the cap (c) to Eq. (5). The relation between these three equations is again given by Penrose's wire bending duality: in a basis, bending a wire corresponds to changing a bra to a ket, and vice versa, allowing one to translate between Eqs. (3), (4), and (5) at will.

The contraction of two tensor indices diagrammatically amounts to appropriately joining open wires. Given tensors $T^i{}_{jk}$, $A^l{}_n$, and B^m_q we form a contraction by multiplying by $\delta^j_l \delta^q_k$ resulting in the tensor

$$T^i{}_{jk} A^j{}_l B^k{}_m := \Gamma^i{}_{lm} \tag{6}$$

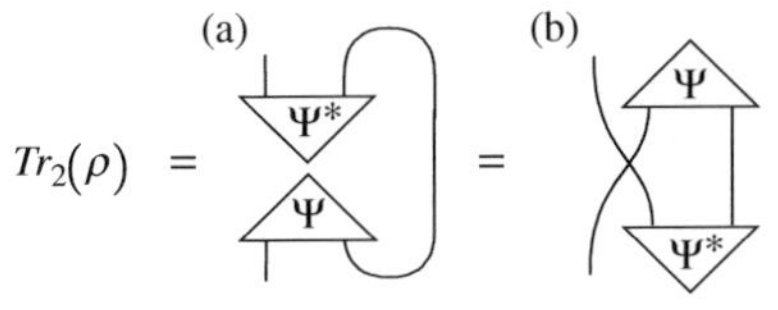

Figure 2. Graphical representation of a partial trace. The network (b) is due to Penrose. Note that Penrose used reflection across the page to represent adjoint, yet we have placed a star on Ψ to represent conjugate for illustrative purposes.

where we use the Einstein summation convention (repeated indices are summed over) and the tensor ${\Gamma^i}_{lm}$ is introduced per definition to simplify notation. In quantum physics notation, this is typically expressed in equational form as

$$\Gamma = \sum_{ilm} {\Gamma^i}_{lm} \left| lm \right\rangle \left\langle i \right| = \sum_{ijklm} {T^i}_{jk} {A^j}_l {B^k}_m \left| lm \right\rangle \left\langle i \right| \tag{7}$$

Remark 3 (Graphical trace—Penrose [10]) *Graphically the trace is performed by appropriately joining wires. The following depiction in Fig. 2b is Penrose's representation of a reduced density operator, where the stars on the Ψs represents complex conjugation.*

We have now presented the key tensor network building blocks used in this study. In practice, tensor networks contain an increasing number of tensors, making it difficult to form expressions using (inherently one-dimensional) equations. The two-dimensional diagrammatic depiction of tensor networks can simplify such expressions and often reduce calculations. Here, we tailor this graphical tool to depict internal structure and lend insight into presenting a unified theory of entanglement evolution. A key component of this unified view relies on the *natural equivalence* induced by the so-called *snake equations*, which we will review next.

The snake equation [10] amounts to considering the transformation of raising or lowering an index. One can raise an index and then lower this index, or vice versa, which amounts essentially to the net effect of doing nothing at all. This is captured diagrammatically by the so-called *snake* or *zig-zag equation*:

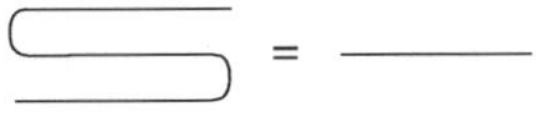

together with its vertical reflection across the page. In tensor index notation, it is given by $\delta_{ji}\delta^{ik} = \delta_j^{\,k}$.

As a tool to aid in our analysis of entanglement evolution, we have noticed a succinct form of map state duality that works by considering the natural equivalence found by the snake equations inside an already connected diagram. We expect that this formulation and its interpretation will have applications outside of the area of entanglement evolution. To date, the snake equation has already

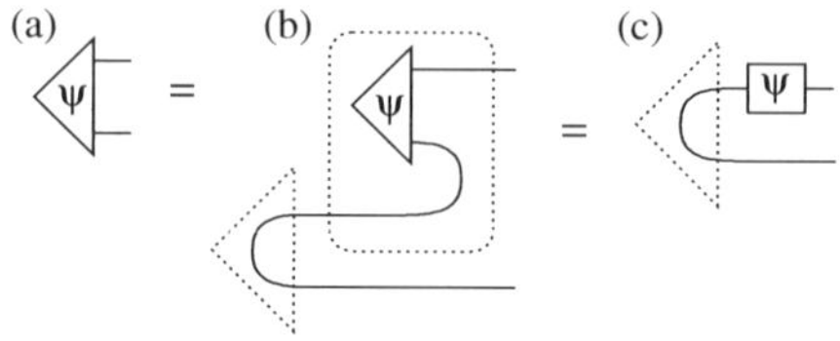

Figure 3. Graphical channel state duality of a state acted on by the identity channel. (a) The state $|\psi\rangle$ can be thought of as being acted on by the trivial or identity channel on both sides. (b) From the snake equation, an identity wire can be equivalently replaced with its S-shaped deformation, which can then be thought of as a map $[\psi]$ (enclosed by a box with dashed lines on its edges) acting on a Bell state (enclosed with a triangle with dashed lines on its edges). (c) An equivalent expression found by redrawing the shape of the tensor representing $|\psi\rangle$, into a box.

been used to model quantum teleportation [19,20], and so the interpretation here is a different one.

Lemma 1 (Graphical map-state duality) *Every bipartite quantum state $|\psi\rangle$ is naturally equivalent to a single-sided map $[\psi]$ acting on the Bell state $|\Phi^+\rangle$ with graphical depiction as follows.*

We will make a few remarks related to Lemma 1. To prove the result using equations (instead of diagrams, which was already done), we proceed as follows. To consider $|\psi\rangle$ as a map $[\psi]$, we act on it with a Bell state as $[\psi] := (\mathbf{1} \otimes \langle\Phi^+|)\,|\psi\rangle$. (Notice that the net effect is to turn kets of one of the spaces into bras, and vice versa, i.e., $|0\rangle \otimes |0\rangle$ becomes $|0\rangle\,\langle 0|$.) We then note that $|\psi\rangle = ([\psi] \otimes \mathbf{1})\,|\Phi^+\rangle$ and hence that $|\psi\rangle = ([(\mathbf{1} \otimes \langle\Phi^+|)\,|\psi\rangle] \otimes \mathbf{1})\,|\Phi^+\rangle$. The result seems more intuitive by considering the graphical rewrites.

To explain the diagrammatic manipulations, begin with Fig. 3a, representing the state $|\psi\rangle$. We can think of this (vacuously, it would seem) as a state being acted on by the identity operation. Application of the snake equation to one of the outgoing wires allows one to transform this into the equivalent diagram (b). We can think of (b) as a Bell state (left) being acted on by a map found from coefficients of the state $|\psi\rangle$. We have illustrated this map acting on the bell state in (b) by a light dashed line around $|\psi\rangle$. We then rewrite the figure, capturing the duality and arriving at (c).

Remark 4 (Properties of the map ψ) *The map $[\psi] := (\mathbf{1} \otimes \langle\Phi^+|)\,|\psi\rangle$ evidently has close properties with the quantum state $|\psi\rangle$. For instance, $Tr[\psi][\psi]^\dagger = \langle\psi|\psi\rangle$. The state $|\psi\rangle$ is invariant under the permutation group on two elements iff $[\psi] = [\psi]^\top$. The state $|\psi\rangle$ is SLOCC equivalent to the generalized Bell state iff $\det[\psi] \neq 0$ and the operator rank of $[\psi]$ gives the Schmidt rank of $|\psi\rangle$. Rank is a natural number and although it is invariant under action of the local unitary group, it is*

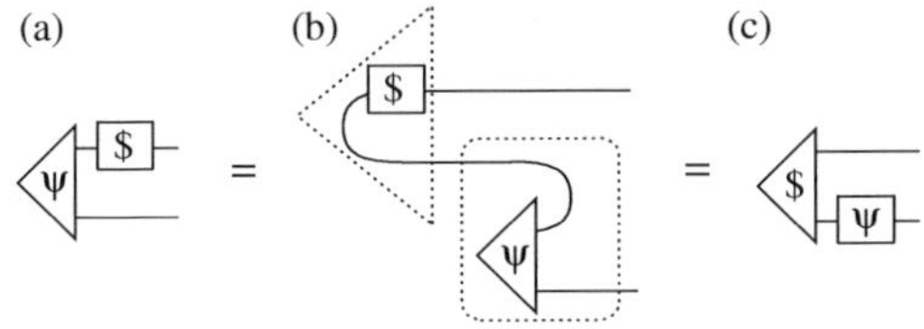

Figure 4. Graphical formulation of channel-state duality. (a) The diagram representing $(\$ \otimes \mathbf{1}) |\psi\rangle$. (b) The tensor network of the same state and the application of the natural equivalence induced by the snake equation per Lemma 1. (c) We then think of the state ψ as a map $\sum \psi_{ij} |i\rangle \langle j|$ acting on the state $\big|\$\big\rangle$.

not what is called a polynomial invariant or algebraic invariant because rank is not expressed in terms of a polynomial in the coefficients of the state. The eigenvalues of the map $[\psi]$ *are all that is needed to express any function of the entanglement of the bipartite state* $|\psi\rangle$.

Remark 5 (Extending Lemma 1) *One can readily extend Lemma 1 and consider instead the scenario given in Fig. 4.*

III. ENTANGLEMENT EVOLUTION

Remark 6 (Background reading) *For more details on standard results in the area of open systems theory, see for instance [21]. For results unifying several pictures of open systems evolutions see Ref. [15].*

Quantum operations are maps acting on quantum states ρ such that they preserve the positivity and boundedness of the corresponding density operator [21]. Such maps are known as *completely positive*. They can be used to represent the action of unitary maps or, more generally, decoherence and measurement or combinations thereof. In general a completely positive map **$**, called a *superoperator*, may be described in terms of Kraus operators $\boldsymbol{A}_k$,

$$\$\,[\rho] = \sum_k \boldsymbol{A}_k \rho \boldsymbol{A}_k^\dagger \tag{8}$$

The map **$** is said to be trace-preserving iff $\mathrm{Tr}(\rho) = \mathrm{Tr}(\$\rho)$ for all positive operators ρ and it can be shown that this is true iff $\sum_k \boldsymbol{A}_k \boldsymbol{A}_k^\dagger = \mathbf{1}$. Such maps faithfully preserve the normalization of ρ.

We consider bipartite states $\rho \in \mathcal{H}_A \otimes \mathcal{H}_B$. Because such states may be entangled, one is often particularly concerned with how entanglement evolves or changes under completely positive maps. Thus, given an entanglement measure $\mathcal{C}$ and a *local* completely positive map $\$ = \$_A \otimes \$_B$ acting as $\$[\rho] = \sum_k (A_k \otimes B_k)\rho(A_k^\dagger \otimes B_k^\dagger)$, we will want to compute $\mathcal{C}[(\$_A \otimes \$_B)\rho]$ for some entanglement measure $\mathcal{C}$. An elementary, but still critically important, example of such maps are one-sided maps of the form $\$ = \$_A \otimes \mathbf{1}_B$.

Entanglement evolution under one-sided maps has been characterized in the literature. In Ref. [1], Konrad et al. showed that for pure bipartite qubit states

$$\mathcal{C}\big[(\$_A \otimes \mathbf{1})\,|\psi\rangle\,\langle\psi|\big] = \mathcal{C}\left[(\$_A \otimes \mathbf{1})|\Phi^+\rangle\langle\Phi^+|\right] \cdot \mathcal{C}\big[|\psi\rangle\,\langle\psi|\big] \tag{9}$$

and so they factored the function $\mathcal{C}[\$, |\psi\rangle]$ of two variable arguments into a product of functions, one depending only on the starting state and the other only on the completely positive map characterizing the one-sided evolution. Thus, calculating the evolution of entanglement for a single initial state, $|\Phi^+\rangle$, allows us to compute it for all other pure initial states. As will be seen later, this provides an upper bound on the quantity of evolved entanglement for all initial mixed states. Here,

$$|\Phi^+\rangle = \frac{1}{\sqrt{2}}\left(|00\rangle + |11\rangle\right) \tag{10}$$

is again the typical Bell state and $\mathcal{C}$ is the concurrence. For mixed states, they found the upper bound to be given by the product [1]

$$\mathcal{C}\big[(\$_A \otimes \mathbf{1})\rho\big] \leq \mathcal{C}\left[(\$_A \otimes \mathbf{1})|\Phi^+\rangle\langle\Phi^+|\right] \cdot \mathcal{C}[\rho] \tag{11}$$

In Ref. [2] the analogous result was obtained for higher-dimensional systems, but where $\mathcal{C}$ was replaced by the $\mathcal{G}$-concurrence, defined for d-dimensional pure states

$$|\psi\rangle = \sum_{i,j=1}^{d} A_{ij}\,|i\rangle \otimes |j\rangle \tag{12}$$

with $A := [\psi]$ as

$$\mathcal{C}_d[|\psi\rangle] = d \cdot \left[\det(A^\dagger A)\right]^{1/d} \tag{13}$$

For mixed states the $\mathcal{G}$-concurrence is defined through the convex roof construction [22]

$$\mathcal{G}_d[\rho] = \inf \sum_i p_i \mathcal{C}_d\left[|\psi_i\rangle\right] \tag{14}$$

where the infimum is taken over all convex decompositions of ρ into pure states, that is, all decompositions of the form

$$\rho = \sum_i p_i |\psi_i\rangle \langle\psi_i| \tag{15}$$

with $p_i > 0$, $\sum_i p_i = 1$ and the $|\psi_i\rangle$ need not be orthogonal. Hence, we note in particular that it is not sufficient to consider only the spectral decomposition.

The results on factorization of entanglement evolution were further extended to the multipartite case in Ref. [4], where Gour showed that the same factorization result holds for all SL-invariant entanglement measures. These are defined as

Definition 1 (SL-invariant measures [4,22]) *Let G be the group $G = \mathrm{SL}(d_1, \mathbb{C}) \otimes \mathrm{SL}(d_2, \mathbb{C}) \otimes \ldots \otimes \mathrm{SL}(d_n, \mathbb{C})$. The groups $\mathrm{SL}(d_k, \mathbb{C})$ are the special linear groups represented by all $d_k \times d_k$ matrices with determinant 1. A SL-invariant measure of multipartite entanglement $\mathcal{C} : \mathcal{B}(\mathcal{H}) \to \mathbb{R}^+$ maps the bounded operators on the Hilbert space $\mathcal{H}$ (not necessarily normalized density matrices) to non-negative real numbers. $\mathcal{C}$ must satisfy the following properties:*

1. *For every $g \in G$ and $\rho \in \mathcal{B}(\mathcal{H})$, we have that $\mathcal{C}[g\rho g^\dagger] = \mathcal{C}[\rho]$.*
2. *It is homogenous of degree 1, $\mathcal{C}[r\rho] = r\mathcal{C}[\rho]$.*
3. *For all valid mixed states, it is given in terms of the convex roof construction:*

$$\mathcal{C}[\rho] = \min \sum_i p_i \mathcal{C}\big[|\psi_i\rangle\langle\psi_i|\big] \tag{16}$$

Here, the minimum is taken over all convex decompositions of ρ (i.e., all decompositions into $\rho = \sum_i p_i |\psi_i\rangle \langle\psi_i|$ where $\langle\psi_i|\psi_j\rangle$ is not necessarily zero).

Now we will show how these results, both for mixed and pure states and generally for qudits, can be described through the use of Penrose graphical calculus and the Choi–Jamio lkowski isomorphism. More precisely, we will show that

$$\mathcal{C}[(\$_A \otimes \mathbf{1}) |\psi\rangle \langle\psi|] = \mathcal{C}[(\$_A \otimes \mathbf{1}) \left|\Phi^+\right\rangle \left\langle\Phi^+\right|] \cdot \mathcal{C}[|\psi\rangle \langle\psi|] \tag{17}$$

The factorized form of the entanglement evolution is desirable because it simplifies and unifies calculations of entanglement once a one-sided map has been applied. To arrive at the factorization formula we first require one further result.

Given an $n \times n$ matrix $\boldsymbol{A}$ with $\det(\boldsymbol{A}) \neq 0$ and an SL-invariant n-dimensional entanglement measure $\mathcal{C}$, we find that

$$\mathcal{C}\big[\boldsymbol{A}\rho\boldsymbol{A}^\dagger\big] = \mathcal{C}\big[|\det(\boldsymbol{A})|^{2/n} g\rho g^\dagger\big] = \left|\det(\boldsymbol{A})^{2/n}\right| \mathcal{C}[\rho] \tag{18}$$

where we have defined $\boldsymbol{A} := \det(\boldsymbol{A})^{1/n} g$, with $g \in \mathrm{SL}(n)$.

As will be seen, this calculation (18) will be applied to maps $\boldsymbol{A}$ that arise under wire-bending duality from states. Consider

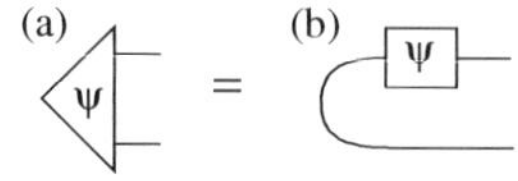

and by Lemma 1 we establish equality of (a) and (b). Hence, every bipartite quantum state $|\psi\rangle$ can be represented as the generalized Bell state $|\Phi^+\rangle$ acted on by a single-sided map $[\psi]$. Let $\boldsymbol{A} := [\psi]$, then the rank of $\boldsymbol{A}$ determines entanglement properties of $|\psi\rangle$. $\boldsymbol{A}$ is full rank iff $|\psi\rangle$ is SLOCC equivalent to $|\Phi^+\rangle$. In fact, $\boldsymbol{A}$ is full rank iff $\det(\boldsymbol{A}) \neq 0$ or equivalently iff $\det(\boldsymbol{A}^\dagger \boldsymbol{A}) > 0$. This will soon become relevant as we will use this duality to view $|\psi\rangle$ as a single-sided map and rely on the determinant to evaluate the entanglement measure of interest.

Now that we have established a factorization of the formula (18) we will show that the factor $\det(\boldsymbol{A})^{2/n}$ is nothing other than $\mathcal{C}[(\$ \otimes \mathbf{1}_B)\,|\phi^+\rangle\,\langle\phi^+|]$. Let us then consider evolution of the bipartite state $|\psi\rangle$ under the superoperator \$, with a tensor network given as

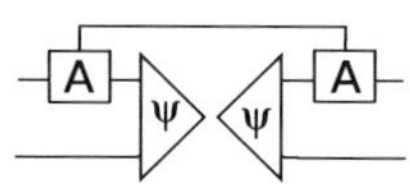

We then apply Lemma 1 to both sides and arrive at

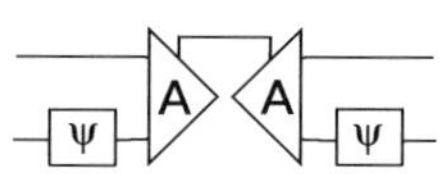

So $|\psi\rangle$ becomes a superoperator acting on the density operator

$$\rho_\$ = \sum_k |\boldsymbol{A}_k\rangle\,\langle\boldsymbol{A}_k| \tag{19}$$

which can be equivalently expressed in Fig. 5c where the remaining subfigures (a, b) summarize the duality behind entanglement evolution. Hence, Lemma 1 allows us to show the equivalence of the states (a) and (c) in Fig. 4 and from this it follows that

$$(\$ \otimes \mathbf{1})\,|\psi\rangle\,\langle\psi| = (\mathbf{1} \otimes \$_\psi)\,\rho_\$ \tag{20}$$

which is alternatively expressed as $(\mathbf{1} \otimes [\psi]) \sum_k |\boldsymbol{A}_k\rangle\,\langle\boldsymbol{A}_k|$.

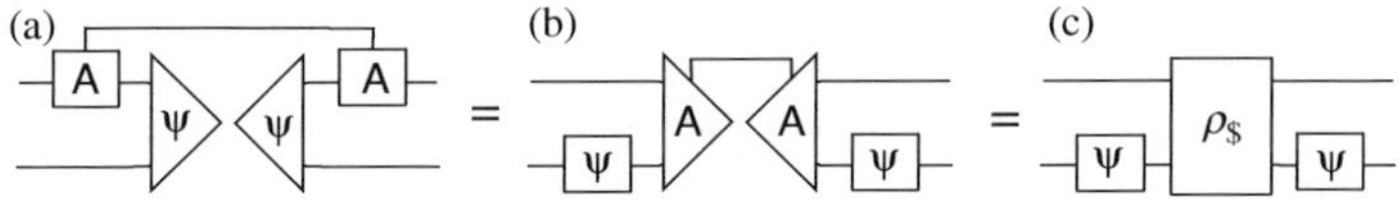

Figure 5. Tensor network summary of the crucial duality behind the theory of entanglement evolution. (a) State $|\psi\rangle$ acted on by a single-sided superoperator \$. (b) Application of Lemma 1 establishes a duality mapping the scenario in (a) into view where $|\psi\rangle$ becomes a single-sided superoperator $[\psi]$ that acts on the bipartite mixed state $\rho_\$ = \sum_k |A_k\rangle\langle A_k|$. (c) An equivalent expression for the valence four tensor $\rho_\$$.

Furthermore, notice that any state $|\psi\rangle$ may be written using the Schmidt decomposition as $|\psi\rangle = \sum_k \sqrt{\omega_k}\,|kk\rangle$. Thus, the map $\$_\psi$ may be considered simply as an action of a matrix

$$\boldsymbol{A} = \sqrt{n}\sum_k \sqrt{\omega_k}\,|k\rangle\langle k| \tag{21}$$

where the factor of $\sqrt{n}$ enters because of normalization. Because of the purity of $|\psi\rangle$ we arrive at only a single Kraus operator. Thus, we find that

$$\mathcal{C}[(\$ \otimes \mathbf{1})\,|\psi\rangle\langle\psi|] = \det(\boldsymbol{A})^{2/n}\mathcal{C}[\rho_\$] \tag{22}$$

Notice, that $\det(\boldsymbol{A})^{2/n} = n\prod_{k=1}^{n}\omega_k^{1/n}$. This is exactly the definition of $\mathcal{G}$-concurrence, reflecting the fact that $\mathcal{G}$-concurrence is the only SL-invariant measure of entanglement for bipartite qudits. The second factor in Eq. (22) is simply equivalent to

$$\mathcal{C}[\rho_\$] = \mathcal{C}\big[(\$ \otimes \mathbf{1})|\Phi^+\rangle\langle\Phi^+|\big] \tag{23}$$

where

$$|\Phi^+\rangle = \frac{1}{\sqrt{n}}\sum_{k=1}^{n}|kk\rangle \tag{24}$$

We can therefore rewrite the equation as

$$\mathcal{C}[(\$ \otimes \mathbf{1})\,|\psi\rangle\langle\psi|] = \mathcal{C}\big[(\$ \otimes \mathbf{1})|\Phi^+\rangle\langle\Phi^+|\big]\cdot\mathcal{C}\big[|\psi\rangle\langle\psi|\big] \tag{25}$$

$$= \mathcal{C}\big[(\$ \otimes \mathbf{1})|\Phi^+\rangle\langle\Phi^+|\big]\cdot n\prod_{k=1}^{n}\omega_k^{1/n} \tag{26}$$

The term $\mathcal{C}\big[(\$ \otimes \mathbf{1})|\Phi^+\rangle\langle\Phi^+|\big]$ still needs to be evaluated. However, its value tells us how concurrence evolves under $\$ \otimes \mathbf{1}$ for all pure states. It thus only needs to be computed once. Here, the entanglement measure $\mathcal{C}$ is now understood to be the $\mathcal{G}$-concurrence. The equation factorizes concurrence into a part entirely due to evolution dynamics and a part entirely due to the initial state.

Having established the evolution equation, Eq. (25), we will now consider an upper bound. Because the measure $\mathcal{C}$ is given by a convex roof construction for mixed states, it is a convex measure, that is,

$$\mathcal{C}\Big[\sum_k p_k \rho_k\Big] \leq \sum_k p_k \mathcal{C}[\rho_k] \tag{27}$$

We can use the convexity of the measure $\mathcal{C}$ to find an upper bound on the entanglement evolution for mixed states. This upper bound is given by

$$\mathcal{C}\big[(\$ \otimes \mathbf{1})\rho\big] \leq \mathcal{C}\big[(\$ \otimes \mathbf{1})|\Phi^+\rangle\langle\Phi^+|\big] \cdot \mathcal{C}[\rho] \tag{28}$$

IV. CONCLUSIONS

We have shown how entanglement evolution can be understood and even unified in terms of a tailored variant of the Penrose graphical calculus. We developed a key graphical tool to simplify the derivation of several important recent results in quantum information theory—that of entanglement evolution of bipartite quantum states. The result was originally obtained through the Choi–Jamio lkowski isomorphism, which we have shown, in the diagrammatic language, reduces to bending a contracting wire into a snake. We introduced the necessary tensor networks and showed how the evolution of $\mathcal{G}$-concurrence, the only SL-invariant measure of bipartite entanglement, can be obtained from these network diagrams. We thus developed a tensor network approach that, through the use of Penrose duality, recovers the known theory of entanglement evolution. This goal necessitated the introduction of several new methods that could prove useful in other areas. In particular, we anticipate the existence of other problems that could be simplified through the use of the diagrammatic variant of channel-state duality we have developed for our purposes here. This cross-pollination could lead to further insights into the strengths and limitations of the approach presented here.

As for the entanglement evolution itself, it is now fairly well understood for pure initial states and one-sided channels; few results are known for more general maps and for mixed initial states. Progress in this direction through the use of the Choi–Jamio lkowski technique is inhibited by the fact that the dual equivalent of a mixed state is a super operator rather than a simple matrix. Mathematically, given a mixed state in its spectral decomposition $\rho = \sum_k p_k \,|\psi_k\rangle\,\langle\psi_k|$, its wire bent dual is the map (see Fig. 5a)

$$\$_\rho = \sum_k p_k \boldsymbol{A}_k \cdot \boldsymbol{A}_k^\dagger \tag{29}$$

where A_k is the wire bent dual of the state $|\psi_k\rangle$. With pure states we have the fortune that $\$_\rho$ reduces to a single matrix and we can thus greatly simplify the calculation. The fact that it is now a genuine super operator limits the applicability of this approach. One would thus expect that a qualitatively different approach is needed to find a closed form expression for entanglement evolution of mixed states. And in fact, the very existence of such an expression is not guaranteed. However, finding it would immediately also yield a closed form expression for the evolution under maps acting on more than one subsystem $\$_1 \otimes \$_2$. To see this, notice that $\$_1 \otimes \$_2 = (\$_1 \otimes \mathbf{1}) \cdot (\mathbf{1} \otimes \$_2)$ thus treating the map as a sequence of two maps acting on only one subsystem.

As with the purely algebraic approach, the drawback of the tensor network approach is that it relies on the simplifying properties of bipartite SL-invariant measures of entanglement, the key contribution being a conceptual aid and exposure of the underlying mathematical structure at play. Namely, the property acts on a single subsystem of a quantum state with a nonsingular matrix as $(\boldsymbol{A} \otimes \mathbf{1})\,\rho\,(\boldsymbol{A}^\dagger \otimes \mathbf{1})$ multiplies an SL-invariant entanglement measure with a function of $\det(A)$. Other entanglement measures may not satisfy this property, and indeed the usual concurrence only satisfies it in two dimensions.

Acknowledgments

We would like to thank Michele Allegra, Dieter Jaksch, Stephen Clark, Mark M. Wilde, James Whitfield, and Zoltan Zimboras.

REFERENCES

1. T. Konrad, F. de Melo, M. Tiersch, C. Kasztelan, A. Aragao, and A. Buchleitner, *Nat. Phys.* **4**(4), 99 (2008).
2. M. Tiersch, F. de Melo, T. Konrad, and A. Buchleitner, *Quant. Inf. Process.* **8**(2), 523–534 (2009).
3. Z.-G. Li, S.-M. Fei, Z. D. Wang, and W. M. Liu, *Phys. Rev. A* **79**, 024303 (2009).
4. G. Gour, *Phys. Rev. Lett.* **105**(19), 190504 (2010).
5. B. Bellomo, R. Lo Franco, and G. Compagno, *Phys. Rev. Lett.* **99**, 160502 (2007).
6. X.-Z. Yuan, H.-S. Goan, and K.-D. Zhu, *Phys. Rev. B* **75**, 045331 (2007).
7. J. Dajka, M. Mierzejewski, and J. Luczka, *Phys. Rev. A* **77**(4), 042316 (2008).
8. T. Yu and J. H. Eberly, *Opt. Commun.* **283**(5), 676–680 (2010).
9. D. Cavalcanti, R. Chaves, L. Aolita, L. Davidovich, and A. Acin, *Phys. Rev. Lett.* **103**, 030502 (2009).
10. R. Penrose, *Combin. Math. Appl.* 221–244 (1971).
11. J. C. Baez and M. Stay, *Physics, Topology, Logic and Computation: A Rosetta Stone. ArXiv e-prints*, 2009.
12. J. D. Biamonte, S. R. Clark, and D. Jaksch, *AIP Advances* **1**(4), 042172 (2011).
13. V. Bergholm and J. D. Biamonte, *J. Phys. A* **44**(24), 245304 (2011).

14. S. J. Denny, J. D. Biamonte, D. Jaksch, and S. R. Clark, *J. Phys. A* **45**, 015309 (2011).
15. C. Wood, J. Biamonte, and D. Corey, *Tensor Networks and Graphical Calculus for Open Quantum Systems. ArXiv e-prints*, 2011.
16. F. Verstraete, V. Murg, and J. I. Cirac, *Adv. Phys.* **57**, 143–224 (2008).
17. J. I. Cirac and F. Verstraete, *J. Phys. A Math. Gen.* **42**, 4004 (2009).
18. T. H. Johnson, S. R. Clark, and D. Jaksch, *Phys. Rev. E* **82**(3), 036702 (2010).
19. L. H. Kauffman, *Opt. Spectrosc.* **99**, 227–232 (2005).
20. S. Abramsky, *Temperley-Lieb Algebra: From Knot Theory to Logic and Computation via Quantum Mechanics. ArXiv e-prints*, 2009.
21. I. Bengtsson and K. Życzkowski, *Geometry of Quantum States*, Cambridge University Press, Cambridge, 2006.
22. G. Gour. *Phys. Rev. A* **71**(1) (2005).

AUTHOR INDEX

Numbers in parentheses are reference numbers and indicate that the author's work is referred to although his name is not mentioned in the text. Numbers in italic show the page on which the complete references are listed.

Advances in Chemical Physics, Volume 154: Quantum Information and Computation for Chemistry, First Edition. Edited by Sabre Kais.

SUBJECT INDEX

Advances in Chemical Physics, Volume 154: Quantum Information and Computation for Chemistry, First Edition. Edited by Sabre Kais.

Printed and bound by CPI Group (UK) Ltd, Croydon, CR0 4YY